T0351940

# Advanced Concrete Technology

# Advanced Concrete Technology

Zongjin Li
Xiangming Zhou
Hongyan Ma
Dongshuai Hou

WILEY

This book is printed on acid-free paper. ♾



Published by John Wiley & Sons, Inc., Hoboken, New Jersey

Published simultaneously in Canada



For general information about our other products and services, please contact our Customer Care Department within the United States at (800) 762-2974, outside the United States at (317) 572-3993 or fax (317) 572-4002.

Wiley publishes in a variety of print and electronic formats and by print-on-demand. Some material included with standard print versions of this book may not be included in e-books or in print-on-demand. If this book refers to media such as a CD or DVD that is not included in the version you purchased, you may download this material at http://booksupport.wiley.com. For more information about Wiley products, visit www.wiley.com.

***Library of Congress Cataloging-in-Publication Data Applied for***

ISBN: 9781119806257 (Hardcover)
ISBN: 9781119806202 (ePDF)
ISBN: 9781119806196 (ePUB)

Cover Design: Wiley
Cover Image: © graemenicholson/Getty Image

SKY10035587_080422

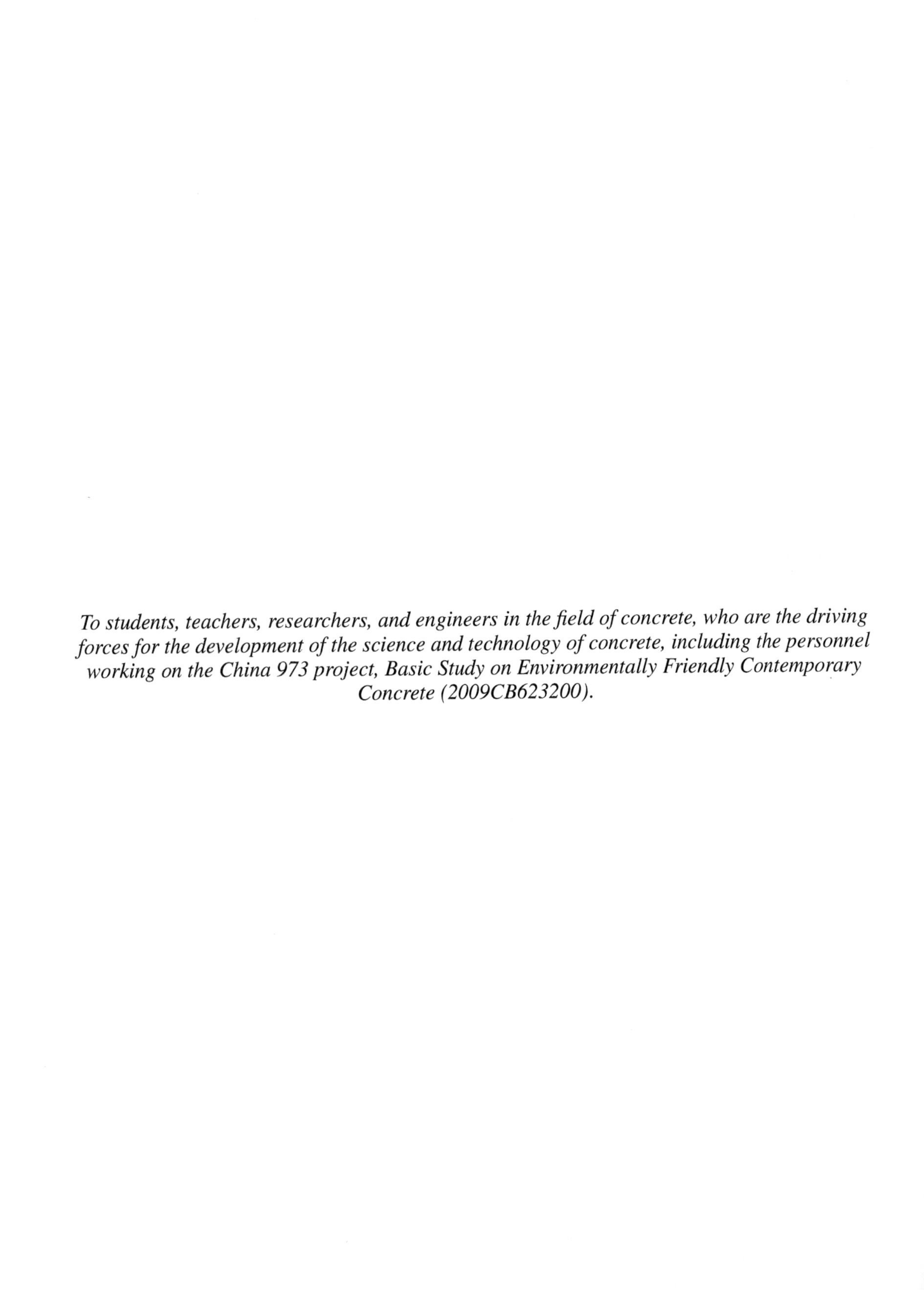

*To students, teachers, researchers, and engineers in the field of concrete, who are the driving forces for the development of the science and technology of concrete, including the personnel working on the China 973 project, Basic Study on Environmentally Friendly Contemporary Concrete (2009CB623200).*

# CONTENTS

# PREFACE

Portland cement concrete is the most widely used building material in the world. It plays an important role in infrastructure and private buildings construction. Due to the fast development of construction worldwide over the past decade, concrete technology has advanced significantly. To reflect the frontier development of concrete technology, the second edition of *Advanced Concrete Technology* now appears.

The second edition of *Advanced Concrete Technology* delivers a state-of-the-art exploration of contemporary and advanced concrete technologies developed over the last decade. It follows the principles and chapter division of the first edition, emphasizes the fundamental and scientific exploration of materials structures of concrete, and clarifies the essential concepts of concrete. The book combines the theories of concrete with practical examples of material design, explains the correlations among the composition, processing, characterization, properties, and performance of concrete and stresses the constraint of end-users on materials development.

The new book is divided into nine chapters, including new chapters on descriptions of the most recent advances in concrete technology. Chapter 1 gives a thorough introduction to concrete, including its definition and its historical evolution as a material used in engineering and construction. Chapter 2 provides in-depth explorations of the materials for making concrete with the addition of limestone and calcined clay cement, artificial sand, sea sand, and recycled aggregates. Chapter 3 discusses the concrete mix design methods and properties of fresh concrete, including its workability and rheology, as well as the methods for manufacturing, delivery, placing, compaction, finishing, and curing of concrete. Chapter 4 focuses on the material structure of concrete at different scales, in particular, the calcium silicate hydrate (C–S–H), the most important hydration product, at the atomic-to-nanometer scale. The structure of C–S–H is discussed with the results obtained by multi-scale simulation, including the quantum chemical method, the molecular potential-based method, and the coarse-grain Monte Carlo method over the past decade. Chapter 5 covers the properties of hardened concrete, including its strength, the stress-strain relationship, dimension stability, and durability. Chapter 6 provides updated knowledge on various advanced cement-based composites, including sea sand and seawater concrete and 3D-printed concrete, in addition to self-consolidation concrete, ultra-high strength concrete, engineered cementitious composites, etc. Chapter 7 introduces fulsome treatments of concrete fracture mechanics and explores the application of fracture mechanics in the design code of concrete structures, in addition to the double-K criterion. Chapter 8 covers essential knowledge of non-destructive testing in concrete engineering, including wave reflection and refraction theories, detecting principles and measurement methodologies for different non-destructive techniques. The new edition also includes an innovative magnetic corrosion detection transducer. Chapter 9 discusses the future and development trends of concrete technology. The new edition also introduces the carbon capture, utilization, and storage (CCUS) technologies, the application of nanotechnology, and data science and artificial intelligence in concrete technology.

This book perfectly fits the teaching needs of undergraduate and graduate students studying civil or materials engineering—especially those taking classes in the disciplines of *Properties of Concrete* or *Concrete Technologies*. It also provides the necessary knowledge and sufficient guidelines for practical engineers in the concrete industry.

During the process of writing the second edition of the book, the authors received enthusiastic help and invaluable assistance from many people, which is deeply appreciated. The authors would like to express their special thanks to Yunjian Li, Zhaoyang Sun, Qing Liu, Xing Ming, Guotao Qiu, Hongda Guo and Ms. Jianyu Xu.

Finally, we would like to thank our families for their love, understanding and support.

*Zongjin Li, University of Macao*
*Xiangming Zhou, Brunel University London*
*Hongyan Ma, Missouri University of Science and Technology*
*Dongshuai Hou, Qingdao University of Technology*

CHAPTER 1

# INTRODUCTION TO CONCRETE

## 1.1 CONCRETE DEFINITION AND HISTORICAL DEVELOPMENT

Concrete is a man-made building material that looks like stone. The word "concrete" is derived from the Latin *concretus*, meaning "to grow together." Concrete is a composite material composed of coarse granular material (the aggregate or filler) embedded in a hard matrix (cement or binder) that fills the space among the aggregate particles and binds them together. Alternatively, we can say that concrete is a composite material that consists essentially of a binding medium in which are embedded particles or fragments of aggregates. The simplest definition of concrete can be written as

$$\text{concrete} = \text{filler} + \text{binder} \tag{1-1}$$

Depending on the types of binder used, concrete can be named in different ways. For instance, if concrete is made with nonhydraulic cement, it is called nonhydraulic cement concrete; if concrete is made of hydraulic cement, it is called hydraulic cement concrete; if concrete is made of asphalt, it is called asphalt concrete; if concrete is made of polymer, it is called polymer concrete. Both nonhydraulic and hydraulic cement need water to mix in and react. They differ here in the ability to gain strength in water. Nonhydraulic cement cannot gain strength in water, while hydraulic cement can.

Nonhydraulic cement concretes are the oldest concrete used in human history. As early as around 6500 BC, nonhydraulic cement concretes were used by the Syrians and spread through Egypt, the Middle East, Crete, Cyprus, and ancient Greece. However, it was the Romans who refined the mixture's use. The nonhydraulic cements used at that time were gypsum and lime. The Romans used a primal mix for their concrete. It consisted of small pieces of gravel and coarse sand mixed with hot lime and water, and sometimes even animal blood. The Romans were known to have made wide usage of concrete for building roads. It is interesting to learn that they built some 5300 miles of roads using concrete. Concrete is a very strong building material. Historical evidence also points out that the Romans used pozzolana, animal fat, milk, and blood as admixtures for making concrete. To trim down shrinkage, they were known to have used horsehair. Historical evidence also shows that the Assyrians and Babylonians used clay as the bonding material. Lime was obtained by calcining limestone with a reaction of

$$CaCO_3 \xrightarrow{1000^\circ C} CaO + CO_2 \tag{1-2}$$

When CaO is mixed with water, it can react with water to form:

$$CaO + H_2O \xrightarrow{\text{ambient temperature}} Ca(OH)_2 \tag{1-3}$$

and is then further reacted with $CO_2$ to form limestone again:

$$Ca(OH)_2 + CO_2 + H_2O \xrightarrow{\text{ambient temperature}} CaCO_3 + 2H_2O \tag{1-4}$$

The Egyptians used gypsum mortar in construction, and the half-water gypsum was obtained by calcining two-water gypsum with a reaction of:

$$2CaSO_4 \cdot 2H_2O \xrightarrow{107-130^\circ C} 2CaSO_4 \cdot \frac{1}{2}H_2O + 3H_2O \quad (1\text{-}5)$$

When mixed with water, half-water gypsum could turn into two-water gypsum and gain strength:

$$2CaSO_4 \cdot \frac{1}{2}H_2O + 3H_2O \xrightarrow{\text{ambient temperature}} 2CaSO_4 \cdot 2H_2O \quad (1\text{-}6)$$

The Egyptians used gypsum instead of lime because it could be calcined at much lower temperatures. As early as about 3000 BC, the Egyptians used gypsum mortar in the construction of the Pyramid of Cheops. However, this pyramid was looted long before archeologists knew about the building materials used. Figure 1-1 shows a pyramid in Giza. The Chinese also used lime mortar to build the Great Wall in the Qin dynasty (220 BC) (see Figure 1-2).

A hydraulic lime was developed by the Greeks and Romans using limestone containing argillaceous (clayey) impurities. The Greeks even used volcanic ash from the island of Santorini, while the Romans used volcanic ash from the Bay of Naples to mix with lime to produce hydraulic lime. It was found that mortar made of such hydraulic lime could resist water. Thus, hydraulic lime mortars were used extensively for hydraulic structures from the second half of the first century BC to the second century AD. However, the quality of cementing materials declined throughout the Middle Ages. The art of burning lime was almost lost and siliceous impurities were not added. High-quality mortars disappeared for a long period. In 1756, John Smeaton was commissioned to rebuild the Eddystone Lighthouse off the coast of Cornwall, England. Realizing the function of siliceous impurities in resisting water, Smeaton conducted extensive experiments with different limes and pozzolans, and found that limestone with a high proportion of clayey materials produced the best hydraulic lime for the mortar to be used in water. Eventually, Smeaton used a mortar

**Figure 1-1** Pyramid built with gypsum mortar, Giza, Egypt

**Figure 1-2** The Great Wall built in the Qin dynasty (Photo provided by Tongbo Sui)

prepared from a hydraulic lime mixed with pozzolan imported from Italy. He made concrete by mixing coarse aggregate (pebbles) and powdered brick and mixed it with cement, very close to the proportions of modern concrete. The rebuilt Eddystone Lighthouse lasted for 126 years until it was replaced with a modern structure.

After Smeaton's work, hydraulic cement developed very fast. James Parker of England filed a patent in 1796 for a natural hydraulic cement made by calcining nodules of impure limestone containing clay. Vicat of France produced artificial hydraulic lime by calcining synthetic mixtures of limestone and clay. Portland cement was patented by Joseph Aspdin of England in 1824. The name Portland was coined by Aspdin because the color of the cement after hydration was similar to that of limestone quarried in Portland, a town in southern England. Portland cement was prepared by calcining finely ground limestone, mixing it with finely divided clay, and calcining the mixture again in a kiln until the $CO_2$ was driven off. This mixture was then finely ground and used as cement. However, the temperature claimed in Aspdin's invention was not high enough to produce true Portland cement. It was Isaac Johnson who first burned the raw materials to the clinkering temperature in 1845 to produce modern Portland cement. After that, the application of Portland cement spread quickly throughout Europe and North America. The main application of Portland cement is to make concrete. It was in Germany that the first systematic testing of concrete took place in 1836. The test measured the tensile and compressive strength of concrete. Aggregates are another main ingredient of concrete, which include sand, crushed stone, clay, gravel, slag, and shale. Plain concrete made of Portland cement and aggregate is usually called the first generation of concrete. The second generation of concrete refers to steel bar-reinforced concrete. François Coignet in France was a pioneer in the development of reinforced concrete. (Day and McNeil, 1996). Coignet started experimenting with iron-reinforced concrete in 1852 and was the first builder ever to use this technique as a building material (Encyclopaedia Britannica, 1991). He decided, as a publicity stunt and to promote his cement business, to build a house made of *béton armé*, a type of reinforced concrete. In 1853, he built the first iron reinforced concrete structure anywhere; a four-story house at 72 Rue Charles Michels in St. Denis (Sutherland et al., 2001). This location was near his family cement plant in St. Denis, a commune in the northern suburbs of Paris. The house was designed by local architect Theodore Lachez (Collins, 2004).

Coignet had an exhibit at the 1855 Paris Exposition to show his technique of reinforced concrete. At the exhibit, he forecast that the technique would replace stone as a means of construction. In 1856, he patented a technique of reinforced concrete using iron tirants. In 1861, he published his techniques of reinforced concrete.

Reinforced concrete was further developed by Hennebique at the end of the 19th century, and it was realized that performance could be improved if the bars could be placed in tension, thus keeping the concrete in compression. Early attempts worked, with the beams showing a reduced tendency to crack in tension, but after a few months, the cracks reopened. A good description of this early work is given in Leonhardt (1964). The first reinforced concrete bridge was built in 1889 in the Golden Gate Park in San Francisco, California.

To overcome the cracking problem in reinforced concrete, prestressed concrete was developed and was first patented by a San Francisco engineer named P.H. Jackson as early as 1886. Prestressed means that the stress is generated in a structural member before it carries the service load. Prestressed concrete was referred to as the third generation of concrete. Prestressing is usually generated by the stretched reinforcing steel in a structural member. According to the sequence of concrete casting, prestressing can be classified as pretensioning or post-tensioning. Pretensioning pulls the reinforcing steel before casting the concrete and prestress is added through the bond built up between the stretched reinforcing steel and the hardened concrete. In the post-tensioning technique, the reinforcing steel or tendon is stretched after concrete casting and the gaining of sufficient strength. In post-tensioning, steel tendons are positioned in the concrete specimen through prereserved holes. The prestress is added to the member through the end anchorage. Figure 1-3 shows the sequence of the pretensioning technique for prestressed concrete.

Prestressed concrete became an accepted building material in Europe after World War II, partly due to the shortage of steel. North America's first prestressed concrete structure, the Walnut Lane Memorial Bridge in Philadelphia, Pennsylvania, was completed in 1951. Nowadays, with the development of prestressed concrete, long-span bridges, tall buildings, and ocean structures have been constructed. The Barrios de Lura Bridge in Spain is currently the longest-span prestressed concrete, cable-stayed bridge in the world, with a main span of 440 m, while the Shibanpo Yangtze

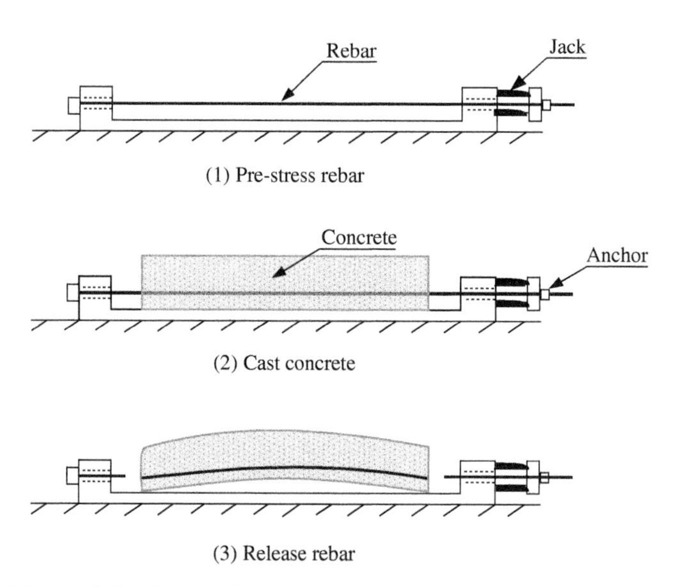


**Figure 1-3** Pretensioning sequence for prestressed concrete

Bridge is the world's longest prestressed concrete girder bridge with the main span of 330 m. In Canada, the prestressed Toronto CN tower reaches a height of 553 m.

As a structural material, the compressive strength at an age of 28 days is the main design index for concrete. There are several reasons for choosing compressive strength as the representative index. First, concrete is used in a structure mainly to resist the compression load. Second, the measurement of compressive strength is relatively easier. Finally, it is thought that other properties of concrete can be related to its compressive strength. Pursuing high compressive strength has been an important direction of concrete development. As early as 1918, Duff Adams found that the compressive strength of concrete was inversely proportional to the water-to-cement (*w/c*) ratio. Hence, a high compressive strength could be achieved by reducing the *w/c* ratio. However, to keep concrete workable, there is a minimum requirement on the amount of water; hence, the *w/c* ratio reduction is limited, unless other measures are provided to improve concrete's workability. For this reason, progress in achieving high compressive strength was very slow before the 1960s. At that time, concrete with a compressive strength of 30 MPa was regarded as high-strength concrete. Since the 1960s, the development of high-strength concrete has made significant progress due to two main factors: the invention of water-reducing admixtures and the incorporation of mineral admixtures, such as silica fume, fly ash, and slag. Water-reducing admixture is a chemical admixture that can enable concrete with good workability under a very low *w/c* ratio; the latter are finer mineral particles that can react with a hydration product in concrete, calcium hydroxide, to make concrete's microstructure denser, hence improving concrete's properties. Silica fume also has a packing effect to further improve the matrix density. In 1972, the first 52-MPa concrete was produced in Chicago for the 52-story Mid-Continental Plaza. In 1972, a 62-MPa concrete was produced, also in Chicago, for Water Tower Place, a 74-story concrete building, the tallest in the world at that time (see Figure 1-4). In the 1980s, the industry was able to produce a 95-MPa concrete to supply to the 225 West Wacker Drive building project in Chicago, as shown in Figure 1-5. The highest compressive strength of 130 MPa was realized in the 220-m-high, 58-story building, the Union Plaza, constructed in Seattle, Washington (Caldarone, 2009).

Concrete produced after the 1980s usually contains a sufficient amount of fly ash, slag, or silica fume as well as many different chemical admixtures, so its hydration mechanism, hydration products, and other microstructure characteristics are very different from the concrete produced without these admixtures. Moreover, the mechanical properties are also different from the conventional concrete; hence, such concretes are referred to as contemporary concretes.

There have been two innovative developments in contemporary concrete: self-compacting concrete (SCC) and ultra-high-performance concrete (UHPC). SCC is a type of high-performance concrete (HPC). High-performance concrete is a concept developed in the 1980s. It is defined as a concrete that can meet special performance and uniformity requirements, which cannot always be achieved routinely by using only conventional materials and normal mixing, placing, and curing practices. The requirements may involve enhancement of the characteristics of concrete, such as placement and compaction without segregation, long-term mechanical properties, higher early-age strength, better toughness, higher volume stability, or longer service life in severe environments.

Self-compacting concrete is a typical example of high-performance concrete that can fill in formwork in a compacted manner without the need of mechanical vibration. SCC was initially developed by Professor Okamura and his research group in Japan in the late 1980s (Ozama et al., 1989). At that time, concrete construction was blooming everywhere in Japan. Since Japan is in an earthquake zone, concrete structures are usually heavily reinforced, especially at beam-column joints. Hence, due to low flowability, conventional concrete could hardly flow past the heavily reinforced rebars, leaving poor-quality cast concrete and leading to poor durability. Sometimes, the

**Figure 1-4** Water Tower Place in Chicago, USA (Photo provided by Xiaojian Gao)

reinforcing steel was exposed to air immediately after demolding. To solve the problem, Professor Okamura and his research group developed concrete with very high flowability. With the help of the invention of the high-range water reducer or plasticizer, such highly flowable concrete was finally developed. They were so excited that they called this concrete "high-performance concrete" in the beginning. It was corrected later to SCC, as HPC covers broader meanings. Durability is a main requirement of HPC. It has been found that many concrete structures could not fulfill the service requirement, due not to lack of strength, but to lack of durability. For this reason, concrete with high performance to meet the requirement of prolonging concrete service life was greatly needed.

In the 1990s, a new type of "concrete" with a compressive strength higher than 200 MPa was developed in France. Due to the large amount of silica fume incorporated in such a material, it was initially called reactive powder concrete and later on the name changed to ultra-high-strength (performance) concrete (UHSC), due to its extremely high compressive strength (Richard and Cheyrezy, 1995). The ultra-high-strength concrete has reached a compressive strength of 800 MPa with heating treatment. However, it is very brittle, hence, incorporating fibers into UHSC is necessary. After incorporating fine steel fibers, a flexural strength of 50 MPa can be reached. The first trial application of UHSC was a footbridge built in Sherbrooke, Canada (Aitcin et al., 1998).

**Figure 1-5** The 225 West Wacker building in Chicago, USA (Photo provided by Xiaojian Gao)

## 1.2 CONCRETE AS A STRUCTURAL MATERIAL

In this book, the term concrete usually refers to Portland cement concrete, if not otherwise specified. For this kind of concrete, the compositions can be listed as follows:

Portland **cement**
+ water (& admixtures) → cement **paste**
+ fine aggregate → **mortar**
+ coarse aggregate → **concrete**

Here it should be noted that admixtures are used in almost all modern practice and thus have become an essential component of contemporary concrete. Admixtures are defined as materials other than aggregate (fine and coarse), water, and cement that are added into a concrete batch immediately before or during mixing. The use of admixtures is widespread, mainly because many benefits can be achieved by their application. For instance, certain chemical admixtures can modify the setting and hardening characteristics of cement paste by influencing the rate of cement hydration. Water-reducing admixtures can plasticize fresh concrete mixtures by reducing the surface tension of the water. Air-entraining admixtures can improve the durability of concrete, and mineral admixtures such as pozzolans (materials containing reactive silica) can reduce thermal cracking. A detailed description of admixtures is given in Chapter 2.

Concrete is the most widely used construction material in the world, and its popularity can be attributed to two aspects. First, concrete can be used for many different structures, such as dams, pavements, building frames, or bridges, much more than any other construction material. Second, the amount of concrete used is much more than any other material. Its worldwide production exceeds that of steel by a factor of 10 in tonnage and by more than a factor of 30 in volume.

In a concrete structure, there are two commonly used structural materials: concrete and steel. A structural material is a material that carries not only its self-weight, but also the load passing from other members.

Steel is manufactured under carefully controlled conditions, always in a highly sophisticated plant; the properties of every type of steel are determined in a laboratory and described in a manufacturer's certificate. Thus, the designer of a steel structure need only specify the steel complying with a relevant standard, and the constructor needs only to ensure that the correct steel is used and that connections between the individual steel members are properly executed (Neville and Brooks, 1993).

On the other hand, concrete is produced in a cruder way and its quality varies considerably. Even though the quality of cement, the binder of concrete, is guaranteed by the manufacturer in a manner similar to that of steel, the quality of concrete is hardly guaranteed because of many other factors, such as aggregates, mixing procedures, and the skill of the operators of concrete production, placement, and consolidation.

It is possible to obtain concrete of specified quality from a ready-mix supplier, but, even in this case, it is only the raw materials that are bought for a construction job. Transporting, placing, and, above all, compacting greatly influence the quality of a cast concrete structure. Moreover, unlike the case of steel, the choice of concrete mixes is virtually infinite and therefore the selection has to be made with a sound knowledge of the properties and behavior of concrete. It is thus the competence of the designer and specifier that determines the potential qualities of concrete, and the competence of the supplier and the contractor that controls the actual quality of concrete in the finished structure. It follows that they must be thoroughly conversant with the properties of concrete and with concrete making and placing.

In a concrete structural element, concretes mainly carry the compressive stress and shear stress while the steel carries the tension stress. Moreover, concrete usually provides stiffness for structures to keep them stable.

As a type of structural material, concrete has been widely used to build various structures. High-strength concrete has been used in many tall building constructions. In Hong Kong, grade 90 concrete (i.e., compressive strength of 90 MPa) was used in the columns of the tallest building in the region, i.e., the 108-story International Commerce Centre (see Figure 1-6), which was built in 2010 and stands 484 m (1588 ft) tall.

Concrete has also been widely used in bridge construction. Figure 1-7 shows the Sutong Bridge that crosses the Yangtze River in China between Nantong and Changshu, a satellite city of Suzhou, in Jiangsu province, east China. It is a cable-stayed bridge with the third-longest main span, 1088 meters, in the world after the Russky bridge (1104 m) and the Hutong Yangtze River bridge (1092 m). Its two side spans are 300 m (984 ft) each, and there are also four small cable spans.

**Figure 1-6** International Commerce Centre, Hong Kong. (Source: Wing1990hk / Wikimedia Commons / CC BY-SA 3.0)

Dam construction is another popular application for concrete. The first major concrete dams, the Hoover Dam and the Grand Coulee Dam, were built in the 1930s and they are still standing. The largest dam ever built is the Three Gorges Dam in Hubei province, China, as shown in Figure 1-8. The total concrete used for the dam was over 22 million $m^3$.

Concrete has also been used to build high-speed railways. Shinkansen, the world's first contemporary high-volume (initially 12-car maximum), "high-speed rail," was built in Japan in 1964. In Europe, high-speed rail was first introduced during the International Transport Fair in Munich in June 1965. Nowadays, high-speed rail construction is blooming in China. With 37,900 kilometers of lines by 2021, China has the world's largest network of high-speed railways. According to planning, 70,000 km of high-speed rail will be built in China by 2035. Figure 1-9 shows a high-speed rail system in China.

In addition, concrete has been widely applied in the construction of airport runways, tunnels, highways, pipelines, and oil platforms. Up to now, the annual world consumption of concrete has reached a value such that if the concrete were edible, every person on the Earth would have 2000 kg per year to "eat." You may wonder why concrete has become so popular.

**Figure 1-7** Sutong Bridge in Suzhou, Jiangsu, China (Photo provided by Xiaoyan Liu)

**Figure 1-8** Three Gorges Dam, Hubei, China (Photo provided by Zhen He)

**Figure 1-9** High speed rail in China (Photo provided by Guotang Zhao)

## 1.3 CHARACTERISTICS OF CONCRETE

### 1.3.1 Advantages of Concrete

(a) *Economical*: Concrete is the most inexpensive and the most readily available material in the world. The cost of production of concrete is low compared with other engineered construction materials. The three major components in concrete are water, aggregate, and cement. Compared with steels, plastics, and polymers, these components are the most inexpensive, and are available in almost every corner of the globe. This enables concrete to be produced worldwide at a very low cost for local markets, thus avoiding the transport expenses and associated carbon emission for most other materials.

(b) *Ambient temperature-hardened material*: Because cement is a low-temperature bonded inorganic material and its reaction occurs at room temperature, concrete can gain its strength at ambient temperature. Normally, no high temperature curing is needed.

(c) *Ability to be cast*: Fresh concrete is flowable like a liquid and hence can be poured into various formworks to form different desired shapes and sizes right on a construction site or a precast plant. Hence, concrete can be cast into many different configurations. One good example to show concrete castability is the Baha'i Temple located in Wilmette, Illinois, USA, as shown in Figure 1-10. The very complex configurations of the different shapes of flowers in the wall and roof were all cast by concrete.

(d) *Energy-efficient*: Compared to steel, the energy consumption of concrete production is low. The energy required to produce plain concrete is only 450–750 kWh/ton and that of reinforced concrete is 800–3200 kWh/ton, while structural steel requires 8000 kWh/ton or more to make.

(e) *Excellent water resistance*: Unlike wood (timber) and steel, concrete can harden in water and can withstand the action of water without serious deterioration, which makes concrete an ideal material for building structures to control, store, and transport water, such as

**Figure 1-10** Baha'i Temple in Wilmette, Illinois (Photo provided by Xiaojian Gao)

**Figure 1-11** Pipeline under construction (Photo provided by Zhulin Zhang)

pipelines (Figure 1-11), dams, and submarine structures. A typical example of a pipeline application is the Central Arizona Project, which provides water from the Colorado River to central Arizona. The system contains 1560 pipe sections, each 6.7 m long, 7.5 m outside diameter, and 6.4 m inside diameter. Contrary to popular belief, water is not deleterious to concrete, even to reinforced concrete; it is the chemicals dissolved in water, such as chlorides, sulfates, and carbon dioxide, that cause deterioration of concrete structures.

**(f)** *High-temperature resistance*: Concrete conducts heat slowly and can store considerable quantities of heat from the environment. Moreover, the main hydrate that provides binding to aggregates in concrete, calcium silicate hydrate (C–S–H), will not be completely dehydrated until 910°C. Thus, concrete can withstand high temperatures much better than wood and steel. Even in a fire, a concrete structure can withstand heat for 2–6 hours, leaving sufficient time for people to be rescued. This is why concrete is frequently used to build up protective layers for a steel structure.

**(g)** *Ability to consume waste*: With the development of industry, civilization, and urbanization, more and more by-products or waste have been generated, which causes a serious environmental pollution problem. To solve the problem, people have to find a way to consume such wastes. It has been found that many industrial wastes can be recycled as a substitute (replacement) for cement or aggregate, such as fly ash, slag (GGBFS = ground granulated blast-furnaces slag), waste glass, and ground vehicle tires in concrete. Production of concrete with the incorporation of industrial waste not only provides an effective way to protect our environment, but also sometimes leads to the better performance of a concrete structure. Due to a large amount of concrete produced annually, it is possible to consume most of the industrial waste in the world in concrete production, provided that suitable techniques for individual waste incorporation are available.

**(h)** *Ability to work with reinforcing steel*: Concrete has a similar value to steel for the coefficient of thermal expansion (steel $1.2 \times 10^{-5}$; concrete $1.0-1.5 \times 10^{-5}$). Concrete produces good protection to steel due to the existence of CH and other alkalis (this is for normal conditions). Therefore, while steel bars provide the necessary tensile strength, concrete provides a perfect environment for the steel, acting as a physical barrier to the ingress of aggressive species and giving chemical protection in a highly alkaline environment (pH value is about 13.5), in which black steel is readily passivated.

**(i)** *Less maintenance required*: Under normal conditions, concrete structures do not need coating or painting as protection for weathering, while for a steel or wooden structure, it is necessary to have a protective layer. Moreover, the coatings and paintings have to be replaced every few years. Thus, the maintenance cost for concrete structures is much lower than that for steel or wooden structures.

### 1.3.2 Limitations

**(a)** *Quasi-brittle failure mode*: The failure mode of materials can be classified into three categories: brittle failure, quasi-brittle failure, and ductile failure, as shown in Figure 1-12. Glass is a typical brittle material. It will break as soon as its tensile strength is reached. Materials exhibiting a strain-softening behavior (Figure 1-12b) are called quasi-brittle materials. Both brittle and quasi-brittle materials fail suddenly without giving a large deformation as a warning sign. Ductile failure is a failure with a large deformation that serves as a warning before the collapse, such as low-carbon steel. Concrete is a type of quasi-brittle material with low fracture toughness. Usually, concrete has to be used with steel bars to form so-called reinforced concrete, in which steel bars are used to carry tension and the concrete compression loads. Moreover, concrete can provide a structure with excellent stability. As elaborated previously, reinforced concrete is realized as the second generation of concrete.

**(b)** *Low tensile strength*: Concrete has different values in compression and tension strength. Its tensile strength is only about 1/10 of its compressive strength for normal-strength concrete, even lower for high-strength concrete. To improve the tensile strength of concrete, fiber-reinforced concrete and polymer concrete have been developed.

**(c)** *Low toughness (ductility)*: Toughness is usually defined as the ability of a material to consume energy. Toughness can be evaluated by the area of a load–displacement curve. Compared to steel, concrete has very low toughness, with a value only about 1/50 to 1/100 of that of steel, as shown in Figure 1-13. Adding fibers is a good way to improve the toughness of concrete.

**(d)** *Low specific strength (strength/density ratio)*: For normal-strength concrete, the specific strength is less than 20, while for steel it is about 40. There are two ways to increase concrete specific strength: one is to reduce its density and the other is to increase its strength. Hence, lightweight concrete and high-strength concrete have been developed.

**(e)** *Formwork is needed*: Conventional fresh concrete is in a liquid state and needs formwork to hold its shape and to support its weight. Formwork can be made of steel or wood, as shown

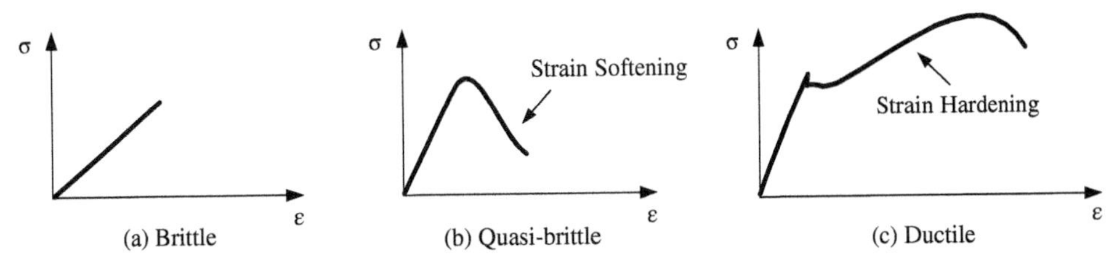


**Figure 1-12** Three failure modes of materials

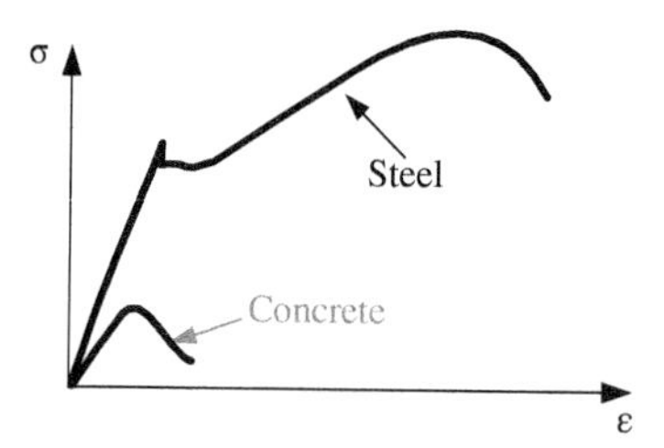


**Figure 1-13** Stress-strain curves of steel and concrete

**Figure 1-14** Formwork for concrete casting

in Figure 1-14. The formwork is expensive because it is labor-intensive and time-consuming to build and demold it. To improve efficiency, precast, extrusion, and 3D printing techniques have been developed.

**(f)** *Long curing time*: The design index for concrete strength is the 28-day compression strength. Hence, full strength development needs a month at ambient temperature. The improvement measure that is able to reduce the curing period is steam curing or microwave curing.

**(g)** *Working with cracks*: Even for reinforced concrete structure members, the tension side has a concrete cover to protect the steel bars. Due to the low tensile strength, the concrete cover cracks. To solve the crack problem, prestressed concrete is developed, and it is also realized as a third-generation concrete. Most reinforced concrete structures have existing cracks on their tension sides while carrying the service load.

## 1.4 TYPES OF CONCRETE

### 1.4.1 Classification in Accordance with Unit Weight

According to the unit weight of concretes, they can be classified into four categories, as shown in Table 1-1. Ultra-lightweight concrete can only be used to build up nonstructural members. Lightweight concrete can be used to build both nonstructural and structural members,

**Table 1-1** Classification of concrete in accordance with unit weight

| Classification | Unit weight (kg/m$^3$) |
|---|---|
| Ultra-lightweight concrete | $< 1{,}200$ |
| Lightweight concrete | $1{,}200 < UW < 1{,}800$ |
| Normal-weight concrete | $\approx 2{,}400$ |
| Heavyweight concrete | $> 3{,}200$ |

**Table 1-2** Concrete classified in accordance to compressive strength

| Classification | Compressive strength (MPa) |
|---|---|
| Low-strength concrete | $< 20$ |
| Moderate-strength concrete | $20 \sim 50$ |
| High-strength concrete | $50 \sim 150$ |
| Ultra high-strength concrete | $> 150$ |

depending on its specified composition. Normal-weight concretes are commonly used concretes in the construction of infrastructures and buildings. Heavyweight concrete is used to build some special structures, such as laboratories, hospital examination rooms, and nuclear plants, where radioactive protection is needed to minimize its influence on people's health.

The main component that makes a concrete unit weight (UW) difference is the aggregate. As discussed in Chapter 2, the four types of concrete differentiated by UW correspond to four different types of aggregates, i.e., ultra-lightweight aggregates for ultra-lightweight concretes, lightweight aggregates for lightweight concretes, normal-weight aggregates for normal-weight concretes while heavyweight aggregates for heavyweight concretes.

### 1.4.2 Classification in Accordance with Compressive Strength

According to its compressive strength, concrete can be classified into four categories, as listed in Table 1-2. Low-strength concrete is mainly used to construct mass concrete structures, subgrades of roads, and partitions. Moderate-strength concretes are the most commonly used concretes in buildings, bridges, and similar structures. High-strength concretes can be used to build tall building columns, bridge towers, and shear walls. Ultra-high-strength concretes have not yet been widely used in structural constructions. Only a few footbridges and some structural segments, such as girders, have been built using such concretes.

### 1.4.3 Classification in Accordance with Additives

According to the materials other than cement, aggregate, and water that are added into concrete mixtures as additives, concretes can be classified into different categories. Four examples are shown in Table 1-3. Fiber-reinforced concrete (FRC) is a type of concrete with fibers incorporated. Many different fibers have been used to produce fiber-reinforced concrete, including steel, glass, polymerics, and carbon. The purpose of incorporating fibers into concrete includes toughness enhancement, tension property improvement, shrinkage control, and decoration. Detailed information regarding FRC can be found in Chapter 6. Macro-defect-free (MDF) is a cement-based composite that incorporates a large amount of water-soluble polymer, produced in a twin-roll

**Table 1-3** Concrete classifications in accordance to additives

| Classification | Additives |
|---|---|
| MDF | Polymers |
| Fiber reinforced concrete | Different fibers |
| DSP concrete | Large amount silica fume |
| Polymer concrete | Polymers |

**Figure 1-15** 3D printing technology (Photo provided by Zhendi Wang)

mixing process. It was developed to enhance the tensile and flexural properties of concrete. Concrete that has been densified with small particles is called DSP concrete which has incorporated a large amount of silica fume, a mineral admixture with very small particles. DSP concrete has excellent abrasion resistance and is mainly used to produce machine tools and industrial molds. Three methods have been developed to incorporate polymers into concrete: (1) using the polymer as a binder; (2) impregnating normal Portland cement concrete members with the polymer; and (3) using the polymer as an admixture in normal Portland concrete. MDF, DSP, and polymer in concrete are discussed in detail in Chapter 6.

### 1.4.4 Classification in Accordance with Construction Methods

Apart from the common casting method, there are some special types of concrete made by different construction methods. Associated with these different construction methods, these concretes possess specified properties and thus require special design. Some include 3D printable concrete, sprayable concrete, and roller-compacting concrete.

3D concrete printing (3DCP) (Figure 1-15) is a new construction technology developed in recent years. It adopts the layer-by-layer deposition method to construct the desired structure as per computer-aided design. Compared with the common casting method, 3D concrete printing has various advantages, such as less waste generated (formwork-free characteristic), less labor required (automatic characteristic), and higher construction efficiency (automatic characteristic). In addition, it facilitates the construction of structures with complex geometry through customized design and automatic printing. 3DCP can largely reduce construction costs.

In the development of 3D printable concrete, the multi-level material design should be adopted, which considers requirements of the mixture design, rheology, the printing process, and the composite structure. A successful 3D printable concrete material cannot be "too stiff," otherwise it cannot be delivered through the pump and hose; meanwhile, it cannot be "too soft," otherwise it cannot support the weight of the printed structure. In other words, the fresh concrete needs to possess good pumpability for delivery and good buildability for deposition. Hence, the requirement for its fresh state mainly focuses on the careful tailoring of its flow characteristic. In other words, tailoring its rheological properties is the key focus. The details of 3D concrete printing are introduced in Chapter 3 of this book.

Spray concrete technology is a conventional construction method, which is also known as shotcrete. It refers to the pumping of sprayable concrete through a hose, and the concrete can be pneumatically projected at a high velocity onto the substrate to build the desired structure. The common practice of spraying is to construct structures on vertical and overhead surfaces, where it is difficult to apply casting in these applications. It is also applied to reinforce caves or tunnels as the support to surrounding rocks, and it is one of the key methods in the repair and retrofit of structures with defects. Compared to concrete that suits casting, the sprayable concrete has enhanced coherence to prevent pressure-induced bleeding in the pumping process. Meanwhile, it has enhanced adherence to guarantee the bond strength between the concrete-substrate interface. In addition, sprayable concrete is required to possess good pumpability for the delivery of the material.

Roller-compacting concrete is another conventional construction method that specifically is applicable to the construction of road pavements. Compared to the concrete that suits casting, roller-compacting concrete has enhanced anti-abrasion properties. Fresh roller-compacting concrete is very stiff and hard to flow. Hence, road rollers are applied to compact the fresh material for it to be uniformly distributed and leveled to the same height.

### 1.4.5 Classification in Accordance with Non-Structural Functionality

As a structural material, concrete has the primary function of load-bearing. Thus, mechanical properties, such as compressive strength, flexural strength, elastic modulus, are strongly emphasized during the development of concrete technology. Accordingly, new types of concrete, such as high-strength concrete, ultra-high-strength concrete, have been developed and utilized. Meanwhile, multi-functional concretes have been developed, in which, in some circumstances, the non-structural functionality attracts more attention.

**(a)** *Decoration function*: Due to being in close contact with people in their daily lives, plain, gray, and boring concrete structures need to be turned into beautiful decorative elements. Concretes for decoration purposes have been developed, such as architectural concrete/colored concrete/polished concrete/stained concrete/stamped concrete.

**(b)** *Abrasion-resistant function*: Abrasion-resistance is very important in dams, diversion channels, water channels, collecting pipes, floors, pavements, etc. A unique selection of the binder (i.e., high calcium aluminate ferrite phase), the stronger fine/coarse aggregates, and the particular mix proportion design ensure this functionality.

**(c)** *Shielding function*: The ability to shield from micro-waves, radiation, acoustic noises, heat, etc., is an essential feature governing the safety of people's daily environment. The unique designs of concrete density, pores, compositions, etc., have been explored to endow the shielding features of concrete structures.

With the development of nanotechnology, the multi-functionalities of concrete structures have been innovatively realized in recent years. By taking advantage of nanomaterials (such as nano$TiO_2$, nano$SiO_2$, nano$CaCO_3$, nanoclay, carbon nanotubes, graphene, etc.), novel functionalities, such as photocatalysis, self-cleaning, rheology modifying, and heat conduction, have been developed to endow specific features and improved performances of traditional concrete.

## 1.5 FACTORS INFLUENCING CONCRETE PROPERTIES

### 1.5.1 *w/c* Ratio (or *w/b* or *w/p* Ratio)

One very important factor dictating the properties of concrete is the water/cement (*w/c*) ratio. In contemporary concrete, the *w/c* is frequently replaced with the *w/b* (water/binder) or the *w/p* (water/powder), since Portland cement is not the only binding material in concrete. The *w/c* or *w/b* ratio is one of the most important factors influencing concrete properties, such as compressive strength, permeability, and diffusivity. A lower *w/c* ratio will generally lead to stronger and more durable concrete. The influence of *w/c* on the concrete compressive strength has been known since the early 1900s (Abrams, 1927), leading to Abrams's law:

$$f_c = \frac{A}{B^{1.5(w/c)}} \tag{1-7}$$

where $f_c$ is the compressive strength, $A$ is an empirical constant (usually 97 MPa or 14,000 psi), and $B$ is a constant that depends mostly on the cement properties (usually 4). It can be seen from the formula that the higher the *w/c* ratio, the lower the compressive strength. Another way to show the influence of the *w/c* ratio on the compressive strength of concrete can be written as

$$f_c = Af_{ce}\left(\frac{c}{w} - B\right) \tag{1-8}$$

where $f_c$ is the compressive strength, $A$ and $B$ are empirical constants that depend on the aggregate, and $f_{ce}$ is the compressive strength of a specified type of cement at 28 days. *c/w* is the reverse of *w/c*.

### 1.5.2 Cement Content

When water is added to a concrete mixture, cement paste will be formed. Cement paste has three functions in concrete: (1) binding; (2) coating; and (3) lubricating. Cement paste provides binding to individual aggregates, reinforcing bars, and fibers and glues them together to form a unique material. Cement paste also coats the surface of the aggregates and fibers during the fresh stage of concrete. The rest of the paste after coating can make the movement of the aggregates or fibers easier, rather like a lubrication agent. The cement content influences the concrete workability in the fresh stage, the heat release rate in the fast hydration stage, and the volume stabilities in the hardened stage. The range of the amount of cement content in mass concrete is usually 160–200 kg/m$^3$, in normal strength concrete, it is less than 400 kg/m$^3$, and in high strength concrete, it is 400–600 kg/m$^3$.

### 1.5.3 Aggregate

**(a)** *Maximum aggregate size*: The maximum coarse aggregate size mainly influences the cement paste requirement in the concrete. For the same volume of aggregates, those with a large aggregate size will lead to a small total surface area and a lower amount of cement paste coating. Hence, if the same amount of cement is used, concrete with a larger maximum aggregate size will have more cement paste left as a lubricant and the fluidity of concrete can be enhanced, as compared to concrete with a smaller maximum aggregate size. For normal-strength concrete, at the same *w/c* ratio and with the same cement content, the larger the maximum sizes, the better the workability; at the same workability, the larger the maximum sizes, the higher the strength. However, a larger aggregate size has some drawbacks. First, a larger aggregate size may make the concrete appear nonhomogeneous. Second, a larger aggregate size may lead to a large interface that can influence the concrete transport properties and mechanical properties.

Generally, the maximum size of coarse aggregate should be the largest that is economically available and consistent with the dimensions of the structure. In choosing the maximum aggregate size, the structural member size and spacing of reinforcing steel in a member have to be taken into consideration. In no circumstances should the maximum size exceed one-fifth of the narrowest dimension in the sizes of the forms, one-third of the depth of slabs, or three-quarters of the minimum clear spacing between reinforcing bars.

**(b)** *Aggregate grading*: Aggregate grading refers to the size distribution of the aggregates. The grading mainly influences the space-filling or particle packing. The classical idea of particle packing is based on the Apollonian concept, in which the smaller particles fit into the interstices left by the large particles. Well-defined grading with an ideal size distribution of aggregate will decrease the voids in the concrete and hence the cement content. As the price of the aggregate is usually only one-tenth that of cement, well-defined grading not only will lead to higher compressive strength and low permeability, but also is more economical at a lower cost.

**(c)** *Aggregate shape and texture*: The aggregate shape and texture can influence the workability, bonding, and compressive strength of concrete. At the same *w/c* ratio and with the same cement content, aggregates with an angular shape and rough surface texture result in lower workability, but lead to a better bond and better mechanical properties. On the other hand, aggregates with spherical shape and smooth surface texture result in higher workability, but lead to a lower bond and lower mechanical properties.

**(d)** *Sand/coarse aggregate ratio*: The sand/coarse aggregate ratio will influence the packing of the concrete. It also influences the workability of the concrete in the fresh stage. Increasing the sand to coarse aggregate ratio can lead to an increase in cohesiveness, but reduces the consistency. Of all the measures for improving the cohesiveness of concrete, increasing the sand/coarse aggregate ratio has been proven to be the most effective one.

**(e)** *Aggregate/cement ratio*: The aggregate/cement ratio has an effect on the concrete cost, workability, mechanical properties, and volume stability. Due to the price difference between the aggregate and cement, increasing the aggregate/cement ratio will decrease the cost of concrete. From a workability point of view, an increase of the aggregate to cement ratio results in a lower consistency because of less cement paste for lubrication. As for mechanical properties, increasing the aggregate/cement ratio can lead to higher stiffness and compressive strength if proper compaction can be guaranteed. Increasing the aggregate/cement ratio will definitely improve the concrete's dimension stability due to the reduction in shrinkage and creep.

### 1.5.4 Admixtures

Admixtures (chemical admixtures and mineral admixtures) are important and necessary components for contemporary concrete technology. The concrete properties, both in fresh and hardened states, can be modified or improved by admixtures. For instance, concrete workability can be affected by air-entraining agents, water reducers, and fly ash. Concrete strength can be improved by silica fume. More details regarding the effects of admixtures on concrete properties can be found in Chapter 2.

### 1.5.5 Mixing Procedures

Mixing procedures refer to the sequence of putting raw materials into a mixer and the mixing time required for each step. Mixing procedures directly influence the workability of fresh concrete and indirectly influence some mature properties of concrete.

Currently, ready-mixed concrete (RMC) is becoming popular. RMC can be manufactured in three ways:

**(a)** *Central-mixing*: The concrete is mixed thoroughly in a stationary mixer and then discharged into transporting equipment such as a truck agitator, a truck mixer operating at agitation speed, or a no-agitating truck.

**(b)** *Shrink-mixing*: The concrete is partially mixed in a stationary mixer and partially in a concrete truck mixer to complete mixing during its travel to the work site.

**(c)** *Truck-mixing*: The concrete is mixed completely in a concrete truck mixer. The mixing constituents are batched at a plant, loaded into the transporting truck, and mixed as the truck travels to the work site.

The mixing time depends on the power of a mixer, the slump requirement, and the properties of the raw materials. It can range from 30 seconds to a few minutes.

The following special mixing procedure can be used to obtain very good workability with a good coating on the coarse aggregate to prevent alkali-aggregate reaction.

**Step 1:** Coarse aggregate + 50% water + 50% cement: mixing for 30 sec to 1 min.
**Step 2:** Adding 50% cement + 25% water + superplasticizer + fine aggregate: mixing for 2 min.
**Step 3:** Adding 25% water: mixing for 3 min.

### 1.5.6 Curing

Curing is defined as the measures for taking care of fresh concrete right after casting. The main principle of curing is to keep favorable moist conditions under a suitable temperature range during the fast hydration process for concrete. It is a very important stage for the development of concrete strength and in controlling early volume changes. Fresh concrete requires considerable care, just like a baby. Careful curing will ensure that the concrete is hydrated properly, with a good microstructure, proper strength, and good volume stability. On the other hand, careless curing always leads to improper hydration with defects in the microstructure, insufficient strength, and unstable dimensions. One of the common phenomena resulting from careless curing of concrete is plastic shrinkage, which usually leads to an early-age cracking that provides a path for harmful ions and agents to get into the concrete body easily and cause durability problems. Curing is a simple measure to achieve a good quality of concrete. However, it is often ignored on construction sites.

Some methods could be helpful in curing:

**(a)** Moisten the subgrade and forms.
**(b)** Moisten the aggregates.
**(c)** Erect windbreaks and sunshades.
**(d)** Cool the aggregates and mixing water.

**(e)** Fog spray.
**(f)** Cover.
**(g)** High temperature (70–80°C) steam curing.
**(h)** Use shrinkage compensating admixtures.

Recently, a new technique called internal curing has been developed, which uses the saturated porous aggregate to form a reservoir inside the concrete and provide water for concrete curing internally. Details of the relevant curing methods and the effects on the properties of concrete are explained in Chapter 3.

## 1.6 APPROACHES TO STUDY CONCRETE

The scope of materials, including concrete, research, design and development can be explained by the Chinese word, 材料, which is pronounced as *tsai liao* and means material. The first character, 材(pronounced tsai), is composed of two parts, 木 (pronounced mu) and 才 (pronounced tsai). The first part, 木, means wood and is real, while the second part, 才, means properties or performance and is virtual. The two parts 木 and 才 represent the hardware and software of materials research, development and design. Similarly, the second character in the Chinese word for material, 料 (pronounced liao), is also composed of two parts, 米 (pronounced mi) and 斗 (pronounced dou). The first part 米 means rice and is real and the second part 斗 means container and is virtual. The two parts 米 and 斗 also represent the hardware and software of materials research, development and design. Materials research, design and development involve two aspects: hardware and software. The hardware includes material composition, microstructure, synthesis/processing. The software includes characterization, measurement, properties, and performance.

As a structural material, the fundamental approach in materials study also applies to concrete. About 25 years ago, a pyramid diagram was used to describe the philosophy in materials research, as shown in Figure 1-16. In this pyramid, the top is performance and the base is a triangle formed by three points: properties, microstructure, and processing. The philosophy behind this pyramid is that the processing, microstructure, and properties of material should be designed, developed, or investigated according to its performance requirement.

This concept was changed in 1999. The U.S. National Research Council has developed a new pyramid for materials research, as shown in Figure 1-17 (National Research Council, 1999). In this pyramid, the top is changed to end-use needs/constraints and the base is changed to a square with processing, properties, microstructure, and performance at each corner. The processing of concrete includes raw materials selection, mixing, placing, compacting, and curing. The properties of concrete include load-carrying capabilities, such as compressive strength, flexural strength, and fatigue

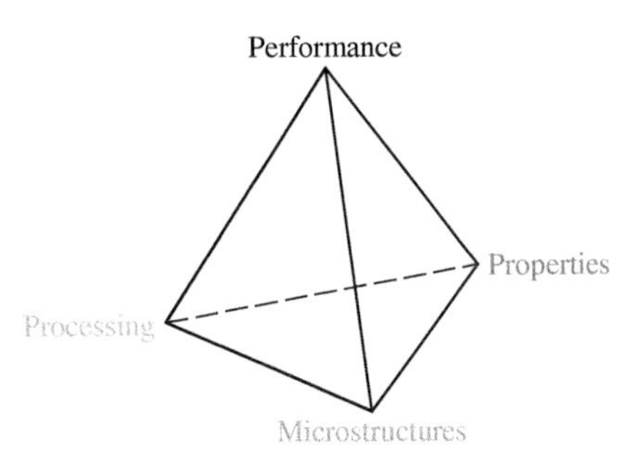


**Figure 1-16** Fundamental approach of materials research – 1

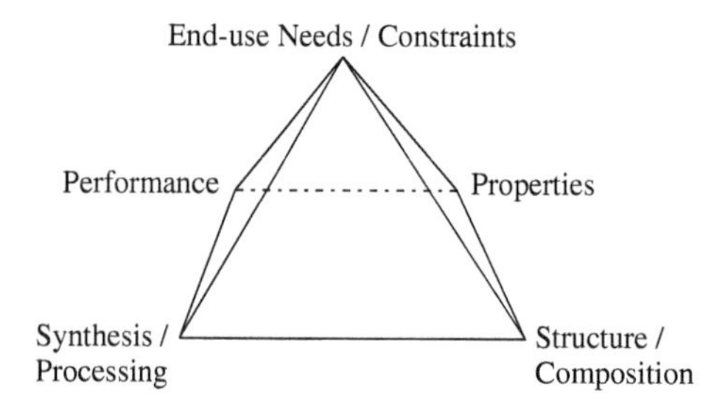


**Figure 1-17** Fundamental approach of materials research – 2. (Source: U.S. National Research Council, 1999)

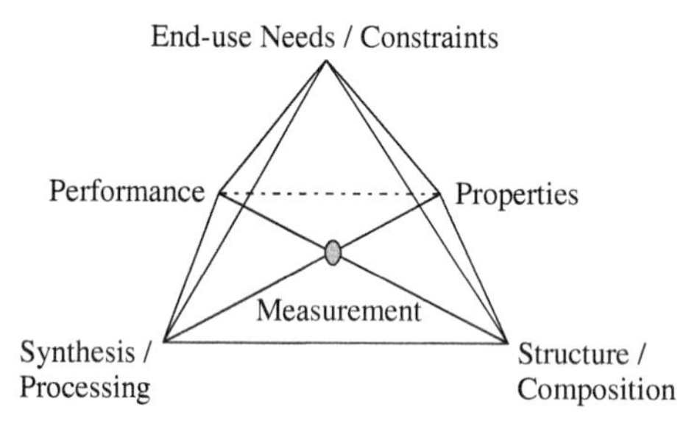


**Figure 1-18** Measurement is an essential part of materials science and engineering

strength, dimensional stability, such as shrinkage and creep, and stress-strain or load-deformation relationship. The structure of concrete consists of different phases with different amounts, sizes, and special arrangements. It covers the nanoscale, microscale, and millimeter-scale, a typical multi-phase composite. The performance of concrete includes safety, durability, and serviceability.

The philosophy behind this pyramid is that the microstructure, processing, properties, and performance of material have to be designed, investigated, or developed comprehensively to meet the requirements of the end-use. In other words, it is the end-use that governs the design, research, or development of a material, not the material itself. The design and construction of the Eddystone Lighthouse by John Smeaton form a good example of an end-use constraint. Since the structure to be built was a lighthouse, which must be able to withstand watery conditions, Smeaton designed the building materials according to the constraints of such end needs, in their composition, processing, properties, and performance, which turned out to be hydraulic cement. The end requirements of a tall building are very different from those of a hydraulic dam. Hence, the materials' design and development must be very different for a tall building and a hydraulic dam.

By adding measurement and characterization at the center of the base of the pyramid shown in Figure 1-17, the essential portion of materials science and engineering, the scope of materials design and development is complete, as demonstrated in Figure 1-18. With measurement and characterization, the material's structure/composition can be quantified, the properties can be specified, and synthesis/processing can be identified. Moreover, with measurement, the four corners of the base can be connected.

A close look at the historic development of concrete shows that concrete has been applied in practice for more than 150 years without a systematic scientific background. Most practices followed empirical formulae and observations. Attention was paid mainly to the properties of concrete, especially compressive strength. Very limited understanding of the material structure of concrete emerged. Concrete is a typical multiscale material, and its material structure covers the nanometer scale, the micrometer scale, and the millimeter scale. The concrete phases in the nanometer scale mainly constitute calcium silicate hydrate (C–S–H). It is believed that C–S–H contributes most to the binding strength, and understanding the nature of the C–S–H gel is key to revealing the behavior of concrete. The structure of C–S–H on the atomic level determines the nature of the mechanical properties, transport mechanism, and dimensional stability of hydrated cement paste. However, due to the limitations in experimental techniques and computer simulation in the past, studies on C–S–H structure are very limited. Nowadays, with the fast development of microstructure measurement technology and powerful computer simulation methods, it is possible to study and develop concrete technology more scientifically at the C–S–H level. The research activities aimed at understanding the nature of concrete hydrates at the nanometer-scale structure are developing very fast. A revolutionary breakthrough in concrete science is very likely. With such an understanding, it will be possible to design or develop concrete structures/compositions, properties, processing methods, and performance with solid knowledge to meet the need of every different end-use.

## DISCUSSION TOPICS

Why is concrete so popular?
What are the weaknesses of concrete?
What are the factors influencing concrete properties?
Give some examples of concrete applications.
Can you list a few topics for concrete research?
When you do a structural design, which failure mode should be applied?
How would you like to improve concrete workability (fluidity or cohesiveness)?
How can you enhance concrete compressive strength?
Which principles are you going to follow if you are involved in concrete research?

## REFERENCES

Abrams, D. A. (1927) "Water–cement ratio as a basis of concrete quality, " *ACI Journal* 23(2), pp. 452–457.
Aitcin, P. C., Lachemi, M., Adeline, R., and Richard, P. (1998) "The Sherbrooke reactive powder concrete footbridge, " *Structural Engineering International* 8(2), 140–144.
Caldarone, M. A. (2009) *High-strength concrete: a practical guide*, London: Taylor & Francis.
Collins, P. (2004) *Concrete: the vision of a new architecture*, Montreal: McGill-Queen's University Press.
Day, L., and McNeil, I. (1996) *Biographical dictionary of the history of technology*, New York: Routledge.
Encyclopaedia Britannica, Inc. (1991) *The new encyclopaedia Britannica*, 15th edition, Chicago, IL: Encyclopaedia Britannica.
Leonhardt, F. (1964), *Prestressed concrete: Design and construction*, Berlin: Ernst & John.
National Research Council (1999) *Materials science and engineering: forging stronger links to users*, Washington, DC: The National Academies Press.
Neville, A. M., and Brooks, J. J. (1993) *Concrete technology*, Harlow, England: Longman Scientific and Technical.
Ozama K., Mekawa, K., Kunishima, M., and Okamura, H. (1989) "Development of high performance concrete based on durability design of concrete structure," in *Proceedings of the second East-Asia and Pacific conference on structural engineering and construction (EASEC-2)*, 445–450.
Richard, P. and Cheyrezy, M. (1995) "Composition of reactive powder concretes," *Cement and Concrete Research* 25(7), 1501–1511.
Sutherland, J., Humm, D., and Chrimes, M. (2001) *Historic concrete: background to appraisal*, London: Thomas Telford.

CHAPTER 2

# MATERIALS FOR MAKING CONCRETE

Concrete is one of the most versatile and widely produced construction materials in the world (Penttala, 1997). Its worldwide annual production exceeds 12 billion metric tons, i.e., more than two metric tons of concrete was produced each year for every person on Earth in 2007. The global ever-increasing population, improving living standards, and economic development lead to an increasing demand for infrastructure development and hence concrete materials. As a composite material, concrete is composed of different graded aggregates or fillers embedded in a hardened matrix of cementitious material. The properties of major constituents of concrete mixtures, such as aggregates, cementitious materials, admixtures, and water, should be understood first to better learn the properties and performance of concrete.

## 2.1 AGGREGATES FOR CONCRETE

Aggregates constitute the skeleton of concrete. Approximately three-quarters of the volume of conventional concrete is filled by aggregate. It is inevitable that a constituent occupying such a large percentage of the mass should contribute important properties to both the fresh and hardened product. Aggregate is usually viewed as an inert dispersion in the cement paste. However, strictly speaking, aggregate is not truly inert because physical, thermal, and, sometimes, chemical properties can influence the performance of concrete (Neville and Brooks, 1990).

### 2.1.1 Effects of Aggregates

**(a)** *Aggregate in fresh and plastic concrete*: When concrete is freshly mixed, the aggregates are suspended in the cement–water–air bubble paste. The behavior of fresh concrete, such as fluidity, cohesiveness, and rheological behavior, is largely influenced by the amount, type, surface texture, and size gradation of the aggregate. The selection of aggregate has to meet the requirement of the end use, i.e., what type of structure is to be built.

**(b)** *Aggregate in hardened concrete*: Although there is little chemical reaction between the aggregate and cement paste, the aggregate contributes many qualities to the hardened concrete. In addition to reducing the cost, aggregate in concrete can reduce the shrinkage and creep of cement paste. Moreover, aggregates have a big influence on stiffness, unit weight, strength, thermal properties, bond, and wear resistance of concrete.

### 2.1.2 Classification of Aggregates

Aggregates can be divided into several categories according to different criteria, such as size, source, and unit weight.

**(a)** In accordance with size

*Coarse aggregate*: Aggregates predominately retained on a No. 4 (4.75-mm) sieve are classified as coarse aggregate. Generally, the size of coarse aggregate ranges from 5 to

**Figure 2-1** Different sizes of coarse aggregates

150 mm. For normal concrete used for structural members, such as beams and columns, the maximum size of coarse aggregate is about 25 mm. For mass concrete used for dams or deep foundations, the maximum size can be as large as 150 mm. Figure 2-1 shows some examples of coarse aggregates.

*Fine aggregate (sand)*: Aggregates passing through a No. 4 (4.75 mm) sieve and predominately retained on a No. 200 (75 μm) sieve are classified as fine aggregate. River sand and artificial sand are two major sources of fine aggregate. Usually, river sand is the most commonly used fine aggregate. However, in recent years, there has been a world shortage of the supply of river sand, so the applications of artificial sand in concrete become much more popular. In China, more than 60% of fine aggregates have been replaced by artificial sand. Usually, the finish of concrete with artificial sand or crushed rock fines is not as good as that with river sand. Figure 2-2 shows the profile of river sand.

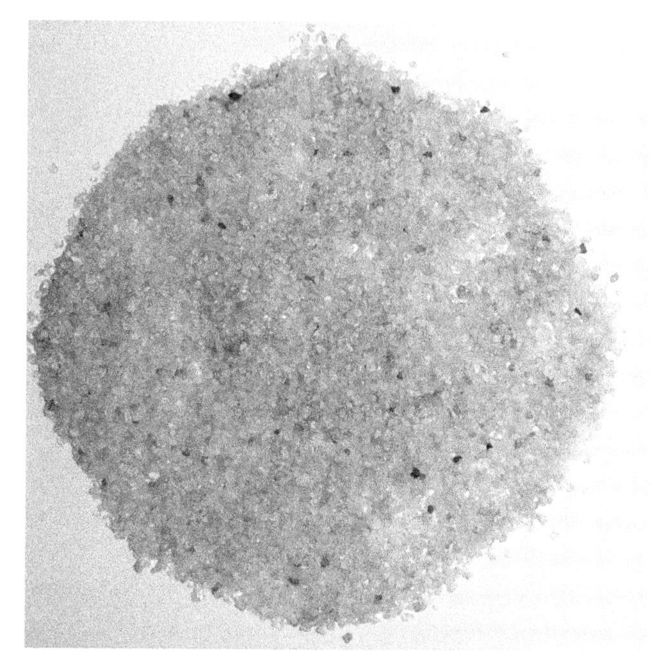

**Figure 2-2** Profile of sand

**(b)** In accordance with source

*Natural aggregates*: This kind of aggregate, such as sand and gravel, is taken from natural deposits without changing the nature during production.

*Manufactured (synthetic) aggregates* (see Figure 2-3): These kinds of aggregate are man-made materials, resulting from products or by-products of industry. Some examples are blast furnace slag and lightweight aggregate. In recent years, more and more artificial sands have been produced to meet the construction needs of infrastructure and buildings. Artificial sand will be discussed in a separate section later.

**(c)** In accordance with unit weight

*Ultra-lightweight aggregate*: The unit weight of such aggregates is less than 500 kg/m$^3$, including expanded perlite and foam plastic. The concrete made of ultra-lightweight aggregates has a bulk density from 800–1100 kg/m$^3$, depending on the volume fraction of aggregate. Such a concrete can be used only as nonstructural members, like partition walls.

*Lightweight aggregate*: The unit weight of such aggregates is between 500 and 1120 kg/m$^3$. Examples of lightweight aggregates include cinder, blast-furnace slag, volcanic pumice, and expanded clay. The concrete made of lightweight aggregate has a bulk density between 1200 and 1800 kg/m$^3$. Such concrete can be either a structural member or nonstructural member, depending on what type of aggregate is used.

(a) Foam plastic

(b) Expanded volcano rock

(c) Expanded clay

(d) Expanded fly ash

**Figure 2-3** Synthetic aggregates

*Normal-weight aggregate*: An aggregate with a unit weight of 1520–1680 kg/m$^3$ is classified as normal-weight aggregate. Sand, gravel, and crushed rock belong to this category and are the most widely used. Concrete made with this type of aggregate has a bulk density of 2300–2400 kg/m$^3$. It is the main concrete used to produce important structural members.

*Heavy-weight aggregate*: If the unit weight of aggregate is greater than 2100 kg/m$^3$, it is classified as heavy-weight aggregate. Materials used as heavy-weight aggregate are iron ore, crushed steel pieces, and magnesite limonite. The bulk density of the corresponding concrete is greater than 3200 kg/m$^3$ and can reach 4000 kg/m$^3$. This kind of concrete has special usage, such as radiation shields in nuclear power plants, hospitals, and laboratories. It can also be used as sound-shielding material.

### 2.1.3 Properties of Aggregates

#### 2.1.3.1 Moisture Conditions

The moisture condition defines the presence and amount of water in the pores and on the surface of the aggregate. There are four moisture conditions, as demonstrated in Figure 2-4.

**(a)** *Oven dry (OD)*: This condition is obtained by keeping the aggregate in an oven at a temperature of 110°C long enough to drive all water out from the internal pores and hence reach a constant weight.

**(b)** *Air dry (AD)*: This condition is obtained by keeping the aggregate at ambient temperature and ambient humidity. Under such a condition, pores inside the aggregate are partly filled with water. When the aggregate is under either the OD or AD condition, it will absorb water during the concrete mixing process until the internal pores are fully filled with water.

**(c)** *Saturated surface dry (SSD)*: In this situation, the pores of the aggregate are fully filled with water and the surface is dry. This condition can be obtained by immersing the coarse aggregate in water for 24 h followed by drying of the surface with a wet cloth. When the aggregate is under the SSD condition, it will neither absorb water nor give out water during the mixing process. Hence, it is a balanced condition and is used as the standard index for concrete mix design.

**(d)** *Wet (W)*: The pores of the aggregate are fully filled with water and the surface of the aggregate has a film of water. When aggregate is in a wet condition, it will give out water to the concrete mix during the mixing process. Since sand is usually obtained from a river, it is usually in a wet condition.

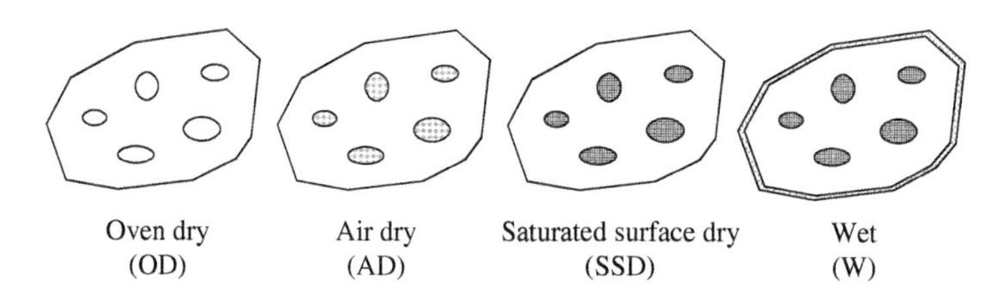


**Figure 2-4** Moisture condition of aggregates

#### 2.1.3.2 Moisture Content (MC) Calculations

The moisture content of aggregates can be calculated with respect to either the OD or SSD condition.

**(a)** For the oven dry condition

$$\text{MC(OD)} = \frac{W_{\text{stock}} - W_{\text{OD}}}{W_{\text{OD}}} \times 100\% \tag{2-1}$$

where $W_{\text{stock}}$ is the weight of aggregate in the stock condition, and $W_{\text{OD}}$ the weight of oven-dried aggregates. It can be seen that $\text{MC}_{\text{OD}}$ is a nonnegative value.

**(b)** For the saturated surface dry condition

$$\text{MC(SSD)} = \frac{W_{\text{stock}} - W_{\text{SSD}}}{W_{\text{SSD}}} \times 100\% \tag{2-2}$$

where $W_{\text{SSD}}$ is the weight of aggregate in the SSD condition. As $W_{\text{stock}}$ can be greater than, equal to, or less than $W_{\text{SSD}}$, $\text{MC}_{\text{SSD}}$ can be greater than, equal to, or less than zero.

**(c)** Absorption capacity

$$\text{AC} = \frac{W_{\text{SSD}} - W_{\text{OD}}}{W_{\text{OD}}} \times 100\% \tag{2-3}$$

The absorption capability of an aggregate is defined as the total amount of water that can be taken by an aggregate from the OD to the SSD condition.

It should be noted that in designing a concrete mix, the moisture content usually uses the SSD condition as a reference, because it is an equilibrium condition at which the aggregate will neither absorb water nor give up water to the paste. Thus, if the $\text{MC}_{\text{SSD}}$ value is positive, it means that the aggregate is under a surface moisture condition. If it is negative, it means that the pores in the aggregate are only partly filled with water. The amount of water used for mixing concrete has to be adjusted according to the $\text{MC}_{\text{SSD}}$ value in order to keep a correct *w/c* ratio, especially for a high-strength concrete in which a small *w/c* ratio is used, and the amount of adjusted water involved in MC can easily be a large portion of the total amount of water in the mixture.

#### 2.1.3.3 Density and Specific Gravity

Since aggregates are porous materials, even a single piece of aggregate contains both solid material volume and pores volume. Hence, two types of aggregate density are defined.

Density ($D$) is defined as the weight per unit volume of solid material only, excluding the pores volume inside a single aggregate:

$$D = \frac{\text{weight}}{V_{\text{solid}}} \tag{2-4}$$

Bulk density (BD) is defined as the weight per unit volume of both solid material and the pores volume inside a single aggregate:

$$\text{BD} = \frac{\text{weight}}{V_{\text{solid}} + V_{\text{pores}}} \tag{2-5}$$

where BD can be either $\text{BD}_{\text{SSD}}$ or $\text{BD}_{\text{AD}}$ according to the moisture condition of the aggregate when it is weighed.

Specific gravity (SG) is a ratio of density or bulk density of aggregate to density of water. Or SG is the mass of a given substance per unit mass of an equal volume of water. Depending on the definition of volume, the specific gravity can be divided into absolute specific gravity (ASG) and bulk specific gravity (BSG).

$$\text{ASG} = \frac{\dfrac{\text{weight of aggregate}}{V_{\text{solid}}}}{\text{density of water}} = \frac{D}{\rho_{\text{w}}} \tag{2-6}$$

$$\text{and} \quad \text{BSG} = \frac{\dfrac{\text{weight of aggregate}}{V_{\text{solid}} + V_{\text{pores}}}}{\text{density of water}} = \frac{\text{BD}}{\rho_{\text{w}}} \tag{2-7}$$

In practice, the BSG value is the realistic one to use since the effective volume that an aggregate occupies in concrete includes its internal pores. The BSG of most rocks is in the range of 2.5–2.8. Similar to BD, BSG can be either $\text{BSG}_{\text{SSD}}$ or $\text{BSG}_{\text{AD}}$ according to the moisture condition of the aggregate. The BSG can be determined using the so-called displacement method. In this method, Archimedes' principle is used. The weight of aggregate is first measured in air, e.g., under the SSD condition, and is denoted as $W_{\text{SSD}}$ in air. Then, the weight of the sample is measured in water, denoted as $W_{\text{SSD}}$ in water. Thus, we have

$$\text{BSG}_{\text{SSD}} = \frac{W_{\text{SSD in air}}}{W_{\text{displacement}}} = \frac{W_{\text{SSD in air}}}{W_{\text{SSD in air}} - W_{\text{SSD in water}}} \tag{2-8}$$

where $W_{\text{displacement}}$ is the weight of water displaced by the aggregates.

#### 2.1.3.4 Unit Weight (UW)

The unit weight is defined as the weight per unit bulk volume for bulk aggregates. In addition to the pores inside each single aggregate, the bulk volume also includes the space among the particles. According to the weight measured at different conditions, the unit weight can be divided into UW(SSD) and UW(OD):

$$\text{UW(SSD)} = \frac{W_{\text{SSD}}}{V_{\text{solid}} + V_{\text{pores}} + V_{\text{spacing}}} \tag{2-9}$$

$$\text{and} \quad \text{UW(OD)} = \frac{W_{\text{OD}}}{V_{\text{solid}} + V_{\text{pores}} + V_{\text{spacing}}} \tag{2-10}$$

The percentage of spacing (voids) among the aggregates can be calculated as

$$\text{Spacing (void)} = \frac{\text{BD} - \text{UW}}{\text{BD}} \times 100\% \tag{2-11}$$

#### 2.1.3.5 Measurement of Moisture Content

Once the $BSG_{SSD}$ is obtained for a type of aggregate, the moisture content of the aggregate under different moisture conditions can be conveniently determined using the following equation:

$$MC(SSD) = \frac{W_{stock} - \dfrac{W_{water} \times BSG_{SSD}}{BSG_{SSD} - 1}}{\dfrac{W_{water} \times BSG_{SSD}}{BSG_{SSD} - 1}} \tag{2-12}$$

where $W_{stock}$ is the weight of the sample under the stockpile condition, and $W_{water}$ is the short form of $W_{SSD}$ in water.

If AC is known for the aggregate, MC(SSD) can also be calculated using the absorption capability of aggregates as

$$MC(SSD) = \frac{W_{stock} - W_{OD}(1 + AC)}{W_{OD}(1 + AC)} \tag{2-13}$$

### 2.1.4 Grading Aggregates

#### 2.1.4.1 Grading and Size Distribution

The particle size distribution of aggregates is called grading. Grading determines the paste requirement for a workable concrete since the amount of voids among aggregate particles requires the same amount of cement paste to fill out in the concrete mixture. To obtain a grading curve for an aggregate, sieve analysis has to be conducted. The commonly used sieve designation is listed in Table 2-1.

As shown in Figure 2-5, five size distributions are generally recognized: dense, gap graded, well-graded, uniform graded, and open graded. The dense and well-graded types are essentially the wide size ranges with smooth distribution. They are the desired grading for making concrete. The dense graded is for coarse aggregate and well-graded for fine aggregate. Gap grading is a kind of grading that lacks one or more intermediate size; hence, a nearly flat horizontal region appears in the grading curve. For uniform grading, only a few sizes dominate the bulk materials, and the grading curve falls almost vertically at the dominating size. Open grading is defined as being under compact conditions, the voids among the aggregate are still relatively large. In open

**Table 2-1** Commonly used sieve designation and the corresponding opening size

| Sieve designation | Nominal size of sieve opening |
|---|---|
| 3 in. | 75 mm |
| 1.5 in. | 37.5 mm |
| 3/4 in. | 19 mm |
| 3/8 in. | 9.5 mm |
| No. 4 | 4.75 mm |
| No. 8 | 2.36 mm |
| No. 16 | 1.18 mm |
| No. 30 | 600 μm |
| No. 50 | 300 μm |
| No. 100 | 150 μm |
| No. 200 | 75 μm |

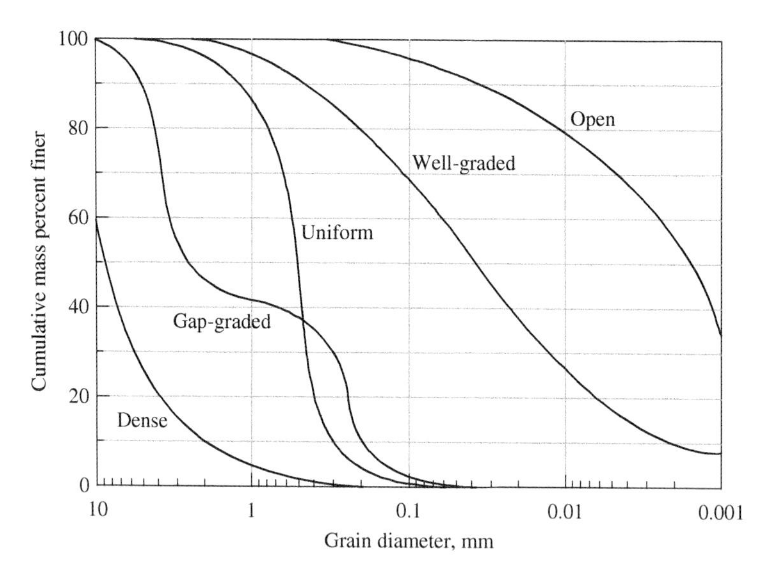


**Figure 2-5** Five types of aggregate gradation

grading, usually the smaller size of aggregate dominates the bulk and can easily be disturbed by a small cavity. Open-grade material is not suitable to be used for subgrade construction of a road.

### 2.1.4.2 Fineness Modulus

To characterize the overall *coarseness* or *fineness* of an aggregate, the concept of a fineness modulus is developed. The fineness modulus is defined as

$$\text{fineness modulus} = \frac{\sum \text{(cumulative retained percentage)}}{100} \tag{2-14}$$

It can be seen from the formula that calculation of the fineness modulus requires that the sum of the cumulative percentages retained on a definitely specified set of sieves be determined, and the result divided by 100. The sieves specified to be used in determining the fineness modulus are No. 100, No. 50, No. 30, No. 16, No. 8, No. 4, 3/8", 3/4", 1.5", 3", and 6". It can be seen that the size of the opening in the above sieves has a common factor of 0.5 in any two adjacent ones. They are called full-size sieves. Any sieve size between full-size sieves is called half-size.

The fineness modulus for fine aggregate should lie between 2.3 and 3.1. A small number indicates a fine grading, whereas a large number indicates a coarse material. The fineness modulus can be used to check the constancy of grading when relatively small changes are expected, but it should not be used to compare the gradings of aggregates from two different sources. The fineness modulus of fine aggregates is required for the mix proportion since sand gradation has the largest effect on workability. A fine sand (low fineness modulus) has much higher paste requirements for good workability.

### 2.1.4.3 Fineness Modulus for Blending of Aggregate

Blending of aggregates is undertaken for a variety of purposes, for instance, to remedy deficiencies in grading. The desired value of the fineness modulus can be calculated if the characteristics of the component aggregate are known. If two aggregates, designated as A and B, are mixed together, having fineness modulus of $FM_A$ and $FM_B$, respectively, the resultant blend will have the following fineness modulus:

$$FM_{Blend} = FM_A \times \frac{P_A}{100} + FM_B \times \frac{P_B}{100} \tag{2-15}$$

where $P_A$ and $P_B$ are the percentages, by weight, of aggregate A and B in the blend.

#### 2.1.4.4 Maximum Granular Size

Increasing the maximum granular size usually leads to an increase of the compressive strength of concrete. The highest compressive strength was exhibited by the concrete mixture having a maximum granular size of ~20 mm compared to the other mixes (Meddah et al., 2010). Several studies have mentioned that the maximum granular sizes are not the only predominant factor regarding the compressive strength of concrete but also the grain size distribution of the granular system has a great impact on the compressive strength development (Donza et al., 2002).

### 2.1.5 Shape and Texture of Aggregates

#### 2.1.5.1 Shape of Aggregates

The aggregate shape affects the workability of concrete due to the differences in surface area caused by different shapes. Sufficient paste is required to coat the aggregate to provide lubrication. The typical shapes of aggregates are shown in Figure 2-6. Among these, spherical, cubical, and irregular shapes are good for applications in concrete because they can benefit the strength. Flat, needle-shaped, and prismatic aggregates are weak in load-carrying ability and are easily broken. Besides, the surface-to-volume ratio of a spherical aggregate is the smallest.

#### 2.1.5.2 Texture of Aggregates

The surface texture of aggregates can be classified into six groups: glassy, smooth, granular, rough, crystalline, and honeycombed. The surface texture of aggregates has a significant influence on the fluidity of fresh concrete and the bond between aggregate and cement paste of hardened concrete. According to experimental statistics, the relative effects of the shape and surface texture of aggregates on concrete strength are summarized in Table 2-2 (Waddall and Dobrowolski, 1993).

### 2.1.6 Artificial Sand

Amounts of good quality natural sand have continued to decrease in the past 10–20 years. Environmental concerns are also being raised against the uncontrolled extraction of natural sand. As a result, artificial sand has to be used as a replacement.

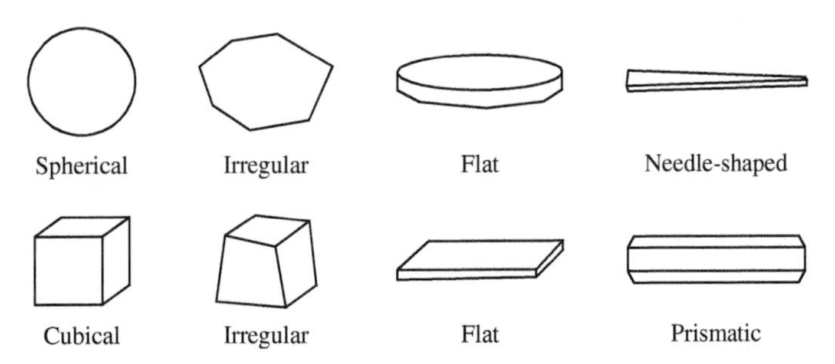


**Figure 2-6** Different basic shapes of aggregates

**Table 2-2** Effects of aggregate shape and surface texture on concrete strength

| Affected strength | Relative effect (%) of | |
|---|---|---|
| | Shape | Surface texture |
| Compressive | 22 | 44 |
| Flexural | 31 | 26 |

Artificial sand, also called manufactured sand or mechanical sand, refers to crushed rocks, mine tailings, or industrial waste granules with a particle size of less than 4.75 mm. It is processed by mechanical crushing, grinding, and sieving (Adinkrah-Appiah et al., 2016). A standard artificial sand making process is shown in Figure 2-7a.

**Figure 2-7** Artificial sand (a) sand-making process; (b) production line of artificial sands in western China

The particle size of artificial sand is more evenly distributed than natural sand and the surface area of artificial sand has a wider range than natural sand (Danielsen and Ørbog, 2000). Generally, artificial sands contain high fines content. The fines are composed of rock dust with a particle size less than 75 μm. Rather than silts and clay impurity in natural sand, the rock dusts have the same chemical and mineral composition as the parent rocks. Hence, the fine rock dusts have a micro-aggregate effect and a proper amount (5–10%) of dust incorporation can improve the packing density in making concrete. Nevertheless, an excessive amount of fines can significantly increase the water demand and weaken the workability of fresh concrete. Additionally, the crushing process tends to produce angular, sharp-edged particles. Rough angular particles yield a granular packing of lower density, lower small-strain stiffness, and higher critical state friction angle when compared with more rounded natural sands. The shape and texture of crushed sand particles could lead to improvements in the strength of concrete due to better interlocking between particles.

In recent years, the manufacture of artificial sands has developed rapidly in China. Many large-scale production lines have been constructed in new building material companies. Figure 2-7b shows a production line manufacturing over 10 million tons of sand per year in the western region of China. Also, artificial sands have been used in the construction of many hydropower systems, such as the Yellow River Xiaolangdi Project, the Three Gorges Project, and other cross-sea bridge projects in China. These projects usually took raw materials for artificial sand production locally to avoid environmental problems and high transportation expenses.

### 2.1.7 Sea Sand

As the demand for concrete increased rapidly, the consumption of the primary materials of concrete, particularly river sand, has raised serious environmental problems. The lack of river sand led many countries to use sea sand to replace river sand as an alternative of arterial sand, especially in coastal regions. Nowadays, many infrastructure projects have used sea sand as a raw material in construction, such as the city expansion in Singapore, the airport in Hong Kong, and the vast reclamation projects in Dubai and Saudi Arabia (Xiao et al., 2017).

Sea sand has various advantages such as:

- **(a)** It is rounded or cubical like river sand.
- **(b)** Being a natural deposit, the grading is generally good and consistent.
- **(c)** It contains no organic contaminant or silt.
- **(d)** It is abundantly available.
- **(e)** It can be mined at a low cost.

The main reason for the rejection of sea sand in concrete construction is the presence of sea shells and the high chloride content in sea sand. Sea sand should be appropriately treated before being used in concrete to avoid corrosion problems of steel in reinforced concrete. The process of reducing the amount of salt (the chloride ions) in sea sand is usually called desalting. For this effect, the rain or other freshwater is used to wash out the chloride content of the sea sand. However, the traditional desalting method can only wash out the free chloride ions on the sand particle surface. The physically or chemically bonded chloride ions in the sea sand particles cannot be removed.

In addition, the desalting process usually leads to an extra construction cost. The freshwater for washing sea sand is also in short supply. Thus, the direct use of sea sand without the desalting process in concrete production has been considered in many projects with other measures to prevent the steel corrosion. The existing research and the application results show that the effect of sea sand on the workability of concrete is minimal. The use of sea sand is likely to accelerate the strength development of concrete in the early hydration ages due to the high salt content. The concrete made with sea sand usually shows a higher 7-day compressive strength and a similar long-term compressive strength compared to the concrete made without sea sand. The corrosion problem of reinforcing steel can be completely eliminated by using fiber reinforced plastic as a reinforcement instead of steel.

### 2.1.8 Recycled Aggregates

Recycled aggregates (RA) are developed from the reprocessing of materials that have been originally used in construction, including construction waste and demolished masonry and concrete structures. A procedure of reprocessing is consisted of pretreatment, crushing, sieving, and separation. Post-treatment is usually required for the aggregates to ensure they meet legal requirements (Rao et al., 2007). Recycled aggregates and natural aggregates are shown in Figure 2-8. The particle size distribution, surface texture, and particle shape of RA are dependent largely on the parent concrete composition, the recycling technique, and the number of crushing cycles. Multi-stage crushing of concrete rubble results in RA with a larger quantity of finer particles, compared to RA produced with a single stage crushing process. Regarding the surface texture, the coarse recycled concrete aggregates are as rough as the crushed natural aggregates (Nedeljković et al., 2021). The RA's physicochemical characteristics depend on the original design and utilization history of the concrete as well as the recycling technique and storage conditions. These physicochemical characteristics have an influence on the performance of the RA concrete mix in the fresh stage as well as in the hardened stage.

Recycled aggregates (RA) derived from construction and demolished waste are normally of inferior quality when compared to virgin aggregates due primarily to the presence of old cement mortar coated on virgin aggregate. It made the RA multiphase and multi-scale in nature. At the macroscale, the RA may be considered a two-phase material: virgin aggregate core and surface glued cement mortar. The old cement mortar usually has a porous structure, which leads to high water absorption, shrinkage and creep for concrete made with RA instead of natural aggregates. At the microscale, there is a third phase in the RA, which is composed of the interfaces between the virgin aggregates and the coated cement mortar. When RA is used to produce a concrete, a new interface will be formed between the RA and new mortar. These interfaces influence the properties of the concrete, such as the range of elastic behavior, water absorption capability, and mechanical properties.

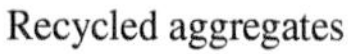
Recycled aggregates

Natural aggregates

**Figure 2-8** Recycled aggregates and natural aggregates

It should be emphasized that there exists a noticeable difference between RA and natural aggregate in the water adsorption (WA) ability. The reported WA values for RA vary between 4.28 and 13.1%, with an average of 8.4%, while the reported WA values for natural fine aggregate vary with an average of only 1.1%. This can be attributed to the complementary effect of the high content of open pores and the rougher surface texture of the RA particles. In addition, the RA has lower density than natural sand (on average, RA: 2295 kg/m$^3$; NA: 2637 kg/m$^3$) (Le and Bui, 2020). The high water absorption capacity of RA directly affects the effective water-to-cement ratio of paste in cement-based materials, giving poor fresh state consistency. The general mechanical properties of the concrete made of recycled aggregates are lower than that of natural aggregates. The compressive strength of concrete made with RA decreases with the increase of the replacement ratio of RA to natural aggregate. When the replacement ratio is less than 30%, the strength reduction is not obvious. When the replacement ratio reaches 100%, a 10–20% compressive strength decrease can be observed in RA concrete according to different researchers' results (Xiao et al., 2012).

Recently, research on how to improve the properties of RA has gained increasing interest. Several methods have been reported in the literature along these lines, mainly focused on reducing the water absorption through residual mortar removal or surface treatments by chemicals. The means to remove the mortar include the use of thermal expansion, mechanical grinding, acid dissolution and ultrasonic cleansing (Katz, 2004; Akbarnezhad et al., 2011). The surface treatment methods include the use of lime or silica fume, the addition of a saline type of chemicals and polymers, and a two-stage mixing method with the aim of improving the interfacial bond between the new cement matrix and the attached mortar on the RA. However, the above-mentioned methods are either energy-intensive or have cumbersome processing procedures, making them difficult to implement in practice.

Considering that a significant amount of old cement mortar adheres to the surface of virgin aggregate and the major hydration products of the paste are calcium hydroxide and calcium silicate hydrates, $CO_2$ treatment of the RA would cause the carbonation of the calcium hydroxide and C–S–H in the old cement paste according to the following reactions:

$$Ca(OH)_2 + CO_2 = CaCO_3 + H_2O \tag{2-16}$$

$$\text{C-S-H} + CO_2 = CaCO_3 + SiO_2.nH_2O \tag{2-17}$$

It is well known that the above reactions will result in increases in the solid volume and density, and improvement in the strength of the cement paste. Thus, it is anticipated that $CO_2$ treatment on RA will improve its properties and the properties of concrete prepared with RA (Kou et al., 2004). The other important side benefit is that the carbonation process could permanently fix (sequestrate) a significant amount of $CO_2$ emitted from other industries. It is anticipated that the technique for pre-treating RA with $CO_2$ will not only improve the properties of RA, but also will provide an economic means of capturing and storing $CO_2$.

The application of RA has many benefits, including alleviating the shortage of high-quality raw materials, such as natural rock, gravel and river sand, eliminating construction waste, saving disposal in landfills and reducing environmental maintenance costs. However, currently, no guidelines or regulations are available for optimized application of RA in making high quality concrete with the same or reduced cement contents as the normal concrete. Therefore, further research is needed to enhance the properties and reduce the variability of RAs for large-scale introduction and optimal use of RAs in new concrete production and construction.

## 2.2 CEMENTITIOUS BINDERS

### 2.2.1 Classification of Binders

Based on the composition, the binder can be classified as either organic or inorganic. An organic binder can easily be burned and thus cannot stand fire. Polymer and asphalt are two commonly used organic binders. Polymers consist of random chains of hydrocarbons and can be classified into thermoplastics, thermosets, and elastomers (or rubbers). Carbon atoms form the skeleton of the polymer chain. Along each chain, there are typically 1000–100,000 carbon atoms, held together by covalent bonds. The polymer chain is therefore very stiff and strong. The overall properties of a polymer, however, are governed by the interaction of individual polymer chains with one another. In thermoplastics, the chains interact with one another through weak van der Waal forces (Figure 2-9). In other words, while there is strong bonding along the polymer chain, making it very difficult to deform, there is very weak bonding between the chains, allowing easy relative movement of one chain from the other. The stiffness values of thermoplastics are therefore very low, and range from 0.15 to 3.5 GPa at room temperature (see examples in Table 2-3). In thermosets, the individual chains interact through van der Waal force as well as occasional cross-links. The cross-links are also hydrocarbon chains whose ends are bonded to the main polymer chains. Due to their presence, the stiffness of thermosets is higher than that of thermoplastics. At room temperature, it ranges from 1.3–8 GPa. Elastomers or rubbers are thermosets with a small number of cross-links. Also, they have very low glass transition temperatures, which means that the van

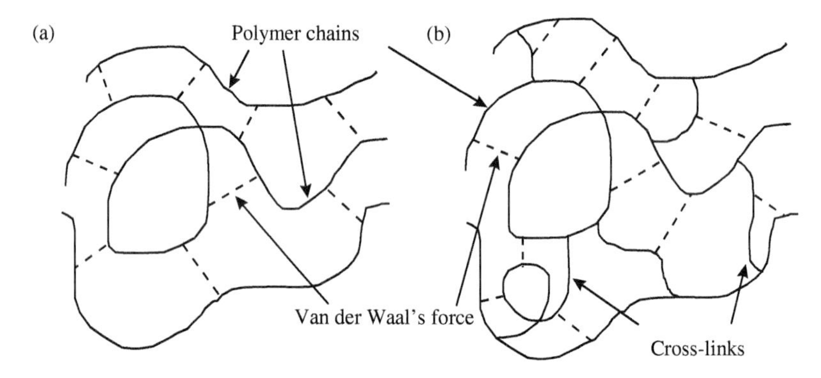


**Figure 2-9** Difference between (a) thermoplastics and (b) thermosets

**Table 2-3** Typical values of Young's modulus ($E$ at room temperature), tensile strength, and glass transition temperature ($T_g$) for various polymers

| Category | Material | $E$ (GPa) | $\sigma_T$ (MPa) | $T_g$ (K) |
|---|---|---|---|---|
| Thermoplastic | Polyethylene (PE), low density | 0.15–0.24 | n/a | 270 |
| | Polyethylene, high density | 0.55–1.0 | 20–37 | 300 |
| | Polypropylene (PP) | 1.2–1.7 | 50–70 | 253 |
| | Polyvinyl chloride (PVC) | 2.4–3.0 | 40–60 | 350 |
| Thermosets | Polyesters | 1.3–4.5 | 45–85 | 340 |
| | Epoxies | 2.1–5.5 | 40–85 | 380 |
| Elastomers | Polyisoprene | 0.002–0.1 | ~10 | 220 |
| | Polybutadiene | 0.004–0.1 | n/a | 171 |

der Waals force has disappeared at room temperature. Therefore, rubbers have very low stiffness values within the range of 0.002 to 0.1 GPa.

In concrete technology, the most widely used polymers are epoxy and latex. Epoxy is composed of resin and hardener and can be used to bind aggregate together or to repair cracks. Latex is mainly used to modify the concrete properties for repair.

Asphalt cement is mainly obtained from the distillation of crude oil. With very high molecular weight, asphalt is generally hard and relatively solid in its original form. Thus, it is easily distributed. To use it in practice, asphalt needs to be softened by heating. To reduce the need for heating, asphalt can be modified by the addition of volatile components or emulsifying agents. This will produce liquid asphalts (or cutbacks) and asphalt emulsions.

Depending on the added component, there are three different types of liquid asphalts: (1) rapid-curing (RC) cutback, a mixture of asphalt and gasoline; (2) medium-curing (MC), a mixture of asphalt and kerosene; and (3) slow-curing (SC), a mixture of asphalt and diesel. Curing here refers to the hardening of the asphalt due to evaporation of the volatile component. When gasoline is used, evaporation occurs at the fastest rate and the curing is rapid. Also, in RC cutbacks, most of the volatile component vaporizes while in MC and SC, some stays with the asphalt. As a result, RC cutbacks produce a harder material, which is appropriate for hot weather. MC cutbacks give a softer material, which is less brittle under cold weather. SC cutbacks, which would produce too soft a material, are used only for dust binding. Asphalt can be used as binder to glue the aggregate together to form asphalt concrete for road construction. In asphalt concrete, the content of asphalt is only 4–6%.

Inorganic binders are usually made of different natural minerals. The inorganic binders can be further classified into nonhydaulic cement and hydaulic cement. However, nonhydraulic here does not mean that it does not need water. In fact, all inorganic binders need water for mixing and reacting to form bonds. Nonhydaulic cement also needs water for mixing. Nonhydraulic means only that such cement cannot harden and thus gain strength in water. Typical examples of nonhydraulic cement are gypsum and lime. They have been used since 6000 BC, as mentioned in Chapter 1. Gypsum is a soft mineral composed of calcium sulfate dihydrate, called two-water gypsum. Below a temperature of 130°C, two-water gypsum can change to half-water gypsum and release some water:

$$2\mathrm{CaSO_4 \cdot 2H_2}O \xrightarrow{(130°C)} 2\mathrm{CaSO_4} \cdot \frac{1}{2}H_2O + 3\mathrm{H_2}O \tag{2-18}$$

When half-water gypsum is mixed with water, it can return to two-water gypsum and form bonds. Lime is the product of calcination of limestone under 1000°C, and consists of the oxides of calcium:

$$\mathrm{CaCO_3} \xrightarrow{(1{,}000°\mathrm{C})} \mathrm{CaO + CO_2} \tag{2-19}$$

When CaO is mixed with water again, the following reactions occur:

$$\mathrm{CaO + CO_2 + H_2O \rightarrow Ca(OH)_2 + CO_2 \rightarrow CaCO_3 + H_2O} \tag{2-20}$$

It can be seen that lime returns to limestone and forms bonds. Differing from nonhydraulic cement, hydraulic cement can harden and gain strength in water. The main difference in composition between the two types of inorganic cements is that the hydraulic cement contains some amounts of clayey impurities (silicate composition). Examples of hydraulic cement include hydraulic lime, pozzolan cement, and Portland cement. Hydraulic lime is composed of lime and clayey impurities. The pozzolan cement contains lime and volcanic rock powders. The name pozzolan originated from the Romans, who used some volcanic tuff from Pozzuoli village, near

Mt. Vesuvius. Since the history of the development of hydraulic cement has been discussed in Chapter 1, only the important events are briefly mentioned here.

In 1756, John Smeaton (the first person to style himself a civil engineer) was commissioned to rebuild the Eddystone Lighthouse off the coast of Cornwall, England). Recognizing that the normal lime mortars would not stand the action of water, Smeaton carried out an extensive series of experiments with different limes and pozzolans. He found that the high proportion of clayey materials could increase the water-resistance properties. He was the first person to control the formation of hydraulic lime. He used a mortar prepared from a hydraulic lime mixed with pozzolan from Italy to build the Lighthouse, which was to last 126 years.

Portland cement (PC) concrete is the most popular and widely used building material, due to the availability of the basic raw materials all over the world, and its ease of use in preparing and fabricating all sorts of shapes. The applications of concrete in the realms of infrastructure, habitation, and transportation have greatly promoted the development of civilization, economic progress, stability, and quality of life. Nowadays, with the occurrence of high-performance concrete (HPC), the durability and strength of concrete have been improved greatly. However, due to the restriction of the manufacturing process and the raw materials, some inherent disadvantages of Portland cement are still difficult to overcome. There are two major drawbacks with respect to sustainability: (1) about 1.5 tons of raw materials is needed in the production of every ton of PC, while, at the same time, about 1 ton of carbon dioxide ($CO_2$) is released into the environment during the production. The world's cement production has increased from 1.4 billion tons in 1995 to almost 3 billion tons by the year 2009. Therefore, the production of PC is an extremely resource- and energy-intensive process; and (2) concrete made of PC deteriorates when exposed to harsh environments, under either normal or severe conditions. Cracking and corrosion have significant influence on service behavior, design life, and safety.

To overcome these problems, other different cementitious materials have been developed recently. Two of them to be discussed in this chapter are geopolymer and magnesium phosphate cement (MPC). Compared with Portland cement, the properties of these two cements have some advantages to the sustainable development of modern society.

### 2.2.2 Portland Cement

The development of Portland cement can be summarized as follows:

| | | | |
|---|---|---|---|
| 1796 | James Parker | England | Patent on a natural hydraulic cement |
| 1813 | Vicat | France | Artificial hydraulic lime |
| 1824 | Joseph Aspdin | England | Portland cement |

Portland cement was developed by Joseph Aspdin in 1824, so named because its color and quality are similar to a kind of limestone, Portland stone (Portland, England).

#### 2.2.2.1 Manufacture of Portland Cement

Portland cement is made by blending an appropriate mixture of limestone and clay or shale together, and by heating them to 1450°C in a rotary kiln. Currently, the capability of a rotary kiln can reach 10,000 metric tons daily. Figure 2-10 shows a model production line of the China Hai Luo Cement Company.

The sequence of operations is shown in Figure 2-11, in which the preliminary steps are a variety of blending and crushing operations. The raw feed must have a uniform composition and be of fine enough size that reactions among the components can be completed in the kiln.

**Figure 2-10** A production line of 10,000 metric tons per day (Hai Luo Cement Company, China)

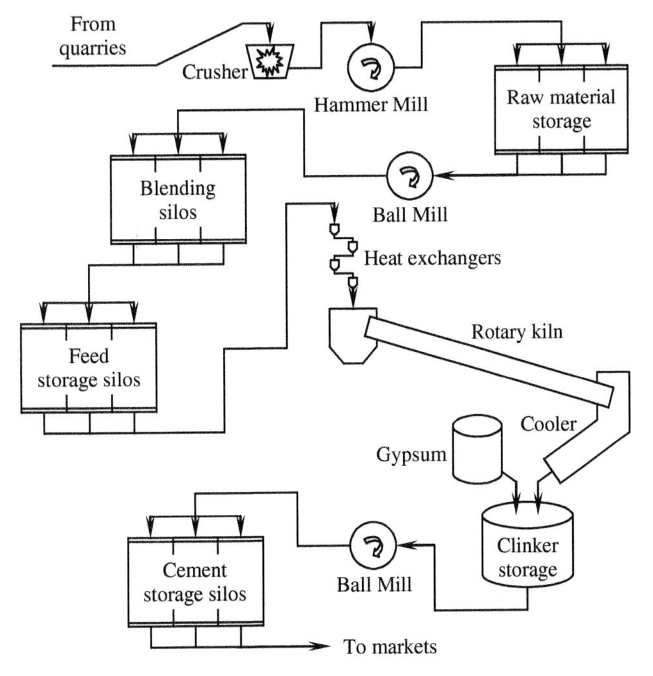


**Figure 2-11** Manufacturing process of Portland cement

Subsequently, the burned clinker is ground with gypsum to form the familiar gray powder known as Portland cement. The basic raw materials used for manufacturing Portland cement are limestone, clay, and iron ore. The primary reactions during the calcination process are listed below.

**(a)** Clay mainly provides silicates ($SiO_2$) together with small amounts of $Al_2O_3$ and $Fe_2O_3$. The decomposition of clay happens at a temperature around 600°C:

$$\text{Clay} \xrightarrow{(600^\circ\text{C})} SiO_2 + Al_2O_3 + Fe_2O_3 + H_2O \tag{2-21}$$

**(b)** Limestone ($CaCO_3$) is mainly providing calcium (CaO) and is decomposed at 1000°C:

$$CaCO_3 \xrightarrow{(1{,}000^\circ\text{C})} CaO + CO_2 \tag{2-22}$$

**(c)** Iron ore and bauxite provide additional aluminum and iron oxide ($Fe_2O_3$), which help the formation of calcium silicates at low temperature. They are incorporated into a row mix.

**(d)** There are different temperature zones in a rotary kiln. At various temperatures between 1000 and 1450°C, different chemical compounds are formed. The initial formation of $C_2S$ occurs at a temperature of around 1200°C. $C_3S$ is formed around 1400°C.

$$\left.\begin{array}{l}\text{Limestone}\\ \text{Clay}\\ \text{Iron ore, Bauxite}\end{array}\right\} \xrightarrow[\text{High temperature}]{(1{,}450^\circ\text{C})} \left\{\begin{array}{l}3CaO\cdot SiO_2\\ 2CaO\cdot SiO_2\\ 3CaO\cdot Al_2O_3\\ 4CaO\cdot Al_2O_3\cdot Fe_2O_3\end{array}\right. \tag{2-23}$$

**(e)** The final product from the rotary kiln is called clinker. Pulverizing the clinker into small sizes (<75 μm) with the addition of 3–5% gypsum or calcium sulf produces Portland cement. Gypsum is added to control fast setting caused by $3CaO \cdot Al_2O_3$.

The majority of cement particle sizes are from 2–50 μm. Plots of typical particle size distribution data analysis are given in Figure 2-12.

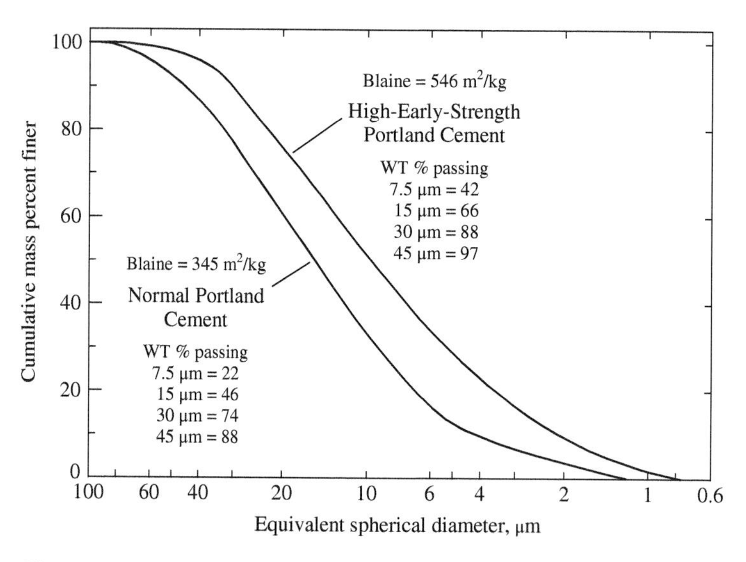


**Figure 2-12** Typical particle size distributions of Portland cement

#### 2.2.2.2 Chemical Composition

(a) *Abbreviations*: In the cement chemistry, the following abbreviations are adopted. Please note that they are not consistent with other types of chemistry, although these notations are also frequently used in ceramic chemistry.

$$CaO = C \quad SiO_2 = S \quad Al_2O_3 = A \quad Fe_2O_3 = F$$
$$H_2O = H \quad SO_3 = \overline{S} \quad MgO = M$$

Thus, we can write $Ca(OH)_2 = CH$, $3CaO = C_3$, and $2CaO \cdot SiO_2 = C_2S$.

(b) *Major compounds*: The major compounds of ordinary Portland cement are listed in Table 2-4. They are $C_3S$, $C_2S$, $C_3A$, and $C_4AF$. $C_3S$ is called tricalcium silicate; $C_2S$, dicalcium silicate; $C_3A$, tricalcium aluminate; and $C_4AF$, tetracalcium aluminoferrite. In addition, $C_3S$ has the nickname alite; $C_2S$, belite; and $C_4AF$, ferrite. It should be noted that $C_3S$ and $C_2S$ occupy 68–75% of Portland cement. Since the primary constituents of Portland cement are calcium silicates, we can define Portland cement as a material that combines CaO and $SiO_2$ in such a proportion that the resulting calcium silicate will react with water at room temperature and normal pressure.

The typical oxide composition of a general-purpose Portland cement can be found in Table 2-5. From the weight percentage of these oxides, the weight percentage of $C_3S$, $C_2S$, $C_3A$, and $C_4AF$ in Portland cement can be calculated using an equation initially developed by Bogue and adopted by ASTM C150 as follows:

$$C_3S(\%) = 4.071C - 7.600S - 6.718A - 1.450F - 2.852\overline{S}$$

$$C_2S(\%) = 2.867S - 0.754C_3S$$

**Table 2-4** Major compounds of ordinary Portland cement

| Compound | Oxide composition | Color | Common name | Weight percentage |
|---|---|---|---|---|
| Tricalcium silicate | $C_3S$ | White | Alite | 50 |
| Dicalcium silicate | $C_2S$ | White | Belite | 25 |
| Tricalcium aluminate | $C_3A$ | white/grey | n/a | 12 |
| Tetracalcium aluminoferrite | $C_4AF$ | Black | Ferrite | 8 |

**Table 2-5** Typical oxide composition of a general purpose Portland cement

| Oxide | Shorthand notation | Common name | Weight percent |
|---|---|---|---|
| CaO | C | lime | 64.67 |
| $SiO_2$ | S | silica | 21.03 |
| $Al_2O_3$ | A | alumina | 6.16 |
| $Fe_2O_3$ | F | ferric oxide | 2.58 |
| MgO | M | magnesia | 2.62 |
| $K_2O$ | K | alkalis | 0.61 |
| $Na_2O$ | N | alkalis | 0.34 |
| $SO_3$ | $\overline{S}$ | sulfur trioxide | 2.03 |
| $CO_2$ | $\overline{C}$ | carbon dioxide | – |
| $H_2O$ | H | water | – |

$$C_3A(\%) = 2.650A - 1.692F$$
$$C_4AF(\%) = 3.043F \quad (2\text{-}24)$$

where C, S, A, F, and $\overline{S}$ are weight percentages of corresponding oxide in a Portland cement such as what is listed in Table 2-5.

It should be noted that the above equations are valid only when $A/F \geq 0.64$. Fortunately, most Portland cements satisfy that condition.

**(c)** *Minor components of Portland cement*: The most important minor components of cement are gypsum, MgO, and alkali sulfates. Gypsum ($2CaSO_4 \cdot 2H_2O$) is added in the last procedure of grinding the clinker to produce Portland cement. The reason for adding gypsum cement is to avoid the flash setting caused by a fast reaction of $C_3A$, because it can react with $C_3A$ and form a hydration product called ettringite on the surface of $C_3A$ to prevent further reaction of the $C_3A$ as a barrier. The normal percentage of gypsum added to cement is about 4–5%. Only when gypsum is more than 3% in a Portland cement, can the formation of ettringite be guaranteed. When the percentage of gypsum is between 1 and 3%, both ettringite and monosulfoaluminate will be formed. When the percentage of gypsum is less than 1%, only monosulfoaluminate will be formed.

Alkalis (MgO, $Na_2O$, and $K_2O$) can increase the pH value of concrete up to 13.5, which is good for reinforcing steel protection. However, a high alkaline environment can also cause some durability problems, such as alkali aggregate reaction and leaching.

#### 2.2.2.3 Hydration

Hydration of cement is the reaction between the cement particles and water, including chemical and physical processes. The properties of fresh concrete, such as setting and hardening, are the direct results of hydration. The properties of hardened concrete are also influenced by the process of hydration. Hence, to understand the properties and behavior of cement and concrete, some knowledge of the chemistry of hydration is necessary.

**(a)** *Hydration of pure cement compounds*: The mechanism of hydration of cement as a whole is very complex and has not yet been fully understood. So far, the only approach to studying the hydration of Portland cement is to investigate the reaction mechanism of individual compounds separately. This assumes that the hydration of each compound takes place independently and no interaction occurs. Although this assumption is not valid completely, it helps to understand the chemistry of hydration.

**(b)** *Calcium silicates*: Taylor (1997) and Mindess et al. (2003) have given good descriptions of the hydration of the calcium silicate phases. The hydrations of two calcium silicates are stoichiometrically similar, differing only in the amount of calcium hydroxide formed, the heat released, and the reaction rate:

$$2C_3S + 11H \rightarrow C_3S_2H_8 + 3CH \quad (2\text{-}25)$$

$$2C_2S + 9H \rightarrow C_3S_2H_8 + CH \quad (2\text{-}26)$$

The principal hydration product is $C_3S_2H_8$, calcium silicate hydrate, or C–S–H (nonstoichiometric). C–S–H occupies about 50% of the structural component in a cement paste and forms directly on the surface of cement particles. In addition, the size of C–S–H is quite small. It is believed that C–S–H is the major strength provider for Portland cement concrete due to its amount and small size. The structure of C–S–H is in the nanometer scale and is not a well-defined compound. The formula $C_3S_2H_8$ is only an approximate description. In an X-ray diffraction (XRD) investigation, C–S–H does not show sharp peaks and has been considered an amorphous structure. C–S–H is usually called a glue gel binder. Recently, much research has been conducted to gain an understanding of the structure of C–S–H and thus the nature of binding of concrete. The progress is introduced in detail in Chapter 4.

Another product is CH, calcium hydroxide. This product is a good crystalline with a plate shape in most cases. CH is formed in solution by crystallization and occupies about 25% of the structural component of cement paste. CH can bring the pH value to over 12 and it is good for th corrosion protection of steel. From a durability of concrete point of view, CH may lead to leaching due to its solubility, carbonation due to its reaction with carbon dioxide, alkali aggregate reaction due to its high pH value, or sulfate attack due to its reaction with sulfate. Hence, in contemporary concrete technology, there has been a trend to reduce the amount of CH in concrete as much as possible. However, a minimum amount of CH is needed to keep the high alkali environment in concrete.

One more thing that needs to be mentioned is that both C–S–H and CH are very "gentle" and pass around something blocking their way or stopping their formation, as there is no free space for them to occupy. It should be pointed out that although $C_3S$ and $C_2S$ produce the same hydration products, their reaction rates are very different; $C_3S$ reacts very fast at the early stage, releases more hydration heat, and contributes most to early age strength of concrete. On the other hand, $C_2S$ reacts very slowly, releases less heat, and contributes minimally to early age strength of concrete. $C_2S$ contributes the most to the long-term strength of concrete.

**(c)** *Tricalcium aluminate and ferrite phase*: The primary initial reaction of $C_3A$ with water in the presence of a plentiful supply of gypsum is

$$C_3A + 3\left(C\,\overline{S}H_2\right) + 26H \rightarrow C_6A_3H_{32} \tag{2-27}$$

The 6-calcium aluminate trisulfate-32-hydrate is usually called calcium sulfoaluminate hydrate and more commonly *ettringite*, which is the name of a naturally occurring mineral of the same composition. The nickname of ettringite is AFt. The formation of ettringite is right on the surface of the particles of $C_3A$. It can slow down the hydration of $C_3A$ because it acts as a diffusion barrier around $C_3A$, analogous to the behavior of C–S–H during the hydration of the calcium silicates. Thus, it can avoid a $C_3A$ flash setting. Ettringite is a needle-shaped crystal with a large volume expansion. Moreover, ettringite is very aggressive and will make space to grow if there is no free space left. The effect of ettringite on concrete strength can be evaluated in two cases. In case 1, ettringite is formed before the paste has hardened and gained strength due to the hydration of $C_3S$. It will contribute to the early strength development of concrete since the needle-shaped crystals can work as a reinforcement for the surrounding C–S–H, and the expansion is not so significant. In case 2, if ettringite is formed after the concrete has hardened and free space has been occupied by other hydration products, it will make its space to grow by breaking the hardened hydration products and hence create cracks and volume instability.

The ettringite is stable only when there is an ample supply of sulfate available and at a temperature lower than 60°C. If all the sulfate is consumed before the $C_3A$ has completely hydrated or the temperature rises to above 60°C, it can be broken down during the hydration of the conversion to monosulfoaluminate:

$$2C_3A + C_6A_3H_{32} + 4H \rightarrow 3C_4A\overline{S}H_{12} \tag{2-28}$$

Monosulfoaluminate is also called tetracalcium aluminate monosulfate-12-hydrate. Its nickname is AFm. When monosulfoaluminate is brought into contact with a new source of sulfate ions, ettringite can be formed again:

$$C_4A\bar{S}H_{12} + 2C\bar{S}H_2 + 16H \rightarrow C_6A\bar{S}_3H_{32} \quad (2\text{-}29)$$

If there is no gypsum, $C_3A$ will react with water very quickly:

$$2C_3A + 21H \rightarrow C_4AH_{13} + C_2AH_8 \quad (2\text{-}30)$$

The hydrates can be further converted to

$$C_4AH_{13} + C_2AH_8 \rightarrow 2C_3AH_6 + 9H \quad (2\text{-}31)$$

The reaction occurs so fast that it causes flash set of concrete. The hydration products of $C_4AF$ are similar to those of $C_3A$. However, the reaction rate of $C_4AF$ is slower than that of $C_3A$. When reacting with gypsum, the following equation applies:

$$C_4AF + 3\left(C\bar{S}H_2\right) + 21H \rightarrow C_6(A,F)\bar{S}_3H_{32} + (F,A)H_3 \quad (2\text{-}32)$$

In the equation, the expression $C_6\,(A,F)\,\bar{S}_3H_{32}$ indicates that iron oxide and alumina occur interchangeably in the compound. The order of symbols in the brackets implies the order of richness of the corresponding element in the compound. $C_6\,(A,F)\,\bar{S}_3H_{32}$ can further react with $C_4AF$ and water:

$$C_4AF + C_6(A,F)\bar{S}_3H_{32} + 7H \rightarrow 3C_4(A,F)\bar{S}H_{12} + (F,A)H_3 \quad (2\text{-}33)$$

**(d)** *Kinetics and reactivities*: The rate of hydration during the first few days is in the order of $C_3A > C_3S > C_4AF > C_2S$. The rate of hydration can be observed in Figure 2-13. Figure 2-14 shows the strength development of different minerals. It can be seen that $C_3S$ has a high early strength and $C_2S$ develops its strength slowly in the early age but very fast later on. It can be deduced that $C_3S$ contributes the most to early strength and $C_2S$ to the long-term strength of Portland cement. On the other hand, the contribution to the strength of Portland cement

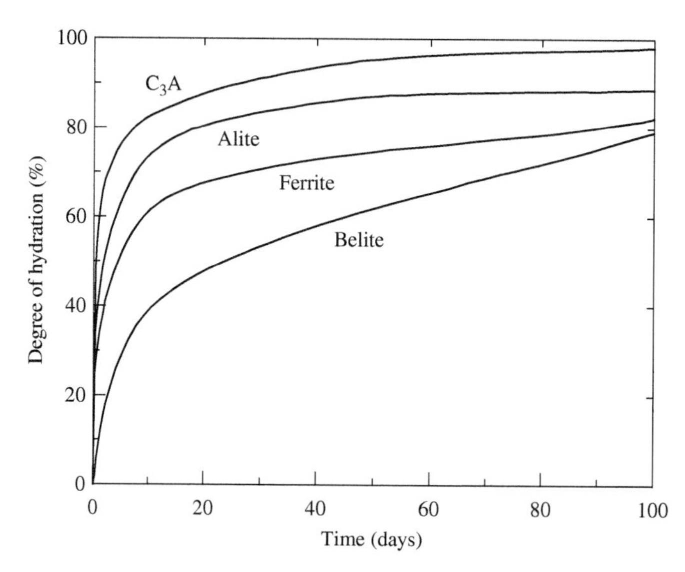


**Figure 2-13** Hydration process of primary constituents of Portland cement

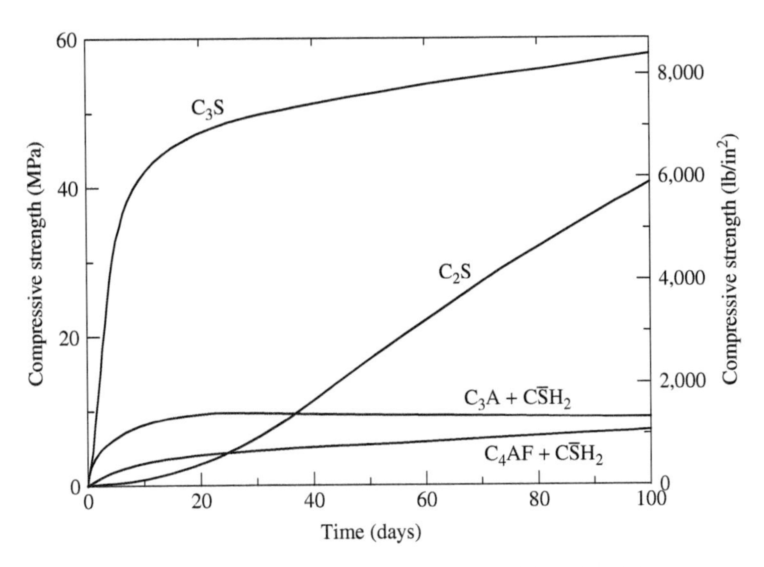


**Figure 2-14** Strength development of primary constituents of Portland cement

from $C_3A$ and $C_4AF$ is not significant. Moreover, the strength development and kinetics of different minerals are somehow but not very closely related.

**(e)** *Calorimetric curve of Portland cement*: In discussing the hydration of individual major compounds of Portland cement, it is assumed that there is no interaction among the reaction of each compound. In fact, this is not entirely true and there are some mutual influences in the reaction of different compounds. For instance, all the hydration products consume CaO and there may be some competition in catching $Ca^{2+}$ from other compounds. In addition, both $C_3A$ and $C_4AF$ compete for sulfate ions and this will cause different reaction rates and change their reactivity. Moreover, C–S–H can incorporate some sulfate, alumina, and iron during the hydration of cement that leads to less calcium sulfoaluminates in the hydration of cement than for pure compounds. Hence, the hydration of Portland cement is far from being fully understood and more research is needed.

To study the dynamics of hydration of Portland cement, heat release rate measurement is frequently adopted because the process is exothermal. A typical calorimetric curve of Portland cement is shown in Figure 2-15. Several characteristic points can be observed in

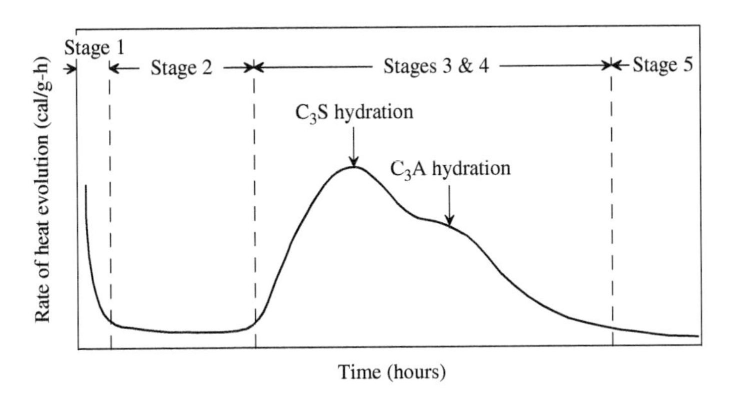


**Figure 2-15** A typical calorimetric curve of Portland cement

**Table 2-6** Kinetics of reaction, chemical processes and relevance to concrete of the different reaction stages of cement

| Reaction stage | Kinetics of reaction | Chemical processes | Relevance to concrete |
|---|---|---|---|
| 1. Initial hydrolysis | Chemical control; rapid | Initial hydrolysis; dissolution of ions | n/a |
| 2. Induction period | Nucleation control; slow | Continued dissolution of ions | Determines initial set |
| 3. Acceleration | Chemical control; rapid | Initial formation of hydration products | Determines final set and rate of initial hardening |
| 4. Deceleration | Chemical and diffusion control; slow | Continued formation of hydration products | Determines rate of early strength gain |
| 5. Steady state | Diffusion control; slow | Slow formation of hydration products | Determines rate of later strength gain |

Figure 2-15. The curve decreases initially and then keeps flat for a while after reaching the lowest value of heat evolution rate. Then it goes up in a very steep manner. After the first crest is reached, the curve goes down and rises up again to form the second peak. After that the curve decreases gradually. The two peaks in the curve represent the dominant effect of $C_3S$ or $C_3A$ correspondingly and their order of occurrence can be reversed.

Based on the characteristic points on the curve, the hydration process of Portland cement can easily be distinguished in five stages: (1) dissolution; (2) dormant; (3) acceleration; (4) deceleration; and (5) steady state. It should be pointed out that the curve of heat evolution of Portland cement is very similar to that of $C_3S$ as $C_3S$ has a dominant effect in cement. Thus, the five stages traditionally defined in cement chemistry can be explained by using the reaction process of $C_3S$, as detailed in Table 2-6.

On first contact with water, calcium ions and hydroxide ions are rapidly released from the surface of each $C_3S$ grain; the pH values rise to over 12 within a few minutes. This hydrolysis slows down quickly but continues throughout the induction period. The induction (dormant) period is caused by the need to achieve a certain concentration of ions in solution, before the crystal nuclei form, from which the hydration products grow. At the end of the dormant period, CH starts to crystallize from the solution with the concomitant formation of C–S–H, and the reaction of $C_3S$ again proceeds rapidly (the third stage, acceleration, begins). CH crystallizes from the solution, while C–S–H develops on the surface of $C_3S$ and forms a coating covering the grain. When the first peak of the rate of heat evolution is reached, the deceleration stage is started. As hydration continues, the thickness of the hydrate layer increases and forms a barrier through which water must flow to reach the unhydrated $C_3S$ and through which ions must diffuse to reach the growing crystals. Eventually, movement through the C–S–H layer determines the rate of reaction, and hydration becomes diffusion-controlled and moves into the fifth stage, the steady-state stage.

**(f)** *Setting and hydration*: Traditionally, it is believed that the initial set of cement corresponds closely to the end of the induction period, 2–4 h after mixing. The initial set indicates the beginning of gel formation. It is controlled primarily by the rate of hydration of C3S. The final set occurs 5–10 h after mixing, which indicates that sufficient hydration products are formed and the cement paste is ready to carry some external load.

It should be noted that the initial and the final set have a physical importance. However, there is no fundamental change in the hydration process for these two different sets. The hydration process of fresh cement paste is schematically illustrated in Figure 2-16.

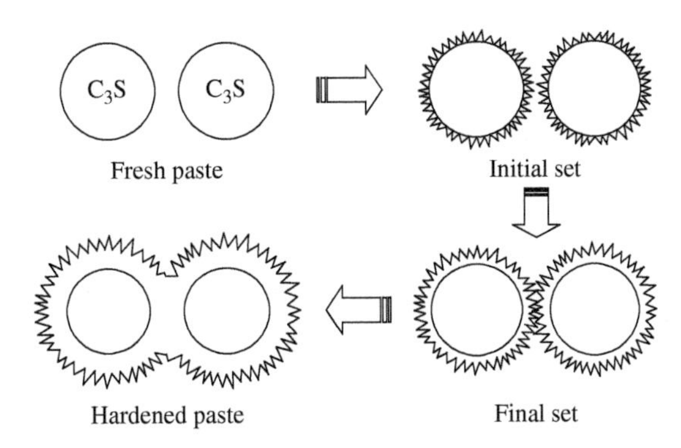


**Figure 2-16** Setting of fresh cement paste

The rate of early hardening, which means a gain in strength, is primarily determined by the hydration of $C_3S$, and the strength gain is roughly proportional to the area under the heat peak in the calorimetric curves of Portland cement. The strength development is mainly derived from the hydration of silicates.

#### 2.2.2.4 New Understanding of the Dynamics of Cement Hydration

Cement hydration is a complex physical–chemical process. During the process, a cement–water mixture is changed from a fluid state to a porous solid state. An adequate understanding of the mechanism of cement hydration is necessary for a full appreciation of cement concrete properties.

The cement hydration process was traditionally studied using the calorimetric method. The hydration stages were identified by heat liberation measurement and the hydration mechanism was explained based on heat evolution. Some limitations (Gartner et al., 2002) were pointed out, such as that the liberated heat content was simply proportional neither to the degree of cement hydration nor to the development of the physical properties. This method provides only an approximation of the understanding of cement hydration and leaves room for cement scientists and engineers to explore more accurate ways.

Chemically, cement hydration involves ion dissolution and the formation of new chemical compounds. Physically, cement hydration involves microstructure formation and a porosity decrease process. When water is added to cement, the soluble ions in the cement dissolve in the water. The ions in the pore solution are conductive and can form an electrical current under a certain electrical field. The conductivity of a cement mix depends mainly on the concentration of the ion solution and microstructure, especially the porosity and pore connectivity of the cement paste. Therefore, the electrical conductivity or resistivity of a cement mixture can be used to interpret the cement hydration process.

Taylor and Arulanandan (1974), Christensen et al. (1994), Tashiro et al. (1994), McCarter et al. (1981, 2003) and Gu et al. (1985) observed the electrical responses of the cementitious hydration systems using alternating current impedance spectroscopy. However, since the experiments were carried out with electrodes, the accuracy and consistency of the test results were largely affected by contact problems, such as the interface gap between the electrodes and the shrunken cement paste mixes.

Recently, the dynamics of the cement hydration process have been studied with an advanced monitoring technique, noncontact electrical resistivity measurement (Li and Li, 2003; Li and Zhu, 2003; Wei and Li, 2005), as shown in Figure 2-17. It was realized long ago that resistivity of cement paste is a fingerprint of its hydration. However, the inaccurate nature of resistivity measurement due to the contact of electrodes in fresh concrete made it difficult to reach any useful conclusion.

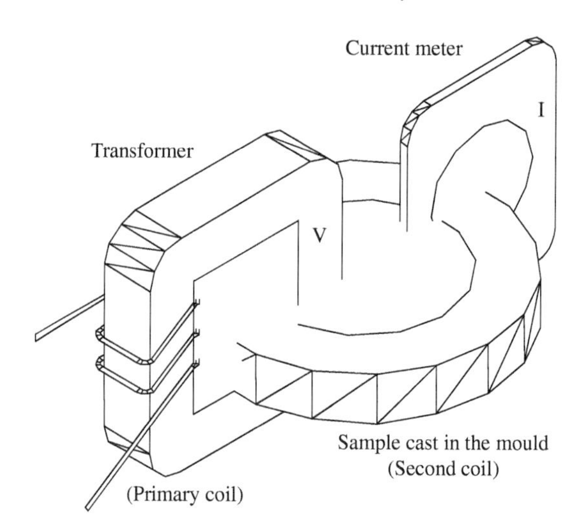


**Figure 2-17** Electrical resistivity measurement setup

The newly developed technique adopts a transformer principle and eliminates electrodes in the resistivity measurement; thus, it completely eliminates the contact problems and provides more reliable results. For the purpose of obtaining a clear understanding of the relationship between electrical resistivity development and the hydration process, microstructure analysis has been conducted and the results are correlated to each stage of the hydration process.

A typical electrical resistivity development curve with time ($\rho(t) - t$) and its differential curve ($d\rho(t)/dt - t$), which represents the rate of resistivity development, are shown in Figure 2-18. The minimum point $M$ and the level point $L$, which is at the end of a relatively flat region after point $M$, are marked on the curve of $\rho(t) - t$. Peak points $P1$ and $P2$ are marked on the curve of $d\rho(t)/dt - t$. The times at which the characteristic points occurred and the paste setting time are listed in the table in Figure 2-18. According to the characteristic points, the hydration process is divided into five stages: (1) dissolution; (2) dynamic balance; (3) setting; (4) hardening; and (5) hardening rate deceleration stage.

To investigate the cement hydration process and understand what happens at each characteristic point on the resistivity and its differential curves, microstructural investigations have been conducted on the samples corresponding to the characteristic point occurrences using scanning electron microscopy (SEM), XRD, differential thermal analysis (DTA), and Fourier transform infrared spectroscopy (FTIR). The results are shown in Figures 2-19, 2-20, and 2-21, respectively.

The XRD patterns in Figure 2-19 show the presence of crystalline substances in anhydrous cement and in the hydrated samples at various ages. At a very early hydration age, the ettringite (AFt) phase starts to form with the consumption of sulf, which is detected in the sample at 0.73 h. The intensity of the ettringite peaks decreases and monosulf (AFm) appears, as shown in the hydrated sample for 12.8 h in comparison with that in the 6.7 h sample. The amorphous hydrates of C–S–H cannot be detected by the XRD technique. The relative density of the peaks of the crystal calcium hydroxide (CH) become significant with time due to the increase of CH content in the hydrated samples.

The DTA has been performed for characterization of the hydration products and the quantification of calcium hydroxide in the samples, as shown in Figure 2-20. The existence of four endothermic peaks is shown in the detected temperature range. Peak 1 and peak 2, located at 85 and 128°C, mainly characterize gypsum. The shift of the first peak from 85°C to around 92°C and the reduced intensity of the second peak with hydration indicate that gypsum is consumed and the AFt formation is at the expense of the gypsum peaks. The third peak, observed at about 165°C,

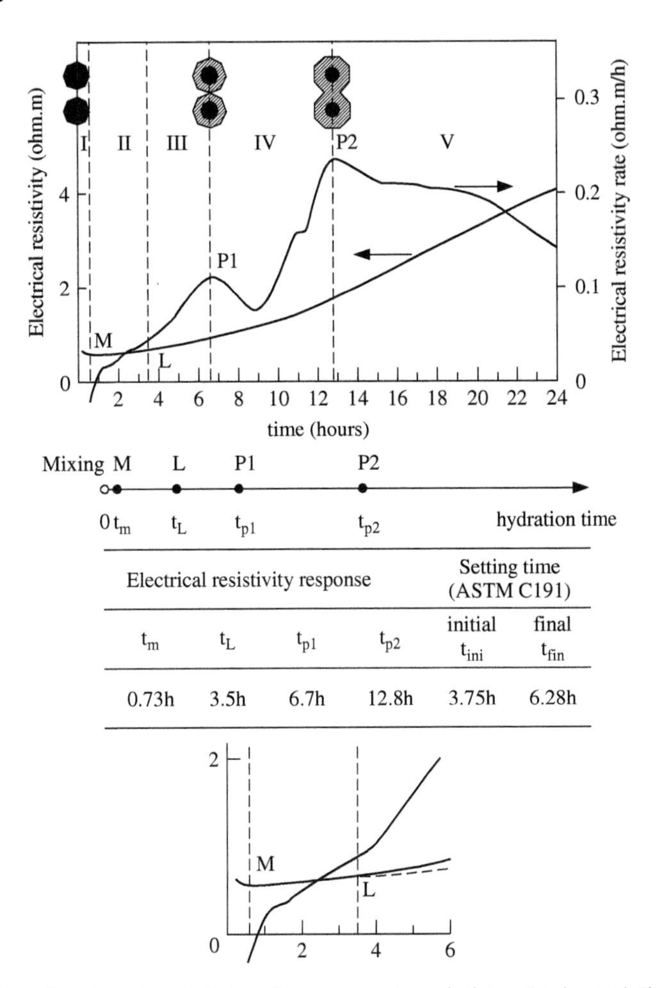


**Figure 2-18** Electrical resistivity development and the rate in resistivity of the cement paste during the first 24 h

represents the decomposition of AFm in the samples hydrated for 12.8 h and 24 h. The fourth peak, at 465–470°C, is due to the CH. The shifts of the CH peaks to the right with age increase show that CH crystalline become denser and larger, and increasing temperature is needed for the decomposition of the CH crystalline. The index of the formed CH content, estimated by determining the endothermic area of the CH peaks, as an indication of the hydration product contents, is marked in Figure 2-20.

The FTIR spectra of the samples at various ages, as shown in Figure 2-21, provide information on chemical bonding. Calcium hydroxide gives a peak at 3643 $cm^{-1}$; ettringite has a strong peak at 3422 $cm^{-1}$ and a weaker one at 3545 $cm^{-1}$, marked A in Figure 2-21; and monosulfate has bands at 3260, 3333, 3402, 3458, and 3527 $cm^{-1}$, which are separated from a broad band 3200–3600 $cm^{-1}$, and the broad band centered at 3402 $cm^{-1}$, marked in Figure 2-21.

Bands $\nu 3$ ($SiO_4$) and $\nu 4$ ($SiO_4$) in the anhydrous cement are located at 924 and 523 $cm^{-1}$. The hydration of the silicate phases causes a shift in band $\nu 3$ ($SiO_4$) from 924 to 978 $cm^{-1}$ over 24 h. There is an obvious shift occurring in band $\nu 3$ ($SiO_4$) from 6.7 to 12.8 h, corresponding to new Si–O bonding formation during the period. The shift is likewise observed in band $\nu 4$ ($SiO_4$), from a wave number of 523 $cm^{-1}$ in the anhydrous cement to a lower wave number 518 $cm^{-1}$ in the hydrated sample at 0.73 h, characterizing $C_3S$ initial hydration.

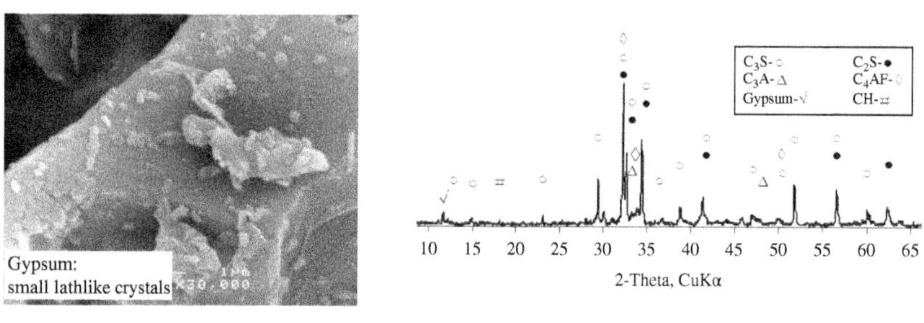


**(a) t = 0, anhydrous cement**

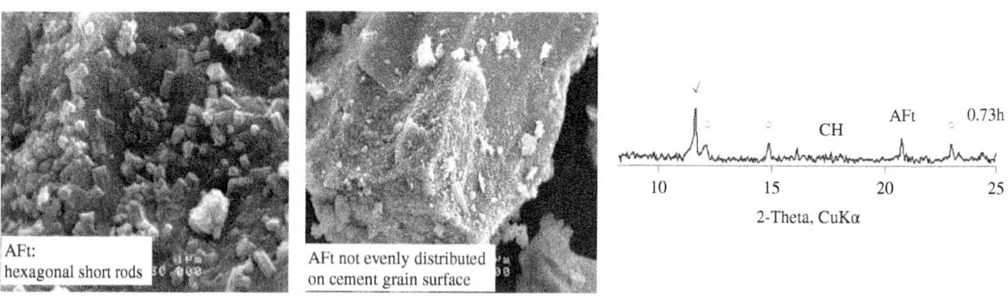


**(b) t = 0.73h (at point M), $C_3A$ rapid reaction**

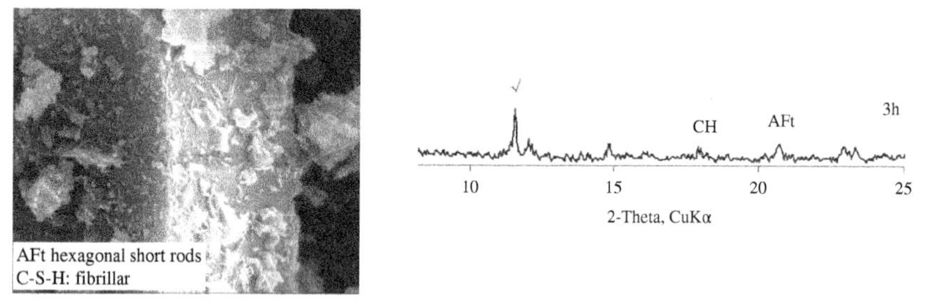


**(c) t = 3h (close to point L)**

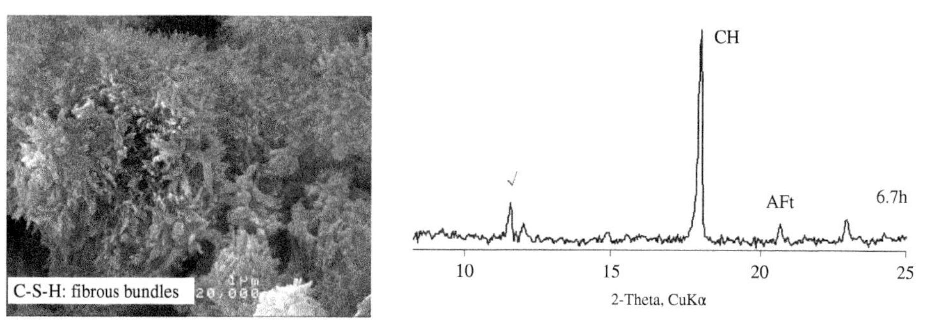


**(d) t = 6.7h (at point P1), close to final setting**

**Figure 2-19** Cement and hydrated pastes at the ages of 0.73, 3, 6.7, 12.8, and 24 h from SEM and XRD

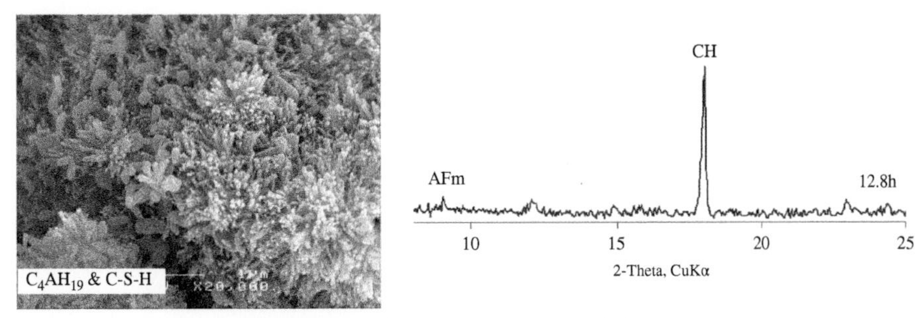


**(e) t = 12.8h (at point P2), AFt transfers to AFm**

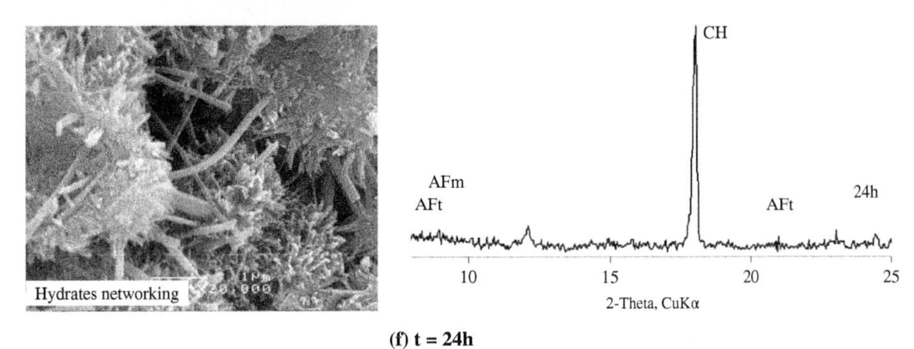


**(f) t = 24h**

**Figure 2-19** *(Continued)*

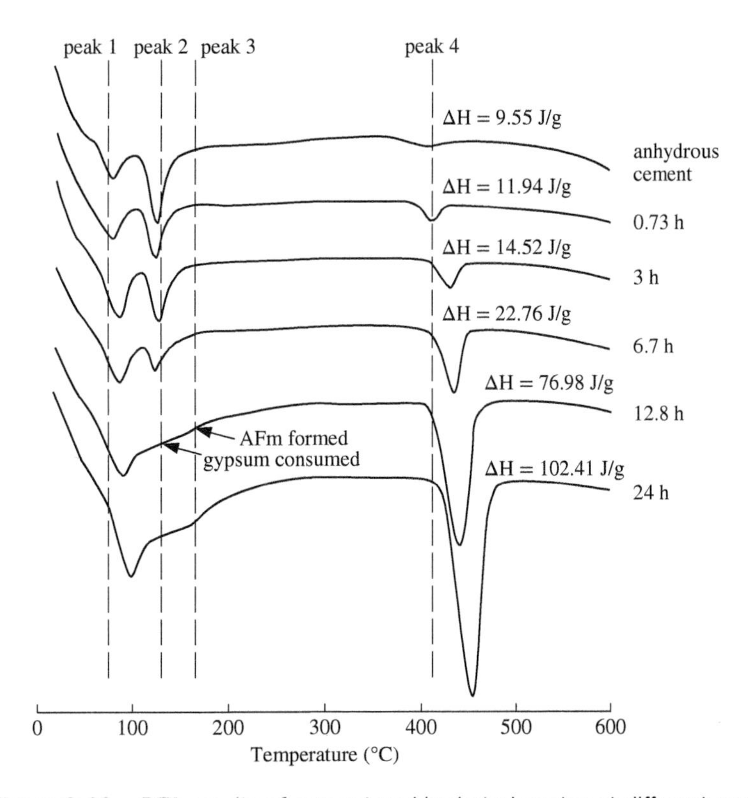


**Figure 2-20** DTA results of cement and hydrated pastes at different ages

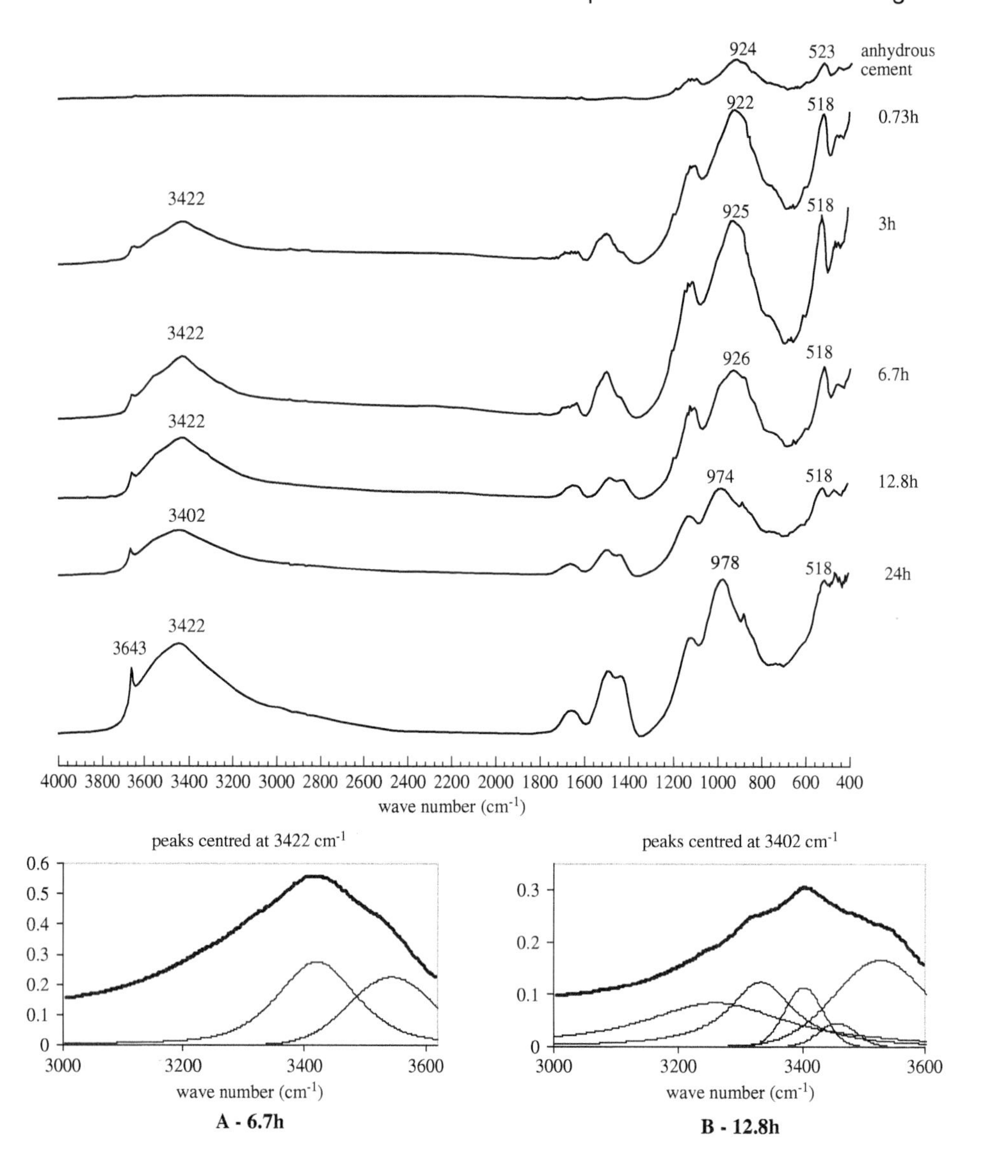


**Figure 2-21** FTIR spectra of anhydrous cement and hydrated samples at different ages

Microstructure investigations can help to understand the relationship between electrical resistivity development and the hydration process. The correlation of the microstructure analysis to the hydration stages is respectively described as follows.

(I) *Dissolution stage*: From the start of mixing to point M (0.73 h) the resistivity ρ(t) of the mixture decreases. When cement is mixed with water, the mobile ions in the cement, such as potassium ($K^+$), sodium ($Na^+$), calcium ($Ca^{2+}$), hydroxyl ions ($OH^-$), and sulfate ($SO_4^{2-}$), dissolve in the water and form an electrolytical solution. The dissolving process of the ions causes the resistivity to decrease in the hydration system.

An immediate increase in resistivity is observed right after point *M*, and it is inferred that some hydration products have been formed. This point has been verified by SEM, XRD, DTA, and FTIR results.

$C_3A$ is the most active chemical compound in cement, and it reacts with sulfate ions in the solution dissolved from the gypsum. As shown in Figure 2-19a, only raw materials appear in the SEM photo at $t = 0$. The hydration product, ettringite crystal, as short rods, is formed directly on the surface of the cement grains, as can be observed in the SEM picture in Figure 2-19b, which corresponds to the time of the M point at $t = 0.73$ h. The picture with a magnification of 5000 shows that the AFt crystals are not evenly distributed on the cement surface; one possibility is that the AFt forms just on the surface $C_3A$.

At same time, a supersaturation point of $Ca^{2+}$ is reached, and the CH precipitates from the solution, as detected by the DTA technique in the hydrated sample for 0.73 h, which is shown in Figure 2-20. The DTA results show that the index of CH (peak 4) content in the hydrated sample for 0.73 h ($\Delta H = 11.94$ J/g) is larger than that of the CH formed from free lime by chemisorbed water in the anhydrous cement ($\Delta H = 9.55$ J/g) during storage, indicating that new CH is formed at that moment.

The shift of the band $\nu 4$ $SiO_4$ over 0.73 h indicates a $C_3S$ hydration occurrence. It has been reported that a gelatinous layer on the surface of the cement was observed in an undried specimen soon after mixing by using an environmental cell, and identified as an amorphous colloidal product (Ménétrier et al., 1979; Jennings and Pratt, 1980).

Therefore, during this period, the resistivity decrease implies that the process is dominated by the ion dissolution and is identified as the dissolution stage until the *M* point. The formation of hydration products AFt, a gelatinous substance, and CH in the solution leads to the resistivity increase from point *M* and signals the end of the dissolution stage.

(II) *Dynamic balance stage*: From point M (0.73 h) to point L (3.5 h), the electrical resistivity slightly increases, by approximately 0.03 Ωm from point *M* to point *L*. Point *L* is the transition point from a linear to a curved shape. In this stage, the hydration products continue to form and break the balance condition, then ions continue to dissolve to recover to the saturation point. Thus, a competitive balance between the ion dissolution and CH, and other hydrates precipitation is dynamically kept. This stage is identified as a dynamic balance stage.

The C–S–H formation can be observed in the dried sample as small acicular shapes (diamond type I, fibrillar morphology) among AFt rods in the sample hydrated at $t = 3$ h, from the SEM picture shown in Figure 2-19c. The DTA results show that the CH content index $\Delta H$ increases from 11.94 to 14.52 J/g, and the FTIR shows that the $\nu 3$ bands ($SiO_4$) have a peak shift from 922 to 925, which confirms C–S–H formation during this period. During this stage, the hydration products are increasing to reach a point at which they start to contact each other at the end of the stage. The mix maintains its fluidity in this stage due to the hydration products not being joined together.

(III) *Setting stage*: From the end of the dynamic balance stage, the hydration products start to join together. Concrete starts to lose its plastic behavior and proceeds to the setting period. As can be seen from Figure 2-18, correspondingly, from point *L* (3.5 h) to point *P*1 (6.7 h), the resistivity increases at a fast rate, indicating a rapid change in the microstructure during this period. It can be seen in the sample hydrated for 6.7 h, as shown in Figure 2-19d, that a great amount of C–S–H is formed in the bundles, compared with 3 h, and the cement particles are surrounded by the C–S–H bundles. The solid phase in the hydration system is closely connected, resulting in a percolation of the solid phase. The initial setting time and the final setting time of the paste are 3.75 and 6.28 h, respectively, obtained by a Vicat needle penetration. The initial setting time is close to the end of the dynamic balance period, and the final setting time is close to the time point of *P*1. Thus, this period is identified as the setting stage. The XRD result at 6.7 h shows a significant increase of the CH peak intensity.

The setting of cement paste appears to be a consequence of $C_3S$ hydration and the formation of sufficient C–S–H and CH hydrates. The $C_3S$ hydration dominates the setting period and the C–S–H formations lead to a normal cement paste set.

(IV) *Hardening stage*: From point $P1$ (6.7 h) to point $P2$ (12.8 h), the resistivity continuously increases at a reduced rate and then the rate increases again. When the hydration proceeds to 12.8 h, gypsum is significantly consumed, the AFt peak disappears, and AFm appears in the XRD pattern in Figure 2-19e. This indicates the transformation of phase AFt to phase AFm during the period of 6.7–12.8 h. When there is sufficient $SO_4^{2-}$ in the pore solution, the ettringite forms and remains stable. After $SO_4^{2-}$ is consumed, the ettringite is transformed into monosulfate (AFm). The DTA results at 6.7 and 12.8 h, as shown in Figure 2-20, confirm the phase transformation occurrence by the disappearance of the gypsum peak (peak 2) and the appearance of the AFm peak (peak 3) at 12.8 h. Additionally, the FTIR pattern, as shown in Figure 2-21, from 6.7 to 12.8 h, and the shift of the band $SO_4^{2-}$ from wave number of 3422 to 3402 support the observation.

The transformation of AFt to AFm (El-Enein Abo et al., 1995) leads to an increasing trend in ion concentration due to the releasing of $SO_4^{2-}$ and $Ca^{2+}$, and a decreasing trend in the solid volume fraction due to phase AFm possessing a higher density than phase AFt, while the reaction in Equation 2-26 (Hewlett, 1998) leads to the latter effect only. No matter which way it takes place, the consequence is that there is a decreasing trend in resistivity. However, the $C_3S$ hydration continuously progresses and the formation of hydrates C–S–H largely increases the solid phase in the hydration system, which compensates the decreasing resistivity trend caused by the phase transformation. The phase transformation slows down the rate of resistivity development, but it does not change the increasing trend of the resistivity. This corresponds to the resistivity increase at a lowered rate ($d\rho(t)/dt$). When the balance between the phase transformation and the AFt formation process, by releasing and consuming ions $Ca^{2+}$ and $SO_4^{2-}$, is achieved, the factors leading to the $d\rho(t)/dt$ decrease, become minor, or disappear, and eventually a larger increase in C–S–H and the other hydrate contents dominates the resistivity development, and the rate ($d\rho(t)/dt$) increases again.

When there is insufficient gypsum, there is also a trend of forming hexagonal plate $C_4AH_{19}$, as shown in Figure 2-19e (Hewlett, 1998), which belongs to the broad group of AFm. This process leads to a similar effect on electrical resistivity development as the formation of phase AFm. The SEM picture shows that C–S–H becomes denser at 12.8 h than at 6.7 h. The FTIR spectra of the samples at 6.7 and 12.8 h show new C–S–H hydrate formation, as described earlier. With the thickness of hydration products increasing, the solution in the pores must go through the thick layer of the products for further hydration and the rate of the chemical reaction then decreases, which corresponds to a decreased rate $d\rho(t)/dt - t$ from point $P2$.

(V) *Hardening rate deceleration stage*: After point P2 (from 12.8 h onward), the resistivity increases at a decreasing rate. The decreasing rate $d\rho(t)/dt$ describes the slowdown in the chemical reaction rate, and the process is controlled by ion diffusion, with the increase of hydrate thickness on the cement particle surface (Mindess et al., 2003).

When hydration proceeds up to the age 24 h, shown in Figure 2-19f, the C–S–H bundles and long rods of AFt intersect each other to form solid networking for strength gain. Meanwhile, the free water space decreases, porosity decreases, and tortuosity increases for conduction paths. Hydration product contents increase with hydration time and can also be represented by the increase in CH hydrate content as the endothermic heat of CH in Figure 2-20.

The new hydration process stages are summarized in Table 2-7. From a macro point of view, fresh concrete changes from a fluid to a solid state. From the micro point of view, a series of physical and chemical reactions are occurring and the consequences lead to microstructure formation. Microstructure formation and concrete property time dependence are essentially correlated. Electrical resistivity can dynamically record the microstructure change process and is related to the concrete properties. Therefore, the electrical resistivity response $\rho(t)$ can provide an indication of setting behavior and strength development as a nondestructive test method.

**Table 2-7** New understanding of the stages of the hydration process

| Hydration stage | Kinetics of hydration | Main chemical phenomena | Chemical reaction |
|---|---|---|---|
| I. Dissolution (mixing to M) | Ion dissolution dominating | Initial rapid chemical reaction of $C_3A$ | $C_3A + 3\left(C\bar{S}H_2\right) + 26H \rightarrow C_6A\bar{S}_3H_{32}$ |
| II. Dynamic balance (M to L) | A competition process of dissolution and precipitation | CH nucleation | $Ca^{2+} + OH^- \rightarrow Ca(OH)_2$ |
| III. Setting (L to P1) | The formation of hydration product C-S-H dominating | Chemical reaction control of $C_3S$ | $C_3S + 11H \rightarrow C_3S_2H_8 + 3CH$ |
| IV. Hardening (P1 to P2) | Continuous formation of hydration products | $C_3S$ hydration; phase transfer from Aft to AFm | $C_3S + 11H \rightarrow C_3S_2H_8 + 3CH$<br>$2C_2S + 9H \rightarrow C_3S_2H_8 + CH$<br>$C_6A\bar{S}_3H_{32} \rightarrow C_4A\bar{S}H_{12} + 2\left(CaSO_4\right)$<br>$C_6A\bar{S}_3H_{32} + 2C_3A + 4H \rightarrow 3C_4A\bar{S}H_{12}$ |
| V. Hardening deceleration (P2 onwards) | Chemical reaction slows down; diffusion control | The second reaction of C3A, C3S and other components | $C_3A + 3\left(C\bar{S}H_2\right) + 26H \rightarrow C_6A\bar{S}_3H_{32}$<br>$C_3S + 11H \rightarrow C_3S_2H_8 + 3CH$<br>$2C_2S + 9H \rightarrow C_3S_2H_8 + CH$ |

#### 2.2.2.5 Types of Portland Cements

According to the ASTM standard, there are five basic types of Portland cement:

| | |
|---|---|
| Type I | regular cement, general use. |
| Type II | moderate sulfate resistance, moderate heat of hydration |
| Type III | increase $C_3S$, high early strength |
| Type IV | low heat |
| Type V | high sulfate resistance |

In BSI, four basic Portland cements are standardized: ordinary Portland cement (OPC), rapid-hardening Portland cement (RHPC), low-heat Portland cement (LHPC), and sulfate-resistant Portland cement (SRPC). OPC is equivalent to type I in ASTM, and RHPC, type III; LHPC, type IV; and SRPC, type V. There is no Portland cement similar to type II in BSI.

The typical chemical compositions of five types of Portland cement in ASTM are given in Table 2-8. Type I is usually used as a reference, which contains 50% $C_3S$, 25% $C_2S$, 12% $C_3A$, 8% $C_4AF$, and 5% gypsum. Compared to type I, type III has more $C_3S$ (60%) and less $C_2S$ (15%). Moreover, type III has a larger fineness number than type I. As a result, the early strength of type III at 1 day is doubled as compared to that of type I. Meanwhile, the heat released by type III increases to 500 J/g (type I is 330 J/g). On the other hand, type IV has less $C_3S$ (25%) and more $C_2S$ (50%). Hence, the early strength of type IV at 1 day is only half of that of type I. However, the heat released by type IV greatly decreases to 210 J/g, and thus is called-low heat Portland cement. As for type V, its sum of $C_3A$ and $C_4AF$ is only 14% and much less than the 20% of type I. Since these two compounds readily react with sulfate, the lower content gives it less opportunity to be attacked by sulfate ions. In addition, Figures 2-22 and 2-23 show the strength and temperature rise for the different types of cement, which are consistent with the information in Table 2-8.

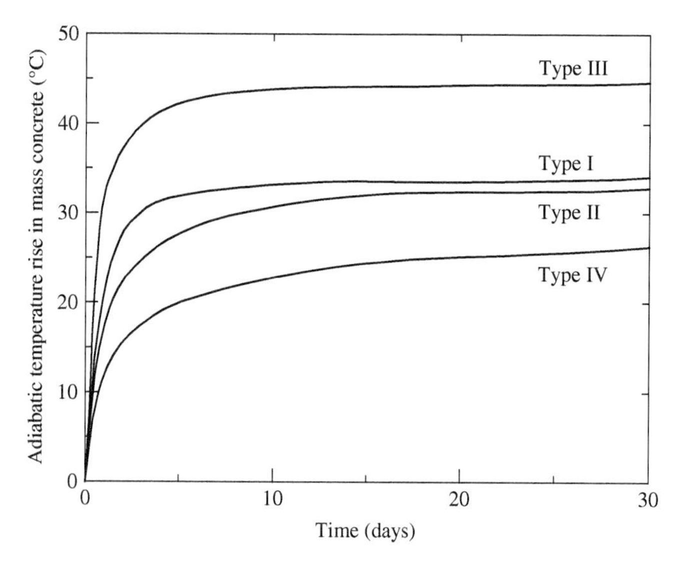


**Figure 2-22** Adiabatic temperature rise in mass concretes with different types of cement

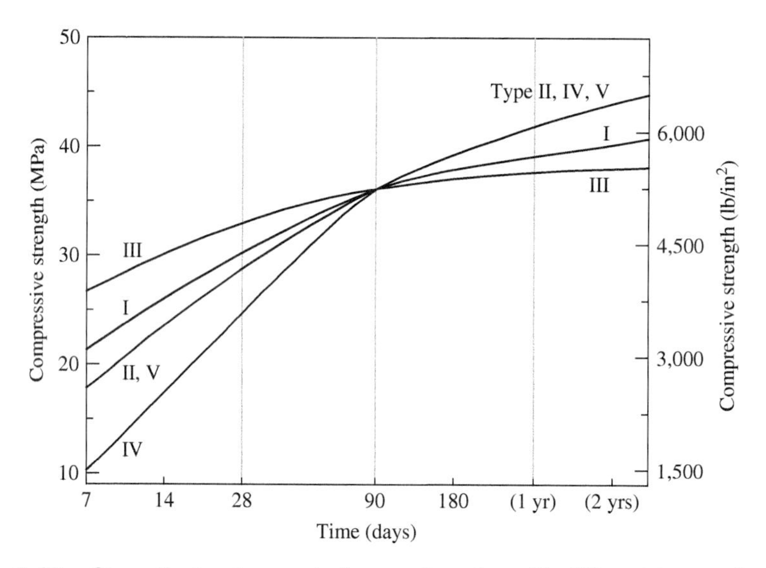


**Figure 2-23** Strength development of cement pastes with different types of cement

From the information provided in Table 2-8, we can evaluate the behavior of each type of cement. The various behaviors provide the basic justification in selecting cement for engineering practice. For instance, for massive concrete structures, hydration heat is a big consideration because too much heat will cause a larger temperature gradient, thermal stress, and cracking. Hence, type IV cement should be the first candidate and type III should not be used. For a marine structure, high sulfate resistance and lower ettringite are needed; thus, type V should be selected. If high early strength is needed, type III will be the best choice. Generally, type I is the most popular cement used in civil engineering.

**Table 2-8** Chemical compositions and physical properties of different Portland cements

| Chemical compositions and physical properties | Portland cement type | | | | |
|---|---|---|---|---|---|
| | I | II | III | IV | V |
| $C_3S$ | 50 | 45 | 60 | 25 | 40 |
| $C_2S$ | 25 | 30 | 15 | 50 | 40 |
| $C_3A$ | 12 | 7 | 10 | 5 | 4 |
| $C_4AF$ | 8 | 12 | 8 | 12 | 10 |
| $CSH_2$ | 5 | 5 | 5 | 4 | 4 |
| Fineness (Blaine, $m^2/kg$) | 350 | 350 | 450 | 300 | 350 |
| Compressive strength (1-day, MPa [psi]) | 7 [1,000] | 6 [900] | 14 [2,000] | 3 [450] | 6 [900] |
| Heat of hydration (7-day, J/g) | 330 | 250 | 500 | 210 | 250 |

#### 2.2.2.6 The Role of Water

Of course, water is necessary for the hydration of cement. However, the water added in the mix is usually much higher than what the chemical reaction needs due to the fluidity requirement of concrete for placing. Thus, we can distinguish the three kinds of water in cement paste according to their roles: chemically reacted water, absorbed water, and free water. The chemically reacted water or chemically bonded water is the water that reacts with C, S, A, F, and $\overline{S}$ to form a hydration products such as C–S–H, CH, and AFt. This type of water is difficult to remove from cement paste and a complete decomposition happens at a temperature about 900°C. Absorbed water is the water molecules inside the layers of C–S–H gel. The loss of absorbed water causes shrinkage, and the movement or migration of absorbed water under a constant load affects the creep. Free water is the water outside the C–S–H gel. It behaves as bulk water and creates capillary pores when evaporated, and can influence the strength and permeability of concrete.

Porosity is a major component of the microstructure that is mainly caused by loss of water. The size of the capillary pores formed due to the loss of free water is in the range of 10 nm to 10 μm. The size of the gel pores involved in absorbed water is in the range of 0.5 to 10 nm. A knowledge of porosity is very useful since porosity has such a strong influence on strength and durability. According to experiments, the gel porosity for all normally hydrated cements is a constant, with a value of 0.26. The total volume of the hydration products (cement gel) is given by

$$V_g = 0.68\ \alpha\ \text{cm}^3\text{g of original cement} \tag{2-34}$$

where α represents the degree of hydration. The capillary porosity can then be calculated by

$$P_c = \frac{w}{c} - 0.36\ \alpha\ \text{cm}^3\text{g of original cement} \tag{2-35}$$

where $w$ is the original weight of water, $c$ is the weight of cement, and $w/c$ is the water to cement ratio. It can be seen that with an increase of $w/c$, the capillary pores increase. The gel/space ratio ($X$) is defined as

$$X = \frac{\text{volume of gel (including gel pores)}}{\text{volume of gel} + \text{volume of capillary pores}} = \frac{0.68\alpha}{0.32\alpha + w/c} \tag{2-36}$$

The gel/space ratio reflects the percentage of solid materials in a cement paste. The higher the ratio, the more solid the materials and hence the higher the compressive strength. It can be seen from

Equation (2-36) that the gel/space ratio is inversely proportional to the *w/c*. It can be deduced that a higher *w/c* leads to a low compressive strength of cement paste or concrete.

The minimum *w/c* ratio for complete hydration is usually assumed to be 0.36 to 0.42. It should be noted that complete hydration never happens and that residual anhydrate cement is beneficial for attaining a high ultimate strength. The space requirements for the cement gel are less than the requirements of water plus cement particles so that when the available water is used up, the cement paste will self-desiccate.

#### 2.2.2.7 Basic Tests of Portland Cement

Portland cement concrete is the most widely used material in the world. The quality of Portland cement plays an important role in assuring the quality of construction and hence requires strict quality control. In this section, the basic tests for checking the quality of Portland cement are introduced.

**(a)** *Fineness (= surface area/weight)*: The fineness of Portland cement is an important quality index. It represents the average size of the cement grains. The fineness controls the rate and completeness of hydration due to the exposure surface of cement particles to the water. The finer the cement particles, the more rapid the reaction, the higher the rate of heat evolution, and the higher the early strength. However, finer cement particles can lead to high hydration heat, high possibility of early age cracking, and possible reduced durability. The fineness of Portland cement can be measured by different methods. One is the Blaine air permeation method defined in ASTM C204. In this method, cement particles are placed on a porous bed and then a given volume of fluid (air) is passed through the bed at a steady diminishing rate. After all the air is passed, the time ($t$) for the process is recorded. The specific surface of the cement can be calculated using

$$S = K\sqrt{t} \tag{2-37}$$

where $K$ is a constant. In practice, $S$ (or $K$) can be determined by comparing the sample to the known surface area issued by the U.S. National Institute of Standards and Technology. Examples of surface measurement are the surface area and pore size analyzer, which utilizes the adsorption of nitrogen on the particle surface. The ultimate sensitivity of the equipment is sufficient to detect less than 0.001 $cm^3$ of desorbed nitrogen from 30% $N_2$/He mixture. The BET (Brunauer, Emmett and Teller) test is commonly used to analyze the specific surface area of particles and pores, which is shown in Figure 2-24. In the BET test, the amount of a monolayer adsorbate gas on the surface is measured by a volumetric or continuous flow method, which can determine the specific surface area of a particle or a pore. In Figure 2-24a, a bare sample surface begins to adsorb gas molecules at a low gas pressure, Figure 2-24b shows the number of adsorbed gas molecules increases until a monolayer is formed (Figure 2-24c), after that, more gas molecules fill the hole until the pores are fully filled (Figures 2-24d and 2-24e). Thus, an adsorption peak can be illustrated by the instrument. The specific surface area can be calculated from the adsorption peak according to the BET equation (Dollimore et al. 1976).

Another method to determine the surface area of cement particles is the Wagner turbidimeter method. In this method, cement particles are placed in a tall glass container that is filled with kerosene and then parallel rays of lights are passed through the container onto a photoelectric cell. The cross-sectional area of the particles intersecting the beam can be determined by measuring the light intensity. The radius of the particles, assuming spheres,

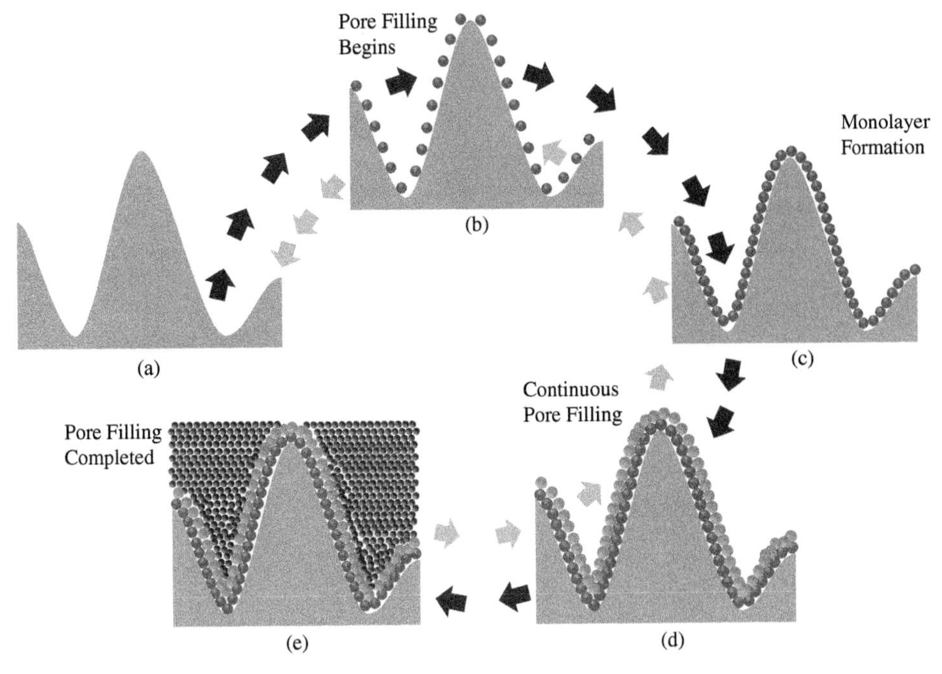


**Figure 2-24** The BET (Brunauer, Emmett and Teller) test method

can be obtained as

$$a = \sqrt{\frac{9V\eta}{2g(D_1 - D_2)}} \tag{2-38}$$

where $V$ is the velocity of the cement particle falling in the viscous medium (kerosene), $\eta$ is the viscosity coefficient, $g$ is the acceleration of gravity, $D_1$ is the density of the particles, and $D_2$ is the viscous medium. A similar instrument is the particle size analyzer, which uses a laser beam as the measurement light and has a revolution of 0.04–2000 μm, the working principle is shown in Figure 2-25.

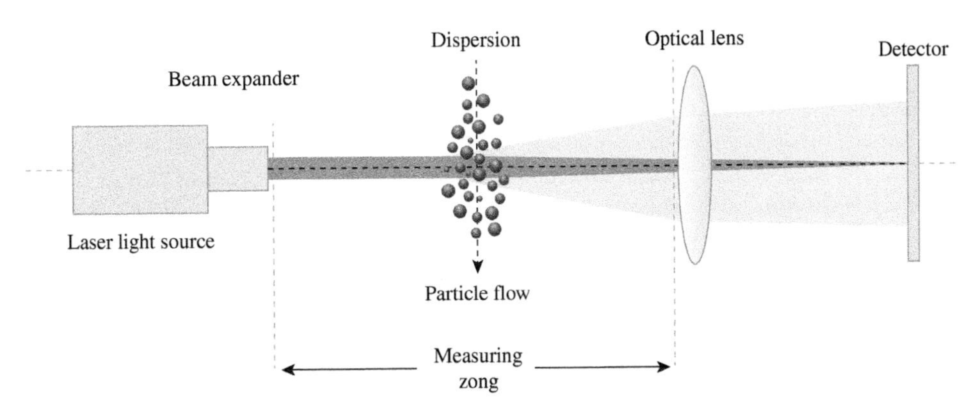


**Figure 2-25** The working principle of laser particle size analyzer

There is also a simple method to determine the fineness through sieving (ASTM C184-94). This method involves the following steps: Place a 50-g sample of cement on a clean and dry sieve having a hole size of 0.15 mm with the pan attached to it. While holding the sieve and the uncovered pan in both hands, sieve with a gentle wrist motion until most of the fine materials have passed through and the residue looks fairly clean. Place the cover on the sieve and remove the pan. Now, holding them firmly in one hand, gently tap the side of the sieve. The dust adhering to the sieve will be dislodged and the underside of the sieve may now be swept clean. Empty the pan and thoroughly wipe it out with a cloth or napkin. Replace the sieve in the pan and carefully remove the cover. Return any coarse material in the cover caught during tapping of the sieve. Continue the sieving as described earlier, without the cover, depending on the condition of cement. Continuously rotate the sieve along with gentle wrist motion, taking care not to spill any cement. Do this for about 9 min. Replace the cover and clean, following the same procedure as described earlier. If the cement is in a proper condition, there should now be no appreciable dust remaining in the residue or adhering to the sieve and the pan. Hold the sieve in one hand, with the pan and the cover attached, in a slightly inclined position and move it backward and forward in the plane of inclination. At the same time, gently strike the side about 150 times per minute against the palm of the other hand on the upstroke. Perform the sieving over a sheet of white paper. Return any material escaping from the sieve or pan and collecting on the paper. The fineness of cement can then be calculated as

$$F = 100 - \left(R_t / W\right) \times 100 \tag{2-39}$$

where $F$ = fineness of cement expressed as the percentage passing through the 0.15-mm sieve, $R_t$ = weight remaining in the 0.15-mm sieve, and $W$ = total weight of the sample in grams.

**(b)** *Normal consistency test*: This test is undertaken to determine the water requirement for the desired cement paste plasticity state required by the setting and soundness test for Portland cement. The normal consistency test is regulated in ASTM C187. The procedures are listed below:

- Secure 300 g of cement.
- Mix cement with measured quantity of clean water.
- Mold the cement paste into the shape of a ball. With gloved hands, toss the ball six times through a free path of about six inches from one hand to the other.
- Press the ball into the larger end of a Vicat ring and completely fill the ring with paste.
- Remove excessive paste without compressing samples and locate the ring under the plunger of the Vicat apparatus.
- Place the plunger with a needle of 10-mm diameter under a load of 300 g in contact with the top of the paste and lock. Set the indicator on the scale to zero.
- Release the plunger and record the settlement of the plunger in mm after 30 sec.
- Repeat the process with the trial paste with varying percentages of water until normal consistency is observed, i.e., when a Vicat plunger penetrates 10 ± 1 mm in 30 sec.

The *w/c* for a normal consistency of Portland cement is 0.24 to 0.33.

**(c)** *Time of setting*: This test is undertaken to determine the time required for the cement paste to harden. The initial set cannot be too early due to the requirements of mixing, conveying, placing, and casting. Final setting cannot be too late owing to the requirement of strength development.

Time of setting is measured by the Vicat apparatus with a 1-mm-diameter needle. The initial setting time is defined as the time at which the needle penetrates 25 mm into the cement paste. The final setting time is the time at which the needle does not sink visibly into the cement paste.

**(d)** *Soundness*: Unsoundness in cement paste results from excessive volume change after setting. Unsoundness in cement is caused by the slow hydration of MgO or free lime. The reactions are

$$MgO + H_2O \rightarrow Mg(OH)_2 \tag{2-40}$$

$$\text{and} \quad CaO + H_2O \rightarrow Ca(OH)_2 \tag{2-41}$$

Another factor that can cause unsoundness is the later formation of ettringite. Since these reactions are very slow processes, taking several months and even years to finish, and their hydration products are very aggressive, their crystal growth pressure will crack and damage the already hardened cement paste and concrete. The soundness of the cement must be tested by an accelerated method due to the slow process. One test is called the Le Chatelier test (BS 4550), and is used to measure the potential for the volumetric change of the cement paste. The Le Chatelier test is used mainly for free lime detection. The main procedures are as follows:

- Fill the cylinder-shaped container with cement paste of normal consistency as shown in Figure 2-26.
- Cover the container with glass plates.
- Immerse in water (20°C) and measure the distance of the indicator at the top of the apparatus.
- Boil the specimen for 1 h and measure the distance again after cooling.
- Expansion should less than 10 mm for acceptable cement quality.

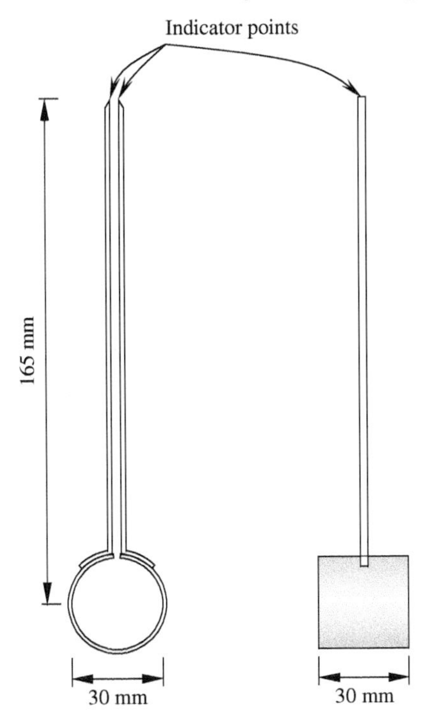


**Figure 2-26** Le Chatelier test apparatus

Another method is the autoclave expansion test (ASTM C151), which uses an autoclave to increase the temperature to accelerate the process. The procedures are as follows: Mold a cement paste with normal consistency into a container and cure normally for 14 h. Then remove from the mold, measure the size of the specimen, and place the specimen into autoclave. Raise the temperature in the autoclave so that the stem pressure inside can reach 2 MPa in 45 to 75 min. Maintain the pressure of 2 MPa for 3 h. Cool the autoclave down so that the pressure is released in 1.5 h. Cool the specimen in water to 23°C in 15 min. After another 15 min, measure the size of the specimen again; the expansion must be less than 0.80% to be acceptable. Autoclave testing can test both excess free lime and excess MgO.

**(e)** *Strength*: The strength of cement is measured on mortar specimens made of cement and standard sand (silica). Compression testing is carried out on a 50-mm cube with an *S/C* ratio of 2.75:1 and *w/c* ratio of 0.485, for Portland cements. The specimens are tested wet, using a loading rate that can let the specimen fail in 20 to 80 sec. A direct tensile test is carried out on a specimen shaped like a dog's bone, and the load is applied through specifically designed grips. Flexural strength is measured on a 40 × 40 × 160-mm prism beam test under a center-point bending.

**(f)** *Heat of hydration test* (BS 4550: Part 3: Section 3.8 and ASTM C186): Hydration is a heat-release process. The heat of hydration is usually defined as the amount of heat released during the setting and hardening at a given temperature, measured in J/g. The experiment is called the heat of solution method. Basically, the heat of solution of dry cement is compared to the heats of solution of separate portions of the cement that have been partially hydrated for 7 and 28 days. The heat of hydration is then the difference between the heats of solution of the dry and partially hydrated cements for the appropriate hydration period. Useful information is the accumulated percentage of the heat release. It is verified by experiments that the heat release is 50% for 1–3 days, 75% for 7 days, 83–91% for 6 months.

The hydration heat can raise the temperature in concrete to 50–60°C and thus cause microcracks. The hydration heat should be taken care of in massive concrete constructions.

**(g)** *Other experiments, including sulfate expansion and mortar air content*: These tests are more meaningful for concrete and hence we discuss them in Chapter 5 in detail. Cement S.G. and U.W. The S.G. for most types of cements is 3.15, and the U.W. about 1000–1600 kg/m$^3$.

### 2.2.3 Supplementary Cementitious Materials

#### 2.2.3.1 Silica Fume

Silica fume is a by-product of induction arc furnaces in the silicon metal and ferrosilicon alloy industries. Reduction of quartz to silicon at temperatures up to 2000°C produces $SiO_2$ vapors, which oxidize and condense in the low temperature zone to very fine spherical particles consisting of noncrystalline silica. Hence, silica fume is also called condensed silica fume or microsilica. The material is collected by filtering the outgoing gases in bag filters. The typical size distributions of silica fume, ordinary Portland cement, and fly ash are compared in Figure 2-27.

More accurately, the size distribution of a typical silica fume product is provided as: 20% below 0.05 micron, 70% below 0.10 micron, 95% below 0.20 micron, and 99% below 0.50 micron. As most of the silica fume particles are less than 100 nanometers, they can be treated as nano-particles. Moreover, the typical chemical composition of a silica fume is shown in Table 2-9. It can be seen that silica dominates (> 92%) in the material.

Micro silica has a surface area around 20 m$^2$/g and an average bulk density of 586 kg/m$^3$. Compared to normal Portland cement and typical fly ashes, silica fume sizes are two orders of

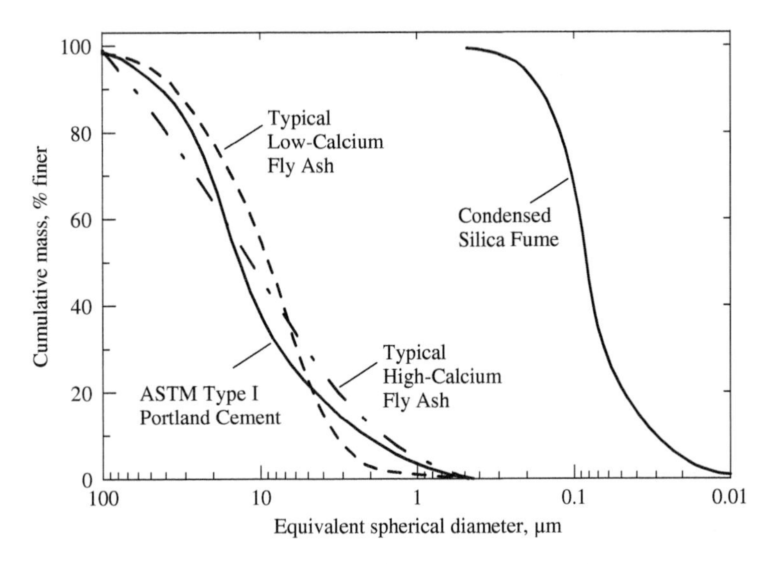


**Figure 2-27** Size distribution of typical Portland cement, fly ash, and silica fume

**Table 2-9** Typical chemical compositions of silica fume

| Chemical compositions | Typical contents by weight (%) | Standard deviation | Minimum (%) | Maximum(%) |
|---|---|---|---|---|
| $SiO_2$ | 92.9 | 0.60 | 92.0 | 94.0 |
| $Al_2O_3$ | 0.69 | 0.10 | 0.52 | 0.86 |
| $Fe_2O_3$ | 1.25 | 0.46 | 0.74 | 2.39 |
| CaO | 0.40 | 0.09 | 0.28 | 0.74 |
| MgO | 1.73 | 0.31 | 1.23 | 2.24 |
| $K_2O$ | 1.19 | 0.15 | 1.00 | 1.53 |
| $Na_2O$ | 0.43 | 0.03 | 0.37 | 0.49 |
| C | 0.88 | 0.19 | 0.01 | 0.03 |
| Cl | 0.02 | 0.01 | 0.01 | 0.03 |
| S | 0.20 | – | 0.10 | 0.30 |
| P | 0.07 | – | 0.03 | 0.12 |
| LOI | 1.18 | 0.26 | 0.79 | 0.73 |
| Moisture | 0.30 | 0.09 | 0.09 | 0.50 |

magnitude finer. Thus the material is highly pozzolanic. However, it creates handling problems and increases the water requirement in concrete appreciably, unless superplasticizer is used.

Silica fume is supplied in two forms: powder and slurry. The slurry of silica fume is adopted to avoid possible health issues caused by breathing in the fine particles of silica fume when people work with it. Before the mid-1970s, silica fume was discharged into the atmosphere. After environmental concerns necessitated the collection and landfilling of silica fume, it has been applied in concrete production since the 1980s and now has become valuable material for high performance concrete. The normal percentage of silica fume used to replace Portland cement is from 5–15%. Silica fume benefits concrete properties in two ways: particle packing and pozzolanic reaction. Since the size of the silica fume is two magnitudes of order smaller than that of Portland cement, it can easily fill the spaces between the cement particles. Subsequently, a denser concrete microstructure can be achieved and a high compressive strength can be reached. Pozzolanic reaction is defined as:

$$\text{Pozzolan} + \text{calcium hydroxide} + \text{water} = \text{calcium silicate hydrate (secondary)} \qquad (2\text{-}42)$$

Silica fume is a very active pozzolan material and readily reacts with CH and water to form secondary C-S-H. Through pozzolanic reaction, silica fume can consume a large amount of CH and generated C-S-H also fills the capillary voids. The process can further reduce the porosity and permeability in concrete as well as the possibility of chemical reaction of CH with other ions to form harmful products. Thus, the durability of concrete can be greatly enhanced. Li, Mu and Peng (1999) have shown that silica fume can greatly enhance the resistance of high performance concrete to alkali aggregate reaction and chloride diffusion. Nowadays, silica fume is widely used in producing high performance and ultrahigh performance concrete worldwide.

#### 2.2.3.2 Fly Ash

Fly ash (pulverized fuel ash) is a by-product of an electricity-generating plant using coal as a fuel. During combustion of powdered coal in modern power plants, as coal passes through the high-temperature zone in the furnace, the volatile matter and carbon are burned off, whereas most of the mineral impurities, such as clays, quartz, and feldspar, will melt at the high temperature. The fused matter is quickly transported to lower-temperature zones, where it solidifies as spherical particles of glass. Some of the mineral matter agglomerates, forming bottom ash, but most of it flies out with the flue gas stream and thus is called fly ash. This ash is subsequently removed from the gas by electrostatic precipitators.

Fly ash can be divided into two categories according to the type of coal burned (ASTM C618): class F fly ash is obtained by burning anthracite or bituminous coal and class C by burning lignite or sub-bituminous coal. The chemical compositions of class F and class C are listed in Table 2-10. It can be seen from Table 2-10 that class F contains less than 10% CaO and class C 15–30% of CaO. Thus, sometimes class F is called low-calcium fly ash and class C is called high-calcium fly ash. Usually, high-calcium fly ash is more reactive because it contains most of the calcium in the form of reactive crystalline compounds, such as $C_3A$ and CS.

The morphology of fly ash is compared with that of silica fume in Figure 2-28. It shows that most of the particles in fly ash occur as solid spheres of glass, but sometimes a small number of hollow spheres, called cenospheres (completely empty) and plerospheres (packed with numerous

**Table 2-10** Chemical compositions of fly ashes (%)

| Oxide | Class F | Class C |
|---|---|---|
| $SiO_2$ | 49.1 | 53.79 |
| $Al_2O_3$ | 16.25 | 15.42 |
| Fe2O3 | 22.31 | 5 |
| TiO2 | 1.09 | 1.68 |
| CaO | 4.48 | 18 |
| MgO | 1 | 3.4 |
| Na2O | 0.05 | 0.5 |
| K2O | 1.42 | 0.5 |
| SO3 | 0.73 | 1.44 |
| LOI | 2.55 | 0.8 |

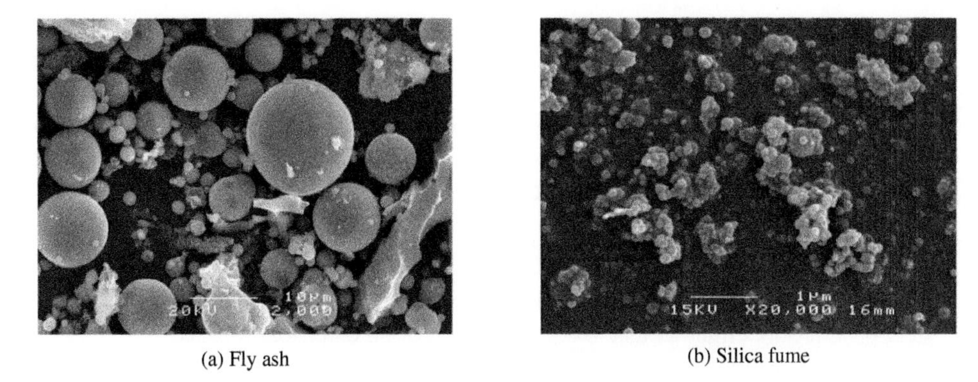


(a) Fly ash (b) Silica fume

**Figure 2-28** Morphology of fly ash and silica fume

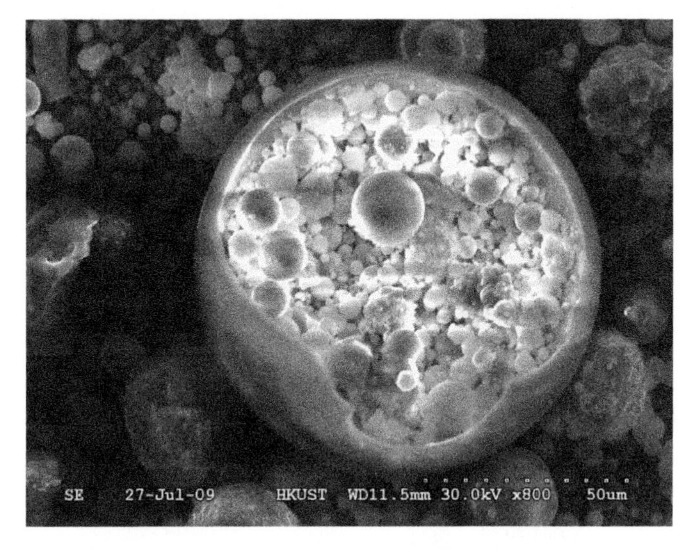


**Figure 2-29** Fly ash particle having plerospheres

small spheres), are present, see Figure 2-29. The size distributions of fly ash are slightly smaller than those of Portland cement with more than 50% under 20 μm.

Fly ash can be used to produce a so-called blended cement. In this case, fly ash is milled with clinker and gypsum in the last procedure in the cement production. Fly ash can also be used to produce modified concrete by adding it during the mixing process. Research carried out in the past 40 years or so shows that the incorporation of fly ash into concrete has certain advantages and some disadvantages. Incorporation of fly ash into concrete can improve the workability due to the spherical shape and glassy surface of the fly ash particles. By replacing cement with fly ash, the cost of concrete can be reduced, since fly ash costs less than cement. Because fly ash is an industrial by-product, it lowers the energy demand in producing concrete. In addition, incorporating fly ash into concrete can reduce the hydration heat of fresh concrete and is good for mass concrete structures.

The disadvantages of fly ash concrete are low early age strength and longer initial setting time due to the low reactivity of fly ash. Much research has been carried out to improve the reactivity by chemically activating it (Xie and Xi, 2000). One method is to add an alkali activator such as

1 or 2% NaOH or KOH into concrete mix. Another is to use lime to mix with the fly ash for a few days before incorporating the fly ash into the concrete mix. Recently, some studies using a mechanical method to wet grind or mill fly ash to "activate" its reactivity have been conducted and some promising results have been obtained (Blanco et al., 2005). The normal replacement of cement by fly ash is around 25–30% by weight. However, in high-volume fly ash concrete, fly ash content up to 60% is reached (Langley et al., 1989).

#### 2.2.3.3 Slag

Slag is a by-product of iron or steel production. For iron production, blast furnaces are used and for steel production, either a basic oxygen furnace or an electrical furnace. The slag produced in iron production is different from that from steel production. Currently, the slag used in the concrete industry is mainly the slag from iron production, and its full name is ground, granulated blast furnace slag or GGBS.

In the production of cast iron or pig iron, the liquid slag is usually quenched from a high temperature to ambient temperature rapidly by either water or a combination of air and water. Due to the very fast process, most of the lime, magnesia, silica, and alumina in slag are held in a noncrystalline or glassy state. The water-quenched product is called granulated slag due to the sand-size particles, while the slag quenched by air and a limited amount of water, which is in the form of pellets, is called palletized slag. Normally, the former contains more glass; however, when ground to powder with a fineness of 400–700 $m^2/kg$ Blaine, both products can develop satisfactory cementitious properties. Figure 2-30 shows the slag particles observed with SEM. The production and use of ground, granulated blast-furnace slag have a long history of more than 100 years. The typical chemical composition of slag is shown in Table 2-11. It can be seen from Table 2-11 that CaO and $SiO_2$ are the two main components, and the CaO content in slag is quite close to that of high-calcium fly ash.

Ground granulated blast-furnace slag and high-calcium fly ashes are similar in mineralogical character and reactivity. Besides a similar content of CaO, both are essentially noncrystalline, and their high-calcium glassy phases have a similar order of reactivity. Compared to low-calcium fly

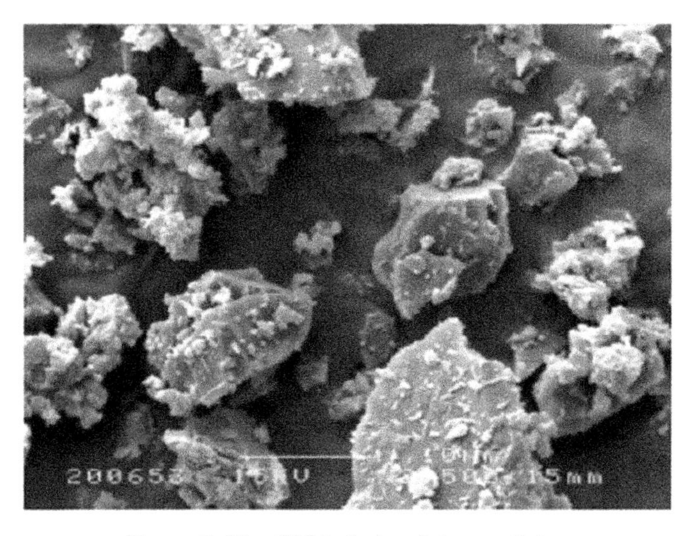

**Figure 2-30** SEM photo of slag particles

**Table 2-11** Chemical compositions of slag (%)

| CaO | $SiO_2$ | $Al_2O_3$ | $Fe_2O_3$ | MgO | $SO_3$ | $K_2O$ | $Na_2O$ | LOI |
|---|---|---|---|---|---|---|---|---|
| 30~45 | ~30 | 10~15 | 1~2 | < 6 | < 6 | 0.4~1.5 | 0.05~0.5 | 0.2~1 |

ash, which usually does not make any significant contribution to the strength of Portland cement concrete until after about 4 weeks of hydration, the strength contribution by high-calcium fly ash or granulated blast-furnace slag may become apparent as early as 7 days after hydration. Although the particle size characteristics, composition of glass, and glass content are the primary factors determining the activity of fly ashes and slag, the reactivity of the glass itself varies with the thermal history of the material. Glass chilled from a higher temperature and at a faster rate will have a more disordered structure and will therefore be more reactive.

Slag particles of less than 10 μm contribute to early strength in concrete up to 28 days; particles of 10–45 μm contribute to later strength, but particles coarser than 45 μm are difficult to hydrate. Since the slag obtained after granulation is very coarse and moist, it is dried and pulverized to a particle size mostly under 45 μm, which corresponds to approximately a 500-$m^2$/kg Blaine surface area.

Similar to fly ash, slag can also be used in both cement and concrete production. In cement production, slag can be milled with clinker and gypsum to produce Portland–slag blended cement. In concrete production, up to 50% of cement can be replaced by slag. In slag-modified concrete, the products of slag hydration form a mixture of C–S–H and AFm. The C–S–H formed through slag has a lower C/S ratio, as expected. Slag-modified concrete shows an improved slump retention as well as both early and long-term strength (Yao et al., 1998). However, reactivities can vary from slag to slag even if they have similar composition. Experimental verification is necessary to identify the reactivity of specific types of slag in practice.

#### 2.2.3.4 Calcined Clay and Limestone Calcined Clay Cement

A coupled addition of calcined clay and limestone is used to substitute part of the clinker in a blended cement, which develops a new type of cement: limestone calcined clay cement (LC3). This new type of cement is usually produced with low-grade clay and dolomite-rich limestone, which is considered a traditional production waste. Firing takes place at half the clinkerization temperature. Because of this low temperature, the $CO_2$ emissions can be reduced by up to 40% in the LC3 production process.

LC3 has a composition of 50% clinker, 30% calcined clay, 15% limestone, and 5% gypsum. The limestone usually provides $CaCO_3$ to the blended system. The calcined clay supplies additional alumina with a further reaction to form alumina phases (Antoni et al., 2012). The maximum dosage of pozzolan should < 35% in a normal blended cement system. If the content of pozzolan is kept at ~35%, the clinker can be offset by additional limestone in the blended system. The ratio of calcined clay:limestone can affect their substitution rate in the clinker. The total substitution of calcined clay and limestone can increase 15%, and reach up to 50% in the clinker when calcined clay:limestone is 2:1. The performance of this ternary system is usually better than common binary clinker systems with those pozzolan admixtures, although there is a large amount of substitute species in it (Scrivener and Favier, 2015). This ternary system usually shows a higher early strength compared with traditional binary systems. This phenomenon can be attributed to the high hydration reactivity of alumina phases at an early age, eliminating the main shortcoming of pozzolanic cement.

Manufacture of LC3 involves two stages: calcination and grinding. For calcination, a normal rotary kiln is needed, similar in operating principle to rotary kilns for clinkerization. Two technologies can be used at ~700 to ~850 °C for clay calcination: stationary and flash calcination. During the process of calcination, the clay remains a layered structure to some extent. The chemical bonds between the layered clay structures are driven off and substantial reorganization of the basic building units results in a highly amorphous material. The calcined clay platelets contribute to increased specific surface of the cementitious blend which may result in a slightly higher water demand in comparison to pure Portland systems. Additionally, grinding can also be carried out with conventional equipment. Due to the multi-component nature of LC3, having ingredients with different hardness, separate grinding may be preferable.

High-reactivity metakaolin (MK) is one of the recently developed supplementary cementing materials for high-performance concrete. It is produced by calcining purified kaolinite clay in a specific temperature range (650–800°C) to drive off the chemically bound water in the interstices of the kaolin and destroy the crystalline structure, which effectively converts the material to the MK phase, an amorphous aluminosilicate. Unlike industrial by-products, such as silica fume (SF), fly ash, and blast-furnace slag, MK is carefully refined to lighten its color, remove inert impurities and control its particle-size. This well-controlled process results in a highly reactive white powder that is consistent in appearance and performance. The particle size of MK is generally less than 2 μm, which is significantly smaller than cement particles, though not as fine as SF, see Figure 2-31. It is typically incorporated into concrete to replace 5~20 % of cement by mass.

The typical chemical composition of MK is shown in Table 2-12. It can be seen that silicon dioxide and aluminum oxide are the two main components in MK and little CaO exists in MK. Similar to silica fume, metakaolin improves the concrete performance by the packing effect and reacting with calcium hydroxide to form secondary C-S-H. However, MK was found to improve concrete properties while offering good workability. It has been found that concrete modified by MK requires less water-reducing admixture than that modified by SF to achieve a comparable fluidity (Caldarone et al., 1997). It has also been demonstrated that MK is particularly effective in reducing the rate of diffusion of sodium and chloride ions. The diffusion coefficient can be reduced

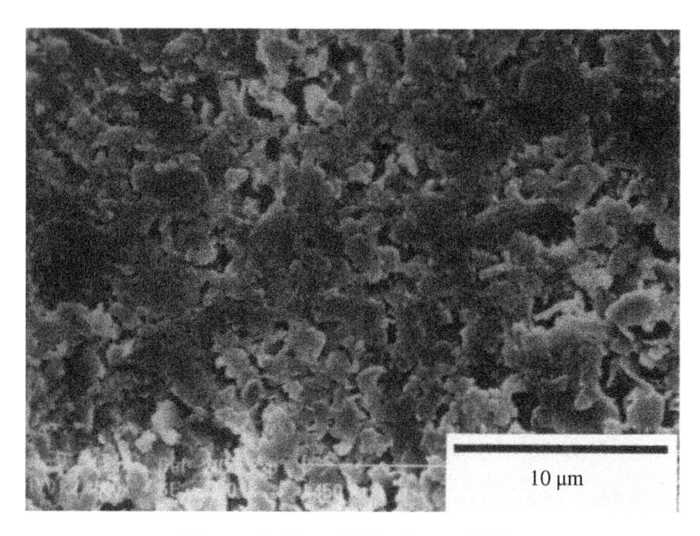


**Figure 2-31** SEM photo of MK

**Table 2-12** Chemical composition of MK (%)

| $SiO_2$ | $Al_2O_3$ | $Fe_2O_3$ | CaO | LOI |
|---|---|---|---|---|
| 51.34 | 41.95 | 0.52 | 0.34 | 0.72 |

by about 50% when 8% MK is added into the concrete mixture (Thomas et al., 1997). Ding and Li (2002) systematically studied the effect of MK and SF on the workability, strength, shrinkage, and resistance to chloride penetration of concrete. They found that MK and SF had similar functions in improving the strength of concrete, reducing free drying shrinkage and chloride diffusion rate. In addition, because of its white color, high-reactivity MK does not darken concrete as SF typically does (the white-colored SF is very limited in tonnage), which makes it suitable for color matching and other architectural applications. However, the cost of MK is now much lower than that of SF. Thus, MK has a more potential for application in concrete structures.

According to recent research and application results, LC3 generally exhibits better durability characteristics than OPC. Components made of LC3 have higher resistance to moisture ingress, gas permeation, chloride ion ingress, and sulfate attack. On the contrary, LC3 shows lower carbonation resistance and poor workability. These shortcomings can be overcome by adding a shrinkage reducing admixture.

#### 2.2.3.5 Alternative Supplementary Cementitious Materials

Supplementary cementitious materials (SCM) such as fly ash, silica fume and blast furnace slag have been widely used in concrete technology. These industrial by-products are the main source of the mineral components of SCM. Because of their pozzolanic activity in cement hydration, the performance of cement-based materials can be improved significantly. However, the demand for concrete is so massive that the available good quality SCM cannot satisfy the developing market (Sobolev, 1993; Scrivener and Kirkpatrick, 2008). Some alternative supplementary cementitious materials including off-spec products and less-investigated by-products (Gokmenoglu and Sobolev, 2002; Sobolev and Arikan, 2002; Sobolev, 2003; 2005) have become the alternative SCM of choice. It should be noted that the long-term performance of these alternative SCMs has not been very clear, thus these alternative SCMs cannot be widely used. The strength, durability, and the unwanted substances, such as heavy metals, should be considered carefully in the alternative SCM applications. The current economic conditions make recycling feasible for only a limited number of by-products. To increase the rates of recycling, cement and concrete manufacturers need a uniform supply of quality waste. The classification and identification technology of industrial waste should be further developed. In the previous sections, silica fume, fly ash and slag were discussed as the SCM. In this section, some alternative supplementary cementitious materials are introduced:

**(a)** *Steel slag*: The steel industry provides valuable raw materials in the form of different slags. Because slag replaces nonrenewable mineral resources and reduces the consumption of natural materials in construction, its use is a sustainable approach. Steelmaking slag includes blast furnace slag (BFS) from cast iron production, electric arc furnace (EAF), basic oxygen furnace (BOF/BOS) and ladle steel slag types (Heidrich, 2002). The EAF mills produce steel by remelting scrap steel (Azom, 2012) and integrated mills produce steel by combining molten cast iron with scrap steel in the BOF. Steel manufactured in an EAF or BOF is further refined in a ladle furnace and this process further produces different grades of steel ladle slag (Azom, 2012).

The EAF/BOS slag from steel mills can be recycled into a range of useful products for building roads and pavements, and to substitute for virgin quarried materials, mainly aggregates for asphalt and Portland cement concrete. Some common applications for EAF/BOS slag are:

- base course and top course for asphalt roads;
- sub-base material for rigid pavements;
- anti-skid surfacing for roads on accident-prone intersections and curves;
- low strength concrete;
- controlled low strength material for backfill and trench stabilization;
- raw material for production of Portland cement clinker;
- control of $CO_2$ emissions.

Steel slag has been used in construction since at least the mid-nineteenth century. It is used in all industrialized countries. The German steel industry conducted extensive research on the application of BOF, EAF, and blast furnace slags. Modern steel slags are well characterized and evaluated for long-term performance. These are commonly used as aggregates for road construction (e.g., asphaltic or unbound layers), as armor stones for hydraulic engineering construction (e.g. shore stabilization), and as a fertilizer for agriculture.

**(b)** *Sugarcane bagasse ash*: Sugarcane bagasse ash (SCBA) is obtained from the boilers of the sugar industries as a combustion by-product. After crushing of sugarcane in sugar mills and extraction of the juice from the prepared cane by milling, the discarded fibrous residual matter of cane is called bagasse. Bagasse is very commonly used as a fuel in boilers in the sugar mills for cogeneration processes. After burning in the cogeneration boiler, bagasse ash is collected in a baghouse filter and is disposed of locally, which causes severe environmental problems. Because of the rapid increase in cogeneration plants in the sugar industries, bagasse ash generation is also increasing significantly in India and other major sugarcane-producing countries, such as Brazil and Thailand. SCBA is mainly composed of silica and can be used as a supplementary cementitious material in concrete instead of being disposed of. Several studies have reported that sugarcane bagasse ash can be used as a pozzolanic material (Chusilp et al., 2009; Cordeiro et al., 2009; Frias et al., 2011; Montakarntiwong et al., 2013; Bahurudeen et al., 2015; Bahurudeen et al., 2016).

**(c)** *Waste glass*: Waste glass comes from various sources: glass containers (bottles and jars), construction glass (windows), and electrical equipment (lamps, old style monitors, and TVs). Most (89%) of the waste glass comes from various containers (Sobolev et al., 2004). Generally, recovered glass containers are recycled into new containers, other portions are used in newly emerging sectors such as fiberglass insulation, abrasives, light-weight aggregates, yet some quantities are used for Portland cement and asphalt concrete as an aggregate (Stewart, 1986; Apotheker, 1989; Geiger, 1994). Recycling of waste glass is attractive to the glass manufacturers because it reduces the costs associated with raw materials and the technological process; it lowers energy consumption; and also eliminates the need to dump waste glass in landfills. However, to recycle waste glass effectively within the glass industry, it must be glass of similar composition, which has been separated from contaminants which could reduce the quality of the new glass products. The following contaminants affect the recycling of waste glass (Stewart, 1986; Rodriguez, 1995; Mayer 2000; Sobolev et al., 2004; Gunalaan and Kanapathy, 2013):

- glass of fluctuating composition or color (compared with the mainstream);
- ceramics (dishware, porcelain, pottery, bricks, concrete);

- metals (including container lids or seals);
- organics (paper, plastics, cork, wood, plants, food residue, especially sugar).

Utilization of glass powder (GLP) from mixed color waste packaging glass comprising soda-lime glass was reported by Shayan and Xu (2006). It was demonstrated that incorporating different proportions (0%, 20%, and 30%) of GLP as a cement replacement in 40 MPa concrete can result in some improvements with respect to drying shrinkage and alkali reactivity. Incorporation of GLP into a concrete matrix decreases chloride ion penetration, thereby reducing the risk of chloride corrosion of steel rebars in concrete. The results of reported research proved that up to 30% of GLP can be used as a supplementary cementitious material to replace cement without loss in concrete performance.

#### 2.2.3.6 Benefits of Using SCMs in Concrete

For modern or advanced concrete technology, various mineral admixtures have been widely used in concrete construction. The aims in using these SCMs in concrete are to gain economic benefits, protect the environment, improve the workability of fresh concrete, enhance the strength, and especially the durability of hardened concrete, and decrease hydration heat. The benefits of using various SCMs in modern concrete construction are shown in Tables 2-13 and 2-14.

**Table 2-13** Benefits of using mineral admixtures in fresh concrete

| | | | | Workability of fresh concrete | | |
|---|---|---|---|---|---|---|
| **Mineral Admixtures** | **Usual Dosage[a] (%)** | **Economic Benefits** | **Protecting Environment** | **Increasing Fluidity** | **Increasing Cohesiveness** | **Decreasing Segregation** |
| PFA or FA | 10–40 | ✓ | ✓ | ✓ | ✓ | ✓ |
| BFS or slag | 20–60 | ✓ | ✓ | ✓ | ✓ | ✓ |
| Silica fume | 5–15 | | ✓ | | ✓ | ✓ |
| Other pozzolans | 10–30 | ✓ | ✓ | | | ✓ |

[a]Replacement of Portland cement.

**Table 2-14** Benefits of using mineral admixtures in hardened concrete

| **Mineral Admixtures** | **Usual Dosage[a] (%)** | **Increasing Strengths** | **Decreasing Heat** | **Decreasing Cracking** | **Impermeability** | **Anti-chemical Attack[d]** |
|---|---|---|---|---|---|---|
| PFA or FA | 10–40 | ✓[b] | ✓ | ✓ | ✓ | ✓ |
| BFS or slag | 20–60 | ✓[c] | ✓ | | ✓ | ✓ |
| Silica fume | 5–15 | ✓[c] | | ✓ | ✓ | ✓ |
| Other pozzolans | 10–30 | ✓[b] | ✓ | ✓ | ✓ | ✓ |

[a]Replacement of Portland cement.
[b]Increasing strength at long term age.
[c]Increasing strength at early age and long-term age, especially using silica fume.
[d]Anti-chemical attack may involve salt scaling, sulfates or seawater attack, resistance to alkali–aggregate reaction (AAR), resistance to corrosion of rebar, and so on.

### 2.2.4 Alternative Binders

#### 2.2.4.1 Alkali Triggered Binders (Geopolymers)

Alkali triggered binders, also called "geopolymers" or "inorganic polymers" based on the physio-chemical properties of the source materials and the type of alkali activator (Duxson et al. 2007), are inorganic binding materials that are produced through the reaction of an alkali source and aluminosilicates. This section uses the term of geopolymers and introduces their advantages, development, reaction mechanism, and applications.

**(a)** *Advantages of Geopolymers*:

Compared to ordinary Portland cement, newly developed inorganic binder geopolymers possess the following characteristics:

*Abundant raw material resources*: Any pozzolanic compound or source of silicates or alumino-silcates that is readily dissolved in alkaline solution will suffice as a source for the production of a geopolymer.

*Energy saving and environment protection*: Geopolymers do not require large energy consumption. A large amount of $CO_2$ is emitted during the production of Portland cement, which is one of the main reasons for global warming. Studies have shown that one ton of carbon dioxide gas is released into the atmosphere for every ton of Portland cement made anywhere in the world. In contrast, geopolymer cement is manufactured in a different way from Portland cement. It does not require extremely high-temperature treatment of the limestone. Only low-temperature processing of naturally occurring or direct man-made alumino-silicates (kaoline or fly ash) provides suitable geopolymeric raw materials. This leads to a significant reduction in energy consumption and $CO_2$ emission. Thermal processing of natural alumino-silicates at a relatively low temperature (600–800°C) provides suitable geopolymeric raw materials, resulting in much less energy consumption than for Portland cement. In addition, only a small amount of $CO_2$ is emitted. It was reported by Davidovits (1994b) that about 60% less energy is required, and 80–90% less $CO_2$ is generated in the production of geopolymers than for Portland cement. Thus the development and application of geopolymer cement are of great significance in environmental protection.

*Simple preparation technique*: Geopolymers can be synthesized simply by mixing alumino-silicate-reactive materials and strongly alkaline solutions, then curing at room temperature. In a short period, reasonable strength will be gained. This is very similar to the preparation of Portland cement concrete.

*Good volume stability*: Geopolymers have 80% lower shrinkage than Portland cement.

*Reasonable strength gain in a short time*: Geopolymers can obtain 70% of the final compressive strength in the first 4 hrs of setting.

*High fire resistance and low thermal conductivity*: Geopolymer cement possesses excellent high temperature resistance up to 1200°C and can endure 50-kW/m$^2$ fire exposure without sudden property degradation. In addition, no smoke is released after the extended heat flux. The heat conductivity of geopolymers varies from 0.24 to 0.3 w/m-k, and compares well with that of lightweight refractory bricks (0.3 to 0.438 w/m-k).

**(b)** *Development of Geopolymers*

Since French scientist Davidovits invented geopolymer materials in 1978 (Davidovits, 1993), great interest in the development of geopolymers has been voiced around the world.

More than 28 international scientific institutions and companies have presented updated research and published their results. These works mainly focus on the following aspects.

**(i)** *Solidification of toxic waste and nuclear residues*: Davidovits et al. (1994c) first began to investigate the possibilities of heavy metal immobilization by commercial geopolymeric products in the early 1990s. The leachate results for geopolymerization on various mine tailings showed that over 90% of heavy metal ions included in the tailings can be tightly solidified in a 3D geopolymer framework. In the middle of the 1990s, Van Jaarsveld and Van Deventer et al. (Davidovits et al., 1990; Van Jaarsveld and Van Deventer, 1997, 1999; Van Jaarsveld et al., 1998) also set out to study the solidification effectiveness of geopolymers manufactured from fly ash. The bond mechanism between heavy metal ions and the geopolymer matrix is also simply explained on the basis of the XRD, infrared spectroscopy,and magic-angle spinning nuclear magnetic resonance and leaching results. The European research project GEOCISTEM (Van Jaarsveld et al., 1999) successfully tested geopolymerization technology in the context of the East German mining and milling remediation project, carried out by WISMUT. Another research project into the solidification of radioactive residues was jointly carried out by Cordi-Geopolymer and Comrie Consulting Ltd. (European R&D project BRITE-EURAM BE-7355-93, 1997).

**(ii)** *Fire resistance*: The Federal Aviation Administration (FAA), USA, and the Geopolymer Institute of Cordi-Geopolymere SA, France (Comrie Consulting Ltd, 1988), jointly initiated a research program to develop low-cost, environmentally-friendly, fire-resistant matrix materials for use in aircraft composites and cabin interior applications. The flammability requirement for new materials is that they withstand a 50-kW/m$^2$ incident heat flux characteristic from a fully developed aviation fuel fire penetrating a cabin opening, without propagating the fire into the cabin compartment. The goal of the program is to eliminate cabin fire as a cause of death in aircraft accidents. As with this program, the fire-resistance properties of geopolymers reinforced by various types of fiber, such as carbon fiber, glass fiber, and SiC fiber, were tested and the fireproof mechanics were also analyzed. In addition, comparisons were made among geopolymer composites and carbon-reinforced polyesters, vinyl, epoxy, bismaleinide, cyanate ester, polyimide, phenolic, and engineering thermoplastic laminates. The test results showed that these organic large molecular polymers ignited readily and released appreciable heat and smoke, while carbon-fiber-reinforced geopolymer composites did not ignite, burn, or release any smoke even after extended heat flux exposure. On the basis of these fireproof studies, some nonflammable geopolymer composites for aircraft cabin and cargo interiors were produced on November 18, 1998, in Atlantic City, NJ, USA.

**(iii)** *Archeological research*: In the 1970s, Davidovits proposed a controversial theory documented in a book by Lyon (1994) that has since gained widespread support and acceptance. He postulated that the great pyramids of Egypt were not built by natural stones, but that the blocks were cast in place and allowed to set, creating an artificial zeolitic rock using geopolymerization technology. He collected a great amount of evidence from ancient Egyptian literature and samples in sites to confirm his geopolymerization theory. From then on, many experts began to focus their concerns on geopolymer studies. Some related papers (Davidovits, 1987; Campbell and Folk, 1991; Morris, 1991; Folk and Campbell, 1992; Mckinney, 1993) and patents were also published.

**(c)** *Reaction Mechanism of Geopolymers*

Many studies on the formation mechanism have been made since the invention of geopolymers, but only one formation mechanism was proposed by Davidovits. He believed that geopolymer synthesis consists of three steps—dissolution of alumino-silicate under a

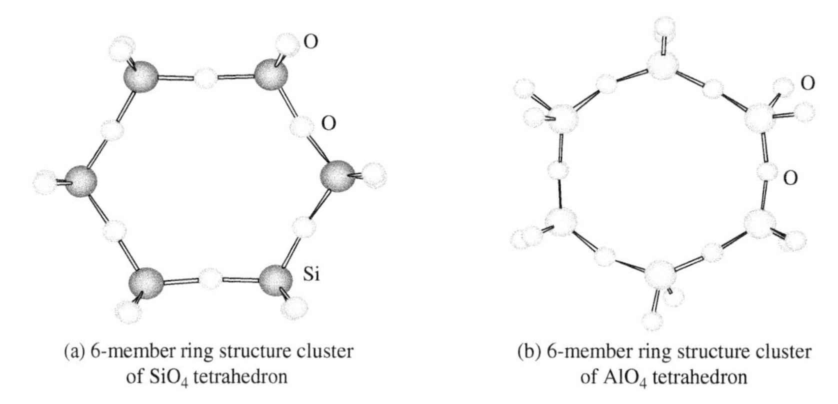


(a) 6-member ring structure cluster of $SiO_4$ tetrahedron

(b) 6-member ring structure cluster of $AlO_4$ tetrahedron

**Figure 2-32** Molecular structure representing a model of metakaolinite

strong alkali solution, reorientation of the free ion clusters, and polycondenzation—but that each step includes many pathways. The dissolution step, for example, includes 8 pathways, according to thermodynamics. Different pathways can create different ion clusters that directly determine the final properties of a geopolymer. Thus, it is important to understand the actual pathway when producing a geopolymer in order to gain insights into the mechanism of geopolymerization. However, until now, these studies have not been carried out. Because the forming rate of a geopolymer is very rapid, these three steps take place almost at the same time, which make the kinetics of the three steps interdependent. Thus, it is impossible to separate these steps in experimental studies, which leads to the use of molecular simulation to solve these problems.

In the studies, two 6-membered-ring molecular structural models, to represent the chemical structure of metakaolinite (main raw material for synthesizing geopolymer), were established to quantitatively analyze the formation process of a geopolymer, as shown in Figure 2-32. Based on these two 6-membered ring models, all possible dissolution pathways of metakaoline under a strongly alkali environment were numerically simulated using quantum mechanics, quantum chemistry, computation chemistry, and thermodynamics theories. All possible pathways, Equations 2-43–2-50, involved in the formation process of the geopolymer were analyzed, and the enthalpies of each possible pathway were also calculated, as listed in Table 2-15. As a result, the optimum theoretical pathways in the geopolymerization process were determined.

$$\left(\mathrm{Si(OH)_2O}\right)_6 + 3\mathrm{NaOH} \rightarrow$$
$$\mathrm{(OH)_3Si} \equiv \left(\mathrm{Si(OH)_2O}\right)_3 - \mathrm{Si(OH)_3} + \mathrm{OH} - \mathrm{Si} - 3\mathrm{(ONa)_3} + \mathrm{H_2O}\Delta\mathrm{E1} \tag{2-43}$$

$$\left(\mathrm{Si(OH)_2O}\right)_6 + 3\mathrm{KOH} \rightarrow$$
$$\mathrm{(OH)\,3Si} - \left(\mathrm{Si(OH)_2O}\right)_3 - \mathrm{Si(OH)_3} + \mathrm{HO} - \mathrm{Si} - 3\mathrm{(OK)_3} + \mathrm{H_2O}\Delta\mathrm{E2} \tag{2-44}$$

$$\left(\mathrm{Si(OH)_2O}\right)_6 + 4\mathrm{NaOH} \rightarrow$$
$$\mathrm{(OH)_3Si} - \left(\mathrm{Si(OH)_2O}\right)_3 - \mathrm{Si(OH)_2} - \mathrm{ONa} + \mathrm{HO} - \mathrm{Si} - 3\mathrm{(ONa)_3} + 2\mathrm{H_2O}\Delta\mathrm{E3} \tag{2-45}$$

**Table 2-15** Reaction heats of single 6-member ring structure models under strongly alkaline solution

(a) single 6-member ring of $SiO_4$ tetrahedral

| The molecular structural unit | Formation enthalpy | Reaction enthalpy (kJ/mol) | | | |
|---|---|---|---|---|---|
| | | ΔE1 | ΔE2 | ΔE3 | ΔE4 |
| $(Si(OH)_2O)_6$ | −1491.45 | | | | |
| $(OH)_3Si{-}(Si(OH)_2O)_3{-}Si(OH)_3$ | −1294.65 | | | | |
| $(OH)_3Si{-}(Si(OH)_2O)_3{-}Si(OH)_2{-}ONa$ | −1385.43 | | | | |
| $(OH)_3Si{-}(Si(OH)_2O)_3{-}Si(OH)_2{-}OK$ | −1370.88 | | | | |
| $HO{-}Si{\equiv}(ONa)_3$ | −500.94 | −5.49 | 12.12 | −36.23 | −19.23 |
| $HO{-}Si{\equiv}(OK)_3$ | −437.85 | | | | |
| NaOH | −119.30 | | | | |
| KOH | −104.14 | | | | |
| $H_2O$ | −59.25 | | | | |

(b) single 6-member ring of $AlO_4$ tetrahedral

| The molecular structural unit | Formation enthalpy | Reaction enthalpy (kJ/mol) | | | |
|---|---|---|---|---|---|
| | | ΔE1 | ΔE2 | ΔE3 | ΔE4 |
| $(Al^-(OH)_2O)_6$ | −619.67 | | | | |
| $(OH)_3Al^-{-}(Al^-(OH)_2O)_3{-}Al^-(OH)_3$ | −776.45 | | | | |
| $(OH)_3Al^-{-}(Al^-(OH)_2O)_3{-}Al^-(OH)_2{-}ONa$ | −839.39 | | | | |
| $(OH)_3Al^-{-}(Al^-(OH)_2O)_3{-}Al^-(OH)_2{-}OK$ | −810.59 | | | | |
| $HO{-}Al^-{\equiv}(ONa)_3$ | −441.66 | −299.80 | −245.79 | −302.69 | −235.03 |
| $HO{-}Al^-{\equiv}(OK)_3$ | −342.17 | | | | |
| NaOH | −119.30 | | | | |
| KOH | −104.14 | | | | |
| $H_2O$ | −59.25 | | | | |

$$(Si(OH)_2O)_6 + 4KOH \rightarrow (OH)_3Si - (Si(OH)_2O)_3 - Si(OH)_2 - OK + HO - Si - 3(OK)_3 + 2H_2O\,\Delta E4 \tag{2-46}$$

$$(Al^-(OH)_2O)_6 + 3NaOH \rightarrow (OH)_3Al^- - (Al^-(OH)_2O)_3 - Al^-(OH)_3 + HO - Al^- - 3(ONa)_3 + H_2O\,\Delta E5 \tag{2-47}$$

$$(Al^-(OH)_2O)_6 + 3KOH \rightarrow (OH)_3Al^- - (Al^-(OH)\,2O)_3 - Al^-(OH)_3 + HO - Al^- - 3(OK)_3 + H_2O\,\Delta E6 \tag{2-48}$$

$$(Al^-(OH)_2O)_6 + 4NaOH \rightarrow (OH)_3Al^- - (Al^-(OH)_2O)_3 - Al^-(OH)_2 - ONa + HO - Al^- - 3(ONa)_3 + 2H_2O\,\Delta E7 \tag{2-49}$$

$$(Al^-(OH)_2O)_6 + 4KOH \rightarrow (OH)_3Al^- - (Al^-(OH)_2O)_3 - Al^-(OH)_2 - OK + HO - Al^- - 3(OK)_3 + 2H_2O\,\Delta E8 \tag{2-50}$$

**(d)** *Microstructure Characterization of Geopolymers*

The structural characteristics of the products directly determine the final mechanical and durability properties. The case is also true for geopolymers. Many researchers have investigated the microstructure using different advanced techniques. Because a geopolymer is a type of amorphous 3D material with complex composition, it is very difficult to quantitatively measure the exact arrangement and chemical atmosphere of different atoms in geopolymers. If we want to solve this difficulty, we have to turn to statistical theories for establishing its molecular model. Unfortunately, until now, these studies have not yet been done. Therefore, the structural nature of geopolymers is not understood thoroughly.

The relationship between geopolymers and the corresponding zeolites has been investigated and the intertransformation between geopolymers and zeolites can be realized under specified conditions. On the basis of these results, the microstructure of geopolymers can clearly be characterized: a geopolymer is an amorphous 3D alumino-silicate material, which is composed of $AlO_4$ and $SiO_4$ tetrahedral lined alternatively by sharing all oxygen atoms. Positive ions ($Na^+$, $K^+$) are present in the framework cavities to balance the negative charge in a fourfold coordination. In addition, 3D statistical models of geopolymers, as shown in Figure 2-33, were simulated according to the decomposition results of MAS-NMR spectra.

**(e)** *Applications of Geopolymers*:

Geopolymers are an abundant raw resource, and have low $CO_2$ emission, less energy consumption, low production cost, high early strength, and fast setting. These properties make geopolymers suitable for applications in many fields of industry, such as civil engineering, the automotive and aerospace industries, nonferrous foundries and metallurgical industries, the plastics industries, waste management, art and decoration, and retrofitting of buildings. Several areas of application are described below.

**(i)** *Toxic waste treatment*: Immobilization of toxic waste may be one of the major areas where geopolymers can impact significantly on the status quo. The molecular structure of a geopolymer is similar to that of zeolites or feldspathoids, which are known for their excellent abilities to adsorb and solidify toxic chemical wastes, such as heavy metal ions and nuclear residues. It is the structures that make a geopolymer a strong candidate for immobilizing hazardous elemental wastes. Hazardous elements that are present in waste materials mixed with geopolymer compounds are tightly locked into the 3-D network of the geopolymer bulk matrix.

**(ii)** *Civil engineering*: Geopolymer binders behave similarly to Portland cement, and can set and harden at room temperature, and gain reasonable strength in a short period of time. Some geopolymer binders have been tested and proven to be successful in construction, transportation, and infrastructure applications. They yield synthetic mineral products with properties such as high mechanical performance, hard surface (Davidovits 1994a; Lyon et al., 1997), thermal stability, and high acid resistance. Any current building component, such as bricks, ceramic tiles, and cement, could be replaced by geopolymers.

**(iii)** *Automotive and aerospace*: The merits of high-temperature resistance allow geopolymers to have great advantages in the automotive and aerospace industries. At present, some geopolymer products are being used in aircraft to avoid cabin fires in aircraft accidents.

(a) Statistical structure model of K-PS geopolymer: the Si ions with a symbol of ⊕ corresponding for $Si^4(2Al)$, ⊖ for $Si^4(4Al)$, • for $Si^4(4Si)$

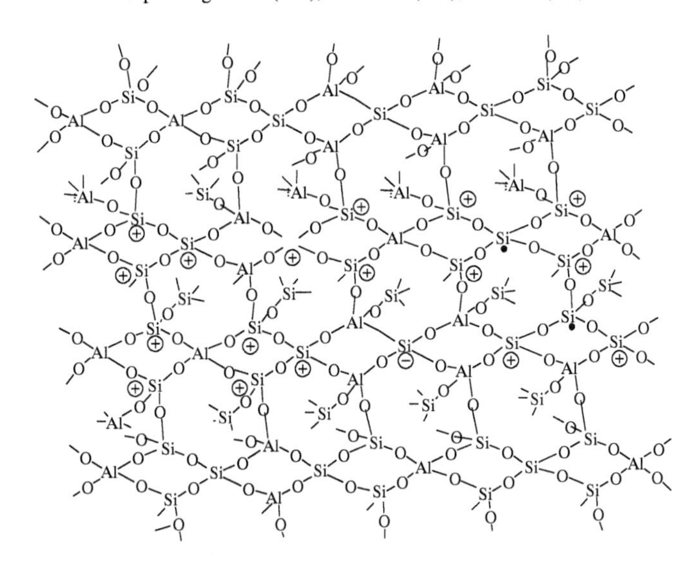

(b) Statistical structure model of K-PSDS geopolymer

(c) Statistical structure model of K-PSS geopolymer

**Figure 2-33** Statistical structure models of K-PSS geopolymer

#### 2.2.4.2 Magnesium-Based Binder

The typical magnesium-based binder includes magnesium phosphoric cement (MPC) and magnesium oxychloride cement (MOC).

**(a)** *Magnesium Phosphoric Cement (MPC)*:

MPC is a type of artificial stone made from an acid-base reaction of magnesia and phosphates. They possess some properties that Portland cements do not possess according to the previous studies. Therefore, they can be used in fields in which Portland cements are not suitable (Kingery, 1950; Yoshizake et al., 1989; Seehra et al., 1993; Singh et al., 1997; Wagh et al., 1999; Yang et al., 2002). The applications of MPCs include the following:

**(i)** Due to its rapid setting and high early strength, MPC has been used in rapid repair of concrete structures, such as highways, airport runways, and bridge decks, for many years. It can save a lot of idle time and cost caused by the long disruptive period of other materials. If the interruption period is too long for a busy highway, airport runway, or bridge, etc., it will cause losses of millions of dollars. By using MPC materials, the interruption time for transportation can be greatly shortened. Therefore, valuable time and resources can be saved.

**(ii)** MPC can be incorporated with nontoxic industrial waste, such as class F fly ash (FA) and be converted into useful construction materials. The addition of FA to MPC can be over 40% by mass of MPC, about two times that of PC. In addition, MPC can combine with the FA that is not suitable in PC because of its high carbon content and other impurities. Besides FA, acid blast furnace slag, red mud (the residue of the aluminum industry), and even tails in gold mines can also be used in MPC in large amounts. These wastes are difficult to use in PC concrete in appreciable amounts.

**(iii)** Due to the high alkali environment of PC (pH over 12.5), when they are used with fiber reinforcement, some components such as natural fibers, notably lignin, and hemicellulose will be susceptible to degradation. However, the lower alkalinity of MPC matrices (pH value 10 to 11) makes them potentially better suited to fiber reinforcement obtained from vegetation. Furthermore, the sugar in some natural fibers, such as sugarcane and cornstalks, can prohibit the setting of PC, which weakens the bonding between the Portland cement and the fiber. The set of MPC is not influenced by sugar.

**(iv)** MPC can be used in the management and stabilization of toxic and radioactive wastes, including solids and liquids. The waste can be micro- and/or microencapsulated and chemically bonded by MPC to form a strong, dense, and durable matrix that stores the hazardous and radioactive contaminants as insoluble phosphates, and microencapsulates insoluble radioactive components. The waste forms are not only stable in groundwater environments, but are also nonignitable and hence safe for storage and transportation.

**(v)** MPCs are very suitable for repairing deteriorated concrete pavements in cold regions. MPC can develop strength at low temperature due to its exothermic hydration and low water-to-binder ratio. At the same time, MPCs possess a higher deicer scaling property than Portland cement.

**(vi)** The raw material of MPC is hard burnt magnesia, and is, in fact, a refractory material. Therefore, MPC can be designed to be fireproof and/or as a cold setting refractory, according to practical need.

Phosphate bonding has been known about for about a century, since the advent of dental cement formulations. In the refractory industry, the properties of cold-setting and heat-setting compositions were used as chemically bonded refractory material. According to the comprehensive studies of Kingery in 1950, phosphate bonding can be classified as (1) zinc–phosphate bonding; (2) silicate–phosphoric acid bonding; (3) oxide–phosphoric acid bonding; (4) acid phosphate bonding; and (5) metaphosphate–polyphosphate bonding (Kingery, 1950). The oxides, such as magnesium, aluminum, and zirconium, will react with phosphoric acid or acid phosphate at room temperature, forming a coherent mass, which sets quickly and gives high early strength. The hydration system, based on magnesia and ammonia phosphate (Kingery, 1950; Yoshizake et al., 1989; Seehra et al., 1993; Singh et al., 1997; Wagh et al., 1999; Yang et al., 2002) attracted the most attention in the past.

From the 1970s, many patents using the reaction of magnesia and acid ammonia phosphate had been granted for rapid repair of concrete. The variation in patents arises from the use of different raw materials, inert materials to reduce cost, and retarders to control the reaction rate. Most claims are supported by a few examples cited in the patents without a systematic scientific approach. From the middle of the 1980s, systematic studies on magnesia and ammonia phosphates were made by researchers (Yoshizake et al., 1989; Seehra et al., 1993; Yang et al., 2002). The hydration products, setting process, and strength development were the main thrusts of these investigations. Only a very few papers focused on the durability of the system (Yoshizake et al., 1989; Seehra et al., 1993; Yang et al., 2002). By the mid-1990s, it was found that MPC can be incorporated with industrial waste, producing a solid to bond toxic waste (Davidovits, 1993; Singh et al., 1997; Wagh et al., 1999). Therefore, MPC became a leading candidate for sustainable development. The environmental benefits arise from two aspects: (1) the nontoxic industrial waste can be recycled to useful building materials, and (2) many toxic and radioactive wastes are difficult to treat with traditional processes, but can easily be treated by MPC. This function ensures more promising uses of MPC in the future, especially for the sustainable development of modern society.

Concerning the durability of MPCs, research that has been done by other investigators mainly includes superior durability topics, such as freezing–thawing and scaling resistance, protection of steel from corrosion, better bonding properties with waste organic materials, the transfer of noncontaminated industrial wastes into useful construction materials, and the stabilization of toxic or radioactive wastes.

The deterioration of concrete pavements is mainly caused by frost action in cold regions, and is severely amplified by the use of deicer chemicals. The repair material must possess high frost/deicer resistance, and MPC has very high deicer–frost resistance (Yoshizake et al., 1989; Yang et al., 2002). Scaling does not occur on the surfaces of MPC materials until after 40 freeze–thaw cycles. The regime of freeze–thaw cycling was achieved with a cooling rate of about 0.5°C/min, freezing for 4 h at -20 ± 2°C and then thawing for 4 h at 20 ± 5°C. A 3% NaCl solution was used as the deicer solution. The studies showed that the freezing/thawing resistance of MPCs was basically equal to the well-air-entrained PC concrete in general.

Steel corrosion in PC concrete is a serious problem, and MPC is an inhibitor of steel corrosion, forming an iron phosphate film on the surface of the steel. The pH of hardened MPC mortar is 10 to 11, and this may contribute a little to the inhibition of reinforcing steel corrosion. In addition, the ratio of permeability of MPC to PC concrete is 47.3%, or more than double the resistance to permeation (Yoshizake et al., 1989). Abrasion resistance testing has shown that MPC mortar possesses approximately double the abrasion resistance compared with slab-on-grade floor concrete, and nearly equal that of pavement concrete (Yoshizake et al., 1989; Seehra et al., 1993). With respect to chemical corrosion resistance, in the case of continuous immersion of specimens in sulfate solutions and potable water, results indicate that MPC mortar patches will remain durable under sulfate and moist conditions.

A wide range of waste particle sizes can be used when producing structural products using MPC. Styrofoam materials are candidates for optimal results. Styrofoam articles can be completely coated with a thin, impermeable layer of MPC. The uniform coating of the Styrofoam particles not only provides structural stability but also confers resistance to fire, chemical attack, humidity, and other weathering conditions. The Styrofoam insulation material provides superior *R* values. Furthermore, wood waste (suitable size range from 1 to 5 mm long, 1 mm thick, and 2 to 3 mm wide) can be bonded with MPC to produce chipboard having good flexural strength. For example, samples containing 50 wt.% of wood and 50 wt.% of binder achieve approximately 10.4 MPa in flexural strength. Samples containing 60 and 70 wt.% of wood exhibit flexural strength of 2.8 and 2.1 MPa, respectively. Once the wood and binder are thoroughly mixed, the samples are subjected to pressurized molding to the order of approximately 18.3 MPa, for approximately 30–90 min.

With the progress of modern civilization, living conditions have been greatly improved; at the same time, however, a large amount of industrial waste (including toxic and nontoxic) has been produced. MPC can bind lots of nontoxic industrial waste to useful construction materials. If the wastes are toxic, MPC can solidify and stabilize them. It is important to recycle or stabilize waste, especially when natural resources are becoming more and more scarce. The waste is in various forms, such as aqueous liquids, inorganic sludge, particles, heterogeneous debris, soils, and organic liquids. However, only a few parts of the total waste can be recycled, such as fly ash and red mud, which can be blended with Portland cement. Most of the wastes need to be solidified and stabilized. Because of the diverse nature of the physical and chemical composition of these wastes, no single solidification technology can be used to successfully treat and dispose of them. For example, the low-level wastes contain both hazardous chemical and low-level radioactive species (Singh et al., 1997). To stabilize them requires that the two kinds of contaminants be immobilized effectively. Generally, the contaminants are volatile compounds and hence cannot be treated effectively by high-temperature processes.

In a conventional verification or plasma hearth process, such contaminants may be captured in secondary waste streams or off-gas particulates that need further low-temperature treatment for stabilization. Also, some of these waste streams may contain pyrophoric that will ignite spontaneously during thermal treatment and thus cause hot spots that may require an expensive control system and equipment with demanding structural integrity. Therefore, there is a critical need for a low-temperature treatment and stabilization technology that will effectively treat the secondary wastes generated by high-temperature treatment processes and wastes that are not amenable to thermal treatment. Now those wastes can be successfully solidified by magnesia phosphate cement or chemically bonded phosphate ceramics (CBPC) (Singh et al., 1997). Other forms of waste, such as ash, liquid, sludge, and salts can be also solidified by MPC.

MPC is extremely insoluble in groundwater which will protect groundwater from contamination by the contained waste. Long-term leaching tests conducted on magnesium phosphate systems have shown that these phosphates are insoluble in water and brine. The radiation stability of MPC is excellent (Wagh et al., 1999). No changes in the mechanical integrity of the materials were detected even after gamma irradiation to a cumulative dosage of $10^8$ rads.

**(b)** *Magnesium Oxychloride Cement (MOC)*:

Magnesium oxychloride cement (MOC), also known as Sorel cement (Sorel, 1867), is a type of nonhydraulic cement. It is formed by mixing powdered magnesium oxide (MgO) with a concentrated solution of magnesium chloride ($MgCl_2$). Magnesium oxychloride cement has many superior properties as compared to ordinary Portland cement (Bensted and Barnes, 2002). It has high fire resistance, low thermal conductivity, and good resistance to abrasion, and is unaffected by oil, grease, and paint. It also has high early strength and is suitable for use with all kinds of aggregates in large quantities, including gravel, sand, marble flour, asbestos, wood particles, and expanded clays. The lower alkalinity of magnesium oxychloride (pH of 10–11), compared to the higher alkalinity of ordinary cement (pH of 12–13), makes it suitable for use with glass fiber by eliminating aging problems.

Magnesium oxychloride cement has drawn much research interest due to the ever-increasing awareness of the need for environmental protection (Li et al., 2003). One of the important issues is to recycle waste wood in light of producing cement-based wood composites. However, lignin compounds and some other adverse chemicals contained in wood significantly retard the hydration of ordinary Portland cement. To solve the problem, magnesium oxychloride cement provides an excellent substitute for the binder (Odler, 2000). Some other major commercial applications of magnesium oxychloride cement are industrial flooring, fire protection, and grinding wheels. Due to its resemblance to marble, it is also used for rendering wall insulation panels, stuccos with revealed aggregates, and decorative purposes (de Henau and Dupas, 1976).

Magnesium oxide, or calcined magnesia, is normally obtained by calcinations of magnesite ($MgCO_3$) at a temperature of around 750°C. The quality or reactivity of the formed magnesium oxide powder is largely affected by its thermal history (calcination temperature and duration) and particle size (Haper 1967; Sorrel and Armstrong 1976; Matkovic et al., 1977). This in turn influences both the reaction rate and the properties of the reacted products of magnesium oxychloride cement. The setting and hardening of the magnesium oxychloride cement takes place in a through-solution reaction (Urwongse and Sorrell, 1980). The four main reaction phases in the ternary MOC system are $2Mg(OH)_2{\cdot}MgCl_2{\cdot}4H_2O$ (phase 2), $3Mg(OH)_2{\cdot}MgCl_2{\cdot}8H_2O$ (phase 3), $5Mg(OH)_2{\cdot}MgCl_2{\cdot}8H_2O$ (phase 5), and $9Mg(OH)_2{\cdot}MgCl_2{\cdot}5H_2O$ (phase 9). Of these, phases 3 and 5 may exist at ambient temperature, whereas phases 2 and 9 are stable only at temperatures above 100°C (Cole and Demediuk, 1955). Another possible reaction product with a suitable reaction environment is magnesium hydroxide or brucite, $Mg(OH)_2$.

Therefore, optimum formation of phase 5 crystals in the hydrated MOC cement is desirable, as it is widely reported that the crystals provide the best mechanical properties. The formation mechanism of the four magnesium oxychloride phases has sparked vigorous discussion (Bilinski, et al., 1984; Ménétrier-Sorrentino et al., 1986; Deng and Zhang 1999). Theoretically speaking, phase 5 can be obtained from a molar ratio of $MgO/MgCl_2$ of 5 along with the water required by the stoichiometry. The mechanical strength developed largely depends on the MOC phases produced and consequently on the appropriate proportions of the starting materials. Nevertheless, the correct or theoretical proportions of the starting materials alone are not sufficient to ensure the formation of phase 5 crystals, since the reactivity of MgO can have an influence.

For normal practice, it is believed that the chemical reactions in the system of $MgO{-}MgCl_2{-}H_2O$ are not complete and many unreacted MgO particles are expected to be left in the final reaction products. While unreacted MgO particles can be treated as a filler, surplus chloride ions are troublesome, as they cause corrosion problems (Maravelaki-Kalaitzaki and Moraitou, 1999) when they are involved in reinforcing steels. Besides, a higher water content is usually required as a lubricant for the required workability of a mixture. Therefore, excess magnesium oxide and water are suggested to be used in producing magnesium oxychloride cement to ensure the formation of phase 5, while keeping the free chloride ions to a minimum. From the practical point of view, the optimal molar ratios between different components of the ternary system $MgO{-}MgCl_2{-}H_2O$ would be the most important thing to know before further formulation for commercial products with various additives and fillers (Ji, 2001; Deng, 2003).

#### 2.2.4.3 Carbonated Binders

Carbonated binders are those that harden by carbonation reactions. They are obtained by carbonating reactive magnesia, calcium hydroxide, high-calcium fly ash, etc. in well-controlled (temperature and pressure) $CO_2$ curing conditions. The carbonated binder is environmentally-friendly due to the significant consumption of $CO_2$ during production. In this section, the reactive magnesia cement is mainly introduced.

Reactive MgO cement formulations were developed and patented just over a decade ago by the Australian scientist John Harrison (2003) and are blends of PC and reactive MgO (sometimes also incorporating fly ash) in different proportions, depending on the intended application, ranging from structural concrete to porous masonry units. They have been developed with strong emphasis on a range of sustainability advantages over PC and have received significant publicity. Advantages include: (1) manufacture at a much lower temperatures (650–800°C); (2) potential complete recyclability of the MgO; (3) uptake of significant quantities of $CO_2$; and (4) insensitivity to impurities.

Most MgO commercially produced from the calcination of magnesite is likely to have been calcined at temperatures at the higher end of the reactive MgO range of ~1000 °C hence their reactivity is likely to be close to that of hard burned MgOs. The MgO content was found to vary between 60% and 99.6% where the synthetically produced (seawater and chemical precipitation) were at the higher purity end. The main impurities of the calcined MgOs were CaO and $SiO_2$, which are common minerals of rock, while those of the synthetic MgOs were CaO, Cl and $SO_3$. The CaO content varied between 0.15% and 6.9% and the loss of ignition (LOI) between 0.8% and 7%. The surface areas measured were between 2 and 148$m^2$ /g and the average particle size between 1.8 and 35mm. The reactivity of these MgOs, using the citric acid test, varied significantly from as little as 9 seconds (extremely reactive) to >2.5 hours (highly unreactive). The equilibrium pH of the MgOs also varied significantly between 10.0 and 12.5, mainly related to the CaO content. A degree of hydration of 40–80% was observed with a limit of 80–95%; the degree and rate of hydration were enhanced by the presence of certain hydration agents similar to those used with PC.

The hydration and carbonation reaction mechanisms of MgO have been studied. In the presence of water and under sufficient $CO_2$ in the curing atmosphere (5 ~ 20%), MgO hydrates to form $Mg(OH)_2$, or brucite; and brucite further carbonates to form one or more hydrated magnesium carbonates:

$$\begin{aligned} &MgO + H_2O \rightarrow Mg(OH)_2 \\ &Mg(OH)_2 + CO_2 + 2H_2O \rightarrow MgCO_3 \cdot 3H_2O(\text{nesquehonite}) \end{aligned} \tag{2-51}$$

and/or

$$5Mg(OH)_2 + 4CO_2 + H_2O \rightarrow Mg_5(CO_3)_4(OH)_2 \cdot 5H_2O(\text{dypingite}) \tag{2-52}$$

and/or

$$5Mg(OH)_2 + 4CO_2 \rightarrow Mg_5(CO_3)_4(OH)_2 \cdot 4H_2O(\text{hydromagnesite}) \tag{2-53}$$

These hydrated magnesium carbonates form well-ramified networks of massive dense crystals, with different morphologies depending on the conditions of their formation, with a very effective binding ability. They are metastable compounds and can undergo transformation to less hydrated forms depending on the conditions they are exposed to, including elevated temperature, $CO_2$ partial pressure, pH, and water activity.

The reactive MgO cements, which are blends of MgO and PC, have emerged as a more sustainable alternative to PC and with anticipated superior technical performance. Compared to PC, the water demand of reactive MgO is quite high in the region of 0.45–0.70. In PC–MgO paste blends, both PC and MgO were found to hydrate mainly independently forming a blend of their own hydration products, typical of their own raw materials, but did interact such that the setting time was prolonged compared to that of the individual materials, due to the common ion effect (Masse et al., 1993). As the MgO content in the blends increased, the ability to obtain densely packed mix decreased. With MgO content of up to 15% in PC–MgO pastes, the MgO hydration rate was found to vary in the range of 50–95%, between MgOs from different sources and different calcination conditions, and seems to correlate reasonably well with the reactivity and surface area. Some strength enhancement, up to 5%, was observed for some of the MgOs used and this was usually in the MgO content range of ~4–7% (Li, 2012). Additionally, reactive MgO, are also used as activators for slags. Reactive MgO, in 2.5–20% content, showed effective activation of a range of slags, with the performance being primarily and strongly affected by the slag used and the reactivity of the MgO (Li, 2012; Yi et al., 2012). The early age degree of hydration of MgO in slag–MgO pastes was found to be ~50% higher than in PC–MgO pastes (Li, 2012). Dense microstructure was observed generally supporting high compressive strengths observed for the pastes of up to 35 MPa with 5% MgO content. It should be noted that the expansive MgO reactions have been found, in the appropriate dosages, to compensate for shrinkage of cement-based materials. Hard burned MgO has been used for shrinkage compensation in concrete dams in China for more than half a century.

The performance of reactive MgO has been investigated in a number of different applications including porous blocks, both in the laboratory as well as commercial full-scale trials, concrete, ground improvement and in a range of aggressive and extreme environments. As discussed in the use of PC, reactive MgO additives and admixtures play a major role in significantly enhancing the performance of the cement. Hence the applicability of the typical range of such additives and admixtures that are commonly used with PC is currently being investigated with reactive MgO.

#### 2.2.4.4 Special Cements

Compared to the normal cement mentioned above, special cements are different from those cementitious materials that have special properties, special functions, or special application. In this section, two types of special cements, aluminate cement and phosphoaluminate cement, are introduced.

**(a)** *Aluminate cement*: Calcium aluminate (CA) cements are similar to the more familiar Portland cements in that they both require water for hydration, they both form concretes that set in about the same time, and they both require similar mix designs and placing techniques. There are, however, important differences between the two cements.

First, Portland cements are made by reacting limestone and clay to produce calcium silicates, while calcium aluminate cements (also called high-alumina cements) are made by reacting a lime-containing material with an aluminous material to produce calcium aluminates.

Second, calcium aluminate cements, when mixed with suitable aggregates, are generally used for special applications where advantage can be taken of their unique properties. Calcium aluminate cements are rarely used for cast-in-place structural work, except for emergency repairs and foundation construction. Some of the purposes for which CA cement concretes may be specified include: cold weather work; resistance to high temperatures; resistance to mild acids and alkalis; resistance to sulfates, sea water, and pure water; and rapid hardening. It is important to differentiate between rapid setting and rapid hardening. Calcium aluminate cement concretes are not rapid setting. They are, however, rapid hardening; that is, they will develop as much strength in 24 hours as Portland cement concrete will achieve in 28 days.

Calcium aluminate cement concretes should be cured for at least 24 hours, using a water spray or fog, ponding, wet burlap or a curing membrane. When working with calcium aluminate cement concrete for the first time, one must remember that is should be handled much the same way as Portland concrete. For satisfactory results, however, two points should be emphasized. First, keep the water-cement ratio below 0.4 and use mechanical vibration to place the concrete. Second, these concretes develop heat much more rapidly than Portland cement concrete does, and good curing for 24 hours after placement is mandatory for satisfactory strength gain.

**(b)** *Phosphoaluminate cement*: Phosphoaluminate cement (PAC) is an anionic cementitious material including [P-O] and [Al-O] groups, which possesses its characteristic mineral composition during sintering, phase L, ternary phosphoaluminate compounds, modified phase CA and CxP, and so on (Liu et al., 2011). Early-high strength property of PAC is due to the replacement between [$P^{5+}$] and [$Al^{3+}$], which produces lots of defects and increases the hydration activity. Portland cements (PC), as the commonest cementitious materials used in construction engineering, have low-early strength which is nearly 20 to 30 MPa.

Main hydration products of PC are calcium hydroxide (CH), calcium silicate hydrate (C-S-H), and C-A-H. CH easily reacts with sulphate and produces expansive products which can destroy the structures. So it is not suitable for any special projects and the environment. PAC has many special properties compared to PC. The main hydration products of PAC are calcium phosphorus aluminates hydrate (C-A-P-H), calcium phosphate hydrate (C-P-H), calcium aluminates hydrate (C-A-H), the corresponding hydration microcrystal as well as gels. No CH and calcium sulphoaluminate hydrate (AFt) are formed during the process of the hydration.

## 2.3 ADMIXTURES

Historically, an admixture is almost as old as concrete itself. The Romans used animal fat, milk, and blood to improve their concrete properties. Although these were added to improve workability, blood was a very effective air-entraining agent and might well have improved Roman concrete durability. In more recent times, calcium chloride was often used to accelerate the hydration of cement. The systematic study of admixtures began with the introduction of air-entraining agents in the 1930s, when it was accidentally found that cement ground with beef tallow (grinding aid) had more resistance to freezing and thawing than a cement ground without beef tallow. Nowadays, as we mentioned earlier, admixtures are important and necessary components for modern concrete technology. The concrete properties, both in fresh and hardened states, can be modified or improved by admixtures. The benefits of admixtures to concrete are listed in Table 2-16. Today, almost all the contemporary concretes contain one or more admixtures. It is thus important for civil engineers to be familiar with commonly used admixtures.

### 2.3.1 Definition and Classifications

An admixture is defined as a material other than water, aggregates, cement, and reinforcing fibers that is used in concrete as an ingredient, and added to the batch immediately before or during mixing. Admixtures can be roughly divided into the following groups:

**(a)** *Air-entraining agents (ASTM C260)*: This kind of admixture is used to improve the frost resistance of concrete.

**(b)** *Chemical admixtures (ASTM C494 and BS 5075)*: A chemical admixture is any chemical additive to the concrete mixture that enhances the properties of concrete in the fresh or hardened state. The general-purpose chemicals include those that reduce the water demand for a given workability (called *water reducers*), and those chemicals that control the setting time and strength gain rate of concrete (called *accelerators* and *retarders*). Apart from these chemicals, there are others for special purposes—viscosity-modifying agents, shrinkage-reducing chemicals, and alkali–silica reaction-mitigating admixtures.

**Table 2-16** Beneficial effects of different kinds of admixtures on concrete properties

| Concrete property | Admixture type | Category of admixture |
|---|---|---|
| Workability | Water reducers | Chemical |
| | Air-entraining agents | Air entraining |
| | Inert mineral powder | Mineral |
| | Pozzolans | Mineral |
| | Polymer latexes | Miscellaneous |
| Set control | Set accelerators | Chemical |
| | Set retarders | Chemical |
| Strength | Pozzolans | Mineral |
| | Polymer latexes | Miscellaneous |
| Durability | Air-entraining agents | Air entraining |
| | Pozzolans | Mineral |
| | Water reducers | Chemical |
| | Corrosion inhibitors | Miscellaneous |
| | Shrinkage reducer | Miscellaneous |
| Special concrete | Polymer latexes | Miscellaneous |
| | Silica fume | Mineral |
| | Expansive admixtures | Miscellaneous |
| | Color pigments | Miscellaneous |
| | Gas-forming admixtures | Miscellaneous |

**(c)** *Mineral admixtures*: This kind of admixture consists of finely divided solids added to concrete to improve its workability, durability, and strength. Slag and pozzolans are important categories of mineral admixtures.

**(d)** *Miscellaneous admixtures* include all those materials that do not come under the above-mentioned categories, such as latexes, corrosion inhibitors, and expansive admixtures.

### 2.3.2 Chemical Admixtures

This class of admixtures encompasses the total spectrum of soluble chemicals that are added to concrete for the purpose of modifying setting times and reducing the water requirements of concrete mixes. Figure 2-34 shows some typical chemical admixtures in liquid form.

#### 2.3.2.1 Water-Reducing Admixtures

Water-reducing admixtures are used to reduce the water content of a concrete mixture while maintaining a given constant workability. Application of a water-reducing admixture can achieve different purposes. The resultant effect of reduced water content is increased strength and durability of concrete. However, water reducers may also be employed to "plasticize" the concrete, i.e., to make concrete flowable. In this case, the water content (or water-to-cement ratio) is held constant, and the addition of the admixtures makes the concrete flow better, while the compressive strength (which is a function of the water-to-cement ratio) is not affected. Another use of water reducers is to lower the amount of cement (since water is proportionately reduced) without affecting strength and workability. This makes the concrete cheaper and more environmentally friendly, as less cement is consumed.

Water reducers are classified broadly into two categories: (1) normal and (2) high range. The normal water reducers are also called *plasticizers*, while the high-range water reducers are called *superplasticizers*. While the normal water reducers can reduce the water demand by 5–10%, the

**Figure 2-34** Commonly used admixtures: (a) corrosion inhibitor; (b) set-retarding admixture; (c) air entraining agent; and (d) high-range water-reducing admixture (superplasticizer)

high-range water reducers can cause a reduction of 15–40%. Lignosulfonate salts of sodium and calcium are an example of normal water reducers. Lignosulfonates are derived from neutralization, precipitation, and fermentation processes of the waste liquor obtained during production of paper-making pulp from wood. Sulfonated naphthalene formaldehyde (SNF) and polycarboxylic ether (PCE) are examples of the high-range water reducers. SNF is produced from naphthalene by oleum or $SO_3$ sulfonation; subsequent reaction with formaldehyde leads to polymerization and the sulfonic acid is neutralized with sodium hydroxide or lime. Polycarboxylic ether is manufactured by a polymerization process in which a free radical mechanism with peroxide initiators is used.

Lignosulfonates are generally regarded as first-generation water reducers, while the sulfonated naphthalene formaldehyde condensates are second-generation water reducers and polycarboxylic ether is a third-generation water reducer. The introduction of polycarboxylate dispersants into the concrete industry in the last decade has made it possible to develop water-reducer molecules in specific and tailored ways to influence the performance of the material. This is a tremendous technological advancement for the concrete industry as it enables the use of molecules developed for the sole purpose of dispersing Portland cement, whereas previous dispersants were mainly by-products of other industries. Polycarboxylates are classified as *comb* polymers. The name itself implies much about the structure of these molecules in that they consist of a backbone with pendant side chains, much like the teeth of a comb, as shown in Figure 2-35. For these molecules to be effective as dispersants, they must be attracted to the surface of a cement particle. The backbone of the polycarboxylate molecules typically serves two functions: as the location

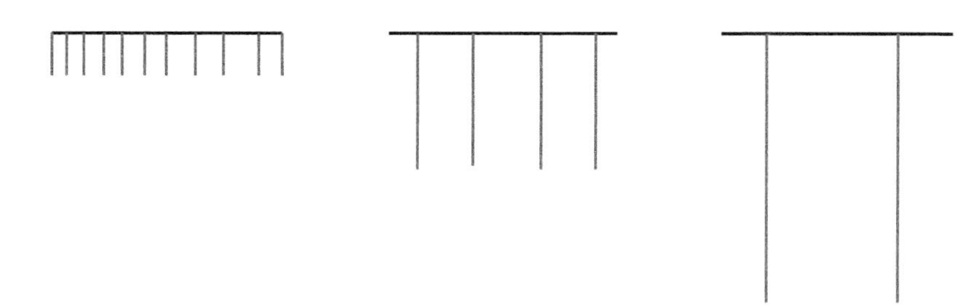

**Figure 2-35** Structure of PCE

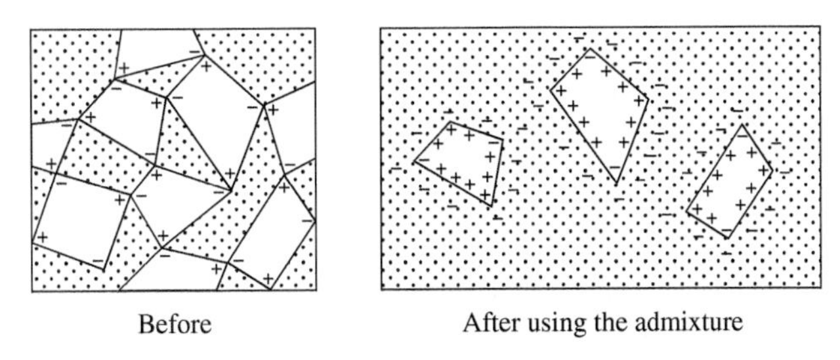


**Figure 2-36** Mechanism of water-reducing admixture

of binding sites (to the surface of the cement particle) and to provide anchoring sites for the side chains of the molecule. The pendant side chains serve as a steric, or physical, impediment to reagglomeration of the dispersed cement grains.

The mechanism of water reduction is different for different water reducers. Lignosulfonates (normal and sugar-refined), and SNF-based water reducers work on the mechanism of lowering zeta potential, which leads to electrostatic repulsion to separate cement particles from flocculation, thus releasing the entrapped water by cement particle clusters, as shown in Figure 2-36. Polycarboxylic ether-based water reducers are polymers with backbone and side chains. The backbone gets adsorbed on the surface of the cement grains, and the side chains cause dispersion of cement grains by steric hindrance (Uchikawa et al., 1997). This phenomenon relates to the separation of the admixture molecules from each other due to the bulky side chains, demonstrated in Figure 2-37. Steric hindrance is a more effective mechanism than electrostatic repulsion. The side chains, primarily of polyethylene oxide extending on the surface of cement particles, migrate in water and the cement particles are dispersed by the steric hindrance of the side chains.

Water-reducing chemicals are generally supplied in two types: powder and liquid. In a liquid form, the active solids content in the liquid is in the range of 30–40%. The dosages of water reducer for the liquid form can refer to either the solid content or liquid mass and is expressed as a weight percentage of the binder. However, referring to solid content is preferred because it is consistent to the dosage of water reducer in powder form and it is more scientific since, the reducer in liquid form has a different amount of solid content. For example, if a concrete mix uses 350 kg cement and 100 kg fly ash as well as 1% of water reducer in 1 m3 concrete, the solid content of water

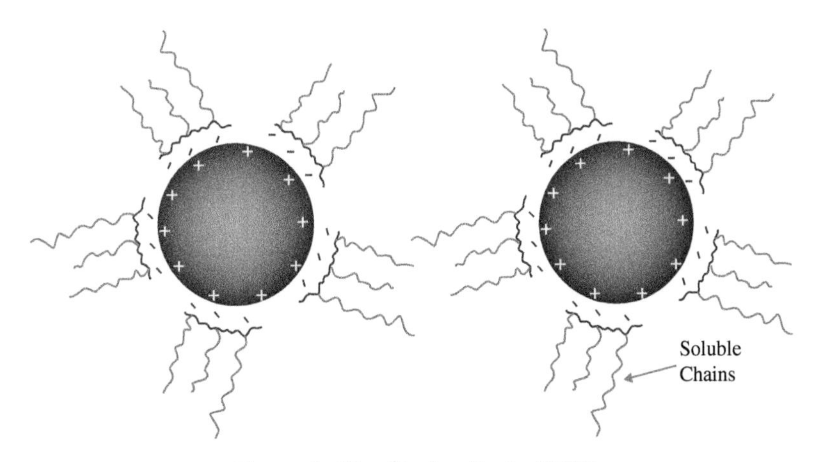


**Figure 2-37** Steric effect of PCE

reducer used should be $0.01 \times (350 + 100) = 4.5$ kg. If a liquid form of water reducer is used and there is 30% solid in the liquid, the amount of liquid is $4.5/0.3 = 15$ kg. Thus, the additional water added into concrete mix is 10.5 kg, that has to be deducted from the free mixing water.

Superplasticizers (SPs) are used mainly for two main purposes: (1) to produce high-strength concrete at a *w/c* ratio in a range of 0.23–0.3; and (2) to create "flowing" concrete with high slump flow in the range of 500 to 600 mm, the self-compacting concrete. Another benefit is that a lower *w/c* ratio would lead to better durability and lower creep and shrinkage. The major drawbacks of superplasticizers are: (1) they retard setting (especially in large amounts); (2) they cause more bleeding; and (3) they entrain too much air.

The most common problem in the application of water reducers in concrete is incompatibility, which refers to the abnormal behavior of a concrete due to the superplasticizer used. Common problems include flash setting, delayed setting, rapid slump loss, and improper early-age strength development. These issues in turn affect the hardened properties of concrete, primarily strength and durability. Compatibility between cements and superplasticizers is affected by many factors, including cement composition, admixture type and dosage, and concrete mixture proportions.

The $C_3A$ content or, more specifically, the $C_3A$ to $SO_3$ ratio has a profound effect on the compatibility between cement and SPs. When the $C_3A$ content of cement is high and the sulfate availability is low, superplasticized concretes experience high rates of slump loss. When there is less $C_3A$ available, higher amounts of SPs tend to adsorb on $C_3S$ and $C_2S$, resulting in a reduction in the rate of strength development.

SP molecules with sulfonate functional groups have an affinity for the positively charged aluminates. They compete with the sulfate released from gypsum for the aluminate reaction sites (Jolicoeur et al., 1994; Ramachandran, 2002). When the solubility of the calcium sulfate is low, the SP molecules tend to get adsorbed first on the aluminate compounds, thus preventing the normal setting reaction. It must be also noted that the solubility of sulfates would decrease in the presence of SPs with sulfonate functional groups, thus affecting the normal setting process of the cement. To avoid such problems, cements usually contain sufficient amounts of quickly soluble alkali sulfates.

The type and dosage of admixture have major effects on the cement–admixture compatibility issues. Primarily, SPs cause the slowing down of the dissolution of $Ca^{2+}$ and inhibit ettringite crystallization (Prince et al., 2002). The adsorption of the admixture on the surface of cement particles leads to a reduction of the chemical in solution. Thus, when adsorption levels are higher, more admixture is required to obtain a given fluidity. The surface adsorption of the admixture increases with the molecular weight of the polymer, and the presence of calcium ions promotes this adsorption. In the case of the lignosulfonates (Rixom and Mailvaganam, 1999), the presence of low-molecular-weight ingredients causes excessive air entrainment and leads to loss of strength. In addition, the high sugar content of these admixtures could cause unnecessary retardation, especially at high dosages. SNF-based admixtures are most prone to rapid loss of workability, particularly at low water-to-cement ratios, which are the norm for most special concretes today. Another common problem with SNF admixtures is excessive retardation, which may be caused because of the blending of these chemicals with lignosulfonates in commercial formulations. Similar to lignosulfonates, the presence of moderate- to high-molecular-weight chain fractions leads to a better performance for SNF admixtures. The low-molecular-weight fractions cause excessive retardation by covering reactive sites on the cement surface and inhibiting reactions.

The factors influencing compatibility between SP and cement are broad and the mechanism is very complicated. For construction practice, a simple method is needed for quick selection of the proper SP as well as its dosage for concrete proportion. Traditionally, there are many methods to evaluate SP from a rheological point of view, such as the flow-table method of ASTM C230-90, the mini-slump cone test (Kantro, 1981), and the Marsh cone test (Aitcin et al., 1994). However, these methods cannot provide information on the effect of SP on setting and strength development

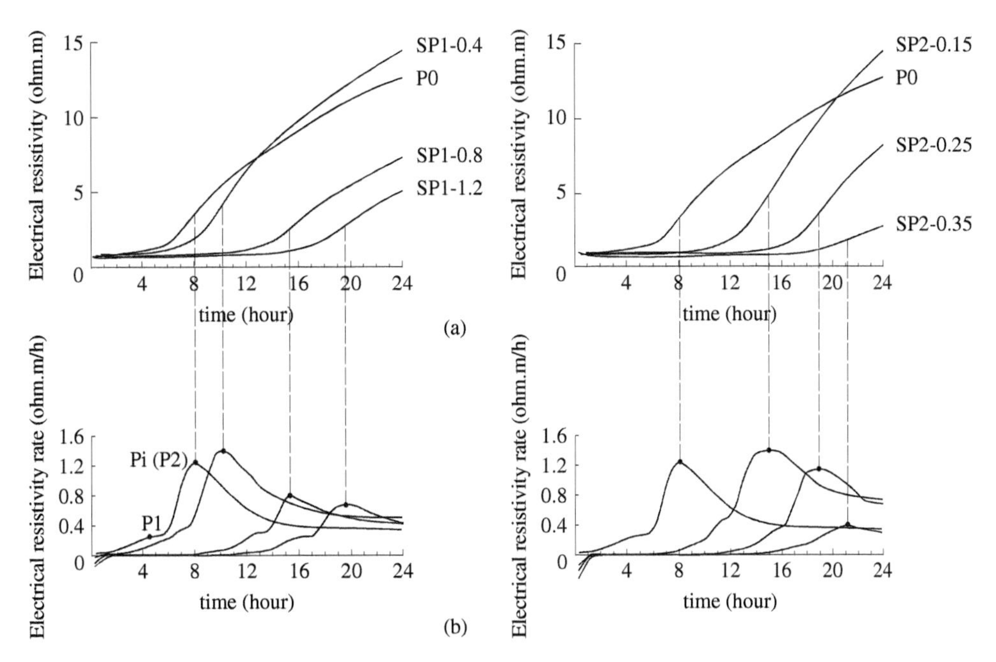


**Figure 2-38** Typical resistivity curves and their differential curves

of the concrete. Recently, a new method has been proposed for selection of SP using noncontact resistivity measurement (Xiao et al., 2007). Typical resistivity curves and the differential curves of the cement pastes with or without SPs are shown in Figure 2-38.

Figure 2-38 shows that for the same SP, the paste with the larger dosage demonstrated a longer flat region, which implies a more effective retarding influence and slow strength development. With a suitable dosage, such as 0.4% for SP1 and 0.15% for SP2, the resistivity of corresponding pastes increases quickly after the dynamic balance period and exceeds the one of the reference paste without SP addition. In this method, two parameters are proposed for selecting SPs:

$$K_t = \frac{t_{i,SP}}{t_{i,PO}} \tag{2-54}$$

where $t_{i,SP}$ is the inflection point time for a specimen with SP, and $t_{i,PO}$ is the inflection point time for the reference specimen without SP.

$$K_r = \frac{\rho_{24,SP}}{\rho_{24,PO}} \tag{2-55}$$

where $\rho_{24,SP}$ is the value of resistivity for a specimen with SP at 24 h, and $\rho_{24,PO}$ is the value of resistivity for reference specimen without SP at 24 h.

A typical plot of $K_t$ and $K_r$ is shown in Figure 2-39. Since both $t_{i,SP}$ and $t_{i,PO}$ are related to the setting time of the cement paste, the parameter $K_t$ is in fact, reflecting the change of the setting time for the specimen with SP. If $K_t$ is smaller than 1, it implies an accelerated setting due to the addition of SP. If $K_t$ is greater than one, it means a retarded setting due to the addition of SP. Hence, a value of $K_t$ close to 1 is preferred. On the other hand, both $\rho_{24,SP}$ and $\rho_{24,PO}$ are related to the microstructure development, especially the porosity. A higher $\rho_{24}$ value implies smaller porosity and hence a higher compressive strength. If $K_r$ is smaller than 1, it implies a slow strength development and a lower compressive strength at 1 day due to the addition of SP. If $K_r$ is greater

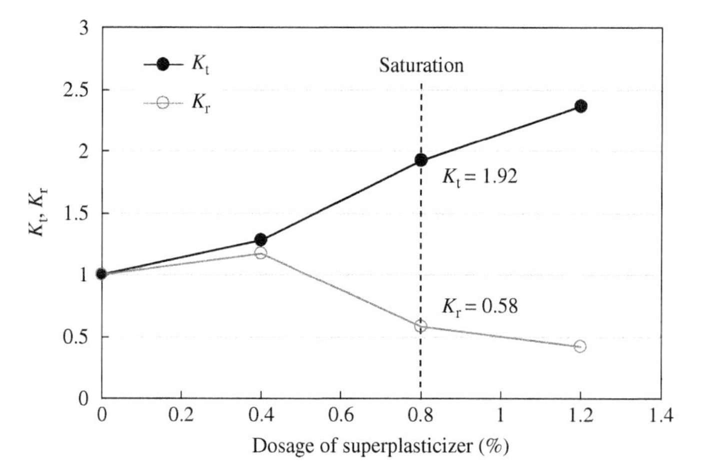


**Figure 2-39** Typical plot of $K_t$ and $K_r$

than one, it means a fast strength development and a higher compressive strength at 1 day that implies the addition of SP benefits the strength development. From Figure 2-39, it can be seen that for this type of SP, the optimum dosage is 0.4%.

#### 2.3.2.2 Setting-Control Admixture

Setting-control admixtures are used to either extend or shorten the plastic stage of concrete to meet the special requirements of the construction of concrete structures. If the admixture is used to extend the plastic period, it is called a retarder. If the admixture is used to shorten the plastic period, it is called an accelerator.

**(a)** *Mechanism*: The setting phenomenon of Portland cement paste signals the end of the plastic stage of concrete. It is a result of the progressive crystallization of the hydration products. Thus, to change the length of the plastic period equally means to change the setting rate. To influence the rate of setting, one has to change the rate of the crystallization. One way to do it is to add certain soluble chemicals to the cement–water system to influence the ion dissolution rate.

To understand the mechanism of acceleration or retardation, it is helpful to consider a hydrating Portland cement paste as being composed of certain anions (silicate and aluminate) and cations (calcium), the solubility of each being dependent of the type and concentration of the acid and base ions present in the solution. Usually, the setting will be speeded up when the dissolution rates of cations and anions are higher. On the other hand, the setting will be slowed down when the dissolution rates of cations and anions are lower. Thus, an accelerating admixture must promote the dissolution of cations (calcium ions) and anions from the cement, especially the silicate ions, which have the lowest dissolving rate, during the early hydration period. On the other hand, a retarding admixture must impede the dissolution of cations (calcium ions) and anions from the cement, especially the anions, which have the highest dissolving rate during the early hydration period (e.g., aluminate ions).

It should be noted that most chemical admixtures used for setting control purposes have both complementary and opposing effects. For instance, the presence of monovalent cations in solution (i.e., $K^+$ or $Na^+$) reduces the solubility of $Ca^{2+}$ ions but promotes the solubility of silicate and aluminate ions. Also, the presence of certain monovalent anions, such as $Cl^-$ or $NO_3^-$, reduces the solubility of silicates and aluminates but tends to promote the solubility

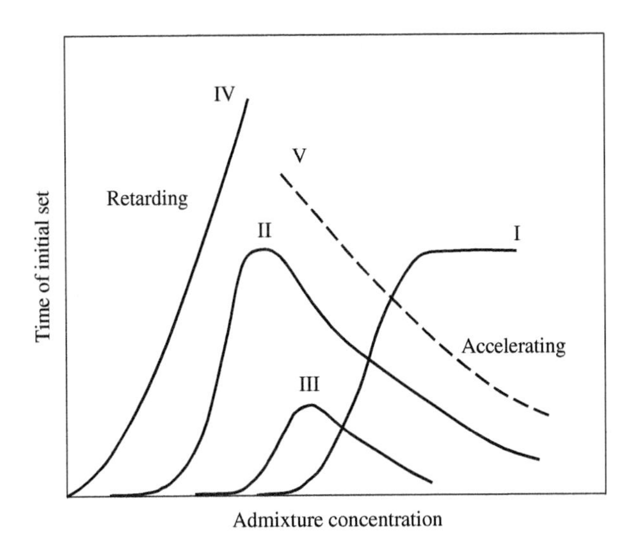


**Figure 2-40** Different types of setting-control admixtures

of calcium ions. Moreover, the effects of promotion and reduction are a function of concentrations or dosage of these chemicals. For example, for a small dosage of the monovalent cations in solution (i.e., $K^+$ or $Na^+$), the dominant effect is in reducing the solubility of $Ca^{2+}$ ions and thus acting as a retarder. For a large dosage, the dominant effect is promoting the silicate and aluminate ions and thus acting as an accelerator. The monovalent anions, such as $Cl^-$ or $NO_3^-$, have the same phenomena.

There is also a monotonic retarding agent, which works with a different mechanism. These kinds of retarders reduce the solubility of the anhydrous constituents from cement by forming insoluble and impermeable products around the particles, or delaying the bond formation among hydration products. Once insoluble and dense coatings are formed around the cement grains, further hydration slows down considerably. Sugar and carbonated beverages belong to this category. There is some usage of sugar to retard the concrete inside the drum of the transporting truck when it has a problem in traffic or operation. Figure 2-40 demonstrates the different types of setting control admixtures: monotonic retarding admixture, monotonic accelerating admixture, and dual-role setting control admixture. It can be seen from Figure 2-40 that the function of a dual-role setting admixture largely depends on its concentration.

**(b)** *Applications*: The major applications for setting-control admixtures are as follows.

**(i)** *Retarding admixtures*: Mainly used (1) to offset fast setting caused by ambient temperature, particularly in hot weather; (2) as a setting control of large structural units to keep concrete workable throughout the entire placing period; (3) to meet the requirement of long transportation time from the concrete plant to the construction site. For example, for the construction of a pier for a bridge built up in the ocean, the concrete has to be transported by ferry from land to the site. In this case, 5 to 6 h or even longer may be required.

**(ii)** *Accelerators*: Accelerators have been widely used in civil engineering. The applications include plugging leaks in swimming pools, water tanks, and pipelines; emergency repairs for highways, bridges, airport runways, and tunnels; and winter construction in cold regions. Soluble inorganic salts, such as calcium chloride, are by far the best-known and most widely used accelerators. A side effect of using chloride, however, is that it induces corrosion of the reinforcement in concrete structures.

#### 2.3.2.3 Shrinkage-Reducing Admixture

Shrinkage-reducing admixtures (SRA) provide a significant technical approach to reduce the drying shrinkage of concrete. SRA was invented in 1982, and Goto et al. (1985) applied for their patent in 1985. Subsequently, many researchers performed detailed studies on SRA. SRA is a liquid organic compound consisting of a blend of propylene glycol derivatives. Some literature has indicated that SRA can reduce long-term drying shrinkage by 50%, and there is a significant improvement in restrained shrinkage performance. Even for concrete with proper curing at which the drying shrinkage would reduce to minimum, there is still a substantial reduction in drying shrinkage due to effect of SRA (Berke et al., 1999). The main mechanism of SRA in reducing drying shrinkage of concrete is that the SRA lowers the surface tension of the pore solution and subsequently reduces the stresses in the pore solution that are directly proportional to the surface tension. With the reduction of the driving stress, the drying shrinkage can be reduced. However, in addition to reducing the surface tension, SRA also influences the dynamic process of cement hydration, as studied by He et al. (2006).

#### 2.3.2.4 Waterproof Admixtures for Durability-Enhancement

Although, in a broad sense, water reducers and air-entraining admixtures can enhance the durability of concrete structures, the durability enhancement ability of waterproofing admixtures must be mentioned.

Water tightness is a key performance criterion for concrete structures designed for the containment, treatment, or transmission of freshwater, seawater, wastewater, or other fluids. From a concrete perspective, water tightness can be achieved by using well-proportioned, low permeability, workable concrete that is appropriately consolidated and cured. Waterproofing admixtures can reduce the permeability as well as the water absorption of concrete and hinder the transition of harmful ions and other species without changing the water/cement ratio.

Many kinds of waterproofing admixture have been used for decades, such as fatty acid, petroleum products, and extractives from plants and vegetable waxes. These kinds of waterproofing are usually hydrophobic, which can reduce the contact angle between water and concrete surfaces. In past research, the waterproofing admixture does not change the properties of fresh concrete significantly and can improve the workability of concrete (Hamid et al., 2018).

### 2.3.3 Air-Entraining Admixtures

Air-entraining admixtures entrain air in the concrete. An air-entraining admixture contains surface-active agents that have two poles: one is hydrophobic and the other hydrophilic. The agents are concentrated at the air–water interface, the hydrophilic side with water and the hydrophobic side with air (see Figure 2-41). The surface tension is lowered so that bubbles can form more readily and then stabilize once they are formed. Commonly, carboxylic acid or sulfonic acid groups are used to achieve the hydrophilicity, while aliphatic or aromatic hydrocarbons are used for hydrophobility. In summary, the entrained air is produced by admixtures that cause the mixing water to foam, and the foam is locked into the paste during hardening.

Entrained air voids are different from entrapped air voids. Entrained air voids are formed on purpose, while an entrapped air void is formed by chance when the air gets into the fresh concrete during mixing. Entrapped air voids may be as large as 3 mm; entrained air voids usually range from 50 to 200 μm. The size distribution of the solids and pores in a hydrated cement paste is given in Figure 2-42.

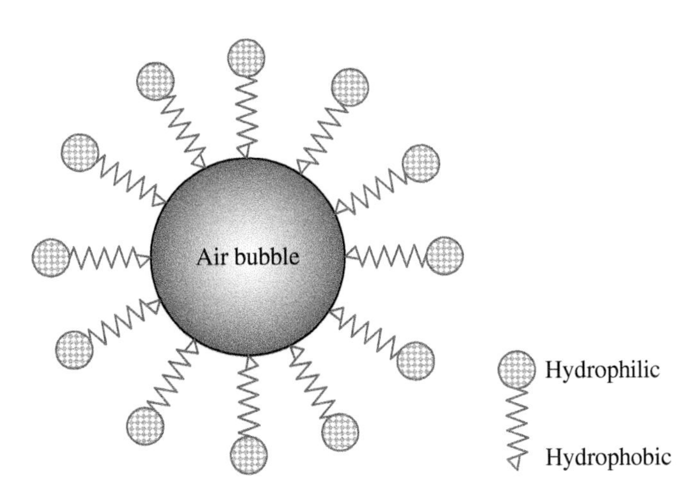


**Figure 2-41** Mechanism of air entraining

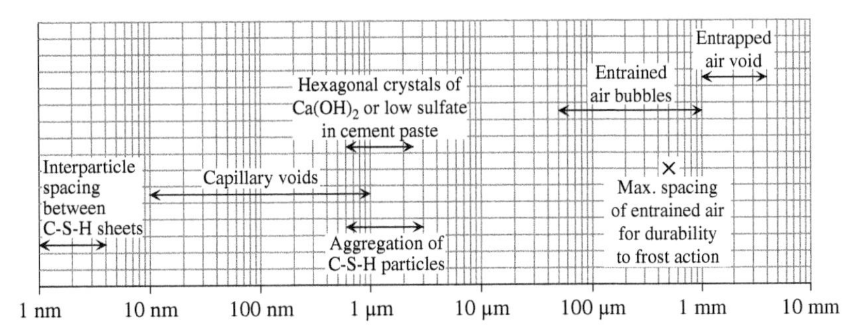


**Figure 2-42** Size distribution of solids and pores in hydrated cement paste

Various advantages can be obtained by adding air-entraining admixtures into concrete. First, the workability of concrete can be improved because air bubbles act as lubricants in a fresh concrete. Subsequently, the water amount can be reduced for a targeted workability. Second, the ductility of concrete can be improved, since the air bubbles generated by the air-entraining agent provide more room for deformation to occur. Third, the permeability of concrete can be improved due to the effect of air-entraining agent in enclosing the air bubbles. Fourth, the impact resistance of concrete, can be improved as the air bubbles provide more deformation. Finally and most importantly, the durability, especially the ability to resist freezing and thawing of concrete, can be significantly improved by adding an air-entraining agent. This is because, in addition to the improved permeability, the small air bubbles in concrete provide spaces to release the pressure generated during the ice formation in the freezing process, which can prevent concrete from cracking and damage.

The spacing among the air bubbles also plays an important role in determining durability. As shown in Figure 2-43, the smaller the spacing factor (which is defined as the average distance from any point in the paste to the edge of a nearest entrained air bubble), the more durable the concrete. For the case of a spacing factor greater than 0.3 mm, the air entrained has a small or negligible effect on durability. Although the entrained air bubbles can decrease the capillary porosity due to the pore enclosing effect and the ability for increasing the workability of the entrained air, generally, a strength loss of 10–20% can be anticipated for most air-entrained concrete, as shown in Figure 2-44.

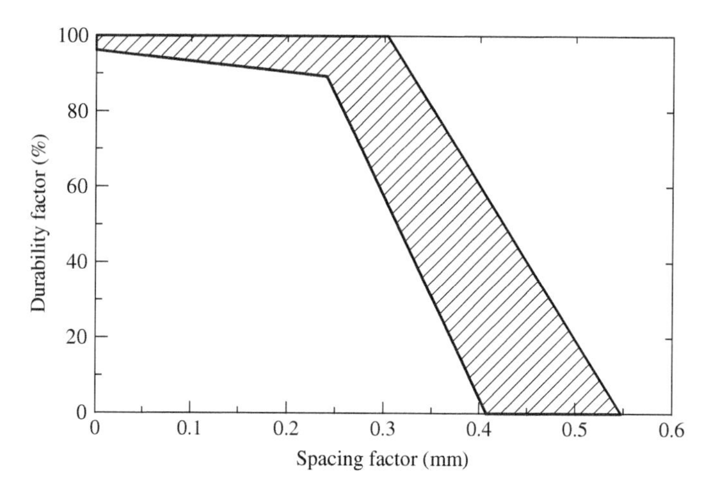


**Figure 2-43** Effects of spacing factor on durability factor

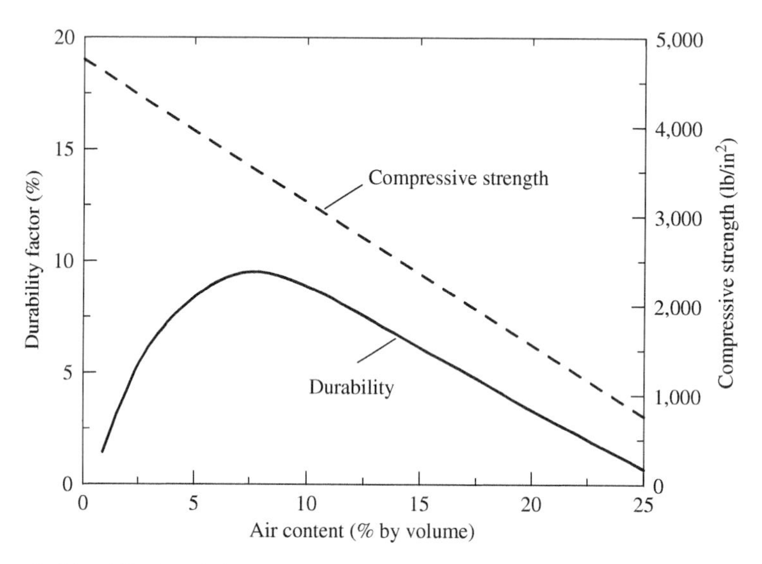


**Figure 2-44** Effects of air content on durability factor and compressive strength

The volume of air required to give optimum durability has been found to be about 4–8% by volume of concrete, as observed from the above figures. The actual fraction depends on the maximum size of aggregate: the larger the aggregate size, the lower requirement for air content. This is because more paste is required to provide similar workability for concrete with a smaller size of coarse aggregate due to the surface coating requirement.

The formula used to calculate the gel space ratio has to be modified if entrained air in cement paste is considered:

$$X = \frac{\text{volume of gel (including gel pores)}}{\text{volume of gel} + \text{volume of capillary pores} + \text{entrained air}} = \frac{0.68a}{0.32a + w/c + \text{entrained air}} \tag{2-56}$$

The percentage of the entrained air is based on the volume of concrete, and not the volume of cement paste.

### 2.3.4 Rheology Modifier

Controlling the rheological properties, expressed in practical terms as workability, of fresh concretes is important as it will determine the efficiency of the casting process during construction (Marchon et al., 2018). In the case of novel applications like digital fabrication, it will determine its success or failure. The workability is better described through fundamental rheological parameters, primarily yield stress and plastic viscosity. Yield stress is the stress above which flow initiates, or below which flow terminates. It also determines the suspension stability and accounts for the results of stoppage tests, such as slump or slump flow. Plastic viscosity describes the resistance to flow, which increases with increasing shear rate––an important parameter for robustness. In addition, fresh cement-based materials exhibit thixotropy, which is a time-dependent property. It can be generally defined as the continuous decrease of apparent viscosity with time under shear and subsequent recovery at rest.

In general, the rheology of concretes is affected by the mix design, including the volume fraction of the binding system, its composition, and the attributes of the aggregates. Admixtures can be used as an effective tool to tailor the rheology. Superplasticizers, which act as dispersants to lower yield stress, and viscosity modifying agents, which increase plastic viscosity, are now regularly used in concretes to control workability. Clays can also be used as thixotropy modifiers.

**(a)** *Superplasticizers*: Dispersing agents and superplasticizers, such as lignosulfonate (LS), polynaphthalene sulfonate (PNS), and polycarboxylate ether (PCE), are polymeric dispersants used in cementitious materials to reduce yield stress and viscosity at a constant solids content. Details of superplasticizers are discussed in Section 2.3.2.1.

**(b)** *Viscosity modifiers*: Viscosity modifying agents (VMAs) are commonly used in industrial practice to control water transport and porous structure in both the fresh and hardened state (Khayat and Mikanovic, 2012). The most commonly used VMAs in cementitious systems are cellulose-ether derivatives, such as welan gum and diutan gum. Anionic polyacrylamides are also used in the construction field for their high flocculating efficiency. Such admixtures are essential to control the risk of segregation of highly flowable concretes, improving the water retention, and in some cases for promoting particle flocculation, such as for shotcrete applications. VMAs can enhance the yield stress of cementitious materials and effectively reduce material deformation under their weight, which is critical for additive manufacturing concrete.

**(c)** *Clays and inclusions as thixotropic agents*: Inorganic additions, namely clays, and organic inclusions, such as polymeric fibers, can also be effective rheological modifiers. Clays themselves exhibit shear-thinning (i.e., decrease of the apparent viscosity by increasing the shear rate) due to their opposing surface charges, which gives rise to a house of cards structure that can form at rest and break down under shear (Van Olphen, 1964). They have a wide range of applications as rheological modifiers, including paints and drilling fluids. Among the different families of clays, attapulgite (or palygorskite), bentonite (montmorillonite-based), kaolinite, sepiolite, and contaminated clays have been studied as rheological modifiers for cement-based materials.

For successful additive manufacturing, it will be key to control rheology through the use of suitable admixtures, likely in combination. Most of them are considered to be chemically inert as they alter the rheological properties through physical effects.

## 2.4 WATER

Water is an important ingredient of concrete, and a properly designed concrete mixture, typically with 15–25% water by volume, will possess the desired workability for fresh concrete and the required durability and strength for hardened concrete. The roles of water have been discussed earlier and are known as hydration and workability. The total amount of water in concrete and the water-to-cement ratio may be the most critical factors in the production of good-quality concrete. Too much water reduces concrete strength, while too little makes the concrete unworkable. Because concrete must be both strong and workable, a careful selection of the cement-to-water ratio and total amount of water are required when making concrete (Popovics, 1992).

### 2.4.1 Fresh Water

#### 2.4.1.1 Mixing water

Water can exist in a solid form as ice, a liquid form as water, or in a gaseous form as vapor. The mixing water is the free water encountered in freshly mixed concrete. It has three main functions: (1) it reacts with the cement powder, thus producing hydration products; (2) it acts as a lubricant, contributing to the workability of the fresh mixture; and (3) it secures the necessary space in the paste for the development of hydration products. The amount of water added for adequate workability is always greater than that needed for complete hydration of the cement in practice.

Unlike other raw materials, the raw water supply varies significantly in quality, both from one geographical region to another and from season to season. Water derived from an upland surface source, for instance, usually has a low content of dissolved solids and is relatively soft, but has a high concentration of organic contamination, much of it colloidal. By contrast, water from an underground source generally has a high content of dissolved solids and a high hardness level but a low organic content. Seasonal variations in water quality are most apparent in surface waters. During the autumn and winter months, dead leaves and decaying plants release large quantities of organic matter into streams, lakes and reservoirs. As a result, the degree of organic contamination in surface waters reaches a peak in spring, and falls to a minimum in summer. Excessive impurities in mixing water not only may affect setting time and concrete strength, but also may cause efflorescence, staining, corrosion of reinforcement, volume instability, and reduced durability.

There is a simple rule concerning the acceptability of mixing water: if water is potable, that is, fit for human consumption, with the exception of certain mineral waters and water containing sugar, it is also suitable for concrete making. In other words, if water does not have any particular taste, odor or color, and does not fizz or foam when shaken, then there is no reason to assume that such water will hurt the concrete when used properly as mixing water. On the other hand, some water unsuitable for drinking is still satisfactory for concrete making.

#### 2.4.1.2 Impurities in Water

The unique ability of water in dissolving, to some extent, virtually every chemical compound and supporting practically every form of life means that raw water supplies contain many contaminants. Water impurities can be either dissolved in the water or present in the form of suspensions. Water should be avoided if it contains large quantities of suspended solids, excessive amounts of dissolved solids, or appreciable amounts of organic materials. The major categories of impurities found in raw water include: (1) suspended solids; (2) dissolved solids; and (3) dissolved organic material.

Suspended solids in water include silt, clay, pipe work debris, organic matter and colloids. Usually up to about 2,000 ppm of suspended clay or silt can be tolerated (Mindess et al., 2003). Higher amounts may increase water demand, increase drying shrinkage, or cause efflorescence. Muddy water should be allowed to clear in settling basins before use. Colloidal particles, either organic or inorganic, are not truly in solution or suspension and give rise to haze or turbidity in the water. The degree of colloidal contamination can be determined by a fouling index test or by turbidimetry. In the fouling index, test raw water is passed through a standard filter and the rate of blockage is measured. The greater the rate of blockage, the greater the amount of colloidal contamination. The turbidimetric method determines the total suspended solids content of the raw water by passing a beam of light through the water and measuring the proportion of light scattered from the suspended particles. Since the mix water quickly becomes a highly alkaline solution, organic materials may dissolve during mixing and subsequently retard setting and strength development by interfering with cement hydration. Organic impurities may also entrain excessive amounts of air, thereby reducing strength, or conversely, they may interfere with the action of air-entraining agents.

Total dissolved solids (TDS) are the residue in ppm obtained by the traditional method of evaporating a water sample to dryness and heating at 180°C. This residue includes colloids, non-volatile organic compounds and salts which are stable at this temperature. It can be measured directly or estimated by multiplying the conductivity of the water in μS/cm at 25°C by 0.7. Water containing less than 2,000 ppm of dissolved solids can in most instances be used safely, although this depends on the nature of the dissolved material. As little as 100 ppm of sodium sulfide may cause problems. At the other extreme, seawater, which contains about 34,000 ppm (3.4%) of dissolved salts, can be used to make satisfactory concrete if certain precautions are taken. Indeed, soluble salts may be added deliberately as admixtures and the most common example being $CaCl_2$, which is used as an accelerating agent. Soluble carbonates and bicarbonates can promote rapid setting; large quantities of carbonates and sulfates may cause a reduction in 28-day strength or long-term strength. Some soluble inorganic salts may retard the setting and hardening of concrete. Salts of zinc, copper, lead, and, to a lesser extent, manganese and tin fall into this category, as well as phosphates, arsenates, and borates. Soluble inorganic salts of up to 500 ppm can generally be tolerated in mixing water. Acidic waters can be used in concrete making; the pH of the water may be as low as 3.0, at which level there are more problems surrounding the handling of the water than will occur in the concrete. Organic acids may affect the setting and hardening of concrete. Alkaline waters, containing sodium or potassium hydroxide, may cause quick setting and low strengths at concentrations above 500 ppm.

Organic impurities in water usually arise from the decay of vegetable matter, principally humic and fulvic acids, and from farming, paper making and domestic and industrial waste. These include detergents, fats, oils, solvents and residues from pesticides and herbicides. In addition, water-borne organics may include compounds leached from pipe lines, tanks and purification media. A water purification system can also be a source of impurities and so must be designed not only to remove contaminants from the feed water, but also to prevent additional recontamination from the system itself. Colored natural waters generally indicate the presence of dissolved organic material, mostly tannic and humic acids, which may retard the hydration of cements. Many organic compounds that occur in industrial wastes may also severely affect the hydration of cement or entrain excessive amounts of air. Wastes from the pulp and paper industries, the tanning industries, and food-processing industries have been used as a source of chemicals for the formulation of set-retarding or air-entraining admixtures. Thus, untreated industrial wastewaters should be treated with caution, but if they have passed through a sewage treatment process, the organic matter will be reduced to safe levels. The degree of organic contamination can be measured by the oxygen-absorption (OA) test using potassium permanganate solution or the chemical oxygen demand (COD) test. Increasingly, however, total organic carbon (TOC) analyzers are being used because of the ease of interpretation of the results and their sensitivity in detecting low levels of organic compounds in water samples.

Furthermore, oxygen and carbon dioxide are the two gases most commonly found in natural waters. Carbon dioxide behaves as a weak anion and is removed by strong exchange resins. Dissolved oxygen can also be removed by degassing or by anion exchange resins in the sulfite form, and the level of dissolved oxygen in the feed water can be monitored with oxygen-specific electrodes.

#### 2.4.1.3 Water for Curing and Washing

The requirements for curing water are less stringent than those discussed above, mainly because curing water is in contact with the concrete for only a relatively short time. Such water may contain more inorganic and organic materials, sulfuric anhydride, acids, chlorides, and so on, than acceptable mixing water, especially when slight discoloration of the concrete surface is not objectionable. Nevertheless, the permissible amounts of the impurities are still restricted. In cases of any doubt, water samples should be sent to a laboratory for testing and recommendations. Water for washing aggregates should not contain materials in quantities large enough to produce harmful films or coatings on the surface of aggregate particles. Essentially the same requirement holds when the water is used for cleaning concrete mixers and other concreting equipment. Chemical limitations for the impurities in wash water are specified in ASTM C94.

### 2.4.2 Seawater

The main components of sea water are NaCl, $MgCl_2$, $MgSO_4$, $K_2SO_4$, and $Na_2CO_3$ et al. The use of seawater usually has a negative effect on the workability, strength development, and durability of concrete. Table 2-17 exhibits the components that are often added to freshwater to simulate seawater (Cotruvo, 2005; Cwirzen et al., 2014; Bazli et al., 2020).

The main reactive ions in seawater to affect the properties of sea water concrete and potentially threaten the durability of concrete structures include $Cl^-$, $SO_4^{2-}$ and $Mg^{2+}$. The high concentration of $Cl^-$ can accelerate the corrosion of steel, whereas the presence of $SO_4^{2-}$ accelerates the formation of gypsum and ettringite which may result in the expansion and cracking of concrete (Wegian and Falah, 2010; Shi et al., 2018). The $Mg^{2+}$ in the pore solution of cement-based materials can react with calcium hydroxide to produce brucite, thus decreasing the salinity of pore solution and destabilizing the C-S-H gels. Nevertheless, the chemical ions in seawater may also interact with each other or with other ions in the system to affect the solid phases within concrete (Al-Amoudi et al., 1994). The combined effects of $Cl^-$ and $SO_4^{2-}$ or other cation ions such as $Na^+$, $K^+$, $Ca^{2+}$ and $Mg^{2+}$ on microstructure and durability of concrete need to be considered when seawater with different chemical compositions is applied as mixing water.

Although it is widely believed that seawater is not suitable for the concrete system, there have been many successful applications of seawater in structural concrete. The problems of steel corrosion and concrete durability can be solved by using non-reinforced concrete structures or replacing rebars with other materials such as FRP (fiber reinforced polymer).

**Table 2-17** Chemical composition of substitute seawater (mg/L)

| Region | $Cl^-$ | $Na^+$ | $SO_4^{-2}$ | $Mg^{2+}$ | $Ca^{2+}$ | $K^+$ |
|---|---|---|---|---|---|---|
| Red Sea | 22,219 | 14,255 | 3078 | 742 | 225 | 210 |
| Arabian Gulf | 23,000 | 15,850 | 3200 | 1,765 | 500 | 460 |
| Eastern Mediterranean | 21,200 | 11,800 | 2950 | 1403 | 423 | 463 |
| Northern Baltic Sea | 3,000 | 1,800 | 410 | 240 | 98 | 67 |
| Brighton beach | 20,700 | 11,940 | 3420 | 1430 | / | 622 |

**Table 2-18** The main requirements of prEN1008 in comparison with those of ASTM C94

| Parameter | $Cl^-$ content | $SO^{4-}$ content | Solid material content | Comparative samples strength |
|---|---|---|---|---|
| prEN 1008 | <= 600 mg/L (concrete/grout) $\leq$1200 mg/l (reinforced concrete) | $\leq$2000 mg/L | <= 1% w.t. | The mean 7-days and 28-days compressive strength of the mortar or concrete samples wash water must be at least 90% of the mean strength of the control samples (prepared with distilled or tap water) |
| ASTM C94 | $\leq$400 mg/l (prestressed concrete) $\leq$1200 mg/l (reinforced concrete) | <=3000 mg/l (Optional) | $\leq$50000 mg/l | The mean 7-days and 28-days strength of the concrete samples prepared with prepared with wash water must be at least 90% of the mean strength of the control samples (prepared with distilled or city water) |

### 2.4.3 Recycled Water

With the great demand for concrete and the current water shortage worldwide, it is necessary to find an alternative source of potable water for concrete production. Recycled water is usually obtained from municipal sources, wells, streams, and other sources that are not potable. In the early 1970s, there was an initial need for recycled water for concrete production in California. The use of recycled water can solve some environmental problems.

The use of recycled waste wash water for the production of new concrete is dealt with by prEN 1008 standard: mixing water must meet severe requirements in composition and may be used only if concrete exhibits 7- and 28-day compressive strengths higher than 90% of the value exhibited by samples prepared with distilled water (in Table 2-18, the main requirements of prEN1008 are shown in comparison with those of ASTM C94). Despite the requirements that have been achieved so far by prEN 1008, building companies often fear the use of ready-mix truck wash water in the production of fresh concrete, also a limited investigation has been carried out on the effects on physical-mechanical properties and microstructure as a function of the characteristics of wastewater used. The results have shown that mortar and concrete prepared with recycled water exhibit 28-day mechanical strength in no way lower than 96% of the reference materials (90% is the minimum allowed in prEN 1008) and, in some cases, even better. Moreover, the use of wash water in concrete leads to a reduction of the concrete capillary water absorption and mortar microporosity, which surely improves the durability of the material. This effect can be ascribed to the filling action of the fines present in the wash water and to the slight reduction of the actual water/cement ratio.

## DISCUSSION TOPICS

How are aggregates classified?

Give some examples of aggregates.

Why is the gradation of aggregate important?

What are the important physical properties of aggregates?

Why can the moisture content influence concrete properties?

What are the differences between organic and inorganic binders?
What are the differences between hydraulic and nonhydraulic cement?
What are main chemical components of Portland cement?
What are main hydration products of Portland cement? State their functions in concrete.
How many stages are there in Portland cement hydration?
What happens chemically during each stage of cement hydration?
What are the differences among five types of Portland cement? Give some suitable applications for each type of cement in practice.
Why can application of $LC_3$ reduce the carbon dioxide emissions?
What are the main advantages of a geopolymer? What are its main applications?
What are the main advantages of MPC? What are its main applications?
What are the main advantages of MOC? What are its main applications?
How are admixtures classified?
Why does modern concrete technology stress the importance of admixtures?
What is the mechanism for water reducers, air entraining agents, and setting control admixtures?
Compare the chemical composition for four main mineral admixtures.
What are the major influences of fly ash on concrete's properties?
What are the main functions of silica fume?
What is the common reaction of a mineral admixture in concrete?
What should be concerned when seawater is used in concrete preparation?

## PROBLEMS

1. A sample of sand weighs 490 g in stock and 475 g in OD condition, respectively. If absorption capability of the sand is 1.1%, calculate the MC(SSD) for the sand.
2. The sieve results of a batch of aggregate are as follows: retained on 1.5" sieve: 0.3 kg; retained on 1" sieve: 1.2 kg; retained on 0.75" sieve: 7.6 kg; retained on 0.5" sieve: 6.2 kg; retained on 3/8" sieve: 4.2 kg; retained on No. 4 sieve: 0.8 kg; retained on No. 8 sieve: 0.1 kg. Calculate the fineness modulus for the aggregate. Is it coarse or fine aggregate?
3. A 1000 g sample of coarse aggregate in the SSD condition in air weighed 633 g when immersed in water. Calculate the BSG of the aggregate.
   If a sample from the same batch of aggregate after being exposed to air dry condition for some time weighed 978 g in air and weighed 630 g after immersed in water for 2 hours, calculate the moisture content, MC(SSD), of the air dried aggregate at that time.
4. A mixture of 1800 g of gravel with an absorption of 1.3% and 1200 g of sand with a surface moisture of 2.51% was added into a concrete mix. Compute the adjustment of water that must be made to maintain a constant *w/c* ratio.
5. Demonstrate that the specific gravity of a blend of n aggregates with specific gravities of $SG_1$, $SG_2$, …, SGn can be given by:
   **(a)** SG = SG1 P1 + SG2 P2 + … + SGn Pn if P1, P2, … , Pn are the volume fractions of each of the n aggregates
   **(b)**
   $$SG = \frac{1}{\frac{p_1}{SG_1} + \frac{P_2}{SG_2} + \cdots + \frac{P_n}{SG_n}}$$
   If $P_1$, $P_2$, …, $P_n$ are the weight fractions of each of the n aggregates.

6. The material ratio for concrete mix is 1:1.5:2 (C:Sand:Coarse Aggregate by weight). The BSG for cement is 3.15, for sand is 2.5 and for coarse aggregate is 2.7. Air content is 4.8%. The gel/space ratio is 0.72. Calculate the water cement ratio for $\alpha = 0.8$.
7. If one ton Portland cement contains 620 kg CaO, how much $CO_2$ is produced during cement production due to decomposition of limestone?

## REFERENCES


Adinkrah-Appiah, K., Kpamma, E Z., Nimo-Boakye, A., et al. (2016) "Annual consumption of crushed stone aggregates in Ghana," *Journal of Civil Engineering and Architecture Research*, 3(10), 1729–1797.

Aitcin, P., Jolicoeur, C., and MacGregor, J. (1994) "A look at certain characteristics of superplasticizer and their use in the industry," *Concrete International*, 16, 45–52.

Akbarnezhad, A., Ong, K.C.J., Zhang, M.H., Tam, C.T., and Foo, T.W.J. (2011) "Microwave-assisted beneficiation of recycled concrete aggregates," *Construction and Building Materials*, 25, 3469–3479.

Al-Amoudi, O.S.B., et al. (1994) "Influence of chloride ions on sulphate deterioration in plain and blended cements," *Magazine of Concrete Research*, 46(167), 113–123.

Antoni M., Rossen J., Martirena, F., et al. (2012) "Cement substitution by a combination of metakaolin and limestone," *Cement and Concrete Research*, 42(12), 1579–1589.

Apotheker S. (1989) "Glass processing: the link between collection and manufacture," *Resource Recycling*, 7, 38.

Azom (2012) "Recycling EAF slag waste in commercial products."

Bahurudeen, A., Kanraj, D., Gokul Dev, V., and Santhanam, M. (2015) "Performance evaluation of sugarcane bagasse ash blended cement in concrete," *Cement and Concrete Composites*, 59. 77–88.

Bahurudeen, A., Wani, K., Basit, M., and Santhanam, M. (2016) "Assessment of pozzolanic performance of sugarcane bagasse ash," *ASCE Journal of Materials in Civil Engineering*, 28(2).

Bazli, M., Zhao, X.L., Raman, R.K., Bai, Y., and Al-Saadi, S.(2020). "Bond performance between FRP tubes and seawater sea sand concrete after exposure to seawater condition," *Construction and Building Materials*, 265.

Bensted, J. and Barnes, P. (2002) *Structure and performance of cements*, 2nd ed., London: Spon Press.

Berke, N.S., Dallaire, M.P. and Hicks, M.C. (1999) "New development in shrinkage-reducing admixtures," ACI SP-173-48, Hormington Hills, Michigan: American Concrete Institute, pp. 973–998.

Bilinski, H., Matkovic, B., Mazuravic, C., and Zunic, T.A. (1984) "The formation of magnesium oxychloride phases in the system of $MgO-MgCl_2-H_2O$ and $NaOH-MgCl_2-H_2O$," *Journal of the American Ceramic Society*, 67 266–269.

Blanco, F., Garcia, M.P. and Ayala, J. (2005) "Variation in fly ash properties with milling and acid leaching", *Fuel*, 84(1), 89–96.

Caldarone, M.A., Gruber, K.A., and Burg, R.G. (1997) "High-reactivity metakaolin: a new generation mineral admixture," *Concrete International*, 16(11), 37–40.

Campbell, D.H. and Folk, R.L. (1991) "The ancient pyramids – concrete or rock?," *Concrete International: Design & Construction*, 13(8), 28–39.

Christensen, B.J., Coverdale, R., Olson T., Ford, R.A., S.J., Garboczi, E.J, Jennings, H.M. and Mason, T.O. (1994) "Impedance spectroscopy of hydrating cement-based materials: measurement, interpretation, and application," *Journal of the American Ceramic Society*, 77(11), 2789–2804.

Chusilp, N., Chai, J., and Kraiwood, K. (2009) "Utilization of bagasse ash as a pozzolanic material in concrete," *Construction and Building Materials*, 23(11), 3352–3358.

Cole, W. F. and Demediuk, T. (1955) "X-ray, thermal and dehydration studies on magnesium oxychloride," *Australian Journal of Chemistry*, 8, 234–237.

Comrie Consulting Ltd. (1988) *Comrie Preliminary Examination of the Potential of Geopolymers for Use in MineTailings Management*, Mississauga, Ontario, Canada: D. Comrie Consulting Ltd.

Cordeiro, G.C., Filho, R.D., Tavares, L.M., and Fairbairn, EM. (2009) "Effect of calcination temperature on the pozzolanic activity of sugar cane bagasse ash," *Construction and Building Materialss*, 23, 3301–3303.

Cotruvo, J. (2005) "Water desalination processes and associated health and environmental issues," *Water Conditioning and Purification*, 47, 13–17.

Cwirzen, A., Sztermen, P., and Habermehl-Cwirzen, K. (2014) "Effect of Baltic seawater and binder type on frost durability of concrete," *Journal of Materials in Civil Engineering*, 26, 283–287.

Danielsen, S.W. and Ørbog, A. (2000) "Sustainable use of aggregate resources through manufactured sand technology," *Quarry Management*, 27(7), 19–28.

Davidovits, J. (1987) "Ancient and modern concretes: what is the real difference?" *Concrete International: Design & Construction*, 9(12), 23–35.

Davidovits, J. (1993) "Geopolymer cement to minimize carbon-dioxide greenhouse warming," *Ceramic Transactions*, 37, 165–182.

Davidovits, J. (1994a) "Recent progresses in concretes for nuclear waste and uranium waste containment," *Concrete International*, 16(12), 53–58.

Davidovits, J. (1994b) "Properties of geopolymer cements," *Alkaline Cements and Concretes*, Kiev, Ukraine, p. 9.

Davidovits, J. (1994c). "Geopolymers: inorganic polymeric new materials", *Journal of. Materials Education*, 16, 91–139.

Davidovits, J., Comrie, D.C., Paterson, J.H. and Ritcey, D.J. (1990) "Geopolymeric concretes for environmental protection," *Concrete International: Design & Construction*, 12(7), 30–40.

de Henau, P. and Dupas, M. (1976) "Study of the alternation in acropolis monuments," *Proceedings of Second International Symposium on the Deterioration of Building Stone, Athens*, pp. 319–325.

Deng, D. (2003) "The mechanism for soluble phosphates to improve the water resistance of magnesium oxychloride cement," *Cement and Concrete Research*, 33, 1311–1317.

Deng, D. and Zhang, C. (1999) "The formation mechanism of the hydrate phases in magnesium oxychloride cement," *Cement and Concrete Research*, 29, 1365–1371.

Ding, J. and Li, Z. (2002) "Effects of metakaolin and silica fume on properties of concrete," *ACI Materials Journal*, 99(4), 393–398.

Dollimore, D., Spooner, P., and Turner, A. (1976) "The BET method of analysis of gas adsorption data and its relevance to the calculation of surface areas," *Surface Technology*, 4, 121–160.

Donza, H., Cabrera, O., and Irassar, E.F. (2002) "High-strength concrete with different fine aggregate," *Cement and Concrete Research*, 32(11), 1755–1761.

Duxson, P., Provis, J.L., Lukey, G.C., and van Deventer, J.S.J.(2007) "The role of inorganic polymer technology in the development of 'green concrete'," *Cement and Concrete Research*, 37(12), 1590–1597.

El-Enein, A. S.A, Kotkata, M.F. and Hannan, G.B. (1995) "Electrical resistivity of concrete containing silica fume," *Cement and Concrete Research*, 25(8), 1615–1620.

European R&D project BRITE-EURAM BE-7355-93 (1997) *Cost-Effective Geopolymeric Cement for Innocuous Stabilization of Toxic Elements (GEOCISTEM)*. Final Report, April 1997.

Folk, R.L. and Campbell, D.H. (1992) "Are the pyramids built of poured concrete blocks?" *Journal of Geological Education*, 40, 25–34.

Frias, M., Ernesto, V., and Holmer S. (2011) "Brazilian sugar cane bagasse ashes from the cogeneration industry as active pozzolans for cement manufacture," *Cement and Concrete Composites*, 33, 490–496.

Gartner, E.M., Young, E.M., Damidot, D.A., and Jawed, I. (2002) "Hydration of Portland cement." In: Bensted J. and Barnes P. (Eds), *Structure and performance of cements*, London: Spon Press, pp. 83–84.

Geiger, G. (1994) "Environmental and energy issues in the glass industry," *American Ceramics Society Bulletin*, 73(2), 32–37.

Gokmenoglu, Z. and Sobolev, K. (2002) "Solid waste-cement composite construction materials," In: *Creating the future—second FAE international symposium*, Lefke, Cyprus, pp. 153–158.

Goto et al., (1985) United States Patent Number 4547223, Oct, 15.

Gu, P., Xie, P. and Fu, Y. (1985) "Microstructural characterization of cementitious materials: conductivity and impedance methods," In: Skalny, J. and Mindess, S. (Eds), *Materials Science of Concrete IV* . Cincinnati, OH: American Ceramics Society, pp. 94–124.

Gunalaan, V. and Kanapathy, S.G. (2013) "Performance of using waste glass powder in concrete as replacement of cement," *American Journal of Engineering Research (AJER)*, 2(12), 175–181.

Hamid, N.B., Ramli, M.Z., Sanik, M.E., et al. (2018) "Effect of waterproofing admixtures on concrete," *AIP Conference Proceedings.* AIP Publishing LLC, 2030(1): 020239.

Haper, F. C. (1967) "Effect of calcinations temperature on the properties of magnesium oxides for use in magnesium oxychloride cements," *Journal of Applied Chemistry*, 17, 5–10.

Harrison, J. (2003) "New cements based on the addition of reactive magnesia to Portland cement with or without added pozzolan," *Proceedings of the CIA Conference: Concrete in the Third Millenium*, CIA. Brisbane, Australia, pp. 24–25.

He, Z., Li, Z., Chen, M., and Liang, W. (2006) "Properties of shrinkage-reducing admixture-modified pastes and mortar," *Materials and Structures RILEM*, 39(4), 445–453.

Heidrich, C. (2002) "Slag—not a dirty word," conference paper, Geopolymers 2002, Melbourne.

Hewlett, P.C. (1998) *Lea's Chemistry of Cement and Concrete*, 4th ed, London: Arnold.

Jennings, H.M. and Pratt, P.L. (1980) "On the reactions leading to calcium silicate hydrate, calcium hydroxide and ettringite during the hydration of cement," *Proceedings of the 7th international congress on the chemistry of cement*, Vol. II. Editions Septima, Paris, 141–146.

Ji, Y. (2001) "Study of the new type of light magnesium cement foamed material," *Materials Letters*, 50, 28–31.

Jolicoeur, C., Nkinamubanzi, P.C., Simard, M.A., and Piotte, M. (1994) "Progress in understanding the functional properties of superplasticizer in fresh concrete," *ACI* SP, 148, 63–88.

Kantro, D. (1981) "Influence of water-reducing admixtures on properties of cement paste – a miniature slump test," *Research and Development Bulietin*, RD079.01T, Portland Cement Association.

Katz, A. (2004) "Treatments for the improvement of recycled aggregate," *Journal of Materials in Civil Engineering*, 16(6), 597–603.

Khayat, K.H., and Mikanovic, N. (2012) "Viscosity-enhancing admixtures and the rheology of concrete," in *Understanding the rheology of concrete*, Cambridge: Woodhead Publishing, pp. 209–228.

Kingery, W.D. (1950) "Fundamental study of phosphate bonding in refractories (I): literature review," *Journal of the American Ceramic Ssociety*, 33(8), 239–250.

Kou, S.C., Zhan, B.J., and Poon, C. S. (2004). "Use of a $CO_2$ curing step to improve the properties of concrete prepared with recycled aggregates," *Cement and Concrete Composite*, 45, 22–28.

Langley, W., Carette, G., and Malhotra, V. (1989) "Structural concrete incorporating high volume of class F fly ash," *ACI Materials Journal*, 86(5), 507–514.

Le, H.B., and Bui, Q.B. (2020). "Recycled aggregate concretes–a state-of-the-art from the microstructure to the structural performance," *Construction and Building Materials*, 257, 119522.

Li, X. (2012) "Mechanical performance and durability of reactive magnesia cement concrete," PhD thesis, University of Cambridge.

Li, Z. and Li, W. (2003) "Contactless, transformer-based measurement of the resistivity of materials," United States Patent 6639401.

Li, Z., Mu, B., and Peng, J. (1999a) "The combined influence of chemical and mineral admixtures upon the alkali-silica reaction," *Magazine of Concrete Research*, 51(3), 163–169.

Li, Z., Peng, J., and Ma, B. (1999b) "Investigation of chloride diffusion for high-performance concrete containing fly ash, microsilica, and chemical admixtures," *ACI Materials Journal*, 96(3), 391–396.

Li, Z., Wei, X., and Li, W. (2003) "Preliminary interpretation of Portland cement hydration process using resistivity measurements," *ACI Material Journal*, 100(3) 253–257.

Li, Z. and Zhu, D. (2003) "Property improvement of Portland cement by incorporating with metakaolin and slag," *Cement and Concrete Research*, 33(4), 579–584.

Liu, P., Yu, Z.W., Chen, L.K., et al. (2011) "Mechanical property of phosphoaluminate cement," *Advanced Materials Research*, 150: 1754–1757.

Lyon, R.E. (1994) Technical Report DOT/FAA/CT-94/60.
Lyon, R.E., Foden, A., Balaguru, P.N., Davidovits, M. and Davidovits, J. (1997) "Fire resistant alumino-silicate composites," *Journal Fire and Materials*, 21(1), 67–73.
Maravelaki-Kalaitzaki, P. and Moraitou, G. (1999) "Sorel's cement mortars decay susceptibility and effect on pentelic marble," *Cement and Concrete Research*, 29, 1929–1935.
Marchon, D., Kawashima, S., Bessaies-Bey, H., Mantellato, S., and Ng, S. (2018). "Hydration and rheology control of concrete for digital fabrication: Potential admixtures and cement chemistry," *Cement and Concrete Research*, 112, 96–110.
Masse, S., Zanni, H., Lecourtier, J., et al. (1993) "29Si solid state NMR study of tricalcium silicate and cement hydration at high temperature," *Cement and Concrete Research*, 23(5), 1169–1177.
Matkovic, B. et al. (1977) "Reaction products in magnesium oxychloride cement pastes system $MgO-MgCl_2-H_2O$," *Journal of the American Ceramic Society*, 60, 504–507.
Mayer, P. (2000) "Technology meets the challenge of cullet processing," *Glass Industry*, 2.
McCarter, W.J., Chrisp, T.M., Starrs, G., and Blewett, J. (2003) "Characterization and monitoring of cement-based systems using intrinsic electrical property measurements," *Cement and Concrete Research*, 33(2), 197–206.
McCarter, W.J., Whittington, H.W., and Forde, M.C. (1981) "The conduction of electricity through concrete," *Magazine of Concrete and Research*, 33(114), 48–60.
McKinney, R.G. (1993) "Comments on the work of Harrell and Penrod," *Journal of Geological Education*, 41, 369.
Meddah, M.S., Zitouni, S., and Belaabes, S. (2010) "Effect of content and particle size distribution of coarse aggregate on the compressive strength of concrete," *Construction & Building Materials*, 24(4), 505–512
Ménétrier, D., Jawed, I., Sun, T.S., and Skalny, J. (1979) "ESCA and SEM studies on early $C_3S$ hydration," *Cement and Concrete Research*, 19(4), 473–482.
Ménétrier-Sorrentino, D., Barret, P., and Saquat, S. (1986) "Investigation in the system $MgO-MgCl_2-H_2O$ and hydration of Sorel cement," *Proceedings of the 8th ICCC, Rio de Janeiro*, 4, 339–343.
Mindess, S., Young, J.F., and Darwin, D. (2003) *Concrete*, 2nd ed., Upper Saddle River, NJ: Prentice Hall.
Montakarntiwong, K., Chusilp, N., Tangchirapat, W., and Jaturapitakkul, C. (2013) "Strength and heat evolution of concretes containing bagasse ash from thermal power plants in sugar industry," *Mater Design*, 49, 414–420.
Morris, M. (1991) "The cast-in-place theory of pyramid construction," *Concrete International: Design & Construction*, 13(8), 39–44.
Nedeljković, M., Visser, J., Šavija, B., Valcke, S., and Schlangen, E. (2021). "Use of fine recycled concrete aggregates in concrete: A critical review," *Journal of Building Engineering*, 102196.
Neville, A.M. and Brooks, J.J. (1990) *Concrete Technology*, Harlow: Longman Group.
Odler, J. (2000) *Special inorganic cements*, London: E & FN Spon.
Penttala V. (1997) "Concrete and sustainable development," *ACI Materials Journal*, 94(5), 409–416.
Popovics, S. (1992) *Concrete materials: properties, specifications, and testing*, Park Ridge, NJ: Noyes Publications.
Prince, W., Edwards-Lajnef, M. and Aitcin, P.C. (2002) "Interaction between ettringite and polynapthalene sulfonate superplasticizer in cementitious paste," *Cement and Concrete Research*, 32(1), 79–85.
Ramachandran, V.S. (2002) *Concrete admixtures handbook*, New Delhi: Standard Publishers.
Rao, A., Jha, K.N., and Misra, S. (2007) "Use of aggregates from recycled construction and demolition waste in concrete," *Resources, Conservation and Recycling*, 50(1), 71–81.
Rixom, R. and Mailvaganam, N. (1999) *Chemical admixtures for concrete*, London: E & FN Spon.
Rodriguez, D. (1995) "Application of differential grinding for fine cullet production and contaminant removal," *Ceramic Engineering and Science Proceedings,* 16(2), 96–100.
Scrivener, K.L. (2014) "Options for the future of cement," *Indian Concrete Journal*, 88(7), 11–21.
Scrivener, K.L. and Favier. A. (2015) "Calcined clays for sustainable concrete" *RILEM*, 17–18.
Scrivener, K.L. and Kirkpatrick, R.J. (2008) "Innovation in use and research on cementitious material," *Cement and Concrete Research,* 38,128–136.

Seehra, S.S., Gupta, S. and Kumar, S. (1993) "Rapid setting magnesium phosphate cement for quick repair of concrete pavements – characterization and durability aspects," *Cement and Concrete Research*, 23(2), 254–266.

Shayan, A. and Xu, A. (2006) "Performance of glass powder as a pozzolanic material in concrete: a field trial on concrete slabs," *Cement and Concrete Research,* 36, 457–468.

Shi, L., Liu, J., Liu, J.P., Gao, X.L., Zhang, W.N., and Jiang, Q. (2018), "Transmission and binding behavior of chloride ion in concrete under sulfate ion coupling effect in marine environment," *Concrete*, 348, 67–71.

Singh, D., Wagh, A., Cunnane, J. and Mayberry, J. (1997) "Chemically bonded phosphate ceramics for low-level mixed-waste stabilization," *Journal of Environmental Science and Health, A*, 32(2), 527–541.

Sobolev, K. (1993) "High-strength concrete with low cement factor," PhD dissertation, Chemical Admixtures Lab., Research Institute of Concrete and Reinforced Concrete, Moscow, Russia.

Sobolev, K. (2003) "Effect of complex admixtures on cement properties and the development of a test procedure for the evaluation of high-strength cements," *Adv. Cem. Res.,* 15(1), 67–76.

Sobolev, K. (2005) "Mechano-chemical modification of cement with high volumes of blast furnace slag," *Cement and Concrete Composites,* 27(7–8), 848–853.

Sobolev, K. and Arikan, M. (2002) "High-volume mineral additive ECO-cement," *American Ceramic Society Bulletin,* 81(1), 39–43.

Sobolev, K., Türker, P., Yeginobali, A., and Erdogan, B. (2004) "Microstructure and properties of eco-cement containing waste glass," In: International Conference on Sustainable Waste Management and Recycling: Challenges and Opportunities, UK.

Sorel, S. (1867) "On a new magnesium cement," *Comptes Rendus, Hebdomadaires des Séances de l'Académie des Sciences*, 65, 102–104.

Sorrel, C. A. and Armstrong, C. R. (1976) "Reactions and equilibria in magnesium oxychloride cements," *Journal of the American Ceramic Society*, 59, 51–59.

Stewart, G. (1986) "Cullet and glass container manufacture," *Resource Recycling*, 2.

Tashiro, C., Ikeda, K. and Inome, Y. (1994) "Evaluation of pozzolanic activity by the electric resistance measurement method," *Cement and Concrete Research*, 24(6), 1133–1139.

Taylor, H.F.W. (1997) *Cement chemistry*, 2nd ed., London: T. Telford.

Taylor, M.A. and Arulanandan, K. (1974)"Relationships between electrical and physical properties of cement pastes," *Cement and Concrete Research*, 4(6), 881–897.

Thomas, M.D.A., Gruber, K.A. and Hooton R.D. (1997) "*The use of high-reactivity metakaolin in high-performance concrete*," High Strength Concrete, Proceedings of the First International Conference, A. Azizinamini, D. Darwin, and C. French, eds, ASCE, Kona, Hawaii, pp. 517–530.

Uchikawa, H., Hanehara, S., and Sawaki, D. (1997) "The role of steric repulsive force in the dispersion of cement particles in fresh paste prepared with organic admixture," *Cement and Concrete Research*, 27, 37–50.

Urwongse, L. and Sorrell, C. A. (1980) "The system $MgO-MgCl_2-H_2O$ at 23°C," *Journal of the American Ceramic Society*, 63, 501–504.

Van Jaarsveld, J.G.S. and van Deventer, J.S.J. (1997) "The potential use of geopolymeric materials to immobilize toxic metals: part I. theory and applications," *Minerals Engineering*, 10(7), 659–669.

Van Jaarsveld, J.G.S. and van Deventer, J.S.J. (1999). "The effect of metal contaminants on the formation and properties of waste-based geopolymers," *Cement and Concrete Research*, 29(12), 1189–1200.

Van Jaarsveld, J.G.S., van Deventer, J.S.J. and Lorenzen, L. (1998) "Factors affecting the immobilization of metals in geopolymerized fly ash," *Metallurgical and Materials Transactions B*, 29B(1), 283–291.

Van Jaarsveld, J.G.S., van Deventer, J.S.J., and Schwartzman, A. (1999) "The potential use of geopolymeric materials to immobilize toxic metals: part II. material and leaching characteristics," *Minerals Engineering*, 12(1), 75–91.

Van Olphen, H. (1964). "An introduction to clay colloid chemistr,," *Soil Science*, 97(4), 290.

Vogel, A. I. and Jeffery, G. H. (1989) *Vogel's textbook of quantitative chemical analysis*, New York: John Wiley & Sons.

Waddall, J.J. and Dobrowolski, J.A. (1993) *Concrete construction handbook*, 3rd edn, New York: McGraw-Hill.

Wagh, A., Strain, R., Jeong, S., Reed, D., Krouse T., and Singh, D. (1999) "Stabilization of rocky flats Pu-contaminated ash within chemically bonded phosphate ceramics," *Journal of Nuclear Materials*, 265(3), 295–307.

Wegian, L. and Falah, M. (2010) "Effect of seawater for mixing and curing on structural concrete," *The IES Journal Part A: Civil & Structural Engineering*, 235–243.

Wei, X. and Li, Z. (2005) "Study on hydration of Portland cement with fly ash using electrical measurement," *Materials and Structures*, 38(277), 411–417.

Xiao, J., Li, W., Fan, Y., and Huang, X. (2012). "An overview of study on recycled aggregate concrete in China (1996–2011)," *Construction and Building Materials*, 31, 364–383.

Xiao, J., Qiang, C., Nanni, A., et al. (2017) "Use of sea-sand and seawater in concrete construction: Current status and future opportunities," *Construction and Building Materials*, 155, 1101–1111.

Xiao, L., Li, Z. and Wei, X. (2007) "Selection of superplasticizer in concrete mix design by measuring the early electrical resistivities of pastes," *Cement and Concrete Composites*, 29, 350–356.

Xie, Z.H., and Xi, Y.P, (2000) "Hardening mechanism of an activated class F fly ash," *CCR*, 31, 1245–1249.

Yang, Q.B., Zhang, S.Q., and Wu, X.L. (2002) "Deicer-scaling resistance of phosphate cement-based binder for rapid repair of concrete," *Cement and Concrete Research*, 32(1), 165–168.

Yao, W., Li, Z., and Wu, K.R. (1998) "Effects of slag upon the flowability and mechanical properties of high strength concrete," *Proceedings of Sixth CANMET/ACI International Conference on Fly Ash, Silica Fume, Slag and Natural Pozzolans in Concrete*, Bangkok, Thailand, pp. 413–424.

Yi, Y.L., Martin, L., and Abir, A.T. (2012) "Initial investigation into the use of GGBS-MgO in soil stabilization," *Grouting and Deep Mixing*, 444–453.

Yoshizake, Y., Ikeda, K., Yoshida, S., and Yoshizumi, A. (1989) "Physicochemical study of magnesium-phosphate cement," *MRS International Meeting. on Advanced Materials*, 13, 27–38.

CHAPTER 3

# FRESH CONCRETE

## 3.1 INTRODUCTION

Fresh concrete is defined as a fully mixed concrete in a rheological state that has not lost its plasticity. The fresh concrete stage covers the cement hydration stages I and II. The plastic state of fresh concrete provides a time period for transportation, placing, compaction, and surface finishing. The properties of fresh concrete have a large influence on the construction speed and decision making.

Having discussed the constituents of concrete, we can now examine the properties of freshly mixed concrete. The properties of fresh concrete affect the choices of handling, consolidation, and construction sequence. They may also affect the properties of the hardened concrete. The properties of fresh concrete are short-term requirements in nature, and should satisfy the following requirements:

1. It must be easily mixed and transported.
2. It must be uniform throughout a given batch, and between batches.
3. It must keep its fluidity during the transportation period.
4. It should have flow properties such that it is capable of completely filling the forms.
5. It must have the ability to be fully compacted without segregation.
6. It must set in a reasonable period of time.
7. It must be capable of being finished properly, either against the forms or by means of troweling or other surface treatment.

Compaction plays an important role in ensuring the long-term properties of the hardened concrete, as proper compaction is vital in removing air from concrete and in achieving a dense concrete structure. Subsequently, the compressive strength of concrete can increase with an increase in the density. Traditionally, compaction is carried out using a vibrator. Nowadays, the newly developed self-compacting concrete can reach a dense structure by its self-weight without any vibration.

## 3.2 WORKABILITY AND RHEOLOGY

### 3.2.1 Workability

#### 3.2.1.1 Definition

Workability of concrete is defined in ASTM C125 as the property determining the effort required to manipulate a freshly mixed quantity of concrete with minimum loss of homogeneity (uniform). The term manipulate includes the early-age operations of placing, compacting, and finishing. Mindess et al. (2003) defined the workability of fresh concrete as "the amount of mechanical work, or energy, required to produce full compaction of the concrete without segregation."

The effort required to place a concrete mixture is determined largely by the overall work needed to initiate and maintain flow, which depends on the rheological properties of the cement paste and the internal friction between the aggregate particles, on the one hand, and the external friction between the concrete and the surface of framework, on the other hand. Workability of fresh concrete consists of two aspects: consistency and cohesiveness. Consistency describes how easily fresh concrete flows, while cohesiveness describes the ability of fresh concrete to hold all the ingredients together uniformly. Traditionally, consistency can be measured by a slump-cone test, the compaction factor, or a ball penetration compaction factor test as a simple index for fluidity of fresh concrete. Cohesiveness can be characterized by a Vebe test as an index of both the water-holding capacity (the opposite of bleeding) and the coarse-aggregate-holding capacity (the opposite of segregation) of a plastic concrete mixture. The flowability of fresh concrete influences the effort required to compact concrete. The easier the flow, the less work is needed for compaction. A liquid-like self-compacting concrete can completely eliminate the need for compaction. However, such a concrete has to be cohesive enough to hold all the constituents, especially the coarse aggregates in a uniform distribution during the process of placing.

Workability is not a fundamental property of concrete; to be meaningful it must be related to the type of construction and methods of placing, compacting, and finishing. Concrete that can be readily placed in a massive foundation without segregation would be entirely unworkable in a thin structural member. Concrete that is judged to be workable when high-frequency vibrators are available for consolidation, would be unworkable if hand tamping were used.

The significance of workability in concrete technology is obvious. It is one of the key properties that must be satisfied. Regardless of the sophistication of the mix design procedures used and other considerations, such as cost, a concrete mixture that cannot be placed easily or compacted fully is not likely to yield the expected strength and durability characteristics.

#### 3.2.1.2 Measurement of Workability

Unfortunately, there is no universally accepted test method that can directly measure the workability as defined earlier. The difficulty in measuring the mechanical work defined in terms of workability, the composite nature of the fresh concrete, and the dependence of the workability on the type and method of construction makes it impossible to develop a well-accepted test method to measure workability. The most widely used test, which mainly measures the consistency of concrete, is called the slump test. For the same purpose, the second test in order of importance is the Vebe test, which is more meaningful for mixtures with low consistency. The third test is the compacting factor test, which attempts to evaluate the compactability characteristic of a concrete mixture. The fourth test method is the ball penetration test that is somewhat related to the mechanical work.

**(a)** *Slump test*: The equipment for the slump test is indeed very simple. It consists of a tamping rod and a truncated cone, 300 mm in height, 100 mm in diameter at the top, and 200 mm in diameter at the bottom, see Figure 3-1. To conduct a slump test, first moisten the slump test mold and place it on a flat, nonabsorbent, moist, and rigid surface. Then hold it firmly to the ground by foot supports. Next, fill 1/3 of the mold with the fresh concrete and rod it 25 times uniformly over the cross-section. Likewise fill 2/3 of the mold and rod the layer 25 times, then fill the mold completely and rod it 25 times. If the concrete settles below the top of the mold, add more. Strike off any excessive concrete. Remove the mold immediately in one move. Measure and record the slump as the vertical distance from the top of the mold to average concrete level. The sequence of a slump test is shown in Figure 3-2.

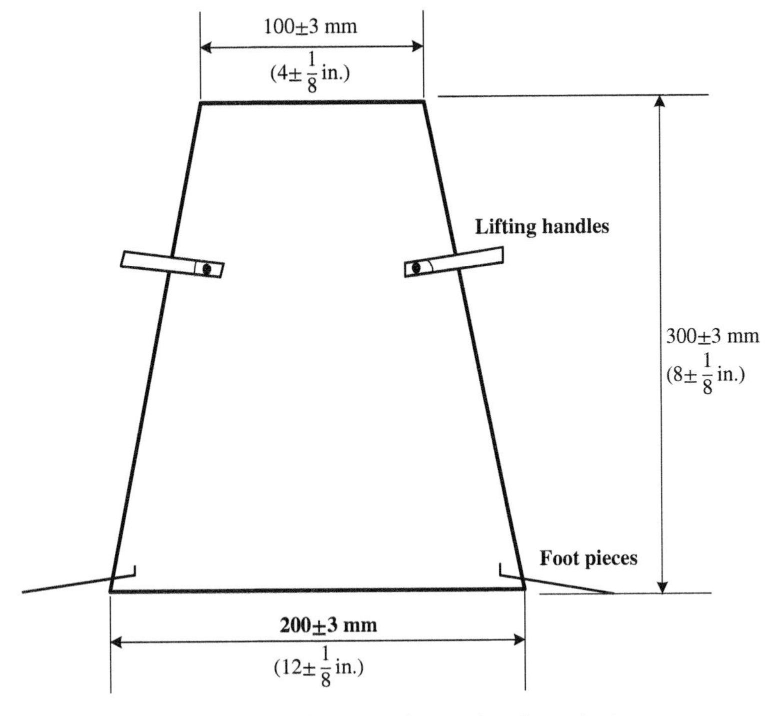


**Figure 3-1** Truncated cone for slump test

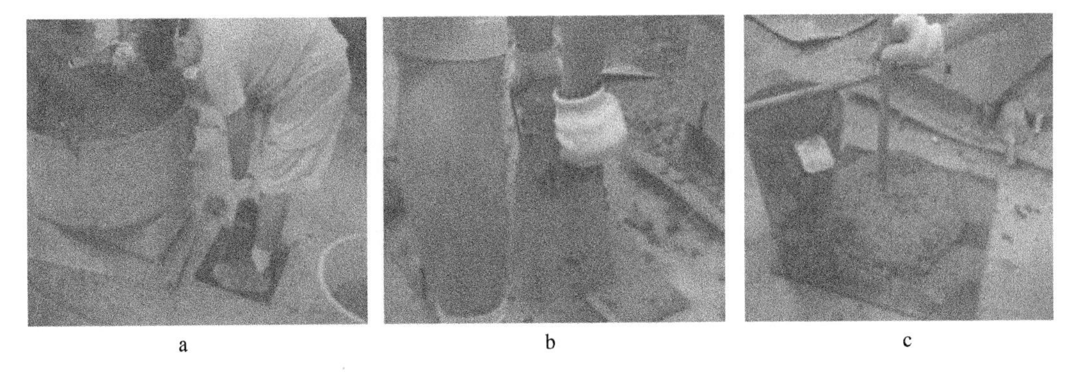


**Figure 3-2** Experimental sequence of slump test

If slumping occurs evenly all around, it is regarded as a true slump. If one-half of the cone slides down along an inclined plane, it is regarded as shear slump. Shear slump is caused by insufficient cohesiveness and the concrete proportions should be adjusted. Mixes of very stiff consistency have zero slump, so that in the rather dry range no slump can be detected between mixes of different workability. There is no problem with rich mixes, their slumps are sensitive to variations in workability; however, in a lean mix with a tendency to harshness, a true slump can easily be changed to the shear type, or even collapse with a nonuniform distribution of aggregates, especially coarse aggregates, and widely different values of slump can be obtained in different samples from the same mix. Thus, the slump test is unreliable for lean mixes.

**(b)** *Vebe test*: The test equipment, which was developed by Swedish engineer V. Bahrner, is shown in Figure 3-3. It consists of a vibrating table, a cylindrical pan, a slump cone, and a glass or plastic disk attached to a free-moving rod, which serves as a reference endpoint. The cone is placed in the pan. After it is filled with concrete and any excess concrete is struck off, the cone is removed. Then, the disk is brought into a position on top of the concrete cone, and the vibrating table is set in motion. The time required for the concrete cone to shorten and change from the conical to a cylindrical shape, until the disk on the top is completely covered with concrete, is the index of workability and is reported as the number of Vebe seconds.

The Vebe test is a good laboratory test, particularly for very dry mixes. This is in contrast to the compacting factor test where error may be introduced by the tendency of some dry mixes to stick in the hoppers. The Vebe test also has the additional advantage that the treatment of concrete during the test is comparatively closely related to the method of placing in practice. Moreover, the cohesiveness of concrete can easily be distinguished by the Vebe test through the observation of distribution of the coarse aggregate after vibration.

**(c)** *Compaction factor*: Figure 3-4 shows the compaction factor test apparatus. It consists of two hoppers and one cylindrical mold stacked in three levels. To perform an experiment, the upper hopper is first fully filled with fresh concrete. Then the hinged door is slid open and hence the concrete will fall into the lower hopper by gravity. Next, the hinged door of the power hopper is slid open and the concrete free falls into the 150 × 300 mm cylindrical mold. After the excess concrete is struck off on the top of the mold, the weight of the cylinder is measured and noted as $M_p$, representing a partially compacted cylinder mass. Another cylinder is made with same concrete by three layers with 25 times rodding on each layer and striking off any excess concrete. The weight of the cylinder is measured as $M_f$, representing the fully compacted cylinder mass. The compaction factor is defined as

$$\text{compaction factor} = \frac{M_p}{M_f} \tag{3-1}$$

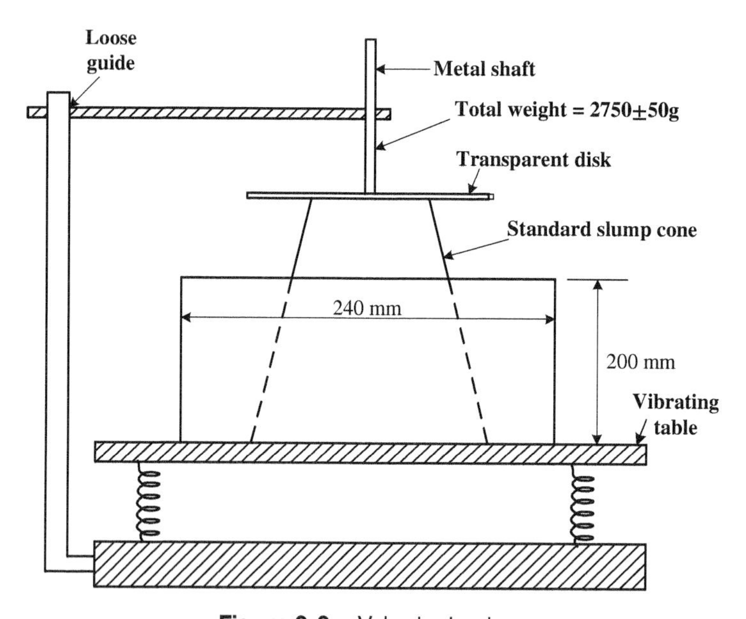


**Figure 3-3** Vebe test setup

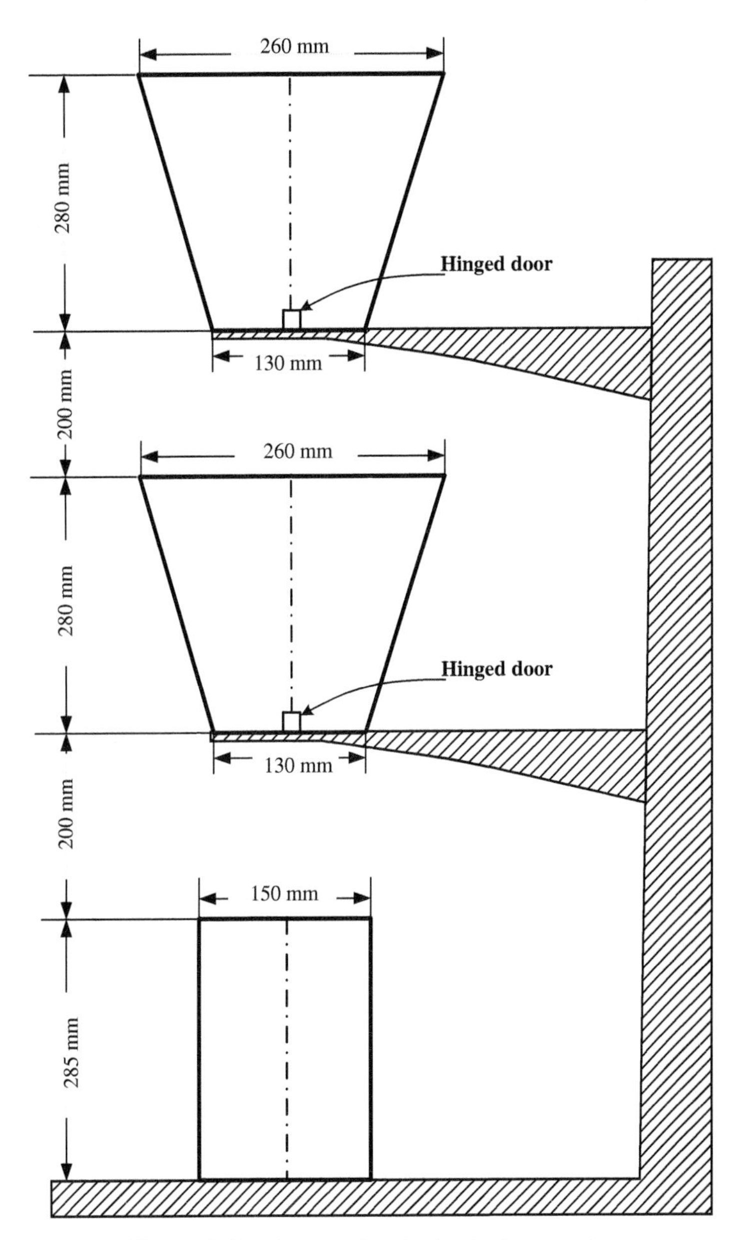


**Figure 3-4** Compaction factor test apparatus

Usually, the range of compaction factor is from 0.78 to 0.95 and concrete with high fluidity has a higher compaction factor.

**(d)** *Ball penetration test*: ASTM C360 covers the Kelly ball penetration test. The test setup is shown in Figure 3-5. A 152-mm-diameter hemisphere hammer of weight 13.6 kg is connected to a handle with a ruler. The hammer is fixed on a box container through a pin. When taking measurements, the box is placed on the top of the concrete to be tested with the surface of the hammer touching the concrete. When the pin is removed, the hammer will sink into

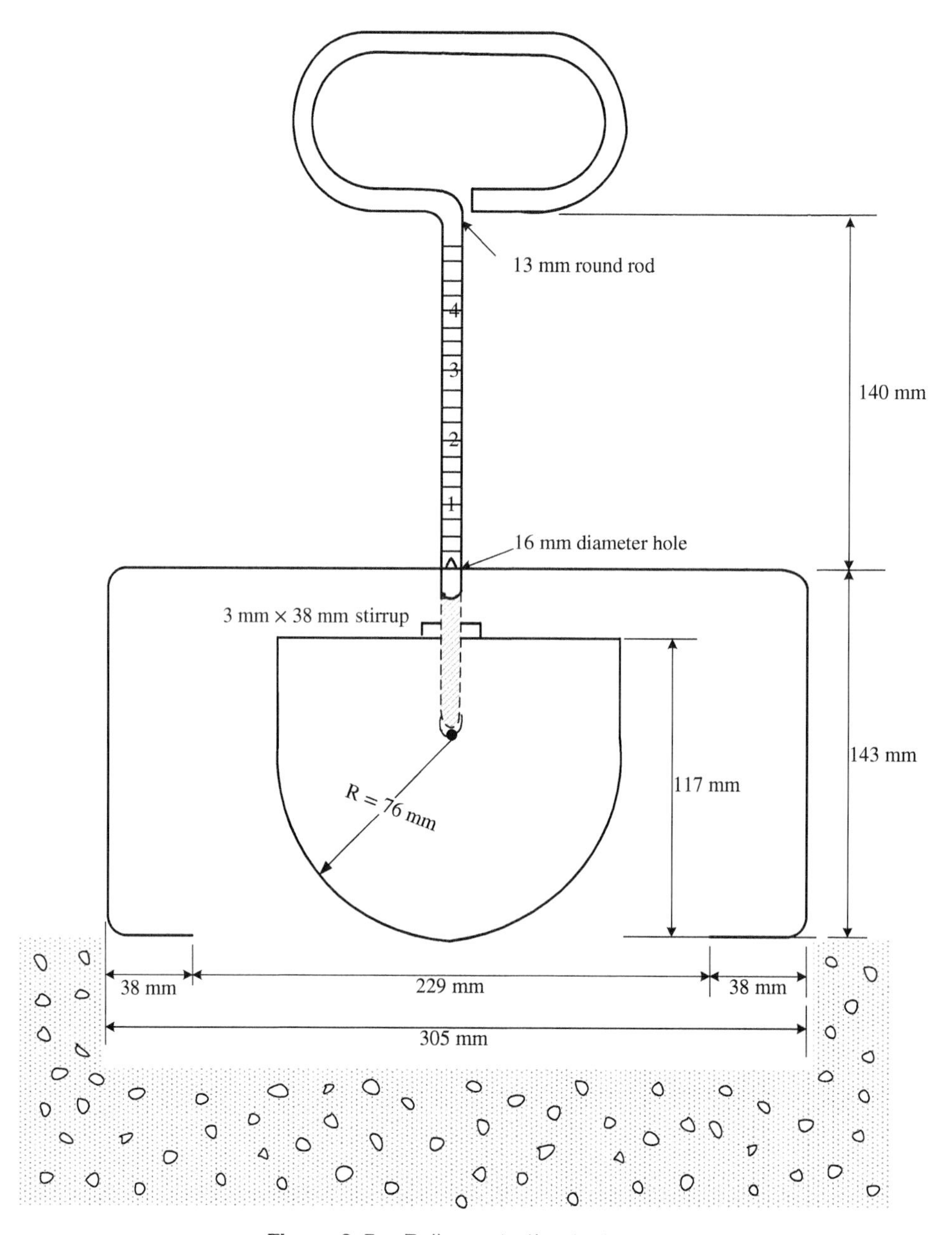


**Figure 3-5** Ball penetration test setup

the fresh concrete by its own weight. The depth of the hammer penetration can be read from the ruler and is used as an index of workability. A concrete with higher consistency leads to a deeper ball penetration. Since the measurement is related to the work done by the hammer penetration, ball penetration measurement is close to the definition of workability given by Mindess et al. (2003). The penetration test is usually very quick and can be done on site, right in the formwork, provided it is wide enough. The ratio of slump value to penetration depth is from 1.3 to 2.0.

#### 3.2.1.3 Factors Affecting Workability

The workability of concrete contains two aspects: consistency and cohesiveness. Due to the different requirements and characteristics of the two aspects, the influence of a factor on workability may be opposite. In general, through the influence on consistency and/or cohesiveness, the water content, the cement content, the aggregate grading, and other physical characteristics, and admixtures can affect the workability of concrete mixtures.

**(a)** *Water content*: Water content is regarded as the most important factor influencing the workability of concrete. After adding water to a concrete mix, the water is absorbed on the surface of the particles of the cement and aggregates. Additional water fills the spaces among the particles and "lubricates" the particles by a water film. Decreasing the water content will result in a low fluidity. If the water content is too small, the concrete will become too dry to mix and place. Increasing the amount of water will increase the amount of water for lubrication and hence improve the fluidity and make it easy to be compacted. However, too much water will reduce cohesiveness. This not only leads to segregation and bleeding, but also reduces the concrete strength. The water content in a concrete is determined by *w/c* or *w/b* and cement or binder content.

**(b)** *Cement content*: Cement content influences the workability of concrete in two ways. First, for a given *w/c* ratio, the greater the cement content, the higher the total water amount in the concrete; hence, the consistency of concrete will be enhanced. Second, cement paste itself plays the roles of coating, filling, and lubrication for aggregate particles. In normal concrete, a considerably low cement content tends to produce a harsh mixture, with poor consistency and, subsequently, poor finishability. High cement content implies that more lubricant is available for consistency improvement. Finally, with an increase of the cement content at a low *w/c* ratio, both consistency and cohesiveness can be improved. Under the same *w/c* ratio, the higher the cement content, the better the workability.

Increasing the fineness of the cement particles will decrease the fluidity of the concrete at a given *w/c* ratio, but will increase the cohesiveness. Concretes containing a very high proportion of cement or very fine cement show excellent cohesiveness but tend to be sticky.

**(c)** *Aggregate characteristics*: Aggregates can influence the workability of concrete through their need for surface coating and their friction and mobility during mixing, placing, and compaction. Maximum aggregate size, aggregate/cement ratio, fine aggregate/coarse aggregate ratio, and aggregate shape and texture are four aspects influencing the workability of concrete.

The particle size of coarse aggregates influences the paste requirement for coating through the surface area. The larger aggregates have a smaller surface area than smaller aggregates with the same volume. Subsequently, the amount of the paste available for lubrication is increased for concrete with large aggregates, and consistency is improved. Hence, for a given w/c ratio, as the maximum size of aggregate increases, the fluidity increases. Moreover, very fine sands or angular sands will require more paste for a given consistency; alternatively, they will produce harsh and unworkable mixtures at water contents that might have been adequate with coarser or well-rounded particles. In general, to get a similar consistency of concrete, more water is needed when crushed sand is used instead of natural sands.

The aggregate/cement ratio influences the paste requirement. A higher aggregate/cement ratio implies more aggregates and less cement paste. Thus, the concrete consistency decreases with aggregate/cement ratio increase due to less cement paste being available for lubrication.

Fine aggregate/coarse aggregate ratio also affects the cement paste requirement. With an increase of the fine aggregate/coarse aggregate ratio, concrete contains more fine aggregates and less coarse aggregates. Thus, the total surface area of the aggregates increases, which leads to a higher demand on the cement paste for surface coating. As a result, the consistency of concrete decreases and the cohesiveness improves. Increasing the fine aggregate/coarse aggregate ratio is the most effective measure to increase the cohesiveness of concrete.

The shape and texture of aggregate particles can affect the workability of concrete through the influence on paste requirement, particle moving friction, and moving ability. Cubical, irregular, granular, and rough aggregates require more coating cement paste and have higher friction than spherical, glassy, and smooth aggregates. As a general rule, the more spherical the particles, the more workable is the concrete.

**(d)** *Admixtures*: Both chemical and mineral admixtures can influence the workability of concrete. Their effects have been discussed in Chapter 2. For instance, an air-entraining agent increases the paste volume and improves the consistency of concrete for a given water content through the entrained air. The entrained air also increases cohesiveness by reducing bleeding and segregation. Improvement in consistency and cohesiveness by air entrainment is more pronounced in harsh and unworkable mixtures, such as in mass concrete, which has low cement content. Water-reducing admixtures can improve the fluidity of concrete due to the dispersing effect on cement particles and the releasing of entrapped water by cement clusters. Similarly, when the water content of concrete mixtures is held constant, the addition of water-reducing admixtures (plasticizer) will increase the consistency.

Different mineral admixtures have different effects on workability, although they all tend to improve the cohesiveness of concrete. Fly ash, when used as a partial replacement for cement, generally increases the consistency at a given water content due to the spherical shape and glassy surface. When silica fume is used to replace part of the cement, it tends to reduce the amount of water used for lubrication, due to its very large surface area and hence the need for a water film coating.

**(e)** *Temperature and time*: Freshly mixed concrete stiffens with time due to evaporation of the mixing water, particularly when the concrete is directly exposed to sun or wind, absorption by the aggregate, and consumption in the formation of hydration products. The stiffening of concrete is effectively measured by a loss of workability with time, known as slump loss, which varies with richness of the mix, type of cement, temperature of the concrete, and initial workability. A high temperature reduces the workability and increases the slump loss because the hydration rate is higher and the loss of water is faster at a higher temperature. In practice, when the ambient conditions are unusual, it is best to perform actual site tests to determine the workability of the mix.

### 3.2.2 Rheology

#### 3.2.2.1 Definition

Rheology is the science that deals with the deformation and flow of materials, primarily in a liquid state, but also as "soft solids" that respond with plastic flow. Rheology study is originated from experiments, and it is treated as one of the most suitable methods to investigate the fresh behavior of cement-based materials. In general, the fluid can be divided into two groups: Newtonian fluid and non-Newtonian fluid. For the cement-based materials, due to the fact that they are composed of a series of different materials with various particle size distributions, the cement-based materials commonly present a typical non-Newtonian fluid characteristic. When the mineral admixtures or superplasticizers are added to the cementitious system, the thickening or thinning of the tested paste can be observed during the shearing process. These thinning or thickening phenomena have a close relationship with the flowability, viscosity and other fresh properties of concrete (Schwartzentruber and Cordin, 2006). Viscosity and rheology are often confused in the literature because they both deal with fluid flow. However, they are different concepts. Viscosity of a fluid is the measure of its resistance to deformation at a given rate. The key difference between rheology and viscosity is that rheology is the study of the flow of matter, whereas viscosity is a measure of its resistance to deformation.

#### 3.2.2.2 Rheology Models of Concrete

By far, there are four rheological models for cement-based materials, which are presented as follows (Hu et al., 1996; Ferraris et al., 2001; Senff et al., 2009; Ouyang et al., 2016; Bruno et al., 2016).

**(a)** *Bingham model*: Many researchers treated the cement-based materials as a plastic fluid, which could be described by the Bingham model, as shown in Equation (3-2):

$$\tau = \tau_0 + \mu\dot{\gamma} \tag{3-2}$$

where $\tau$ is shear stress (Pa), $\gamma$ is share rate ($s^{-1}$), $\tau_0$ is yield stress for the tested paste, and $\mu$ means the plastic viscosity (Pa·s). In this model, the tested paste can flow when the share stress exceeds a critical value $\tau_0$.

**(b)** *Modified Bingham model*: In practice, it is noted that the relationship between the stress and strain of the tested paste in a low share rate mode is not linear. Hence, a quadratic term ($c\gamma^2$) is added to the Bingham model for further modification, as shown in Equation (3-3):

$$\tau = \tau_0 + \mu\gamma + c\gamma^2 \tag{3-3}$$

where $c$ is correction index.

**(c)** *Casson model*: The Casson fluid model is based on two hypotheses: (1) there are interaction forces between the solid particles that suspended in a Newtonian fluid system; and (2) when the shear stress is relatively small, the particles can agglomerate together and generate a rigid rod, and the length of the rigid rod could inversely proportionally decrease with an increase in the shear stress. The Casson fluid model can be shown in Equation (3-4):

$$\tau = \tau_0 + \mu_\infty\gamma + 2\left(\sqrt{\mu_\infty\tau_0}\right)\sqrt{\gamma} \tag{3-4}$$

where $\mu_\infty$ is ultimate viscosity (Pa·s).

**(d)** *Herschel-Bulkley model*: In the Herschel-Bulkley model, the paste can flow only when the shear stress exceeds a critical shear stress, and the shear rate will increase based on a power-law when the shear stress is increasing. The Herschel-Bulkley model can be presented as Equation (3-5):

$$\tau = \tau_0 + m\gamma^n \tag{3-5}$$

where $m$ is consistency coefficient (Pa·s), $n$ means rheological behavior index. It is clear that the "n" can reflect the difference between the Newtonian fluid and non-Newtonian fluid. When $n = 1$, this model is similar to that shown in the Bingham model; when $n<1$, the thinning of the tested paste can be observed during the shearing process; when $n>1$, the thickening of the tested paste can be observed during the shearing process. In the Herschel-Bulkley model, the fluid viscosity can be expressed by Equation (3-6):

$$\mu = \frac{3m}{n+2}\gamma_{max}^{n-1} \tag{3-6}$$

where $\gamma_{max}$ is the maximum shear rate during the rheological testing process.

#### 3.2.2.3 Factors Affecting Rheology

Concrete is a multiphase material, involving aggregates, cement, water, and mineral and chemical admixtures. The rheology properties of concrete are primarily dependent on the quality of its constituents and their interactions.

**(a)** *Paste*: Paste plays a critical role in concrete, which coats aggregates, fills the spaces between aggregates, and provides workability. One of the most important factors affecting rheological behavior is the volume of paste. Generally, the interactions between particles or fluid and particle exhibit hydrodynamic processes in concrete with a relatively high paste volume fraction. On the contrary, rheological behavior is governed by the direct frictional contacts between aggregate particles at a relatively low paste volume fraction. In order to quantify the effect of paste volume on rheological behavior, Reinhardt et al. (2006) proposed a factor named the thickness of the excess paste, which can be calculated by the volume of excess paste, total paste, and voids between the aggregates, as well as the mass, density, and loose bulk density of the aggregates. With an increase in the thickness of the excess paste, a certain decline in both the yield stress and plastics viscosity values can be witnessed. Apart from the volume of the paste, cementitious materials compositions also directly affect the rheological behavior of fresh concrete. For example, the paste made of cement and fly ash reduced the plastic viscosity of concrete due to its spherical geometry and smooth surface compared with the paste that includes plain cement.

**(b)** *Aggregate*: Aggregate accounts for more than 75% of the total volume of normal concrete. From the rheological point of view, the physical properties of aggregate affect the plastic viscosity and yield stress significantly. With regard to the geometric parameters, i.e., shape and roughness of aggregate, the factors influence the workability of concrete by packing density and surface morphology. For instance, aggregate with a smooth surface of rounded aggregate contributes to a reduced inter-particle friction, thus, resulting in a decrease in the plastic viscosity and yield stress. On the contrary, the introduction of crushed stone aggregate usually increases the specific surface area and reduces the dosage of free water due to its angular shape and rough surface. Besides, the angular shape hinders the flow of particles and increases the inter-particle friction resistance. As a result, the yield stress and plastic viscosity increased significantly (Hu and Wang, 2007; Aissoun et al., 2016). Similarly, the flat and elongated aggregate will increase the collision due to its shape. Therefore, concrete mixtures containing flat and elongated aggregate suffer a sharp increase in plastic viscosity and yield stress. Apart from the geometric parameter of aggregate, the relative volume fraction of coarse and fine aggregates also affects the rheological behavior of concrete. In general, yield stress and plastic viscosity increase along with the volume fraction of coarse aggregate, while increasing the fine aggregate volume raises the yield stress but lowers the plastic viscosity of the concrete. The addition of coarse aggregate increases the paste coating the surface aggregate, and reduces the friction among aggregates. With the increase of fine aggregate volume fraction, the surface area of solid particles and water demand rise, leading to an increase in yield stress and a reduction in plastic viscosity. In addition, the gradation and maximum particle size of aggregate play predominant roles in the rheological properties of concrete. To be specific, the larger the maximum particle size, the smaller the specific surface area, and the less the amount of paste requirement needed, thus the lower the values of rheological parameters of concrete (Feys et al., 2009).

**(c)** *Chemical admixtures*: Superplasticizer is a widely used chemical admixture in concrete because of its economic and technical benefits. Generally, with the inclusion of superplasticizer, the action of electrostatic and/or steric hindrance effects could decrease the yield stress and viscosity of the cement-based materials dramatically. With the increase of superplasticizer, the bridging distance between the particles increases, the colloidal interaction between the particles decreases, and therefore the yield stress and the degree of flocculation decrease.

An air-entraining agent, a chemical admixture typically used in the improvement of resistance to freezing and thawing damage, also influences the rheological properties of the entrained air bubbles. The increase in yield stress can be attributed to the fact that air bubbles are attracted to cement particles and form bubble bridges, which enhance the bonding between particles. Once the bubble bridges are broken and paste flow again, the air bubbles act as a lubricant to reduce the plastic viscosity.

Besides, viscosity modifying agents (VMA) have been used to alter the rheology of the concrete. Usually, the introduction of a VMA contributes to high plastic viscosity and yield stress, which brings benefits in reducing the risk of separation of the heterogeneous concrete. By far, concrete with VMA is promising in achieving stable concrete for underwater repair, curtain and deep foundation walls.

### 3.2.3 Segregation and Bleeding

#### 3.2.3.1 Segregation

In discussing the workability of concrete, it has been pointed out that cohesiveness is an important characteristic of the workability. A proper cohesiveness can ensure concrete to hold all the ingredients in a homogeneous way without any concentration of a single component, and even after the full compaction is achieved. An obvious separation of different constituents in concrete is called segregation, as shown in Figure 3-6. Thus, segregation can be defined as concentration of individual constituents of a heterogeneous (nonuniform) mixture so that their distribution is no longer uniform. In the case of concrete, it is the differences in the size and weight of particles (and sometimes in the specific gravity of the mix constituents) that are the primary causes of segregation, but the extent can be controlled by the concrete proportion, choice of suitable grading, and care in handling.

#### 3.2.3.2 Bleeding

Bleeding is a form of local concentration of water in some special positions in concrete, usually the bottom of the coarse aggregates, the bottom of the reinforcement, and the top surface of the concrete member, as shown in Figure 3-7. During placing and compaction, some of water in the mix tends to rise to the surface of freshly placed concrete. This is caused by the inability of the solid constituents of the mix to hold all the mixing water when they settle downward due to the lighter density of water. Bleeding can be expressed quantitatively as the total settlement (reduction in height) per unit height of concrete, and bleeding capacity as the amount (in volume or weight) of water that rises to the surface of freshly placed concrete.

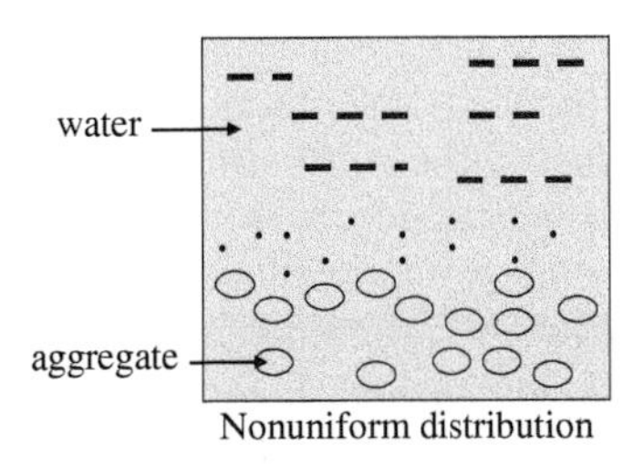


**Figure 3-6** Segregation of concrete mixture

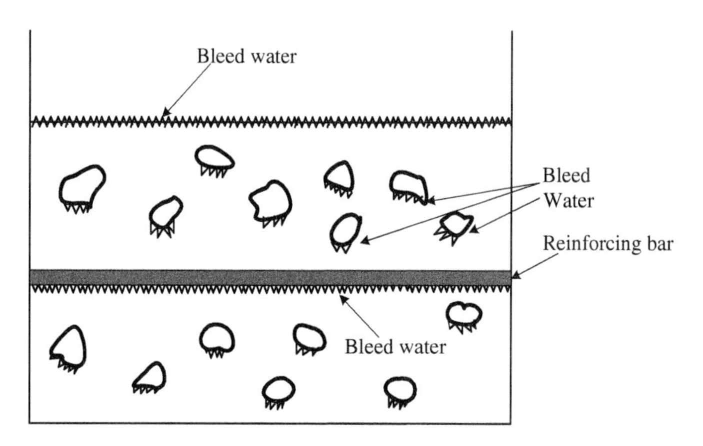


**Figure 3-7** Bleeding phenomenon

As a result of bleeding, an interface between aggregates and bulk cement paste is formed, and the top of every lift (layer of concrete placed) may become too wet. If the water is trapped by the superimposed concrete, a porous and weak layer of nondurable concrete may result. If the bleeding water is remixed during the finishing process of the surface, a weak wearing surface can be formed. This can be avoided by delaying the finishing operations until the bleeding water has evaporated, and also by the use of wood floats and avoidance of overworking the surface. On the other hand, if evaporation of water from the surface of the concrete is faster than the bleeding rate, plastic shrinkage cracking may be generated.

### 3.2.4 Slump Loss

Slump loss can be defined as the loss of consistency in fresh concrete with elapsed time. Slump loss is a normal phenomenon in all concretes because it results from gradual stiffening and setting of hydrated cement paste, which is associated with the formation of hydration products such as ettringite and calcium silicate hydrate. Slump loss occurs when the free water from a concrete mixture is removed by hydration reactions, by absorption on the surface of hydration products, and by evaporation. Slump loss should be controlled to an acceptable value, especially for concrete transported with a long delivery time, to ensure that it is still placeable and compactable when shipped to the construction site. Slump loss can be minimized by using a setting retarder.

### 3.2.5 Setting of Concrete

#### 3.2.5.1 Definition

Setting of concrete is distinguished as the initial and final setting of cement paste. The initial setting is defined as the loss of plasticity or the onset of rigidity (stiffening or consolidating) in fresh concrete. The final setting is defined as the onset point of strength. It is different from hardening, which describes the development of useful and measurable strength. Setting precedes hardening, although both are controlled by the continuing hydration of the cement. The measurement of the setting time for concrete is very different from that of cement paste. For cement paste, it uses the samples made of the water amount needed for consistency. For concrete, it uses the sieved mortar from a concrete with different water/cement or water/binder ratios. Moreover, for cement paste, it measures the penetration depth of the Vicat needle, 1 mm in diameter, under a constant weight. For concrete, it measures the resistance of the mortar to a rod under an action of the load. A setup for concrete setting time measurement is shown in Figure 3-8.

**Figure 3-8** Measurement setup of concrete mixture setting time

It can be seen from Figure 3-8 that the container is full with the sieved mortar from a concrete. A steel rod is fixed on a load cell. The rod and load cell can be moved downward and penetrate into the mortar. The resistance of the rod encountered during the penetration process can be measured by a load cell and displayed on a meter. ASTM C 403 (1995) defines two points as the initial setting time and the final setting time, corresponding approximately to the point at which the concrete will no longer be plastic during vibration, and a concrete strength gain of about 0.7 MPa. Initial setting and final setting of concrete are determined by the times at which the penetration resistance reaches 3.5 and 27.6 MPa, when a designated rod penetrates 25.4 mm into the mortar, sieved from the fresh concrete.

### 3.2.5.2 Abnormal Setting

**(a)** *False setting*: If a concrete stiffens rapidly in a short time right after water is added and restores its fluidity by remixing and sets normally, it is called false setting. The main reason causing the false setting is crystallization of gypsum. In the last procedure in the process of cement production, gypsum is milled with a clinker by intergrounding. During grinding, due to clinkers that may be still very hot and heat generated by friction, the temperature can rise to about 120°C, thus causing the reaction

$$C\bar{S}H_2 \xrightarrow{120\ \mathrm{deg\ C}} C\bar{S}H_{1/2} \tag{3-7}$$

$C\bar{S}H_{1/2}$ is called half-water gypsum or plaster. During mixing, when water is added, the plaster will rehydrate back to two-water gypsum, form a crystalline matrix quickly, and make the concrete stiffen. However, due to the small amount of plaster in the mix, very little strength will actually develop and the fluidity can easily be restored by further mixing to break the plaster set. Hence, it is not real setting, but false setting.

**(b)** *Flash setting*: Flash setting is caused by the formation of large quantities of monosulfoaluminate or other calcium aluminate hydrates due to quick reactivities of $C_3A$, without the presence of gypsum. This is a rapid set and thus is a more severe condition than false setting. However, as mentioned before, flash setting has been largely eliminated by the addition of 3–5% gypsum to the cement, which can react with $C_3A$ and water to form AFt as a barrier layer of $C_3A$ to prevent further reaction of $C_3A$, as discussed in Chapter 2.

#### 3.2.5.3 New Method for Determining Concrete Setting Time

Concrete setting time is a crucial parameter for construction progress and concrete quality control. When the setting time is known, the times of mixing, transporting, casting, and finishing can be regulated and the effectiveness of various set-controlling admixtures can be decided. In addition, the setting time also influences the time of demolding. As discussed earlier, the initial and final setting times of concrete are measured by a penetration method that has been standardized in ASTM C 403 (1995). This method utilizes a sieved mortar for penetration testing. However, it is hard work to obtain the mortar fraction through a 4.75-mm sieve from fresh concrete, especially for concretes with poor fluidity. Additionally, for concrete with the incorporation of a retarder, it probably takes an operator more than ten hours to finish the test, at regular intervals. Moreover, the test results can be largely affected by the skill of the operators. For these reasons, there have been some attempts to use alternative techniques, such as the impact-echo method (Pessiki and Carino, 1988) and ultrasonic measurement (Lee et al., 2004; Subramaniam et al., 2005; Voigt et al., 2005) to determine the setting time for concretes. The impact-echo method defines the setting time of concrete as when the wave velocity begins to increase. However, it is hard to impact on early-age concrete and is thus not practical. The ultrasonic method is based on generation, transmission, and reflection of mechanical waves in concrete. Preliminary studies have shown good correlation between the measured wave reflection factor and the hydration process.

The noncontact electrical resistivity method has been used to determine the setting time of concrete based on the characteristic points on the resistivity measurement curve of concrete (Li et al., 2007). The electrical resistivity of concrete is measured by a noncontacting electrical resistivity apparatus, which is introduced in detail in Chapter 8. The transformer principle was adopted in this apparatus. Three typical examples of the bulk electrical resistivity development with time ($\rho$–$t$) for plain concrete, concrete containing superplasticizers, and concrete with a higher water cement ratio are plotted in Figure 3-9. Figure 3-9a shows $\rho$ change with time up to 400 min and the characteristic point at the minimum resistivity, $P_m$, is marked as solid dots on the curves. Figure 3-9b shows the $\rho$ change up to 1440 min on a logarithmic scale and the second critical point $P_t$, is identified on the curves.

These two characteristic points have been utilized to relate to the setting times of the concrete measured by the penetration method. The relationships (Equations 3-8 and 3-9) have been developed using regression methods:

$$t_t = 1.8807t_m + 0.4429t_t \left(R^2 = 0.8950\right) \quad (3\text{-}8)$$

$$t_f = 0.9202t_t + 0.2129 \left(R^2 = 0.9895\right) \quad (3\text{-}9)$$

Equation 3-8 is for the initial setting time and Equation 3-9 for the final setting time. Figure 3-10 shows the comparison of the predicted initial and final setting using Equations 3-8 and 3-9 and

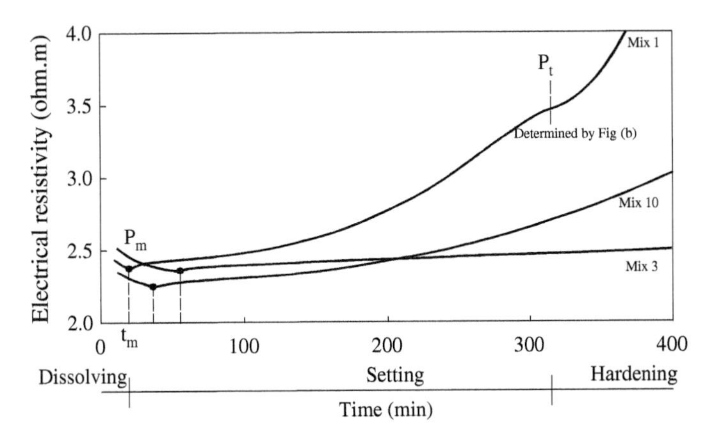


(a) Minimum point $P_m$ and 3 development periods on curves -t

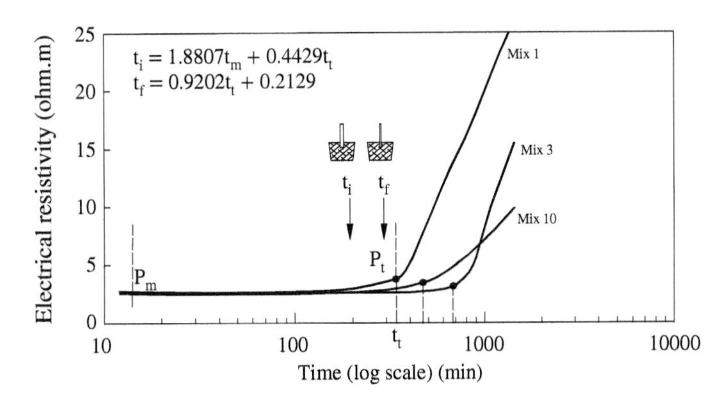


(b) Transition point $P_t$ on curves -t (log scale)

**Figure 3-9** The electrical resistivity response of concretes

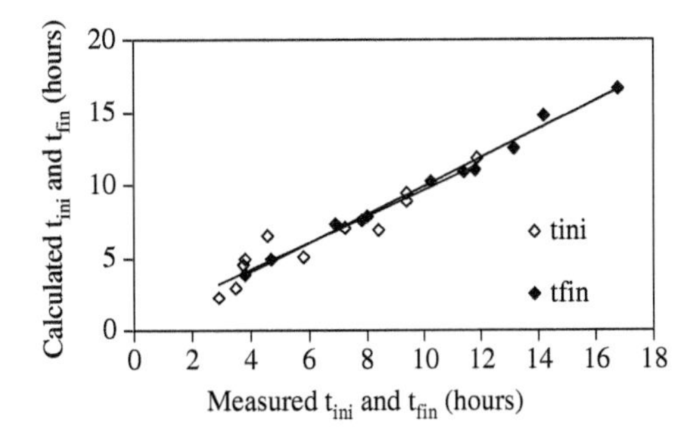


**Figure 3-10** Comparison of the predicted and measured initial and final setting time

that measured using the penetration test. Good agreement is found, as shown in Figure 3-10. This proves that it is possible to use the electrical resistivity measurement method to judge the setting time of concrete, especially the final setting time. Figure 3-10 plots the measured and calculated setting times and a good correlation can be observed.

The electrical resistivity method (ERM) provides practical advantages over the conventional penetration resistance method because it can continuously measure the data immediately after mixing and can be conducted directly on concrete rather than on sieved mortar. ERM needs only about 0.0016 $m^3$ of fresh concrete, but the penetration method needs at least 0.01 $m^3$ of mortar sieved from fresh concrete. ERM is a practical method to assess the setting time and strength development trend, and gives the in situ information to determine the demolding time.

## 3.3 MIX DESIGN

The mix design of concrete is the process of deciding what type of raw material and how much of each raw material needs to be selected to make concrete that can meet prerequisites such as strength, durability, and workability. The required properties of hardened concrete are specified by the designer of the structure and the properties of fresh concrete are governed by the type of construction and by the techniques of placing and transporting. These two sets of requirements are the main factors that determine the composition of the mix, also taking account of the construction experience on site. Mix design can, therefore, be defined as the processes of selecting suitable ingredients and determining their relative quantities, with the purpose of producing an economical concrete that has certain minimum properties, notably workability, strength, and durability. It should be pointed out that the mix design of concrete is frequently done by trial and error. Hence, mix design of concrete is an art, not a science. This means that the mix design of concrete in the strict sense is not possible: the materials used vary in a number of respects and their properties cannot be assessed truly quantitatively, so that we are really making no more than an intelligent guess at the optimum combinations of the ingredients on the basis of relationships established in the earlier sections. It is not surprising, therefore, that to obtain a satisfactory mix, we must check the estimated proportions of the mix by making trial mixes and, if necessary, make appropriate adjustments to the proportions until a satisfactory mix has been obtained (Neville and Brooks, 1994).

### 3.3.1 Principal Requirements for Concrete

The main purpose of the mix design is to obtain a product that will perform according to predetermined requirements. These requirements include the following concrete properties.

**(a)** *Quality (strength and durability)*: Strength and permeability of hydrated cement paste are mutually related through the capillary porosity that is controlled by the *w/c* ratio and degree of hydration. Since the durability of concrete is controlled mainly by its permeability, there is a relationship between strength and durability. Consequently, routine mix design usually focuses on strength and workability only. When the concrete is exposed to special environmental conditions, provisions on durability (e.g., limit on *w/c* ratio, minimum cement content, minimum cover to steel reinforcement) will also be considered.

**(b)** *Workability*: As mentioned earlier, workability is a complicated concept for fresh concrete and embodies various properties, including consistency and cohesiveness. There is still not a single test method that can fully reflect workability. Since the slump represents the ease with which the concrete mixture will flow during placement, and the slump test is simple and quantitative, most mix design procedures rely on slump as a crude index of workability. Sometimes, the Vebe test may be employed.

**(c)** *Economy*: Among all the constituents of the concrete, the admixture has the highest unit cost, followed by cement. The cost of aggregates is about one-tenth that of cement. Admixtures are often used in small amounts, or they are required to achieve certain properties. To minimize the cost of concrete, the key consideration is the cement cost. Therefore, all possible steps should be taken to reduce the cement content of a concrete mixture without sacrificing the desirable properties, such as strength and durability. The scope for cost reduction can be enlarged further by replacing a part of the Portland cement with cheaper materials, such as fly ash or ground blast-furnace slag.

As mentioned earlier, under normal conditions, it is sufficient to consider workability and strength for concrete design. For special conditions, additional considerations on dimensional stability and durability have to be taken into account.

### 3.3.2 Weight Method and Absolute Volume Method

There are two approaches for concrete mix design; the weight method and the absolute volume method. In the weight method, the unit weight of fresh concrete is known from previous experience for the commonly used raw materials and is used to calculate the weight of the last unknown component of concrete, usually the sand. If the unit weight of fresh concrete (wet concrete) is known, we have

$$W_{\text{wet concrete}} = W_{\text{cement}} + W_{\text{water}} + W_{\text{aggregate}} + W_{\text{sand}} + W_{\text{admixtures}} \tag{3-10}$$

If the weights of cement, water, coarse aggregate, and admixtures have been determined, then the weight of sand can be obtained from above equation. The unit weight of wet concrete usually ranges from 2300 to 2400 kg/m$^3$.

In the absolute volume method, the total volume (1 m$^3$) is equal to the sum of volume of each ingredient (i.e., water, air, cement, and coarse aggregate). Thus, we have

$$\frac{W_{\text{cement}}}{\rho_{\text{cement}}} + \frac{W_{\text{water}}}{\rho_{\text{water}}} + \frac{W_{\text{aggregate}}}{\rho_{\text{aggregate}}} + \frac{W_{\text{sand}}}{\rho_{\text{sand}}} + \frac{W_{\text{admixture}}}{\rho_{extadmixture}} + \text{volume\% (air)} = 1 \tag{3-11}$$

Again, when the weights of cement, water, coarse aggregate, and admixture have been determined, their corresponding volumes can be calculated, with their densities known. With the volume of air determined during concrete mix design, the volume of sand (usually the last unknown component) can be calculated using the above equation. Since the weight of each ingredient is easier to measure than the volume, the design proportion of concrete is usually expressed as a weight ratio. Hence, the proportion obtained in the volume method has to be converted to weight units by multiplying the volume with the density of the material.

### 3.3.3 Factors to Be Considered

In the previous sections, the various factors that influence the properties of concrete have been discussed in detail. In this section, the technical and economic factors that need to be considered in concrete mix design are addressed. As always, the projected strength of concrete has to be considered first. The projected strength is usually specified by the structural designer. Normally, the strength at 28 days is used as the design index for structural purposes, but other considerations may dictate the strength at other ages, e.g., formwork demolding time. By adopting quality-control techniques, the variability of strength can be minimized so that the projected strength can be achieved. Nowadays, most concrete is supplied by commercial concrete operators. In these designed mixes, specifications for a range of properties must be satisfied. These properties include the maximum water/cement ratio, minimum cement content, projected strength, projected workability, maximum size of aggregate, and air content.

#### 3.3.3.1 Water/Cement Ratio

Although the *w/c* ratio can be estimated from Abram's law, based on projected strength of concrete, in concrete design, the *w/c* ratio required to produce a given mean compressive strength is usually determined from previously established relations for mixes made from similar ingredients, or by carrying out tests using trial mixes made with the actual ingredients to be used in the construction, including admixtures. These relations can be summarized into tables or graphs. Tables 3-1 and 3-2 are two examples. They may be used to estimate the approximate *w/c* ratio for the cements listed for each set of values as a starting point.

It is important that the *w/c* ratio selected on the basis of strength is also satisfactory for the durability requirements. Moreover, this *w/c* ratio for durability should be established prior to the commencement of the structural design because, if it is lower than necessary from structural considerations, the advantage of the use of a higher strength of concrete can be taken in the design calculations.

When supplementary cementitious materials are used in concrete, the water/binder ratio by mass has to be considered. With supplementary cementitious materials, the ACI 211.1-81 approach treats the water/binder ratio as equivalent to the *w/c* ratio of a Portland cement mix. With the

**Table 3-1** Relation between w/c and average compressive strength of concrete, according to ACI 211.1-91

| Average compressive strength at 28 days[a] (MPa) | Effective water/cement ratio (by Mass) | |
|---|---|---|
| | Non-air-entrained concrete | Air-entrained concrete |
| 45 | 0.38 | — |
| 40 | 0.42 | — |
| 35 | 0.47 | 0.39 |
| 30 | 0.54 | 0.45 |
| 25 | 0.61 | 0.52 |
| 20 | 0.69 | 0.60 |
| 15 | 0.79 | 0.70 |

[a]Measured on standard cylinder. The values given are for a maximum size of aggregate of 19 to 25 mm, for concrete containing not more than the percentage of air shown in Table 3-8, and for ordinary Portland (type I) cement.

**Table 3-2** Relation between w/c and specified compressive strength of concrete, according to ACI 318-83

| Specific compressive strength at 28 days[a] (MPa) | Absolute water/cement ratio (by mass) | |
|---|---|---|
| | Non-air-entrained concrete | Air-entrained concrete |
| 30 | 0.40 | — |
| 25 | 0.50 | 0.39 |
| 20 | 0.60 | 0.49 |
| 17 | 0.66 | 0.54 |

[a]Measured on standard cylinder. Applicable for cements: ordinary Portland (types I & IA), modified Portland (types II an IIA) cement, rapid-hardening Portland (types III and IIIA), sulfate-resisting Portland (type V); also Portland blast furnace (types IS, IS-A) and Portland pozzolan (types IP, P, I(PM), IP-A), including moderate sulfate-resisting cement (MS). *Note:* The use of admixtures, other than air-entraining, or of low-density aggregate is not permitted. The values of absolute water/cement ratio are conservative and include any water absorbed by the aggregates. Hence, with most materials, the water/cement ratio will provide average strengths that are greater than the specified strength.

mass method, water/binder ratio is equal to the *w/c* ratio of the Portland-cement-only mix. In the absolute volume method, however, because supplementary cementitious materials usually have a lower specific gravity than Portland cement does, the volume replaced by the supplementary cementitious materials is greater than that of Portland cement. With the volume method, the mass of the cementitious material is smaller than that of the cement in the Portland-cement-only mix so that the water/cementitious material ratio is greater than in the Portland-cement-only mix. Whichever approach is used, a partial replacement of cement by pozzolan generally reduces the strength at an early age. For this reason, the ACI 211.1-81 mix design is used mainly for mass concrete, in which the reduction of the heat of hydration is of paramount importance and the early strength is of lesser significance.

#### 3.3.3.2 Durability

Severe exposure conditions require a stringent control of the *w/c* ratio because it is the fundamental factor determining the permeability and diffusivity of the cement paste and, to a large extent, of the resulting concrete. In addition, adequate cover to embedded reinforcing steel is essential. However, the *w/c* ratio can be assessed indirectly through the workability of the mix, the cement content, and strength. If the *w/c* ratio is determined due to durability requirements, the cement content can be reduced by the use of a larger-size aggregate.

The requirements on the water/binder ratio and the minimum cover thickness for reinforced concrete by ACI 318-83 are given in Tables 3-3 and 3-4. It must be remembered that air entrainment is essential under conditions of freezing and thawing or exposure to deicing salts (see Table 3-5), although entrained air does not protect concrete containing coarse aggregate that undergoes disruptive volume changes when frozen in a saturated condition.

#### 3.3.3.3 Workability

As discussed earlier, the workability of concrete consists of two aspects: flowability and cohesiveness. Two factors have to be taken into consideration when determining the workability. One is the

**Table 3-3** Requirements of ACI 318-83 for Water/Cement Ratio and Strength for Special Exposure Conditions

| Exposure condition | Maximum water/cement ratio, normal-density aggregate concrete | Minimum Ddesign strength in MPa, low-density aggregate concrete |
|---|---|---|
| Concrete intended to be watertight | | |
| Exposed to fresh water | 0.50 | 25 |
| Exposed to brackish or seawater | 0.45 | 30 |
| Concrete exposed to freezing and thawing in a moist condition | | |
| Curbs gutters, guardrails, or thin sections | 0.45 | 30 |
| Other elements | 0.50 | 25 |
| In presence of deicing chemicals | 0.45 | 30 |
| For corrosion protection of reinforced concrete exposed to deicing salts, brackish water, seawater, or spray from these sources | 0.40[a] | 33[a] |

[a]If minimum cover required in Table 3-4 is increased by 10 mm, water/cement ratio may be increased to 0.45 for normal density concrete or design strength reduced to 30 MPa for low-density concrete.

**Table 3-4** Requirements of ACI 318-83 for minimum cover for protection of reinforcement

| Exposure condition | Minimum cover in mm | | |
|---|---|---|---|
| | Reinforced concrete cast in suit | Precast concrete | Prestressed concrete |
| Concrete cast against, or permeability exposed to, earth | 70 | — | 70 |
| Concrete exposed to earth or weather | | | |
| Wall panels | 40–50 | 20–40 | 30 |
| Slabs and joists | 40–50 | — | 30 |
| Other members | 40–50 | 30–50 | 40 |
| Concrete not exposed to weather or in contact with earth | | | |
| Slabs, walls, joists | 20–40 | 15–30 | 20 |
| Beams, columns | 40 | 10–40 | 20–40 |
| Shells, folded plate members | 15–20 | 10–15 | 10 |
| Nonprestressed reinforcement | — | — | 20 |
| Concrete exposed to deicing slats, brackish water, seawater, or spray from these sources | | | |
| Walls and slabs | 50 | 40 | — |
| Other members | 60 | 50 | — |

*Note*: Ranges of cover quoted depend on the size of steel used.

**Table 3-5** Recommended air content of concretes containing aggregates of different maximum size, according to ACI 201.2R-77 (reaffirmed **1982**)

| Maximum size of aggregate (mm) | Recommended total air content of concrete (%) for level of exposure | |
|---|---|---|
| | Moderate[a] | Severe[b] |
| 10 | 6.0 | 7.5 |
| 12.5 | 5.5 | 7.0 |
| 20 | 5.0 | 6.0 |
| 25 | 4.5 | 6.0 |
| 40 | 4.5 | 5.5 |
| 50 | 4.0 | 5.0 |
| 70 | 3.5 | 4.5 |
| 150 | 3.0 | 4.0 |

[a]Cold climate where concrete will be occasionally exposed to moisture prior to freezing, and where no deicing salts are used, e.g., exterior walls, beams, slabs not in concrete with soil.
[b]Outdoor exposure in cold climate where concrete will be in almost continuous contact with moisture prior to freezing or where deicing salts are used, e.g., bridge decks, pavements, sidewalls, and water tanks.

geometry of the member to be cast, including size of cross-section and the amount and spacing of reinforcement. The other is the compaction method, including the equipment for compacting and duration of consolidation.

It is clear that when the cross-section of the member to be cast is narrow and complicated in shape, the concrete must have a high fluidity so that full compaction can be achieved. The same applies when the member is heavily reinforced with steel bars that make placing and compaction difficult. Moreover, it is important to choose proper compacting equipment, such as a plate-type vibrator or a sticker-type vibrator, and a compaction duration to ensure that concrete can be fully compacted during the entire progress of construction. A guide to workability for different types of construction is given in Tables 3-6 and 3-7.

After choosing the workability, the water content of the mix (mass of water per unit volume of concrete) can be estimated by considering the workability requirement. ACI 211.1-91 gives the water content for various maximum sizes of aggregate and slump value (as an index of workability), with and without air entrainment (see Table 3-8). The values apply for well-shaped coarse aggregates and, although the water requirement is influenced by the texture and shape of the aggregate, the values given are sufficiently accurate for a first estimate.

**Table 3-6** Workability, slump, and compacting factor of concretes with 19 or 38 mm maximum size of aggregate

| Degree of workability | Slump | | Compacting factor | Use for which concrete is suitable |
|---|---|---|---|---|
| | (mm) | (in.) | | |
| Very low | 0–25 | 0–1 | 0.78 | Roads vibrated by power-operated machines. At the more workable end of this group, concrete may be compacted in certain cases with hand-operated machines. |
| Low | 25–50 | 1–2 | 0.85 | Roads vibrated by hand-operated machines. At the more workable end of this group, concrete may be manually compacted in roads using aggregate of rounded or irregular shape. Mass concrete foundations without vibrated or lightly reinforced sections with vibration. |
| Medium | 25–100 | 2–4 | 0.92 | At the less workable end of this group, manually compacted flat slabs using crushed aggregate. Normal reinforced concrete manually compacted and heavily reinforced sections with vibration. |
| High | 100–175 | 4–7 | 0.95 | For sections with congested reinforcement. Not normally suitable for vibration. |

Source: Building Research Establishment, Crown copyright.

**Table 3-7** Recommended values of slump for various types of construction as given by ACI 211.1-91

| Type of construction | Range of slump[a] | |
|---|---|---|
| | (mm) | (in.) |
| Reinforced foundation walls and footings | 25–75 | 1–3 |
| Plain footings, caissons and substructure walls | 25–75 | 1–3 |
| Beams and reinforced walls | 25–100 | 1–4 |
| Building columns | 25–100 | 1–4 |
| Pavements and slabs | 25–75 | 1–3 |
| Mass concrete | 25–75 | 1–2 |

[a]The upper limit of slump may be increased by 20 mm for compaction by hand.

**Table 3-8** Approximate requirement for mixing water and air content for different workabilities and nominal maximum sizes of aggregates, according to ACI 211.1-91

| Workability or air content | Water content (kg/m³) of concrete for indicated maximum aggregate size in mm | | | | | | | |
|---|---|---|---|---|---|---|---|---|
| | **9.5** | **12.5** | **19** | **25** | **37.5** | **50** | **75** | **150** |
| | Non-air entrained concrete | | | | | | | |
| Slump: | | | | | | | | |
| 25–50 mm | 207 | 199 | 190 | 179 | 166 | 154 | 130 | 113 |
| 75–100 mm | 228 | 216 | 205 | 193 | 181 | 169 | 145 | 124 |
| 150–175 mm | 243 | 228 | 216 | 202 | 190 | 178 | 160 | - |
| Approximate entrapped air content (%) | 3 | 2.5 | 2 | 1.5 | 1 | 0.5 | 0.3 | 0.2 |
| | Air-entrained concrete | | | | | | | |
| Slump: | | | | | | | | |
| 25–50 mm | 181 | 175 | 168 | 160 | 150 | 142 | 122 | 107 |
| 75–100 mm | 202 | 193 | 184 | 175 | 165 | 157 | 133 | 119 |
| 150–175 mm | 216 | 205 | 197 | 184 | 174 | 166 | 154 | - |
| Recommended average total air content (%): | | | | | | | | |
| Mild exposure | 4.5 | 4.0 | 3.5 | 3.0 | 2.5 | 2.0 | 1.5[a] | 1.0[a] |
| Moderate exposure | 6.0 | 5.5 | 5.0 | 4.5 | 4.5 | 4.0 | 3.5[a] | 3.0[a] |
| Extreme exposure[b] | 7.5 | 7.0 | 6.0 | 6.0 | 5.5 | 5.0 | 4.5[a] | 4.0[a] |

[a]For concrete containing large aggregate which will be wet-screened over 40 mm sieve prior to testing of air content, the percentage of air expected in the material smaller than 40 mm should be as tabulated in the 40 mm column. However, initial proportioning calculations should be based on the air content as a percentage of the whole mix.
[b]These values are based on the criterion that a 9% air content is needed in the mortar of concrete.
*Notes*: Slump values for concrete containing aggregate larger than 40 mm are based on slump test made after removal of particles larger than 40 mm by wet-screening.
Water contents for nominal maximum size of aggregate of 75 mm and 150 mm are average values for reasonable well-shaped coarse aggregates, well graded from coarse to fine.

#### 3.3.3.4 Cement Type and Content

The properties of the different types of cement were discussed in Chapter 2. The choice of the types of cement depends on the required hydration rate and strength development, the likelihood of chemical attack, and thermal considerations. Although all have been discussed earlier, it is worth reiterating the need for a cement with a high rate of heat of hydration developed for cold-weather concreting, and with a low rate of heat of hydration for mass concreting, as well as for concreting in hot weather. In the latter case, it may be necessary to use a lower *w/c* ratio to ensure a satisfactory strength at early ages.

Because cement is more expensive than aggregate, it is desirable to reduce the cement content as much as possible, provided it can satisfy the strength, durability, and workability requirements. Moreover, low to moderate cement content confers the technical advantage of a lower hydration heat as well as cracking potential in the case of mass concrete, where the heat of hydration needs to be controlled, and in the case of structural concrete where shrinkage cracks should be minimized.

In technical terms, the cement content is obtained from the water/binder ratio and the mixing water requirement from the workability requirement. However, the cement content has to meet the minimum requirement by specification from the durability considerations.

#### 3.3.3.5 Major Aggregate Properties and Aggregate Content

Many parameters have to be determined in choosing an aggregate. Usually, the maximum size of aggregate is determined first, as it has a significant influence on concrete properties. In reinforced concrete, the maximum size of an aggregate is governed by the geometry of the member and the spacing of the reinforcement. Generally, the maximum aggregate size has to be smaller than 1/4 to 1/5 of the smallest size of the cross-section of a member and 3/4 of the net spacing distance of reinforcement. With this proviso, it is generally considered desirable to use as large a maximum size of aggregate as possible. However, the improvement in the properties of concrete with an increase in the size of aggregate does not extend beyond about 40 mm. For high-strength concrete, the maximum aggregate size is limited to 20 mm. For mass concrete utilized in dam construction, the maximum aggregate size can be as large as a few hundred millimeters.

The choice of the maximum size of aggregate may also be governed by the availability of material and by its cost. For instance, when various sizes are screened from a pit, it is generally preferable not to reject the largest size, provided this is acceptable on technical grounds.

Another important parameter of an aggregate is its grading. In all cases, dense-graded or well-graded aggregate is preferred, and uniformity has to be achieved. In the case of a coarse aggregate, uniformity can be obtained relatively easier by the use of separate stockpiles for each size fraction. For mass concrete with a maximum size of aggregate larger than 40 mm, ACI 211.1-81 recommends a combination of coarse aggregate fractions to give maximum density and minimum voids. In the case of fine aggregate, however, considerable care is required in maintaining the uniformity of grading of fine aggregate as the sand is usually obtained from river beds directly and can vary from place to place. This is especially important when the water content of the mix is controlled by the mixer operator on the basis of a constant workability: a sudden change toward finer grading requires additional water for the workability to be preserved, and this means a lower strength of the batch concerned. An excess of fine aggregate may also make full compaction impossible and thus lead to a drop in strength.

Table 3-9 gives the idealized combined grading for the 150 and 75 mm nominal maximum sizes of the aggregate. To demonstrate the proportioning of fractions of crushed coarse aggregate so as to obtain the ideal combined grading of the first column of Table 3-9, consider four size fractions: 150 to 75 mm, 75 to 37.5 mm, 37.5 to 19 mm, and 19 to 4.76 mm. The gradings of these

**Table 3-9** "Ideal" combined grading for coarse aggregate of nominal maximum size of 150 and 75 mm

| | Cumulative percentage passing for nominal size of aggregate in mm | | | |
|---|---|---|---|---|
| | 150 mm | | 75 mm | |
| **Sieve size (mm)** | **Crushed** | **Rounded** | **Crushed** | **Rounded** |
| 150 | 100 | 100 | — | — |
| 125 | 85 | 89 | — | — |
| 100 | 70 | 78 | — | — |
| 75 | 55 | 64 | 100 | 100 |
| 50 | 38 | 49 | 69 | 75 |
| 37.5 | 28 | 39 | 52 | 61 |
| 25 | 19 | 28 | 34 | 44 |
| 19 | 13 | 21 | 25 | 33 |
| 9.5 | 5 | 9 | 9 | 14 |

fractions are given in Table 3-10. In concrete mix design, the aggregate content has to be determined carefully. As the aggregate occupies 65 to 75% of the total volume of concrete, it plays an important role in determining the concrete properties and cost.

The parameters for aggregate content in a concrete mix design that need to be decided include the total aggregate-to-binder ratio and the fine aggregate-to-coarse aggregate ratio. Table 3-11 provides the dry bulk volume of coarse aggregate per unit volume of concrete, which is expressed as a function of both fineness modules of the fine aggregate and the maximum size of aggregate. The mass of the coarse aggregate can then be calculated from the product of the dry bulk volume and the density (or unit weight) of the dry coarse aggregate.

The fine aggregate content per unit volume of concrete can be then estimated using either the mass method or the volume method. In the former, the sum of the masses of cement, coarse aggregate, and water is subtracted from the mass of a unit volume of concrete, which is often known from previous experience with the given materials. However, in the absence of such information, Table 3-12 can be used as a first estimate; adjustment is made after trial mixes.

**Table 3-10** Example of grading of individual coarse aggregate fractions to be combined into an "ideal" grading for mass concrete

| | Cumulative percentage passing for fraction (%) | | | |
|---|---|---|---|---|
| **Sieve size (mm)** | **150–75 mm** | **75–37.5 mm** | **37.5–19 mm** | **19–4.76 mm** |
| 175 | 100 | — | — | — |
| 150 | 98 | — | — | — |
| 100 | 30 | 100 | — | — |
| 75 | 10 | 92 | — | — |
| 50 | 2 | 30 | 100 | — |
| 37.5 | 0 | 6 | 94 | — |
| 25 | 0 | 4 | 36 | 100 |
| 19 | 0 | 0 | 4 | 92 |
| 9.5 | 0 | 0 | 2 | 30 |
| 4.76 | 0 | 0 | 0 | 2 |

**Table 3-11** Dry bulk volume of coarse aggregate per unit volume of concrete as given by ACI 211.1-91

| Maximum size of aggregate (mm) | Dry bulk volume of rodded coarse aggregate per unit volume of concrete for different fineness modulus of sand | | | |
|---|---|---|---|---|
| | 2.40 | 2.60 | 2.80 | 3.00 |
| 9.5 | 0.50 | 0.48 | 0.46 | 0.44 |
| 12.5 | 0.59 | 0.57 | 0.55 | 0.53 |
| 19 | 0.66 | 0.64 | 0.62 | 0.60 |
| 25 | 0.71 | 0.69 | 0.67 | 0.65 |
| 37.5 | 0.75 | 0.73 | 0.71 | 0.69 |
| 50 | 0.78 | 0.76 | 0.74 | 0.72 |
| 75 | 0.82 | 0.80 | 0.78 | 0.76 |
| 150 | 0.87 | 0.85 | 0.83 | 0.81 |

*Note*: The values will produce a mix with a workability suitable for reinforced concrete construction. For less workability concrete, e.g. that used in road construction, the values may be increased by about 10 per cent. For more workable concrete, such as may be required for placing by pumping, the values may be reduced by up to 10 per cent.

**Table 3-12** First estimate of density (unit weight) of fresh concrete, as given by ACI 211.1-81

| Maximum size of aggregate (mm) | First estimate of density (unit weight) of fresh concrete (kg/m³) | |
|---|---|---|
| | Non-air entrained | Air entrained |
| 9.5 | 2,280 | 2,200 |
| 12.5 | 2,310 | 2,230 |
| 19 | 2,345 | 2,275 |
| 25 | 2,380 | 2,290 |
| 37.5 | 2,410 | 2,350 |
| 50 | 2,445 | 2,345 |
| 75 | 2,490 | 2,405 |
| 150 | 2,530 | 2,435 |

A more precise estimate can obtained from the following equation:

$$\rho = 10\gamma_a (100 - A) + C\left(1 - \frac{\gamma_a}{\gamma}\right) - W(\gamma_a - 1)\ (\mathrm{kg/m^3}) \tag{3-12}$$

where

$\rho$ = density (unit weight) of fresh concrete, kg/m$^3$
$\gamma a$ = weighted average bulk specific gravity (SSD) of combined fine and coarse aggregate; clearly, this needs to be determined from tests
$A$ = air content, %
$C$ = cement content, kg/m$^3$
$\gamma$ = specific gravity of cement (generally 3.10 for Portland cement)
$W$ = mixing water requirement, kg/m$^3$

The volume method is an exact procedure for calculating the required amount of fine aggregate. Here, the mass of fine aggregate, $A_f$, is given by

$$A_f = \gamma_f\left[1000 - \left(W + \frac{C}{\lambda} + \frac{A_c}{\gamma_c} + 10A\right)\right]\ (\mathrm{kg/m^3}) \tag{3-13}$$

where

$A_c$ = coarse aggregate content, kg/m$^3$
$\gamma f$ = bulk specific gravity (SSD) of fine aggregate
$\gamma_c$ = bulk specific gravity (SSD) of coarse aggregate

### 3.3.4 Approaches for Concrete Mix Design

By far, various approaches in designing concrete have been proposed, which can be classified into three main categories: the prescriptive method, the statistical method, and the close packing-based method.

#### 3.3.4.1 The Prescriptive Method

The process of traditional mixture design is decided upon mixture proportions via either prescriptive or performance-based design methods, for example, the ACI method and the UK method. However, there is no fundamental difference between these methods. Thus, it is sufficient to introduce one method. Here, the method proposed by American Institute of Concrete (ACI 211.1-91) is introduced.

Before starting concrete mix design, basic information on raw materials and background data should be collected, including

**(a)** Sieve analysis results and fineness modulus of fine and coarse aggregate
**(b)** Dry-rodded density (unit weight) of coarse aggregate
**(c)** Bulk specific gravity of each raw material
**(d)** Absorption capacity or moisture content of the aggregates
**(e)** Variation of the approximate mixing water requirement with slump, air content, and grading of the available aggregates
**(f)** Relationships between strength and water/cement ratio for available combinations of cement and aggregate
**(g)** Job specifications, if any, e.g., maximum water/cement ratio, minimum air content, minimum slump, maximum size of aggregate, and strength at early ages (normally, 28-day compressive strength is specified)

Regardless of whether the concrete characteristics are prescribed by the specifications or left to the mix designer, the batch weights in per kilograms cubic meter of concrete can be computed in the following sequence:

**Step 1:** *Choice of slump.* If the slump value is not specified, an appropriate value for the particular work can be selected from Table 3-7 according to the type of structure to be built.

**Step 2:** *Choice of maximum size of aggregate.* For the same volume of coarse aggregate, using a large maximum size of well-graded aggregate will produce less void space than using a smaller size, and this will have the effect of reducing the mortar requirement in a unit volume of concrete. Generally, the largest size of coarse aggregate economically available should be selected as long as it can meet the general requirement mentioned earlier. In no event should the maximum size exceed one-fifth of the narrowest dimensions between the size of the forms, one-third the depth of slabs, or three-fourths of the minimum clear spacing between reinforcing bars.

**Step 3:** *Estimation of mixing water and the air content.* The quantity of water per unit volume of concrete required for a given slump value depends on the maximum particle size, shape, and grading of the aggregates, as well as on the amount of entrained air. If data based on experience with the given aggregates are not available, assuming normally shaped and well-graded particles, an estimate of the mixing water, with or without air entrainment, can be obtained from Table 3-8 for the purpose of deriving the trial batches.

**Step 4:** *Selection of water/cement ratio.* Since different aggregates and cements generally produce different strengths at the same *w/c*, it is highly desirable to develop the relationship between strength and *w/c* for the materials actually to be used. In the absence of such data, approximate and relatively conservative values for the concretes made with type I Portland cement can be selected from Table 3-1. Since the selected *w/c* must satisfy both the strength and the durability criteria, the values of *w/c* (*w/b*) should conform to the values in Table 3-3.

**Step 5:** *Calculation of cement content.* The required cement content is equal to the mixing water content obtained in step 3 divided by the *w/c* determined in step 4.

**Step 6:** *Estimation of coarse aggregate content.* To reduce the cost of concrete, aggregates should be used as much as possible. Statistics on a large number of tests have shown that for properly graded materials, the finer the sand and the larger the size of the particles in a coarse aggregate, the more the volume of coarse aggregate that can be used to produce a concrete mixture with satisfactory workability. It can be seen from Table 3-11 that, for a suitable degree of workability, the volume of coarse aggregate in a unit volume of concrete depends only on its maximum size and fineness modulus of the fine aggregate. It is assumed that differences in the amount of mortar required for workability with different aggregates, due to differences in particle shape and grading, are compensated for automatically by differences in dry-rodded void content.

**Step 7:** *Estimation of fine aggregate content*. At the completion of step 6, all the ingredients of the concrete have been estimated except the fine aggregate. The amount of fine aggregate can be determined by either the *weight* method or *volume* method.

According to the weight method, if the unit weight of the wet fresh concrete is known from previous experience, then the required weight of fine aggregate is simply the difference between the unit weight of concrete and the total weight of water, cement, and coarse aggregate. In the absence of a reliable estimate of the unit weight of concrete, Table 3-12 can be used to as a guide to choose the unit weight of fresh concrete. Experience shows that even a rough estimate of the unit weight is adequate for the purpose of producing trial concrete.

$$U_m = 10G_a(100 - \overline{A}) + C_m\left(1 - \frac{G_a}{G_c}\right) - W_m(G_a - 1)\ \text{kg/m}^3 \tag{3-14}$$

where $U_m$ is the weight of fresh concrete, kg/m$^3$; $G_a$ is the weighted average bulk specific gravity (SSD) of combined fine aggregate and coarse aggregate, assuming reasonable weight proportions; $G_c$ is the specific gravity of cement; $A$ is the air content, %; $W_m$ is the mixing water content, kg/m$^3$; $C_m$ is the cement content, kg/m$^3$.

In the absolute volume method, the total volume displaced by the known ingredients (i.e., water, air, cement, and coarse aggregate) is subtracted from the unit volume of concrete to obtain the required volume of fine aggregate. This in turn is converted to weight units by multiplying by the density of fine aggregates.

**Step 8:** *Adjustment of amount of free water*. The mix proportions determined by steps 1 to 7 assume that aggregates are in SSD condition. Generally, however, the stock aggregates are not in a balanced condition, i.e., the SSD condition. They are either in an air dry or wet condition that will either absorb mixing water or given up extra water to the mix during mixing process. Moreover, when admixtures in liquid form are used, extra water will be supplied. If water related to these sources is not taken into account for the adjustment of mixing water, the actual *w/c* of the trial mix will be inaccurate. Hence, the moisture content of aggregates and the extra water in liquid admixtures have to be carefully calculated and the amount of mixing water should be adjusted. The procedures will be demonstrated in the sample computations.

**Step 9:** *Trial mixes*. Due to so many assumptions underlying the foregoing theoretical calculations, the mix proportions for the actual concrete have to be checked and adjusted by means of laboratory trials consisting of small batches (e.g., 0.02 m$^3$ or 50 kg of concrete). Fresh concrete should be tested for slump, cohesiveness, finishing properties, and air content, as well as for unit weight. The specimens of hardened concrete cured under standard conditions should be tested for strength at specified ages. If any property cannot meet the design requirement, adjustments to the mix proportions have to be conducted. For example, lack of cohesiveness can be corrected by increasing the fine aggregate content at the expense of the coarse aggregate content. The rules of thumb for other adjustments are as follows:

**(a)** If the correct slump is not achieved, the estimated water content is increased (or decreased) by 6 kg/m$^3$ for every 25 mm increase (or decrease) in slump.

**(b)** If the desired air content is not achieved, the dosage of the air-entraining admixture should be adjusted to produce the specified air content. The water content is then increased (or decreased) by 3 kg/m$^3$ for each 1% decrease (or increase) in air content.

**(c)** If the estimated density (unit weight) of fresh concrete by the mass method is not achieved and is of importance, the mix proportions should be adjusted, with allowance being made for a change in air content.

**(d)** If the projected strength cannot be met, w/b should be reduced at a rate of 0.05 for every 5 MPa.

**Step 10:** *Mix proportion adjustments.* After several trials, when a mixture satisfying the desired criteria of workability and strength is obtained, the mix proportions of the laboratory-size trial batch can be fixed and scaled up for producing large amounts of field batches.

## Examples

### Example I Concrete Mix Design

Concrete is required for a column that will be moderately exposed to freezing and thawing. The cross-section of the column is 300 × 300 mm. The smallest spacing between reinforcing steel is 30 mm. The specified compressive strength of concrete at 28 days is 40 MPa with a slump of 80 to 100 mm. The properties of materials are as follows:

**(a)** Cement used is type I Portland cement with a specific gravity of 3.15.

**(b)** The available coarse aggregate has a maximum size of 20 mm, a dry-rodded unit weight of 1600 kg/m$^3$, a bulk specific gravity (SSD) of 2.68, absorption capacity of 0.5%, and moisture content (OD) of 0.25%.

**(c)** The fine aggregate has a bulk specific gravity (SSD) of 2.65, absorption capacity of 1.3%, a moisture content (SSD) of 3%, and a fineness modulus of 2.60. The aggregates conform to the ASTM C33-84 requirements for grading.

With the given information, the mix design will be carried through in detail, using the sequence of steps outlined.

**Step 1:** *Choice of slump.* The slump is given and consistent with Table 3-7.

**Step 2:** *Maximum aggregate size.* The maximum aggregate size is 20 mm, which meets the limitations of 1/5 of the minimum dimension between forms and 3/4 of the minimum clear space.

**Step 3:** *Estimation of mixing water and air content.* The concrete will be exposed to freezing and thawing; therefore, it must be air entrained. From Table 3-8, the recommended mixing water amount is 180 kg/m$^3$, and the air content recommended for moderate exposure is 5.0%.

**Step 4:** *Water/cement ratio (w/c).* According to both Table 3-1 and Table 3-3, the estimate of the required *w/c* ratio to give a 28-day compressive strength of 40 MPa is 0.35.

**Step 5:** *Calculation of cement content.* Based on the steps 3 and 4, the required cement content is 180/0.35 = 514 kg/m$^3$.

**Step 6:** *Estimation of coarse aggregate content.* From Table 3-11, for fineness modulus of the fine aggregate of 2.60, the volume of dry-rodded coarse aggregate per unit volume of concrete is 0.64. Therefore, there will be 0.64 $m^3$ coarse aggregate in per volume concrete. And, the OD weight of the coarse aggregate is $0.64 \times 1600 = 1.024$ kg. The SSD weight is 1024 $\times$ 1.005 = 1029 kg.

**Step 7:** *Estimation of fine aggregate content.* The fine aggregate content can estimated by either the weight method or the volume method.

**(a)** Weight method. From Table 3-12, the estimated concrete weight is 2280 kg/$m^3$. Although for a first trial it is not generally necessary to use the more exact calculation based on Equation (3-14), this value will be used here:

$$U_m = (10)(2.67)(100 - 5) + 514(1 - 2.67/3.15) - 180(2.67 - 1)$$
$$= 2314 \ \text{kg/m}^3$$

Based on the already determined weights of water, cement, and coarse aggregate, the SSD weight of the fine aggregate is 2314 − 180 − 514 − 1,029 = 591kg.

**(b)** *Volume method.* Based on the known weights and specific gravity of water, cement, and coarse aggregate, the air volume, the volumes per $m^3$ occupied by the different constituents can be obtained as follows:

$$\text{Water: } \frac{180}{1000} = 0.180 \ \text{m}^3$$

$$\text{Cement: } \frac{514}{1000 \times 3.15} = 0.163 \ \text{m}^3$$

$$\text{Coarse aggregate (SSD): } \frac{1029}{1000 \times 2.68} = 0.384 \ \text{m}^3;$$

$$\text{Air: } 0.05 \ \text{m}^3;$$

Therefore, the fine aggregate must occupy a volume of 1 − (0.180 + 0.163 + 0.384 + 0.05) = 0.223 $m^3$. The required SSD weight of the fine aggregate is $0.223 \times 2.65 \times 1000 = 591$ kg.

**Step 8:** *Adjustment for moisture in the aggregate.* Since the aggregates will be neither SSD nor OD in the field, it is necessary to adjust the aggregate weights for the amount of water contained in the aggregate. Since absorbed water does not become part of the mix water, only surface water needs to be considered. For the given moisture contents, the adjusted aggregate weights become

Coarse aggregate (stock): From

$$W(\text{stock}) = W(\text{OD})[1 + \text{MC}(\text{OD})]$$

$$\text{Get: } \quad W(\text{stock}) = 1024 \times 1.0025 = 1026 \text{kg}$$

The extra water needed for coarse aggregate absorption is

$$\text{W(SSD)} - \text{W(stock)} = 1029 - 1026 = 3 \text{kg}$$

$$\text{Fine aggregate (stock): } 591 \times 1.03 = 609 \ \text{kg/m}^3$$

$$\text{Extra water provided by fine aggregate: } 609 - 591 = 18 \text{kg}$$

$$\text{The mixing water is then: } 180 + 3 - 18 = 165 \text{kg.}$$

Thus, the estimated batch weights per $m^3$ are as follows: water, 165 kg; cement, 514 kg; coarse aggregate, 1026 kg; fine aggregate, 609 kg; total, 2314 kg.

**Step 9:** *Trial mixes*. Trial mixes should be carried out using the proportions calculated. The properties of the concrete in the trial mix must be compared with the desired properties, and the mix design must be corrected as described.

### Example II Raw Material Calculation for a Given Concrete Mix Ratio

A concrete mix has a proportion of 1:0.4:1.8:2.5 (P:W:S:A). In the powder content, 90% is Portland cement and 10% is solid silica fume (SF). The concrete also uses 0.25% retarder and 1.5% superplasticizer. The absorption for sand is 1.2% and for aggregate is 0.9%. The MC (OD) is 2.5% for sand and 0.4% for coarse aggregate. The solid content is 35% in retarder and 40% in superplasticizer. The SF used is a slurry with 50% of water. To cast 5 beams (100 × 100 × 500 mm) and 18 cylinders (100 × 200 mm), how much of each individual ingredient should be used (consider 5% extra amount)?

Solution

1. The total volume of concrete specimen is

$$5 \times 0.1 \times 0.1 \times 0.5 + 18 \times 3.14159 \times (0.1/2)^2 \times 0.2 = 0.053274\ \text{m}^3$$

2. The total weight is (assuming the unit weight of concrete is 2400 kg/m$^3$)

$$2400 \times 0.053274 \times (1 + 0.05) = 134.2492\ \text{kg}$$

(Note that P : W : S : A refers to the SSD states)

3. The weight of cement and SF is

$$W_{\text{C+SF}} = 134.2492/\,(1 + 0.4 + 1.8 + 2.5 + 0.0025 + 0.015) = 23.480\ \text{kg}$$

$$W_{\text{C}} = 23.48 \times 0.9 = 21.132\ \text{kg}$$

4. For sand

$$W_{\text{SSD}} = 1.8 \times 23.480 = 42.265\ \text{kg}$$

$$\text{Absorption capability (AB)} = \left(W_{\text{SSD}} - W_{\text{OD}}\right)/W_{\text{OD}}$$

$$\text{MC (OD)} = \left(W_{\text{stock}} - W_{\text{OD}}\right)/W_{\text{OD}}$$

$$\text{So}\ W_{\text{stock}}/W_{\text{SSD}} = (1 + \text{MC (OD)})\,/\,(1 + \text{AB})$$

$$\text{Thus,}\ W_{\text{stock}} = (1 + 0.025)\,/\,(1 + 0.012) \times 42.265 = 42.808\ \text{kg}$$

$$\text{Water provided by sand} = 42.808 - 42.265 = 0.543\ \text{kg}$$

5. For aggregate

$$W_{\text{SSD}} = 2.5 \times 23.48 = 58.701\ \text{kg}$$

$$W_{\text{stock}} = (1 + 0.004)\,/\,(1 + 0.009) \times 58.701 = 58.410\ \text{kg}$$

$$\text{Water required by aggregate} = 58.410 - 58.701 = -0.291\ \text{kg}$$

6. For retarder

$$W_{\text{solid}} = 0.0025 \times 23.48 = 0.059\ \text{kg}$$

$$W_{\text{solution}} = 0.059/0.035 = 0.168\ \text{kg}$$

$$\text{Water provided by retarder solution} = 0.168 - 0.059 = 0.109\ \text{kg}$$

**7.** For superplasticizer

$$W_{solid} = 0.015 \times 23.48 = 0.352 \text{ kg}$$
$$W_{solution} = 0.352/0.4 = 0.881 \text{ kg}$$
$$\text{Water provided by superplasticizer solution} = 0.881 - 0.352 = 0.528 \text{ kg}$$

**8.** For SF

$$W_{solid} = 0.1 \times 23.48 = 2.348 \text{ kg}$$
$$W_{solution} = 2.348/0.5 = 4.696 \text{ g}$$
$$\text{Water provided by superplasticizer solution} = 4.696 - 2.348 = 2.348 \text{ kg}$$

**9.** For water

$$\text{Water} = 0.4 \times 23.48 - (0.543 - 0.291 + 0.109 + 0.528 + 2.348) = 6.155 \text{ kg}$$

This example provides the method that can be used to make adjustments to the mixing water in practice.

#### 3.3.4.2 The Statistical Method

To satisfy the multiple and rigorous requirements in concrete design, tremendous efforts have been made to improve the forecasting accuracy and efficiency, as well as saving time and resources. Therefore, statistical methods and approaches to infer the nature of the object, optimize the performance of the object and even predict the future of the object by searching, sorting, analyzing and describing data, have the advantage of objectivity and comprehensiveness (Hayslett and Murphy, 1981).

Here, the response surface methodology (RSM) and artificial neural network (ANN) are introduced.

**(a)** *The response surface methodology (RSM)*: The introduction of the RSM begins with the clarification of basic terms since the section is designed to provide information for designers regardless of statistical background.

*Data*, the values obtained by observation, experiment, or calculation, which are used for scientific research, technical design, verification, decision-making, etc.

*Variables*, the independent factors of the optimization problem, which is typically controlled by the decision-maker.

*Response*, the decision factors of the desired outcome for the problem, which is used to seek the optimal results.

*Input*, the specific mixed decision variables.

*Output*, the objectives of the desired outcome.

*Run*, an experiment.

RSM is the most commonly used statistical design method for the modeling and analysis of the problem in which a response of interest is influenced by several variables. The advantages of RSM include the possibility of evaluating factors' interaction within the experimental region, recognition of the optimal response based on different priorities, and the establishment of nonlinear relationships. As illustrated in Figure 3-11, the process of RSM

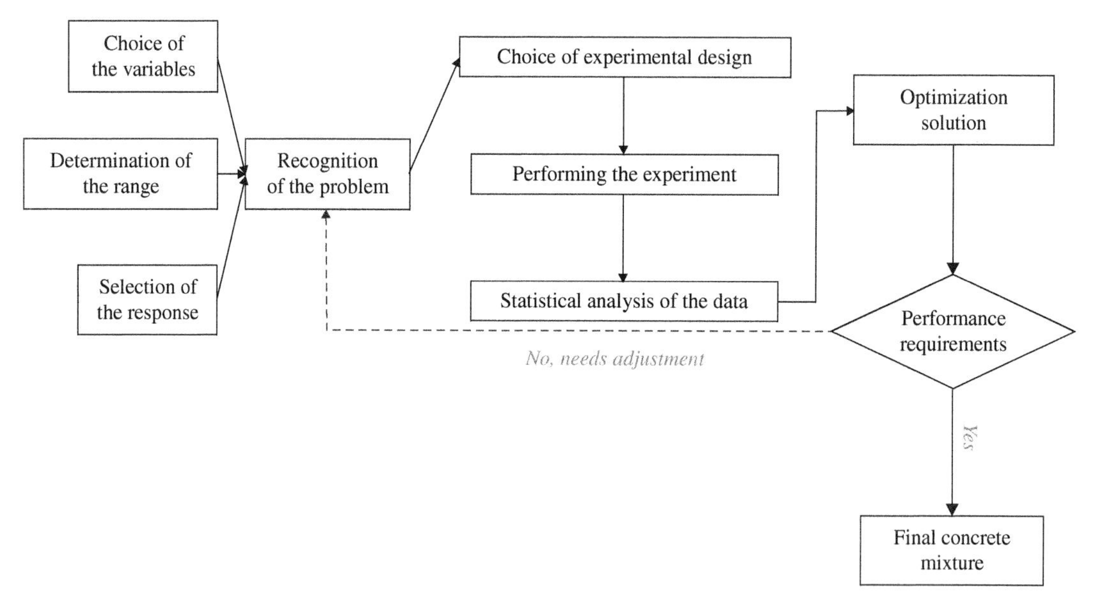


**Figure 3-11** The flow sheet of the statistical method

goes through five main steps, i.e., the formulation of problem, the selection of the design approach, the performing of runs, the statistical analysis of the data, and the determination of optimized solutions.

**Step 1:** *The formulation of problem.* First, the transformation of the complex, practical concrete design problem into the solution of the functional relation. Once the problem has been formulated, designers need to select the variables, the range over which each factor will be varied, the responses, and the evaluation methods. For instance, the water to binder ratio, the composition and property of cementitious materials, the binder to aggregate ratio, the dosage of SP and fiber, the mixing process, the curing schedule that may affect the performance, all can be determined as variables. Correspondingly, the environmental burden, cost, mechanical properties, workability are selected as the response. Besides, the range of constituents needs to be determined

$$i = 1, 2, \cdots, q \qquad 0 \leq L_{\mathrm{i}} \leq x_i \leq U_i \leq 1 \tag{3-15}$$

where $L_{\mathrm{i}}$ and $U_{\mathrm{i}}$ refer to the lower and the upper constraints, respectively.

**Step 2:** *The selection of the design approach.* Once the problem (concrete design) has been transformed to a process of functional relationship solution. For the solution of the functional relationship between water to binder ratio and slump flow, the key lies in the adoption of an appropriate design approach. A second-order model based strategy is effective in solving the complex concrete design problem. An optimal predictor quadratic model used to determine the optimal condition of the responses is shown in Equation (3-16).

$$Y = \beta_0 + \sum \beta_i x_i + \sum \beta_{ii} x_i^2 + \sum \beta_{ij} x_i x_j \tag{3-16}$$

where $Y$ is the response, $x_i$ are the design factors, $\beta$ is unknown parameters that will be estimated from the data in the experiment, $\beta_{ii}$ are the quadratic coefficients, $\beta_{ij}$ corresponds to coefficients of the interaction, and ε is a random error term that accounts for the experimental error in the system being studied.

Once the design approaches have been selected, specific runs are determined to build the model. Taking a RSM, i.e., Central Composite Design (CCD) as an example, specific runs (design points shown in Figure 3-12), including the factorial points in the corners of the cube, the axial points in the center of each face of the cube, and the center points in the center of the cube are selected.

**Step 3:** *Performing runs.* Do the experiments.

**Step 4:** *The statistical analysis of the data.* Once the model has been established, confirmation tests such as analysis of variance (ANOVA) should also be performed to further validate the accuracy of the developed model.

**Step 5:** *The determination of optimized solutions.* After establishing the regression model between mix design variables and responses, all independent variables vary simultaneously and independently in order to optimize the objective functions. The global desirability function developed by Derringer et al. (Derringer and Suich, 1980) is:

$$D = \left(d_1^{r_1} \times d_2^{r_2} \times d_3^{r_3} \times \cdots \times d_n^{r_n}\right)^{1/\sum r_i} = \left[\prod_{i=1}^{n} d_i^{r_i}\right]^{1/\sum r_i} \tag{3-17}$$

where $n$ is the number of responses included in the optimization, and $r_i$ is the relative importance of each individual functions $d_i$. Importance ($r_i$) varies from 1 to 5, from least to most important, respectively. Individual desirability functions ($d_i$) range between 0 (completely undesired response) and 1 (fully desired response). For a value of $D$ close to 1, the response values are near the target values. For each variable and response, four different goals could be assigned, namely maximize, minimize, target, and in range. Besides, a weight factor is defined, ranging between 0.1 and 10, which represents the shape of the desirability for each goal. The values higher than 1 give more emphasis to the goal and those lower than 1

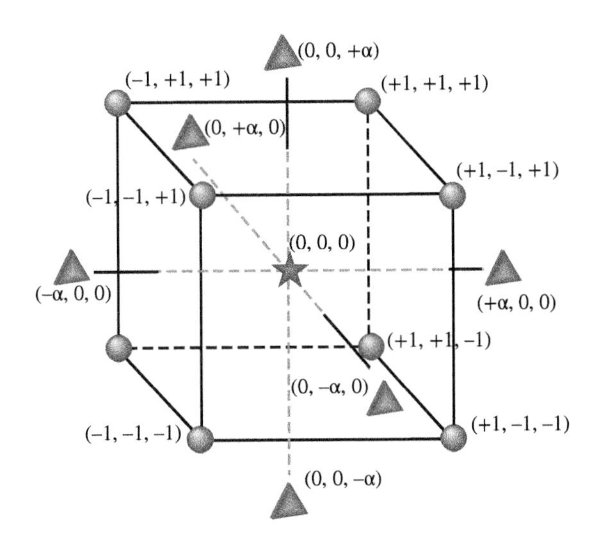


**Figure 3-12** Center, axis and factorial points in CCD

have the opposite effect. The meaning of goal parameters is described as Equations (3-18) and (3-19).

$$d = \begin{cases} 0 & Y_i \leq L \\ \left[\frac{Y_i - L}{U - L}\right]^{wt_i} & L < Y_i < U \\ 1 & Y_i \geq U \end{cases} \tag{3-18}$$

$$d = \begin{cases} 0 & Y_i \leq L \\ 1 & L < Y_i < U \\ 1 & Y_i \geq U \end{cases} \tag{3-19}$$

where $L$ is a lower limit, $U$ is an upper limit for a response, and $wt_i$ is the weight for a response.

After that, solutions of concrete could be calculated, and the theoretical solutions are also confirmed by comparing the predicted value to the ensured value.

**(b)** *The artificial neural network (ANN)*: The artificial neural network (ANN) is a machine learning algorithm whose architecture essentially mimics the learning capability of the human brain to process complex and nonlinear problems. Recently, the ANN has been applied for the prediction and optimization in concrete properties. The ANN has the advantage of modeling the relationship between a large number of decision variables and objectives, as well as avoiding the over-fitting with many decision variables and explanatory terms. However, ANN is not popular in the field of concrete design. A possible reason is the incomprehensible characteristic. As shown in Figure 3-13, the process of ANN is like a black box, the internal structure does not provide any insights on the causal relationships between inputs and outputs.

In fact, ANN includes a multi-layer feed-forward perception network, which presents five basic elements (the inputs, weights, sum function, activation function, and outputs) that resemble the biological ones. By adjusting the weights and thresholds among layers, designers could accomplish movement between the input and the output layer. The weighted sum of the input components can be calculated:

$$net_j = \sum w_{ij} x_i + b_j \tag{3-20}$$

where $net_j$ is the weighted sum of the $j_{th}$ neuron received from the lower layer with $n$ neurons, $w_{ij}$ is the weight between the $i_{th}$ neuron and the $j_{th}$ neuron in the preceding layer, $x_i$ is the

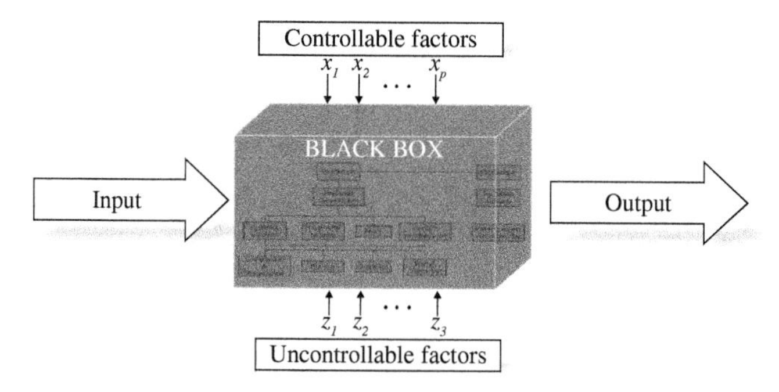


**Figure 3-13** The black box of the complex relationship between the composition of raw materials and the performances

output of the $i_{th}$ neuron, and $b_j$ is a fix value as internal addition. Typical ANN goes through three steps.

**Step 1:** *Data collection.* Designers obtain the data by personally conducting experiments or gathering them from the literature. The data include variables (such as OPC, SF, and alternatives content), and responses (such as slump flow, strength, and energy consumption), which are similar to that of RSM.

**Step 2:** *ANN structure determination.* The basic structure of ANN consists of the input layer, the hidden layer, and the output layer. The number of input and output nodes is determined by the parameters of the research problems or estimated by trials. Then, the activation function in the hidden layer and output layer should be determined. Furthermore, to develop the ANN model to predict the performance of UHPC, a series of trials are carried out in order to determine the number of layers and other parameters of the ANN models by the minimum mean square error (MSE) of the training data, aiming to develop the ANN model with high accuracy.

**Step 3:** *Data processing.* The collected data applied to develop the ANN models should be normalized within the specific limits to eliminate the non-singular data, improve the precision of results, accelerate the convergence speed, and reduce the calculation time. The majority of normalization expressions are linear or logarithmic functions.

#### 3.3.4.3 The Close Packing-Based Design

It is generally accepted that porosity plays a vital role in determining the mechanical properties of cement-based materials. Based on a great number of experimental results, formulas have been developed to uncover the impact of porosity on the strength development of concrete. One representation can be expressed by Balshin's equation.

$$\sigma = \sigma_0 \bullet (1\text{-}P)^A \tag{3-21}$$

where $\sigma_0$ is the compressive strength of concrete at zero porosity, $P$ is the porosity, $\sigma$ refers to the compressive strength at porosity $P$, and $A$ is an experimental constant. It can be concluded that a higher strength of concrete can be generated through the reduction of total porosity. After that, numerous equations such as Ryshkewitch's formula (Ryshkewitch, 1953) have been developed and all equations draw a similar conclusion (higher strength of concrete can be generated through the reduction of total porosity) although with different expressions.

Inspired by the foregoing discussion, various porosity reduction strategies have been employed to meet the increasing mechanical properties requirements. One critical work can be found in Birchall's study (Birchall et al., 1981), where concrete with almost no pores (lately named as macro-defect-free cements (MDF)) has been synthesized via the use of high physical extrusion (4–10 MPa). However, the technology is limited in its field application due to strict requirements. The breakthrough in developing dense concrete from the view of a more practical porosity reduction strategy began with Bache (Bache et al., 1981) who filled the voids of the cement particle with ultrafine particles (such as silica fume) to achieve a concrete with the enhanced mechanical property. The strategy, known as densified with small particles (DSP) theory, which expanded the gradation from aggregate to the matrix consisting of both cementitious powder and aggregate, laid the theoretical foundations of concrete design.

Later, encouraged by the DSP, many approaches or models considering particle packing optimization have been developed. In fact, the particle packing was an old mathematical problem 400 years ago before it attracted the attention of concrete designers in 1981. The earliest particle packing system can be traced back to the Kepler Conjecture in 1611, i.e., how to put shells on the ship most efficiently. Although this ideal model was not proven by Hales' flyspeck project in the field of mathematics until 2017, a lot of researchers have already found that the densest structure was face centered cubic (FCC) or hexagonal closed-packed (HCP) for single size sphere packing in three-dimensional space (see Figure 3-14), and the results have already guided the practical applications (for instance, the most space-saving orange placement in fruit stalls also involves the packing problem). Later, studies focus on the packing of spheres with different particle sizes. It was obvious that the packing regarding two types of spheres contributed to a higher density, which was actually consistent with Bache's theory, i.e., where the cement and SF particles were regarded as spheres with different diameters. On this basis, designers started to realize that the guidelines generated from ideal spheres packing (such as the optimal size ratio of large and small spheres and the interaction, i.e., the wall effect and loosen effect) could provide ideas for solving the packing problem in concrete. So far, various particle packing models (as shown in Figure 3-15, which can be divided into discrete models and continuous models, according to the assumed material size distribution) have been developed to build a close packing structure. Among them, the compressible packing model (CPM) (belonging to the discrete models) and modified Andreasen and Andersen (MAA) model (a typical continuous model) have drawn special attention in the design of self-compacting concrete (SCC) and ultra-high performance concrete (UHPC).

**(a)** *The concrete design based on CPM*: As mentioned above, the CPM can be expressed as Equations (3-22), (3-23) and (3-24), and the index K can be calculated form Equation (3-25).

$$\gamma = \gamma_i = \frac{\beta_i}{1 - \sum_{j=1}^{i-1} \left[1 - \beta_i + b_{ij}\beta_i \left(1 - 1/\beta_j\right)\right] y_j - \sum_{j=i+1}^{n} \left[1 - \alpha_{ij}\beta_i/\beta_j\right] y_j} \tag{3-22}$$

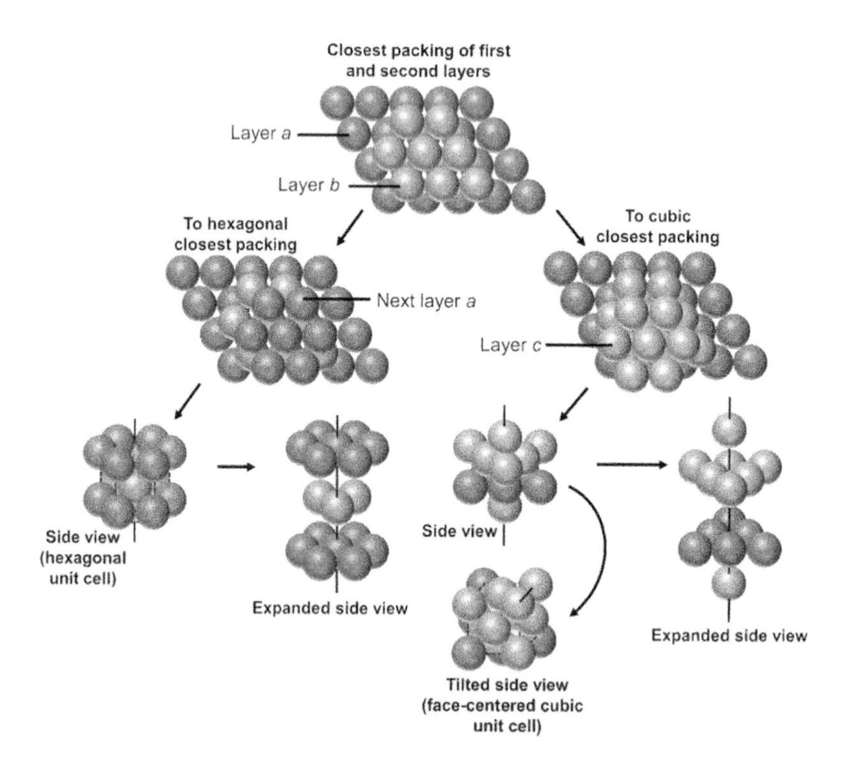


**Figure 3-14** The densest packing structure of same sphere in three-dimensional space

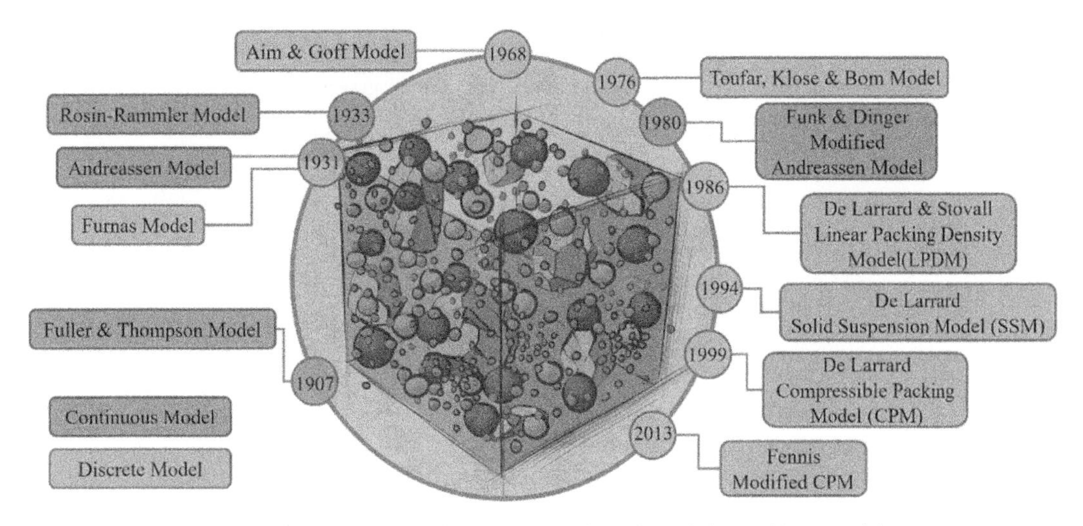


**Figure 3-15** The development timeline of particle packing models

**Table 3-13** K values for different packing processes

| Mixture condition | Packing processes | K values |
|---|---|---|
| Dry | Sticking | 4.5 |
| | Vibration | 4.75 |
| | Vibration + compression | 9 |
| Wet | Smooth thick paste | 6.7 |
| | Proctor test | 12 |

$$\alpha_{ij} = \sqrt{1 - \left(1 - d_j/d_i\right)^{1.02}}, (j = i + 1, \cdots, n) \tag{3-23}$$

$$b_{ij} = 1 - \left(1 - d_i/d_j\right)^{1.50}, (j = 1, \cdots, i - 1) \tag{3-24}$$

$$K = \sum_{i=1}^{n} K_i = \sum_{i=1}^{n} \frac{y_i/\beta_i}{1/\phi - 1/\gamma_i} \tag{3-25}$$

where $\beta_i$ refers to the residual packing density, which can be calculated via the virtual packing density displayed when the class $i$ is isolated and fully packed. For example, the $\beta_i$ of monosized sphere is 0.74. $y_j$ is the volume fractions of the size class $j$ by reference to the total solid volume. The loosening effect $a_{ij}$ and wall effect $b_{ij}$ (as shown in Figure 3-16) have also been considered. $d_i$ and $d_j$ refer to the particle size of the class $i$ and $j$, respectively. With regard to the design of concrete via CPM, the packing density of the mixtures could be predicted via software such as RENE-LCPC, BETONLABPRO, and the MATLAB-based program after the compaction index K, and the input parameters (either from experiments or assumption) are determined. Typically, the selection of K value, based on the different packing process, affects the accuracy of the results directly. The recommended K values for different packing system are list in Table 3-13.

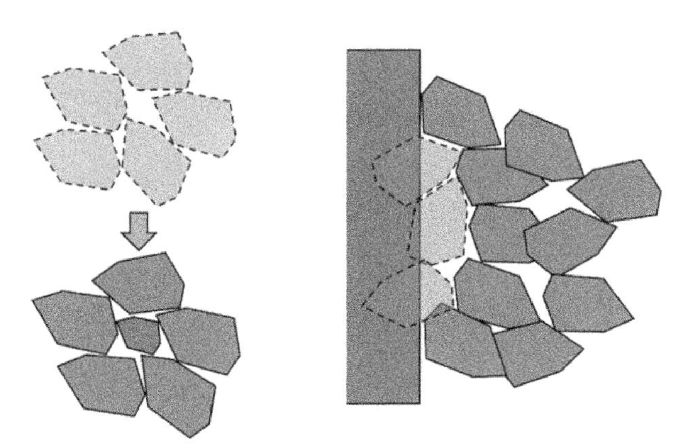

**Figure 3-16** Two perturbations occurring in a dense assembly of particles with at least two populations of particles. Left: the "loosening effect." An enforced fit of a particle larger than the central void can push the larger particles away. Right: the "wall effect." Close to a wall, like the surface of a framework or the surface of a large particle, for instance, the density of the packing of smaller particles is decreasing due to the gap in the contact interface.

**(b)** *The concrete design based on the MAA model*: Unlike the discrete models, the essence of most continuous models is altering the composites' content to approximate the target curve (generated based on the experimental data or empirical statistics). The first continuous model was made by Fuller and Thompson (1907), which focuses on solving the aggregate packing problem. Later, the Fuller curve has been improved by Talbot and Richart (1923) (as illustrated in Equation 3-26), aiming to solve the optimal aggregate gradation more conveniently

$$P(D) = \left(\frac{D}{D_{\max}}\right)^{0.5} \tag{3-26}$$

where $P(D)$ is the proportion of the solid particles less than size $D$, $D$ is the particle size (μm), and $D_{\max}$ is the maximum particle size (μm). In 1930, Andreasen and Andersen developed another continuous particle packing model

$$P(D) = \left(\frac{D}{D_{\max}}\right)^{q} \tag{3-27}$$

where the distribution modulus $q$ is not a constant, and the recommended value of $q$ ranges from 0.33 to 0.50 based on the aggregate gradation. It is noteworthy that the building of the Andreasen and Andersen (AA) model differed from that of the Fuller model although both models exhibited similar formulas and target curves (the comparison results of optimization curves are presented in Figure 3-17). Fundamentally, the target curve of the Fuller model was generated based on the experimental data or empirical statistics, while that of the AA model was built based on the assumption of particle packing similarity (as shown in Figure 3-18), size analysis, and geometry (more closed to the calculation of discrete models). Later, Funk and Dinger proposed the modified Andreasen and Andersen (MAA) model by considering the impact of the minimum particle size, aiming to achieve more accurate prediction results. The theoretically optimal particle packing curve could be expressed as Equation (3-28)

$$P(D) = \frac{D^q - D_{\min}^q}{D_{\max}^q - D_{\min}^q} \tag{3-28}$$

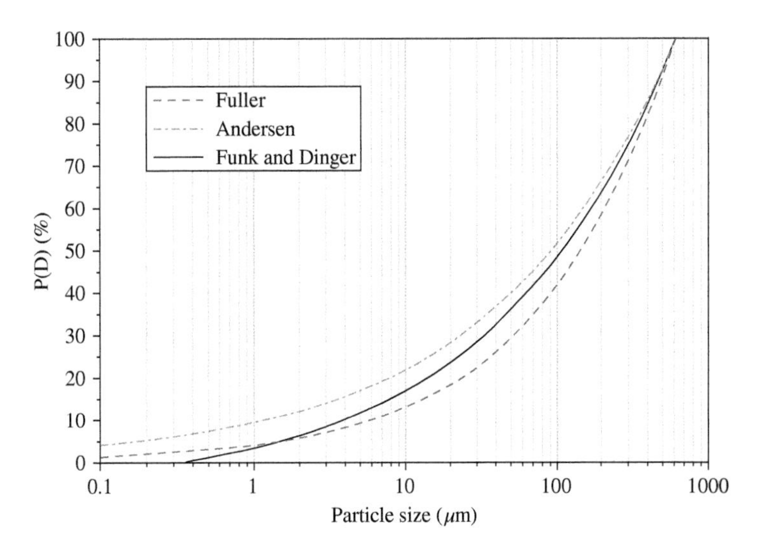


**Figure 3-17** The comparison of the optimization curves of different continuous models

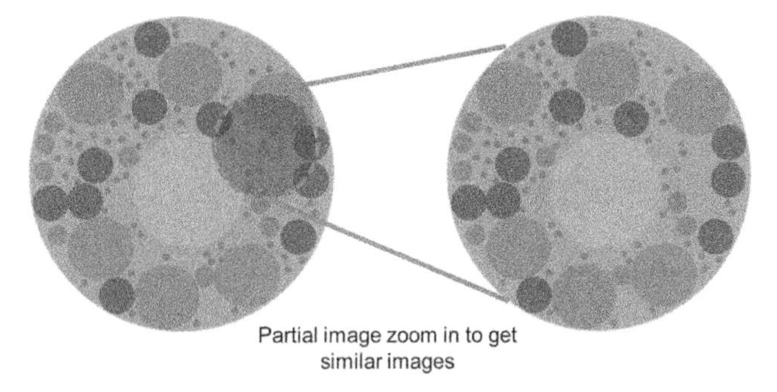


**Figure 3-18** The schematic diagram of the assumption of "particle accumulation similarity" in the AA model

where $D$ refers to the particle size, $P(D)$ is the proportion of the solid particles less than size $D$. $D_{max}$ and $D_{min}$ represent the maximum and minimum particle sizes, and q is the distribution modulus. In the MAA model, higher values of the distribution modulus ($q > 0.5$) lead to coarse mixtures, while lower values ($q < 0.25$) result in concrete mixes which are rich in fine particles (typically UHPC).

As mentioned above, the packing problem solution was a process of adjusting the material proportion to approach the target curve, and it was often solved by an iterative algorithm. In addition, the sum of the squares of the residuals (RSS) value has also been used to evaluate the deviation of the model. Specifically, by adjusting the proportions of each individual material using an optimization algorithm based on the Least Squares Method (LSM), an optimum fit between the composed mixture and the target curves could be reached

$$RSS = \sum_{i=1}^{n} \left[P_{mix}(D_i) - P_{tar}(D_i)\right]^2 \rightarrow \min \qquad (3\text{-}29)$$

where $P_{mix}$ and $P_{tar}$ are the composed mixture and the target grading.

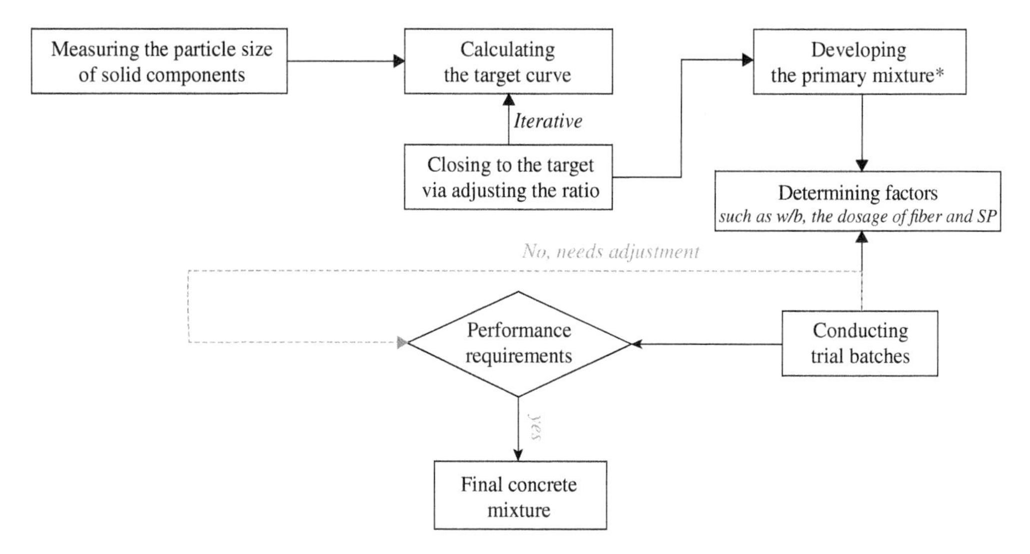


**Figure 3-19** The flowsheet of design of the MAA model-based design

The flowsheet of the MAA method is shown in Figure 3-19. As can be seen from Figure 3-19, four steps (involving measuring the particle size of solid particles, solving the target curve via adjusting the proportion of each material in the mixture, determining the key factors such as *w/b*, the dosage of fiber and SP, and obtaining the final sustainable UHPC mixture after conducting the trial batches) were concluded.

**Step 1:** The target curve is determined according to the MAA model.

$$P(D) = \frac{D^q - D^q_{\min}}{D^q_{\max} - D^q_{\min}} \tag{3-30}$$

where $D$ refers to the particle size (μm), $P(D)$ is the proportion of the solid particles less than size $D$, which is the target curve. $D_{\max}$ and $D_{\min}$ represent the maximum and minimum particle size, and $q$ is the distribution modulus.

**Step 2:** The different particle size combinations are obtained by adjusting the proportion of each material in the mixture. The cumulative distributed probability model is shown as follows

$$y = a_1x_1 + a_2x_2 + \cdots + a_nx_n + \varepsilon \tag{3-31}$$

s.t.

$$\sum_i a_i = 1, \quad lb_i < a_i < \mu b_i$$

where $y$ represents the cumulative probability of particle size distribution, $a_i$ refers to the proportion, $lb_i$ and $ub_i$ are the upper and lower limits, respectively.

**Step 3:** After that, the probability model in step 2 has been transformed into a quadratic programming problem with constraints

$$\min J(A) = -y^T XA + \frac{1}{2}A^T X^T XA \tag{3-32}$$

s.t.

$$A = (a_1, a_2, \cdots, a_n)$$
$$\sum_i a_i = 1, \quad lb_i < a_i < \mu b_i$$

**Step 4:** Finally, Lagrange function methods have been used to solve the constrained optimization problems, and get the optimal solutions

$$L(A, \lambda, \gamma) = \frac{1}{2}A^T GA + (-y^T X) A + \lambda^T (vA - 1) + \gamma^T \begin{pmatrix} AI - \mu b \\ -AI - lb \end{pmatrix} \tag{3-33}$$

where $\lambda$ and $\gamma$ are Lagrange multipliers, $v = (1, 1, \cdots, 1, \cdots, 1)$, $I$ refers to the identity matrix. The optimal solutions that satisfy the constraint condition are obtained by the active set method.

## 3.4 MANUFACTURE OF CONCRETE

Nowadays, concretes are usually produced in two ways, one is manufactured in commercial concrete plant, the other is produced on construction site. Since the last few decades, ready mixed concrete has developed very fast. The advantages of ready mixed concrete are: (1) provides concrete with better quality due to specialized operation; (2) mass production; (3) elimination of storage space for basic materials at site; (4) basic materials can be fully utilized; (5) reduces the labor requirement; (6) reduces noise and dust pollution on the construction site; and (7) reduces the production cost of concrete. ASTM C94 / C94M 09a (Standard Specification for Ready-Mixed Concrete) specifies the requirements for quality control of ready-mixed concrete manufactured and delivered to a purchaser in a freshly mixed and unhardened state. Mixers will be stationary mixers or truck mixers. Except as otherwise specifically permitted, cement, aggregate, and admixtures are measured by mass. Figure 3-20 shows a modern ready-mixed concrete plant. The large pipes in Figure 3-20 are silo containers for raw materials. The truck position is the location of the mixed concrete discharge. Nowadays, stationary mixers are usually used to produce wet concretes and are referred to as wet processing. A dry process just discharges all the raw materials into a truck mixer and the mixing is finished on the truck on the way to the site. The mixing capacity of a stationary type of drum mixer reaches 9 $m^3$. The pan mixer (see Figure 3-21) is usually used at the construction site, precast concrete plant and laboratories. Mostly, it is for small amounts of concrete, as compared to ready-mix concrete plant.

Mixing of cement paste, mortar or concrete is a mechanical action to reduce or eliminate any in-homogeneities present in the mixture. The mixing conditions, especially mixing time and mixing sequence, significantly impact the rheological and workability of the fresh concrete.

Due to insufficient mixing by hand or paddle, agglomerates of cement powder and cement paste remain unbreakable leading to high shear stresses, while the prolonged mixing of concrete significantly affects and alters its rheological behavior. Mixing times are usually longer for high performance concretes and ultra-high performance concretes than for normal concretes. Besides, it is noteworthy that the further increase in mixing time does not increase the homogeneity or the workability of the concrete. As mentioned in Chapter 1, the mixing time depends on the power of a mixer, the slump requirement, and the properties of the raw materials. It ranges from 30 seconds to a few minutes. The selection of the mixing time has to be adjusted on a case-by-case basis.

Apart from the mixing time, when and how to introduce the superplasticizer (SP) into the mix is another important parameter. There are presently two approaches, i.e., adding all the SP in one throw at the beginning of the mixing, or first put two-thirds of the superplasticizer into the mixture at the beginning of the mixing, and add the remaining one-third at the very end of the mixing period. There is no conclusion which approach acts better.

**Figure 3-20** Ready-mix concrete plant

**Figure 3-21** Small pan mixer

## 3.5 DELIVERY OF CONCRETE

The delivery of fresh concrete from the concrete plant to the construction site is usually done by agitators, either truck mixers or truck agitators. The truck is equipped with a rotating drum for agitation, shown in Figure 3-22. The truck mixer receives raw materials from the plant and completely mixes them into workable fresh concrete on the way to the construction site. The advantage of truck mixing is that the water can be stored separately and added into the solid materials for

**Figure 3-22** Delivery of concrete by truck

**Figure 3-23** Delivery of concrete to an ocean construction site using a ferry (Photo provided by Ove Arup, HK)

mixing according to the time of shipping, to avoid slump loss. If the construction site is offshore, a ferry is used to carry a large number of trucks from land to the site (see Figure 3-23). In this case, due to a long shipping period, special care has to be paid to the slump loss. Usually, retarding admixtures have to be used to keep concrete workable for a period of 5 to 6 h, and an initial setting time of 7 to 8 h. In this case, a truck mixer has priority to be selected, if available. A truck mixer has to meet the requirements of environmentally-friendly production nowadays.

## 3.6 CONCRETE PLACING

Placing concrete is a construction process that can be divided into four operations: (1) site preparation to receive the concrete; (2) conveying and placing the concrete into the forms; (3) compacting the concrete; and (4) taking care of the concrete after it has been compacted. Concrete should be placed as close to its final position as possible. To minimize segregation, it should not be moved over too long a distance. After concrete is placed in the formwork, it has to be compacted to remove entrapped air. Compaction can be carried out by hand rodding or tamping, or by the use of mechanical vibrators. In this section, the focus is on preparation, conveying, placing, and compacting.

### 3.6.1 Site Preparation

Before ordering and receiving concrete from commercial concrete plants, the construction site has to be prepared carefully. Site preparations include processing steel reinforcement, setting up formwork, and arranging the handling equipment for concrete. Different structures have different preparations.

**(a)** *Foundations*: Preparation for foundations to receive concrete is complicated due to the large area and excavation involved. Excavation for foundations should extend into sound, undisturbed soil or rock. If a large hole is encountered during excavation, the hole must be backfilled with the material selected having similar stiffness with the surrounding soil to avoid uneven settlement of the foundation. Rock surfaces should be clean and sound. If free water is present, it should be blown out with air jets or removed by other methods. In some cases it may be necessary to provide a sump (outside the form area) into which the water drains, for removal by means of a pump.

Foundations should be free of frost and ice when concrete is placed in winter. During the dry seasons, the earth should be moist but not muddy. Steel shells for cast-in-place piles and shafts for caissons should be inspected to ensure that the way is not blocked.

**(b)** *Construction joints*: For a structural member with large areas or volumes or big heights, it is impossible to cast the whole member at one time. The limitations of concrete supply, construction processes, and time make concrete casting stop at some point, leaving a construction joint to separate the currently cast concrete to the later resumed casting. A construction joint may be horizontal or vertical, depending on the type of structural member and construction process. Locations of construction joints have to be planned carefully according to the structural member size, the amount of concrete that can be provided during the period, the coefficient of thermal expansion, the construction planning, the moment and shear force distribution in the structural member, and the shrinkage properties of concrete. The joint should be made in a plane normal to the main reinforcing bars and in a region with minimum shear force. Construction joints must be made and located so as not to impair the strength of the structure. Where a joint is to be made, the surface of the concrete must be cleaned and laitance removed. Immediately before new concrete is placed, all construction joints must be wetted and standing water removed (ACI 318 Section 6.4). Reinforcing steel is normally continuous across a construction joint to keep the concrete structure unified as whole.

**(c)** *Formworks*: Before casting concrete, a formwork has to be built. The formwork is a temporary mold into which concrete is poured. The formwork supports the dead load of fresh concrete and some construction load until the concrete hardens and can carry the load itself. The formwork can be built with steel plate or timber board. Formworks should be clean, tight, and strong enough to carry the wet concrete and resist the pressure generated by the liquid concrete. The formwork lining should be coated with appropriate oil or parting compound to aid in demolding. Wood forms should be moistened with water before concrete pouring to avoid their absorbing water from the concrete. Figure 3-24 shows a formwork for construction of a transfer block of a tall building. Also shown in Figure 3-24 is the falsework. Falsework is a type of temporary structure used in construction to support spanning or arched structures by holding the component in place until its construction is sufficiently advanced to support itself. Falsework usually includes temporary support structures for formwork, and scaffolding to give workers access to the structure being constructed.

In placing concrete in high, thin walls or similar structural units, it is common practice to provide ports or windows in the forms. If possible, these windows should be made on a surface that will not be exposed to view in the finished structure, such as the back side of a wing wall on a highway structure. When the level of the fresh concrete within the structure approaches the window, the hole should be closed as tightly and neatly as possible. Because of the danger of segregation resulting from a high-velocity stream of concrete entering the form at an angle, and because of the surface blemishes usually resulting in the area where the hole was closed, it is best to avoid using these ports, if at all possible, or to provide a collecting hopper outside the opening.

Permanent formwork is a part of a concrete structure that can be used to hold the fresh concrete, mold it to the required dimensions and remain in place for the life of the structure. The steel tube-reinforced concrete utilizes steel tubing as permanent formwork. The use of permanent formwork can reduce the number of skilled form workers on a construction site, speed up the construction process, improve curing and reduce shrinkage, reduce construction costs, and improve safety by reducing hazards during construction. It also reduces construction waste generation during construction.

**(d)** *Reinforcing steel*: After the formwork is built up, a reinforcing steel cage has to be put into the formwork before concrete casting. The main reinforcing steel bars should be held by steel stirrups in the right position. The reinforcing steel should be free of dirt, paint, oil, grease or other foreign substances. The rust should be removed by wire brushing. During the placing and compacting, all reinforcing steel should be held accurately in the right position. Distances from the forms should be maintained by means of chairs, ties, or hangers. The reinforcing steel cage should also be strong enough to carry the construction load. Figure 3-25 shows the reinforcing steel framework before concrete casting.

**Figure 3-24** Formwork for transfer block of a tall building (Photo provided by Mr. Peter Allen)

**Figure 3-25** Reinforcing steel frame work (Photo provided by Mr. Peter Allen)

(e) *Embedded items*: Many concrete structures have objects and fixtures embedded, including manholes, anchor bolts, pipes, instruments for health monitoring, and tubes for holding prestress tendons. Most of them have to be fixed in place prior to concrete placement by attaching them to the formwork or the reinforcing steel. Adjustments to the steel location to accommodate these items should be made only when the load-carrying ability is not affected.

### 3.6.2 Conveying Concrete

After the formwork, the steel work, and embedded items are ready, a final inspection should be conducted to make sure that the plant and equipment are ready to go. The correct amount of concrete should be ordered and the ready-mix supplier prepared to furnish concrete at the required rate. Transporting equipment, such as pumps, cranes, batch trucks, buckets, conveyors, and helicopters, should be capable of handling the concrete at the required rate. Curing materials should be available.

The methods of placing fresh concrete include direct discharge from a mixer into the forms, crane and buckets, pumps, conveyors, buggies, wheelbarrows, pneumatic placers, small railcars, or a combination of two or more of these methods. Helicopters have been employed for transporting equipment and concrete buckets into especially isolated or difficult sites.

The method used depends on the size of the job, adequacy of space, and availability of equipment. The displacing concrete should be as close as possible to the actual site. Direct discharge is a method of pouring concrete from a transport truck directly into the forms or on the subgrade. Extra lengths of chute on the truck can provide a placing radius of about 5 m from the truck.

For a concrete placing to a lower level, such as a foundation, some sort of chuting arrangement for moving concrete is needed. Chutes should be of rounded cross-section, made of metal or lined with metal, smooth to prevent the concrete from friction and sticking, and of the proper slope for the concrete to slide fast enough to keep the chute clean. As long as there is no segregation or separation, any reasonable slope can be tolerated. A slope of about 1:3 is good for application without segregation. See Figure 3-26.

Buckets are excellent means of conveying concrete. The capacity range is from less than 1 $m^3$ for structural use to 10 $m^3$ for mass concrete. Buckets can be handled by cranes, derricks, trucks, rail cars, helicopters, or cableways. Moving by cranes is the usual method of handling buckets in building construction. Care should be taken to avoid shaking and jarring the concrete, as this causes segregation, especially for relatively high-slump concrete. Figure 3-27 shows concrete being unloaded using a bucket.

**Figure 3-26** Concrete is conveyed into a large foundation construction through chutes (Photo provided by Mr. Peter Allen)

**Figure 3-27** Unloading concrete (Photo provided by Ove Arup, HK)

Concrete is also frequently moved by belt conveyor, wheelbarrows, and buggies. Belts can move concrete long distances horizontally and, to some extent, vertically. Pneumatic-tired wheelbarrows can be used for moving small amounts of concrete for short distances. About 60 m is the maximum horizontal distance for a wheelbarrow. One person with a wheelbarrow can move a maximum of about 1.2 $m^3$ of concrete per hour. Hand-operated carts can carry about 0.12 or 0.2 $m^3$

each, with a maximum haul of about 60 m. A power-driven cart has a capacity of up to 1.2 $m^3$ and can move a maximum of 17 $m^3$ of concrete per hour on a moderate length of haul. Maximum haul should not exceed 300 m. Figure 3-28 shows a small wheelbarrow used on a construction site.

Pumping is another effective method to move concrete to the formwork, see Figure 3-29. Pumped concrete is conveyed under pressure through a rigid pipe or flexible hose. With the development of self-compacting concrete, pumping has become more and more popular in the construction of tall buildings, bridge towers, and tunnels. Nowadays, with the help of special

**Figure 3-28** Wheelbarrow used on construction site

**Figure 3-29** Pumping of fiber reinforced concrete to a height of 306m at Su-Tong Bridge, Suzhou, China (Photo provided by Jinyang Jiang)

high-pressure pumps, concrete can be pumped a distance as far as 1400 m horizontally or as high as 420 m vertically.

The total pressure needed to pump concrete to a height of $H$ can be estimated from the following equation:

$$P_t = \Delta P_f(L + H) + \rho g H + P_l \tag{3-34}$$

where $P_t$ is the total pressure needed, MPa; $\Delta P_f$ is the pressure loss due to friction of the wall of the pipe, MPa/m; $L$ is the total horizontal distance; H is the total vertical distance; ρ is the density of concrete, kg/m$^3$; g is gravity acceleration, m/s$^2$; and $P_l$ is pressure loss due to local effects, such as the bending tube, MPa. Once $P_t$ is obtained, the capacity of the pump can be decided and equipment selected.

### 3.6.3 Depositing Concrete in Forms

Once all the preparation work and equipment are ready, depositing concrete into the formwork can be started. A basic rule is that concrete should be deposited as close as possible to its final location, especially for vertical dropping. As long as the concrete is deposited near its final position without undue segregation, any method is acceptable. If possible, concrete should fall vertically. In most cases, free fall should be limited to 0.9 to 1.5 m to avoid aggregate bouncing off from faces and striking the reinforcing steel, which may increase segregation. However, when the formwork is open and the concrete drop is unobstructed, free fall to a depth of 45 m has proven successful without segregation (Mindess et al., 2003). Usually, a drop chute or pipe is used to guide or protect the concrete during its fall, as shown in Figure 3-30.

For concrete deposited in the formwork of walls, footings, beams, and shear walls, concrete should be placed from the ends or corners toward the center, in horizontal layers not exceeding about 450 mm in depth. Mass concrete in dams and foundations is usually placed in lifts of 1.5 or 2.5 m depth, each lift consisting of several layers. To avoid cold joints, these layers are carried across the form in a series of steps. For a rock foundation, a layer of mortar with a thickness of 12 mm should be placed before depositing the first layer of concrete. The compatibility between

**Figure 3-30** Drop chute-guided concrete fall

the mortar and concrete should be taken into consideration. Subsequent layers, continuing to the full height of the structure, should be placed and consolidated before the underlying layer has hardened.

For slope placing and consolidation, the work should be started from the bottom, allowing gravity to aid, rather than to hinder consolidation. The concrete should be constrained to fall vertically onto the slope. A strict watch should always be kept for segregation during placement, and action should be taken to correct problems as soon as they arise.

A number of occasions present unique problems in handling, placing, and consolidating concrete. Among these are slip forms, underwater placing, and preplaced aggregate concrete.

### 3.6.4 Compacting and Finishing

After depositing concrete into formworks, it should be compacted right away. The purpose of compacting is to remove the air entrapped during concrete placement and to consolidate plastic concrete into all the spaces in the formworks, including the corners and the gaps in the reinforcing steels. Compacting can make concrete denser and stiffer and thus have a good compressive strength and low permeability. Without proper compacting, high-quality concrete cannot be achieved. Many years ago, consolidation was accomplished by laborers wielding a variety of spades, tampers, and similar tools. Now, nearly all concrete is consolidated with high-frequency vibrators. At a construction site, two vibrators are frequently used, an internal and an external vibrator. An internal vibrator consists of a poker, housing an eccentric shaft driven through a flexible drive from a mortar. The poker can be immersed in concrete and vibrates in a harmonic way, exerting pressure to the surrounding concrete. The poker vibrator ranges in size from 20 to 150 mm diameter, with head lengths from 250 to 750 mm. Pokers can generate vibration with a frequency ranging from 70 to 200 Hz, and acceleration from 0.7 to 4 m/s$^2$ at speeds of 5500 to 15,000 vpm (vibrations per minute). The vibration can produce a noise with a level up to 90 dB, and, obviously, poker vibration is not good for people's health. Self-compacting concrete can completely eliminate the vibration and hence is environmentally friendly. Figure 3-31 shows the vibration process of an internal vibrator.

**Figure 3-31** Internal vibration

The external vibrator usually has a flat metal base like a plate, including the surface, and pan or screed vibrators. The mortar on the top of the plate can generate vibration on the plate. It can be placed on the surface of the fresh concrete and is usually suitable for slab and floor member compacting. Slabs up to 200 mm thick can be consolidated adequately. Thicker slabs require additional internal vibration. In addition to consolidating the concrete in the slab, the unit strikes off the surface and prepares it for final finishing. Another type of external vibrator is the form vibrator. Form vibrators are attached to the exterior of the mold or form. They are used in locations where it is difficult to use internal vibrators, such as in tunnel linings or heavily congested forms. They are also used for making pipes, masonry units, and many other types of precast concrete. Pneumatically driven units develop vibration by the rotation of an eccentric weight. The speed can be varied by changing the volume of air supplied. When the surface of the concrete takes on a flattened glistening appearance, the rise of entrapped air bubbles ceases, the coarse aggregate blends into the surface but does not disappear, and vibration can be stopped.

Overvibration sometimes occurs. If so, the coarse aggregate will have sunk below the surface, and the surface may have a frothy appearance. In this case, the slump should first be reduced, and the amount of vibration then has to be adjusted.

A simple finishing on the fresh concrete is usually done by trowel just before initial setting. The purpose of finishing is to make a smooth surface on the concrete member and to achieve a denser, compact, and properly graded surface layer to prevent water evaporation and increase wear resistance. Proper finishing of good-quality concrete can minimize the maintenance cost of a structural member. In addition to simple finishing, many special techniques have been developed to achieve a decoration effect, as shown in Figure 3-32.

**Figure 3-32** Decoration effect of concrete

## 3.7 CURING OF CONCRETE

### 3.7.1 Definition

Curing refers to the procedures for the maintaining of a proper environment in fresh concrete for the hydration reactions to proceed. . In concrete curing, the critical thing is to keep the concrete in a sufficiently moist condition, so that the hydration will not stop. Inadequate curing permits loss of mixed water, thereby leading to the initiation of detrimental processes on the concrete quality and decreasing the service life of a structure in the short and long run. Curing has a significant impact on both the fresh and hardened properties of concrete. The curing should start 3h–18h after the concrete is poured. The curing time and curing regimes depend on the type of concrete and the construction environment. For Portland cement concrete, a minimum period of 7 days of moist curing is generally recommended.

Curing is a simple procedure, and is frequently ignored. However, it is the most important step in producing a strong, durable and watertight concrete.

### 3.7.2 General Curing Methods

This section summarizes the concrete curing regimes as follows:

**(a)** *Standard curing*: Standard curing refers to the curing of concrete in a humid environment or water with a temperature of $20 \pm 2$°C and relative humidity of not less than 95%. The standard curing time is 28d.

**(b)** *Natural curing*: Natural curing refers to the curing measures such as covering, watering and wetting, wind shielding, and heat preservation of concrete under ambient temperature conditions (above 5°C). Natural curing can be divided into two types: covering watering curing and plastic film curing. The former (covering watering curing) usually uses grass curtains, mats, and blankets to cover concrete within 3–12 hours after the concrete is poured. Then water is sprayed on the covering from time to time to keep the moist condition. The plastic film curing refers to the use of plastic film coverings to separate the concrete from the air, so that the moisture no longer evaporates, and the cement is hydrated and hardened by the water in the concrete.

**(c)** *Heat curing*: Heat curing is a curing regime that accelerates the hydration process of concrete by heating the concrete. Commonly used heat curing methods include the warm shed method, steam curing, pressurized steam curing, dry and radiant heat curing. Steam curing refers to the curing of concrete in an environment full of saturated steam or a mixture of steam and air with high temperature and humidity to accelerate the hydration of concrete. The parameters related to the steam curing include the natural curing time before steam curing, the heating and the cooling speeds, the temperature, constant temperature keeping time, and the relative humidity. Steam curing can increase the 28d strength of concrete by 10–40%, and accelerate the construction speed of concrete without being affected by the external environment temperature. Steam curing can be performed either in a curing room, stored chamber or on site with steam guided into a sealed space with a concrete member inside. Pressurized steam curing is performed, based on steam curing by increasing the atmospheric pressure (not less than 8 standard atmospheres) and the curing temperature (>174.5°C) of the curing environment to obtain higher strength concrete. The warming shed method refers to the method of placing the concrete elements in a shed with thermal insulation materials and installing radiators, pipes, electric heaters or stoves and other equipment to heat the air in the shed so that the concrete is cured under a positive temperature environment. The warm shed method is mostly used in underground projects, foundation projects, and masonry structures with tight schedules. Radiant heat curing refers to a curing process by using heat radiation. It can be done either by coating a layer of heat-absorbing objects on the surface of the concrete to collect heat from the sunlight for concrete curing. Or using an electric infrared heater to radiate heat on the concrete, so that the concrete surface absorbs radiant heat and the heat passes into the concrete through heat conduction to increase the internal temperature of the concrete, accelerate the hydration of the cement, and promote the formation of the internal structure of the concrete. The radiant heat curing time is short, the prepared concrete has high compressive strength.

**(d)** *Cold-curing*: In large concrete structures, cooling of the interior (e.g., by circulation of water in embedded pipes) is important, not only to prevent the reduction of concrete strength, but also to avoid thermal cracking as a result of nonuniform heating/cooling of the structure. Cold-curing can also be conducted using liquid nitrogen spray to cover on site concrete structure.

**(e)** *Curing compound method*: A concrete curing compound is a compound that cures concrete by preventing the moisture loss from the concrete. It is done by forming a membrane after it is sprayed on the surface of fresh concrete. The membrane is moisture-tight and hence can keep the moist condition of concrete to ensure proper curing. Waxes, natural resins, synthetic resins and solvents of high volatility all are good examples of curing compounds. Generally pigments having white or gray colors are incorporated into the curing compound to provide heat reflectance and clear margin for the area of curing compound applied. A curing compound is especially useful for concrete members with vertical surfaces such as walls and columns.

### 3.7.3 Internal Curing

As defined by ACI in 2010, internal curing is "supplying water throughout a freshly placed cementitious mixture using reservoirs, via prewetted lightweight aggregates, that readily release water as needed for hydration or to replace moisture lost through evaporation or self-desiccation." In short, internal curing is to cure concrete from the inside out with the water supplied via internal reservoirs.

In sealed conditions, the continued consumption of water due to cement hydration desaturates the pores in the cement paste and results in the formation of the liquid–vapor meniscus. As a result, the internal relative humidity (RH) of the cement paste decreases and self-desiccation occurs. This phenomenon is a remarkable problem in concrete with a low *w/b* ratio. In this case, deep penetration of water is necessary to keep the concrete internally moist. However, traditional curing techniques, such as watering and covering burlap, cannot allow significant penetration of external water into the cement-based materials with low *w/b* due to its denser microstructure and low permeability. Therefore, curing water needs to be internally provided to avoid self-desiccation, referred to as internal curing. Commonly used internal curing materials include super-absorbent polymers (SAP) and lightweight aggregates (LWA), such as natural pumice or zeolite, artificial expanded clay and shale, aggregates produced by fly ash and ground granulated blast furnace slag, coral, waste ceramics, recycled aggregates, and even chemical admixtures, namely polyethylene glycol. Other high-absorption porous materials, such as rice husk ash (RHA), miscanthus combustion ash, hydrogel and cellulose fibers can also be used as internal curing materials.

#### 3.7.3.1 Principle of Internal Curing

The principle of internal curing is to ensure the hydration of the cement inside concrete continues through the sufficient supply of internal water that is not part of the mixing water.

Internal curing can delay the drop in the critical pore size that remains saturated by providing readily available water to fill the voids created by the chemical shrinkage. Although additional pores carried by the internal curing materials may result in a reduction in the compressive strength of the cement-based materials, the benefits regarding the reduction in autogenous shrinkage and cracking control are overwhelming. Minimum autogenous shrinkage and maximum degree of hydration are the main motivations for using internal curing. The main driving force behind shrinkage in the cement paste is capillary tension. As hydration proceeds, smaller and smaller pores are emptied due to self-desiccation. Consequently, capillary pressure increases. On the other hand, as water is removed, the degree of saturation of the cement pastes decreases. At this early stage of hydration, the stiffness of the paste is so low and the viscous behavior so pronounced that the slightest stress acting on the solid skeleton would result in a large shrinkage. However, the autogenous shrinkage of cement paste can be reduced and a higher degree of cement hydration can be achieved by introducing internal curing water. One can explain this result by taking saturated LWA as an example. LWA usually contains water-filled pores much larger than the capillary pores of cement paste. When the paste undergoes self-desiccation, the gradient in the meniscus radius (or capillary pressure) drives the moisture to migrate from the larger pores to the smaller pores. Therefore, the largest pores in LWA empty first. The radius of the liquid–vapor menisci can be effectively increased by the inclusion of saturated LWA. This leads to a reduction in the capillary tension. The resulting shrinkage strain is thus reduced significantly in comparison to the cement paste without saturated LWA.

To ensure the effectiveness of internal curing in cement-based materials, a sufficient amount of internal curing water is not the only requirement. It is of equal importance to ensure that the internal curing water can replenish the surrounding paste in time. The water desorption at the appropriate RH and the distribution of internal curing materials thus need to be considered. Besides, the distance that the water must migrate in the cement paste also affects the results of internal curing.

#### 3.7.3.2 Factors Affecting Internal Curing of Cement-Based Materials

Two main factors influence the effectiveness of internal curing in cement-based materials: (1) the water absorption and desorption properties of the internal curing materials; and (2) the distribution of the internal curing materials.

**(a)** *The water absorption and desorption properties of the internal curing materials:* The amount of water absorption is the direct factor affecting internal curing since the amount of water used for internal curing is proportional to the degree of the water saturation status of the internal curing materials. Usually, the water absorption process of the internal curing materials can be achieved by either pretreatment before mixing or directly mixing based on the type of internal curing materials. For pretreatment, by varying the soaking internal curing materials time for around 24 h (Ghourchian et al., 2013), the pumice with saturated water absorption (77.9%) can be prepared. For the latter approach (direct absorbing of water in the mixing process) (Shen et al., 2016), most lightweight aggregates (LWAs) can not absorb enough water, which is more suitable for the super-absorbent polymers. In contrast to the water absorption property, the amount of water and the time released from internal curing materials into the slurry are critical to effectively mitigate the autogenous shrinkage, where more factors need to be taken into consideration. According to ASTM C 1761/C 1761M, it is suggested that the internal curing materials should release more than 85% of the absorbed water at the RH of 94%. For the time of water desorption, the internal curing water should be released at an early age when the cement paste undergoes early self-desiccation. It should be noted that too early a release is also not good, for example, water release before initial setting, which will change the rheological behavior and reduce the effectiveness of the internal curing (Liu et al., 2020). The water desorption properties of the internal curing materials are closely related to the availability of the internal curing water, the characteristics of the materials (especially), as well as the property of the cement paste. For instance, not all LWAs possess beneficial water desorption properties for internal curing, even though most LWAs can absorb a significant amount of water. An important reason is that only the LWAs with a large and open pore structure can supply water to a hydrating cement paste with a fine pore structure for internal curing. Different from the LWA, super-absorbent polymers (SAPs) are cross-linked hydrophilic networks, and their ability to release water depends on their chemical composition (such as types and concentration of functional groups) and cross-link density. Therefore, the water molecules are attached by hydrophilic groups in SAPs and limited in its cross-linked network, not held by a pure capillary absorption like LWAs bounded in physical form. Apart from the water absorption and desorption properties mainly determined by the pore structure of LWAs and the molecular structure of SAPs, the cement paste property governed by the hydration process also plays an important role in the internal curing. It is well known that water release from the pores in internal curing materials is driven by capillary forces due to the RH differences between the internal curing materials and the cement paste. Thus, a small RH difference at an early age might develop an insufficient driving force to draw the water out of the fine pores in the internal curing materials. Meanwhile, the critical time when the internal curing water leaves the internal curing materials and enters the cement paste is directly dependent not only on the pore structure of the internal curing materials but also on the pore structure of the cement paste. The water will gradually migrate away from the internal curing materials only when the RH in the cement paste drops to a value that no longer maintains the water-filled pores (Li et al. 2020).

**(b)** *The distribution of internal curing materials*: The effect of internal curing in concrete slurry is also correlated with the distribution of the internal curing materials. As the cement hydration progresses, the pore structure in the cement paste is reduced and water may not be able to penetrate further. Thus, the particle size distribution of the internal curing materials thus needs to be carefully considered during mixture proportioning to ensure that the majority of the surrounding paste is within the migration distance of internal curing water. In this case, the cement paste within this distance is protected from self-desiccation. Therefore, the distance of the internal curing materials from the point in the cement paste where the drop of RH takes place can also determine the effectiveness of the internal curing. For a given amount of internal curing materials, the spacing between internal curing materials and the surrounding paste (distribution spacing) can be adjusted by the particle size. Well-dispersed and small particles can effectively increase the internal curing zone because, if the internal curing materials are well distributed within the cement paste, only shorter distances have to be covered. This is especially important for internal curing at later ages when the distance of water migration is limited to hundreds of micrometers due to the depercolation of capillary porosity in the cement paste. In addition, the calculation of the amount of internal curing materials needs a comprehensive consideration of its characteristics. Water absorption and desorption properties should be optimized under the premise of ensuring the uniform distribution of particles to minimize the required amount of internal curing materials and reduce the undesired effects generated by internal curing.

### 3.7.4 Importance of Curing for Concrete

Curing is a simple procedure, and is frequently ignored. However, it is most important in producing a strong, durable and watertight concrete. In concrete curing, the critical thing is to keep a sufficiently moist condition for the concrete, so that the hydration will not stop. Moist curing is

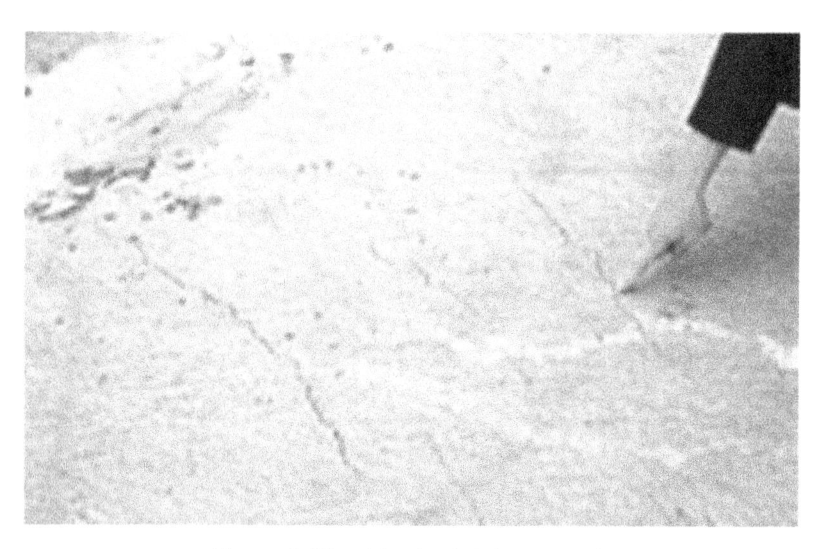

**Figure 3-33** Plastic shrinkage crack

provided by water spraying, ponding or covering the concrete surface with wet sand, plastic sheets, burlaps or mats. Curing compounds, which can be sprayed onto the concrete surface to form a thin continuous film, are also commonly used, especially for vertical surfaces such walls and columns. Loss of water to the surrounding should be minimized. If concrete is cast on a soil subgrade, the subgrade should be wetted to prevent water absorption. In exposed areas (such as a slope), windbreaks and sunshades are often built to reduce water evaporation. For Portland cement concrete, a minimum period of 7 days of moist curing is generally recommended.

If fresh concrete is not properly cured, surface water evaporation is fast and the internal water has almost no change. Plastic shrinkage may occur if the rate of water loss (due to evaporation) exceeds the rate of bleeding. Shrinkage is the reduction in volume due to the loss of water. Such early shrinkage occurs when the concrete is still at the plastic state (not completely stiffened), especially internal concrete, and thus it is called plastic shrinkage. The small amount of volume reduction due to plastic shrinkage is accompanied by the downward movement of the surface layer material. If this downward movement is restrained, by steel reinforcements or large aggregates, cracks will form as long as the low concrete strength is exceeded. Plastic shrinkage cracks often run perpendicular to the concrete surface, above the steel reinforcements. A typical plastic shrinkage crack is shown in Figure 3-33. The presence of plastic shrinkage cracks can affect the durability of the structure, as they allow corrosive agents to easily reach the steel. If care is taken to cover the concrete surface and reduce other water loss (such as absorption by formwork or subgrade), plastic shrinkage cracking can be avoided. If noticed at an early stage, it can be removed by re-vibration.

## 3.8 EARLY-AGE PROPERTIES OF CONCRETE

For concrete at an age less than 7 days, cement paste in the concrete undergoes a fast hydration process. Thus, both the mechanical properties and the pore size distribution of the concrete are different from that of mature concrete. The mechanical properties of concrete at early ages play an important role in determining construction speed and quality, especially for high-rise buildings and nuclear power plants. Better understanding of the mechanical properties of concrete at early ages is essential for engineers to make a right decision in the construction stage on issues such as construction planning. Lew and Reichard (1978) studied the compressive strength, splitting strength, and bond strength for concrete at early stages. They obtained relationships between compressive strength, secant modulus, splitting strength under different temperatures, and degree of hydration. Gardner (1990) studied the property development of young concrete with the incorporation of fly ash and proposed some empirical formulas. In recent years, several studies and conferences have been devoted to various research on concrete at early ages, including the monitoring of determination of modulus of young concrete with the nondestructive method (Jin and Li, 2001). Jin and Li (2000) made tensile strength measurements of young concrete, and determined the differences in mechanical properties for normal and high-strength concretes (Jin et al., 2005).

It has been found that the properties of concrete at early ages are very different from those of mature concretes. Here, a complete stress–strain curve is used as an example for illustration. A complete stress–strain curve represents the comprehensive behavior of concrete under an external force. Many properties can be obtained from the stress–strain curve, such as modulus of elasticity, an important parameter. Curves at ages of 18 h and 1, 2, 3, 7, and 28 days are plotted in Figure 3-34 for NSC and Figure 3-35 for HSC.

In Figures 3-34 and 3-35, there are significant differences in the shape of the compressive stress–strain response at various ages. The slope of the ascending part of the stress–strain curve becomes steeper for the concrete after 7 days for NSC, and after 2 days for HSC, and so does the slope of the descending part. As the compressive strength increases, both the ascending and descending portions of the compressive stress–strain curves become steeper and more linear, which implies that the concrete becomes more brittle as the age increases. Noises can be observed in the later part of curves, which indicate the initiation and propagation of cracks.

It can be seen that the elastic modulus of concrete increases with age. To be quantified, according to ASTM C469-94, the secant modulus at a point with 40% of the maximum stress was used to compare the difference of static elastic modulus at various ages. The values are listed in Table 3-14.

The remarkable difference between the curves of young concrete and those of mature concrete implies their diversity in mechanical properties. More ductile behavior can be observed for

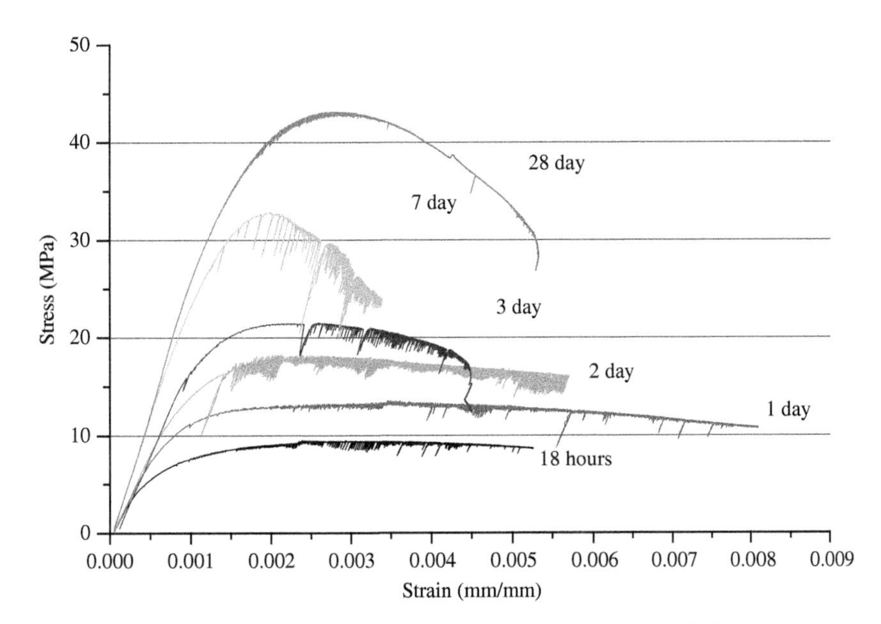


**Figure 3-34** Complete stress-strain curves of NSC

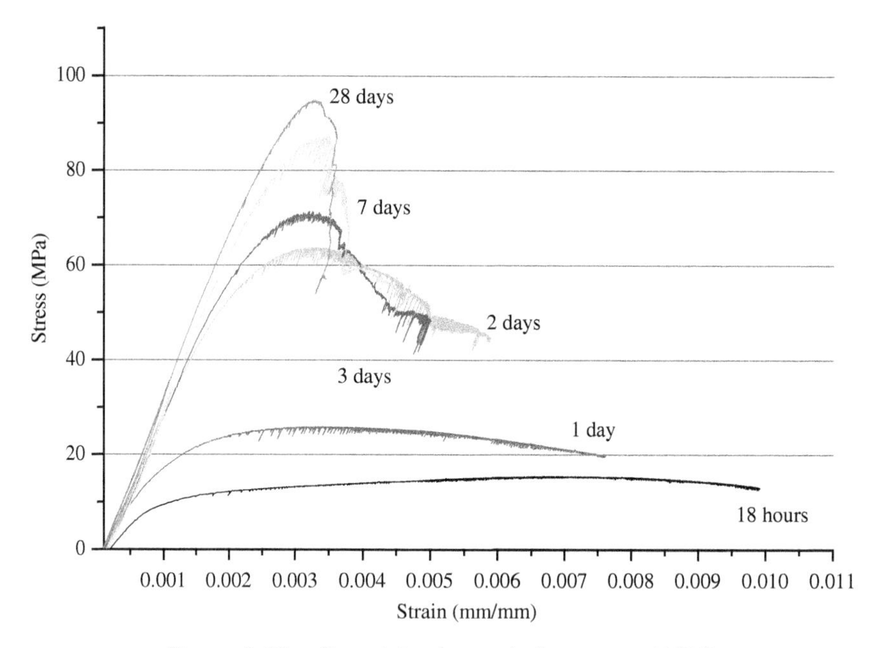


**Figure 3-35** Complete stress-strain curves of HSC

**Table 3-14** Modulus of elasticity of concrete at different ages

| Age | 18h | 1 day | 2 days | 3 days | 7 days | 28 days |
|---|---|---|---|---|---|---|
| NSC | 12.95 | 14.92 | 16.12 | 15.96 | 24.04 | 25.47 |
| HSC | 10.53 | 18.88 | 22.39 | 28.24 | 30.02 | 33.05 |

concrete at early age, compared to a mature concrete's brittleness, due to its viscous characteristics. From the test results for compressive strength, it was found that there was an initial retardation of hydration exhibited for HSC as a result of high dosage of superplasticizer in the mixes. As shown in Table 3-14, after this retardation period, HSC has a higher rate of elastic modulus gain than NSC. Obviously, deducting the influence of the retardation, HSC has higher elastic modulus than NSC.

## DISCUSSION TOPICS

Can you describe the terms workability, consistency, cohesiveness, segregation, and bleeding of freshly concrete?

What are the methods to evaluate the workability of fresh concrete? What are their suitability and limitations?

Discuss the factors affecting the consistency of concrete.

Discuss the factors affecting the cohesion of concrete.

Why does workability decrease with time?

Discuss the role of water in fresh concrete.

What is the purpose of consolidation in concrete construction?

List some methods of curing.

Discuss the effects of curing in ensuring concrete quality.
Discuss the measures that can improve the cohesiveness of concrete.
Discuss the measures that can reduce the bleeding of concrete.
What is the relationship between cohesiveness and segregation?
What is the significance of bleeding in forming microstructures?
Generally, how many common methods are used in a mix design of concrete?
What type of moisture condition in aggregate is assumed in a mix design of concrete?
What is difference between *w/c* and *w/b*?
How is the durability issue considered in a mix design of concrete?
Do you need to justify the moisture content of aggregates? How?
What is the purpose of trial mixes in a mix design o concrete?
What are the main differences in early age properties between HSC and NSC?

## PROBLEMS

1. Use the American method to design a concrete mix that is required to have a specified mean strength of 30 MPa at 28 days. The presence of reinforcement requires a slump of 75 mm and a maximum size of aggregate of 10 mm. The aggregates are of normal weight, and gradings conform to the appropriate standard with a fineness modulus of 2.8. (Assume that absorption is 0.7% and moisture condition of the aggregates is SSD; the bulk density of coarse aggregate is 1600 kg/m$^3$; and there will be extreme exposure condition to freeze-thawing.)
2. Use the American method to design a concrete mix that is required to have a specified mean strength of 25 MPa at 28 days. The presence of reinforcement requires a slump of 30–50 mm and a maximum size of aggregate of 40 mm. The aggregates are of normal weight and gradings conform to the appropriate standard with a fineness modulus of 2.8. (Assume there is negligible absorption and moisture content; a dry-rodded bulk density (unit weight) of coarse aggregate is 1550 kg/m$^3$, and there is a bulk specific gravity (SSD) of 2.70; the fine aggregate has a bulk specific gravity (SSD) of 2.65; and the concrete will be in extreme exposure conditions.)
3. In this problem, we assume that the strength of concrete strictly follows Abram's law as follows: $f_c = \frac{14{,}000}{4^{1.5(w/c)}}$ where, $f_c$ is compressive strength in a dimension of psi (i.e., lb/in$^2$).

   For a concrete mixing with a projected strength of 5000 psi, what value of w/c ratio should be used? If the moisture content of sand (SSD) used for the mixture is 3.5% and the amounts of cement and sand (SSD) used for this mixture are 814 and 1950 lb, respectively, calculate the actual amount of water and sand (stock) that could guarantee the projected strength of the concrete.
4. A concrete mix with a proportion of 1:0.4:1.8:2.5 (B:W:S:A by weight) of 100 kg is prepared. The binder contains 90% Portland cement and 10% silica fume. The silica fume is in a form of slurry with 50% water. The concrete also uses 0.35% retarder and 1.5% superplasticizer. The solid content in the retarder is 35% and in the superplasticizer is 40%. How much water will be brought to concrete by the admixtures?
5. A concrete mix with a proportion of 1:0.4:1.8:2.5 (B: W: S: A) is prepared. In the binder, there is 10% of silica fume. The concrete also uses 0.25% retarder and 1.5% superplasticizer. The absorption for sand is 1.5% and for coarse aggregate is 0.9%. The MC(OD) for sand is 3.5% and for coarse aggregate is 0.4%, respectively. The solid content in the retarder is 35% and in the superplasticizer is 40%. The SF (silica fume) used is a slurry with 50% of water. To cast 5 beams (100 × 100 × 500 mm) and 18 cylinders (100 × 200 mm), how much of each individual ingredient should be used? Suppose that the wet unit weight of the concrete is 2400 kg/m$^3$.

**6.** Two concrete mixes, one is normal strength and other high strength, are considered for use in a tall building column construction. The mix proportion for normal strength concrete is 1:0.6:1.8:2.2 (cement: water: sand: aggregate) while the mix proportion for high strength concrete is 1:0.23:1:2 (binder: water: sand: aggregate, in the binder 70% is cement, 25% is fly ash and 5% is silica fume. Additional 1% of superplasticizer is used.) The column size for normal strength concrete is 2.5 m in diameter and 10 m in height and for high strength concrete is 1.5 m in diameter and 10 m in height. Suppose that the MC(SSD) for sand is 1.5% and for aggregate is -0.3%, the silica fume is in the form of slurry with 50% of water and solid content of superplasticizer is 35%.

**(a)** Calculate the materials usage in each mix;

**(b)** What is difference is cement usage?

**(c)** What is the difference in carbon dioxide emission for two mixes if you only consider the cement? (Suppose the carbon dioxide generated in cement production due to fuel burning is 450 kg/ton and there is 560 kg CaO in 1000 kg cement.)

## REFERENCES


Aissoun, B.M., Hwang, S.D., and Khayat, K.H. (2016) "Influence of aggregate characteristics on workability of superworkable concrete," *Materials & Structures*, 49(1–2), 597–609.

Andreasen, A. H. M. (1930) "Über die Beziehung zwischen Kornabstufung und Zwischenraum in Produkten aus losen Körnern (mit einigen Experimenten)," *Kolloid-Zeitschrift*, 50(3), 217–228.

ASTM (1995) "Standard test method for time of setting of concrete mixtures by penetration resistance," C-403, Philadelphia: ASTM.

Bache, H.H. (1981) "Densified cement/ultra-fine particle-based materials," Second International Conference on Superplastizer Concrete, pp. 1–34.

Birchall, J.D., Howard, A.J., and Kendall, K. (1981) "Flexural strength and porosity of cements," *Nature* , 289(5796), 388–390.

Bruno, L.D., Vanderley, M.J., Björn, L., and Rafael, G.P. (2016) "Viscosity prediction of cement-filler suspensions using interference model: A route for binder efficiency enhancement," *Cement and Concrete Research*, 84, 8–19.

Derringer, G. and Suich R. (1980) "Simultaneous optimization of several response variables," *Journal of Quality Technology*, 12, 214–219.

Ferraris, C.F., Obla, K.H., and Hill, R. (2001) "The influence of mineral admixtures on the rheology of cement paste and concrete," *Cement and Concrete Research*, 21(2), 245–255.

Feys, D., Verhoeven, R., and Schutter, G.D. (2009) "Why is fresh self-compacting concrete shear thickening?" *Cement and Concrete Research*, 39(6), 510–523.

Fuller, W.B. and Thompson, S.E. (1907) "The laws of proportioning concrete," *Transactions of the American Society of Civil Engineers.*

Gardner, N.J. (1990) "Effect of temperature on the early-age properties of type I, type III, and type I/fly ash concretes," *ACI Materials Journal* (American Concrete Institute), 87(1), 68–78.

Ghourchian, S., Wyrzykowski, M., Lura, P., Shekarchi, M., and Ahmadi, B. (2013) "An investigation on the use of zeolite aggregates for internal curing of concrete," *Construction and Building Materials*, 40, 135–144.

Hayslett, H.T., and Murphy, P. (1981) *Statistics*, Oxford: Elsevier, pp. 1–5.

Hu, C., and Larrard, F.D. (1996) "The rheology of fresh high-performance concrete," *Cement and Concrete Research*, 26(2), 283–294.

Hu, J., and Wang, K. (2007) "Effects of size and uncompacted voids of aggregate on mortar flow ability," *ACT*, 5(1), 75–85.

Jin, X. and Li, Z. (2000) "Investigation on mechanical properties of young concrete," *Materials and Structures, RELIM* 33, 627–633.

Jin X. and Li, Z. (2001) "Dynamic property determination for early-age concrete," *ACI Materials Journal*, 98(5), 365–370.

Jin, X., Shen, Y., and Li, Z. (2005) "Behavior of high-and normal-strength concrete at early ages," *Magazine of Concrete Research*, 57(6), 339–345.

Lee, H.K., Lee, K.M., Kim, Y.H., Yim, H., and Bae, D.B. (2004) "Ultrasonic in-situ monitoring of setting process of high-performance concrete," *Cement and Concrete Research*, 34(4), 631–640.

Lew, H.S. and Reichard, T.W. (1978) "Mechanical properties of concrete at early ages," *Journal of the American Concrete Institute*, 75(10), 533–542.

Li, Z., Liu, J., Xiao, J., and Zhong, P. (2020) "Internal curing effect of saturated recycled fine aggregates in early-age mortar," *Cement and Concrete Composites*, 108, 103444.

Li, Z., Xiao, L., and Wei, X. (2007) "Determination of concrete setting time using electrical resistivity measurement," *Materials in Civil Engineering, ASCE* 19(5), 423–427.

Liu, J., Khayat, K.H., and Shi, C.(2020) "Effect of superabsorbent polymer characteristics on rheology of ultra-high performance concrete," *Cement and Concrete Composites*, 103636.

Mindess, S., Toung, J.F., and Darwin, D. (2003) *Concrete*, 2nd ed., Upper Saddle River, NJ: Pearson Education.

Neville, A.M., and Brooks, J.J. (1994) *Concrete Technology*, Harlow: Longman.

Ouyang, J., Han, B., Cao, Y., Zhou, W., Li, W., and Shah, S.P. (2016) "The role and interaction of superplasticizer and emulsifier in fresh cement asphalt emulsion paste through rheology study," *Construction and Building Materials*, 125, 643–653.

Pessiki, S.P. and Carino, N.J. (1988) "Setting time and strength of concrete using the impact-echo method," *ACI Material Journal*, 85, 389–399.

Reinhardt, H.W. and Wuestholz, T. (2006) "About the influence of the content and composition of the aggregates on the rheological behaviour of self-compacting concrete," *Materials & Structures*, 39(7), 683–693.

Ryshkewitch, E. (1953) "Compression strength of porous sintered alumina and zirconia: 9th communication to ceramography," *Journal of the American Ceramic Society*, 36, 65–68.

Schwartzentruber, L.D. and Cordin, R.L.J. (2006) "Rheological behaviour of fresh cement pastes formulated from a self compacting concrete (SCC)," *Cement and Concrete Research*, 36(7), 1203–1213.

Senff, L., Labrincha, J.A. Ferreira, V.M., Hotza, D., and Repette, W.L. (2009) "Effect of nano-silica on rheology and fresh properties of cement pastes and mortars," *Construction and Building Materials*, 23(7), 2487–2491.

Shen, D., Wang, X., Cheng, D., Zhang, J., and Jiang, G. (2016) "Effect of internal curing with super absorbent polymers on autogenous shrinkage of concrete at early age," *Construction Building Materials*, 106(1), 512–522.

Subramaniam, K.V., Lee, J., and Christensen, B.J. (2005) "Monitoring the setting behavior of cementitious materials using one-sided ultrasonic measurement," *Cement and Concrete Research*, 35(5), 850–857.

Talbot, A.N., and Richart, F.E. (1923) "The strength of concrete its relation to the cement aggerates and water," *Bulletin*, 137, 1–118.

Voigt, T., Ye, G., Sun, Z., Shah, S.P., and Van Breugel, K. (2005) "Early age microstructure of Portland cement mortar investigated by ultrasonic shear waves and numerical simulation," *Cement and Concrete Research*, 35(5), 858–866.

CHAPTER 4

# MATERIALS STRUCTURE OF CONCRETE

## 4.1 INTRODUCTION

The type, amount, size, shape, and distribution of different phases present in concrete constitute its materials structure. The structure is multiscale in nature, ranging from the nanometer scale, to the micrometer scale, to the millimeter scale. The elements of the structure above the millimeter scale can readily be seen by the naked eye, whereas the elements below the millimeter scale usually have to be resolved with the help of microscopes, scanning electron microscopy (SEM), transmit electron microscopy (TEM), and atomic force microscopy (AFM). The traditional term *macrostructure* is generally used for the gross structure, visible to the human eye. The limit of resolution of the unaided human eye is approximately one-fifth of a millimeter (200 μm). The traditional term *microstructure* is used for the microscopically magnified portion of a macrostructure. The magnification capability of modern electron optical microscopes is of the order of $10^5$ to $10^6$ times; thus, the application of transmission and scanning electron optical microscopy techniques has made it possible to resolve the structure of materials to a fraction of a micrometer or even a nanometer. Nowadays, the rapid development of advanced experimental tools such as the transmission electron microscope, the high-resolution magic-angle spinning nuclear magnetic resonance spectroscopy, and the atomic force microscope has made it possible to observe the microstructure of the hydration products of concrete at the nanometer scale. For accuracy when discussing the materials structure of concrete, in this book, the terms nanometer-scale structure and micrometer-scale structure are used to describe the materials structure of concrete.

Progress in the field of materials has resulted primarily in the recognition of the principle that the properties of a material originate from its internal structure. In other words, the properties can possibly be modified or improved by making suitable changes in the material structure. Although concrete is the most widely used structural material, its heterogeneous and highly complex structure is still not fully understood. Knowledge on structure–property relationships in concrete is essential before we discuss the factors influencing the important engineering properties of concrete, such as strength, elasticity, shrinkage, creep, cracking, and durability. The highly heterogeneous and dynamic nature of the structure of concrete is the main reason why the theoretical structure–property relationship models, which generally are so helpful in predicting the behavior of engineering materials, are of so little use in the case of concrete. A broad knowledge of the important features of the structure of the individual components of concrete is nevertheless essential for an understanding and control of the properties of such a composite material.

Understanding the behavior of cement-based systems, including traditional concrete, requires that one first should understand its structure, especially its structure in the nanometer and micrometer scales. Concrete and other cement-based materials are typically complex in structure and require investigation at several different levels of magnification to develop an appreciation of their pertinent details. The aims of this chapter are to understand the multiscale nature of concrete, to develop an appreciation of the structure of ordinary Portland cement concrete, and to outline an approach to *nano and microstructural engineering*—methods of modifying the materials structure.

## 4.2 CLASSIFICATION OF MATERIALS STRUCTURAL LEVELS

Concrete is a typical multiscale material. Its structure cannot be approached at a single level of scale, but rather requires examination and documentation over a very large range of magnifications. According to Diamond (1993), the structure of concrete can be examined at five levels: visual, petrographic, intermediate scanning electron microscopy (SEM), high magnification of SEM, and the nanometer level, as shown in Table 4-1. Each level of the structure corresponds to a certain range of the length scale.

The visual level is at the millimeter scale. In this level, aggregates and hardened cement paste can be observed from examination of a cross-section of the concrete with the naked eye. As an example, let us take Figure 4-1. Two phases that can easily be distinguished in Figure 4-1 are the aggregate particles of varying size and shape, as well as the binding medium, composed of an incoherent mass of hydrated cement paste. Details of the size and shape of coarse aggregate are visible. The mineralogy details of the coarse aggregate can be observed and interpreted by an experienced petrographer. At the millimeter scale or visual level, therefore, concrete may be considered to be a two-phase material, consisting of coarse aggregate particles dispersed in a matrix of the cement paste. No details of the hydrated cement paste can be viewed.

The petrographic level falls in the scale of $10^{-5}$ to $10^{-3}$ m, which can be made visible in stereooptical and petrographic microscopes. Optical microscopy is more usefully applied to either plane polished surfaces or specially ground petrographic thin sections, typically 25–30 μm, and transparent to light. With special techniques, large-area thin sections of 150 × 100 mm can be produced. Such sections are often produced by a process that includes impregnation of the empty spaces, such as cracks, air voids, and empty capillary pores, by a fluorescent dye dissolved in low-viscosity epoxy resin. When such specimens are examined in petrographic microscopes equipped for fluorescent light examination, the details of crack patterns and the degree of local inhomogeneity can be revealed. Figure 4-2 shows an area of a concrete examined by SEM at a magnification of 200×. From Figure 4-2, the lower grade of fine aggregates, such as the finer sand of the size of the order of tens of micrometers, entrapped air voids, and unhydrated cement particles can be observed.

**Table 4-1** Levels of microstructure

| Level | Optimal magnification range | Usual method of observation | Structures to be revealed |
|---|---|---|---|
| Visual | 1× ~ 10× | Unaided eye or hand lens | Details of coarse aggregate and air voids |
| Petrographic | 25× ~ 250× | Optical microscope | Fine aggregates, air voids, some paste details and some cracks |
| Intermediate SEM | 250× ~ 2,000× | SEM backscatter mode on plane polished surfaces | Arrangement and juxtaposition of cement paste particles, sand, capillary voids |
| High magnification SEM | 2,000× ~ 20,000× | SEM secondary electron mode on fractured surfaces | Details of the internal structure of individual cement particles and masses |
| Nanostructure | 600,000× ~1,000,000× | AFM, TEM | Some details of C–S–H |

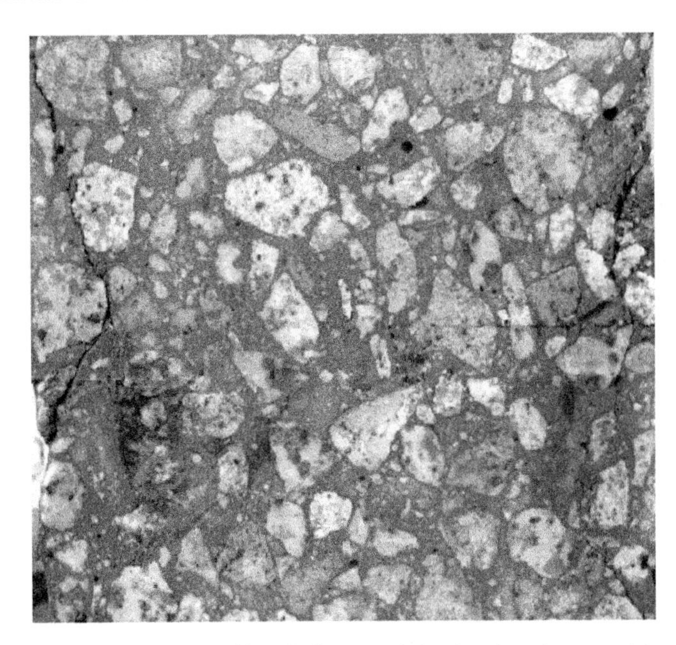

**Figure 4-1** Visual level of concrete's structure (mm scale)

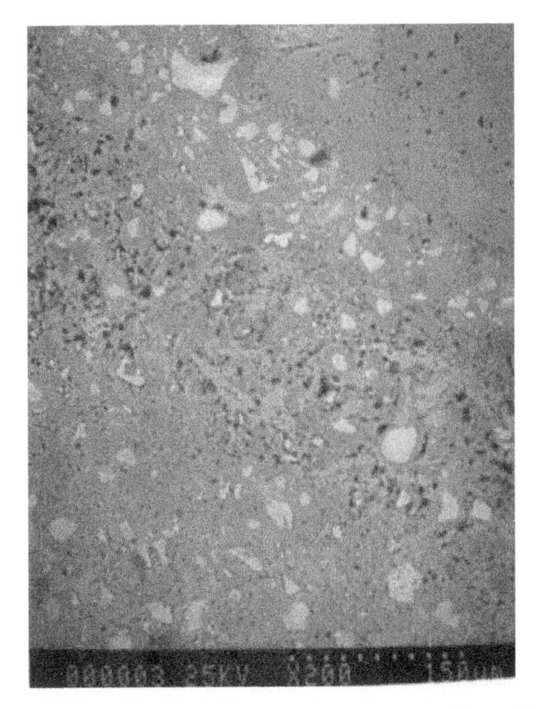

**Figure 4-2** Structure of concrete at the petrographic level (the white ones are unhydrated cement particles; and the black ones are air voids)

**Figure 4-3** Structure of concrete at intermediate SEM level

The corresponding length scale for the intermediate level is $10^{-5}$ m. The technique used for examination at this level is called backscatter electron detector scanning microscopy. It is capable of providing reasonably high resolution and at same time is sensitive to small differences in electron scattering power. In the usual procedure, the specimen is dried and potted in a low-viscosity epoxy preparation, which is then sliced with a diamond saw to get a plane surface. The slice is polished before the SEM test. Figure 4-3 is such an SEM backscattered photo taken at a magnification of 400×. Different gray levels in Figure 4-3 are due to different electron backscatter coefficients of different chemical compositions. The brightest ones are unhydrated cement particles and the darkest areas are pores. From Figure 4-3, the arrangement and juxtaposition of cement particles, capillary voids, and unhydrated cement particles, and the interface between coarse aggregates and matrix can clearly be observed. Moreover, the tiny crack on the surface of the coarse aggregate in the top of Figure 4-3 is readily apparent. Basically, there is no significant difference between the petrographic and intermediate level, and the low end of the intermediate level overlaps with the high end of the range available to the petrographic level.

The high-magnification SEM level corresponds to the micrometer scale. The structure of concrete at this level is basically examined in an SEM using a secondary electron detector. The mechanism of contact formation of the secondary image depends primarily on differences in elevation and the local position of the hydrated cement rather than on differences in the backscattering ability associated with different elemental composition. Thus, the fracture surface has to be used in the secondary mode of an SEM. Figure 4-4 is a typical photo of a secondary mode taken at a magnification of 10,000×. From Figure 4-4, the aggregation of C–S–H, the plate-shaped CH, and the needle-shaped AFt can clearly be distinguished. At this level, the structure of individual cement particles, the details of different hydration products, and the lower grade of the capillary pores can be observed. From Figure 4-4, it can be seen that the structure is neither homogeneously distributed, nor are the products themselves homogeneous. For instance, in some areas, more solid hydrated products appear while other areas are highly porous. For a well-hydrated cement paste,

**Figure 4-4** Structure of concrete at high magnification SEM level

**Figure 4-5** Morphology of C–S–H

the inhomogeneous distribution of solids and voids may be improved. Figure 4-5 shows an SEM photo of the secondary mode taken at a magnification of 20,000×. The morphology of calcium silicate hydrate (C–S–H) can clearly be observed. Figure 4-6 shows a fly ash particle in cement paste. It can be seen that the secondary hydration products are formed on the surface of the fly ash particle. Very fine cracks, about 0.1 micrometer wide, can clearly be seen in the surrounding fly ash particles. The ultimate structure of hydrated cement at the nanometer scale is that of calcium silicate hydrate (C–S–H), which is considered to be responsible for the strength of Portland

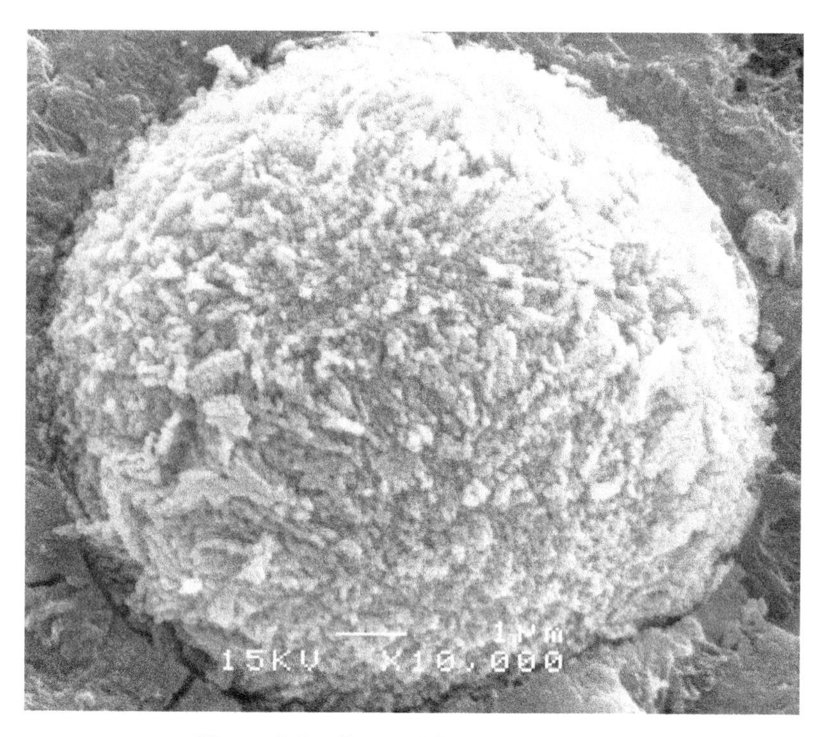


**Figure 4-6** Reacted fly ash in concrete

cement concrete. Understanding the structure of C–S–H plays an important role in revealing the nature of the concrete strength. However, knowledge in this aspect is limited.

Due to the availability of advanced measurement equipment at the nanometer scale, such as atomic force microscopy, nuclear magnetic resonance, and transmission scanning microscopy, as well as powerful computers with powerful software, it is now possible to characterize or simulate some features of C–S–H. Figure 4-7 is a photo taken by transmission scanning microscopy with a magnification of 800,000. It shows a moment during $C_3S$ hydration process. The crystal line of the hydration products of $C_3S$ as well as unhydrated part of $C_3S$ can be seen. Along the edge of the unhydrated part $C_3S$, a layer of hydration product can clearly be distinguished and it should be C–S–H. The crystal line shown in the layer proves the statement: "C–S–H is ordered in short range." Figure 4-8 is a photo of C–S–H taken by AFM. Some aggregation of C–S–H, in an egg shape with a size of 30–50 nanometers, are clearly visible. There is a clear boundary among the egg-shaped grains and a strong trend in orientation can also be observed. Because C–S–H plays an important role in determining concrete properties, it has recently become a hot topic in concrete research. Section 4.3 discusses the modeling of C–S–H structure in detail.

## 4.3 STRUCTURE OF CONCRETE AT NANOMETER SCALE: THE C–S–H STRUCTURE

The structure of calcium silicate hydrate (C–S–H) gel is considered at the nanometer scale. Since C–S–H is the most important hydration product in cement-based materials—it takes up approximately 50–70% of the fully hydrated cement paste and makes the dominant contributions to the mechanical properties of the cement-based material—understanding its structure is critical in revealing concrete properties. Unfortunately, with decades of research and development, even in

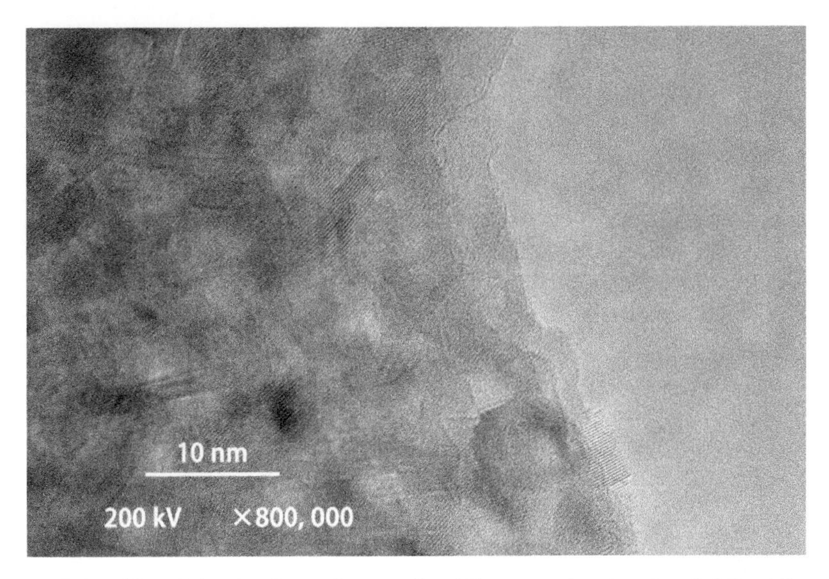


**Figure 4-7** Materials structure of concrete at the nanostructure level: the white particle is unhydrated $C_3S$ and at the edge showing a C–S–H layer

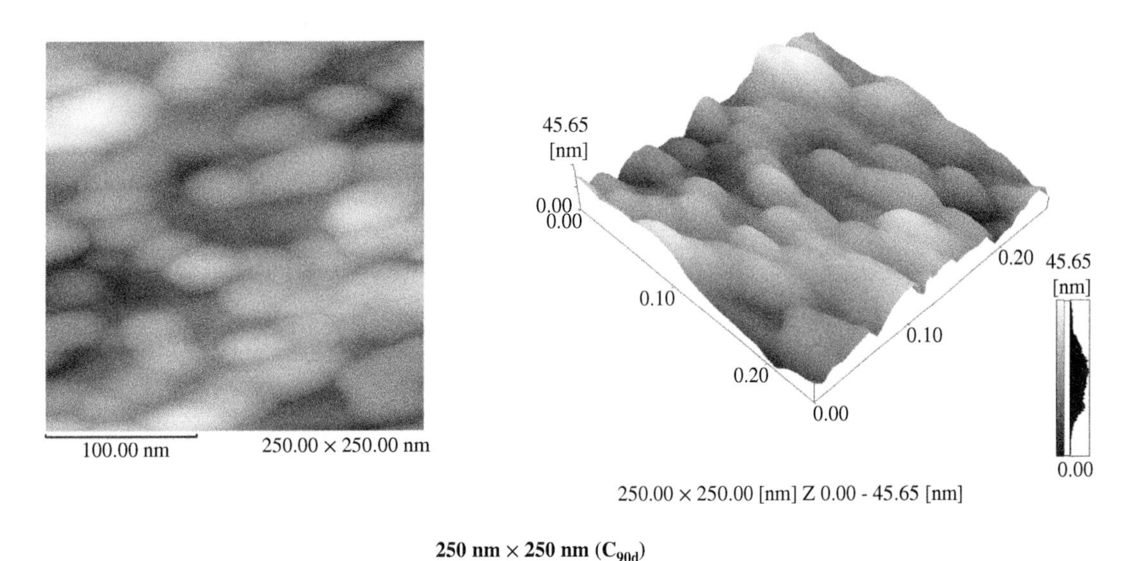


**250 nm × 250 nm ($C_{90d}$)**

**Figure 4-8** AFM photo of C–S–H at nanometer scale

terms of modern advanced technology, the nature of C–S–H gel is still obscure. This is due to not only the intrinsic difficulty in the characterization of the amorphous structure within a single paste and the Ca/Si ratio probably varying from one area to another, but also the complex composition and reaction in the materials, especially when some additives are used. To construct a C–S–H phase model, the following aspects should be considered first.

### 4.3.1 Experimental Characterization of C–S–H Structure

**(a)** *Chemical composition*: A model for the C–S–H structure must be compatible with the widely observed compositional variations, which are contemporarily considered as the most important factors that need to be satisfied. In fact, in recent studies, the models established all start from or concentrate on the chemical composition approach. The Ca/Si ratio in hardened $C_3S$ pastes or pure Portland cements generally has a range of values from 0.7 to 2.3, and a mean value of 1.7 has been experimentally confirmed. Recently a C–S–H composition formula of $(CaO)1.7(SiO_2)(H_2O)1.8$ with a density of 2604 kg/m$^3$ was obtained (Allen et al., 2007) by small-angle X-ray and neutron scattering (SAXS and SANS) methods. It may be used as a starting point to construct a C–S–H model.

**(b)** *Silicate anion structure*: The basic unit of all the silicates is the $SiO_4$ tetrahedron. Different numbers of silicate units can form various silicate structures, as shown in Figure 4-9. It can be seen from Figure 4-9 that each silicate can coordinate with different numbers of silicate units, which is the basis for the characterization of the C–S–H structure by $^{29}Si$ magic-angle spinning nuclear magnetic resonance (MAS-NMR). The $Q_n$ factor ($n = 0, 1, 2, 3, \ldots$) is used to represent the fractional chemical shift of a silicon atom bound to $n$ bridging oxygen ($n$, the coordinate number of a specific silicate unit with other silicate units; 0, single silicate; 1, dimer or end of chains; 2, middle of silicate chains). The silicate anion structure of the C–S–H phase formed in the hydration of Portland cement has been studied most extensively by $^{29}Si$ MAS-NMR or trimethylsililation techniques with some of the results described below.

In $^{29}Si$ MAS-NMR, $Q_1$ and $Q_2$ predominate. The ratio of $Q_1/Q_2$ declines with the hydration progress, indicating that the lengths of silica chains are prolonged at the expense of a decrease of the dimers and an increase of the polymers. However, it has been established that the dimer is the most predominant of all silicate species and the linear pentamer the second most abundant, followed by the octamer, which shows a sequence of silicate chain lengths of $(3n - 1)$. Therefore, for the establishment of the C–S–H gel model, it should

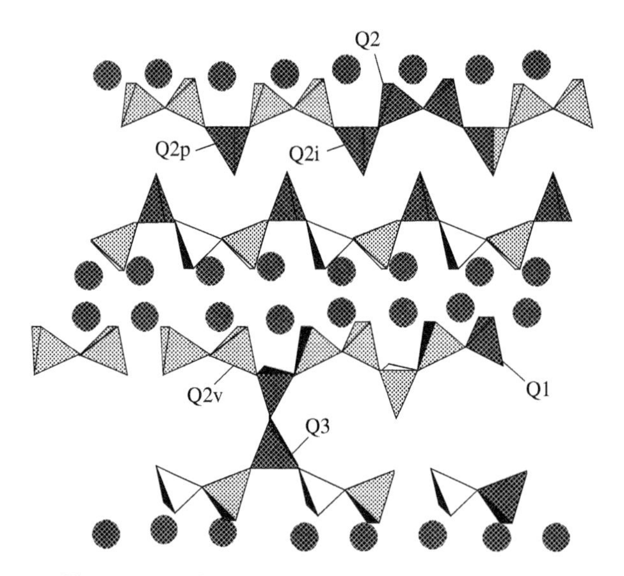


**Figure 4-9** Different types of silicate formation

account for such a sequence of chain length (2, 5, 8, …, $3n - 1$, where $n$ is an integer). According to Richardson and Groves (1992), the mean chain length of silicate chains should also be less than 5. If the dimer and pentamer are mainly used in the model, this requirement can be achieved automatically.

The $Q_3$ and $Q_4$ units cannot form in the hydrated material or in the course of hydration, but they may form in a reaction of the C–S–H phase with $CO_2$ from the atmosphere. For the $Q_0$ unit, its fraction steadily declines with the consumption of nonhydrated $C_3S$, but it may still exist in the late stage of hydration. It has been observed from experiments that in C–S–H, $Q_0$ is about 13%, $Q_1$ 50%, and $Q_2$ 33%.

**(c)** *Morphology*: The C–S–H phase in a mature concrete is not evenly distributed and may exist in a variety of morphologies, such as in the form of fibers, flakes, honeycombs, tightly packed grains, or a seemingly featureless dense material. However, at high resolution, it becomes apparent that all these C–S–H types have similar underlying foil morphology and therefore probably contain only one type of C–S–H at the nanometer level. The morphologies of the C–S–H formed within the boundaries of the original anhydrous grains (called the *inner product*) and in the space originally filled with water (called the *outer product*) are different, as shown in Figure 4-10. The inside is dense with a structure similar to fine foil morphology, while the outside is less dense and more like a fibrillar structure.

### 4.3.2 Classical Models of C–S–H Structure

With more than half a century's development, a large number of C–S–H gel models have been proposed to describe the relative nanostructure. Richardson (2008) has systematically summarized C–S–H models. Tobermorite is one of the earliest models proposed by Taylor and Howison (1956). A tobermorite phase can be classified as tobermorite 14 Å (angstrom), tobermorite 11 Å,

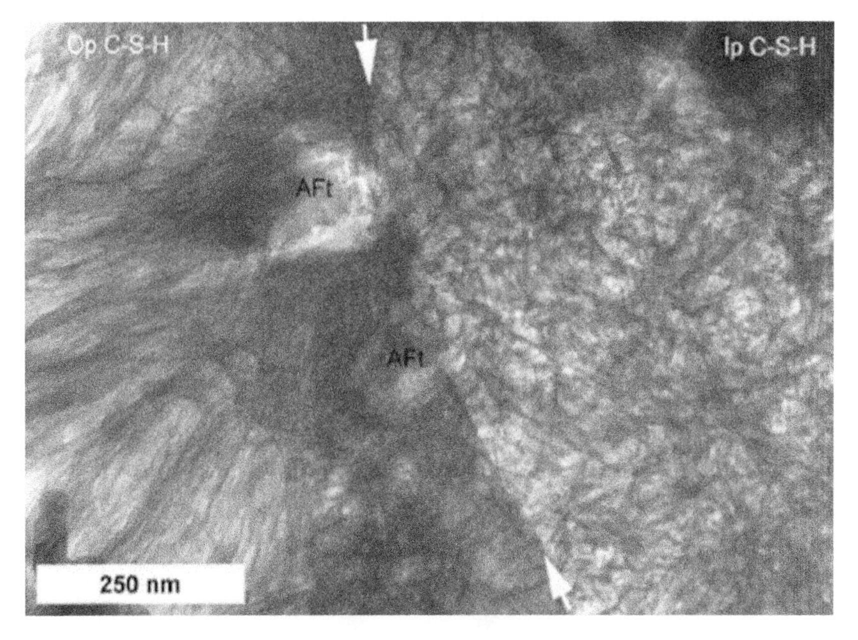


**Figure 4-10** TEM of inner product and outer product. Source: Richardson, I.G., et al, (2004) 34(9), 1733-1777 / with permission of ELSEVIER.

and tobermorite 9 Å, according to their basal spacing. The main difference between these models is the hydration degree, with the order of hydration degree in the order of 14 Å > 11 Å > 9 Å. Its ideal chemical composition is $[Ca_4(Si_3O_9H)_2]Ca \cdot 8H_2O$, according to Taylor (1997). But in other references (Bonaccorisi et al., 2004; Richardson, 2008), some differences exist with respect to the amount of water, e.g., $Ca_5Si_6O_{16}(OH)_2 \cdot 7H_2O$ and $[Ca_4(Si_6O_{16}(OH)_2]Ca \cdot 4H_2O$, indicating the difference in regard to understanding the hydration degree. The widely accepted framework of tobermorite 14 Å can briefly be summarized as follows:

**(a)** The structure of tobermorite 14 Å is basically a layered structure.

**(b)** The central part is a Ca–O sheet (with an empirical formula: $CaO_2$, though the chemical aspect CaO should always be CaO, which implies that the oxygen in $CaO_2$ also includes that of the silicate tetrahedron part).

**(c)** Silicate chains envelope the Ca–O sheet on both sides, with the characteristics mentioned later.

**(d)** Between individual layers, $Ca^{2+}$ and $H_2O$ are filled in the space to balance the charges and determine the layer distance, respectively.

A typical sketch of the structure of tobermorite 14 Å is shown in Figure 4-11. In the chemical formula, this 14 Å tobermorite structure can be determined as shown in Figure 4-12, which is much easier to understand in regard to composition aspects, but is obscure on the stereo structure and water molecules. The subcell chemical composition is $[Ca_4Si_6O_{16}(OH)_2]^{2-}$. In the tobermorite structure, in most cases, the bridging tetrahetra can be removed to increase the Ca/Si ratio, which also agrees with experimental results to some extent.

The Jennite phase is another popular model used to represent the structure of C–S–H. It has the chemical composition of $Ca_9Si_6O_{18}(OH)_6 \cdot 8H_2O$ (Bonaccorisi et al., 2004), which has a much higher Ca/Si ratio of 1.5 than that of tobermorite. The Ca/Si ratio in jennite is closer to the experimentally confirmed value of 1.7 in the C–S–H gel, but its full structure is not well determined due to the poor quality of available crystals. Some information about its structure has been proposed by different researchers, based on jaffeite or metajennite, which are related to the structure of jennite. Until now, the most useful structure for the jennite model was proposed by

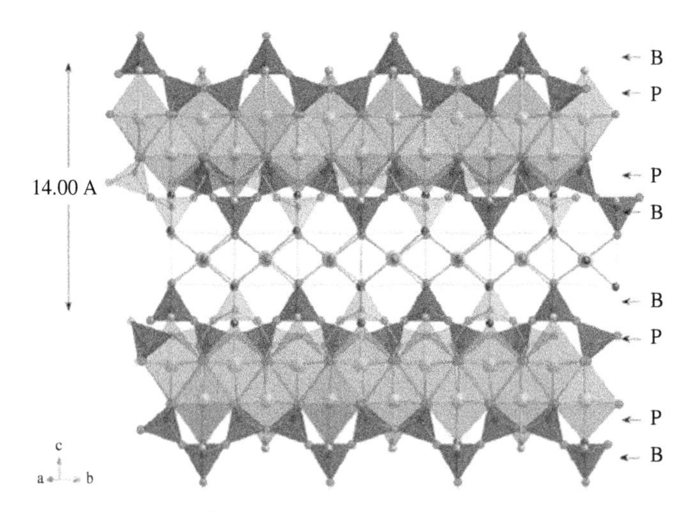


**Figure 4-11** The structure of 14 Å tobermorite. Source: Adapted from Bonaccorsi, et al., 2005.

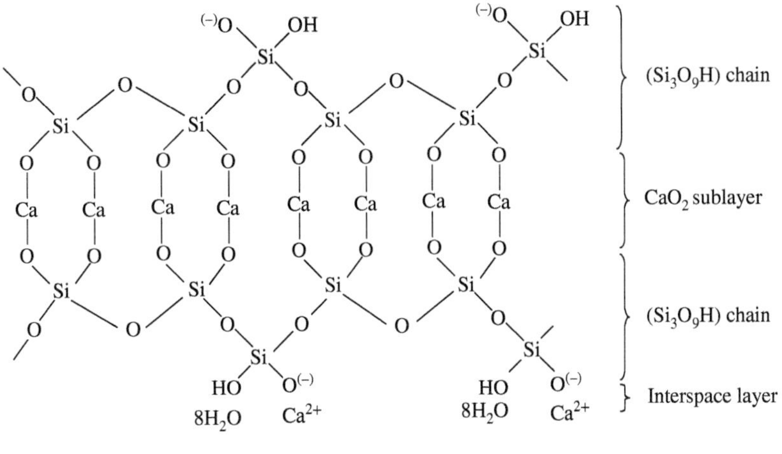

**Figure 4-12** Chemical formula of 14 Å tobermorite

Bonaccorsi et al. in 2004, which agreed well with some experimental results. Basically, jennite is built up by the combination of three modules:

1. Edge-sharing calcium octahedrals.
2. "Dreierkette" form of silicate chains.
3. Additional calcium octahedrals in special positions on the inversion center to connect different layers.

According to Richardson and Groves (1992), a jennite model can be built up from jaffeite using the following procedure:

1. Identify the "tilleyite ribbon" in jaffeite.
2. Remove the ribbon.
3. Insert bridging silicate to form metajennite.
4. Add $Ca^{2+}$ ions and water molecules to form jennite.

In this structure, distinct characteristics of the Ca–O layers can easily be found. In some Ca octahedrals, all the vertices of the octahedrals are shared with tetrahedral or interlayer calcium (see Figure 4-13), while some on both sides of the layers are thoroughly free to combine with water.

The distinct difference between jennite and tobermorite is that in a perfect jennite structure only Ca–OH bonds are present, while in a perfect tobermorite structure only Si–OH bonds are present. Taylor (1997) offered a hypothesis that in the C–S–H model most of the layers were of structurally imperfect jennite and a smaller proportion was related to tobermorite. It implies that the C–S–H model is a combination of tobermorite and jennite. This model is later called the T/J model. The general formula of C–S–H can be expressed as:

$$Ca_xH_{(6n-2x)}Si_{(3n-1)}O_{(9n-2)} \cdot zCa(OH)_2 \cdot mH_2O \tag{4-1}$$

where $Si_{(3n-1)}O_{(9n-2)}$ is an average silicate anion, $Ca_x$ the necessary charge to the balance, not distinguished from those in the interlayers, $H_{(6n-2x)}$ the hydrogen atoms that are directly attached to silicate anions, which are gained by the charge balance and obtained by the following formula:

$$2(9n-2) - 2x - 4(3n-1) = 6n - 2x \tag{4-2}$$

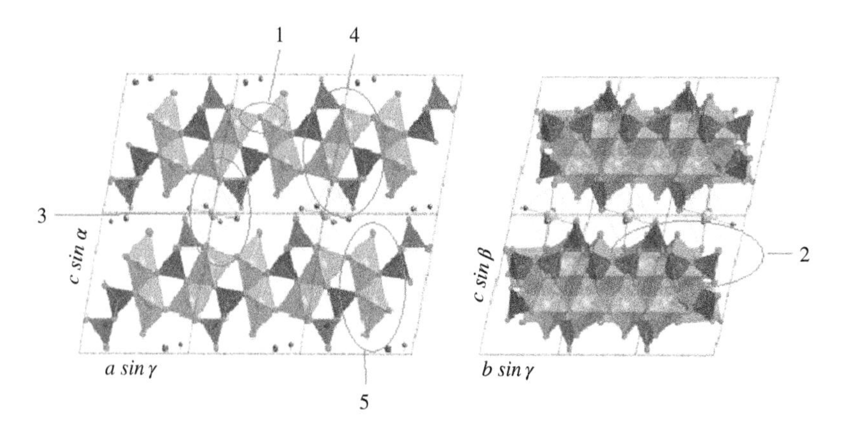


**Figure 4-13** Jennite structure on projection from [010] and [100]. Source: Adapted from Bonaccorsi, et al., 2004

while $zCa(OH)_2$ is the number of $Ca(OH)_2$ units (without considering the structure) with $z = 2n\,(n = 0, 1, 2, \ldots)$, and $mH_2O$ the number of water molecules bound, but not present, as hydroxyl groups.

By combining the general formula with $^{29}$Si NMR measurement, information about the structure can be obtained.

$^{17}$O NMR spectroscopy and $^{29}$Si NMR have demonstrated the existence of the Ca–OH group and Si–OH groups, no matter whether with high Ca/Si ratios or low ones. When Ca/Si < 1.3, the Si–OH group is necessary and dominant, while when Ca/Si > 1.3, the Ca–OH group is necessary and dominant, which is consistent with the jennite structure proposed by Taylor. This point can also be explained by the charge-balance assumption and is consistent with the degree of protonation proposed by Richardson. The charge balance calculation is based on the assumption that the non-bridging oxygen (NBO) is preferentially balanced by $Ca^{2+}$. Therefore, to maintain the electrical neutrality, the following equation has to be satisfied:

$$2C/S - \text{NBO} = 0\,(\text{here NBO represents NBO per tetrahedron}) \tag{4-3}$$

For the case of $2C/S$ – NBO < 0 (*low Ca/Si ratio case*), $Ca^{2+}$ is not sufficient and the residual NBO must be neutralized by the formation of Si–OH; therefore, Si–OH linkage is essential. For the case of $2C/S$ – NBO > 0 (*high Ca/Si ratio case*), $Ca^{2+}$ is surplus and extra $Ca^{2+}$ must receive charge-balance from the formation of Ca–OH; therefore, the Ca–OH linkage is essential. Consequently, the process can be expressed as

$$\text{NBO}\,(\text{Si}-\text{O}-\text{Ca}) + \text{H}_2\text{O} \rightarrow \text{Ca}-\text{OH} + \text{Si}-\text{OH} \tag{4-4}$$

Based on surface chemistry, the interlayers are of electrical neutrality, while the surface can be charged and the thermodynamics analyzed. It is proposed that no matter how C–S–H is synthesized, as long as the equilibrium state is the same, the structure of C–S–H should be the same. Based on some equilibrium constants, the structure of C–S–H can be obtained. Models proposed here are structurally descriptive. Concerning the bulk phase, the structural unit of C–S–H is proposed as follows.

*Low Ca/Si Ratio Case*

Mainlayer: dimeric, charged balance by $Ca^{2+}$ and proton–$Ca_2H_2Si_2O_7$

Two successive dimerics bridged by tetrahedra: $Ca_2H_2Si_2O_7(SiO_2)_x, (x = 1)$

Silanol partially ionized, balanced by Ca (interlayer): $Ca_2H_{2-p}Si_2O_7\left(SiO_2\right)_xCa_{p/2}$ or $Ca_2H_{2-p}$ $Si_2O_7\left(SiO_2\right)_x(CaOH)_{2/p}$

Missing bridging tetrahedra by $Ca(OH)_2$: $Ca_2H_{2-p}Si_2O_7\left(SiO_2\right)_xCa_{p/2}\left(Ca(OH)_2\right)_y$

When the bridging tetrahedra is missing, some Si change should occur, which is not shown in the last formula, and thus is questionable. In the whole process, there is no concern about water, and the process is therefore not well-rounded.

*High Ca/Si Ratio Case*

The steps are the same as those for the low Ca/Si ration, except for the last step, which is:

The missing silicate bridging tetrahedra is replaced by $Ca(OH)_2$—tobermorite-like

Or the nonbridging tetrahedron is replaced by $2OH^-$—jennite like.

Richardson (2008) classified mainly two categories of C–S–H chemistry: T/J and T/CH composition. For the T/J composition, C–S–H is an assembly of tobermorite regions followed by jennite domains, or, briefly speaking, a mixture of tobermorite and jennite, which is widely known. For the T/CH composition, C–S–H is a kind of tobermorite layer sandwiched between calcium hydroxide, or it is based on the structure of $Ca(OH)_2$ to some extent. This kind of structure, according to the literature, was the one most studied, dating back to the 1960s. It did not consider the detailed structural characteristics, but some stoichiometry or compositional ones.

The strongest evidence of this model is the chemical extraction experiment showing that significantly larger amounts of CH can be extracted from the C–S–H gel than that inferred by XRD. Therefore, the CH is considered to be incorporated into the nearly amorphous gel phase (Viehland et al., 1996). If this assumption makes sense, there should be an explicit signal for CH when doing analysis; however, this is not the case for most experimental results when studying common cement-based materials. But, as mentioned by Richardson (2004), the T/CH model is more appropriate for the KOH-activated metakaolin Portland cement due to the observation that CH is always present in the selected area electron diffraction (SAED).

Recently, a realistic model of C–S–H has been proposed by Pellenq et al. (2009) through molecular simulation. The model has been compared with experimental results, showing acceptable agreement. The model is based on the structure of 11 Å tobermorite with the starting Ca/Si ratio of 1. This structure is quite analogous to that of 14 Å tobermorite, having a difference in the distance. However, in this model, it is assumed that no OH group is present, which might not be the case for real C–S–H. The procedures to build up the model are as follows.

**Step 1:** Remove all the water molecules in the tobermorite structure.

**Step 2:** Remove the $SiO_2$ neutral group according to the $^{29}Si$ NMR results to raise the Ca/Si ratio to 1.65, which is the case of the missing tetrahedra in the previously mentioned models. This value is close to the mean value of 1.7 (NMR: $Q_0 = 13\%$, $Q_1 = 67\%$, $Q_2 = 20\%$, no $Q_3$ and $Q_4$, dimers are the predominant ones, monomers are present but not that much).

**Step 3:** Optimize the remaining structure by 0 K energy minimization and relax the whole cell by the core–shell potential model, obtaining a significant distortion of the layer structure.

**Step 4:** Simulate water adsorption by the Grand Canonical Monte Carlo simulation.

**Step 5:** Further relax the whole cell under constant pressure and temperature with the interlayer spacing of 11.9 Å and density 2.45 g/cm$^3$, which is to some extent different from 2.6 g/cm$^3$ (before this step, the obtained density is shown to be 2.56 g/cm$^3$, which is much closer).

The unique observation in this model is the distribution of the water molecules. In previous studies, the water molecules have been considered to be located in the interlayers. However, in this simulation, the water lies not only in the interlayer regions, but also around the silica monomers, which emphasizes the water function in the mechanical properties. Besides, in the optimization of the structure in step 3, a clear distortion structure can be observed. When compared with experimental results, the C–S–H structure can be described as a glass for a short distance range while retained as a layered crystal feature for a long distance range. Here short range and long range refer to intra- and interlayer spacing, respectively.

### 4.3.3 Atomistic Simulation on C–S–H Structure

Investigating the atomic structure of C–S–H generally employs theoretical techniques such as quantum chemistry (QC), molecular dynamics (MD), and some mesoscopic (atomic- to micro-scale) method. The following is a brief description of the methods and their research scopes.

#### 4.3.3.1 Computational Research Methods

The QC methods apply quantum mechanics to solve chemistry problems. The central topic in QC methods is understanding the electronic structure by solving the Schrödinger equations, which is known as determining the electronic structure of the molecule. The physical and chemical properties are implied essentially by the electronic structure of a molecule or crystal. But the exact solution of the Schrödinger equation only exists in a single hydrogen atom system. Almost all of the atomic or molecular systems involve two or more "particles." Their Schrödinger equations cannot be solved exactly, and thus the approximate solutions must be sought. For example, the Born-Oppenheimer approximation is often used to divide the electrons and nuclei. And the local-density approximation used in density functional theory (DFT) converts a multi-electron problem into a single-electron problem. In the QC studies of the C–S–H structure, researchers often focus on the optimal conformations of C–S–H and the basic properties when comparing experimental data. The QC methods are only suitable for the study of the structure of nanometers with less than 100 atoms and femtoseconds process.

The MD method applies statistical mechanics, which is employed to study the physical motion of atoms and molecules. It allows atoms and molecules to interact for fixed times so that the dynamic evolution of the system can be observed. In the commonest method, the trajectory of atoms and molecules is determined by numerically solving the Newtonian equation that is used to determine the interaction of the particle system, where the forces between particles and their potential energy are usually calculated using the interatomic potential or molecular mechanics force field. This method was initially developed in the field of theoretical physics in the late 1950s, but it is currently mainly used in the fields of physical chemistry, materials science, and biology. Different from QC methods, although the accuracy of the MD methods is reduced, the scale range of the simulation system is significantly enlarged. The traditional MD (called MD in this book) methods are suitable for 1∼100 nanometers structure and 1∼100 nanoseconds process.

Due to computational restriction, all-atom MD cannot directly bridge the gap between the molecular level and the micro-level, which is still a great challenge for multi-scale study. These materials range in size between the nanoscale for several atoms (such as a molecule) and materials measuring micrometers. Numerous methods are developed on this scale. This section only pays attention to the scale of "beyond the traditional MD range but below the micron level." One of the most important methods is the coarse-grained molecular dynamics (CGMD) method. Its basic theory, similar to the MD methods, also follows the Newtonian motion equation and force field. Different from a single atom as the basic unit in all-atom molecular dynamics, the CGMD generally

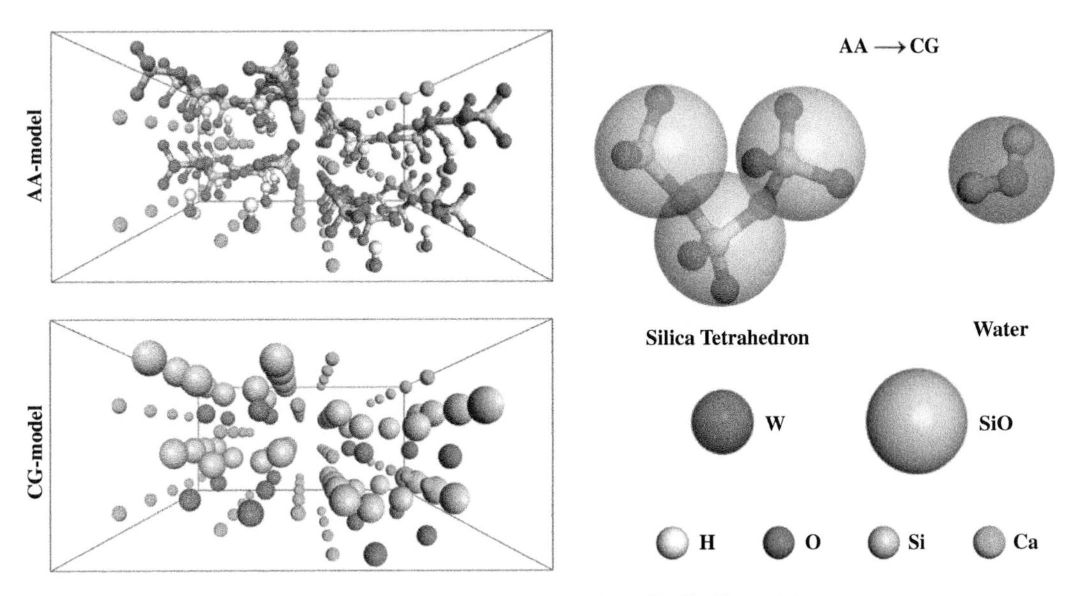


**Figure 4-14** The coarse-grained C–S–H model

defines a group of atoms that remains in a relative position and exhibits specific properties as a basic unit. The C–S–H coarse-grained process is shown in Figure 4-14. Although CGMD simulation has been applied in many materials, it is a very new method in studying the C–S–H system and related research have just emerged in recent years. The CGMD methods can upscale the molecular simulation and allow the study of larger-scale properties, but further sacrifices calculation accuracy, missing many chemical features such as chemical bonds.

How these theoretical techniques are applied to study the nanoscale structure of C–S–H is discussed in Section 4.3.3.2.

#### 4.3.3.2 Nanostructure Simulation of C–S–H Model

The model used today was born out of Taylor's studies (1997), which has ideal chemical composition. At the moment, the most commonly used molecular simulation methods are beginning from the studies of Pellenq et al. (2009), which have proposed a realistic molecular model of calcium silicate hydrates. Both the QC and MD methods are used to generate this C–S–H structure. The presented density functional theory (DFT, one of the QC methods) results of the structure and elastic constants for the tobermorite model illustrate that:

- **(a)** The interlayer distance is appropriate in the tobermorite model so that the Coulomb interactions between interlayers become comparable to the covalent-bond interactions (we can call this an ionic bond).
- **(b)** The existence of interlayer calcium ions and water molecules cannot shield the Coulomb interactions (there is no electrostatic shielding effect). The layers composed of silicon-oxide tetrahedron chains can be connected by the interlayer calcium ions and water molecules.

Such a structure and intra-interactions result in the direction with the weakest mechanical properties in the two inclined regions that form a hinge mechanism, as shown in Figure 4-15. The investigated class of materials are relevant to the chemical composites of C–S–H that is the binding phase of all concrete materials and is the principal source of their strength and stiffness. Moreover,

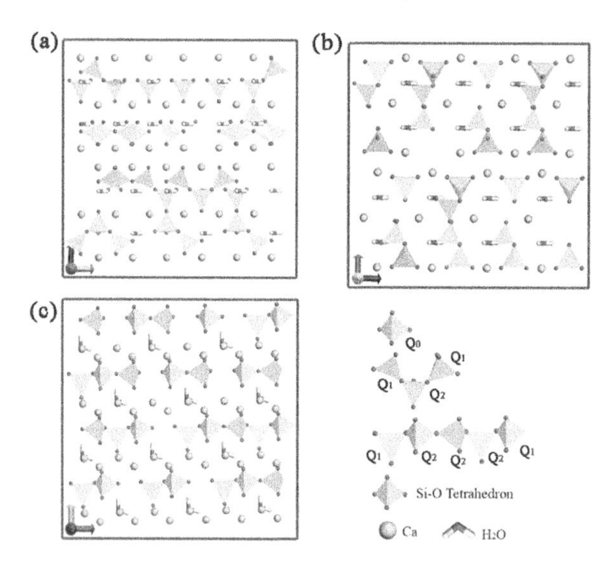


**Figure 4-15** The snapshots of (a) x, (b) y, and (c) z directions for the common C–S–H model. Source: Adapted from Hou et al. (2014)

employing the DFT methods, the composition and mechanical properties of cement clinkers and hydration products were studied by Izadifar et al. (2019). The bulk modulus of all phases was computed and $\gamma$-$C_2S$, jaffeite, and tobermorite C–S–H were combined in isotropic models. The monomer structure of silicate tetrahedra was revealed in these structures. But, due to the scale limitation, the QC results can't directly compare to the experimental properties of C–S–H. The QC methods also are used to evaluate the accuracy of the model. The $^{29}$Si NMR spectra of Rejmak et al. (2012) in cement-based materials are studied through calculations of the isotropic shielding of silicon atoms within the DFT. The widely accepted models based on the observed structures of jennite and tobermorite minerals are compared. The result shows that the NMR spectra for models of the calcium-silicate-hydrate gel based on tobermorite are in better agreement with the experiment than those for jennite-based models. The tobermorite model has been widely accepted for C–S–H modeling. In addition, the first-principles results may serve as a benchmark for validating empirical force fields required for the analysis of complex C–S–H systems.

The accuracy of the atomistic structure model is greatly dependent on the force field that is used to describe the interactions between atoms. The most commonly used force field in the C–S–H system was derived from the ClayFF force field (Cygan et al., 2004), named the C–S–H force field (Shahsavari et al., 2011) that is widely employed for the C–S–H model and used to study concrete hydration products for its good transferability and reliability. But this does not mean that there is only one force field model available. According to the database of Mishra et al. (2017), many force fields have been developed, such as Born–Mayer–Huggins, InterfaceFF (IFF), ClayFF, CSH-FF, CementFF, GULP, ReaxFF, and UFF. The benefits and limitations of these approaches are discussed in the database. They can be accessed on the web link (http://cemff.epfl.ch). Molecular simulations of cementitious minerals, such as tricalcium silicate ($C_3S$), portlandite, and tobermorite are described in this database. By using these force fields, the MD structures are well understood on the atomic scale. Manzano et al. (2007) earlier measured the bulk modulus of tobermorite 9, 11, and 14 Å and jaffeite by using the MD simulation. They adjust the C/S ratio of tobermorite 9 and 14 Å that is equal to 0.8 with the bulk modulus of 68 and 46 GPa, respectively. Although their measuring data is systematically overestimated to the experimental data of C–S–H gels, the sequence of the relative size of their mechanical properties is highly consistent. This difference will disappear after taking into account the finite length of silicate chains. Compared to the DFT result, the MD results showed that the bulk modulus of tobermorite 9 and 14 Å increased by 36% and 43%, respectively. The MD results showed that the bulk modulus increases with the C/S ratio increases.

With further investigation of the C–S–H structure, researchers realized that the small length range (~nm) and short time scale (~ns) of QC and MD models are the main reasons limiting their applications. By simplifying some unimportant atomistic details of MD, the CGMD method could increase the simulation scale as well as explore the nano characteristics at the same time. Assuming that C–S–H is composed of packing colloidal particles, Ioannidou et al. (2016) unraveled how the heterogeneities that developed during the early stages of hydration persist in the structure of C–S–H and impact the mechanical performance of the hardened cement paste. Unraveling such properties of C–S–H facilitates controlling them as that can improve the smarter mix designs of cementitious materials. Further studies by Hou et al. (2020) directly describe the process of C–S–H coarse-graining, as shown in Figure 4-14. However, the current CGMD model has some limitations:

**(a)** The free interlayer structure of C–S–H is not well reflected. Although there is still a layer structure of some studies, they are produced by restraint atom, numerically.

**(b)** The interactions between silicon-oxygen tetrahedron and water are not well reflected. The main reason is that coarse graining has abandoned the direction of interactions and only retained the strength of interactions.

The CGMD is very important, which plays a role in bridging the nano-structure to the macro-performance of C–S–H. Although the CGMD uses the framework of traditional MD methods, the force field should be the new proposal. It is necessary to design an accurate CGMD model that can keep as much atomic information as possible.

### 4.3.4 The Properties of C–S–H at the Nanoscale

In normal concrete, the performance of concrete is usually decided by the properties of cementitious materials. For concrete systems, we mostly care about the mechanical properties and durability, which has to be reflected in the nanoscale. Therefore, the most important precondition for studying the C–S–H properties at the nanoscale is to accurately establish the relationship between the macroscale performance of the cementitious and nanoscale properties of C–S–H. Recently, molecular methods (include QC, MD, and CGMD that were introduced in Section 4.3.3) have been regarded as the most efficient approach for revealing the nanoscale properties. And also, we can roughly divide these research studies of molecular simulation into strength properties and transport properties according to their concerned performance of concrete. The strength properties and transport properties of C–S–H at the nanoscale are often closely related to the mechanical properties and durability at the macroscale, respectively. The following section introduces how the molecular simulation methods are used to investigate the properties of C–S–H at the nanoscale.

#### 4.3.4.1 The Strength Properties of C–S–H

The tobermorite-based C–S–H nano-structure possesses anisotropy characteristic, which contains three directions, as shown in Figure 4-15. In the direction of the silicon-oxygen tetrahedral chains (often called silicate chains), the silicate chains provide a stable backbone that determines the cohesive strength along the *y*-direction. Several silicate chains are connected by ionic bonds between Ca atoms and O atoms on a silicate tetrahedral, which both generate the calcium silicate sheet.

Neighboring calcium silicate sheets are further bridged by the inter-layer Ca atoms and H-bonds from chemically bonded water molecules along the *z*-direction. The mechanical properties of the layered structures of C–S–H are determined by the combination of these chemical bonds.

To characterize the mechanical behaviors of the layered structure, Hou et al. (2014) simulated the uniaxial tensile process of C–S–H along the x, y, and z directions. The constitutive relation between stress and strain is obtained at the nanoscale. Different stress-strain relations of tensile loading in three directions indicate the heterogeneous nature of the layered structures (Figure 4-15). The typical stress-strain of the x, y, and z directions of C–S–H exhibits the large differences as shown in Figure 4-16a. In the x-direction, during the tensile process, stress first increases linearly in the elastic stage and subsequently slowly increases to a maximum value of 6.5 GPa at a strain around 0.17 Å/Å. After the maximum value, the stress directly drops without obvious yield behavior and the stress slowly decreases to zero as the strain reaches 0.8 Å/Å. Stress-strain curves in the x- and y-directions are almost coincident with strain from 0 to 0.5 Å/Å, but the stress in the y-direction slowly reduces from 0.3 to 0.8. The structure of C–S–H gel cannot fracture along the y-direction even as the strain reaches 0.8 Å/Å, implying good plasticity. The deviation illustrates the different deformation mechanisms on the XY plane. The C–S–H gel is more likely to stretch and break in the z-direction and the strain at the fractured state is 0.4 Å/Å, indicating the brittle feature of the interlayer region.

Compared to the C–S–H structure, the strength properties of the original tobermorite structure are significantly larger, as shown in Figure 4-16b. The major changes in the stress-strain curves occur in the y-direction, which is quite different from the x- and z-directions. The high failure strength implies that Si-O bonds grow along the y-direction, which is better in load carrying. It should be emphasized that as the infinitely long silicate chains in the tobermorite crystals transform into the isolated dimer structure in the C–S–H gel, the tensile strength along the y-direction is greatly reduced. The difference in mechanical properties between tobermorite and C–S–H gel confirms that the silicate chain, the skeleton of the calcium silicate phase, plays a vital role in load resistance. In addition, the strength of tobermorite is much greater than the strength in the x and z directions in the C–S–H gel, which also means the mechanical contribution of bridging silicate tetrahedra. Based on the idea of increasing the silicate chain length, some biomimetic methods have been proposed, such as grafting organic groups on the C–S–H mineral sheet layer to improve the mechanical properties of the gelling system (Pellenq et al., 2008).

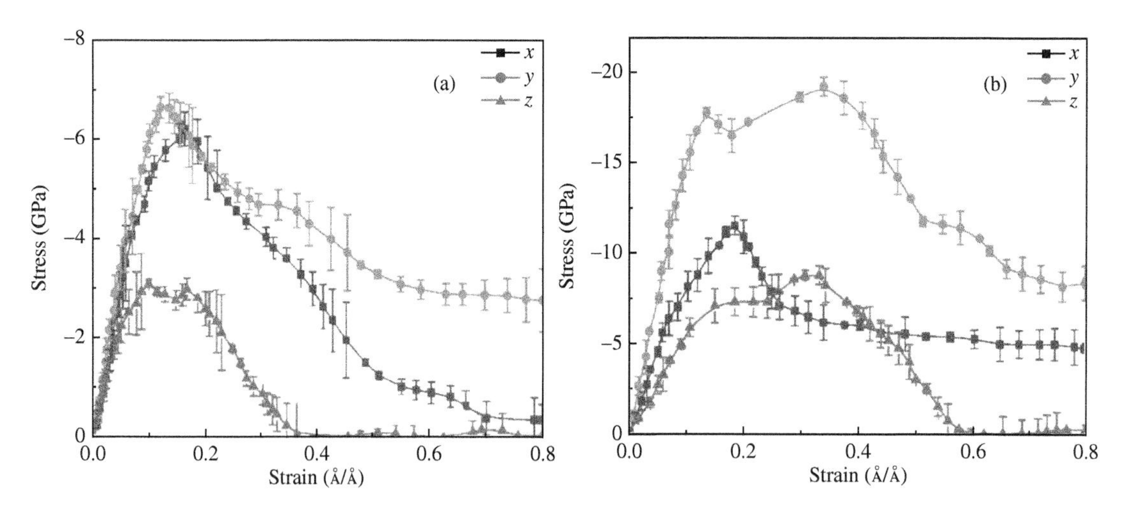


**Figure 4-16** The typical stress-strain curve of (a) C–S–H and (b) tobermorite. Source: Adapted from Hou et al. (2014)

#### 4.3.4.2 The Transport Properties of C–S–H

Many durability problems of concrete material, such as corrosion, carbonation, and sulfate attack are related to water and ions transport in the pores of the material. Unfortunately, despite many decades of studies, the migration mechanism of solution in the gel pores of concrete has not been fully understood. In the cementitious system, the C–S–H gel is a kind of microporous and mesoporous material, which is assembled from ordered and disordered calcium silicate sheets. Water and ions are exchanged in the interlayer space from nanometer to micrometer (Pellenq 2008). The motion of water in the pores relates directly to the cohesion of the C–S–H gel, determining the strength, creep, shrinkage, and chemical and physical properties of the cementitious materials. Experimental techniques can be used to study the water confined in the pores of C–S–H gels, including nuclear magnetic resonance (NMR) relaxation measurements (Bohris et al., 1998), solid-state 1H NMR experiments (McDonald et al., 2005), and quasi-elastic neutron scattering (QENS) (Thomas et al., 2001). QENS divides the water in the Portland cement and the tricalcium silicate into three states: liquid water, chemically bound water, and physically adsorbed water, which are mainly related to the C–S–H in the small nanopores. The water enclosed in the nanochannel exhibits different dynamics, such as residence time in the pores and diffusion coefficient. However, only experimentally studying the structure and dynamics of water is challenged by some limitations, such as material purity and instrument accuracy on the relevant length and time scales. The calculation methods, such as molecular dynamics, help to interpret the experimental results and play a complementary role in understanding the structure and dynamic properties at the molecular level.

Recently, a series of C–S–H gel pore models have been constructed to provide valuable insights into the solid-liquid interface quantitatively. As shown in Figure 4-17, the interfacial model is usually based on the cleaved C–S–H model in the (0 0 1) direction that is terminated with Si-OH, Si-O$^-$ and surface calcium ions. On the (0 0 1) surface of tobermorite, water molecules diffusing in the channel between the silicate chains demonstrate a number of structural water features: large density, good orientation preference, high dipolar moment, ordered interfacial organization, and low diffusion rate (Hou et al., 2014). The glassy nature of the surface water molecules is mainly attributed to the fact that the defective silicate chains provide plenty of non-bridging oxygen sites to form H-bonds with neighboring water molecules. The stable H-bonds connected with oxygen atoms in the silicate chains fix the molecular orientation and restrict the mobility of the channel water molecules. With increasing distance from the calcium silicate surface, the structural and dynamical behavior of the water molecules varies and gradually translates into bulk water properties at distances of 10–15 Å from the liquid-solid interface. In respect of the ions and C–S–H interaction, the cations such as $Ca^{2+}$ and $Na^+$ ions can be strongly captured by the negative silicate chains. On the other hand, secondary adsorption happens in the anions that are weakly adsorbed on the surficial calcium ions by forming ionic pairs or clusters. Based on the accurate interfacial model, the transport behavior of fluid in the nanometer gel pores can be further investigated (Hou 2015). Figure 4-18 is a typical capillary transport model of C–S–H gel. Water gradually penetrates into the gel pores with the evolution of time, penetrated water molecules shows advancing meniscus, orientation preference, and disturbed hydration shell in the vicinity of hydrophilic C–S–H surface due to strong chemical correlation between surface calcium atoms, non-bridging oxygen, and water and ions. More importantly, the water and ions have different capillary transport behavior: the ions migrate slower than water molecules in the C–S–H gel pore. The transport discrepancy between water and ions is more pronounced with the decrease in nanopore size due to the following reasons: (a) The small nanometer channel plays a filtering role to prevent the chloride and sodium ions from entering the larger hydration shell; (b) The ionic pairs such as $Ca^{2+}$ and

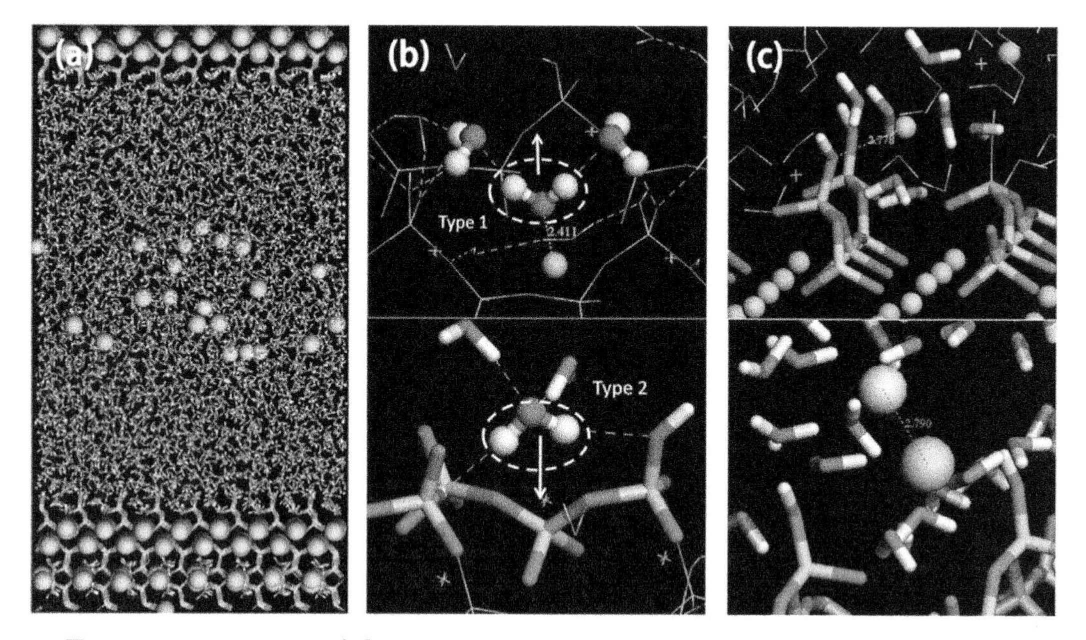


Figure 4-17 The typical C-S-H (a) interfacial model, (b) surface H-bonds, and (c) ions pairs

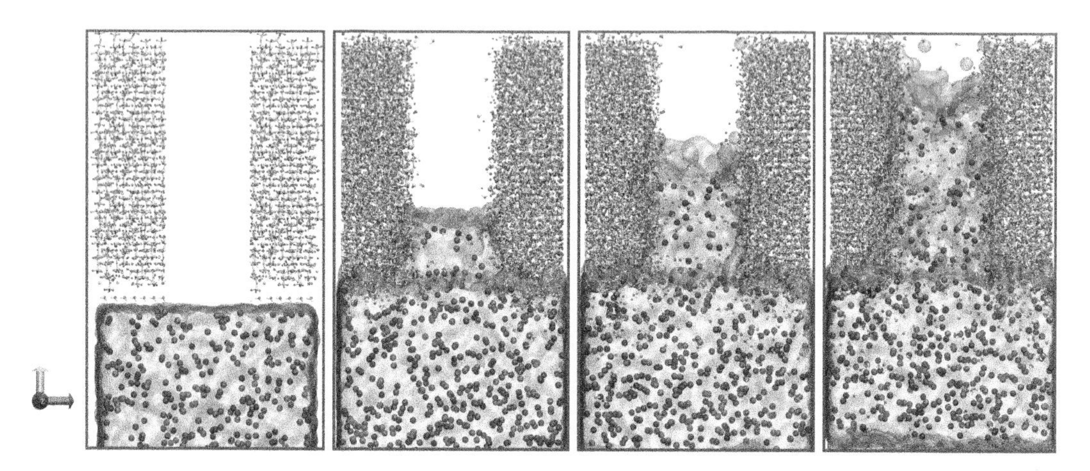

Figure 4-18 The capillary transport model of C-S-H gel

$Cl^-$ are more likely to accumulate to form a cluster, blocking further penetration of other ions; (c) the immobilization effect is quite stronger from the surface calcium ions and non-bridging oxygen atoms to elongate the resident time of ions. These efforts have played a guiding role in exploring the complex C–S–H gel and the interesting findings from the molecular dynamics study are very useful for the durability study of concrete serving in the detrimental environment.

## 4.4 STRUCTURE OF CONCRETE AT THE MICRO-SCALE

Concrete is a very complex heterogeneous composite material, containing microstructures of different length scales, ranging from microscopic hundreds of nanometer scales to macroscopic decimeter scales, over 9 orders of magnitude. The definition of the microscopic scale is so wide that no single experimental method or model can fully reveal the microstructure of concrete. As Wittmann (1983) pointed out, the concrete properties are modeled on three independent simulation levels covering the entire range of the concrete multi-scale system, namely, micro-level, meso-level, and macro-level. When considering the behavior of concrete at the macro level, it can be regarded as a relatively homogeneous material, so continuous methods such as the finite element method or the meshless particle method can be used for structural calculations. It is assumed that all the effects of the internal structure are eliminated by the average parameter. At the meso-level, aggregate and cement slurry can clearly be distinguished. At this scale, both finite element procedures and lattice models are suitable for simulating the behavior of concrete materials, including crack formation. The microscopic level corresponds to the description of the cement slurry in terms of its composition and the level where chemistry and thermodynamics actually play a role, as described in the C–S–H model introduced in Chapter 3. On the nanoscale, C–S–H can be described as a short-ordered structure, and atomic/molecular dynamics simulations can be used to gain insight into the structure based on tobermorite and sphalerite crystals. Recently, it has been recognized that the geometric gap between the micrometer and nanometer scales is too large (it must be reduced by more than three orders of magnitude). Therefore, a sub-micron level should be inserted between the micro-level and the nano-level, where the colloidal or gel-like properties of C–S–H can be studied.

### 4.4.1 The Experimental Measurements on the Micro-Scale

Generally, considering durability issues requires a good understanding of the transport characteristics of concrete. "Concrete transport characteristics" is a term that covers all the characteristics related to the transportation of fluids in concrete. Fluids related to durability include pure water or water containing corrosive ions, carbon dioxide, and oxygen. They can enter and pass through concrete in different ways, but all transmission mainly depends on the microstructure or pore structure of the concrete. The flow of fluid through porous concrete is called penetration. However, the movement of substances in concrete occurs not only through flow but also through (ionic) diffusion and (fluid) adsorption. These are the three main terms that are commonly used in the study of concrete transmission characteristics, namely permeability, diffusivity, and absorption. In addition, since both the diffusivity and permeability can be related to the electrical conductivity of a specific porous material, through the Nernst-Einstein equation and the Katz-Thompson relationship, the electrical conductivity Rate (or sometimes resistivity) has also been widely studied as one of the transmission properties of concrete.

The literature often classifies pores in solids based on their average widths, such as the diameter of cylindrical pores or the distance between two sides of slit-like pores. In this classification, pores are divided into micropores, mesopores, and macropores. The size of the micropores is less than 2 nm, and the mesopores range from 2 to 50 nm. All pores larger than 50 nm are represented by macropores. This classification is officially adopted by the International Union of Pure and Applied Chemistry (IUPAC). In specific technical standards, 2.5 nm and 100 nm are usually selected as boundaries to divide these three types of pores. However, in concrete science and technology, it is more common that pores are divided into gel pores, capillary pores, entrained air, and entrapped air.

The microstructure of cement-based materials is often characterized based on the following available methods, i.e., MIP (mercury intrusion porosimetry), nitrogen adsorption based on the BJH method, and BSE (backscattered scanning electron microscopy). The introduction of these methods is as follows:

**(a)** Pores with a width between 3 nanometers and 100 microns can be measured by the MIP method. A typical MIP measurement process involves pressing mercury into the sample in the chamber by gradually increasing the pressure, and performing a squeezing procedure after the highest pressure. By tracking the pressure and the volume of intrusive or extruded mercury during the test, the volume of the connected pores can be measured. The relationship between the hole diameter and the corresponding pressure is described by the Washburn equation based on an assumed cylindrical hole model.

$$P = -\frac{4\gamma \cos\theta}{d} \tag{4-5}$$

**(b)** Nitrogen adsorption based on the BJH method (Barrett, Joyner, and Halenda, 1951) is one of the most popular techniques for studying the pore structure of microporous and mesoporous materials. The nitrogen adsorption isotherm is usually chosen to analyze the pore structure of cement-based materials, the pore size distribution curve is derived according to the BJH method, and the specific area is calculated according to the Langmuir theory or the BET theory (Brunauer, Emmett, and Teller, 1938). The theory assumes that the pores are cylindrical, which is the same as the basic assumption of the Washburn equation. Based on this assumption, and using nitrogen as the adsorbate, the Kelvin equation can be written as

$$d_K = \frac{-8.3}{\ln\left(\frac{P}{P_0}\right)} \times 10^{-10}\,(m) \tag{4-6}$$

**(c)** BSE is one of the important techniques of image analysis. Image analysis is a method to derive pore size, pore area/volume fraction, and pore size distribution from images. The images used in this method can be obtained by optical microscope, BSE, or computer-based simulation models. Using BSE technology, the pore geometry can be displayed, and the algorithm for measuring pores with random shapes can be stated and modified according to the operator's requirements.

Further, based on these experimental results, some input parameters of the computer model can be determined, such as the porosity of the small capillary. Or these experimental results can be used to verify simulated microstructures at different scales.

### 4.4.2 The Simulation Studies on the Micro-Scale

The microstructure and transmission characteristics of concrete materials change over time. If the composition of the cementitious material is known, its microstructure and performance are directly determined by the degree of hydration reaction, which is a function of time (or age). The relationship between the degree of hydration reaction and time is usually described by the kinetics of hydration reaction. As the basis of the computer model, it is necessary to study the hydration reaction of the cementitious system in contemporary concrete. The output of the simulation first includes the hydration reaction products, and on this basis, the stoichiometric factors of different cementing materials can be determined. There is also a volume stoichiometric factor, also called a hydrate/reaction product volume expansion factor, which is a key parameter in a state-oriented computer model that controls the formation of microstructures in a simulation. The output also includes hydration reaction kinetics, which provides the degree of hydration reaction as a micro-scale state guide to the "state" in the computer model, and helps to calculate the transport properties and time (or age) based on the degree of hydration reaction.

Based on the results of hydration/reaction studies and microstructure characterization, computer models can be used to generate microstructures of different length scales, namely, the nano-scale CSH, the sub-micron-scale hydration product layer, and micro-scale cement slurry. An algorithm based on the random walk (Stauffer and Aharony, 1992) is used to calculate the transmission characteristics of each scale, as shown in Figure 4-19. The lower-scale transmission characteristics are amplified and used in conjunction with the microstructure to calculate the higher-scale effective transmission characteristics.

When considering mortar or concrete, on the meso/macro scale, the material is not a simple mixture of cement paste and aggregate particles. ITZ exists between the cement slurry and the aggregate particles, and the characteristics of ITZ should be studied separately from the so-called "bulk slurry." With the development of computer science, a computer model has been established to characterize ITZ. Its advantage is that it can easily obtain a three-dimensional gradient of porosity, degree of hydration, or calcium hydroxide content that varies with the distance from the aggregate surface. However, these computer models believe that the wall effect is a unique mechanism for forming ITZ, and therefore always choose a simple shape (cube, cuboid, or cylinder) for the calculation space. In this model, other mechanisms that contribute to the formation and evolution of ITZ cannot be simulated, nor can the influence of technical parameters such as aggregate particle size and aggregate volume fraction be effectively considered. In this study, considering the influence of the mix ratio parameters and the degree of hydration, a multi-aggregate method was proposed to simulate the formation and evolution of ITZ in concrete. In this way, the thickness, volume fraction, and microstructure of ITZ in a specific mixture in any state can be determined. A random walk simulation of ITZ microstructure can calculate the effective transmission characteristics of ITZ.

Finally, a three-phase model is used to simulate the microstructure of the mortar/concrete, namely, bulk-paste, ITZ, and aggregate particles. So far, the effective transmission characteristics of bulk-paste and ITZ have been determined, and it can be assumed that the aggregated particles are electrically insulating, non-diffusible to chloride ions, and impermeable. With these inputs,

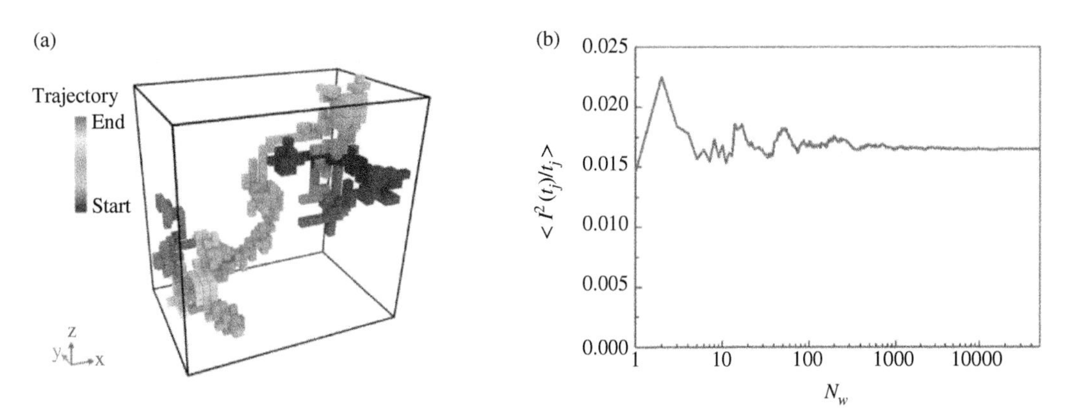


**Figure 4-19** The random walk algorithm (a) trajectories of particles transport in the system and (b) stability analysis with simulation steps

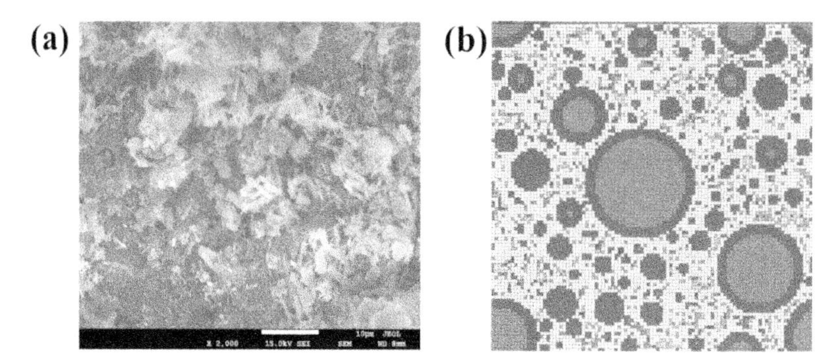


**Figure 4-20** The peridynamics model of cementitious system: (a) experimental reference and (b) numerical correspondence structure

the algorithm proposed by Bentz, Garboczi, and Snyder (1999) for solving the effective diffusion coefficient of the hard-core/soft-shell sphere combination can be used to calculate the effective transmission characteristics of the simulated mortar/concrete.

Additionally, the recently emerging method of peridynamics has received widespread attention. The first peridynamics modeling of concrete structures is proposed by Gerstle et al. (2007), as shown in Figure 4-20. This model is realized by adding pairs of near-field dynamic moments to simulate linear elastic materials with different Poisson's ratios, and is called the "micropolar peridynamics model." The micropolar peridynamics model is placed in a finite element environment to use implicit rather than explicit solving algorithms to effectively apply boundary conditions and efficient calculation solutions, which is suitable for quasi-static simulation of concrete structure damage and cracking. With this new model, a very simple tensile damage mechanism at the microstructure (near field dynamics) level is sufficient to explain the large number of microcracks (damage) and fracture mechanics observed in concrete structures, and the model is computationally efficient.

Although many modeling methods exist on the micro-scale of concrete, there is also a great challenge that researchers have to face: the parameter often comes from the experiment measurement as a top-down approach, which indicates the simulation is an extension of experimental observation. How to get the simulation from the atomic level, and make it a bottom-up approach may be more instructive, but it is also harder to achieve.

## 4.5 THE TRANSITION ZONE IN CONCRETE

In the presence of an aggregate, the structure of the hydrated cement paste in the vicinity of large aggregate particles is usually different from the structure of the bulk paste or mortar in the system. This forms an individual phase different from the aggregate and bulk hardened matrix in concrete. In fact, many aspects of concrete behavior under stress can be explained only when the cement paste–aggregate interface is treated as an individual phase (third phase) of the concrete structure. Thus, the third phase gives unique features to the concrete structure.

Two methods have been used to identify the existence of the third phase. The first method is micro-indentation, which is a method used to detect the hardness of a material. A typical curve of the variation of hardness along a line in the radial direction from the edge of an aggregate is shown in Figure 4-21. It can be seen from Figure 4-21 that an obvious valley appears adjacent to the aggregate, which demonstrates that a relative soft belt exists between the aggregate and the bulk matrix. The second method used to identify the third phase is direct observation of an SEM

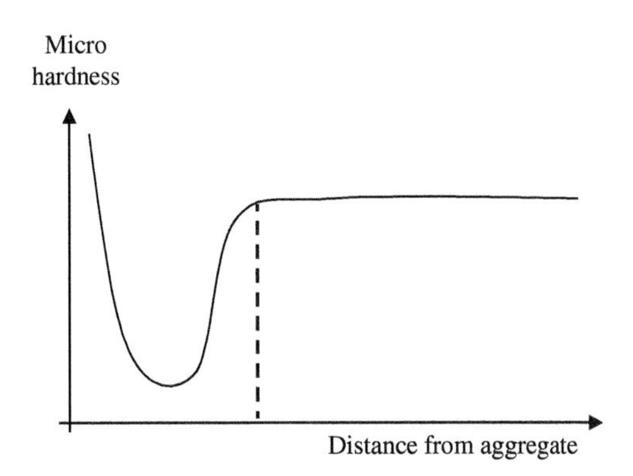


**Figure 4-21** Micro hardness distribution

photo, as shown in Figure 4-3. It can be clearly seen that more dark areas appear in the adjacent region to the aggregate. Since the SEM photo was taken in the backscatter mode, the brightness of the photo is related to the amount of the reflected electrons from the surface examined. The dark area means that less or no electrons are reflected and implies that they are pores on the surface.

The third phase is called the transition zone or interfacial zone in concrete technology, which represents the interfacial region between the coarse aggregate particles and the bulk hydrated cement paste. Existing as a thin shell, typically 10–50 μm thick around the large aggregate, the transition zone is generally weaker than either of the two other phases of concrete. The transition zone exercises a far greater influence on the mechanical behavior of concrete than is reflected by its volume fraction. It should be pointed out that each of the three phases is multiphase in nature. For instance, each aggregate particle may contain several materials, in addition to microcracks and voids. Similarly, both the bulk hydrated cement paste and the transition zone generally contain a heterogeneous distribution of different types and amounts of solid phases, pores, and microcracks, which are described below. Moreover, it should be noted that unlike other engineering materials, the structure of concrete does not remain stable (i.e., it is not an intrinsic characteristic of the material). This is because two components of the structure—the hydrated cement paste and the transition zone—are subject to change with time, environmental humidity, and temperature.

### 4.5.1 Significance of the Transition Zone

The volume fraction of the transition zone in concrete is usually only a few percent, but its influence on concrete properties is far more than such a percentage. It is a fact that many concrete macroscopic properties are sourced in the transition zone. The following examples can support this statement.

**(a)** As shown in Figure 4-22, when tested separately in uniaxial compression, under the same loading level, the aggregate and hardened cement paste (HCP) can remain elastic, whereas concrete itself shows inelastic behavior. The factor responsible for such behavior is the transition zone. Because the transition zone is porous and thus can deform more, its existence will generate higher deformation under the same load than both aggregate and HCP do.

**(b)** Table 4.2 shows the permeability coefficient of the aggregate, HCP, and concrete. It can easily be seen that the permeability of a concrete containing the same aggregate and HCP is the highest, and is an order of magnitude higher than the permeability of the corresponding components. The higher permeability coefficient of concrete can be attributed to its transition zone. Since the transition zone is more porous than the aggregate and bulk HCP, water can more easily flow through the transition zone, which results in a high permeability.

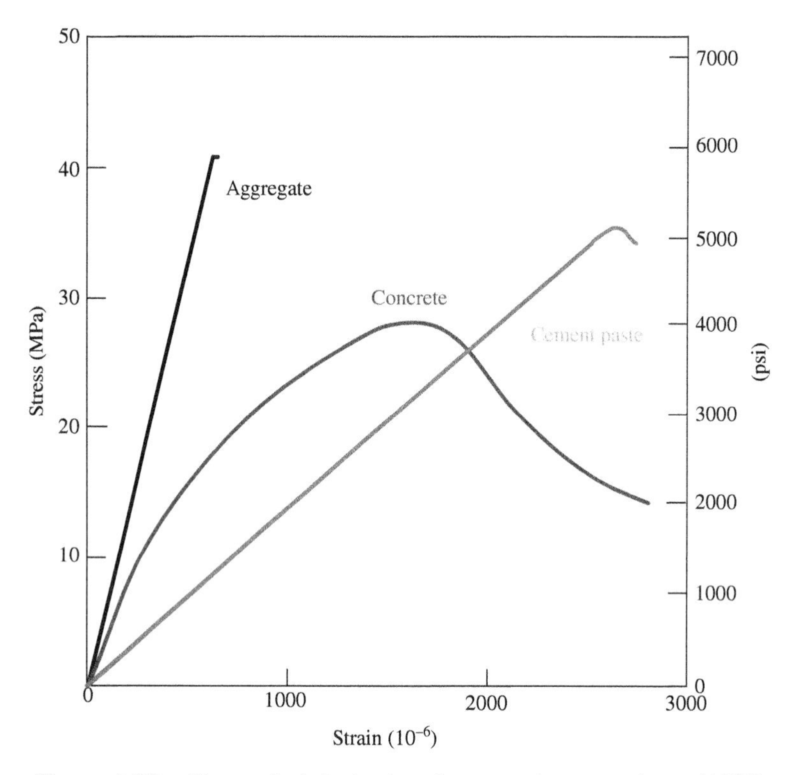


**Figure 4-22** Stress-strain behavior of aggregate, concrete and HCP

**Table 4-2** Permeability coefficients of different materials

| Type | Permeability coefficient |
|---|---|
| HCP | $6 \times 10^{-12}$ cm/sec |
| Aggregate | $1 \sim 10 \times 10^{-12}$ cm/sec |
| Concrete | $100 \sim 300 \times 10^{-12}$ cm/sec |

**(c)** It has been discovered that at a given cement content, water/cement ratio, and age of hydration, both HCP and cement mortar are always stronger than the corresponding concrete. Also, the strength of concrete decreases as the coarse aggregate size is increased. Again, the porous nature of the transition zone in concrete is responsible for the reduction of compressive strength as compared to the corresponding HCP and mortar, as shown in Figure 4-22.

In addition to the examples above, the existence of the transition zone in concrete can be used to explain why concrete is brittle in tension but relatively tough in compression. Why is the compressive strength of a concrete higher than its tensile strength by an order of magnitude?

It can be concluded that although the transition zone is composed of the same elements as the hydrated cement paste, the structure and properties of the transition zone are different from the bulk hydrated cement paste, and its influence on concrete properties is significant. Therefore, it should be treated as a separate phase of the concrete structure.

### 4.5.2 Structure of the Transition Zone

It is not easy to interpret the structure of the transition zone in concrete, even with advanced equipment. The structural characteristics of the transition zone in concrete described here are based on the work of Maso (1980), following the sequence of the development from the time the concrete is placed. In freshly compacted concrete, water films try to form around the large aggregate particles due to hydrophilic behavior. This leads to a higher water/cement ratio closer to the larger aggregates (i.e., in the bulk matrix). Due to the relatively large amount of water, there are fewer cement particles surrounding the aggregates. With the process of hydration, calcium and hydroxyl ions produced by the dissolution of calcium silicate crystallize when a critical saturation is reached and form calcium hydroxide ($Ca(OH)_2$) in the adjacent region of aggregate as in the bulk paste. Owing to the high solution content, the crystalline product of CH in the vicinity of the coarse aggregate can grow into relatively large crystals. Meanwhile, more water evaporation occurs, which forms a more porous framework than in the bulk cement paste or mortar matrix. The platelike calcium hydroxide crystals tend to form in oriented layers. Moreover, with the progress of hydration, poorly crystalline C–S–H and a second generation of smaller crystals of ettringite and calcium hydroxide start to fill the empty space that exists between the framework created by the large ettringite and the calcium hydroxide crystals. This helps to improve the density and hence the strength of the transition zone. A diagrammatic representation of the transition zone in concrete is shown in Figure 4-23.

### 4.5.3 Influence of the Transition Zone on the Properties of Concrete

The transition zone is generally considered the weakest link in the concrete chain. It has a strength-limiting effect in concrete. Because of the presence of the transition zone, concrete fails at a considerably lower stress level than either of the two main components, as demonstrated in

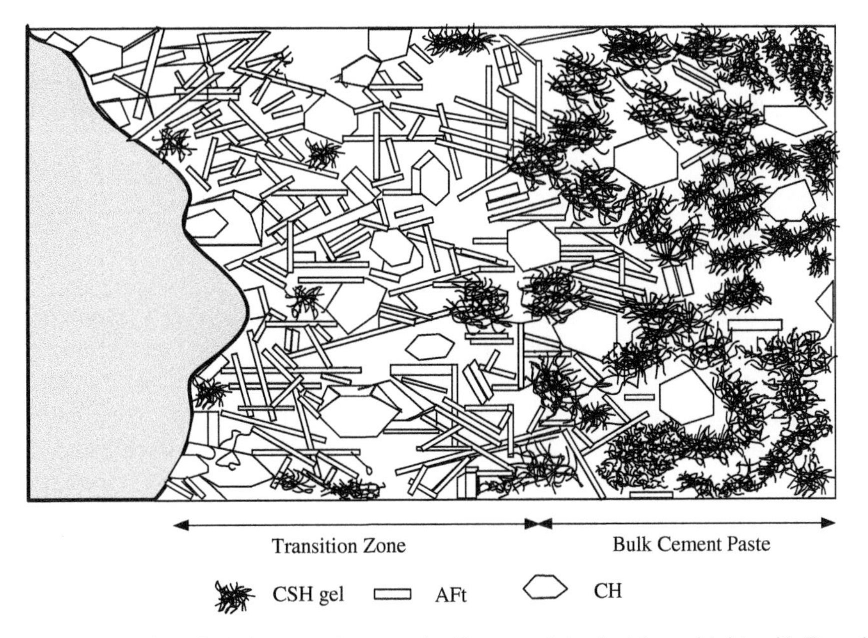


**Figure 4-23** Interfacial zone of concrete. Source: Adapted from Mehta, K. P. and Moteiro, P.J.M. Concrete: Microstructure, Properties, and Materials, Third edition, Mc Graw-Hill Companies Inc, 2006.

Figure 4-22. Since it does not require high energy levels to extend the cracks already existing in the transition zone, even at 40–70% of the ultimate strength, higher incremental strains are obtained per unit of applied stress. This explains the phenomenon that the components of concrete (i.e., aggregate and hydrated cement paste or mortar) usually remain linear elastic until fracture in a uniaxial compression test, whereas concrete itself shows inelastic behavior (Mehta and Monteiro, 2006).

At stress levels higher than about 70% of the ultimate strength, the stress concentrations at large voids in the mortar matrix become large enough to initiate strain localization there. With increasing stress, the matrix cracks gradually spread in the localization zone and join the cracks originating from the transition zone. The crack system then becomes continuous and the material ruptures. Because the crack direction is usually parallel to the compressive load, if the loading platen has little constraint on the concrete specimen, considerable energy is needed for the formation and extension of matrix cracks under a compressive load. On the other hand, under tensile loading, cracks propagate rapidly and at a much lower stress levels. This is why concrete fails in a brittle manner in tension but is relatively tough in compression, and is also the reason why the tensile strength is much lower than the compressive strength of a concrete.

The structure of the transition zone, especially the volume of voids and microcracks present, has a great influence on the stiffness or the elastic modulus of concrete. In a composite material, the transition zone serves as a bridge between the two components: the bulk matrix and the coarse aggregate particles. Even when the individual components are of high stiffness, the stiffness of the composite may be low because of the broken bridge (i.e., voids and microcracks in the transition zone) that hinder stress transfer, as well as larger deformation occurrences due to the porous nature of the interface.

The characteristics of the transition zone also influence the durability of concrete. Prestressed and reinforced concrete elements often fail due to corrosion of the embedded steel. The rate of steel corrosion is greatly influenced by the permeability of concrete. The existence of relatively large numbers of pores and microcracks in the transition zone at the interface with steel and coarse aggregate makes concrete more permeable than the corresponding hydrated cement paste or mortar. Subsequently, oxygen and moisture can penetrate into concrete more easily and lead to corrosion of the steel in the concrete.

## 4.6 NANO- AND MICRO-STRUCTURAL ENGINEERING

### 4.6.1 Overview

A number of aspects of the structure in the nanometer and micrometer scales of cement-based materials are linked to the built-in characteristics of Portland cement. Only limited variation in aspects of cements, such as particle size distribution, relative proportions of the different cement components, and the content and type of gypsum interground with the clinker, are commercially feasible. On the other hand, it is feasible and common to modify various characteristics of the final hydration product by incorporating chemical admixtures, pozzolans, or supplementary cementing components during the mixing process of concretes to improve their properties and performance. Some of these modifications act primarily as processing aids without appreciably changing the internal structure of the final product; most, however, result in definite changes, usually for the better. The changes in structure may be physical, such as an increase in density and reduction in porosity, or chemical, such as producing new reaction products and consumption of some non-useful hydration products. Such changes can be documented with proper advanced instrumentation on a microscopic scale. Initially, improved performance characteristics of the microstructure were "accidentally" achieved in the course of introducing admixtures or additives into concrete to modify its macroscopic properties. Later on, such actions were adopted to deliberately influence the structural development of concrete in the nanometer or micrometer scales to produce desired and measurable changes.

Developing nano- or micro-structural systems with defined and prerequisite properties for concrete is called *nano-micro-structural engineering*. The term "engineering" here, as elsewhere, first requires carefully designed actions with clear objectives in producing a better material. Second, it needs a proper measurement system to measure the changes in the structure as feedback to be used to deliberately improve previous actions; hence, stronger, less permeable, and more durable cement-based materials can be finally produced. In simple terms, nano- micro-structural engineering deals with how to modify the old materials structure at the nanometer or micrometer scale to produce a controlled, engineered microstructure with superior characteristics. It was Diamond and Bonen who proposed this wonderful concept in 1993.

Naturally, the first approach in nano-micro-structural engineering is to reduce the *w*/*c* ratio, as a lower *w*/*c* ratio can lead to a denser microstructure and subsequently to better mechanical behavior. However, the amount of water that can be reduced depends on many factors, such as aggregate characteristics, mixing techniques, cement properties, and the compacting method. As a result, only a limited amount of such reduction is feasible due to the requirement of workability. If water is reduced too much, without other measures, the mix will become too dry to be properly mixed, placed, and compacted. Thus, other measures have to be developed to achieve the objective. Usually these measures consist of the action of chemical and mineral admixtures.

### 4.6.2 Nanostructural Engineering

Nanostructural engineering is an engineered activity to improve the mechanical properties of concrete through modification of its materials structure at the nanometer scale on purpose. Such modification can be realized by introducing nanoparticles such as nano silicate, nano alumina oxide, nanocarbon tubes, graphene, and organic monomer to physically or chemically react with the hydration products, mainly C–S–H. The modified materials structures in the nanoscale are usually denser, more homogeneous, and multi-network connected. Such kind of materials structure can help concrete with its strength, stiffness, and crack resistance. Here we discuss a few examples of nanostructural engineering, including one on how to produce nano particles.

#### 4.6.2.1 Ion Diffusion Method for Fabrication of Nano Particles

At the Hong Kong University of Science and Technology, we developed a new method to produce nano particles, namely ion diffusion method (Liang. 2018). In this method, tricalcium silicate ($Ca_3SiO_5$) was used as an ion generation source. We dispersed tricalcium silicate particles in large amount of water. Upon contact with water, the calcium cations ($Ca^{2+}$) are released from the tricalcium silicate and form calcium hydroxide crystals in the aqueous solution. According to the crystal formation mechanism, temperature is critical in controlling the size of the formed calcium hydroxide. By controlling the temperature to stay at 0°C for the experimental environment, the diffused calcium cations from the tricalcium silicate could be quite uniformly distributed in the surroungding water and the calcium hydroxide spherulites formed of quite a small size, about 5 nm, as can be seen in Figure 4-24, which shows a transmission electron microscopy (TEM) photograph taken on the surroundings of a hydrolyzed tricalcium silicate particle. The left side of Figure 4-24 corresponds to a large block of tricalcium silicate, while the right side of Figure 4-24 show the tiny spherulites with diameters around 5 nm. Energy dispersive spectroscopy analysis confirms that these spherulites contain only Ca and O elements, confirming that they are crystalline calcium hydroxide.

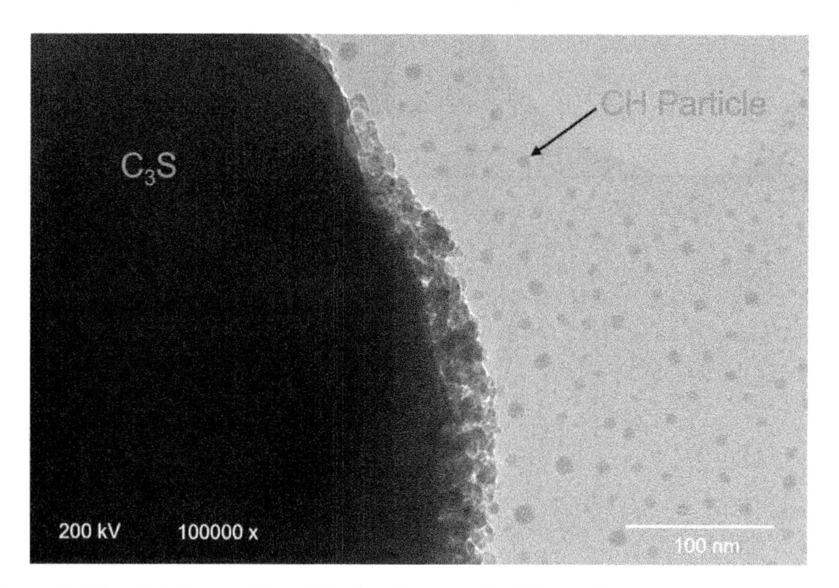


**Figure 4-24** Calcium cation diffusion from a tricalcium silicate generated calcium hydroxide nano particle

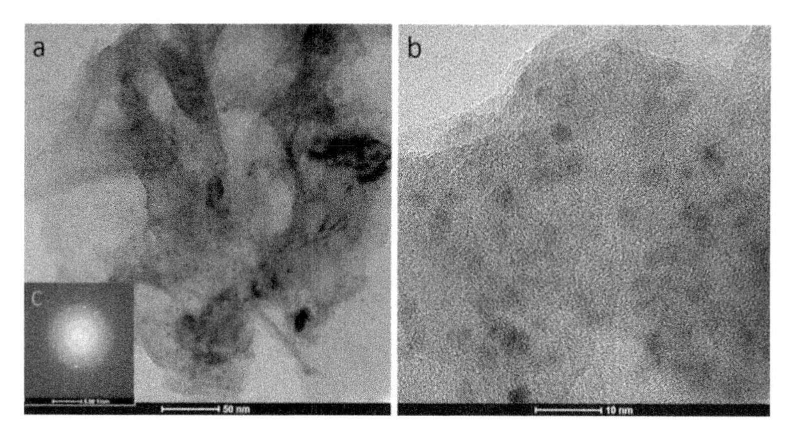


**Figure 4-25** (a) TEM image of magnesium hydroxide nano-spherulites obtained from $MgO_2$, the scale bar is 50nm; (b) zoomed TEM image of (a), shows the diameters of nano particle are ranged from 2 to 4 nm; the scale bar is 10 nm

A similar method was used for magnesium oxide as the precursor. Disperse the precursor with suitable reactivity in 0°C water and maintain for 3 days. The well-dispersed magenesium hydroxide nanoparticles, size ranged from 2 to 4 nm, were obtained in aqueous solution, as shown in Figure 4-25.

#### 4.6.2.2 Development of High Modulus Concrete with Nano Silica

For high-rise buildings the stiffness of the structure is much more critical than its strength. There are two ways to increase the stiffness of a concrete structural member: enhancing the modulus of concrete and increase the cross-section or size of a structural member. Of the two, enhancing the modulus of the concrete is a much more efficient approach. When designing concrete with high modulus, nanostructural engineering provides technical support. Nano-silica is a kind of nanoparticle that has a mean particle size around 12 nm, which is selected to make high modulus concrete. To achieve a uniform dispersion of nano silica, the ultrasonic mixing technique was adopted in concrete mixing. By incorporating only 1% nano-silica, the modulus of concrete can be increased dramatically. As shown in Figure 4-26, the modulus of a concrete with 1% nanomaterials (Fan 2017) reached 53.5 GPa, which is 20% higher than normal concrete in the design code of practice. The stiffness enhancement can be mainly attributed to the secondary C–S–H gel produced by the nano-silica, and the dense packing effect of fine particles. As shown in Figure 4-27, a better and denser interfacial transition zone can be observed in the concrete with the incorporation of nano-silica.

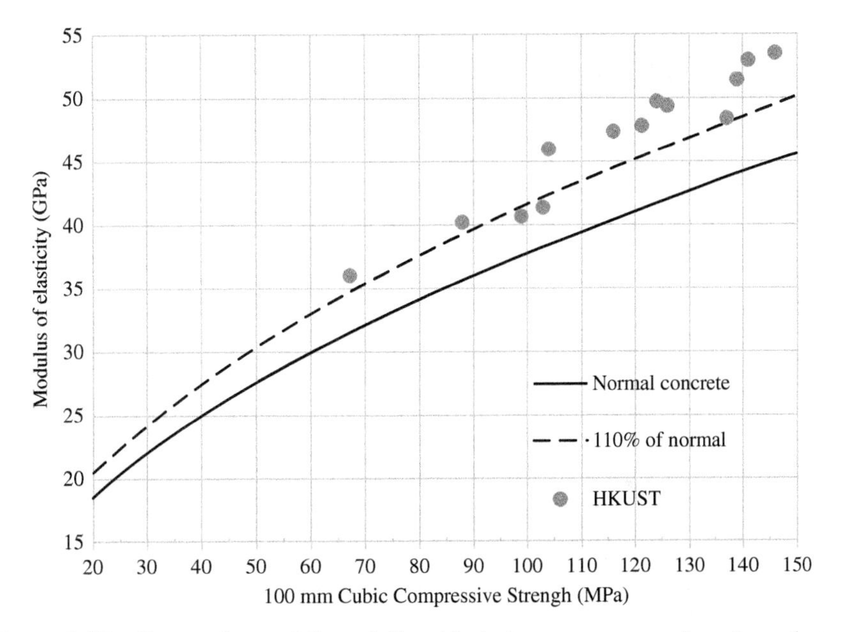


**Figure 4-26** Comparison of the relationship between compressive strength and modulus of elasticity (code of practice and results from Fan's work in HKUST)

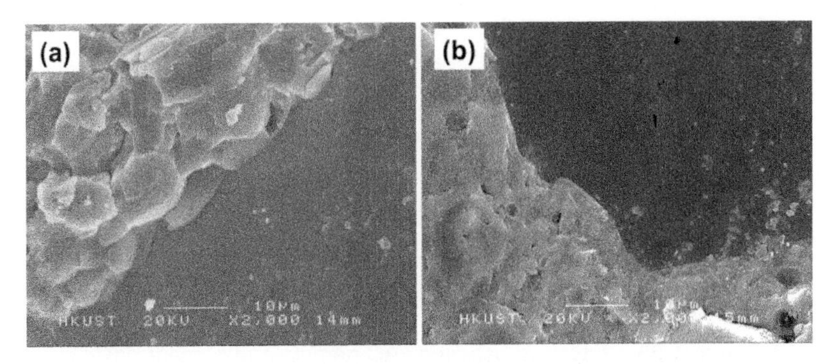


**Figure 4-27** The interfacial transition zone in concrete (a) without nano $SiO_2$ and (b) with nano $SiO_2$

### 4.6.2.3 Development of High-Performance Lightweight Concrete

Another example of nanostructural engineering is the development of high-performance lightweight concrete (LWC) (Hanif 2017). To produce LWC, a lightweight aggregate such as fly ash cenospheres (FAC) is needed. FAC has a structure with a thin wall and a hollow spherical space. The reactivity of the FACs in the cement composites is poor due to their glassy surface, particle size, and chemical composition that lead to a low strength development. To overcome this problem, nano-silica is used to activate the inert FAC particles and enhance the pozzolanic activity of FAC. It can be found in Figure 4-28 that the FAC spherical shells are consumed by hydration or pozzolanic reaction, and the reaction products grow in the interior region of porous FAC. Also, the nano-silica was very efficient in improving the uniformity, compactness, and denseness of the interface between the FAC particle and the surrounding matrix. In Figure 4-29, the dense C–S–H gel packing is found at the nano-silica modified interface, whereas in the matrix without nano-silica, plate-like crystals of CH and needle-shaped ettringite crystals are more pronounced. In the paste with nanomaterials, the gap between FAC and the matrix is bridged by hydration gel, and the particles are tightly held in the paste. The structural enhancement contributes to the early age mechanical performance of LWC. As shown in Figure 4-30, with only 1% nano-silica addition, the 7-day compressive strength of LWC with 10% FAC increased 17% as compared to the specimen without nano-silica. For a mix with 30% FAC and 1% nano silica, the 7-day compressive strength reached 40 MPa with a density of 1500 kg/m$^3$, which is comparable to the strength of normal concrete without FAC (Hanif 2017).

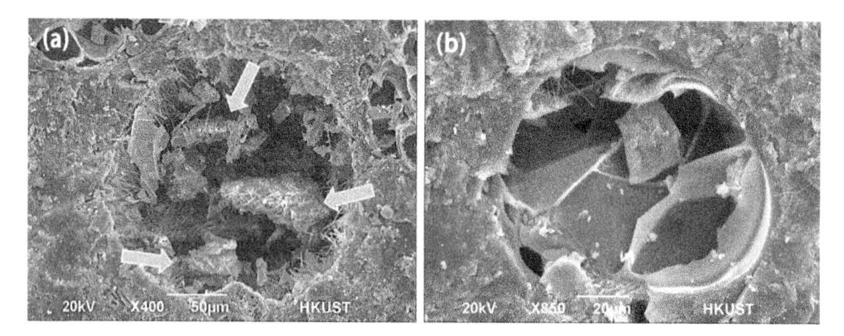


**Figure 4-28** FAC particle behavior in pastes (a) containing NS, and (b) without NS (of 120-day age)

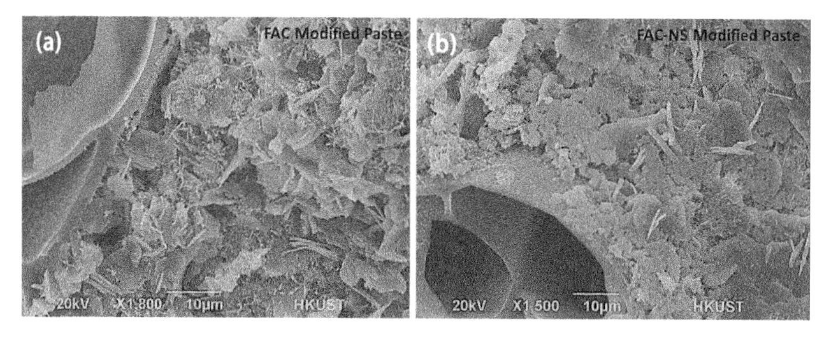


**Figure 4-29** Interface between FAC particle and cement paste (microstructure) at 3-day age; (a) containing NS, and (b) without NS

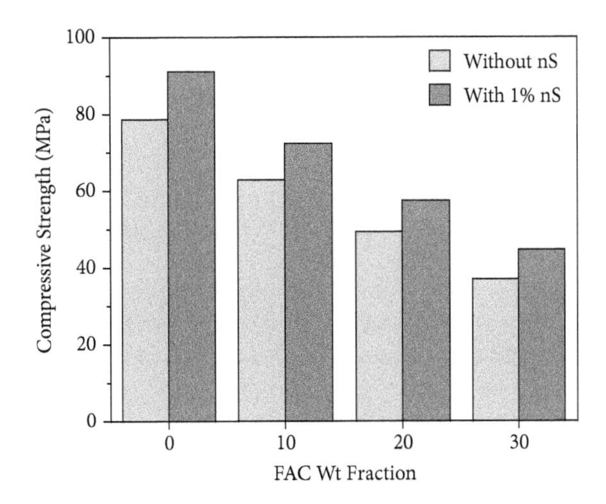


**Figure 4-30** Compressive strength of hardened specimen cubes at various FAC fractions

### 4.6.3 Superplasticizer and Dispersion in Cement Systems

Adding water to normal cement produces a flocculated system. To see the effect of individual flocs, a plain cement paste can be produced at a normal *w/c* ratio, and then diluted with additional water by a factor of ten or above. This produces a suspension in which individual flocs can be seen as they settle rapidly in the graduated cylinder provided. The bed deposited on the bottom of the cylinder will be open-textured, porous, and not visibly size-sorted.

If a sufficient dose of a commercial superplasticizer is added to another cement paste, otherwise identical with the first one, after dilution, it will be seen that no flocs are evident, and the cement grains settle individually. Since the cement grains are much smaller than the flocs, the sedimentation will take much longer. Examination of the bed deposited at the bottom of the cylinder will show dense packing and size stratification, with the largest cement grains on the bottom (Diamond, 1993). Figure 4-31 shows the effect of the superplasticizer in separating the cement particles. Efforts to delineate the boundaries of individual flocs in fresh cement paste, without diluting them, have been universally unsuccessful. The fresh paste at normal *w/c* ratios has to be treated as a single large, flocculated mass.

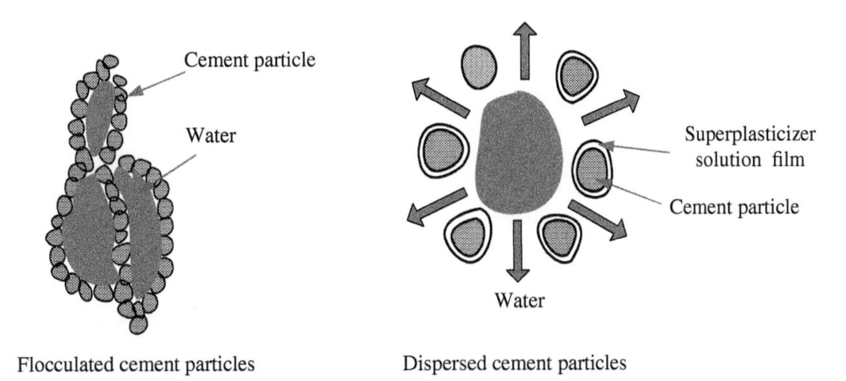


**Figure 4-31** Flocculated cement and dispersed cement

The physical chemistry behind the flocculated condition has been well established. Cement grains develop only modest diffuse double layers in the mix water. The mix water in conventional concretes, in fact, rapidly becomes a high ionic strength solution due to thedissolution of alkali, sulfate, and free lime from the clinker, and of gypsum. In this high ionic strength solution, attractive van der Waals forces vastly exceed any double-layer repulsive forces that may be developed, and the cement grains attract each other. At points of contact they tend to stick together. A significant shearing force exceeding the yield strength must be applied to get the mass to flow, either in a rheological measurement or in practice.

Superplasticizers work primarily by separating cement particles either through imparting a strong negative charge to the surface of the cement particles or in adsorbing the surfaces of cement particles. At a sufficient superplasticizer dosage, the electrostatic repulsive forces between particles due to negative charge overwhelm the attractive forces, and the cement grains try to separate from each other to the extent possible in the dense suspension. Not much separation is possible at practical *w/c* ratios, but the particles slip past each other easily and do not tend to stick to each other at points of contact. As a result, fully dispersed cement does not show yield strength, but acts as a Newtonian fluid when examined by the rheological measurement techniques.

With normal dosages of superplasticizer, concrete with same workability can be mixed and consolidated at lower *w/c* ratio. Thus, the strength of the concrete is improved due to the reduction of the amount of water used, as compared to the concrete mixed using a normal amount of water without superplasticizer. However, for concretes made of the same *w/c* with or without the superplasticizer, the structure of concrete in the nanometer scale and the micrometer scale are the same and so is their strength. The only difference is the workability. In this case, further measures have to be taken to get denser concrete.

### 4.6.4 Silica Fume and Particle Packing

Cement size varies from 2–80 μm, with a mean value of 18–20 μm. It appears that cement particles cannot form a dense packing because the water-filled pockets are roughly the same size as the cement particles that exist throughout the mass, as shown in Figure 4-32. It is obvious that an admixture of much finer particles is needed to pack into the water-filled pockets among the cement grains. Silica fume (or *microsilica*) provides such particles. The mean particle size of commercial silica fume is typically less than 0.1 μm, two magnitudes smaller than cement particles. A denser packing may be ensured when microsilica is added to ordinary cement paste. However, this happy state of affairs does not usually occur if microsilica is simply added to cement. Early experiments of adding microsilica to concrete in Norway led only to increases in water demand, and little or no improvement in concrete properties. To get the dense particle packing, not only must the fine particles be present, but the system must be effectively deflocculated during the mixing process. Only then can the cement particles move around to incorporate the fine microsilica particles. The fine microsilica particles must themselves be properly dispersed so that they can separate from each other and pack individually between and around the cement grains.

Another requirement for best packing is that the mixing should be more effective than the relatively casual mixing done in ordinary concrete production. High-shear mixers of several kinds have been explored. Proper dispersion and incorporation of fine microsilica particles can thus result in a dense local structure of fresh paste with little water-filled space between grains. When the cement hydrates, the overall structure produced in the groundmass is denser, tighter, and stronger. It is also much more difficult to picture in SEM, because it appears massive and lacks distinctive features. Microsilica is an active and effective pozzolan; that is, it reacts readily with the calcium hydroxide produced by the cement around it to generate additional calcium silica hydrate. Also it somehow stimulates the cement to hydrate faster, and its presence modifies the overall chemistry

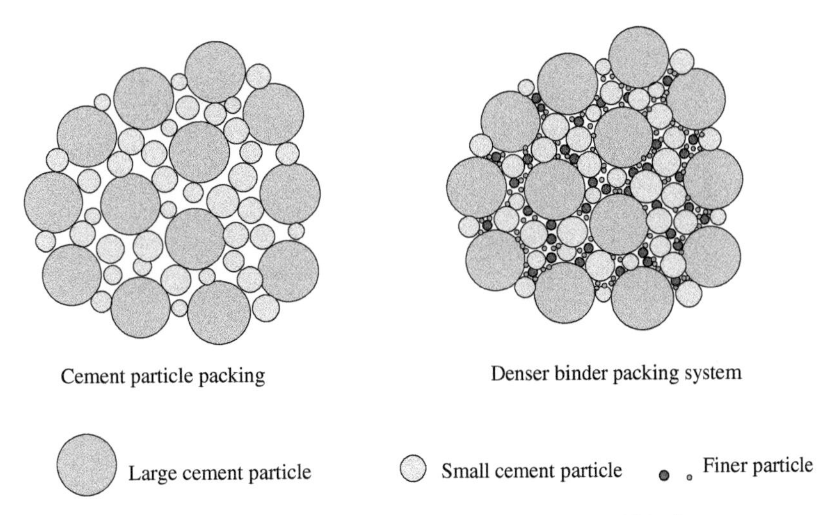


**Figure 4-32** Packing systems of cement and binder

of the "regular" calcium silicate hydrate produced by the cement. Similar benefits can be obtained if carbon black particles of roughly the same fine size as silica fume and fly ash are introduced into a properly dispersed and mixed concrete. Substituting metakaolin for all or part of the silica fume can produce somewhat the same effects in a heavily superplasticized concrete.

It should be pointed out that if only using silica fume, optimized packing may not be achieved. To reach such a goal, different grades of powders have to be adopted, just like aggregates. Along this line, the mixed use of several mineral admixtures is a good solution. Together with metalkaolin with an average particle size of 2 μm and GGBS of 8 μm, adding silica fume into a cement system can lead to an optimal size distribution and denser packing, as shown in Figure 4-32.

### 4.6.5 Transition Zone Improvement

As the transition zone has been identified as a weak link in concrete, it has become customary to ascribe some poor performance characteristics of traditional concrete to the transition zone. From this point of view, microstructural engineering should pay attention to improving the poor structure of the transition zone to enhance the performance of concrete as a whole.

The methodology presented earlier using silica fume to modify the structure of concrete is also effective in improving the structure of the transition zone due to its packing and pozzolanic reaction effect. Bentur and Cohen (1987) prepared two sets of mortar, one was plain mortar and the other silica fume-bearing mortar. Their observation on SEM scans of the interface between the sand grains and the cement paste demonstrated that at least the upper portion of the interface zone of the plain mortar was much more porous than that of the silica fume-modified mortar and obvious CH crystals were visible. In contrast, the silica fume-modified upper interfacial zone was dense and homogeneous, and showed no visible porosity even at a magnification of 200×. Mitsui et al. (1994) have studied the characteristics of the transition zone for different formulations of cement paste and coating methods on coarse aggregate surfaces. Their results also demonstrated that the incorporation of silica fume into cement paste could significantly improve the characteristics of the transition zone and lead to a dense and homogeneous structure. Moreover, together with the surface coating of a slurry of silica fume and cement, the silica fume-modified paste could almost eliminate the transition zone, as shown in Figure 4-33.

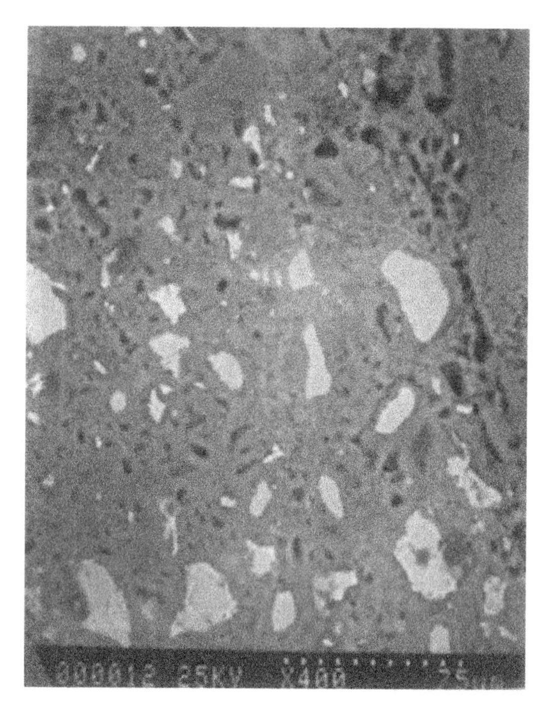


**Figure 4-33** Transition zone of high strength concrete

The image analysis on the SEM photos of the transition zone of different pastes showed that there were clear distinctions between normal paste and silica fume-modified paste in the porosity distribution along the line from the surface of the aggregate, as shown in Figure 4-34. In addition, since the structure of the transition zone has a significant influence on the bond between the aggregate and paste, different transition zone structures have different bonding properties. Figure 4-35 shows the push-out test results for the two specimens mentioned above.

### 4.6.6 Effects of Polymers on Microstructural Engineering

It is possible to significantly change the characteristics of hydrated cement systems by incorporating some polymers into the concrete. Mostly, the effect of the incorporated polymer is physical in nature. In latex-modified concrete, the latex forms a film network among the cement hydration products that helps to improve the bonding properties and to reduce permeability. Moreover, with a combination of effects involving the addition of substantial amounts of soluble polymers, significant reduction of water content, and with special mixing and processing techniques, a special cement system called MDF (macro-defect-free) has been developed. The thin sheet products manufactured by MDF show enormously improved strength levels and enormously modified microstructures. The MDF processing approach works much better with calcium aluminate cement than with Portland cement. In MDF formulation there is only a very small amount of hydration due to very small *w/c* ratio. To some extent, the system behaves as a filled polymer rather than a cement, but there are significant chemical interactions between the hydrated calcium aluminate products and the polymers. This chemical interaction, combined with the extremely dense microstructure induced by the physical processing, produces a virtually void-free and macro defect-free product that can develop enormous compressive and tensile strengths.

Recently, efforts have been made to develop brand new cement systems by inducing chemical reactions into polymer-modified cement systems. This method is to introduce monomer,

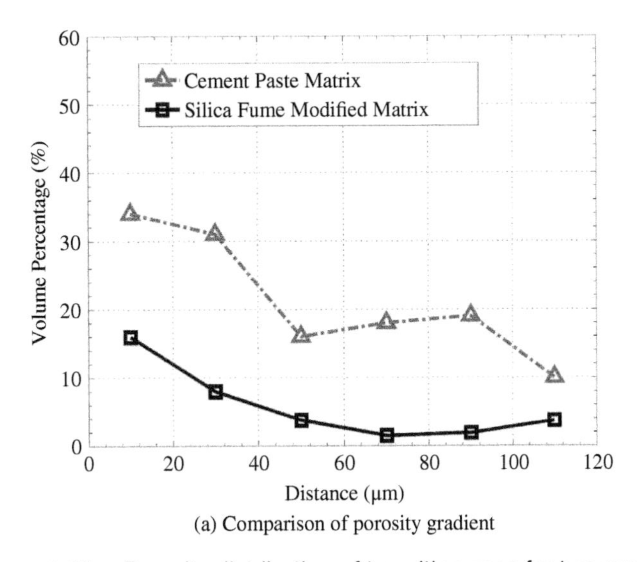

(a) Comparison of porosity gradient

**Figure 4-34** Porosity distribution of transition zone for two pastes

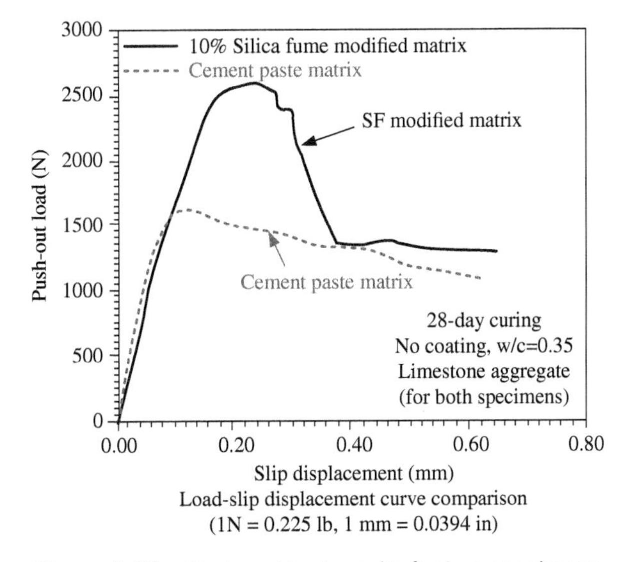

Load-slip displacement curve comparison
(1N = 0.225 lb, 1 mm = 0.0394 in)

**Figure 4-35** Push-out test results for two specimens

initiator, and accelerator into concrete mix. The polymerization of the monomer accompanies the hydration of the cement, forming the second network in concrete additional to the hydration products. One end of the on-site polymerized organic materials can chemically bond with the hydrates of Portland cement, and the other end can form the second network. The monomers that can be used for on-site polymerization include acrylamide, acrylic acid, vinyl acetate, or 2-acrylanmido-2-methylpropanesulfonic acid (AMPS) (Liu 2020) precursors. The parameters, such as raw material amount, ratio, synthesis procedures, synthesis temperature, and additives, are under study for optimization. Figure 4-36 shows a unit that can be used to improve the flexural strength of concrete designed according to the principles explained earlier. Figure 4-37 shows the proof of the chemical bond formed between the polymerized organic materials and the calcium in hydration products of concrete.

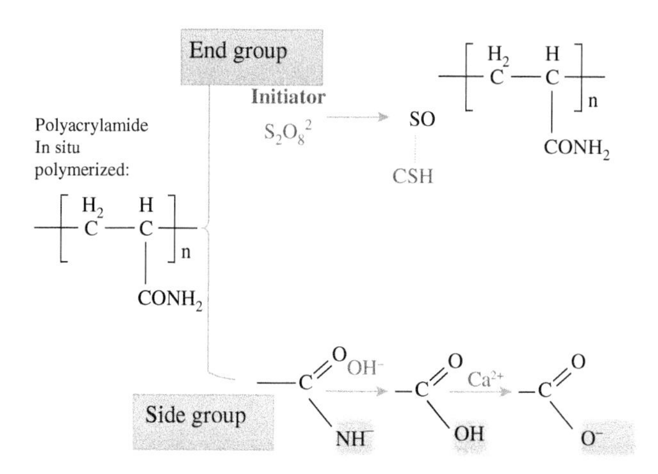


**Figure 4-36** A unit for improving the flexural strength of concrete

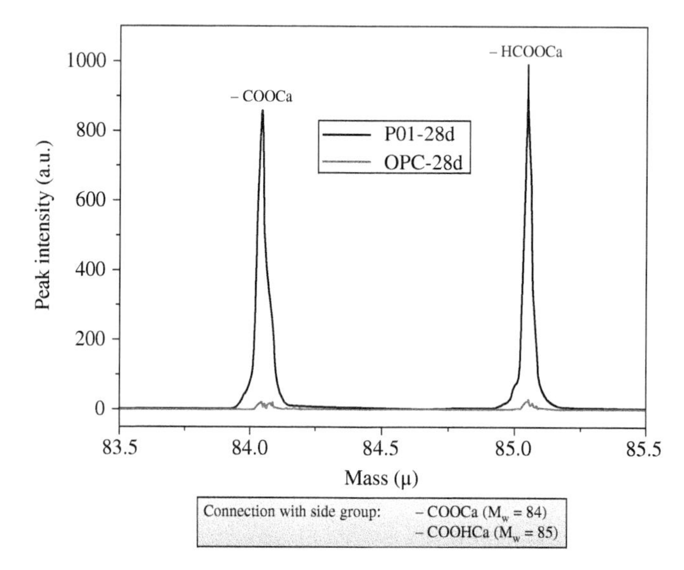


**Figure 4-37** Proof of the chemical bond between in-situ polymerized organic materials with hydration products of concrete by ToF-sims

## DISCUSSION TOPICS

How are macrostructures and microstructures in concrete distinguished traditionally?
How is the interface formed?
What are the important characteristics of the interface?
How can the interface be improved?
What is microstructural engineering?
Can you provide a few methods to improve the interface?
Why must you learn the structure of concrete?

How can you improve concrete compressive strength from the point of view of the microstructure?

Can you eliminate the porous interface in concrete?

How is the microstructure of concrete influenced by a superplasticizer?

Why does modern concrete technology stress the importance of the interface?

Can you list a few common C–S–H structure models?

What do you need to consider if you would like to build a C–S–H model?

Why is the C–S–H structure important?

## REFERENCES


Allen, A.J., Thomas, J.J., and Jennings, H.M. (2007) "Composition and density of nanoscale calcium-silicate-hydrate in cement," *Nature Materials*, 6, 311–316.

Barrett, E.P., Joyner, L.G., and Halenda, P.P. (1951) "The determination of pore volume and area distributions in porous substances. I. Computations from nitrogen isothermsm" *Journal of the American Chemical Society*, 73(1), 373–380.

Bentur, A. and Cohen, M.D. (1987) "Effect of condensed silica fume on the microstructure of the interfacial zone in Portland cement mortars," *Journal of the American Ceramic Society*, 70(10), 738–743.

Bentz, D.P., Garboczi, E.J., and Snyder, K.A. (1999) "A hard core/soft shell microstructural model for studying percolation and transport in three-dimensional composite media." Washington, DC: US Department of Commerce, Technology Administration, National Institute of Standards and Technology.

Bohris, A.J., Goerke, U., McDonald, P.J., Mulheron, M., Newling, B., and Le Page, B. (1998) "A broad line NMR and MRI study of water and water transport in portland cement pastes," *Magnetic Resonance Imaging*, 16(5–6), 455–461.

Bonaccorisi, B. and Merlino, S. (2005) "The crystal structure of tobermorite 14 A (plombierite), a C–S–H phase," *Journal of American Ceramic Society*, 88(3), 505–512.

Bonaccorsi, E., Merlino, S. and Taylor, H.F.W. (2004) "The crystal structure of Jennite, $Ca_9Si_6O_{18}(OH)_6 \cdot 8H_2O$," *Cement and Concrete Research*, 34, 1481–1488.

Brunauer, S., Emmett, P.H., and Teller, E. (1938) "Adsorption of gases in multimolecular layers," *Journal of the American Chemical Society*, 60(2), 309–319.

Colombet, P., Grimmer, A.R., .Zanni, H., and Sozzani, P. (1998) *Nuclear magnetic resonance spectroscopy of cement-based materials*, Berlin: Springer Verlag.

Cygan, R.T., Liang, J.J., and Kalinichev, A.G. (2004) "Molecular models of hydroxide, oxyhydroxide, and clay phases and the development of a general force field," *The Journal of Physical Chemistry B*, 108(4), 1255–1266.

Diamond, S. (1993) "Teaching the materials science, engineering, and field aspects of concrete," in: M.D. Cohen (Ed.), *Microstructure and microstructural engineering*, Evanston, IL: NSF, pp. 46–73.

Diamond, S. and Bonen, D. (1993) "Microstructure of hardened cement paste: a new interpretation," *Journal of American Ceramic Society*, 76(7), 2993–2999.

Fan, T. (2017) "Development of high modulus concrete for tall buildings," doctoral dissertation.

Gerstle, W., Sau, N., and Silling, S. (2007) "Peridynamic modeling of concrete structures," *Nuclear Engineering and Design*, 237(12–13), 1250–1258.

Hanif, A. (2017) "Development and application of high performance lightweight cementitious composite for wind energy harvesting," doctoral dissertation.

Hou, D., and Li, Z. (2014) "Molecular dynamics study of water and ions transport in nano-pore of layered structure: A case study of tobermorite," *Microporous and Mesoporous Materials*, 195, 9–20.

Hou, D., Zhang, W., Sun, M., Wang, P., Wang, M., Zhang, J., and Li, Z. (2020) "Modified Lucas-Washburn function of capillary transport in the calcium silicate hydrate gel pore: A coarse-grained molecular dynamics study," *Cement and Concrete Research*, 136, 106166.

Hou, D., Zhao, T., Ma, H., and Li, Z. (2015) "Reactive molecular simulation on water confined in the nanopores of the calcium silicate hydrate gel: structure, reactivity, and mechanical properties," *The Journal of Physical Chemistry C*, 119(3), 1346–1358.

Hou, D., Zhu, Y., Lu, Y., and Li, Z. (2014) "Mechanical properties of calcium silicate hydrate (C–S–H) at nano-scale: a molecular dynamics study," *Materials Chemistry and Physics*, 146(3), 503–511.

Ioannidou, K., Krakowiak, K.J., Bauchy, M., Hoover, C.G., Masoero, E., Yip, S., Ulm, F.J., Levitz, P., Pellenq, R.J.M. and Del Gado, E. (2016) "Mesoscale texture of cement hydrates," *Proceedings of the National Academy of Sciences*, 113(8), 2029–2034.

Izadifar, M., Königer, F., Gerdes, A., Wöll, C., and Thissen, P. (2019) "Correlation between composition and mechanical properties of calcium silicate hydrates identified by infrared spectroscopy and density functional theory," *The Journal of Physical Chemistry C*, 123(17), 10868–10873.

Liang, R. (2018) "Mechanism and performance study of cement-organic integrated materials," PhD thesis, The Hong Kong University of Science and Technology.

Liu, Q. (2020) "Flexural strength and durability enhancements for cement-based materials by in situ polymerization of monomers," PhD thesis, University of Macao.

Manzano, H., Dolado, J.S., Guerrero, A., and Ayuela, A. (2007) "Mechanical properties of crystalline calcium-silicate-hydrates: Comparison with cementitious C-S-H gels," *Physica Status Solidi (A)*, 204(6), 1775–1780.

Maso, J.C. (1980) "The bond between aggregates and hydrated cement paste," in:*Proceedings of the 7th International Congress on the Chemistry of Cement*, Paris: Editions Septima, I, 3–15.

McDonald, P.J., Korb, J.P., Mitchell, J., and Monteilhet, L. (2005) "Surface relaxation and chemical exchange in hydrating cement pastes: A two-dimensional NMR relaxation study," *Physical Review E*, 72(1), 011409.

Mehta, K.P. and Monteiro, P.J.M. (2006) *Concrete: Microstructure, properties, and materials* (3rd edition). New York: McGraw-Hill.

Mishra, R.K., Mohamed, A.K., Geissbühler, D., Manzano, H., Jamil, T., Shahsavari, R., Kalinichev, A.G., Galmarini, S., Tao, L., Heinz, H. and Pellenq, R. (2017) "CEMFF: A force field database for cementitious materials including validations, applications and opportunities," *Cement and Concrete Research*, 102, 68–89.

Mitsui, K., Li, Z., Lang, D. and Shah, S.P. (1994) "Relationship between microstructure and mechanical properties of the paste-aggregate interface," *ACI Materials Journal*, 91(1), 30–39.

Pellenq, R.M., Kushima, A., Shahsavari, R., Van Vliet, K.J., Buehler, M.J., Yip, S., and Ulm, F.J. (2009) "A realistic molecular model of cement hydrates," *Proceedings of the National Academy of Sciences*, 106(38), 16102–16107.

Pellenq, R.M., Lequeux, N., and Van Damme, H. (2008) "Engineering the bonding scheme in C–S–H: The iono-covalent framework," *Cement and Concrete Research*, 38(2), 159–174.

Rejmak, P., Dolado, J.S., Stott, M.J., and Ayuela, A. (2012) "29Si NMR in cement: A theoretical study on calcium silicate hydrates," *The Journal of Physical Chemistry C*, 116(17), 9755–9761.

Richardson, G. and Groves, G.W. (1992) "Models for the composition and structure of calcium silicate hydrate (C–S–H) gel in hardened tricalcium silicate paste," *Cement and Concrete Research*, 22, 1001–1010.

Richardson, I.G. (2004) "Tobermorite/jennite- and tobermorite/calcium hydroxide-based models for the structure of C–S–H: applicability to hardened pastes of tricalcium silicate, $\beta$- dicalcium silicate, Portland cement, and blends of Portland cement with blast-furnace slag, metakaolin, or silica fume,"*Cement and Concrete Research*, 34, 1733–1777.

Richardson, I.G. (2008) "The calcium silicate hydrates," *Cement and Concrete Research*, 38, 137–158.

Shahsavari, R., Pellenq, R.J.M., and Ulm, F.J. (2011) "Empirical force fields for complex hydrated calcio-silicate layered materials," *Physical Chemistry Chemical Physics*, 13(3), 1002–1011.

Stauffer, D., and Aharony, A. (1992) *Introduction to percolation theory*, New York: Taylor & Francis.

Taylor, H.F.W. (1997) *Cement Chemistry* (2nd edition). London: Thomas Telford Ltd, pp. 142–150.

Taylor, H.F.W. and Howison, J.W. (1956) "Relationship between calcium silicates and clay minerals," *Clay Mineral Bulletin*, 31, 98–111.

Thomas, J.J., FitzGerald, S.A., Neumann, D.A., and Livingston, R.A. (2001) "State of water in hydrating tricalcium silicate and portland cement pastes as measured by quasi-elastic neutron scattering," *Journal of the American Ceramic Society*, 84(8), 1811–1816.

Viehland, D., Li, J.F., Yuan, L.J. and Xu, Z. (1996) "Mesostructure of calcium silicate hydrate (C–S–H) gels in Portland cement paste: Short-range ordering, nanocrystallinity, and local compositional order," *Journal of the American Ceramic Society*, 79(7), 1731–1744.

Wittmann, F H. (1983) "Structure of concrete with respect to crack formation," *Fracture Mechanics of Concrete*, 43(5), 6.

CHAPTER 5

# PROPERTIES OF HARDENED CONCRETE

With the development of hydration, concrete will change from a fluid to a plastic state, and eventually to a solid hardened state. In the hardened state, concrete is ready to support external loads as a structural material. The most important properties of hardened concrete include various strengths, dimension stability, complete stress–strain relationship, various moduli and Poisson's ratio, and durability.

## 5.1 STRENGTHS OF HARDENED CONCRETE

### 5.1.1 Introduction

#### 5.1.1.1 Definitions

To understand the concept of strength, it is necessary to understand what stress and strain mean. Nominal stress is defined as the load divided by the original cross-section area. This stress definition can be expressed as

$$[\sigma] = \frac{F}{A} \tag{5-1}$$

where $F$ is the load (in N) and $A$ is the cross-section area (in $m^2$). The dimension used for stress in the SI system is Pa, with the following definition:

$$\text{Pa} = \frac{\text{N}}{\text{m}^2}, \quad \text{MPa} = 10^6 \frac{\text{N}}{\text{m}^2} = \frac{\text{N}}{\text{mm}^2} \tag{5-2}$$

Conventional strain is defined as the change in length per unit original length, and can be expressed as

$$\varepsilon = \frac{\Delta L}{L_0} \tag{5-3}$$

where $\Delta L$ is the change in length; and $L_0$ is the original length.

Strength is defined as the ability of a material to resist the stress generated by an external force without failure. For concrete, failure is frequently identified with the appearance of cracks. Since the development of a crack is closely related to the development of deformation, in fact, the real criterion of failure for concrete is the limiting strain rather than the limiting stress. The limiting strain for a concrete is different for different loading conditions and different strength levels. For instance, the limiting strain for concrete under uniaxial tension is $100 \times 10^{-6}$ to $200 \times 10^{-6}$, while for uniaxial compression it is $4 \times 10^{-3}$ for a concrete with a strength of 14 MPa, and $2 \times 10^{-3}$ for a concrete with a strength of 70 MPa. For various strengths of concrete, compressive strength is the property generally specified in construction design and quality control. The reasons are as follows: (1) it is relatively easy to measure; and (2) it is believed that other properties can be related to the compressive strength and, thus, can be deduced empirically from compressive strength data.

The 28-day compressive strength of concrete, determined by a standard uniaxial compression test, is accepted universally as a general concrete property index for structural design. To measure different strengths of concrete, various tests have to be conducted with a universal testing machine. As shown in Figure 5-1, a universal testing machine usually consists of frames, loading fixture, actuator, loading cell, and control unit. The loading process can be controlled either by load or deformation. All the tests have to be carried out using proper control methods specified in Section 5.1.1.2.

#### 5.1.1.2 Control Methods for Strength Test

In general, there are two kinds of control methods for strength testing: open-loop or closed-loop control, as shown in Figure 5-2. In open-loop control (OLC), once the prescribed process is set by the controller with the input variables, the process will proceed. The output of the system is not fed back to the controller, and the process depends only on the system input. Hence, there is no chance to correct the process if it deviates from the predetermined directions. Examples of open-loop control include the action of a person throwing a stone to kill a bird or a washing machine washing clothes. In closed-loop control (CLC), the output of the controlled variable is directly monitored by the controller. The current value of the control variable is fed back to the controller and compared with the reference-input signal. The difference between the two signals (i.e., the error) is used to manipulate the control process. Examples of closed-loop control include cruise control of a vehicle and firing missiles. In a strength test of a concrete, a universal testing machine is usually used. The variables that can be controlled in such a system are usually the actuator displacement (or stroke) and the applied load (load cell), which are not significantly affected by the behavior of the test specimen, or the deformation of the concrete, which is significantly affected by the behavior of

**Figure 5-1** The universal testing machine

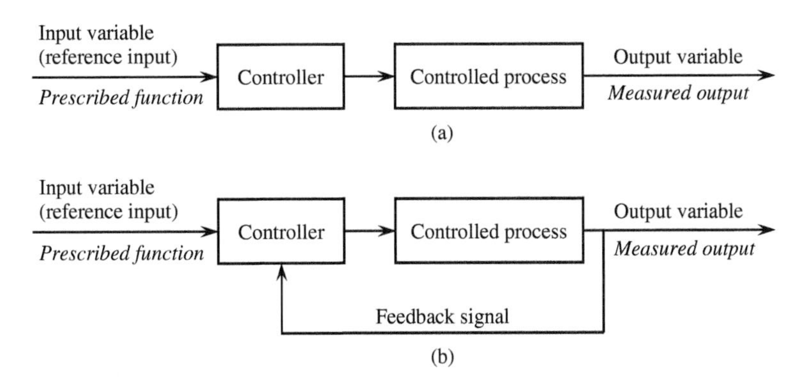


**Figure 5-2** Two basic control methods: (a) open loop control; (b) closed loop control

the test specimen. The reference input in testing machines is provided by a function generator. The feedback signal is normally the measurement of the deformation of a concrete specimen, i.e., the output of a transducer such as an extensometer or linear variable differential transformer (LVDT), which is continuously monitored in analog controllers, and sampled at discrete instants in digital controller.

Obviously, the scope of CLC is greater than that of open-loop control, because of the larger number of variables that can be controlled. Even for the same control variable, the CLC system produces a more accurate output than the open-loop system. However, CLC has a few drawbacks. The most important one, other than the higher initial cost, is that the system requires better operator skills because improper use could make the system unstable. There is always a lag between the actual response and the corrective action of the controller, which may result in the loss of control, overcorrection, or undercorrection. Thus, closed-loop controllers have to be properly designed through modeling and analysis.

#### 5.1.1.3 Calibration of Transducers

To obtain concrete strength or other properties, the parameters of a specimen such as displacement, strain, crack opening, and force have to be recorded to perform analysis. Various transducers are designed for these purposes. However, most transducers are electronic products that directly measure electrical parameters, such as change of resistance, current, or voltage, which may relate to the physical/mechanical variables to be measured. Hence, correlations between electrical variables and physical/mechanical variables are needed for interpreting the experimental results. A process to find such correlations is called calibration.

The purpose of calibration is to build up a linear relationship between the electrical and mechanical parameters. The procedures of calibration include giving a known output of a mechanical variable, from zero to a full range of the transducer to be calibrated, with a constant increment, adjusting the gain of the electronic transducers to the corresponding reading, and building up a linear relationship between the mechanical variables and the transducer readings. Generally, the full working range of a transducer will be calibrated to an output of 10 volts. The calibration line is used later for interpreting the specimen's mechanical behavior by simply converting the electrical signal into mechanical variables.

The commonly used equipment for calibration includes a standard micrometer for displacement transducers and a standard dead weight for load cells. For example, to calibrate a displacement transducer with a full range of 2.5 mm, standard outputs of known displacement values can be provided by a micrometer, as shown in the following case:

| Displacement (mm): | 0.25 | 0.5 | 1.25 | 2.5 |
|---|---|---|---|---|
| Voltage (V): | 1 | 2 | 5 | 10 |

With the displacement of 0.25 mm, the voltage output of transducer is adjusted to 1 volt. Then, 0.5 mm, 2 volts, until full range 2.5 mm, 10 volts. Hence, the relationship between displacement and voltage can be established, as shown in Figure 5-3. During experiments, the reading of the transducer, in voltage, can easily be interpreted into displacement through the equation

$$D = \frac{1}{K}V \tag{5-4}$$

Calibration is an important procedure to ensure accurate measurement of mechanical parameters accuracy.

### 5.1.2 Compressive Strength and Corresponding Tests

#### 5.1.2.1 Failure Mechanism

With a material such as concrete, which contains void spaces of various sizes and shapes in the matrix, and microcracks between the matrix and aggregates or in the matrix, the failure modes under loading are complex and vary with the type of stress. A brief review of the failure modes, however, will be useful in understanding and controlling the factors that influence concrete strength.

In compression, the failure mode is less brittle because considerably more energy is needed to form and to extend cracks in the matrix. It is generally agreed that in a uniaxial compression test on medium- or low-strength concrete, no cracks are initiated in the matrix below about 40–50% of the failure stress. However, at this stage, a stable system of cracks, called shear-bond cracks, already exists in the vicinity of the coarse aggregate. With the increase of the stress level, cracks are initiated in a random manner. Some cracks in the matrix and in the vicinity of the coarse aggregate (shear-bond cracks) eventually join up in a narrow region when the loading reaches about 80% of the ultimate value, while other cracks are closed. Subsequently, major cracks develop in these localized zones in the direction close to the direction of the load, which signals the failure of the compression specimen.

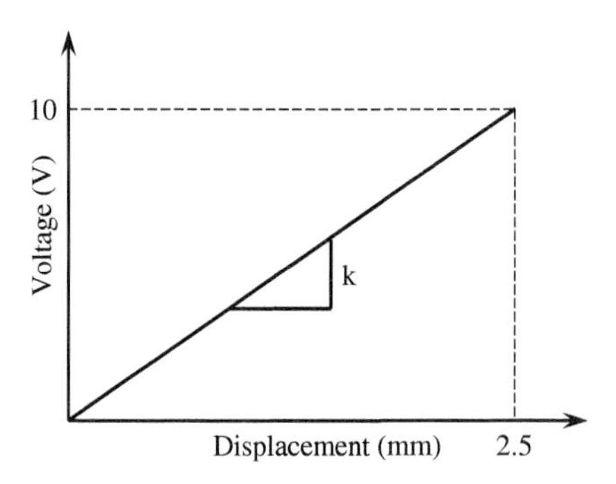


**Figure 5-3** Relationship between displacement and voltage

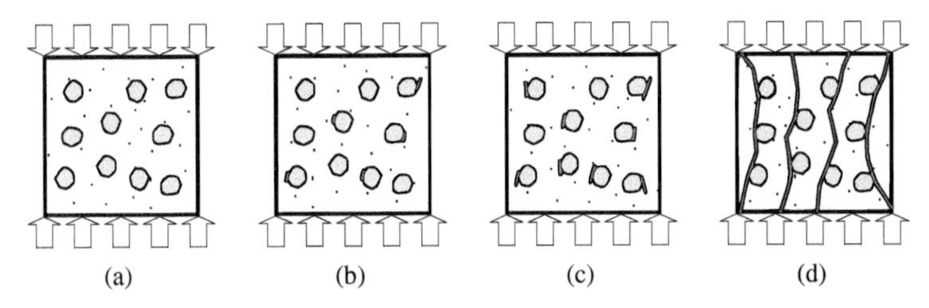


**Figure 5-4** Failure process of a concrete under compression: (a) shear-bond-crack (load less than 40% ultimate value); (b) load-initiated-cracks (load between 40% to 80% ultimate value); (c) crack concentration (load greater than 80% ultimate value); (d) major cracks formed (failure indication)

**Figure 5-5** Typical failure mode of concrete in compression

The process of concrete failure during compressive loading can also be described as shown in Figure 5-4. The development of the vertical cracks causes the expansion of concrete. Final failure is mostly likely splitting (see Figure 5-5), and sometimes failure on the formation of a shear band.

#### 5.1.2.2 Specimen Preparation for Compression Test

There are two types of specimens that can be used for uniaxial compression testing: the cube specimen and the cylinder specimen. The cube specimen method is used in Europe and China, while the cylinder specimen method is used in North America and accepted in Europe too.

**(a)** *Cube specimen (BS EN 12390-3: 2019)*: BS EN 12390 allows the use of a cube, cylinder, or core (from field concrete). The most frequently used standard size of cube specimens is 150 × 150 × 150 mm. However, the nominal size (side length of a cube, or diameter of a cylinder) of specimen should be chosen to be at least 3.5 times the maximum aggregate size. A cube specimen should be prepared by pouring concrete into a cubic mold in at least two layers, but no layer should be thicker than 100 mm. Each layer should be stroked at least 25 times by a hemispherical-tipped steel rod (16 mm diameter) or a steel compacting bar having a square (25 mm × 25 mm) cross-section. An internal vibrator or vibrating table may also be used to compact the specimens. The compacted specimens need to be kept in the mold for at least 16 hours (but no longer than 3 days), protected (often by covering) against dehydration at 20 ± 5°C. After demolding, the specimen should be cured at a temperature of 20 ± 1°C in water or in a chamber at a relative humidity of no less than 95%. For a cube specimen, the height to width ratio is 1 and that at least five surfaces are exposed to the mold, which can generate smooth surfaces on the specimen. Hence, a cube specimen can be tested without surface preparation.

**(b)** *Cylinder specimen (ASTM C39/C39M):* ASTM C39 allows the use of cylindrical specimens of various sizes (length/diameter = 2; and diameter shall be at least 3 times the nominal maximum aggregate size). The most widely adopted standard cylinder size for compression testing is 150 × 300 mm. It should be prepared by pouring concrete into the cylindrical mold in three approximately equal layers, with each layer being stroked 25 times by a hemispherical-tipped steel rod (16 ± 2 mm). After 16–24 hours from casting, the specimens should be demolded and cured at a temperature of 23.0 ± 2.0°C in a water tank or moist conditions (ASTM C192/C192M). The cylinder specimen has a length/diameter ratio of 2. Also, the upper surface of the cylinder is never smooth; hence, grinding or capping is needed to level and smooth the compression surface before a test.

#### 5.1.2.3 Factors Affecting the Measured Compressive Strength

The compressive strength tests appear to be perfectly straightforward. However, the results obtained can be affected considerably by a number of factors. When interpreting the strength values obtained by standard (although arbitrary) procedures, it is necessary to consider in detail how compressive strength is affected by the test parameters.

**(a)** *Loading rate*: In general, the lower the loading rate, the lower the measured compressive strength. This may be attributed to the fact that deformation generated by loading needs time to develop. The slow rates of loading may allow more subcritical crack growth to occur, thus leading to the formation of larger flaws and hence a smaller apparent ultimate load. On the other hand, it may be that slower loading rates allow more creep to occur, which will increase the amount of strain at a given load. When the limiting value of strain is reached, failure will occur. More likely, the observed effect of loading rate is due to a combination of these two, and perhaps other minor factors as well.

So, to make the results of compressive strength tests comparable, a standard load rate has to be followed. For a cylinder specimen, ASTM regulates 0.25 ± 0.05 MPa/sec as the standard loading rate. For a cube specimen, BSI sets 0.2–0.4 MPa/sec as the standard loading rate. In the real situation, the loading rate can be transferred to N/sec by multiplying the area of the specimen under the loading.

**(b)** *End condition*: The compression test assumes a state of pure uniaxial compression. However, this is not really the case, because of friction between the ends of the specimens and the platens of testing machine that make a contact with them. The frictional force arises due to the fact that, because of the differences in moduli of elasticity and Poisson's ratios for steel and concrete, the lateral strain in the platens is considerably less than the lateral expansion of the ends of the specimen if they were free to move. Thus, through friction, the platens act to restrain the lateral expansion of the ends of the specimens and to introduce shear stress that is greatest right at the specimen end and gradually dies out at a distance from each end of approximately $\left(\sqrt{3}/2\right)d$, where $d$ is the specimen diameter or width. The manifestation of this lateral confining pressure is often the appearance of relatively undamaged cones (or

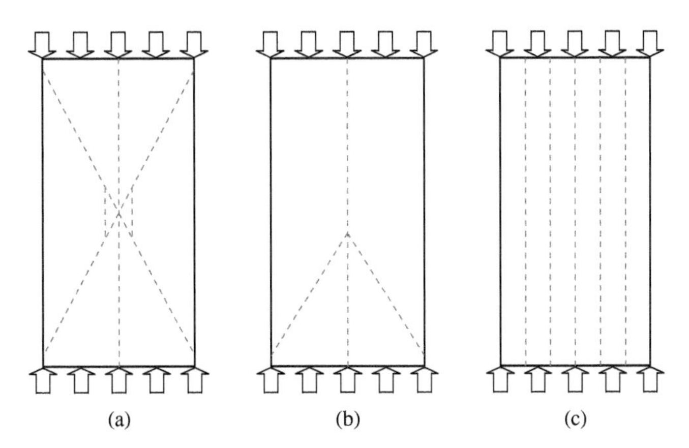


**Figure 5-6** Influence of constraint to failure mode of concrete specimen: (a) with constraint on both ends; (b) with constraint in one end; and (c) without constraint

pyramids) of concrete in specimens tested to failure, as shown in Figure 5-6a. Thus, for a standard cylinder with $l/d = 2$ (where $l$ is the height of specimen), only a small central portion of the cylinder is in true uniaxial compression, the remainder being in a state of triaxial stress. The effect of this type of end restraint is to give an apparently higher strength (i.e., test result) than the true compressive strength of the specimen. For ordinary concrete with a Poisson's ratio approximately equal to 0.2, if the platen-specimen interface is friction-free, lateral tensile strains will occur at fairly low compressive loads, and this could be the cause of failure, as shown in Figure 5-6c. This is probably the natural mode of failure in pure uniaxial compression. The stresses induced due to the end restraints may cause an apparent conical failure of a specimen, as indicated in Figure 5-6a. Since some end restraints cannot be avoided, it is likely that failure occurs through some combination of factors, as indicated schematically in Figure 5-6b. Tensile cracks may not be able to propagate to the areas of the specimen under lateral confining stress.

As mentioned earlier, due to the height-to-diameter or height-to-width ratio being different for cylinder compression and cube compression specimens, the constraints from the platens of a testing machine to the specimens are different. The friction between the platens and the cube specimen ends confines a much greater portion in the specimen than is the case with the cylindrical specimen, as shown in Figure 5-7. This leads to higher strength values when measured on cubes rather than cylinders. Usually, the ratio between the apparent cube strength and apparent cylinder strength is assumed to be 1.25 for normal-strength concrete. However, for high-strength concretes, the ratio will be reduced.

**(c)** *Size effect*: The probability of having large deficiencies, such as void and crack, increases with size. Thus, smaller specimens will give higher apparent strengths. If less-common specimens, e.g., 100-mm cubes or 100 × 200-mm cylinders, are used to measure the compressive strength, the test results for such small specimens may need to be corrected. Generally speaking, a factor of 0.9 could be used to modify the results of small specimens.

### 5.1.3 Uniaxial Tensile Strength and Corresponding Tests

#### 5.1.3.1 Failure Mechanism

A uniaxial tension test is more difficult than a compression test to conduct for three reasons. First, it is difficult to center the loading axis with the mechanical centroid. Second, it is difficult to control

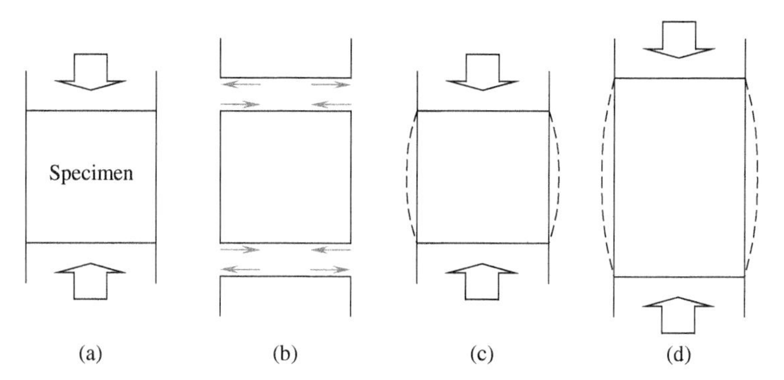


**Figure 5-7** Comparison of the end constraints between cube (a) and (b) and cylinder (c) and (d) specimens

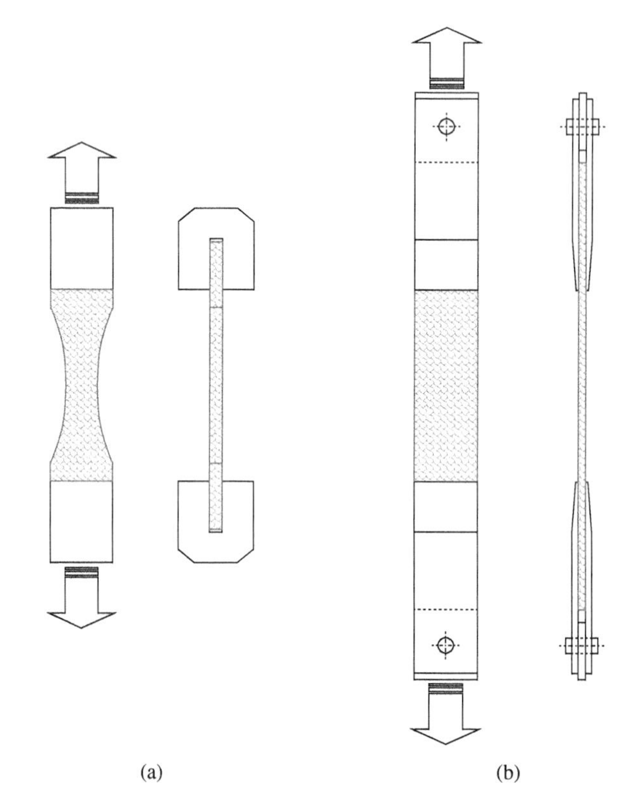


**Figure 5-8** Two conventionally used uniaxial tension methods: (a) grips dog bone test; (b) end plate loading method

the loading process due to the quasi-brittle nature of concrete under tension. Third, the tension process is more sensitive to a sudden change in cross-sectional area, and the specimen-holding devices introduce secondary stress that cannot be ignored. Figure 5-8 shows two conventional test methods for concrete uniaxial tension. One uses a dog bone-shaped specimen to ensure a smooth transfer of the cross-sectional area and the other uses a tapped steel plate glued on the concrete specimen to smoothly transfer the load. In this way, the secondary stresses generated by the holding devices can be minimized.

The processes of failure of a concrete specimen during uniaxial tension can be described as follows: from start of loading to 30% of ultimate value, the response is linear elastic. Microcracks occur randomly when the loading level is higher than 30% of ultimate load value. The phenomenon of strain localization starts at around 80% of peak load. The microcracks start to concentrate in a narrow region and microcracks in other regions close. The major crack is developed along the localization zone shortly after the peak load is reached and keeps propagating, signaling the failure of the specimen, as shown in Figure 5-9.

#### 5.1.3.2 Stress Concentration Factor

The tensile strength of concrete is much lower than its compressive strength, which can be attributed to the stress concentration generated by the defects in the material. Stress concentration can be illustrated by a plate with an elliptical hole, as shown in Figure 5-10, where the hole represents the defect. Under a tension load, the distribution of stress in the cross-section through the center of the ellipse is not uniform. The highest stress occurs at the edge of the ellipse and can be expressed as

$$\sigma_{\max} = \sigma_0 \left(1 + \frac{2a}{b}\right) = K_t \sigma_0 \tag{5-5}$$

where $\sigma_0$ is the remote stress or nominal stress; $a$ is the long radius of the ellipse, and $b$ is the short one; and $K_t$ is the concentration factor. It can be seen that if $a = b$, $K_t = 3$. $K_t$ depends not only on the geometry of the hole but also on the loading pattern. If the loading is pure shear, $K_t$ can reach 4 in the case of $a = b$.

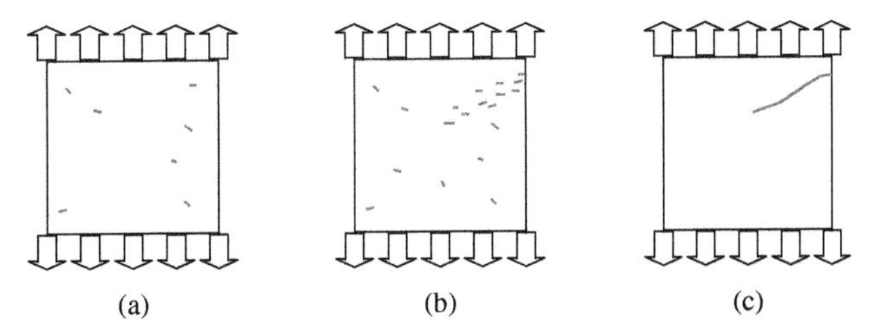


**Figure 5-9** Failure process of concrete specimen under tension: (a) random crack development (after 30% peak load is reached); (b) localization of microcracks (after 80% peak load is reached); and (c) major crack formation and propagation

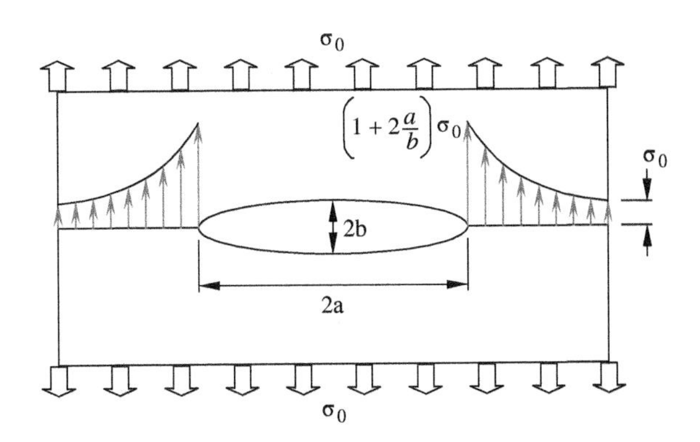


**Figure 5-10** Sketch of stress concentration

#### 5.1.3.3 Relationship Between Compressive Strength and Tensile Strength

It has been pointed out before that other mechanical properties of a concrete can be related to its compressive strength. However, there is no direct proportionality between tensile and compressive strength. As the compressive strength of concrete increases, the tensile strength also increases but at a decreasing rate. In other words, the tensile/compressive strength ratio depends on the general level of the compressive strength, the higher the compressive strength, the lower the ratio. The research work done by Price (1951) showed that the direct (uniaxial) tensile/compressive strength ratio is 10–11% for low-strength, 8–9% for medium-strength, and 5–7% for high-strength concrete.

The relationship between the compressive strength and the tensile/compressive strength ratio seems to be determined by the effect of various factors on the properties of both the matrix and the transition zone in concrete. Not only the curing age but also the characteristics of the concrete mixture, such as water/cement ratio, type of aggregate, and admixtures, affect the tensile/compressive strength ratio to varying degrees. For example, after about one month of curing, the tensile strength of concrete is known to increase more slowly than the compressive strength; that is, the tensile/compressive strength ratio decreases with the curing age. At a given curing age, the tensile/compressive strength ratio also decreases with the decrease in water/cement ratio.

#### 5.1.3.4 Indirect Tension Test (the Split-Cylinder Test or the Brazilian Test)

The indirect tension test is also called the splitting test or Brazilian test. The most frequently used standard specimen for the splitting test is a 150 × 300-mm cylinder (BS EN 12390-6, ASTM C496/C496M). The making and curing requirements are the same as the compression specimen. The splitting test is carried out by applying compression loads along two axial lines that are diametrically opposite, see Figure 5-11a. The loading rate is 0.04–0.06 MPa/sec (i.e., 2.4–3.6 MPa/min) according to BS EN 12390, and 0.7–1.4 MPa/min according to ASTM C496.

Under such a line compression load, the stress along the central diameter will be distributed as shown in Figure 5-11b. It can be seen that the stress distribution as derived from elasticity along

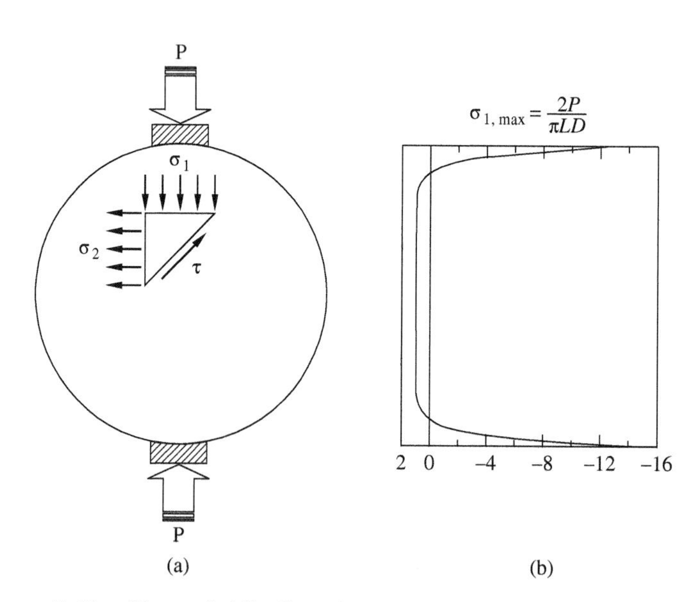


**Figure 5-11** Stress distribution of the specimen under the splitting test

the central part of the diameter is a uniform tensile stress, while only at the edge is compressive stress. The expressions of the stresses are

$$\sigma_{\text{com}} = \frac{2P}{\pi LD}\left(\frac{D^2}{r(D-r)} - 1\right) \tag{5-6}$$

and

$$\sigma_{\text{ten}} = \frac{2P}{\pi LD} \tag{5-7}$$

where $P$ is the applied load, $L$ the cylinder length, $D$ the cylinder diameter, and $r$ the distance from the top of the cylinder. The maximum $\sigma_{\text{ten}}$ calculated from the peak load is the split tensile strength, $f_{\text{st}}$.

According to a comparison of the test results of the same concrete, $f_{\text{st}}$ is about 10–15% higher than the direct tensile strength, $f_{\text{t}}$. In practice, it is difficult to apply a true line load. Usually, a strip of plywood with a width of 25 mm and thickness of 3 mm is used as a bearing to transfer the load from the testing machine to the specimen. It is also allowed to use a cube for a splitting test instead of a cylinder. In this case, the loading is applied to the cube through two hemispherical bars along the center lines of two opposite faces (see Figure 5-12). The formula to obtain the generated tensile stress in the cube specimen subjected to splitting is

$$\sigma_{\text{ten}} = \frac{2P}{\pi a^2} \tag{5-8}$$

where $a$ is the side length of the cube specimen.

### 5.1.4 Flexural Strength and Corresponding Tests

Flexural strength is also called the modulus of rupture (MOR). It can be determined by performing a four-point (or third-point) bending test following the procedures of ASTM C78 or BS EN 12390-5. The specimen for a flexural strength test is a beam, and the minimum cross-sectional dimension of the beam should be selected according to ASTM C31 or BS EN 12390-1. The most widely used cross-sectional dimension is 150 × 150 mm. According to ASTM C78, the beam length should be at least 50 mm greater than 3 times the depth; while according to BS EN 12390, the length should be at least 3.5 times the depth of the beam. The arrangement for the MOR test is shown in Figure 5-13. According to the mechanics of materials, we know that under the four-point bending,

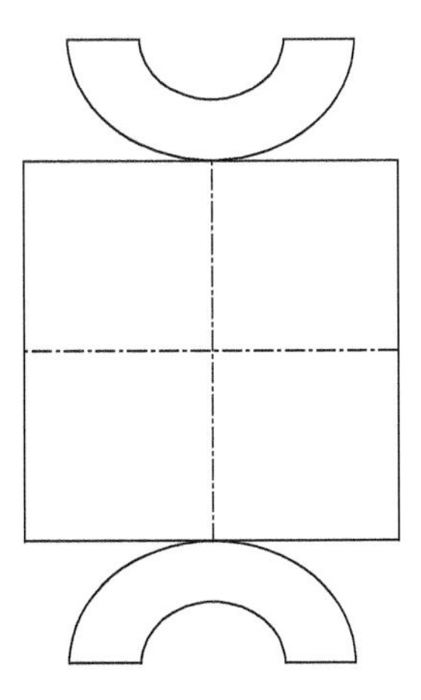

**Figure 5-12** Experimental setup of indirect tension test on a cube specimen

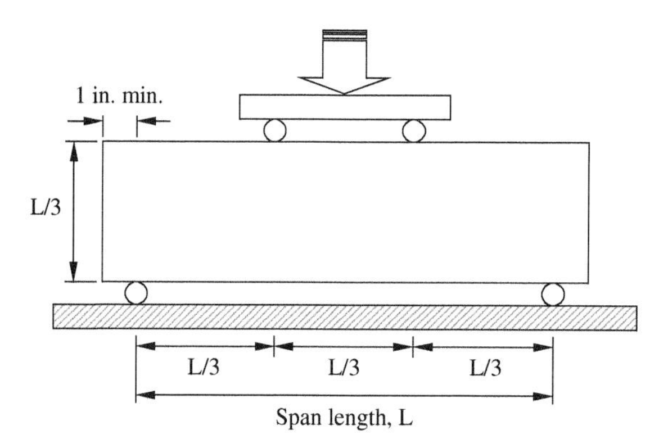


**Figure 5-13** Flexural strength test setup for hardened concrete

the middle one-third portion of the beam is under pure bending. The maximum moment can be calculated by

$$M_{\max} = \frac{P}{2} \times \frac{l}{3} = \frac{Pl}{6} \tag{5-9}$$

If fracture takes place inside the middle one-third, beam theory under pure bending can be applied directly and the bending flexural strength (i.e., maximum tensile stress) can be calculated as

$$f_{\mathrm{bt}} = \frac{M_{\max} y_{\max}}{I} = \frac{\frac{Pl}{6} \times \frac{d}{2}}{\frac{bd^3}{12}} = \frac{Pl}{bd^2} \tag{5-10}$$

However, if fracture occurs outside the middle one-third (pure bending zone), the cross-section is carrying not only bending moment, but also shear force. According to elasticity, if the span to height (of the beam) ratio is greater than 5, beam theory under pure bending can be still applied to calculate the normal stress, with an error of less than 1%. However, for the loading setup regulated by ASTM C78 and BS EN 12390, the span-to-height ratio is only 3. Hence, basically, the formula for calculating normal stress from pure bending cannot be used in this situation. This is why BS EN 12390 suggests discarding such a result. On the other hand, ASTM allows this type of result to be used if the fracture occurs in the tension surface outside of the middle third of the span length by not more than 5% of the span length. In this case, if an average distance between the failure crack and the nearest support is $a$, then the MOR can be computed as

$$f_{\mathrm{bt}} = \frac{M_{\max} y_{\max}}{I} = \frac{\frac{Pa}{2} \times \frac{d}{2}}{\frac{bd^3}{12}} = \frac{3Pa}{bd^2} \tag{5-11}$$

If, however, failure occurs outside of the middle third by more than 5% of the span, the result should be discarded. Although the modulus of rupture is a kind of tensile strength, it is much higher than the results obtained from direct tension because of the support from the inner layers that have not reached their failure criterion.

### 5.1.5 Behavior of Concrete Under Multiaxial Stresses

#### 5.1.5.1 Behavior of Concrete Under Biaxial Stress

The biaxial test can be carried out using a two-dimensional universal testing machine. The loading parameter in two directions can be selected as the ratio of the applied stress to the ultimate compressive strength. The behaviors of three different concretes under different ratios have been plotted in Figure 5-14. It is essential that the level of uniaxial compressive strength of concrete does not affect the shape of the biaxial stress interaction curves. It can be seen that the behavior can be distinguished into four regions. In *region I*, both $\sigma_1/f_c$ and $\sigma_2/f_c$ are negative, which means that both $\sigma_1$ and $\sigma_2$ are in tension. For a concrete specimen under biaxial tension, its strength is similar to the strength obtained under uniaxial tension. In *region II*, the ratio of $\sigma_1/f_c$ is positive and $\sigma_2/f_c$ negative, which means that $\sigma_1$ is in compression and $\sigma_2$ is in tension. Under biaxial compression-tension, the compressive strength decreases almost linearly following the increase of lateral tensile stress. This is because the tensile strain in the lateral direction, generated by compression (as a result of the Poisson's ratio), is enlarged by an additional tensile stress, which speeds up the failure process. In *region III* both $\sigma_1/f_c$ and $\sigma_2/f_c$ are positive, which means that both $\sigma_1$ and $\sigma_2$ are in compression. The test results show that the strength of concrete subjected to biaxial compression is 27% higher than the uniaxial compressive strength. The reason behind the phenomenon is that the tensile strain in the transverse direction, generated by compression, is constrained by the compressive stress in that direction, which slows down the failure process. It is interesting to note that the increase in strength is not proportional to the stress added in the transverse direction. In *region IV* the ratio of $\sigma_1/f_c$ is negative, and $\sigma_2/f_c$ is positive, which means that $\sigma_1$ is in tension and $\sigma_2$ is in compression. The result is similar to that of region II.

The dimensional stability (or maximum strain at failure) of concrete under biaxial stress has different values, depending on whether the stress states are compressive or tensile. In biaxial compression, the average maximum compressive strain is about 3000 με and the average maximum tensile strain varies from 200–400 με. In biaxial tension–compression, the magnitude at failure of both the principal compressive and tensile strain decreases as the tensile stress increases. In biaxial tension, the average value of the maximum principal tensile strain is only about 80 με. It should be pointed out that not only the pure concrete has been tested under biaxial stress conditions, but also reinforced concrete panels. Figure 5-15 shows the setup of such a test at the University of Houston. This universal element tester consists of 40 in-plane jacks of 100 ton each and 20 out-of-plane jacks

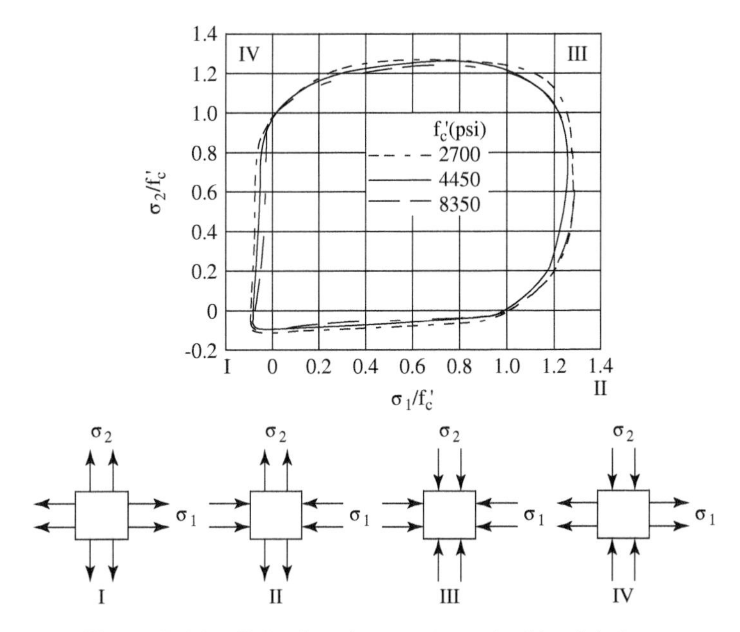


**Figure 5-14** Behavior of concrete under bi-axial stress

**Figure 5-15** Reinforced concrete panel under bi-axial stress. (Source: University of Houston)

of 60 ton each, housed in a giant 5 × 5-m steel frame. It was constructed to test full-size concrete panels of 1.4 $m^2$ and up to 0.4 m thick reinforced with steel bars from 10–25 mm in diameter. Any self-equilibrium loading condition on the four edges of a test panel can be simulated. The panel can be subjected simultaneously to in-plane tension or compression, in-plane and out-of-plane bending, in-plane and out-of-plane shear, as well as torsion.

### 5.1.5.2 Behavior of Concrete Under Triaxial Stress

Triaxial loading experiments can be conducted in two ways according to the shape of specimen. For a cylindrical specimen, the compressive stresses $\sigma_2$ and $\sigma_3$ can be obtained by subjecting the specimen to hydrostatic pressure in a vessel. To prevent the penetration of the pressure fluid into microcracks and pores of the specimen, the cylindrical specimen has to be sealed in a plastic membrane. For a cubic specimen, the lateral pressures, $\sigma_2$ and $\sigma_3$, can be added by flat jacks. The compressive stress $\sigma_1$ can be added through the hydraulic piston as shown in Figure 5-16. The strength of concrete can be significantly increased under triaxial compressive stress because the lateral stress can largely limit the crack development caused by the vertical stress. Roughly, the compressive strength of concrete can be estimated using the following equation:

$$f_{\text{tri-c}} = f_c + Cf_1 \tag{5-12}$$

where $f_{\text{tri-c}}$ is the compressive strength of concrete under triaxial compressive stress; $f_c$ is the compressive strength of concrete under uniaxial compressive stress; $f_1$ is the lateral pressure in triaxial stress stage; and $C$ is an empirical coefficient ranging from 4.5 to 7, or estimated by

$$C = 2 + \frac{1.5}{\sqrt{f_1/f_c}} \tag{5-13}$$

**Figure 5-16** Triaxial testing machine at Dalian University of Technology

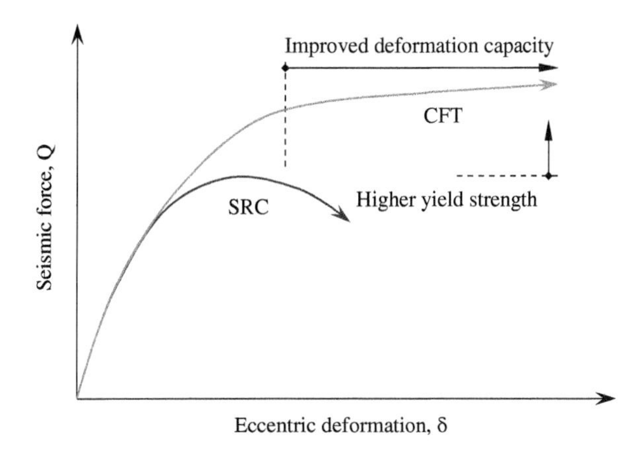


**Figure 5-17** Comparison of the loading behaviors of tube-confined concrete and steel-reinforced concrete

Not only the strength, but also the dimensional stability can be improved by adding the confining pressure. The ultimate strain in the $\sigma_1$ direction can be significantly increased for a tube-confined concrete as compared to a structural steel-reinforced concrete, as shown in Figure 5-17. The philosophy of confinement and triaxial compression has been applied in practice to improve the concrete strength, as well as its ductility. The circular stirrup-reinforced concrete column and tube concrete column are good examples, as shown in Figure 5-18.

The idea of spiral stirrup has been further developed by the Yuntai group in Taiwan. One invention is one-bar confinement, which uses a single reinforcement steel to make a stirrups for the entire cross-section of a structural member in order to hold all the main reinforcing steel. Another example of Yuntai's invention is a group of circular stirrups to hold the main reinforcing steel and provide confinement to the concrete at different zones.

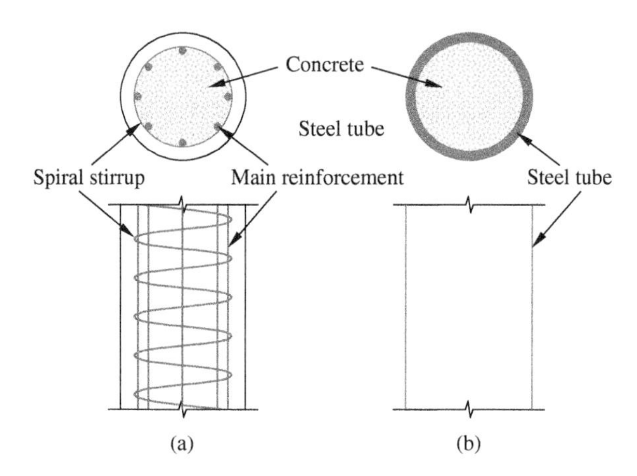


**Figure 5-18** Confinement for concrete column in the form of (a) spiral; and (b) tube

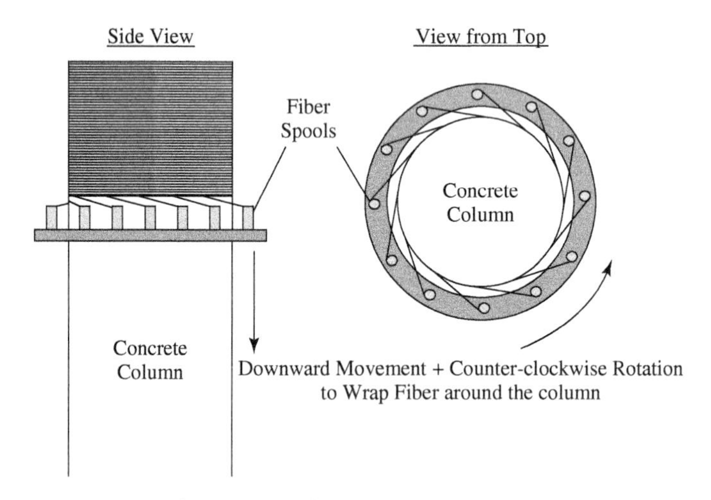


**Figure 5-19** Column retrofitting using the confinement concept

This concept has also been used in column retrofitting by using filament winding, see Figure 5-19. By applying this technique, a fibrous composite shell surrounding the concrete column tightly can be formed by continuous fiber strands. The shell provides a strong circumferential constraint to the concrete column, and hence significantly increases the load-carrying capacity and ductility of the concrete column. Such a strength effect is especially useful for earthquake design.

### 5.1.6 Bond Strength

Bond strength is defined as the shear strength between the aggregate, fiber, reinforcing steel, and cement paste. The bond strength plays an important role in determining the properties of concrete, fiber reinforced concrete, and reinforced concrete. There is considerable evidence to indicate that the interface is the weakest region in concrete. In general, bond failure occurs before failure of either the paste or aggregate. Many people have tried to measure the bond properties and have developed many models to interpret the experimental results. Here, the one developed by Mitsui et al. (1994) is introduced.

Figure 5-20 represents a schematic of the mathematical model for calculating the interface properties. In Figure 5-20, $L$ represents the aggregate embedded length. The aggregate is assumed to be elastic with Young's modulus $E_a$ and cross-sectional area $A$. The bulk matrix is assumed to be rigid except for the interfacial layer, which is idealized as an elastoplastic shear layer. It is assumed that debonding has occurred over a certain length, $a$, starting at $x = L$. Treating the boundary layer as a shear lag and assuming that a constant shear stress is acting at the debonded interface, the following equations can be written:

$$q = \begin{cases} kU(x) & 0 < x < L - a \\ q_f & L - a < x < L \end{cases} \tag{5-14}$$

where $k$ is the stiffness per unit length of the interfacial layer for small deformation, $q$ is the shear force per unit length acting on the aggregate, $q_f$ is the frictional shear force per unit length, and $U(x)$ is the aggregate displacement. Denoting the aggregate push-out force as $P$, the equilibrium equation and the constitutive relationship for the aggregate can be written as

$$\frac{dP}{dx} - q = 0 \tag{5-15}$$

$$\frac{P}{A} = E_a \frac{dU}{dx} \tag{5-16}$$

By introducing Equations 5-14 and 5-16 into Equation 5-15, the following differential equation for $U$ can be obtained:

$$\begin{aligned} &U_{,xx} - \omega^2 U = 0 && 0 < x < (L - a) \\ &U_{,xx} - \frac{q_f}{E_a A} = 0 && (L - a) < x < L \end{aligned} \tag{5-17}$$

in which the subscript comma indicates differentiation. The quantity $\omega$ is defined as

$$\omega = \sqrt{\frac{k}{E_a A}} \tag{5-18}$$

Equation 5-17, together with boundary conditions and continuity conditions, constitute a complete set of equations for the determination of $U(x)$. Solving this set of equations, the following closed form expression for the slip displacement at the loading end, $U^*$, is obtained:

$$U^* = U(L) = \frac{P^* - q_f a}{E_a A \omega} \coth[\omega(L - a)] + \frac{P^* - 0.5 q_f a}{E_a A} a \tag{5-19}$$

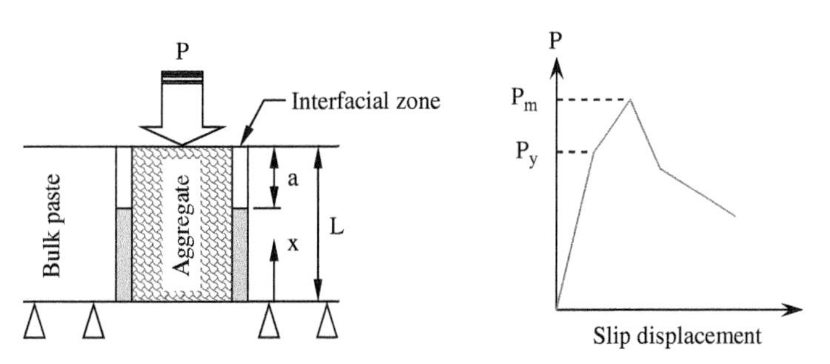


**Figure 5-20** Model for the bond between aggregate and matrix

The relationship between $U$ and $P$ in the elastic stage can be obtained from the above equation by setting $a = 0$. This leads to a form of

$$\frac{P^*}{U^*} = E_a A\omega \tanh\left[\omega\left(L\right)\right] \tag{5-20}$$

The stiffness parameter can be determined from the initial slope of the experimental load–slip displacement curve (see Figure 5-20) for a push-out test using the above equation.

Furthermore, it is shown that the relationship between the push-out force and debonding length, $a$, can be derived from both the shear strength ($q_y$) criterion (material properties represented by $q_y$ and $q_f$) and fracture energy ($\Gamma$) criterion (material properties represented by $\Gamma$ and $q_f$). For the shear strength ($q_y$) criterion the derivation is obtained from the overall equilibrium of the forces acting on the aggregate. For the fracture energy criterion, a differential equation for $P$ is first derived by applying the energy balance concept, and then the expression for $P$ is obtained by solving the differential equation. The expressions for $P$ take the following forms:
For the shear strength criterion

$$P^* = q_f a + \omega \frac{q_y \tanh\left[\omega\left(L - a\right)\right]}{\omega} \tag{5-21}$$

For the fracture energy criterion

$$P^* = q_f a + \left[\frac{q_f}{2\omega} + \sqrt{\left(\frac{q_f}{2\omega}\right)^2 + 2E_a Ap\Gamma}\right] \tanh\left[\omega\left(L - a\right)\right] \tag{5-22}$$

in which $\Gamma$ is the specific fracture energy. Note that three material parameters are needed for either model. They are $\omega$, $q_y$, and $q_f$ for the shear strength criterion, and $\omega$, $\Gamma$, and $q_f$ for the fracture energy criterion. To determine the interfacial yield parameter, $q_y$ (or $\tau_y$, the interfacial yield stress), the interfacial frictional parameter, $q_f$ (or $\tau_f$, the frictional force per unit area), and the specific energy, $\Gamma$, one needs to know the length of the debonded crack, $a$, at the peak load. A method that uses the maximum load, $P^*_{max}$, and the slip displacement corresponding to $P^*_{max}$ is used. The formulas used to calculate $\tau_f$, $\tau_y$, and $\Gamma$ are Equations 5-23, 5-24, and 5-25, respectively:

$$2\pi R\tau_f = \frac{\omega P^*_{max}}{a\omega + \sinh\left[\omega\left(L - a\right)\right]\cosh\left[\omega\left(L - a\right)\right]} \tag{5-23}$$

$$\tau_y = \tau_f \cosh^2\left[\omega\left(L - a\right)\right] \tag{5-24}$$

$$2E_a A\Gamma = \left(\frac{\tau_f}{\omega}\right)^2 \left\{\cosh^4\left[\omega\left(L - a\right)\right] - \cosh^2\left[\omega\left(L - a\right)\right]\right\} \tag{5-25}$$

Note that an additional equation is needed to determine the debonding length, $a$:

$$P^*_{max}\frac{0.5(\omega a)^2 + \cosh^2\left[\omega\left(L - a\right)\right] + \omega a \sinh\left[\omega\left(L - a\right)\right]\cosh\left[\omega\left(L - a\right)\right]}{\omega a + \sinh\left[\omega\left(L - a\right)\right]\cosh\left[\omega\left(L - a\right)\right]} - U^*_{peak}E_a A\omega = 0 \tag{5-26}$$

The procedure of determining the bond properties is as follows: calculate the debonding length, $a$, first; then calculate $\tau_f$; and, finally, calculate $\tau_y$, and $\Gamma$. The test setup for acquiring the required data is shown schematically in Figure 5-21. The specimen is put on a flat, circular plate that is connected with the servo-hydraulic actuator of a material testing system (MTS) machine through a hollow cylinder. The entire specimen fixture can move up with the actuator. A steel rod, which

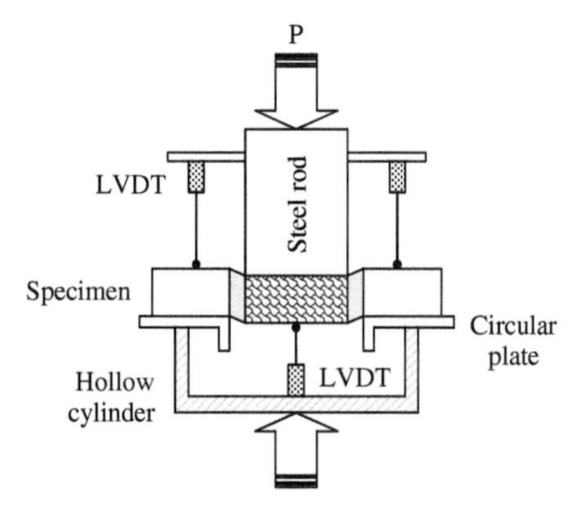


**Figure 5-21** Bond strength test setup

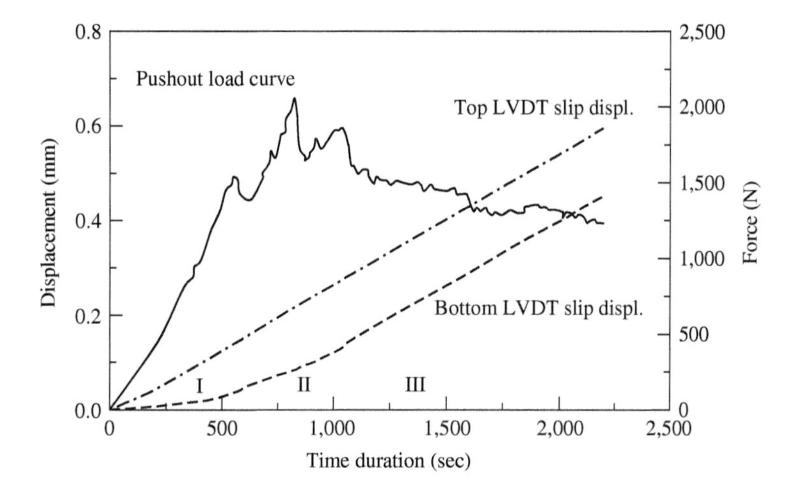


**Figure 5-22** Pushout force and displacement as a function of time

is connected to the load cell, makes contact with the top surface of the aggregate and by resisting the upward movement of the loading fixture pushes the central cylindrical aggregate downward. Two linear variable differential transformers (LVDTs), which are fixed between the circular plate and the rigid wings of the steel rod, are used to measure the slip displacement of the top of the aggregate relative to the surface of the cement matrix annulus. The average output of the LVDTs is used as a feedback signal to control the servo-hydraulic system. The push-out test is performed at a rate of slip displacement of about 1 mm per hour. Slip displacement at the bottom of the aggregate is also measured by another LVDT. Push-out load, slip displacement of aggregate, and stroke of the actuator of MTS are recorded by using a data-acquisition computer.

For a typical push-out test, the four measurements of importance are time, push-out load, average displacement at the top of the aggregate, and slip displacement at the bottom of the aggregate. For a complete analysis of the results, a graph of load vs. top of aggregate displacement and a graph of load, top of aggregate displacement, and bottom of aggregate slip displacement vs. time are produced, as shown in Figure 5-22. Shown in Figure 5-22 are two curves of slip displacement superimposed over the push-out loading curve as a function of time. The displacement at the top of the aggregate is represented by a linearly increasing line, while the curve of the slip displacement at the bottom of the aggregate is expressed as a broken line, composed of three ascending stages. For the measurement of the bottom displacement, the points at which the curve changes its slope signal the transition of the interfacial damage stages. The first portion of the curve represents the elastic deformation of the aggregate–cement interface. The change of the slope, becoming steeper, implies that the deformation rate is increased. This means that the interface becomes less stiff than before, and the elastic bond must have been broken. Hence, the end point of the first portion of the curve marks the initial debonding, and the second stage of the curve could be called the partial debonding stage. The point at which the up measured displacement and bottom measured displacement become parallel signals the start of the third stage. The parallelism of the two lines means that the top of the aggregate and the bottom of the aggregate undergo the same amount of displacement. This corresponds to complete debonding. It should be noted that complete debonding of the interface occurs after the peak load has been achieved, as can be seen from Figure 5-22, by matching the load curve with the second transition point.

The shear bond strength for specimens with different surface areas is plotted in Figure 5-23. It can be seen that the shear bond strength is of the same order of magnitude as the tensile strength of concrete, with slightly lower values around 2.3 to 4.2 MPa. Except for the pushout test, there are several other methods to measure the different properties of the aggregate–cement interface. They are schematically shown in Figure 5-24.

### 5.1.7 Fatigue Strength

There are two terms regarding fatigue in concrete technology. One is called static fatigue and the other cyclic fatigue. The static fatigue describes a failure of concrete under a slowly increasing loading at a peak value slightly lower than the strength obtained in a test at standard loading rate. It can be explained by the effect of loading rate on the load-carrying capacity of concrete, as discussed earlier in Section 5.1.2.3. In other words, the phenomenon of a decreasing load-carrying capacity following the decrease of the loading rate in a wide range is called static fatigue. Under low rates of loading, static fatigue occurs when the stress in concrete exceeds 70%–80% of the defined (short-term) strength. This threshold represents the onset of microcrack localization, which leads to macrocrack development and failure. This stress value also initiates unstable creep development, which is discussed later, so sometimes the static fatigue is also called creep rupture. The influence

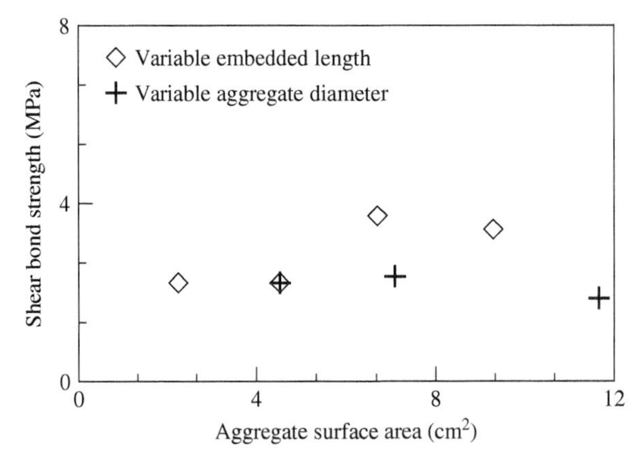


**Figure 5-23** Shear strength as a function of surface area

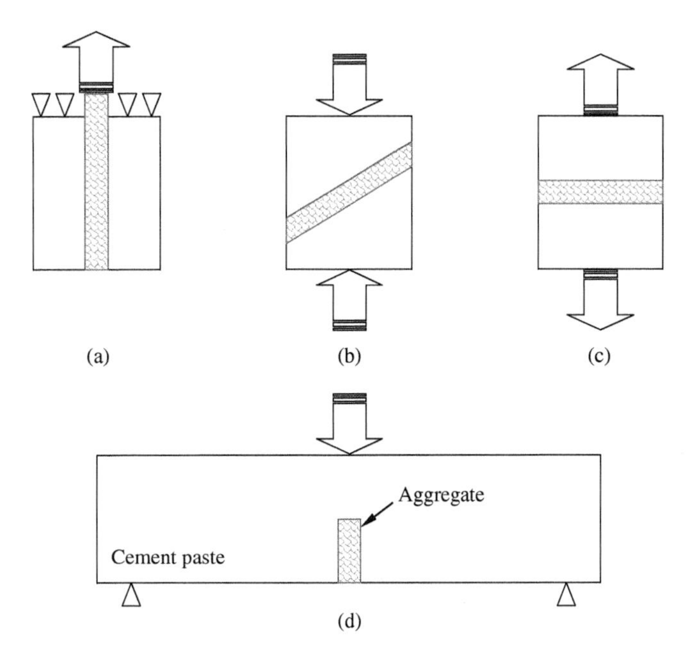


**Figure 5-24** Different types of bond test configuration: (a) aggregate pullout test; (b) shear type test; (c) tension type test; and (d) bending test

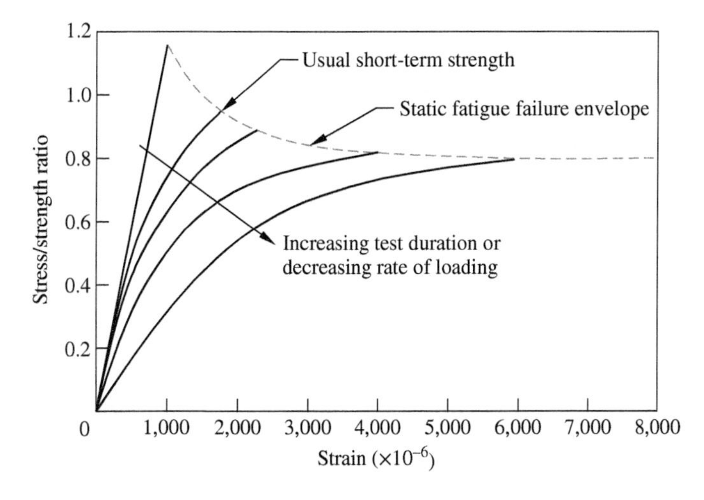


**Figure 5-25** Influence of loading rate on load-carrying capability

of the loading rate on the load-carrying capability, or the phenomenon of static fatigue, can be clearly seen in Figure 5-25.

The cyclic fatigue can be defined as a failure caused by the repeated application of loads that are not large enough to cause failure in a single application. This implies that some internal progressive permanent structural damage accumulates in the concrete under repeated stress. A typical cyclic loading is shown in Figure 5-26. By referring to Figure 5-26, some useful definitions and basic concepts for cyclic loadings can be introduced.

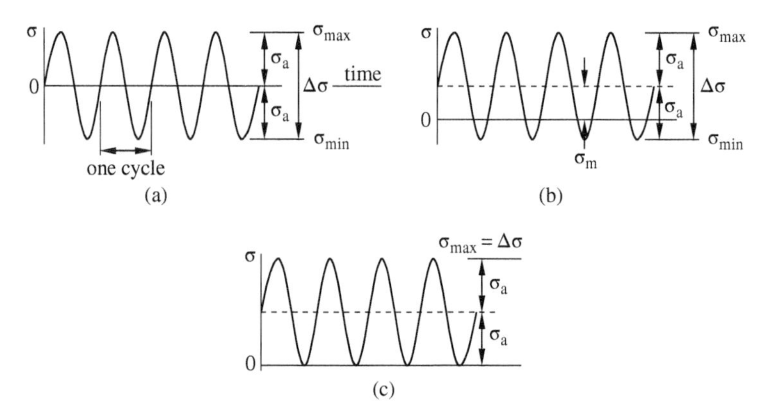


**Figure 5-26** Typical cyclic loading patterns

*Constant amplitude stressing*: cycling between maximum and minimum stress levels that are constant.

*Stress range*, $\Delta\sigma$: the difference between the maximum and the minimum stress values:

$$\Delta\sigma = \sigma_{\text{max}} - \sigma_{\text{min}} \tag{5-27}$$

*Mean stress*, $\sigma_m$: the average of the maximum and minimum stress values:

$$\sigma_{\text{m}} = \frac{\sigma_{\text{max}} + \sigma_{\text{min}}}{2} \tag{5-28}$$

*Stress amplitude*, $\sigma_a$: is half of the stress range:

$$\sigma_{\text{a}} = \frac{\Delta\sigma}{2} = \frac{\sigma_{\text{max}} - \sigma_{\text{min}}}{2} \tag{5-29}$$

*Completely reversed stressing*: means that the mean stress is equal to zero, with constant amplitude.

In cyclic fatigue, the symbol $S$ is usually used to represent a nominal or average stress, which has some difference with the true stress at a point $\sigma$. The nominal stress distribution is determined from the load or moment using mechanics of materials formulas as a matter of convenience, while true stress is determined according to the real materials states (stress concentration, yielding). $S$ is equal to $\sigma$ only in certain situations, see Figure 5-27.

The fatigue strength of a material is largely influenced by the maximum stress applied, the difference between maximum and minimum stress (stress range), and the number of cycles. The fatigue life of a material is usually plotted as nominal stress versus cyclic number on an *S-N* diagram. To get the *S-N* diagram, fatigue tests have to be conducted. Each test deals with a fixed stress amplitude and mean stress. The test continues until the specimen fails at a cyclic number of $N_f$. One experimental result will generate one point in the *S-N* diagram. The *S-N* diagram should have sufficient data points to make the empirical analysis meaningful. To make things simple, usually a completely reversed stressing (i.e., mean stress equal to zero with constant amplitude) is adopted first to build up the *S-N* diagram. For the cases of mean stress unequal to zero, the fatigue life can be estimated by using the *S-N* diagram of completely reversed stress, as stated in the following section.

When sufficient experimental data are obtained from completely reversed stress fatigue tests, an *S-N* diagram can be plotted in linear–linear coordinates, linear–log coordinates, or log–log

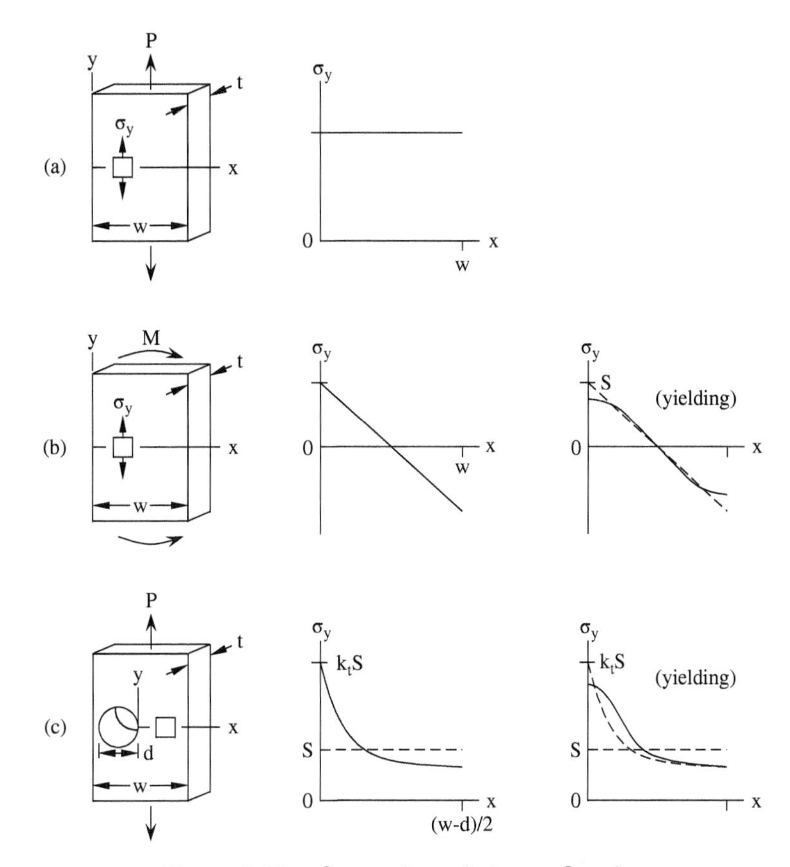


**Figure 5-27** Comparisons between $S$ and $\sigma$

coordinates. Figure 5-28 shows a linear–log plot of an *S-N* diagram. If *S-N* data are found to be a straight line on a log–log plot, the relationship between stress amplitude and fatigue cycles can be written as

$$\sigma_{\mathrm{ar}} = A\left(N_{\mathrm{f}}\right)^{B} \tag{5-30}$$

where $\sigma_{\mathrm{ar}}$ is the stress amplitude for completely reversed stressing ($\sigma_{\mathrm{m}} = 0$) corresponding to $N_{\mathrm{f}}$, $A$ and $B$ are the material constants, and $N_{\mathrm{f}}$ is the number of cycles to failure. For the cases in which $\sigma_{\mathrm{m}} \neq 0$, the relationship between the stress amplitude and the stress amplitude for completely reversed stressing can be expressed by the empirically modified Goodman law,

$$\sigma_{\mathrm{a}} = \sigma_{\mathrm{ar}}\left(1 - \frac{\sigma_{\mathrm{m}}}{\sigma_{\mu}}\right) \tag{5-31}$$

where $\sigma_{\mathrm{a}}$ is the stress amplitude for $\sigma_{\mathrm{m}} \neq 0$ for a given fatigue life, $\sigma_{\mathrm{ar}}$ is the stress amplitude of completely reversed stressing at fatigue failure ($N_{\mathrm{f}}$), $\sigma_{\mathrm{m}}$ is the mean stress, and $\sigma_{\mu}$ is the static strength of the material. This equation provides a base for estimating the fatigue life for a case in which $\sigma_{\mathrm{m}} \neq 0$ by using the *S-N* diagram for completely reversed stress. The fatigue life of any ($\sigma_{\mathrm{m}}$, $\sigma_{\mathrm{a}}$) combination can be estimated from the following procedure. First, substitute Equation 5-30 into 5-31:

$$\sigma_{\mathrm{a}} = \left(1 - \frac{\sigma_{\mathrm{m}}}{\sigma_{\mu}}\right) A N_{\mathrm{f}}^{B} \tag{5-32}$$

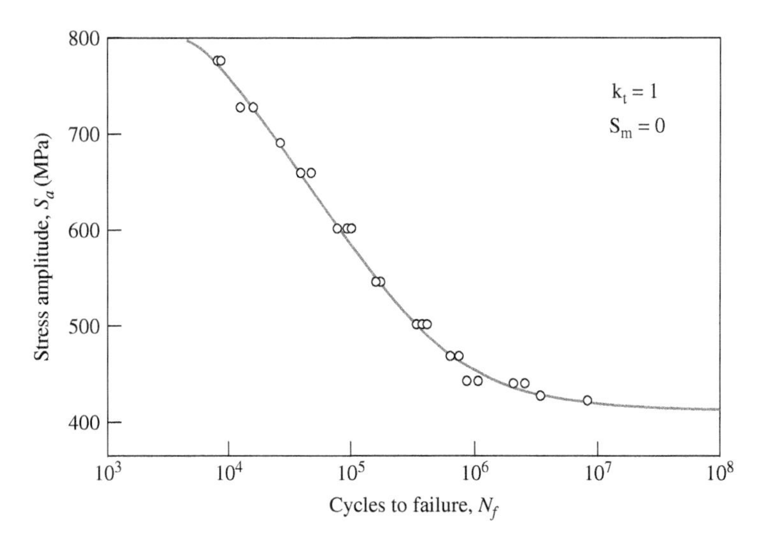


**Figure 5-28** An example of an *S-N* diagram

This equation reduces to $\sigma_a = AN_f^B$ as it should, if $\sigma_m = 0$. On a log–log plot, Equation 5-32 produces a family of *S-N* curves for different values of mean stress, which are all parallel straight lines. In general, let the *S-N* curve be

$$\sigma_a = \left(1 - \frac{\sigma_m}{\sigma_\mu}\right) f\left(N_f\right) \tag{5-33}$$

which is a corresponding family of *S-N* curves. From Equation 5-32,

$$N_f = \sqrt[B]{\frac{\sigma_a \sigma_\mu}{A\left(\sigma_\mu - \sigma_m\right)}} \tag{5-34}$$

If more than one amplitude or mean level occurs in a fatigue test, the fatigue life may be estimated by summing the cycle ratios, called the Palmgren-Miner rule:

$$\sum \frac{N_i}{N_{fi}} = 1 \tag{5-35}$$

where $N_i$ is the number of applied cycles under $\Delta\sigma_i$ or $\sigma_{ai}$, and $N_{fi}$ is the number of cycles to failure under $\Delta\sigma_i$ or $\sigma_{ai}$.

Often, a sequence of variable amplitude loading is repeated a number of times. Under these circumstances, it is convenient to sum cycle ratios over one repetition of the history, and then multiply this by the number of repetitions required for the summation to reach unity.

$$B_f \sum \left(\frac{N_i}{N_{fi}}\right)_{\text{one repetition}} = 1 \tag{5-36}$$

where $B_f$ is the number of repetitions to failure. Another approach for fatigue life prediction involves fracture mechanics concepts. Considering a growing crack that increases its length by an amount $\Delta a$ due to the application of a number of cycles $\Delta N$, the rate of growth with cycling can be characterized by $da/dN$.

Assume that the applied loading is cyclic with constant values of the loads $P_{max}$ and $P_{min}$, the nominal stresses $S_{max}$ and $S_{min}$ will also be constants. For fatigue crack growth, it is conventional to use the nominal stresses that are generally defined based on the gross area to avoid the change of stress values with crack length. The primary variable affecting the growth rate of a crack is the range of the stress intensity factor. This is calculated using the stress range $\Delta S$:

$$\Delta K = F\Delta S\sqrt{\pi a} \tag{5-37}$$

The value of $F$ depends only on the geometry and the relative crack length, $\alpha = a/b$, just as if the loading was not cyclic. Since $K$ and $S$ are proportional for a given crack length, the maximum, minimum, and range for $K$ during a loading cycle are given by

$$K_{max} = FS_{max}\sqrt{\pi a} \tag{5-38}$$

$$K_{min} = FS_{min}\sqrt{\pi a} \tag{5-39}$$

$$\Delta K = K_{max} - K_{min} \tag{5-40}$$

For a given material and set of test conditions, the crack growth behavior can be described by the relationship between cyclic crack growth rate $da/dN$ and the stress intensity range $K$. The empirical curve fitted suggests that the following relationship can be used:

$$\frac{da}{dN} = C(\Delta K)^m \tag{5-41}$$

where $C$ and $m$ are curve-fitting constants (from the log–log plot).

It should be noted that Equation 5-41 is obtained empirically and is valid for an intermediate crack growth rate or $\Delta K$ range. At low growth rates, the curve generally becomes steep and appears to approach a vertical asymptote denoted $K_{th}$, which is called the fatigue crack growth threshold. This quantity is interpreted as a lower limiting value of $K$, below which crack growth does not ordinarily occur. At high growth rates, the curve may again become steep. This is due to rapid unstable crack growth just prior to the final failure of the test specimen. The number of cycles to failure during a fatigue test can be calculated using the following equation:

$$N_f = \int_{a_i}^{a_f} \frac{da}{C(\Delta K)^m} \tag{5-42}$$

where $a_i$ is the initial crack size obtained from inspection (if no crack is found, take $a_i$ as the crack detection threshold), and $a_f$ the final crack size obtained from $K_{max}\left(a_f\right) = K_c$. By substituting the expressions in Equation 5-42, we can obtain

$$N_{if} = \left[\frac{1 - \left(\frac{a_i}{a_f}\right)^{(m/2-1)}}{C\left(F\Delta S\sqrt{\pi}\right)^m\left(\frac{m}{2} - 1\right)}\right]\frac{1}{a_i^{(m/2-1)}} \tag{5-43}$$

where $N_{if}$ is the cycle number for material failure in fatigue from a crack growing from $a_i$ to $a_f$, and $F$ is a geometrical function that depends on the loading pattern and the ratio of crack size to specimen size. In concrete design, a simple diagram or formula is usually used for fatigue strength. For this purpose, the Goodman law can be rewritten as

$$\sigma_a = \sigma_{ar}\left(1 - \frac{\sigma_m}{\sigma_\mu}\right) = \frac{\sigma_{max} - \sigma_{min}}{2} \tag{5-44}$$

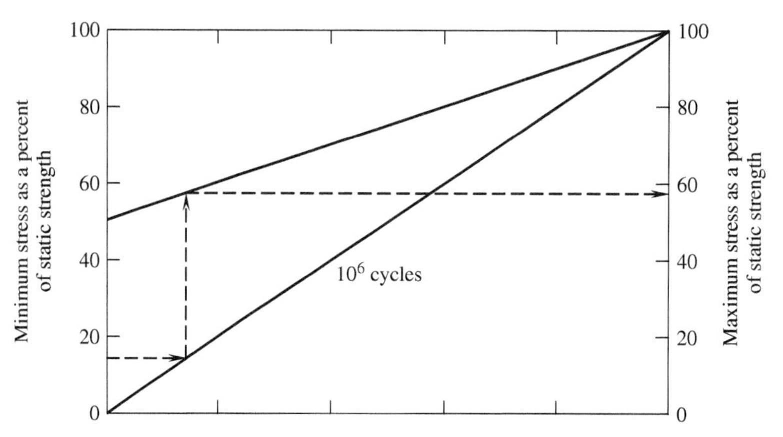


**Figure 5-29** Fatigue design diagram. Source: Adapted from ACI 215R-74

This equation gives a hint that the fatigue life of a concrete structure can generally be governed by the maximum and minimum stress carried by the concrete. Following this idea, the ACI Committee 215 recommended a simple method for use in design for plain concrete in both tension and compression, as shown in Figure 5-29. Figure 5-29 allows the determination of the maximum stress that concrete can withstand for $10^6$ load cycles, for a known minimum stress. For example, for zero minimum stress, the maximum stress for $10^6$ cycles is the value corresponding to the 50% of short-term strength. It provides the maximum stress range. As the minimum stress increases, the stress range reduces, although the maximum stress increases.

## 5.2 STRESS–STRAIN RELATIONSHIP AND CONSTITUTIVE EQUATIONS

### 5.2.1 Methods to Obtain a Stress–Strain (Deformation) Curve

A complete stress–strain curve of concrete includes the post-peak response. To obtain a complete stress–strain curve, a proper loading control mechanism is a must. For a uniaxial compression test, the load cell is used to measure the load, and sensors such as strain gauges, extensometers, or LVDTs are used to measure the deformation. If a very stiff machine is being used, for normal-strength concrete, the stroke control is good enough to obtain the post-peak response. However, for high-strength concrete, the stroke control cannot determine sufficient downward movement as a good feedback signal, and the specimen usually explodes around the peak load. In this case, another control method has to be used to obtain the post-peak behavior. One of the methods is circumferential control by using a roller chain tied on the specimen in a transverse condition, as shown in Figure 5-30. Under uniaxial compression, the circumferential displacement is a monotonic function of the loading process, even in the post-peak response stage. Hence, by using it as feedback in the post-peak response stage, the control is stable and reliable.

For a uniaxial tension test, it is much more difficult to obtain a complete stress–deformation curve. Conventionally, two methods are used to obtain a stable response in the post-peak region for a uniaxial tension test. One is called the loading-sharing method, see Figure 5-31. In this method, an elastic load-sharing system is parallel to the specimen (in the loading direction). In this way unstable failure near the peak load (resulting from the sudden release of stored elastic energy) is avoided. The load shared by the concrete is obtained by subtracting the contribution of the load-sharing system from the total load, implying subtraction of two relatively large numbers to

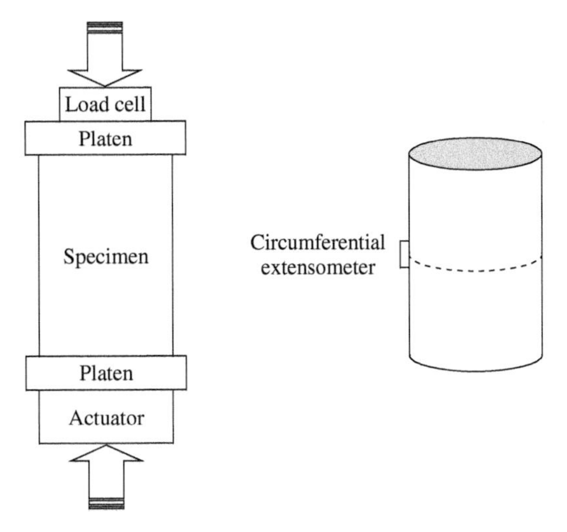


**Figure 5-30** Circumferentially controlled test setup

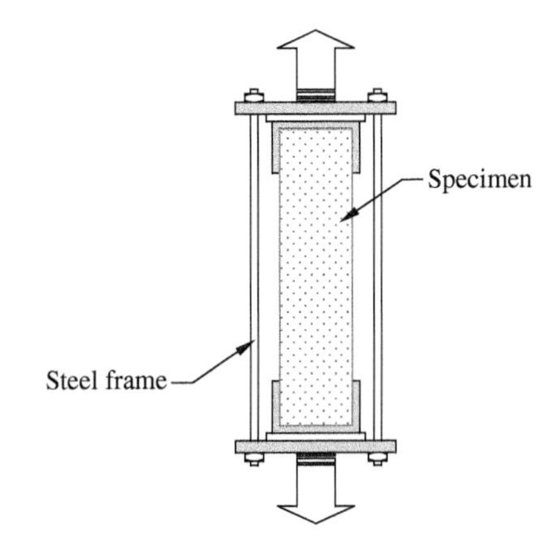


**Figure 5-31** Concrete tension test conducted with loading sharing system

obtain a small number. Since the load carried by the system is calculated from the strain measured by a strain gage attached to the steel bar, any small error in strain gage measurement can lead to a big error in load calculation. Hence, the major drawback of using parallel load-sharing system is poor accuracy, especially in the post-peak region.

Noting the disadvantages of the parallel load-sharing method, attempts were made to perform tensile tests on a specimen with two edge notches using closed-loop control, see Figure 5-32. Tensile failure of concrete is the result of the opening of a single "major" crack. If the opening of the crack is controlled in a closed-loop manner, gradual failure can be obtained, and instability near the peak load can be avoided. Since the cross-section with two edge notches is significantly weaker, a major crack will be forced to form at this position. A pair of displacement sensors is usually used with one on each side of the specimen, to cover the notch, and the average output is used as a feedback control signal. Since the location of the major crack is predetermined by introducing a notch in the specimen, the opening of the notch will always be increasing, and thus a stable control can be obtained. One of the problems associated with a notched specimen is that it forces the major crack to form at a predetermined location. Also, the state of stress in the specimen

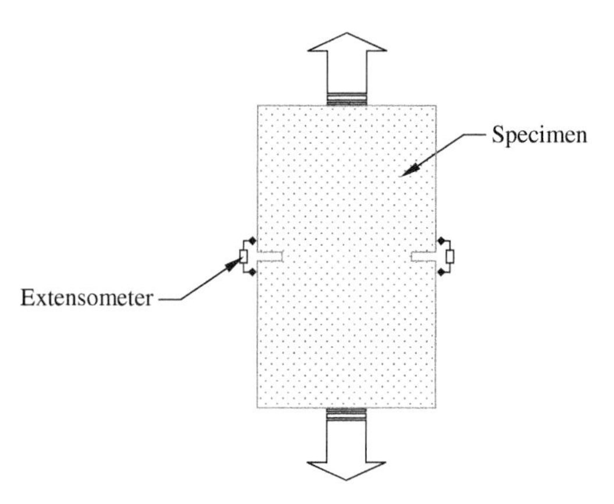


**Figure 5-32** Notched specimen used for uniaxial tension test

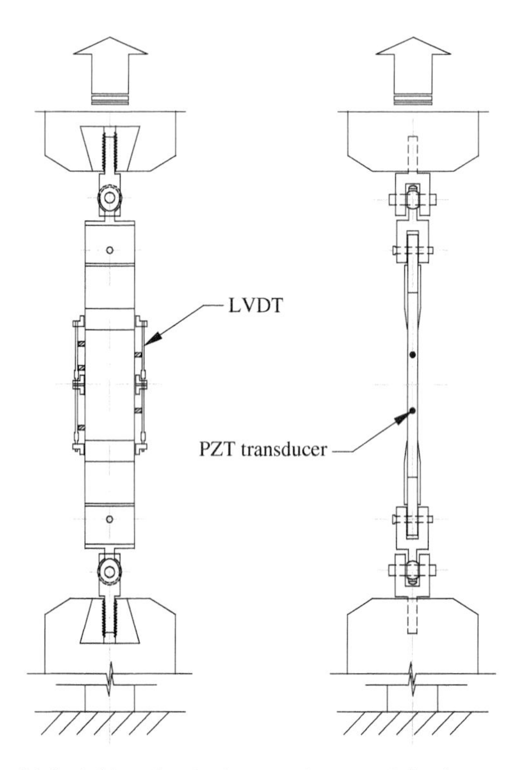


**Figure 5-33** Uniaxial tension test setup for unnotched concrete specimen

is not truly uniform. When a notched specimen is used, it is not possible to study the accumulation of damage and strain localization phenomena, which are of great interest in the failure process.

A new method was developed to obtain the complete stress–deformation response of an unnotched concrete specimen, as shown in Figure 5-33. This method employs a digitally controlled closed-loop-testing machine and five control channels: stroke LVDT and four LVDTs mounted on the specimen. The test portion of the specimen was fully spanned by these four LVDTs. The outputs of these control channels and the load cell were monitored during the test by a computer program, which also enabled quick switching of the mode of control from one LVDT to another. The problem of uncertainty in the location of the major crack was tackled by the said LVDT arrangement and the computer program. It was demonstrated that it is always possible to obtain a stable post-peak response provided one ensures that at any time during the test, the feedback used is the LVDT that exhibits, at that time, the largest slope of the response–time curve. An acoustic emission (AE) measurement system of six channels was also used in the experiments. Monitoring of signals from the AE transducers provides valuable information, which help in making a decision to switch the control.

For the plain concrete specimen tested, a reliable result was obtained and the stress–deformation curve is shown in Figure 5-34. In Figure 5-34, "deformation" refers to displacement measured by the LVDT, which was in control in the post-peak region. The curve can be divided into three parts: the linear elastic part, the nonlinear pre-peak part, and the post-peak part. In the first part, concrete behaves elastically. The linear elastic part is characterized by uniform deformation and "global" behavior of the material. In the second part, due to the damage (indicated by the occurrence of AE events) in the specimen, the modulus of the material starts to reduce and thus nonlinearly appears in the stress–deformation curves. Since damage does not happen uniformly in the specimen, behavior of the material ceases to be "global."

After the peak load, a major crack develops in the specimen. The behavior of the specimen in this region can be explained with the help of fracture mechanics theory. One of the results from this theory is that the stiffness of a specimen (defined as the ratio of load-to-load-point displacement) decreases as a result of the growth of a crack. In the tests described here, the test machine was issued a command to apply load in such a way that the response of the controlling LVDT increased linearly with time. To do this in the post-peak region, the machine must decrease the load on the specimen to maintain equilibrium. Such a reduction in load was explained in earlier hypotheses that the material becomes "soft" in the post-peak region. Today, it is understood that a "softening" response in tension is actually a manifestation of the growth of a single crack.

### 5.2.2 Modulus of Elasticity

The modulus of elasticity can be measured directly from the initial slope of a specially designed stress–strain curve. To get such a curve, the load should be applied gradually, at a rate of 2–3 MPa per second, until the value of the load, P, corresponding to 40% of the peak value is reached. Then

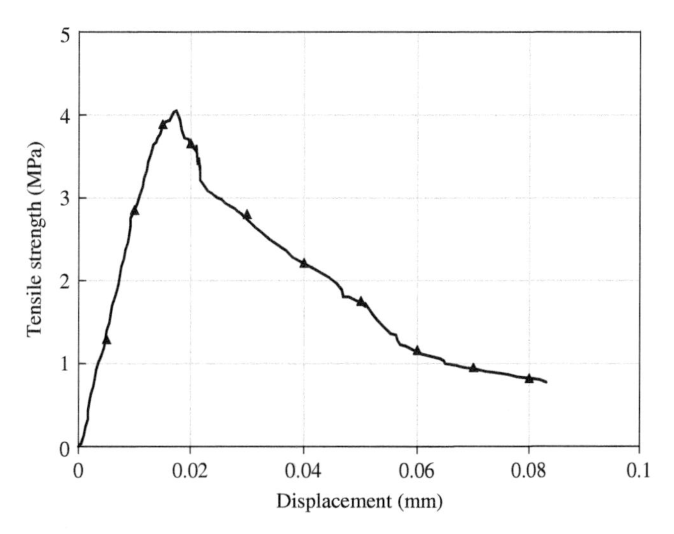


**Figure 5-34** The stress-deformation curve of a unnotched concrete specimen

the load is reduced to near zero at approximately the same rate as it was increased. The above loading/unloading procedure is repeated three times. After finishing three loading cycles, the fourth loading cycle is applied at the aforementioned rate to the value $P_0$, corresponding approximately to a stress value of 5 MPa. This load is maintained for 30 sec. The deformation ($\delta_b$) is measured from the displacement sensors, and then the load is increased to $P$, corresponding to 40% of the peak value, and maintained for 30 sec (see Figure 5-35). Similarly, the corresponding deformation ($\delta_a$) is recorded from the displacement sensor. After calculating the mean value of the difference ($\delta_a - \delta_b$) of the two deformation readings, the result is denoted as $\delta_4$. Loading is then reduced to the base value of $P_0$, as the initial load for the 5th loading cycle. The 5th loading cycle is applied as in the above procedure (i.e., the 4th cycle). The corresponding deformation is recorded and the mean value ($\delta_5$) of the deformation changes recorded by the two gauges is calculated. If the difference between $\delta_4$ and $\delta_5$ is not greater than 0.003 mm, take off the gauges and increase the load at the aforementioned rate until the specimen fails, then the compressive strength ($f_c$) is recorded. The modulus is calculated with the following equation

$$E_h^s = \frac{P - P_0}{A} \times \frac{l}{\delta_n} \tag{5-45}$$

in which, $E_h^s$ is the modulus of elasticity of concrete in compression, $P_0$ is the load corresponding to a stress level of 5 MPa, $P$ is the load corresponding to $0.4\,f_c$, $A$ is the area over which the load is applied, $l$ is the measuring length, and $\delta_n$ is the mean value of the deformation differences in the last loading cycle.

The modulus of elasticity of concrete can also be predicted theoretically using information of the aggregate's modulus and the cement paste's modulus. Here three popular models are introduced.

**(a)** *Parallel model (or isostrain model)*: In the parallel model, it is assumed that all the aggregates are concentrated in the central part of the concrete, in parallel to the loading direction, as shown in Figure 5-36. For this loading pattern, two conditions have to be satisfied: the deformation must be the same in the matrix, the aggregate, and the concrete; and the total force carried by the concrete must be equal to the force carried by the matrix and by the aggregate. Thus, we have

$$\Delta l_c = \Delta l_m = \Delta l_a \tag{5-46}$$

and

$$P_c = P_m + P_a \tag{5-47}$$

where $\Delta l_c$ is the length change of the concrete, $\Delta l_m$ is the length change of the matrix, and $\Delta l_a$ is the length change of the aggregate; $P_c$, $P_m$, and $P_a$ are the loads carried by the concrete,

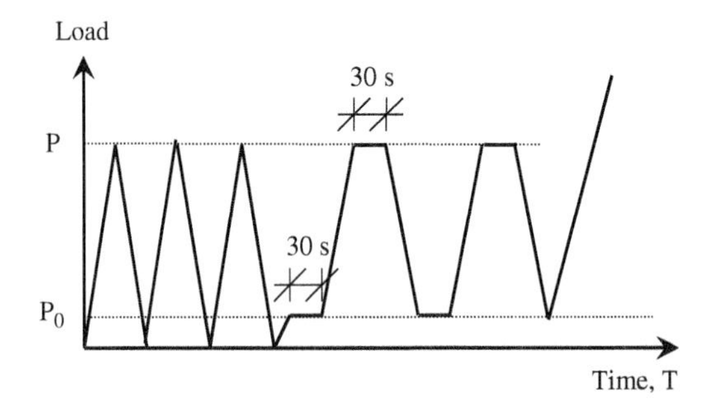


**Figure 5-35** Loading time diagram for evaluation of modulus of elasticity of concrete

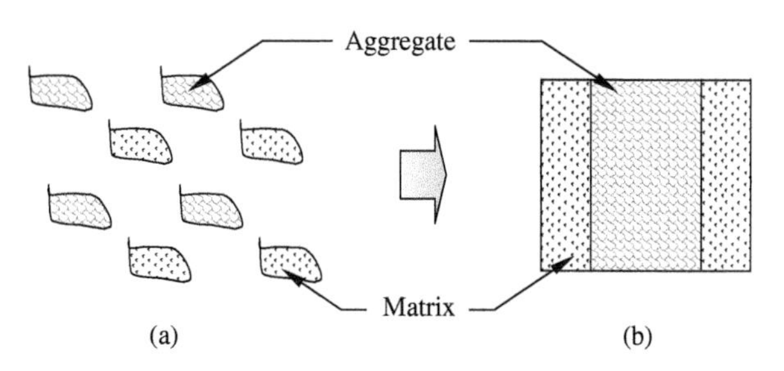


**Figure 5-36** The parallel model for estimation of elastic modulus: (a) the discrete system; and (b) the representative volume

by the matrix, and by the aggregate, respectively. Since all the components in the parallel model have the same original length, from Equation 5-46, we can derive that the concrete, the matrix, and the aggregate have the same strain:

$$\varepsilon_c = \varepsilon_m = \varepsilon_a \tag{5-48}$$

By dividing Equation 5-47 by $A_c$, the cross-sectional area, on both sides, we can get

$$\frac{P_c}{A_c} = \frac{P_m + P_a}{A_c} = \frac{P_m A_m}{A_m A_c} + \frac{P_a A_a}{A_a A_c} \tag{5-49}$$

Based on the definition of stress and the fact that volume fractions of matrix and aggregate equal their area fractions in such a parallel model, we can get the following equation from Equation 5-49:

$$\sigma_c = \sigma_m V_m + \sigma_a V_a \tag{5-50}$$

where $V_m$ and $V_a$ are volume fractions of the matrix and aggregate, respectively. Considering Equation 5-48 and applying Hooke's law, the modulus of elasticity of concrete can be expressed as

$$E_c = E_m V_m + E_a V_a \tag{5-51}$$

This formula is also called the rule of mixtures and is widely accepted in composite mechanics.

**(b)** *Series model (isostress model)*: In this kind of model, it is assumed that the forces in the matrix and aggregate are the same and the deformation of the concrete equals the sum of the deformations in the matrix and aggregate. Starting from

$$P_c = P_m = P_a \tag{5-52}$$

and

$$\Delta l_c = \Delta l_m + \Delta l_a \tag{5-53}$$

we can obtain

$$\frac{1}{E_c} = \frac{v_m}{E_m} + \frac{v_a}{E_a} \tag{5-54}$$

**(c)** *Square-in-square model*: In this kind of model, the representative volume in concrete is idealized as a square, as shown in Figure 5.37b. If we break the system into three slices as shown in Figure 5.37c and consider these three slices to be in series, we can finally obtain the following equation:

$$\frac{1}{E_c} = \frac{1 - \sqrt{v_a}}{E_m} + \frac{\sqrt{v_a}}{\left(1 - \sqrt{v_a}\right) E_m + \sqrt{v_a} E_a} \tag{5-55}$$

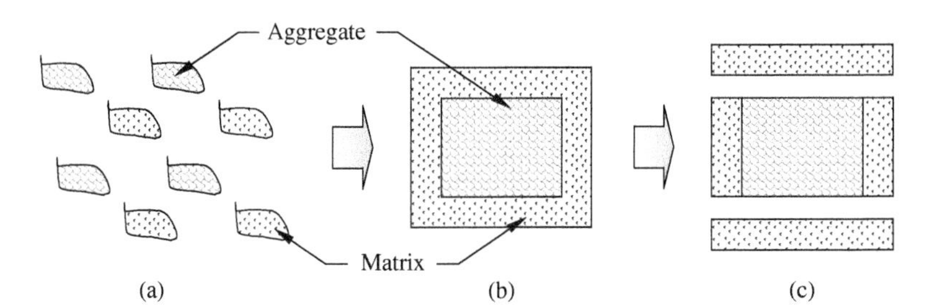


**Figure 5-37** The square-in-square model: (a) the discrete system; (b) the representative volume; and (c) the sliced system

It has been verified by experiments that the results obtained from this equation agree better than lower- and upper-bound equations, i.e., the series and parallel models.

**(d)** *Relationship between compressive strength and modulus of elasticity*: According to the British Standard for the structural use of concrete (BS 8110: part 2; replaced by BS EN 1992-1-1), the modulus of elasticity of concrete can be related to the cube compressive strength by the expression

$$E_c = 9.1 f_c^{0.33} \tag{5-56}$$

when the density of concrete is 2320 kg/m$^3$, i.e., for a typical normal-weight concrete. In the equation, $E_c$ is in GPa and $f_c$ is in MPa. If the density of concrete is between 1400 and 2320 kg/m$^3$, the expression for modulus of elasticity is

$$E_c = 1.7\rho^2 f_c^{0.33} \times 10^{-6} \tag{5-57}$$

where $\rho$ is the density of concrete in kg/m$^3$. According to the ACI Building Code 318, the relationship between the modulus of elasticity and compressive strength for normal density concrete is

$$E_c = 4.70 f_c^{0.5} \tag{5-58}$$

where $f_c$ is the cylinder compressive strength. For concrete with density of 1442–2563 kg/m$^3$ (corresponding to the range of 90–160 lb/ft$^3$ in ACI 318), the relationship changes to

$$E_c = 43\rho^{1.5} f_c^{0.5} \times 10^{-6} \tag{5-59}$$

It should be noted that in all the above equations, MPa is used for strength and stress, and GPa for modulus of elasticity.

### 5.2.3 Constitutive Equations

A constitutive equation is a relation between two physical quantities that describes the response of a material or substance to external functions, such as load, temperature, water flow, or ionic transport. In concrete structural analysis, the most important constitutive equation is the stress–strain relationship that connects applied stress or forces to strain or deformation in concrete. The stress–strain relationship is also called Hooke's law. There are two ways to obtain constitutive equations: the phenomenological method and the upscaling method. The phenomenological method is used to obtain knowledge of constitutive equations through empirical observations of phenomena that are consistent with fundamental theory, but not directly derived from theory. The upscaling method is used to obtain the constitutive equations from the nature of the microstructure of a material and through theoretical derivation based on first principles. For concrete, the constitutive equations are mostly obtained by the phenomenological method, or curve fitting of experimentally obtained stress–strain (deformation) relationships.

As shown in Figure 5-38, a typical stress–strain curve obtained through compressive testing shows several stages: linear elastic, inelastic, and stain-softening or post-peak response (Chen, 1981). Numerous mathematical equations have been developed for the nonlinear constitutive stress–strain relationship of concrete under uniaxial compression. Difficulty arises on how to represent the suspected nature of the deformations of concrete that are generally attributed to the process of progressive microcracking. Moreover, concrete is a highly complex composite, and its deformation response is closely related to its composition, as well as its internal microstructure. It is anticipated that a single mathematical equation is not sufficient to represent the expected wide range of constitutive behavior for different grades of concretes. Yip (1998) has derived a general stress–strain equation for a prismatic concrete specimen, having an aspect ratio of 2.5, under uniaxial compression, as

$$\frac{\sigma_c}{\sigma_{c,u}} = \frac{\varepsilon_c}{\varepsilon_{c,u}} e^{\left(1-\frac{\varepsilon_c}{\varepsilon_{c,u}}\right)} \tag{5-60}$$

where $\sigma_c$ is the compressive stress, $\sigma_{c,u}$ is the ultimate compressive stress, $\varepsilon_c$ is the compressive strain, and $\varepsilon_{c,u}$ is the strain corresponding to $\sigma_{c,u}$. By using the exponential power series expansion, Equation 5-60 can be rewritten as

$$\sigma_c = \frac{2.7182 E_{c,u}\varepsilon_c}{1 + \left(\frac{\varepsilon_c}{\varepsilon_{c,u}}\right) + \frac{1}{2}\left(\frac{\varepsilon_c}{\varepsilon_{c,u}}\right)^2 + \frac{1}{6}\left(\frac{\varepsilon_c}{\varepsilon_{c,u}}\right)^3} \tag{5-61}$$

where $E_{c,u}$ is the modulus of elasticity. It should be pointed out that Equation 5-61 describes not only the ascending branch of the stress–strain curve, but also the descending branch of the curve after the peak stress point. Moreover, the constitutive model presented above is only one typical example in this area. Many other attempts have been made to derive empirical stress–strain equations, including the descending branch, by Wang et al. (1978), Popovics (1973), and Blechman (1992). Recently, with the development of a better understanding of the concrete

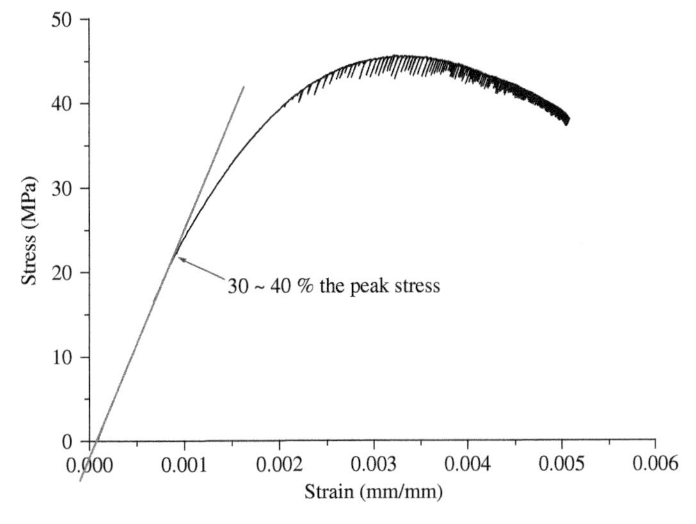


**Figure 5-38** A typical stress–strain curve obtained through the compressive test

structure at the nano- and microscales, attempts to develop constitutive relationships from the microstructure using the upscaling method or multiple scaling techniques have been made. Examples of such new methodologies are atom/molecular simulation-based upscaling (Masoero et al., 2014) and lattice model-based multi-scale modeling (Qian, 2012).

## 5.3 DIMENSIONAL STABILITY—SHRINKAGE AND CREEP

Dimensional stability is defined as the ability of a material to keep its size, shape, or dimension over a long period. The volumetric change of a dimensionally stable material over a long period of time should be sufficiently small that it will not cause any structural problems. For concrete, shrinkage and creep are the two major phenomena that compromise the dimensional stability. Shrinkage and creep are often discussed together because they both originate in the hydrated cement paste (or the hydration process) within concrete. The aggregate in concrete does not exhibit shrinkage and creep. In addition, shrinkage and creep are caused and influenced by many common factors, such as water content, curing conditions, relative humidity, aggregate proportions, and specimen sizes.

### 5.3.1 Shrinkage

Several types of shrinkage can occur in a concrete, including thermal shrinkage, plastic shrinkage, autogenous shrinkage, chemical shrinkage, and drying shrinkage. In this section, we focus on plastic, autogenous, and drying shrinkage.

#### 5.3.1.1 Plastic Shrinkage

Plastic shrinkage is a surface shrinkage that happens at a very early age of concrete (only a few hours after casting) while the concrete underneath is still in the plastic stage. After casting, if the concrete is not properly taken care of with a good curing method, the top surfaces of concrete pours are subjected to evaporation and consequent loss of mix water. The rate of evaporation depends on ambient conditions such as temperature, exposure to sun, wind speed, and relative humidity. The water lost by evaporation on the surface is usually replaced by water rising from the internal regions of the concrete. Once the rate of removal of water from the surface exceeds the rate of immigration of the internal water to the surface, the surface layer's volume starts to show local reductions, or plastic shrinkage. The magnitude of the plastic shrinkage has been shown by L'Hermite (1960) to be over 6,000 microstrain for paste and 2,000 microstrain for concrete; and in extreme cases can be as large as 10,000 microstrain (Troxell et al., 1968). In a plastic state, no great stress is induced in the concrete and further working of the concrete can generally be applied to eliminate consequential cracks.

Plastic shrinkage usually leads to a downward movement of the solid and heavier ingredients in the surface layer. This downward movement may be resisted by the large size of coarse aggregates or by the top layer of reinforcement. In this case, the surface layer of the concrete above the coarse aggregate or reinforcing bar tends to become draped over the aggregates or bars, and hence creates cracks, called plastic shrinkage cracks. The process leading to plastic shrinkage cracking is shown diagrammatically in Figure 5-39. Plastic shrinkage cracks are usually of shallow depth, generally 38 to 50 mm, and 300 to 450 mm long, normally perpendicular to the wind, and typically run parallel to one another. These plastic shrinkage cracks provide a path for water and other chemicals (e.g., chloride ions) to penetrate into the concrete and reach the steel reinforcement, which can greatly affect the durability of concrete structures, for instance, facilitating corrosion of the steel reinforcement.

The most effective ways of preventing plastic shrinkage are by sheltering the surface from the wind and sunshine during construction and by covering the concrete surface immediately after

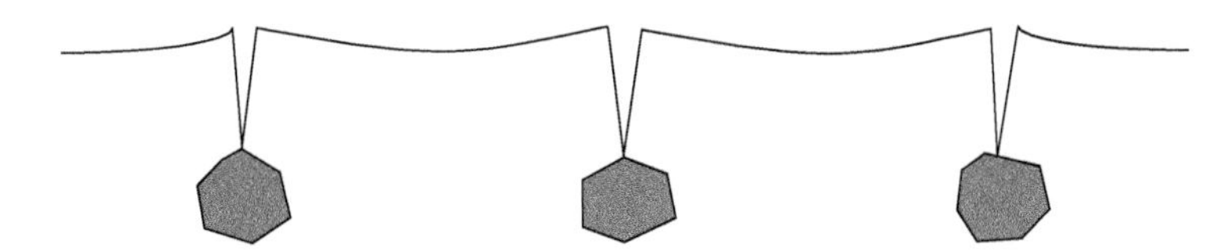

**Figure 5-39** Formation of plastic shrinkage crack

**Table 5-1** Minimum periods of curing and protection

| Type of cement | Ambient conditions after casting | Minimum period of curing and protection — Average concrete surface temperature: 5–10°C | Minimum period of curing and protection — Average concrete surface temperature: $t$ °C (10–25°C) |
|---|---|---|---|
| Portland cement, SRPC | Average | 4 days | $60/(t + 10)$ days |
| | Poor | 6 days | $80/(t + 10)$ days |
| All except Portland cement and SRPC, and all with GGBS or PFA | Average | 6 days | $80/(t + 10)$ days |
| | Poor | 10 days | $140/(t + 10)$ days |
| *All* | Good | No special requirements | |

*Note:* SRPC, sulfate-resisting Portland cement; GGBS, ground granulated blast-furnace slag; PFA, pulverized fuel ash.

finishing, which are all directed toward reducing the rate of evaporation. BS 8110-1: 1997 had recommended minimum periods of curing and protection (see Table 5-1) for fresh concrete to reduce plastic shrinkage and guarantee development of properties. ACI 308R (Guide to External Curing of Concrete) has similar regulations regarding concrete that is maintained above 10 °C. Changes in concrete mix design, and especially the use of air entrainment, may also be helpful in reducing plastic shrinkage. Remedial measures after the cracks have formed usually consist of sealing them against the ingress of water by brushing in cement or low-viscosity polymers (Allen et al., 1993).

### 5.3.1.2 Autogenous Shrinkage

Autogenous shrinkage is defined as the macroscopic volume contraction of concrete at an early age (less than one day after casting) occurring without moisture transfer from the concrete to the surrounding environment. Autogenous shrinkage can be attributed to self-desiccation due to the hydration of cement and is a result of chemical shrinkage (i.e., the total volume of hydration products is smaller than the sum of volumes of water and cement minerals). Autogenous shrinkage was first described in the 1930s by Lyman (1934) as a factor contributing to the total shrinkage. However, in the earlier days, it was noted that autogenous shrinkage occurred only at very low *w/c* ratios, far below the practical *w/c* range, and did not attract much attention. With the development and applications of advanced admixtures such as superplasticizers, *w/c* ratios lower than 0.42 are realized in concrete practice and autogenous shrinkage has become an important issue for contemporary concrete. This issue has been further enlarged by the incorporation of silica fume into concrete mixes.

To better understand the concept of autogenous shrinkage, the chemical shrinkage needs to be examined first. When cement contacts water, the ions dissolve into solutions from cement and react with water to produce hydrates. The hydration leads to the reduction of the total absolute volume of the cement system. This phenomenon is called chemical shrinkage. It was first discovered by

Le Chatelier (1900), who described the basic distinction between the apparent and the absolute volume of cement paste. The apparent volume is essentially the external volume of a sample, which contains the spaces occupied by solid, liquid, and gas phases. The absolute volume excludes the space occupied by the gas phase. Chemical shrinkage can be seen from the volume change before and after a complete reaction of $C_3S$. The stoichiometric equation for fully hydrated $C_3S$ can be expressed as (Damidot et al., 1990):

$$C_3S + 5.2H \rightarrow C_{1.75}SH_{3.9} + 1.3CH \tag{5-62}$$

By substituting the molecular weight and the densities of all products in Equation 5-62, it can be seen that the hydration of $C_3S$ results in a reduction of absolute volume of the system (Nawa and Horita, 2004):

$$\begin{gathered} C_3S + 5.2H \rightarrow C_{1.75}SH_{3.9} + 1.3CH \\ 72.8\,\text{cm}^3 + 93.6\,\text{cm}^3 \rightarrow 95.9\,\text{cm}^3 + 42.9\,\text{cm}^3 \\ 166.4\,\text{cm}^3 \rightarrow 138.8\,\text{cm}^3 \end{gathered} \tag{5-63}$$

Obviously, there is about a 16.5% reduction in volume after hydration if the reactants follow their stoichiometric proportions. It should be pointed out that autogenous shrinkage and chemical shrinkage have some differences. The relationship between autogenous shrinkage and chemical shrinkage is shown schematically in Figure 5-40. When cement paste is in a plastic stage, the apparent volume change or autogenous shrinkage is essentially the same as the reduction of absolute volume or chemical shrinkage. When the hydration products percolate to form a structural skeleton, autogenous shrinkage can be restrained by the skeleton and deviates from the theoretical chemical shrinkage.

Many techniques have been developed to measure the autogenous shrinkage. They can be classified in two basically different ways. One is to measure the volume change and the other is to measure the change in linear length. The volume change of cement paste can be measured by immersing the specimen in a tight rubber balloon and then measuring the amount of water displaced by the specimen. The linear length change can be measured using a strain gauge embedded in the specimen. During the hydration process, the cement paste deforms and generates internal strain that reflects autogenous shrinkage. Autogenous shrinkage has to be measured without any

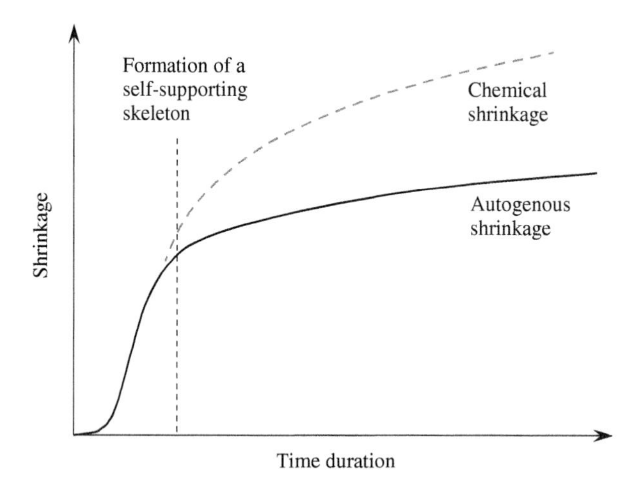


**Figure 5-40** Relationship between autogenous shrinkage and chemical shrinkage

moisture transferred. Hence, the specimen has to be sealed tightly by materials such as a Teflon sheet or a rubber membrane, as shown in Figure 5-41. To pursue high-fidelity measurement or monitoring of such internal strains in concrete, embedded fiber-optic sensors either point sensors (e.g., based on fiber Bragg gratings) (Pei et al., 2014) or distributed sensors (e.g., based on Rayleigh scattering-based optical frequency domain reflectometry system) (Liao et al., 2020; Sun et al., 2019) have been employed widely in the last decade. Recent developments in embedded optical fiber and coaxial cable strain sensors (Du et al., 2017; Huang et al., 2012) have achieved high resolutions at the level of 0.1 micro-strain.

#### 5.3.1.3 Drying Shrinkage

As mentioned earlier, to produce workable concrete, nearly twice as much of the water theoretically needed to hydrate the cement has to be added to the concrete mix if no water-reducing admixture is used for normal concrete. After concrete has been cured and begins to dry, the excessive water that has not reacted with the cement will begin to migrate from the interior of the concrete mass to the surface. As the moisture evaporates, the concrete volume shrinks. The loss of moisture from the concrete varies with distance from the surface. Drying occurs most rapidly near the surface because of the short distance the water must travel to escape, and more slowly from the interior of the concrete because of the increased distance from the surface. The shortening per unit length associated with the reduction in volume due to moisture loss is termed drying shrinkage. A nearly linear relationship exists between the magnitude of the shrinkage and the water content of the mix for a particular value of relative humidity. If the relative humidity increases, the shrinkage of the concrete drops. When concrete is exposed to 100% relative humidity or is submerged in water, it will actually increase in volume slightly as the gel continues to form because of the ideal conditions for hydration.

Three basic mechanisms are responsible for the drying shrinkage of Portland cement concrete. One is the disjointing pressure that is related to the water absorbed on the surface of C–S–H. Water is absorbed in the layers of C–S–H at all relative humidities, and the thickness of the water

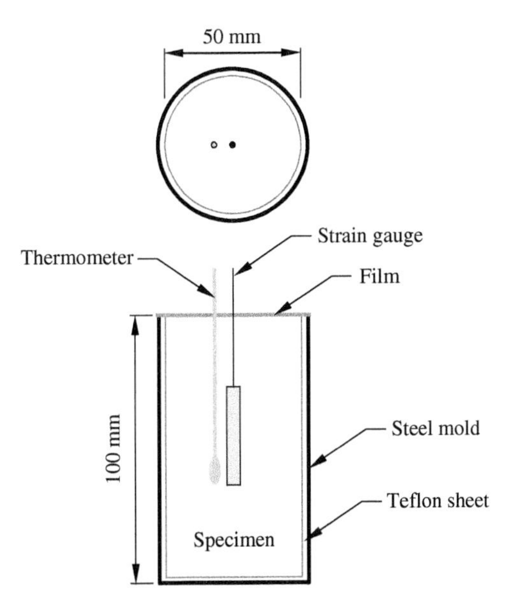


**Figure 5-41** Method of measuring autogenous shrinkage of cement paste

layer increases with increasing humidity. At a relative humidity of more than 10%, the surface bounding in the narrow spaces can form an absorbed film of water, as shown in Figure 5-42. If the distance between two layers is restricted due to the van der Waals force of attraction, the absorbed water molecules between the C–S–H surfaces may generate a pressure, which leads to an expansion. This pressure is termed disjointing pressure. However, if the relative humidity of the system reduces, the disjointing pressure decreases accordingly and the separated surfaces will be brought closer by the van der Waals force again, leading to the reduction of volume or shrinkage. Disjointing pressure plays an important role in shrinkage when the relative humidity is higher than 75% and has no effect when the RH is lower than 45%.

The second mechanism responsible for drying shrinkage is the capillary surface tension effect. With the progress of hydration, the air volume increases (due to chemical shrinkage or the self-desiccation effect) and, subsequently, the interface between the air and water increases in a capillary pore. The interface, or meniscus, creates stress, called capillary stress. Capillary tension effects are due to meniscus formation in the capillary pores. This process results in equal hydrostatic compression in the solid phase, which pulls the voids in the C–S–H body closer (Powers, 1965). The capillary surface tension effect can be viewed by considering the Kelvin and the Laplace equations. The Kelvin equation is given as

$$\ln(\mathrm{RH}) = -\frac{2\sigma}{r}\frac{M}{vRT}\cos\theta \tag{5-64}$$

or

$$\frac{2\sigma}{r}\cos\theta = -\frac{vRT}{M}\ln(\mathrm{RH}) \tag{5-65}$$

where *RH* is the relative humidity, $\sigma$ the surface tension of water in contact with air, *M* the mass of a mole of water, $\theta$ the contact angle of water and solid, *r* the radius of pore, *v* the density of water, *R* the ideal gas constant, and *T* the temperature on the Kelvin scale. The Laplace equation can be written as

$$p_c - p_v = \frac{2\sigma}{r}\cos\theta \tag{5-66}$$

or

$$\Delta p = \frac{2\sigma}{r}\cos\theta \tag{5-67}$$

where $p_c$ is the capillary pressure of water in a pore, $p_v$ is the vapor pressure, and $\Delta p$ is the suction pressure. By substituting the Kelvin equation into the Laplace equation, we get

$$\Delta p = -\frac{vRT}{M}\ln(\mathrm{RH}) = -\frac{\ln(\mathrm{RH})}{K} \tag{5-68}$$

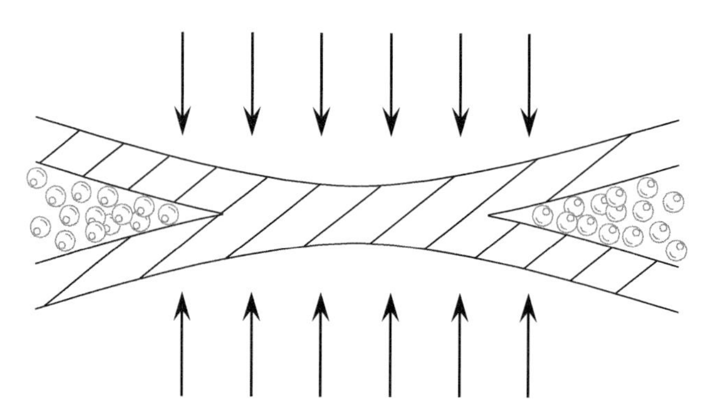

**Figure 5-42** Absorbed water film

From Equation 5-68, it can be seen that suction pressure increases with a decrease in internal humidity. In other words, the reduction of the internal relative humidity will create a larger pressure that leads to the shrinkage of concrete. It cannot exist below about 45% RH, since the disjointing pressure and capillary stress no longer exist, and shrinkage has to be explained by a change of surface energy. As the most strongly adsorbed water (equivalent to one or two molecular layers) is removed, the surface free energy of the solid begins to increase significantly. A liquid droplet is under hydrostatic pressure by virtue of its surface tension (surface energy). This pressure can be described as

$$P_{\text{ave}} = \frac{2\gamma S}{3} \tag{5-69}$$

where $P_{\text{ave}}$ is the mean pressure, $\gamma$ the surface energy in J/m$^2$, and $S$ the specific surface area of solid in m$^2$/g. Since $S$ is large in the case of C–S–H (about 400 m$^2$/g), $P_{\text{ave}}$ can be large and causes compression in the solid phase (Mindess et al., 2003).

Drying shrinkage can create stress inside concrete. Let us take a concrete cylinder as an example. Because the concrete adjacent to the surface of the cylinder dries more rapidly than the interior, shrinkage strains are initially larger near the surface than in the interior. As a result of the differential shrinkage, a set of internal self-balancing forces, i.e., compression in the interior and tension on the outside, is created. The stresses induced by shrinkage can be explained by imagining that the cylindrical core of a concrete cylinder is separated from its outer shell and that the two sections are then free to shrink independently in proportion to the effective quantities of water lost from them. Since deformations must be compatible at the junction between the core and the shell, shear stresses must be created between the core and the shell. If free-body diagrams of the upper half of the cylinder are considered, it is clear that vertical equilibrium requires the shear stresses to induce compression in the core and tension in the shell, see Figure 5-43. The self-balanced shrinkage stress may have some influence on flexural strength and splitting strength measurement. For flexural strength, the existence of a self-balanced stress will reduce a concrete's apparent flexural strength measured by beam bending (i.e., the test result underestimates the strength); contrarily, the self-balanced stress will require a larger peak load to split-break a cylinder (i.e., the test result overestimates the splitting tensile strength of concrete). In addition to the self-balancing stresses set up by differential shrinkage, the overall shrinkage creates stresses if members are restrained in the direction in which shrinkage occurs. Tensile cracking due to shrinkage will take place in any structural element restrained by its boundaries, such as a beam–column joint as shown in Figure 5-44. It must be controlled since it permits the passage of water, is detrimental to appearance, reduces shear strength, and exposes the reinforcement to the atmosphere.

The magnitude of the ultimate shrinkage is primarily a function of the initial water content of the concrete and the relative humidity of the surrounding environment. The shrinkage strain, $\varepsilon_{\text{sh}}$,

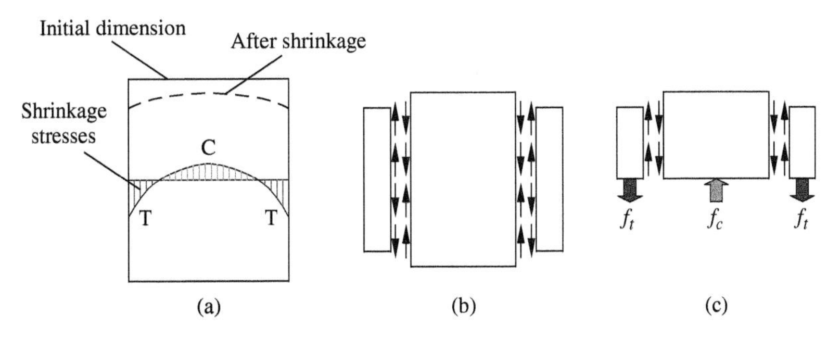


**Figure 5-43** Self-balanced stress generated by drying shrinkage

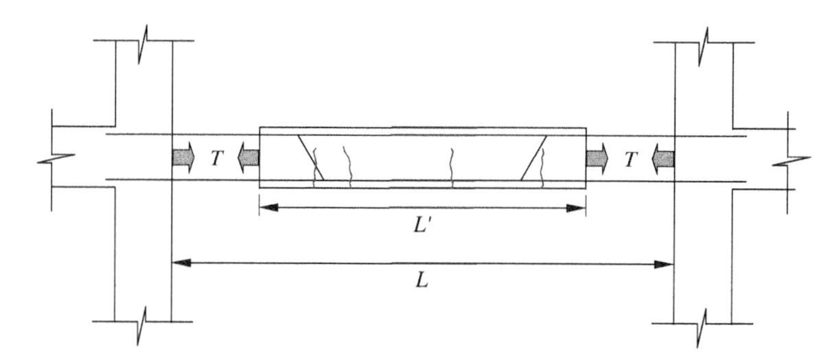


**Figure 5-44** Shrinkage crack generated by restraint

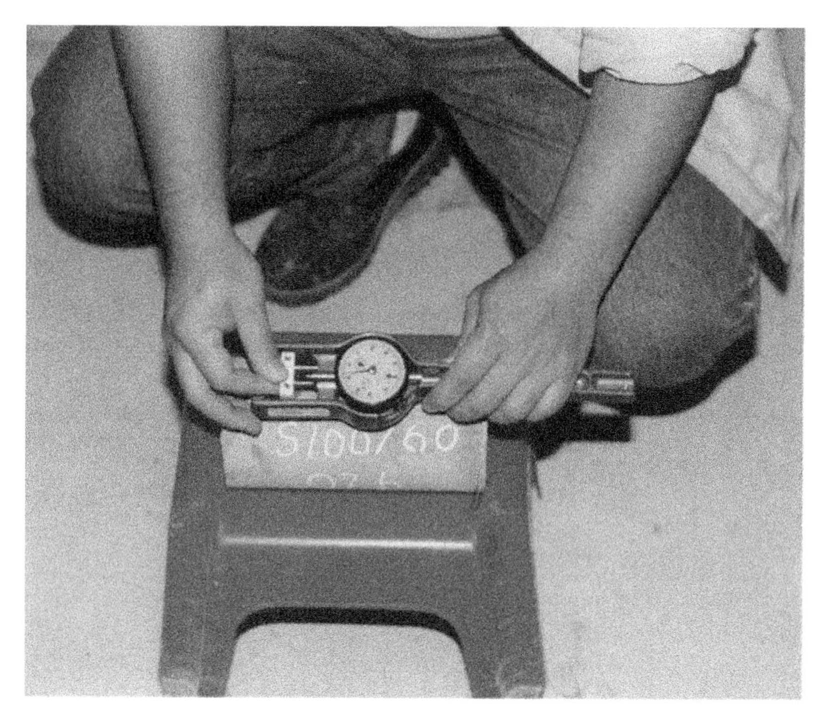

**Figure 5-45** Free drying shrinkage measurement using Demac gage on a cylinder specimen

is time-dependent. Approximately 90% of the ultimate shrinkage occurs during the first year. Both the rates at which shrinkage occurs and the magnitude of the total shrinkage increase as the ratio of the surface to volume increases. This is because the larger the surface area, the more rapidly moisture can evaporate. The value of shrinkage strain for plain concrete members ranges from 0.0004 to 0.0007 for standard conditions. For reinforced concrete members, the shrinkage strain values are between 0.0002 and 0.0003. This means reinforcement is helpful in reducing shrinkage. Free-drying shrinkage can be measured by using a Demec gage on a cylindrical specimen, as shown in Figure 5-45, a prismatic specimen, or a ring-shaped specimen. The specimens for shrinkage tests should be moved into the shrinkage test room after curing to a specified age. The temperature in the room is kept at 23 ± 2°C and relative humidity 50 ± 4% (per ASTM C157/157M and C596). The shrinkage is usually measured at different ages using a Demac gage, such as CT 171M with a maximum displacement measurement of 5 mm.

In addition to the free shrinkage test, restrained shrinkage tests are often required to investigate the cracking sensitivity of concrete. Restrained shrinkage is normally studied using three types of specimens: bars, plates, and rings. The bar-type test can provide uniaxial stress development for specimens with large-size aggregates. However, the difficulties of providing a constant restraint and end conditions often make this type of test complicated. Plate-type testing has been used to evaluate both biaxial and plastic shrinkage and provides a biaxial restraint that depends on geometry and boundary conditions. Ring-type specimens can easily be cast, and the end effects are removed due to the closed loop shape. The ring tests are widely used for restrained shrinkage cracking (Bloom and Bentur, 1995). In such tests, ring-shaped specimens are cast around a rigid inner steel core, as shown in Figure 5-46, to provide a restraint when the specimen has a tendency to shrink. The grade of cracking depends on the restraining conditions and the drying environment. The restraint induces tensile stresses in the concrete ring, which reach a maximal value at the inner surface of the specimen. If the tensile stress developed exceeds the tensile strength of concrete, cracking will be initiated. This method has been standardized as ASTM C1581/1581M to determine age at cracking and induced tensile stresses of mortar/concrete specimens under restrained shrinkage conditions.

Conventional circular ring testing has several disadvantages. First, it is difficult to predict the location of initial cracking due to the equal opportunity of cracking around the circumference of the ring. Second, for a steel ring without adequate stiffness, concrete with a higher toughness cannot generate a visible crack, but more invisible fine cracks, which may result in problems regarding concrete durability (Burrows, 1998). The low cracking sensitivity (mortar or concrete) is associated with geometry and non-stress intensity effects. In a circular ring test, the geometry of the sample does not generate sufficient stress to develop a visible crack at certain locations. The deterioration process in concrete is very complex. Commonly, the deterioration of a concrete can be advanced by early-age cracking. Thus, evaluation of cracking sensitivity of a material is necessary in concrete durability assessment. For the sake of improving the crack propagation rate of mortar or concrete and distinguishing cracking sensitivity of materials in shorter periods, a novel ellipse-type ring for restrained shrinkage test was developed (He and Li, 2005; He et al., 2004). In this setup, the shrinkage diffusive stress development in the direction of the long principal axis is faster than that of the short principal axis, which leads to the crack occurring at a more predictable position. The test apparatus not only can discriminate the extension of mortar or concrete in a short time, but also can lead to an early shrinkage crack. Therefore, the new apparatus is useful to discriminate early-cracking sensitivity of concrete. Moreover, the new method introduced an electrical conduction line along the circumference of the elliptical ring-shaped specimen. Automatic and continuous monitoring on the resistance of the line was able to detect the crack occurrence time exactly, see Figure 5-47.

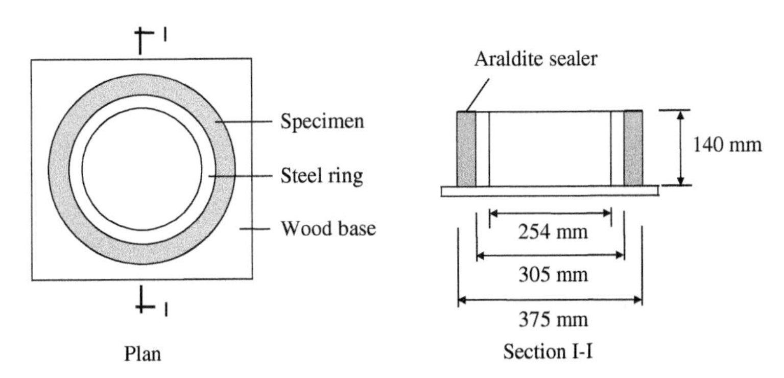


**Figure 5-46** Ring-shaped specimen for restrained shrinkage test

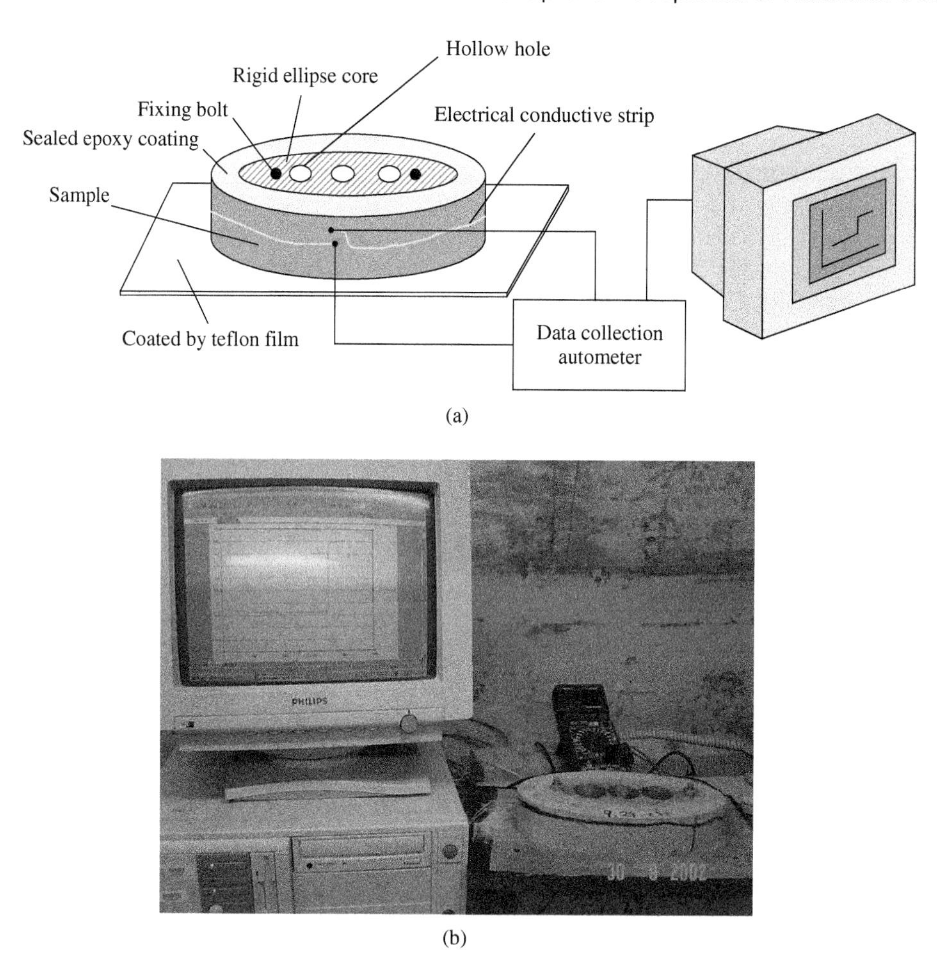


**Figure 5-47** Elliptical ring test for constrained shrinkage: (a) sketch; and (b) photo of equipment

#### 5.3.1.4 Shrinkage Control

Since shrinkage cracking can lead to premature deterioration, which shortens the useful life of concrete structures, many attempts have been made to reduce the shrinkage cracking or to control shrinkage. Following this strategy, using expansive concrete (shrinkage-compensating concrete) is a good solution. The details regarding this special concrete are provided in Chapter 6. Another method to reduce shrinkage is to use reinforcement. Since reinforcement does not shrink, it can restrain the shrinkage of concrete through the mutual bonding. During the restraining process, compression stress will be developed in the reinforcement steel and tension stress in the concrete. The resultant force of the two are self-balanced. In using reinforcement steel to reduce shrinkage, care has to be taken on the magnitude of the tensile stress generated in the concrete to ensure that it does not exceed the tension strength. For this purpose, if the area of the reinforcement steel is fixed, finer steel bars are preferred to thick steel bars. Using fibers to control shrinkage is another popular method. By adding fibers into the concrete, the shrinkage can be reduced due to the restraining effect of the fibers. More importantly, the shrinkage crack width can be controlled due to the bonding between the concrete and fibers (i.e., the bridging effect of the fibers). For early-age shrinkage crack control, polymeric (e.g., polyvinyl alcohol) fibers are frequently used. In recent years, using a shrinkage reducer, a type of chemical admixture consisting of a blend of propylene glycol derivatives, to control shrinkage has gained more and more attention. The shrinkage-reducing admixture (SRA) was invented in 1982, and later patented in 1985 (Goto et al., 1985). Subsequently, many researchers have performed detailed studies on SRA. Some literature has indicated that the reduction of free shrinkage can postpone the time to cracking. Moreover, SRA can reduce the long-term drying shrinkage by 50%, and there is a significant improvement in the restrained shrinkage performance. Even for concrete with proper curing at which the drying shrinkage would reduce to a minimum, there is still a substantial reduction in drying shrinkage due to the effect of SRA (Berke et al., 1999). SRA could also play a vital role in advanced cement-based materials, such as ultra-high performance concrete (UHPC), in which shrinkage represents a critical issue hindering practical applications (Soliman and Nehdi, 2014; Valipour and Khayat, 2018).

### 5.3.2 Creep

#### 5.3.2.1 Phenomenon of Creep

For many materials (e.g., polymers, wood, concrete), the response to stress or strain has a time-dependent component. For example, when a fixed stress is applied, after an instantaneous elastic response, the strain will continue to increase with time. This phenomenon is called creep and is illustrated in Figure 5.48a. Hence, creep is defined as a time-dependent deformation under a constant load. The creep develops in a concrete rapidly at the beginning and gradually decelerates with time. Approximately 75% of the ultimate creep in concrete occurs during the first year. The total deformation of a reinforced concrete specimen consists of the instantaneous deformation, shrinkage deformation, and creep. On the other hand, when a fixed strain is applied (e.g., by stretching a member and then fixing its ends), the stress in the member will decrease with time as shown in Figure 5.48b. This phenomenon is called stress relaxation.

If creep and relaxation are linear (e.g., if the stress is doubled, the strain at a particular time is also doubled), we can define the following two parameters:

$$\text{Creep compliance: } J(t) = \varepsilon(t)/\sigma \tag{5-70}$$

$$\text{Relaxation modulus: } E_r(t) = \sigma(t)/\varepsilon \tag{5-71}$$

Creep compliance can be obtained from a test with a fixed load applied to a specimen. When $J(t)$ is known, the time-dependent behavior of the material under an arbitrary loading history can be obtained from superposition (see Section 5.3.2.5).

For materials exhibiting creep behavior, when a stress is applied, the strain will increase with time. If stress is applied at a slower rate (i.e., over a longer period of time), the resulting strain will be more than that due to a stress applied at a rapid rate. Figure 5-49 shows the

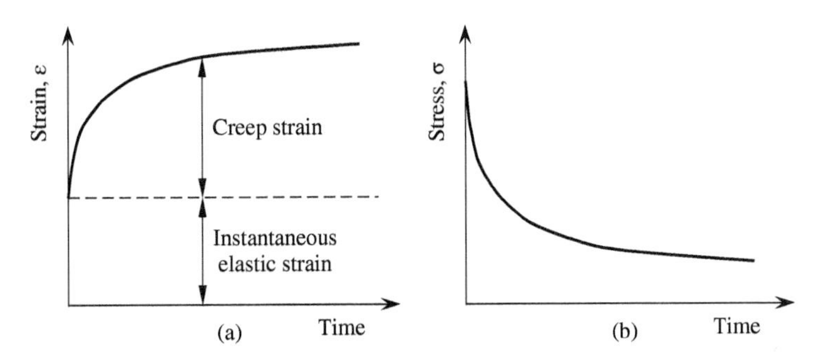


**Figure 5-48** The time-dependent behaviors: (a) creep, and (b) stress relaxation

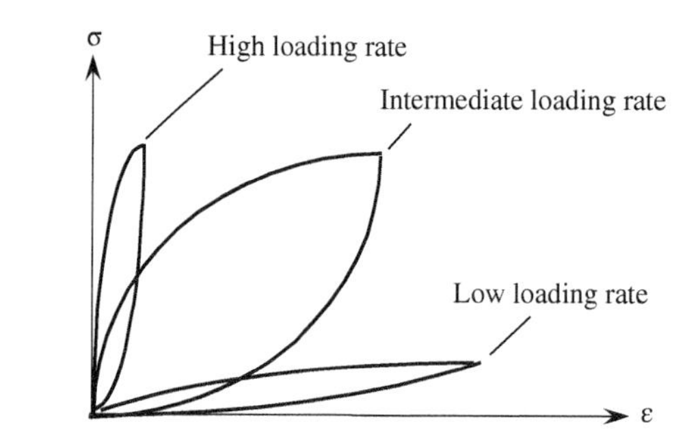


**Figure 5-49** Hysteresis behavior under various loading rates

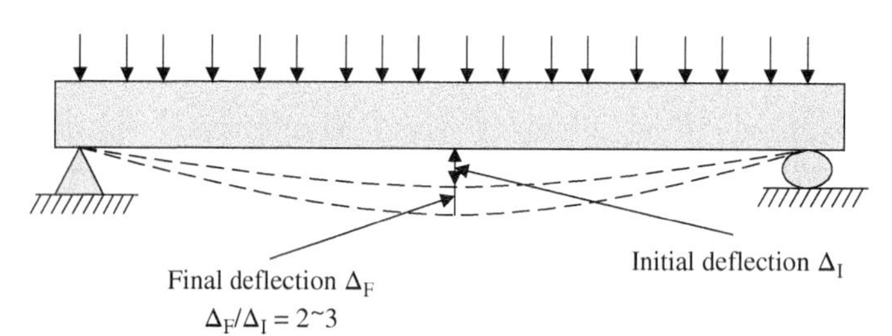


**Figure 5-50** The long-term deformation of a reinforced concrete beam

loading/unloading behavior for three general cases (low, high, and intermediate loading rates). For creeping materials, the loading and unloading curves do not overlap with one another. The area between the two curves (called the hysteresis loop) reflects the energy absorbed by the material over a loading/unloading cycle. This energy absorption varies with loading rate and is the highest at intermediate loading rate.

#### 5.3.2.2 Influence of Creep on Reinforced Concrete

When a material exhibits time-dependent behavior, it will affect the structural behavior in a number of ways. In a reinforced concrete (RC) column supporting a constant load, creep can cause the initial stress in the steel to double or triple with time because steel is noncreeping and thus takes over the force originally carried by the concrete. Creep can influence reinforced concrete in the following aspects:

**(a)** Due to creep effects, the long-term deformation of a reinforced concrete element can be significantly larger than the short-term deflection. For instance, because of the delayed effects of creep, the long-term deflection of a reinforced concrete beam can be 2–3 times larger than the initial deflection (Figure 5-50). Therefore, sufficient stiffness has to be provided during the design process to make sure that the beam deflection meets the long-term requirement. For large structures, the long-term differential creep in different parts of the structure needs to be checked to ensure no mismatching problems will be caused. For a tall building or structure, the persistent shortening of reinforced concrete columns may cause the final height of the building or structure to be significantly shorter, and this has to be taken into consideration in design and construction.

**(b)** The hysteresis loop shown in Figure 5-49 indicates that energy can be absorbed during cyclic loading. The energy absorption results in the damping of a structure as it is set in vibration (e.g., during an earthquake or typhoon). Note that the damping is frequency-dependent, although this is often not considered in civil engineering designs, as damping is difficult to quantify in practice.

**(c)** In prestressed concrete design, the creep of concrete and the relaxation of steel will lead to a loss of prestress. The percentage of the prestress loss due to shrinkage and creep can be as high as 60%. It has to be taken into account in the design of a prestressed structure to make sure that sufficient prestress can be applied. Moreover, in some cases, restressing of the prestressed tendon has to be carried out to compensate the stress caused by creep.

**(d)** For a reinforced concrete column, creep may lead to significant stress redistribution in concrete and in the reinforcing steel. This can be explained by a parallel model of a reinforced concrete column as illustrated in Figure 5-51. As the creep occurs only in concrete, the reinforcing steel will restrain the concrete creep through the bond between the concrete and steel. As a result, the reinforced concrete column will shorten less than a same-sized pure concrete column. This implies that the steel is compressed while the concrete is stretched. During this process, tensile stress will be generated in the concrete and compressive stress in steel. Hence, the original compressive stress level in concrete will be reduced, while in steel it will be increased. Such stress redistribution may cause the final compressive stress in steel to be 2 or 3 times higher than the original stress value. If it is not considered during the design process, overstressing in steel may cause yielding and put the structure in danger.

The misalignment or shortening of a reinforced column due to shrinkage and creep should also be considered during design. Otherwise, the net space of a building on each floor can be compromised. For instance, in Lake Shore Towers in Chicago, combined creep and shrinkage resulted in a shortening of the vertical columns by 2.5 mm per floor (Mindess et al., 2003).

#### 5.3.2.3 Mechanism of Creep in Concrete

For most materials, creep behavior is due to the time needed for atoms or molecules to rearrange themselves under load. For example, when a polymer is under stress, the polymeric chains tend to slide relative to one another. A finite time is required for the chains to go from one state (i.e., a given arrangement) to another. When loaded for a longer time, more movement will occur, leading to creep. For concrete, it is believed that creep is mainly caused by the position rearrangement of water molecules that are absorbed in the layer of C–S–H gel. Creep strains originate from the deformation of a microvolume of paste, named a *creep center*, with higher energy. The creep center will deform when changing from a relatively high energy state to a relatively low energy state through an intermediate state with a certain energy obstacle due to the influence of external sources. The ability of a creep center to cross the intermediate state depends on the height of the

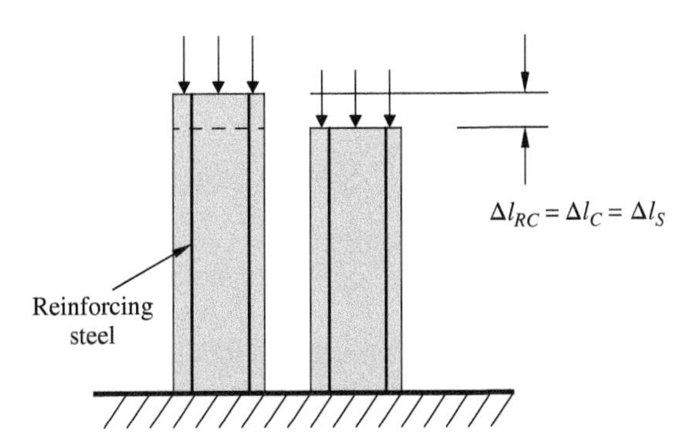


**Figure 5-51** Parallel model for a reinforced concrete column

intermediate state and the energy input from external sources. Such sources may be temperature, stress (strain energy), and variation of moisture concentration. In concrete, the nature of the creep center involves slip between adjacent particles of C–S–H under shear stress. The ease and extent of the slip depend on the process of attraction between particles. If the particles are chemically bonded, no slip can occur. If only van der Waals interaction exists, slip can occur under some conditions. For instance, when there is a sufficient thickness of water layers between the C–S–H particles, the water can reduce the van der Waals forces and cause slippage to occur between the particles (Mindess et al., 2003).

In concrete, creep can also result from the diffusion of adsorbed water in nanoscale pores between C–S–H layers. The thickness of the adsorbed water films that separate C–S–H particles depends on the relative humidity with which the system is in equilibrium. For saturated paste at 100% RH, the equilibrium thickness is about five water molecules thick (about 1.3 nm). If two adjacent C–S–H particles have a distance less than 2.6 nm, the equilibrium water thickness will be attained by pushing the C–S–H particles apart. If the particle positions are fixed, a disjointing pressure can be developed. The equilibrium state of water in nanoscale pores is thus determined by a combination of stress and the thickness. With the application of an external stress, the disjointing pressure exerted on the water is increased. As a result, the thickness of the adsorbed water has to be decreased to maintain equilibrium, and extra water has to be diffused from the pores in the nanoscale to the stress-free capillary pores. This process leads to a bulk deformation or creep. Creep can occur in a saturated specimen at 100% RH as long as the capillary pores can take the diffused water from the nano-pores. The creep occurred in saturated concrete is called basic creep. When water movement occurs due to both diffusion and evaporation as the external RH is reduced to less than 100%, drying creep will develop. Drying creep is much larger than basic creep because the water reduction in pores in nanoscale is much faster by drying plus diffusion than by diffusion only.

While acknowledging the fundamental role of C–S–H in the process of concrete creep, new findings imply that the creep of C–S–H depends only on the packing of low density, high density, and ultra-high density C–S–H gels; and the creep rate depends on the rearrangement of C–S–H nano-granules around their limit packing densities following the free-volume dynamics theory of granular physics (Vandamme and Ulm, 2009). If these conclusions are eventually validated, it is possible to predict the long-term creep behaviors of cement-based materials from short-term nanoscale creep tests (e.g., by nano-indentation), and even from the hardened-state composition of the material, in the future.

In general, the creep rate of concrete (i.e., the rate of strain increases under a given stress) increases with applied stress. Creep behavior is not necessarily linear. For many metals and ceramics, the creep rate at high temperature is proportional to the stress raised to a high power. However, at room temperature and working stress levels, the creep strain of concrete is linearly dependent on stress. In such a case, material behavior can be described by models combining springs and dashpots. The study of these models will constitute the subject matter of the next section.

#### 5.3.2.4 Modeling of Creep at Low Temperature (Viscoelastic Models)

Models with springs and dashpots can be used to describe linear creep behaviors. The spring (Figure 5-52a) is a linear elastic element with direct proportionality between stress and strain. For the dashpot (Figure 5-52b), the rate of strain is directly proportional to the applied stress. This is similar to the behavior of a viscous liquid, the strain rate of which is directly proportional to the applied shear stress. Since the material can be considered as a combination of linear elastic and viscous elements, it is called a linear viscoelastic material. Using one spring and one dashpot, two

**Figure 5-52** Spring and dashpot for the modeling of viscoelastic behavior

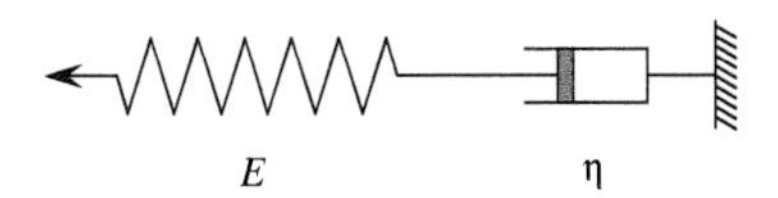


**Figure 5-53** The Maxwell model

different models can be created by putting the elements either in series or in parallel. The behavior of each of these simple models is discussed below.

**(a)** *Maxwell model (spring and dashpot in series)*: In the Maxwell model, shown in Figure 5-53, the material is made up of two parts, a spring and a dashpot, in series. The elastic (time-independent) part is represented by a spring with modulus $E$, and the viscous (time-dependent) part is represented by a dashpot of viscosity $\eta$. The equations regarding the equilibrium, compatibility, and constitutive relationship are as follows. For equilibrium,

$$\sigma_E(t) = \sigma_\eta(t) = \sigma(t) \tag{5-72}$$

where $\sigma_E(t)$ represents the stress in the spring, $\sigma_\eta(t)$ the stress in the dashpot, and $\sigma(t)$ the stress in the system. For compatibility,

$$\varepsilon(t) = \varepsilon_E(t) + \varepsilon_\eta(t) \tag{5-73}$$

where $\varepsilon(t)$ is the strain in the system, $\varepsilon_E(t)$ the strain in the spring, and $\varepsilon_\eta(t)$ the strain in the dashpot. The constitutive relationships read:

$$\sigma_E(t) = E\varepsilon_E(t) \tag{5-74}$$

$$\sigma_\eta(t) = \eta\dot{\varepsilon}_\eta(t) \tag{5-75}$$

Under an applied stress $\sigma$, the strain rate of the spring is given by

$$\frac{d\varepsilon_E}{dt} = \frac{1}{E}\frac{d\sigma}{dt} \tag{5-76}$$

The strain rate in the dashpot is given by

$$\frac{d\varepsilon_\eta}{dt} = \frac{\sigma}{\eta} \tag{5-77}$$

By differentiating Equation 5-73 and substituting Equations 5-76 and 5-77, we get

$$\dot{\varepsilon}(t) = \frac{\dot{\sigma}(t)}{E} + \frac{\sigma(t)}{\eta} \tag{5-78}$$

Now, we will examine the creep behavior and relaxation behavior of the Maxwell model.

1. Creep behavior under constant stress applied from $0 < t < t_1$ (see Figure 5-54) gives

$$\frac{d\sigma}{dt} = 0, \ \frac{d\varepsilon}{dt} = \frac{\sigma}{\eta}, \text{ and } \varepsilon = \frac{\sigma}{\eta}t + \varepsilon(0) \tag{5-79}$$

It takes a finite time for the dashpot to respond to the loading. Therefore, at $t = 0$, $\varepsilon_\eta(0) = 0$ and the dashpot acts as if it is rigid. The initial strain results from the spring alone, and we get

$$\varepsilon(0) = \frac{\sigma}{E}, \varepsilon = \frac{\sigma}{\eta}t + \frac{\sigma}{E} \tag{5-80}$$

At $t = t_1$, the load is completely removed. The spring shortens by an amount equal to $\sigma/E$. The remaining strain is $(\sigma/\eta)t_1$. After load removal, $\sigma = d\sigma/dt = 0$, implying $d\varepsilon/dt = 0$. The strain will stay constant for $t > t_1$. The stress and strain are plotted against time as shown in Figure 5-54.

2. Relaxation behavior (constant strain applied at $t = 0$) under constant strain, $d\varepsilon/dt = 0$, gives the governing equation

$$\frac{1}{E}\frac{d\sigma}{dt} = -\frac{\sigma}{\eta} \tag{5-81}$$

Integrating both sides with respect to $t$, and noting that $\sigma(0) = E\varepsilon$ (the dashpot stays undeformed) at $t = 0$, we have

$$\sigma = E\varepsilon \exp\left(-\frac{Et}{\eta}\right) \tag{5-82}$$

Schematically, the relaxation behavior of the Maxwell model under constant strain is shown in Figure 5-55.

**(b)** *Kelvin-Voigt model (spring and dashpot in parallel)*: For the Kelvin-Voigt model shown in Figure 5-56, the spring and dashpot are arranged in a parallel manner. The equations regarding the equilibrium, compatibility, and constitutive relationship are

$$\text{Equilibrium: } \sigma(t) = \sigma_E(t) + \sigma_\eta(t) \tag{5-83}$$

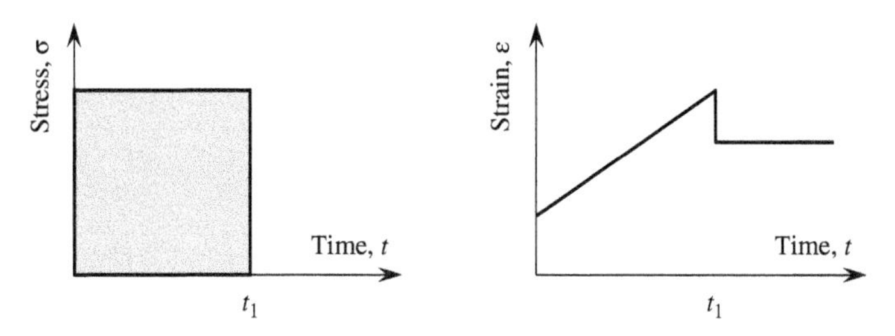


**Figure 5-54** Creep behavior under constant stress for the Maxwell model

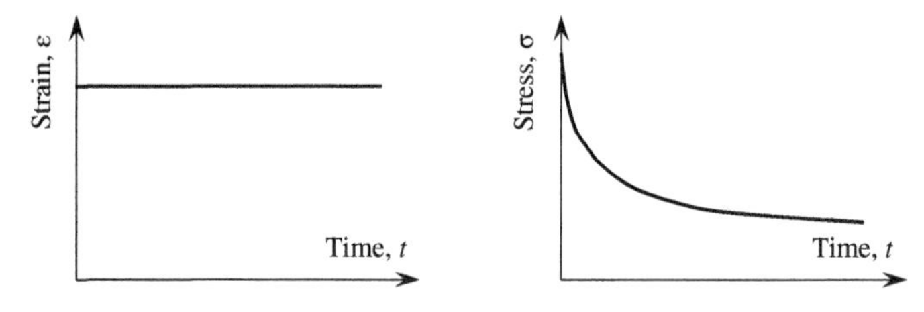


**Figure 5-55** Relaxation behavior of the Maxwell model

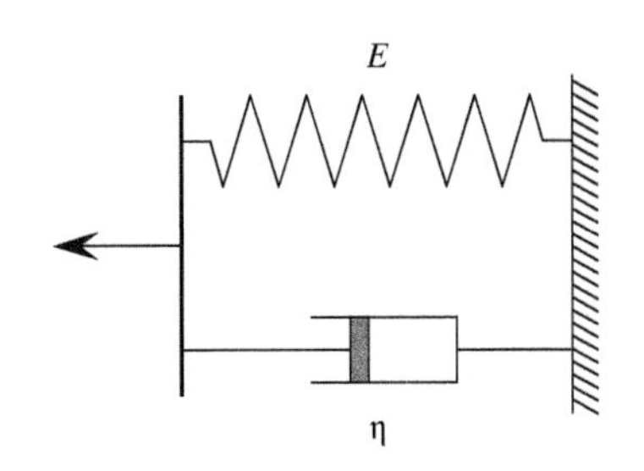


**Figure 5-56** The Kelvin-Voigt model

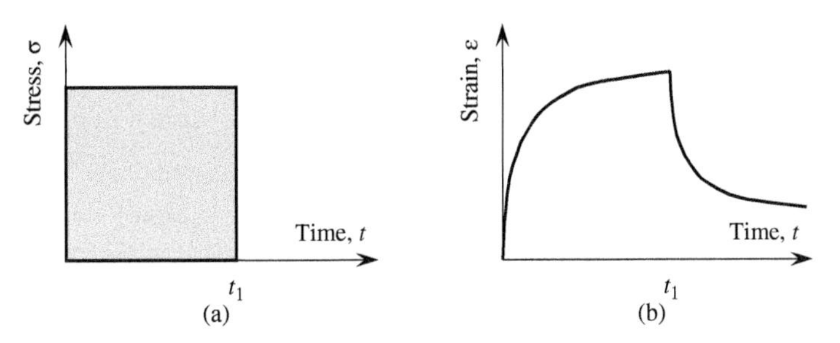


**Figure 5-57** Creep behavior of the Kelvin-Voigt model

$$\text{Compatibility: } \varepsilon(t) = \varepsilon_E(t) = \varepsilon_\eta(t) \tag{5-84}$$

$$\text{Constitutive relationship: } \sigma_E(t) = E\varepsilon_E(t) \tag{5-85}$$

$$\sigma_\eta(t) = \eta\dot{\varepsilon}_\eta(t) \tag{5-86}$$

By substituting constitutive equations into equilibrium, we can get

$$\sigma(t) = \eta\dot{\varepsilon}(t) + E\varepsilon(t) \tag{5-87}$$

Equation (5-87) is the governing equation for the Kelvin-Voigt model and the creep behavior and relaxation behavior of the model will be examined as well.

**1.** Creep behavior under constant stress applied from $0 < t < t_1$ (see Figure 5-57a). The governing equation is a first-order differential equation, which can be solved by the following procedure. Multiplying each side of the governing equation by $\exp(Et/\eta)$, we have

$$\frac{d\varepsilon}{dt} exp\,(Et/\eta) + \frac{E}{\eta}\varepsilon(t)\ exp\,(Et/\eta) = \frac{\sigma}{\eta} exp\,(Et/\eta) \tag{5-88}$$

The left-hand side of the equation can be rewritten as $d[\varepsilon\exp(Et/\eta)]/dt$. Carrying out the integration, and noting that $\varepsilon(0) = 0$ (because the dashpot takes a finite time to respond), the strain is given by

$$\varepsilon(t) = \frac{\sigma_0}{E}\left(1 - e^{(-E/\eta)t}\right) \tag{5-89}$$

This equation can be used to describe the creep process diagram. When load is applied (at $t = 0$), it takes time for the dashpot to react. Thus, initially, the load is taken by the dashpot and the strain is equal to zero. Then the load is gradually transferred to the spring and the strain increases at a slower rate and has an asymptotic value of $\sigma_0/E$.

If the stress is applied at $t = 0$ and removed at $t = t_1$, as shown in Figure 5-57a, then the strain development from $t = 0$ to $t = t_1$ is a rising exponential curve, as shown in the first part of the curve in Figure 5-57b. However, for $t > t_1$, $\sigma = 0$, hence, the governing equation becomes

$$\eta \frac{d\varepsilon}{dt} = -E\varepsilon \tag{5-90}$$

Integrating, with $\varepsilon = \varepsilon\left(t_1\right)$ at $t = t_1$ as the initial condition, gives

$$\varepsilon(t) = \varepsilon\left(t_1\right) e^{-E/\eta t} \tag{5-91}$$

The above behavior is illustrated in the second part of the curve shown in Figure 5-57b.

2. Relaxation behavior (constant strain applied at $t = 0$). At $t = 0$, the dashpot is theoretically rigid. In other words, the strain should be zero. To force the strain to reach a finite value, infinite stress is required. For $t > 0$, the strain is constant, implying $d\varepsilon/dt = 0$. The governing equation gives $\sigma = E\varepsilon$. The relaxation response is shown in Figure 5-58.

In describing the creep/relaxation behavior of real materials, each of the two models above has its own shortcomings. For the Maxwell model, the strain rate is constant, and after stress is removed, there is no time-dependent gradual strain recovery. For the Kelvin-Voigt model, no instantaneous material response is allowed, thus producing infinite stress when a finite strain is suddenly applied. For real materials, the applied stress is always accompanied by an instantaneous response. Subsequently, the strain will increase with time but at a decreasing rate. After the stress is removed, part of the strain is recovered immediately, while another part will be slowly recovered after a period of time. To describe the behavior of real materials, the two simple models can be combined, as shown in Figure 5-59. This combined model is called Burger's body and can be used to describe the time-dependent behavior of both concrete and wood.

#### 5.3.2.5 Strain Response Under Arbitrary Stress History—Superposition

The creep strain for a unit stress, or creep compliance $J(t)$, can be obtained experimentally from a single test (under constant stress). Once the compliance is known, the creep behavior under a nonconstant stress can be obtained by superposition, as illustrated in Figure 5-60. To apply superposition, any increase in stress level is replaced by a new constant stress applied at the time when the stress change takes place. The decrease in stress level is replaced by the removal of a constant stress. In Figure 5-60, the stress is shown to increase or decrease by discrete amounts. For a continuously changing stress, the stress history can be approximated with discrete stress increments occurring over very small-time intervals. This is the same principle of numerical integration.

#### 5.3.2.6 Importance of Applied Stress Level to Creep

Creep in concrete is significantly influenced by the applied stress level. For the case of concrete under a stress level less than 50% of its strength, creep in concrete is linear function of stress.

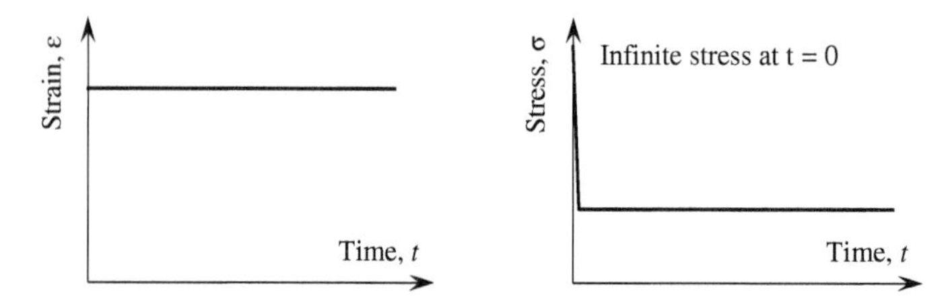


**Figure 5-58** Relaxation behavior of the Kelvin-Voigt model

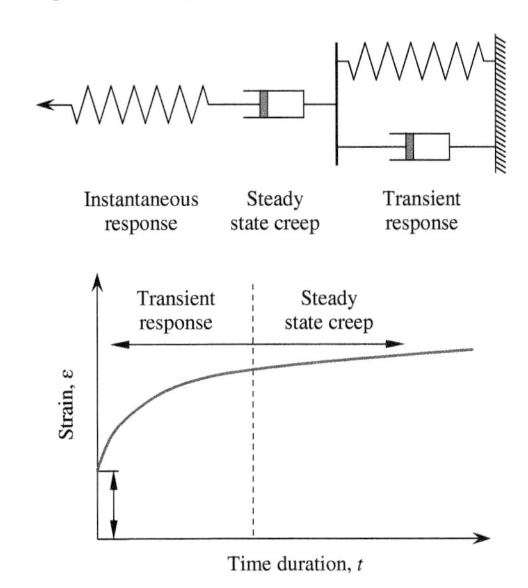


**Figure 5-59** The Burger's body and its response to constant stress

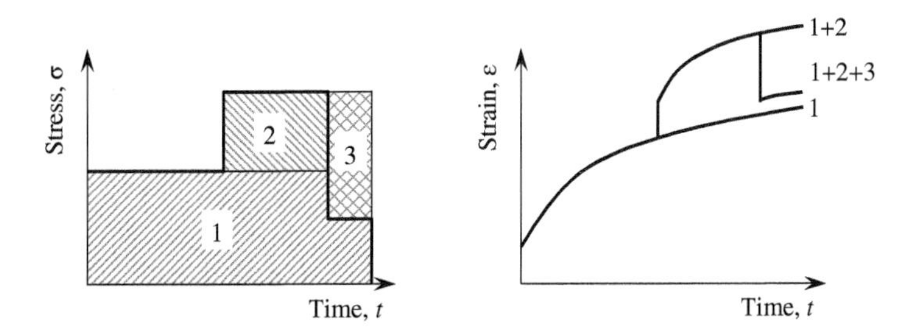


**Figure 5-60** Illustration of the superposition principle

For the case of a stress level higher than 50% but lower than 75% of its strength, the creep is a nonlinear function of stress. For the case of a stress level higher than 75% of a concrete's strength, creep can rapidly increase infinitely and cause structure failure, as shown in Figure 5-61. Thus, it is called unstable rupture creep. When designing a column, it is important to keep the stress level in the concrete sufficiently low, usually less than 40% of its compressive strength.

#### 5.3.2.7 ACI Equation for Predicting Creep

ACI 209R-92 (reapproved 2008) has suggested some equations to predict the creep of concrete. The equations are simple mathematical formulas but provide a level of accuracy good enough for most concrete structures. The basic equation describing the creep–time relationship is

$$v_t = \frac{t^{\varphi}}{d + t^{\varphi}} v_u \tag{5-92}$$

where $v_t$ is the creep coefficient (ratio of creep strain to initial strain) at time $t$ (in days) after being loaded; $v_u$ is the ultimate creep coefficient; $d$ is a constant ranging from 6 to 30 days; and $\varphi$ is a constant from 0.4 to 0.8. ACI 209R provided guidelines on how to select the constants and a series

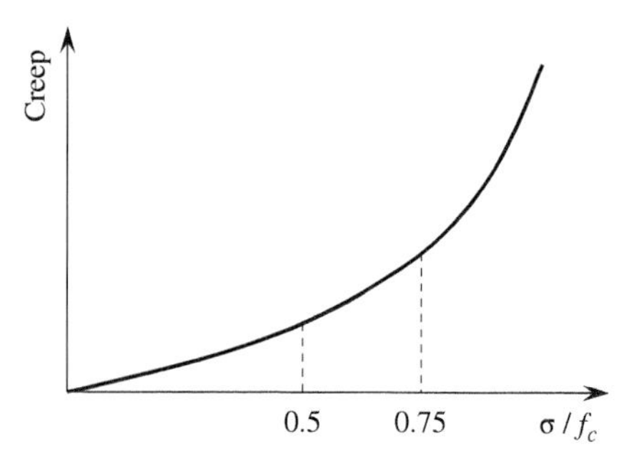


**Figure 5-61** Unstable rupture creep

**Figure 5-62** The experimental setup for measuring creep

of correction factors. For more accurate predictions, it is recommended to determine $v_u$, $d$, and $\phi$ by fitting the data obtained from tests performed in accordance with ASTM C 512.

### 5.3.2.8 Test Method for Creep

The experimental setup for measuring creep is shown in Figure 5-62. It is composed of a frame with four steel bars, together with two thick steel plates. Cylindrical concrete specimens (150 ± 1.5 mm in diameter and at least 290 mm in height, according to ASTM C512/C512M) are placed between two steel plates. The bottom steel plate is supported by a strong spring. A constant load can be applied to the specimens by compressing the spring. The creep under the constant load can be monitored by a gauge. ASTM C512 also provided guidelines on testing creep under various scenarios.

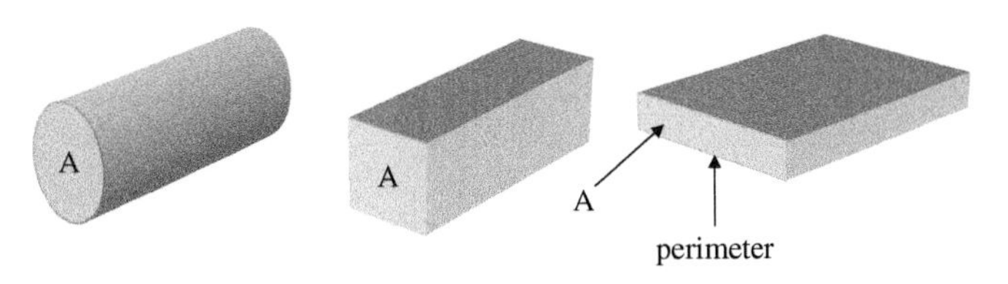


**Figure 5-63** Three shapes with same cross-section area but different perimeter

### 5.3.3 Other Important Factors Affecting Shrinkage and Creep

There are additional factors that influence the shrinkage and creep of concrete.

**(a)** *w/c ratio*: As both shrinkage and creep are sourced at the adsorbed water in C–S–H particles, the higher the *w/c* ratio, the more the adsorbed water and the higher the shrinkage and creep potentials.

**(b)** *Aggregate stiffness (elastic modulus)*: Since the shrinkage and creep are from cement paste, nonshrinking and noncreeping aggregates play a restraining role to the paste. Stiffer aggregates will provide a higher restraining effect and hence lead to smaller shrinkage and creep.

**(c)** *Aggregate fraction*: As aggregates are nonshrinking and noncreeping, as compared to cement paste, a higher aggregate fraction leads to smaller shrinkage and creep due to the restraining effect.

**(d)** *Theoretical thickness*: The theoretical thickness is defined as the ratio of the section area to the semi-perimeter in contact with the atmosphere, as shown in the following equation:

$$t_{\mathrm{TH}} = \frac{2A}{P} \tag{5-93}$$

As can be seen from Figure 5-63, for the same section area, a higher theoretical thickness means less contact surface with the atmosphere. In other words, the distance for water migration from inside to the atmosphere is longer, making the diffusion or migration of water more difficult. Hence, a higher theoretical thickness will lead to smaller creep and shrinkage.

**(e)** *Humidity*: The higher the RH, the lower the shrinkage and creep. This is because an increase in the atmospheric humidity is expected to slow down the relative rate of moisture migration.

## 5.4 DURABILITY

Durability of Portland cement concrete is defined as its ability to resist weathering action, chemical attack, abrasion, or any other processes of deterioration to maintain its original form, quality, and serviceability when exposed to its intended service environment (Mehta and Monteiro, 2006). Durability is most likely to relate to long-term serviceability of concrete and concrete structures. Herein, serviceability refers to the capability of the structure to perform the functions for which it has been designed and constructed after exposure to a specific environment.

### 5.4.1 Why Durability?

Almost universally, concrete has been specified principally on the basis of its compressive strength at 28 days after casting. Reinforced concrete (RC) structures, on the other hand, are almost always designed with a sufficiently high safety factor (i.e., strength is significantly higher than the allowed stress). Thus, it is rare for concrete structures to fail due to lack of intrinsic strength. However, gradual deterioration, caused by the lack of durability, makes concrete structures fail earlier than their specified service lives in ever increasing numbers. The extent of the problem is such that concrete durability was described as a "multimillion dollar opportunity" in the 1980s (Anonymous, 1988), and it has formed a trillion-dollar level market in recent years. For example, in the United States, it was reported that the decks of 2,530,000 concrete bridges had been damaged and needed repair in less than 20 years of operation. In the report of the American Society of Civil Engineers (ASCE) in 2005, it was estimated that US$1.6 trillion was needed in five years to restore the infrastructures to a normal operation condition. ASCE's 2017 Infrastructure Report Card gave a cumulative GPA of D+ to America's infrastructure. To restore a healthy, resilient, and sustainable infrastructure, the U.S. government has been investing trillions of dollars though the Fixing America's Surface Transportation (FAST) Act signed into law by President Obama in 2015, the continuous effort of President Trump to increase infrastructure investment, and potentially the new $3.5 trillion infrastructure plan proposed by President Biden in 2021. These efforts have raised American infrastructure's cumulative GPA to C-, according to ASCE's 2021 Infrastructure Report Card; however, this report also showed that the infrastructure restoring bill has been continuously underpaid as the total investment gap has gone from $2.1 trillion to nearly $2.59 trillion over 10 years. In the United Kingdom there are eight high bridges in the circular expressway in Southern England, and the total initial cost of construction was £28 million. However, the repair cost in 2004 for the bridges reached £120 million, almost 6 times as high as the initial construction cost.

To address the durability problems, many researchers have conducted deep studies, considering single deterioration mechanisms to multiple factors (including the coupling effects of environmental factors and mechanical loading) (Ulm et al., 2000; Sun et al., 2002; Le Bellégo et al., 2003; Kuhl et al., 2004; Nguyen et al., 2007). As a result of durability studies, many countries have proposed durability-based design guidelines (DuraCrete, 2000; CCES, 2004; MDPRC, 2007).

### 5.4.2 Factors Influencing Durability

The causes of concrete deterioration can be grouped into three categories: physical, chemical, and mechanical causes. The factors in the three groups may act alone or, in most cases, in a coupled manner.

Physical causes may include surface wear caused by abrasion, erosion, and cavitation, the effects of high temperature, or the differences in thermal expansion of the aggregate and of the hardened cement paste. Examples are the alternating freezing–thawing cycle and the associated action of deicing salts, and cracking, which is common due to volume changes, normal temperature and humidity gradients, (re-)crystallization of salt in the pores, structural loading, restrained shrinkage, and exposure to fire. Chemical degradation is usually the result of a chemical attack, either internal or external, on the cement matrix. Portland cement is alkaline, so it reacts easily with acids in the presence of moisture, and, in consequence, the matrix may become weakened, and its constituents may be leached out. The most common chemical causes affecting concrete durability are: (1) hydrolysis of the cement paste component; (2) carbonation; (3) cation-exchange reactions; and (4) reactions leading to expansion (such as sulfate expansion, alkali–aggregate expansion, and steel corrosion). Mechanical causes include impact and overloading.

As mentioned earlier, the causes responsible for concrete deterioration are either from the surrounding environment of an exposed structure or the mechanical loading a structure bear. In most cases, the degradation of a concrete structure is a result of the coupling effect of environmental factors and loading. In reality, major durability problems of concrete structures include corrosion of the reinforcing steel, freeze/thaw damage (often coupled with salt scaling), alkali–aggregate reactions, sulfate attack, and so on. The common points of these attacks are that all of them can be initiated due to mass transfer and result in cracking and spalling of the concrete.

#### 5.4.2.1 Transport Properties

The durability of concrete depends, to a large extent, on its transport properties (e.g., permeability and diffusivity). Permeability is defined as the property that governs the rate of flow of a fluid into a porous material under pressure. The permeability of concrete can be measured by determining the rate of water flow through a concrete specimen. The porosity (including pores in bulk cement paste and the interfacial transition zones between aggregate and cement paste, as well as microcracks) in concrete can largely affect the permeability. The flow of water through concrete obeys Darcy's law. In steady state, the flow rate of water at the equilibrium flow condition described by Darcy's law can be written as

$$\frac{dq}{dt} = K\frac{\Delta HA}{L} \tag{5-94}$$

where $dq/dt$ is the flow rate, $K$ the coefficient of permeability, $\Delta H$ the pressure gradient, $A$ the surface area of the specimen, and $L$ the thickness of the specimen. As mentioned earlier, the permeability of concrete is a function of the pores inside the material. This includes two concepts: total percentage of porosity and size distribution of pores or connectivity of pores. Concrete is composed of aggregates, hardened cement paste, and the interface between aggregate and paste. Table 5-2 shows the permeability coefficient of hardened cement paste (HCP), aggregate, and concrete. For matured hardened cement paste, the permeability coefficient is very small even though the total porosity is not that small. The value of the coefficient of matured hardened cement paste is of the same order as that observed in low-porosity aggregate made of rocks. Since a large portion of the porosity of cement paste is from pores less than 0.1 μm and hidden in C–S–H gels, the low permeability coefficient implies that water does not easily move through very small gel pores and permeability is controlled by an interconnecting network of capillary pores, where 0.1 μm is usually regarded as the lower boundary of harmful pores. As hydration proceeds, the capillary network becomes increasingly tortuous as interconnected pores are blocked by the formation of C–S–H. Although concrete is made of HCP and aggregates that both have a small permeability coefficient, its permeability coefficient is one or two orders of magnitude higher than that of HCP and aggregates. This can be attributed to the interface between HCP and aggregates, which is featured by much higher porosity and can easily percolate the whole material at regular aggregate volume fractions (Ma and Li, 2014).

Diffusivity is defined as the rate of migration of ions or moisture in concrete under the action of a concentration gradient, see Figure 5-64. The difference of permeability and diffusivity is that one is driven by pressure difference and the other by concentration difference. Permeability characterizes water flow when pores inside the concrete are filled with water, while diffusivity describes the migration of ions or water vapor under saturated or unsaturated conditions. The two parameters are related to each other and if one of the parameters is known, the other one can be deduced indirectly. The correlation can be described by empirical equations or analytical equations (e.g., the Katz-Thompson equation) (Ma et al., 2014; Ma, 2014). Diffusivity can be described by

**Table 5-2** Typical values of permeability coefficients of concrete materials

| Type | Porosity (%) | Average pore size | Permeability Coefficient |
|---|---|---|---|
| HCP | 20 | 100 nm | $6 \times 10^{-12}$ cm/sec |
| Aggregate | 3–10 | 10 μm | $1–10 \times 10^{-12}$ cm/sec |
| Concrete | 6–20 | nm–mm | $100–300 \times 10^{-12}$ cm/sec |

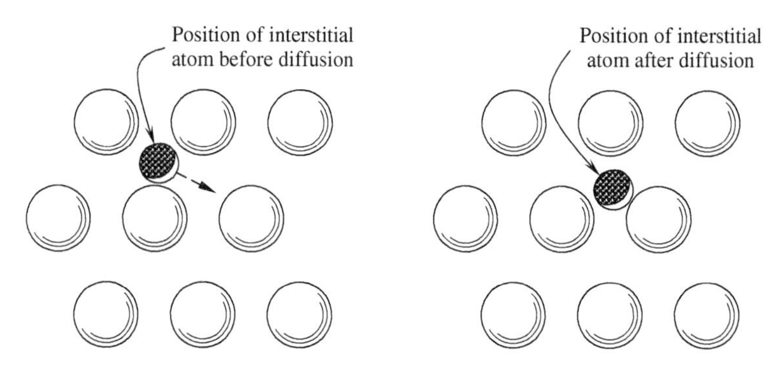


**Figure 5-64** The phenomenon of diffusion

Fick's law. There are two Fick's laws, i.e., Fick's first law and the second law. Fick's first law is normally used to describe steady-state diffusion, and it has the linear form of

$$J = -D\frac{\partial C}{\partial x} \tag{5-95}$$

where $J$ is the diffusion flux or mass transport rate (kg/m$^2$-sec), $D$ the diffusion coefficient (m$^2$/sec), $C$ the concentration of a particular ion or gas (kg/m$^3$), and $x$ the position (m) on the axis pointing low concentrations from the original high concentration point. Fick's second law can describe non-steady state diffusion, and it has the form of

$$\frac{\partial C}{\partial t} = D\frac{\partial^2 C}{\partial x^2} \tag{5-96}$$

The solution of Fick's second law for a semi-infinite plane has the form

$$C(x,t) = C_0\left[1 - \text{erf}\left(\frac{x}{2\sqrt{Dt}}\right)\right] \tag{5-97}$$

where $C(x, t)$ is the ion or gas concentration at distance $x$ and time $t$, $C_0$ the ion or moisture concentration at the higher concentration surface, and erf() is the error function.

The two parameters, permeability coefficient and diffusivity coefficient, apply to different situations. When water is flowing through a piece of concrete or from one part to another part under a hydraulic gradient, permeability is the governing parameter. When ions, moisture, or gases (e.g. oxygen) move through concrete (either dry or wet) or ions (e.g., chloride) move through the pore solution, the process is governed by the diffusion coefficient (or diffusivity). Note that the diffusion coefficient varies for different diffusing substances. Generally speaking, since both the permeability and diffusivity are related to the pore structure of concrete, concrete with low permeability will also possess low diffusivity. Means to reduce permeability and diffusivity (e.g., use lower *w*/*c* ratio to reduce capillary porosity, specification of cement content high enough to ensure sufficient consistency and hence proper compaction, and proper curing to reduce surface cracks) are generally helpful to concrete durability.

#### 5.4.2.2 Measurement of Transport Properties

The objective of a permeability test is to measure the flow rate of a liquid passing right through the test specimen under an applied pressure head. Concrete is a kind of porous material that allows water under pressure to pass slowly through it. There are two common practices for the evaluation of the permeability of concrete using water: the steady-flow method and the penetration-depth method. The steady-flow method is performed on a saturated specimen in which a pressure head is applied to one end of the sample, as shown in Figure 5-65. When a steady-flow condition is reached, the measurement of the outflow enables the determination of the permeability coefficient, by using Darcy's law:

$$k_1 = \frac{(dq/dt)\,L}{\Delta HA} \tag{5-98}$$

where $k_1$ is the permeability coefficient (m/sec), $dq/dt$ the steady flow rate (m$^3$/sec), $L$ the thickness or length of the specimen (m), $\Delta H$ the drop in the hydraulic head across the sample (m), and $A$ the cross-sectional area of the sample (m$^2$).

To achieve a steady-flow state, water has to be absorbed into all pores in the sample so that the pore surfaces do not provide friction or capillary attraction to the passage of water. Negative pressures (induced by self-desiccation during cement hydration) and entrapped air can both affect the permeation test. For example, it has been reported that the presence of 1% air void (by volume) in the pore water could increase the time duration needed to achieve stead-state water flow by one order of magnitude. The steady-state flow conditions can take a great amount of time to be achieved in high-performance contemporary concrete, as it is dense with low porosity. Sometimes, it is impossible to reach such a condition in a year. Hence, the considerable length of time required for testing the concrete and the difficulties of attaining a steady-state outflow can be regarded as disadvantages for the steady-flow method.

To evaluate the permeability of high-performance concrete, the penetration-depth method could be used. In the penetration-depth method, a permeability cell is usually used (Li and Chau, 2000), as shown in Figure 5-66. In the cell, the top and bottom parts are plastic tubes, and the middle is the concrete test specimen. The circumferential surface of the concrete specimen is sealed with epoxy to make sure the 1D flow condition is achieved. The top plastic tube can be filled with water and the bottom plastic tube with air. The pressure of the air in the bottom tube can be tuned through a vacuum process. By putting the specimens in an autoclave as shown in Figure 5-67 and applying pressure, the top end of the unsaturated concrete specimen is subjected to a pressure head through the water in the tube, while the other end is under the tuned normal atmospheric conditions.

The water penetration is achieved either by measuring the volume of water entering the sample or by splitting the cylinder immediately after stopping the test and measuring the average depth of discoloration, due to wetting, which is taken as the depth of penetration. Provided the flow of water is uniaxial, the water penetration depth can be used to calculate the permeability coefficient, which is equivalent to that used in Darcy's law, as derived by Valenta (1969):

$$k_1 = \frac{x^2 v}{2\Delta HT} \tag{5-99}$$

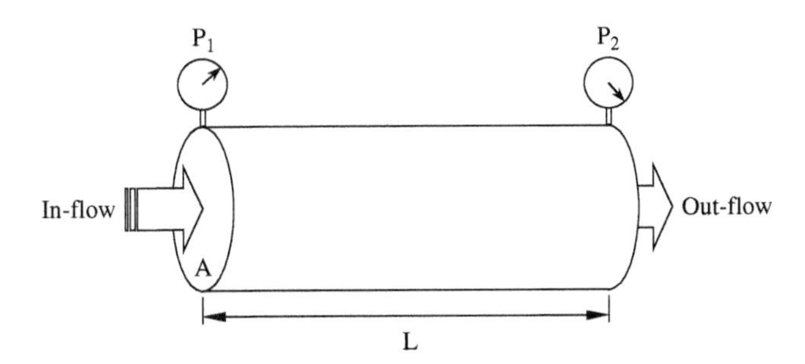


**Figure 5-65** Steady flow permeability test

**Figure 5-66** Permeability cell

where $x$ is the depth of penetration of concrete (m), $v$ the fraction of the volume of concrete occupied by the pores, and $T$ the time duration under pressure (sec). The value of $v$ represents discrete pores, such as air voids, which are not filled with water except under pressure, and can be calculated from the increase in the mass of concrete during the penetration test. Bearing in mind that only the voids in the part of the specimen penetrated by water would be considered, we can write

$$v = \frac{\Delta W}{DAx} \tag{5-100}$$

in which $\Delta W$ is the gain in weight of the specimen during the penetration test, $D$ the density of concrete, and $A$ the cross-sectional area. Based on the penetration-depth method, it is possible to use the depth of penetration of water as a qualitative assessment of concrete permeability. In summary the steady-flow method suits concretes with relatively higher permeability, while the depth of penetration method is most appropriate for concretes with very low permeability. It is important to note that the scatter of the permeability test results on similar concrete at the same age and using the same equipment would be large.

In concrete technology, chloride is usually selected as the medium for diffusion testing. The main reasons for using chloride ions are as follows: (1) chloride causes reinforcing steel corrosion, and (2) the size (radius) of a chloride ion is only $181 \times 10^{-12}$ m, so it is small enough to fit the need for diffusion. There are traditionally two methods for evaluating chloride diffusion in concrete. One is called the rapid permeability test, ASTM C1202-19, and the other is the diffusion cell test method (Li, Peng et al., 1999). In the first method, a standard specimen has a nominal diameter of 100 mm and a thickness of 50 mm, cut from the center of a cylindrical sample (cored from

**Figure 5-67** Penetration depth test with an autoclave

**Table 5-3** Classification of concrete quality using chloride ion penetrability

| Charge passed (Coulombs) | Chloride ion penetrability |
|---|---|
| >4000 | High |
| 2000–4000 | Moderate |
| 1000–2000 | Low |
| 100–1000 | Very low |
| <100 | Negligible |

field concrete or cast in lab). Before mounting the specimen into the test facility, its side surface needs to be sealed by a rapid-setting coating, and the specimen has to be vacuum desiccated for 3 h and vacuum-soaked for 18 ± 2 h. The test facility includes positive and negative terminals made of plexiglass plate with an empty cell and a circular opening of the same diameter as the specimen at one surface. The specimen is mounted with two surfaces connected to the openings of the terminal using a sealing material. Next, a 3% NaCl solution is added to fill the cell on the negative terminal and 0.3 N NaOH is added to fill the cell on the positive terminal. Figure 5-68 shows the experimental setup of the rapid chloride test. A constant voltage of 60 V is then applied for 6 h. The total charge passing through the concrete specimen during testing is taken as the index of the diffusivity of concrete. Table 5-3 shows the classifications of concrete quality using such an index. It is worth noting that the ASTM C1202 method does not provide numerical diffusion coefficients, but qualitative chloride penetrability evaluations.

The latter method is considered a realistic test method for the chloride diffusivity of concrete. In this method, the specimens consisting of $\phi 100 \times 20$ mm ($\phi 3.94 \times 0.79$ in) slices are placed between two chambers and the edges are sealed with an epoxy resin (see Figure 5-69). After the

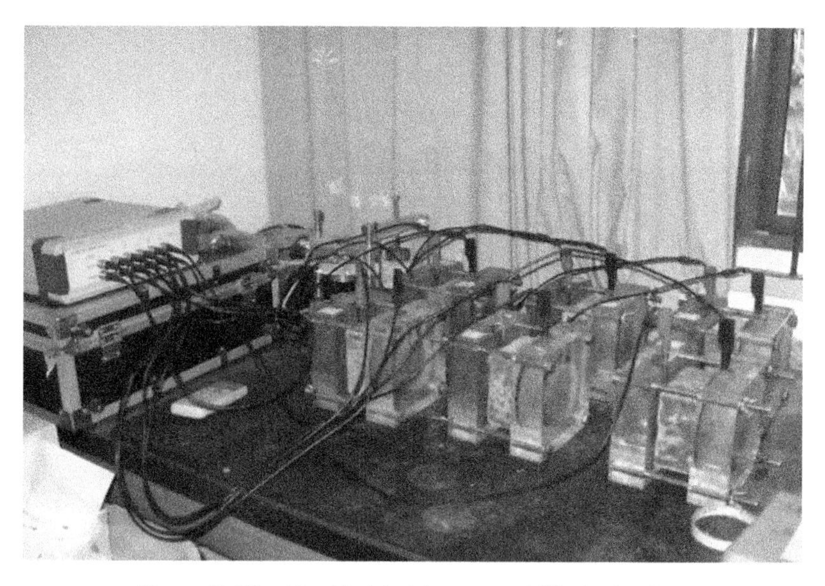

**Figure 5-68** Rapid chloride permeability test setup

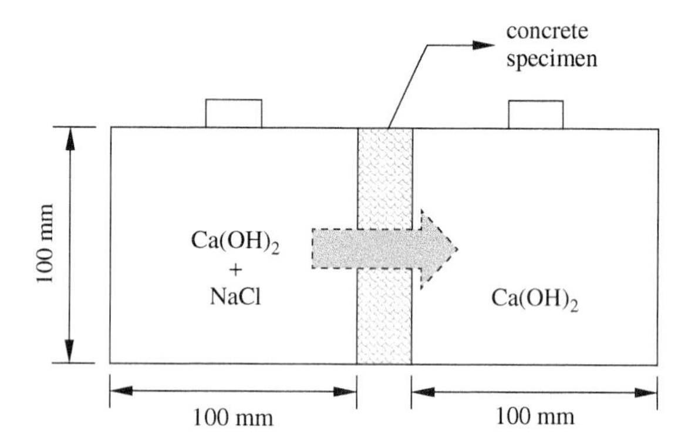


**Figure 5-69** Measurement of diffusion coefficient

epoxy resin is cured, saturated calcium hydroxide solution is poured into the chambers and the specimens are immersed in the solution for 5 days. This procedure is to avoid the anomalous effect due to sorption rather than diffusion of chloride ions. Then a NaCl solution with a concentration of 5 M is added to chamber A to start the chloride diffusion test. The chambers are maintained at $23 \pm 2°C$ and the concentration of the chloride diffused through the specimens in chamber B is measured periodically.

A typical curve of chloride concentration in chamber B obtained from an experiment has a strong nonlinearity between the chloride concentration and time initially. However, after a test time of about 30 weeks, the curve becomes quite linear. This implies that the chloride diffusion has reached a steady state. The linear relation between concentration of chloride ions and time can be expressed as

$$C = kt - A \tag{5-101}$$

where $C$ is the cumulative concentration of chloride ion penetration at time $t$ in chamber B, $k$ the slope of chloride concentration–time curve at steady state, and $A$ constant. In steady state, the flux of the chloride ions through the specimens can be described by Fick's first law of diffusion. For the steady-state case, this law can be written in the following form:

$$C = \frac{DC_A}{l^2}t - A \tag{5-102}$$

where $D$ is the diffusion coefficient in the steady state, $C_A$ the concentration of chloride ions in chamber A, and $l$ the thickness of the slice specimen. Comparing the above two equations, the diffusion coefficient can be expressed as follows:

$$D = \frac{kl^2}{C_A} \tag{5-103}$$

It is clear that once the slope of the chloride concentration–time curve (flux) in the steady state is known, the diffusion coefficient can be calculated. In summary, the latter method is closer to the real situation of chloride penetration in concrete and can be used to obtain the diffusion coefficient. However, it is time- and labor-consuming.

In addition, ASTM C1556 documented a bulk diffusion method to determine the apparent chloride diffusion coefficient of concrete. Cored or molded cylinders or molded cubes are all acceptable for this test. All sides of specimens apart from an exposure surface (e.g., the finished surface) are sealed, and then the specimens are saturated in a calcium hydroxide solution. The pre-processed specimens are placed in a sodium chloride solution (as the exposure liquid) for at least 35 days. After a specified exposure time, thin layers (thickness is dependent on the water-to-cement ratio) are ground off layer by layer in parallel to the exposure surface; the acid-soluble chloride content of each layer is determined according to ASTM C1152. The obtained concentration-depth curve as a function of exposure time can then be fitted to Fick's second law to derive the apparent chloride diffusion coefficient. It can be seen that this bulk diffusion method is also time-consuming.

To save testing time, as well as to evaluate the chloride diffusivity of high-performance concrete which has extremely low diffusivity, a rapid chloride migration (RCM) test has been proposed (Tang and Nilsson, 1992; Tang, 1996), standardized (e.g., NT Build 492), and widely adopted (e.g., in the DuraCrete model code) for service life design of concrete structures. This method uses specimens and devices similar to those of ASTM C1202, but a lower electrical potential (i.e., 30V) is used to drive chloride migration into concrete. After 24-to-96 h test, the specimens are split-broken, and $AgNO_3$ solution is sprayed onto the fracture surface to measure the depth of penetration of chloride so as to calculate a chloride migration coefficient, $D_{RCM}$. It has to be noted that the migration coefficient is not diffusion coefficient, and it is normally larger than the latter. Therefore, if the $D_{RCM}$ is used directly, it will lead to a more conservative design, which is safer (though could be more expensive). If one wants to derive the non-steady state diffusion coefficient from the non-steady state migration coefficient, many factors (e.g., the electrical field parameters and the chloride binding capacity of the material) need to be considered (Ma, 2013; Tang, 1999; Tang and Nilsson, 1999).

#### 5.4.2.3 Cracks in Concrete

Cracking in concrete is another factor that influences concrete durability. The permeability and diffusivity of a concrete with a crack will be significantly higher, and hence the concrete will deteriorate faster. The cracks in concrete may be caused by many different reasons and may range from very small internal microcracks that occur on the application of modest amounts of stress to quite large cracks caused by undesirable interactions with the environment, poor construction practice, or errors in structural design and detailing. In extreme cases, the structural integrity of the concrete may be seriously affected. In many other instances, however, cracks do not affect the load-carrying ability of concrete but may affect the durability of the concrete by providing pathways of easy access to the body of the concrete to aggressive agents that might otherwise not seriously affect the material.

Table 5-4 summarizes the types of cracking that can occur due to interactions in the concrete materials and the surroundings. In most instances, cracking originates internally, forming a network of microcracks gradually throughout the concrete. Internal damage may be considerable before cracks are visible at the exterior surface. In other cases, such as in humidity and temperature changes, localized large cracks may occur in the structure (even on the surface). Cracking may be used to help determine the cause of deterioration of concrete, since in many cases characteristic cracking patterns are produced (see Table 5-5). Concrete that resists cracking under normal environmental conditions may not remain intact under catastrophic conditions, such as fire hazards and earthquakes.

Cracking is best controlled during the design and construction phases. In many instances, cracking may be avoided by proper selection of materials, provided that the potential problem has been anticipated through a careful assessment of the anticipated service environment. For example, unsoundness of cement should never be encountered when ASTM C150, BS EN 197, or GB 175 is adhered to, and proper testing of groundwater should enable severe sulfate attacks to be avoided through the choice of an appropriate binder combination and water/cement ratio.

A chemical attack on concrete involves ingress of moisture, either as a carrier for aggressive agents or as a participant in destructive reactions. Thus, precautions in mix design and construction practice that prevent the entry of water into concrete should improve durability. Concrete of

**Table 5-4** Causes of cracking in concrete due to interaction with surroundings

| Component | Type | Cause of distress | Environmental factor(s) | Variable to control |
|---|---|---|---|---|
| Cement | Unsoundness | Volume expansion | Moisture | Free lime and magnesia |
| | Temperature cracking | Thermal stress | Temperature | Heat of hydration, rate of cooling |
| Aggregate | Alkali-silica reaction | Volume expansion | Supply of moisture | Alkali in cement, composition of aggregate |
| | D-cracking | Hydraulic pressure | Freezing and thawing | Absorption and maximum size of aggregate |
| Cement paste | Plastic shrinkage | Moisture loss | Wind, temperature, relative humidity | Temperature of concrete, protection of surface |
| | Drying shrinkage | Moisture loss | Relative humidity | Mix design, rate of drying |
| | Sulfate attack | Volume expansion | Sulfate ions | Mix design, cement type, admixtures |
| | Thermal expansion | Volume expansion | Temperature change | Temperature rise, rate of change |
| Concrete | Settlement | Consolidation of concrete around reinforcement | | Concrete slump, cover, bar diameter |
| Reinforcement | Electro-chemical corrosion | Volume expansion | Oxygen, moisture | Cover, permeability of concrete |

**Table 5-5** Type of cracking in concrete structures

| Nature of crack | Cause of cracking | Remarks |
|---|---|---|
| Large, irregular, frequently with height differential | Inadequate support, overloading | Slabs on ground, structural concrete |
| Large, regularly spaced | Shrinkage cracking, thermal cracking | Slabs on ground, structural concrete, mass concrete |
| Coarse, irregular "map cracking" | Alkali–silica reaction | Extrusion of gel |
| Fine, irregular "map cracking" (crazing) | Excessive bleeding, plastic shrinkage | Finishing too early, excessive troweling |
| Fine cracks roughly parallel to each other on surface of slab | Plastic shrinkage | Perpendicular to direction of wind |
| Cracks parallel to sides of slabs adjacent to joints (D-cracking) | Excessive moisture content, porous aggregate | Deterioration of concrete slab due to destruction of aggregates by frost |
| Cracks above and parallel to reinforcing bars | Settlement cracking | Structural slabs due to consolidation of plastic concrete around reinforcing bars near upper surface |
| Cracking along reinforcing bar placements, frequently with rust staining | Corrosion of reinforcement | Aggravated by the presence of chlorides |

low permeability can be assured by the use of sufficient (but not excessive) quantities of cementitious materials and low water/cement ratios, proper placement, consolidation, and finishing, and adequate moist curing.

Cracking due to drying shrinkage and thermal expansion is caused by tensile stresses that are created by differential strains that occur under nonuniform drying, temperature rise, or uneven restraint. Thus, shrinkage and thermal cracking resemble flexural cracking and can be controlled by the suitable location of reinforcements and/or adoption of fibers (i.e., distributed micro-reinforcement), which will reduce the amount of cracking and will cause multiple fine cracks rather than a single large crack. The finer the crack, the less likely it is to contribute to durability problems. Crack widths less than 0.10 mm are desirable (as compared to wider cracks) in cases where severe exposure is anticipated.

### 5.4.3 Major Durability Problems

#### 5.4.3.1 Corrosion of Reinforcing Steel

Corrosion of the reinforcing steel is regarded as the most serious durability problem in the world. Originally, in a properly designed, constructed, and used structure, there should be little problem of steel corrosion within the concrete during the designed service life of that structure, because concrete provides a highly alkaline environment, with pH values of 12.5–13.5, in which steel is well passivated. Unfortunately, this highly desirable condition is not always maintained during the service period of concrete structures due to the deterioration of concrete, which results in corrosion of the reinforcement. Apart from normal corrosion of steel exposed to oxygen and water at exposed locations (e.g., large cracks and concrete debonding/spalling), there are two mechanisms that induce and/or accelerate rebar corrosion: carbonation-induced corrosion and chloride-induced/catalyzed corrosion. Carbonation-induced corrosion is caused by a general breakdown of passivity by neutralization of the concrete, and the chloride-induced corrosion is caused by a localized breakdown of the passive film on the steel.

**Carbonation-Induced Corrosion.** Carbonation occurs due to the penetration of carbon dioxide from the atmosphere into the concrete. The chemistry of carbonation is that carbon dioxide molecules penetrate into the concrete and react with calcium hydroxide. Calcium hydroxide is the hydration product in the cement paste that reacts most readily with carbon dioxide. Consequently, $Ca(OH)_2$ carbonates to $CaCO_3$:

$$CO_2 + Ca(OH)_2 \rightarrow CaCO_3 + H_2O \tag{5-104}$$

This is the reaction of main interest, especially for concrete made of Portland cement, though the carbonation of C–S–H is also possible when calcium hydroxide becomes depleted. Other compounds in hydrated cement paste can also be decomposed, such as ettringite, which could also lead to changes in terms of mineralogy. In general, carbonation could cause the concrete to shrink. In the case of Portland cement concrete, carbonation may even result in increased strength (Bertolini et al., 2004). Carbonation itself is not a severe problem; it is the consequences of carbonation on reinforced concrete that causes the major repercussions. The chemical reactions during carbonation result in a drastic decrease in the alkalinity of the concrete, from average values of 12–14 down to 8–9, with the consumption of calcium hydroxide. The reduction in alkalinity destroys the passive environment and leaves the reinforcement in a condition where it is susceptible to corrosion. Therefore, when carbonation penetrates through the rebar cover, and oxygen and moisture are present, corrosion will occur. The depth of carbonation in reinforced concrete is an important factor in the protection of the reinforcement, the deeper the carbonation, the greater the risk of steel corrosion. The second consequence of carbonation is that chlorides bound in the form of calcium chloroaluminate hydrates (i.e., the Friedel's salt) and others bound to the hydrated phases may be liberated due to carbonation, making the pore solution even more aggressive (Birnin-Yauri and Glasser, 1998; Suryavanshi and Swamy, 1996).

The rate of carbonation (expressed as the penetration rate of the carbonation frontier into the concrete) increases with an increase in the concentration of carbon dioxide, especially in concretes with high water/cement ratio. The rate of carbonation also depends on other environmental factors (e.g., relative humidity and temperature) and factors related to the quality of concrete (mainly its alkalinity and permeability). The quality of the concrete has been regarded as the most important parameter controlling the rate of carbonation (Bentur et al., 1997). The quality of the concrete is a function of binder composition (i.e., whether Portland cement or blended cement was used), the water/cement ratio or the water/binder ratio, and the curing conditions. The permeability of concrete has a remarkable influence on the diffusion of carbon dioxide and thus on the carbonation rate. The carbonation penetration can slow down with the decrease in water/cement ratio or water/binder ratio, due to the resultant decrease of the capillary porosity (and, thus, permeability) of the hydrated cement paste. The type of cement also influences the carbonation rate. For blended cement, hydration of pozzolanic materials (e.g., fly ash) or ground granulated blast furnace slag (GGBS) leads to a lower calcium hydroxide content in the hardened cement paste, which may increase the carbonation rate. Consequently, the depth of carbonation is greater for the blended cements than for Portland cement concrete when compared on the basis of equal water/binder ratios (Bentur et al., 1997). Curing condition is also an important factor influencing the rate of carbonation, since a proper curing can densify concrete to slow down permeation or diffusion. If cured properly, the lower alkalinity of cements (a negative factor regarding carbonation) with the addition of fly ash or GGBS can even be compensated by the lower permeability (a positive factor regarding carbonation) of their cement pastes. The rate of carbonation also increases with temperature and varies with the humidity of concrete. In a totally dry or wet environment, there is no carbonation since $CO_2$ diffusion can hardly take place. The carbonation rate may be correlated to the humidity of the environment. The interval of relative humidity most critical for promoting carbonation is 60–70%.

The rate of carbonation-induced steel corrosion is usually expressed as the penetration rate of the corrosion frontier in steel bar and is measured in micrometers per year. The corrosion rate can be considered negligible if it is below 2 μm/year, low between 2 and 5 μm/year, moderate between 5 and 10 μm/year, intermediate between 10 and 50 μm/year, high between 50 and 100 μm/year, and very high for values above 100 μm/year (Bertolini et al., 2004). In high-quality concrete, the rate of carbonation-induced corrosion is negligible for relative humidity (RH) below 80%. It is then assumed that corrosion propagates only while concrete is wet (i.e., RH > 80%). The corrosion rate tends to decrease with time. Besides, corrosion products can reduce the corrosion rate (Alonso and Andrade, 1994). Page (1992) has illustrated the relationship between the corrosion rate in carbonated concrete and the relative humidity of the environment, where the maximum corrosion rates, of the order of 100–200 μm per year, can be reached only in very wet environments with relative humidity approaching 100%. For typical conditions of atmospheric exposure, i.e., RH = 70–80%, the maximum corrosion rates are between 5 and 50 μm per year

**Chloride-Induced Corrosion.** Chloride-induced corrosion in structural concrete is primarily caused by the presence of sufficient free chloride ions in the cement matrix surrounding the steel rebar. Chloride can get into the concrete at the time of mixing, either as an admixture component (e.g., chloride accelerator) or in chloride-contaminated aggregates or mix water (e.g., sea sand or seawater), or penetrate into the hardened concrete later on from external sources, such as seawater, salt spray, or de-icing salt placed on concrete pavements. When sufficient chlorides are present at the time of mixing, the corrosion may start at a very early service stage. For the case of chloride in the service environment penetrating into the concrete, the corrosion of rebar will not start until reaching a certain level of chloride accumulation at the rebar surface. The penetrating chloride ions diffuse through the concrete cover to the rebar surface first, and then sufficient quantities of chloride ions can be accumulated gradually. When the concentration of chloride ions in the concrete reaches a certain level (e.g., 0.6–0.9 kilogram per cubic meter of concrete), they dissolve the protective oxidized passivation film; thus, a localized breakdown of the passive film on the steel is formed where oxidation of iron occurs, and a galvanic cell is created. The local active areas behave as anodes, while the remaining passive areas become cathodes where reduction takes place.

The effect of the separation of the anodes and cathodes has significant consequences for the pattern of corrosion. In the concrete adjacent to the anodic areas, the concentration of positive iron ions increases, causing the pH to fall and allows soluble iron–chloride complexes to form. These complexes can diffuse away from the rebar, permitting corrosion to continue. At some distance from the electrode, where the pH and concentration of dissolved oxygen are higher, the complexes break down, iron hydroxides precipitate, and the chloride is free to migrate back to the anode and react further with the steel. The process thus becomes autocatalytic and deepens the corrosion pits rather than spreading corrosion along the rebar. As the steel increases its state of oxidation, the volume of the corrosion products expands. The unit volume of Fe can be doubled if FeO is formed. The unit volume of the final corrosion product, $Fe(OH)_3{\cdot}3H_2O$, is as large as six and a half times the original Fe. This expansion could create cracking and spalling of the concrete cover, and finally destroy the integrity of the structural concrete and cause failures of buildings and infrastructures.

It can be seen that, chloride-induced corrosion is localized, with corrosion cumulated in a limited area surrounded by non-corroded areas. The threshold concentration of chloride ions, i.e., the critical level of the chloride ion concentration in concrete at which the surface protective layer of reinforcing steel generated in conditions of high alkalinity can be broken, is a function of the pH value (i.e., the hydroxyl ion concentration). In other words, it is the ratio of the free chloride to pH (i.e., $[Cl^-]/[OH^-]$), not merely the free chloride concentration that determines the initiation of corrosion. Hausman (1967) has suggested on the basis of measurements in $Ca(OH)_2$ solutions that the threshold chloride ion concentration is about 0.6 times the hydroxyl ion concentration (i.e., $[Cl^-]/[OH^-] = 0.6$). Gouda (1970) has suggested another relationship in which the threshold chloride ion concentration for a pore solution of a given pH region is expressed in a logarithmic form. Both Hausman's and Gouda's data were derived from experiments in a solution rather than in concrete, and other effects in concrete may influence the threshold value. In fact, most specifications and guidelines specify the total content of chloride in concrete in terms of the percentage by weight of the original cement used. The permitted chloride content in many specifications and recommendations is less than about 0.2% of the cement content of the concrete (Bentur et al., 1997). In the range of 0.2–0.4% there is risk of inducing corrosion, but not always. ACI 318 specified the maximum water-soluble chloride contents (including contributions from various ingredients) by weight of cement determined on the concrete mixture by ASTM C1218 at an age of 28–42 days which are 0.15–0.4% for reinforced concrete, depending on the exposure class; and 0.06% for prestressed concrete. Sometimes, the critical level of chloride is specified in terms of weight of chloride per unit volume of concrete, mostly in the range of 0.6–1.2 $kg/m^3$. This kind of specification is needed when assessing the chlorides in existing structures, where the total chloride content of concrete can be determined experimentally, but not the chloride content by weight of cement. More details regarding the threshold/critical chloride concentration/content can be found in the review by Angst et al. (2009).

Under a normal atmospheric environment, the chloride-induced corrosion rate can vary from several tens of micrometers per year (of steel) to localized values of 1 mm per year (of steel), as the relative humidity rises from 70 to 95% and the chloride content increases from 1% by mass of cement to higher values. High corrosion rates always appear on heavily chloride-containing structures, such as bridge decks, retaining walls, and pillars in seawater. The corrosion rate increases when the temperature changes from lower to higher values. Once the corrosion attack begins in chloride-contaminated structures, it can lead to a relatively short time to an unacceptable reduction in the cross-section of the reinforcement, even under conditions of normal atmospheric exposure. The lower limits of relative humidity near which the chloride-induced corrosion rate becomes negligible are much lower than those that make carbonation-induced corrosion negligible. The influence of temperature and humidity on the corrosion rate is through their influence on the electrochemical reactions at the steel/concrete interface and through their influence on ion transport between the anodes and cathodes. It has long been thought that the concrete resistivity is strongly related to the corrosion rate at moderate or low temperatures (Alonso et al., 1988; Glass et al., 1991). In a given set of conditions in terms of humidity and temperature, and provided corrosion has been initiated, the higher resistivity of blended cements can result in a lower corrosion rate than that of Portland cement.

The signs of corrosion of reinforcing steel can be identified as rust stains and minute cracking over the concrete surface. This can be attributed to the increase in volume associated with the formation of the corrosion products and the leaching of the rust. If repairs are not undertaken at an early stage, corrosion damage will occur, and the corrosion of the steel will proceed further, causing severe damage through the formation of a longitudinal crack in parallel with the underlying reinforcement, delamination and spalling of concrete cover, as well as exposure of the steel and reduction of its cross-section, which may cause a catastrophic collapse. The cost of damage caused by corrosion of reinforcing steel can be very large. For example, a survey by the China Academy of Engineering in 2002 reported that the annual cost due to corrosion of reinforcing steel in China reached CNY100 billion. So, it is desirable to know clearly the mechanism of corrosion of steel and to determine the possible methods for repairing the damage caused by reinforcing steel corrosion.

**Corrosion Mechanisms.** Corrosion of the reinforcing steel in concrete is an electrochemical process, comprising both oxidation and reduction reactions, in which the metallic iron is converted to voluminous corrosion products iron oxides and hydroxides. The process is associated with the presence of anodic and cathodic areas, and a potential difference between the two areas, arising from inhomogeneities in the surrounding liquid medium, or even in the steel itself. The differences in potential are due to the inherent variation in the structure and composition (e.g., porosity and the presence of a void under the rebar or difference in alkalinity due to carbonation) of the concrete cover, and differences in exposure conditions between adjacent parts of the steel (e.g., concrete that is partly submerged in seawater and partly exposed in a tidal zone). In general, corrosion cells are formed due to: (1) contact between two dissimilar materials, such as steel rebars and aluminum conduit pipes; (2) significant variations in surface characteristics, including differences in composition, residual strain due to local cold working, applied stress, etc.; and (3) different concentrations of alkalis, chloride, oxygen, etc. Four components must be present for corrosion to occur in a macro cell: the anode, cathode, electrolyte, and the metallic path. The anode is the electrode at which oxidation (i.e., corrosion) occurs. Oxidation involves the loss of electrons and the formation of metal ions. The cathode is the electrode where reduction occurs. Reduction is the gain of electrons in a chemical reaction. The electrolyte is a chemical mixture, usually a liquid, containing ions that migrate in an electric field. The free electrons travel to the cathode, where they combine with the constituents of the electrolyte, such as water and oxygen, to form hydroxyl ($OH^-$) ions. The metallic path between the anode and cathode is essential for electron movement between the anode and cathode. For the steel corrosion in concrete, the anode, cathode, and metallic path are on the same steel rebar. The electrolyte is the moistened concrete surrounding the steel. The corrosion will stop if any of these components is removed. This provides the basis for corrosion control. The process of corrosion is illustrated in Figure 5-70 with the necessary notations of chemical reactions. At the anode site, the iron dissociates to form ferrous ions and electrons:

$$Fe \rightarrow Fe^{2+} + 2e^- \quad \text{anodic reaction} \tag{5-105}$$

The electrons move through the metal toward the cathodic site, while the ferrous ions are dissolved in the pore solution. At the cathodic site, electrons combine with oxygen and water to form hydroxyl ions:

$$4e^- + O_2 + 2H_2O \rightarrow 4(OH)^- \quad \text{cathode reaction} \tag{5-106}$$

The hydroxyl ions move to the anode through the pore solution, and form corrosion products at the anode:

$$Fe^{2+} + 2(OH)^- \rightarrow Fe(OH)_2 \quad \text{ferrous hydroxide} \tag{5-107}$$

$$4Fe(OH)_2 + 2H_2O + O_2 \rightarrow 4Fe(OH)_3 \quad \text{ferric hydroxide} \tag{5-108}$$

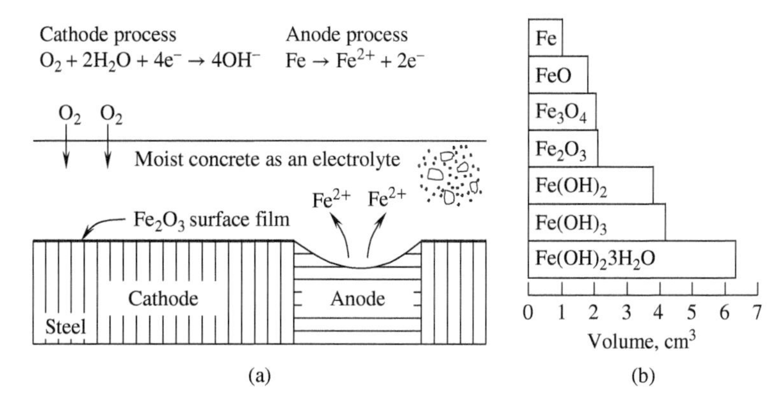


**Figure 5-70** Reinforcing steel corrosion and expansion of corrosion products: (a) corrosion process; and (b) volume expansion of corrosion products

It can be seen from the above equations that oxygen and water are needed for the initiation and propagation of the corrosion. There is no corrosion in a completely dry atmosphere, probably below a relative humidity of 40% (Mailvaganam, 1992). It has been suggested that the optimum RH for corrosion is 70–80%. At higher relative humidity (RH greater than 80%) or under immersion conditions, the diffusion of oxygen is considerably reduced, and the environmental conditions are more uniform along the steel. Consequently, there is little corrosion. When corrosion occurs, ions need to travel through the pores in the surrounding concrete; oxygen needs to diffuse through the concrete cover. The corrosion rate is therefore affected by the electrical resistance, diffusivity, and thickness of the cover concrete. The extent of steel corrosion in concrete depends on the conductivity of the electrolyte, the difference in potential between the anodic and cathodic areas, and the rate at which oxygen reaches the cathode. This controls the velocity of the anodic reaction. For steel in concrete, the strong polarization of the anodic zones under aqueous and highly alkaline conditions raises its potential close to that of the cathode, causing the surface of the steel to be passivated by the formation of an oxide layer (consisting of $Fe_3O_4$, $Fe_2O_3$, and/or $\gamma$-FeOOH) (Montemor et al., 2003). The passivation film prevents further reaction so that the steel remains unaltered over long periods. Another protective effect of concrete on corrosion is that the increased electrical resistivity (following the hydration process) can reduce the flow of electrical currents within the concrete. This is particularly true of high-density concrete (Page, 1992). However, when carbonation reaches the near surroundings of steel bar, the decreased alkalinity will dissolve the passivation film and initiate/resume the corrosion reactions. If chloride ions reach the reinforcement and bypasses the threshold concentration, they can also break down the passivation film, through multiple mechanisms which are still in debate (Montemor et al., 2003). In general, the breakdown of passivation film can be expressed as:

$$Fe^{2+} + 2Cl^- \rightarrow FeCl_2 \tag{5-109}$$

This reaction is faster than the normal anodic reaction; and the dissolved ferrous ions can react with the hydroxyl ions, water, and oxygen to form the corrosion products. It can be seen that chloride ions are not consumed in this process; instead, they serve as a catalyst to facilitate and accelerate the corrosion of steel. Therefore, chloride is a highly harmful contaminant in reinforced concrete.

**Monitoring of Steel Corrosion.** Because serious corrosion is very dangerous to a structure, engineers must attempt to detect or monitor the corrosion to make maintenance decisions. The detection/monitoring methods include visual inspection, electrical/electrochemical methods (e.g., half-cell potential measurement, electrical tomography, and four-electrode configuration) (Rodrigues et al., 2021), and radiographic, fiber-optic, ultrasonic, magnetic perturbation/flux, and acoustic emission techniques. Detailed methodologies can be found in the literature (Agarwala et al., 2000; Zaki et al., 2015). Here, we briefly introduce two techniques: half-cell potential measurement and acoustic emission.

Half-cell potential measurement (ASTM C876) measures the corrosion potential of the reinforcing steel using a standard reference electrode and voltmeters. One connection is made to the reinforcing steel under test and the other is made to a reference electrode, which contacts the concrete surface. The contact area should be moistened before taking the measurement. The probability of corrosion can be inferred according to the potential readings. If the potential is less negative than −200 mV, there is a better than 90% probability of non-corrosion; for a potential between −200 and −300 mV, the probability of corrosion is uncertain; for a potential more negative than −300 mV, there is a higher than 90% probability of corrosion. It should be pointed out that the above values are for copper/copper sulfate electrodes only. If a silver/silver chloride electrode or other type of electrode are used, different potential values have to be used to judge the possibility of corrosion. A big drawback is that half-cell potential is highly dependent on the condition of the concrete at the time of measurement. For instance, moist concrete will yield different measurements from dry concrete.

The application of the acoustic emission (AE) technique to detect corrosion is detailed in Chapter 8. The principle of this application is that under the expansion of the corroded product, microcracks will develop at the interface between the corroded steel bar and the bulk matrix. The energy released by the formation of these microcracks will generate stress waves that propagate along the medium and reach the outer surface. By placing the AE transducer on the surface, the occurrence of microcracks can be detected. Since these microcracks are caused by corrosion products, the signals detected can be used to interpret the corrosion activity (Li et al., 1998; Zhang et al., 2017).

The mathematical model for calculating the stress caused by rebar corrosion at the rebar–concrete interface can be simplified as a shrink-fit model (Li et al., 1998), as shown in Figure 5-71. First, assume that the rebar can freely expand due to corrosion, with an increase of $r$ in the radial direction. Then, we try to put the rebar back into the hole that it occupied before. Due to corrosion expansion, the rebar now is too big to fit freely in the concrete hole. To allow the rebar to fit back into the concrete, pressure has to be applied to both the surrounding concrete and the rebar. Let us consider the surrounding concrete first. This situation can be treated as an internal hole under pressure in an unbounded medium. According to the elasticity solution, the displacement caused by internal pressure $p$ along radius direction is

$$U_{\mathrm{r}}^{\mathrm{c}}(r=a) = \frac{pa}{2\mu_{\mathrm{c}}} \tag{5-110}$$

where $p$ is the pressure, $a$ the radius of hole, and $\mu_{\mathrm{c}}$ the shear modulus of surrounding concrete (interface). For the rebar, it can be treated as an inclusion with pressure on the outside. The displacement along the radial direction can be written as

$$U_{\mathrm{r}}^{\mathrm{s}} = \frac{-p(k-1)a}{4\mu_{\mathrm{s}}} \tag{5-111}$$

where $k$ is the Kolosou constant, and has values of

$$k = \frac{3-v}{1+v} \quad \text{for plane stress} \tag{5-112}$$

$$k = 3-4v \quad \text{for plane strain} \tag{5-113}$$

Compatibility requires that

$$|U_{\mathrm{r}}^{\mathrm{c}}| + |U_{\mathrm{r}}^{\mathrm{s}}| = \Delta a \tag{5-114}$$

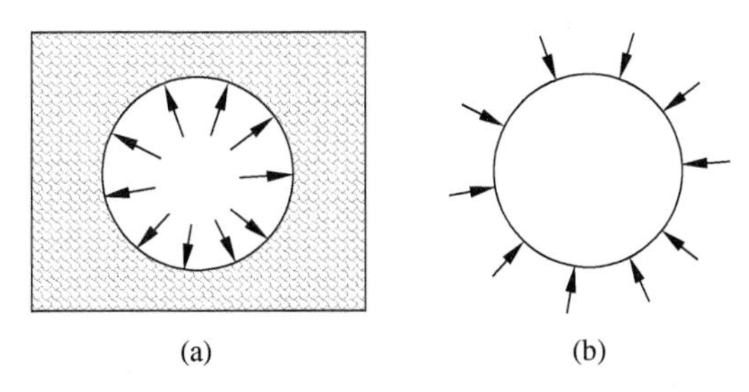


**Figure 5-71** Shrink-fit model for corrosion of steel in concrete: (a) internal pressure on concrete; and (b) external pressure on steel bar

By substituting Equations 5-110 and 5-111 into 5-114, we can obtain the pressure expression as

$$p = \frac{4\mu_c \mu_s}{2\mu_s + (k_s - 1)} \frac{\Delta a}{a} \tag{5-115}$$

The stress produced at surrounding concrete interface is then derived as

$$\sigma_{\theta\theta} = \frac{4\mu_s \mu_c}{2\mu_s + (k_s - 1)\,\mu_c} \frac{\Delta a}{a} = C\frac{\Delta a}{a} \tag{5-116}$$

For steel, the shear modulus is about 81 GPa and the Kolosou constant is around 2. For concrete, the shear modulus is about 12 GPa, thus the value of C is $2.23 \times 10^{10}$. For $\Delta a/a$ equaling 0.0001, the stress produced is 2.23 MPa. Note that the stress is in fact the shear stress at the interface, and this value is large enough to create a microcrack. The stress wave generated by this microcrack can be detected by the acoustic emission transducer. Thus, it is proven that the sensitivity of the detection to rebar corrosion using the AE technique is very high (0.0001 of a radius of 10 mm is only 1 μm).

**Prevention of Steel Corrosion.** Because of the magnitude of the costs of rebar corrosion, significant efforts have been made to protect the steel in recent decades. In most instances, corrosion-control methods can be described as passive. First of all, it is essential to make sure the chloride content in concrete (e.g., through raw materials control) is lower than the threshold values. In this case, the protective layer on the rebar could be kept intact and thus no corrosion will be initiated. Second, a good rebar protection can be obtained by proper design and control of the concrete cover. Such means are usually specified in design codes: minimum concrete cover thickness, the inherent concrete properties (in terms either of design strength or maximum water/cement ratio), and the maximum allowable crack width permitted. Adequate cover thickness and less/narrower cracks can effectively prolong the duration needed for corrosive ingredients (e.g., $CO_2$ and chloride ions) to reach steel rebar. However, as the cover increases, the rebar becomes less effective and the potential for cracking due to tensile stress, shrinkage, and thermal effects increases. Improving the quality of concrete (e.g., smaller water/cement ratio, higher binder content, and/or higher strength) should be considered the primary protection method. On the one hand, this can effectively control the rate of carbonation and chloride ingress; on the other hand, this can cut out one of the four components (i.e., electrolyte) of a corrosion cell. It is obvious that the electrolyte in concrete is made up of the moisture condition and existing air. If we make a denser concrete, there will be less chance for moisture and air to get in, which will reduce the possibility of forming an electrolyte and prevent corrosion. We can also use cathodic protection methods to protect the reinforcing steel. For instance, using zinc as an anode can protect steel, because the corrosion occurs at the zinc anode.

The other protection strategies include coated rebar, cathodic protection, corrosion-resistant rebar, and corrosion inhibitors. Epoxy coating of the reinforcing steel can enhance the durability performance by serving as a barrier preventing the access of aggressive species to the steel surface and providing electrical insulation. Epoxy-coating can be applied in various ways, either as a liquid or as a powder which is fused on the surface. ASTM A775/A775M (Standard Specification for Epoxy-Coated Steel Reinforcing Bars) has addressed the basic requirements for epoxy-coated reinforcing steels by the electrostatic spray method. In general, the performance of epoxy-coated rebars in bridge decks and parking garages in chloride environments, where chloride de-icing salts are applied during winter, has been demonstrated as excellent (Clifton et al., 1975; Satake et al., 1983). However, several notable problems with the corrosion of epoxy-coated bars were reported in substructures in the Florida Keys in the United States (Clear, 1992; Clifton et al.,

1975; Smith et al., 1993). The amount of damage to the epoxy coating prior to concrete casting has been considered the major contributing factor to the poor performance. Thus, training is necessary for the proper production, handling, and applying the coating, and repairing field damage to epoxy-coated bars. Moreover, the bond properties between the concrete and the epoxy-coated rebars are not as strong as those between concrete and conventional rebars, which should be improved. To solve the poor interfacial bonding, enamel coatings have been proposed to compete with epoxy coatings (Tang et al., 2012). However, since the enamels need to be formed under high temperatures at the scale of 800°C, their cost-effectiveness may not be comparable to epoxies. Similar to epoxy coatings, enamels provide only physical barriers, so damage to the coating at the construction stage may induce poor in-service corrosion protectiveness. A new-generation coating could be one that provides not only a physical barrier, but also a surface passivation, so that even though the coating is damaged, the rebar can still be protected by a passivation layer. A couple of chemically bonded phosphate ceramic coatings are being tested (Wagh and Drozd, 2014; Wang et al., 2020), which are expected to provide a dense, ceramic-like barrier and an iron hydrogen phosphate hydrate passivation layer.

Zinc coating of steel (galvanized steel) is also considered a good means for providing corrosion resistance. It acts both as a sacrificial and barrier-type coating. However, one disadvantage is that, like other metal coatings, the zinc coating corrodes over time. The rate of corrosion under the given environmental conditions will determine the loss of coating thickness, and, thus, how long it will be effective. Generally, there is a fairly linear relationship between the metal thickness and the duration of its effective service life for galvanized steel exposed to an industrial atmosphere (Chandler and Bayliss, 1985). The stability of zinc is dependent on the pH of the surrounding solution where the zinc coating is exposed. It has been found that zinc is stable at pH values below about 12.5, but it tends to dissolve at an increasing rate as the pH increases above this level. The corrosion products of zinc may be deposited at the surface of the zinc coating and seal it, thus arresting the evolution of $H_2$ gas and leading to passivation of the zinc coating. However, if galvanized rebars are used with ungalvanized bars, depletion of the galvanized bars will be accelerated. So, if galvanized and ungalvanized bars are used in the same structure, special care should be taken to ensure complete electrical isolation of the two (Broomfield, 1997). Apart from being used as a rebar coating material, zinc is also widely used for cathodic protection methods to protect the reinforcing steel. In these applications zinc serves an anode where corrosion occurs before steel can be corroded.

There have been a few investigations on the use of stainless steel bars. It has been shown that the chloride threshold value for initiation of corrosion in a nonwelded AISI 304 (a kind of stainless steel) rebar is three to five times higher than that of a conventional rebar. However, welding the bar reduced the critical chloride level by 50%. Since the 1990s, there has been burgeoning interest in (carbon, glass, and basalt) fiber-reinforced polymer (FRP) bars as an alternative rebar (Nepomuceno et al., 2021). However, use of stainless/FRP rebars is still a highly expensive solution, and the FRP bars have much more severe high-temperature sensitivities than steel.

Corrosion inhibitors are regarded as useful not only as a preventative measure for new structures but also as a preventative and restorative surface-applied admixture for existing structures. Various corrosion inhibitors can be classified (Trabanelli, 1986) into (1) adsorption inhibitors, which act specifically on the anodic or on the cathodic partial reaction of the corrosion process, or on both reactions; (2) film-forming inhibitors, which block the surface more or less completely; and (3) passivators, which favor the passivation reaction of the steel. The mechanistic action of corrosion inhibitors is thus not against uniform corrosion but against localized or pitting corrosion of a passive metal due to the presence of chloride ions or a drop in the pH value (Bertolini et al., 2004). Corrosion inhibitors added in concrete can act in two different ways: these inhibitors can extend the corrosion initiation time and/or reduce the corrosion rate after depassivation has occurred (Hartt and Rosenberg, 1989). Mixed-in inhibitors are regarded as more reliable since they are easier and more secure to implement. Some laboratory testing has shown that certain corrosion inhibitors do not significantly affect the amount of chloride ions required to initiate corrosion but can reduce the corrosion rate after initiation of corrosion. The field performance of these products has been observed only for a relatively short period and cannot be conclusive in determining their effectiveness.

#### 5.4.3.2 Alkali-Aggregate Reaction

Alkali-aggregate reaction (AAR) is a reaction between alkalis in the pores of cement paste and certain forms of aggregates, which results in excessive expansion of concrete sections, and leads to severe cracking thereafter. Two general types of attacks can occur (Tang, 1987): (1) an alkali attack with dolomitic limestone aggregate (some argillaceous dolomites) is called an alkali–carbonate reaction (ACR), and (2) an alkali attack with siliceous aggregates containing certain forms of amorphous or poorly crystalline silica (such as some chert, flint, opal, tridymite, cristoballite chalcedony, volcanic glasses, and some limestones) is called an alkali–silica reaction (ASR). The alkali content of cement depends on the materials from which it is manufactured and also to a certain extent on the details of the manufacturing process, but it is usually in the range of 0.4–1.6%. In concrete mixes, there may be a contribution to the alkali from other cementitious materials, such as pulverized fly ash or ground granulated blast-furnace slag, which is present. It is well known that $Na_2O$ (sodium oxide) and $K_2O$ (potassium oxide) are present in the cement clinker in small amounts. It is thus conventional to express the results of chemical analysis of cement in terms of these oxides. Furthermore, the alkali content in cement is generally expressed as an equivalent percentage of $Na_2O$ by mass of cement. Since the molecular weights of $Na_2O$ and $K_2O$ are respectively 62 and 94.2 (where 62/94.2 = 0.658), the equivalent percentage of $Na_2O$ is calculated with the formula

$$\%Na_2O_{eq} = \%Na_2O + 0.658\%K_2O \tag{5-117}$$

In the cement paste, $Na_2O$ and $K_2O$ form hydroxides and raise the pH level from 12.5 to 13.5. The concentration of these hydroxides increases as $Na_2O_{eq}$ increases. In such highly alkaline solutions, under certain conditions, the silica can react with the alkalis. The alkali–silica reaction is the most widely spread and best understood. In this reaction, alkali hydroxides in the hardened cement paste pore solution attach to the silica to form an unlimited swelling gel, which draws in any free water from osmosis and expands, disrupting the concrete matrix. The expanding gel products exert internal stress within the concrete, causing characteristic map cracking of unrestrained surfaces (see Figure 5-72), but the cracks may be directionally oriented under the conditions of restraint imposed by the reinforcement, prestressing, or loading. Cracking resulting from alkali–silica reactions can lead to a loss of structural integrity.

The degree of AAR is affected by (1) the presence of water—if there is no water, there is no swelling of the reaction products due to water absorption; (2) the alkali content—if the alkali content ($Na_2O$ and $K_2O$) of binder is less than 0.6%, there is no reaction, and concrete containing more than 3 kg/m$^3$ of alkali can be considered to have a high alkali content; and (3) the concrete porosity—the internal stress may be relieved in concrete with high porosity. ASR can occur only in a moist environment: it has in fact been observed that in environments with a relative humidity below 80–90%, the alkali and reactive aggregate can coexist without causing any damage. With low effective alkali content in the concrete, i.e., when the equivalent content of $Na_2O$ in the concrete is less than 3 kg/m$^3$, deleterious AAR can be prevented. The expansive effect, hence, the ASR, can be negligible. The extent of reaction depends on the amount of reactive silica present in the aggregate mix while the reactivity of the silica minerals depends on their crystal structure and composition. The porosity, permeability, and specific surface of the aggregates and the presence of

**Figure 5-72** Cracks caused by AAR. (Source: Thomas and Folliard, 2007, 247-281 / with permission of ELSEVIER.)

Fe- and Al-rich coatings may influence the kinetics of the alkali–silica reaction. Blended cements containing pozzolana, fly ash, or blast furnace slag give a resulting alkaline solution of slightly lower pH, and the addition of silica fume leads to the lowest pH. Hence, the use of pozzolana materials such as fly ash and GGBS can even prevent damage caused by ASR. These mineral additions reduce the concentration of $OH^-$ ions in the pore solution of the cement paste. This is because hydroxyl ions are consumed by the pozzolanic reaction occurring during hydration. Furthermore, the alkali transport is slowed down because of the lower permeability of pozzolanic and blast-furnace cement paste, which helps in reducing the ASR. Finally, the hydration products of the mineral additives bind alkali ions to a certain extent, preventing them from taking part in the reaction with the silica. It has also been found that by adding calcium nitrite into the concrete mix, the concrete's resistance to ASR can be significantly improved. However, the mechanism is not clear (Li, Peng et al., 1999; Li, Mu et al., 2000).

Temperature also influences the alkali–silica reaction. Normally, the ASR increases as temperature increases. AAR can be deleterious to concrete due to the expansion and possible cracking of the concrete associated with the reaction. What's more, development of the alkali–silica reaction may be very slow, and its effects may show even after long periods (up to several decades). Consequently, cracking caused by an alkali–silica reaction usually takes many years and is often preceded by pop-outs on the concrete surface. Consideration must always be given to the effects of deeply penetrating cracks on the durability of the reinforcement and to the self-stress induced by the expansive reactions caused by alkali reactivity. This may be advantageous in confined sections of normally reinforced concrete members but could be catastrophic in the case of prestressed structures.

AAR was first identified in 1940 by Thomas Stanton of the California State Division of Highways (USA) (Mehta and Monteiro, 2006), but only a limited number of examples were observed in practice until fairly recently. AAR causes deleterious expansion and cracking of the concrete and reduces the tensile strength, which may have consequences for the structural capacity. The significance of AAR or ASR for concrete structures has led to a surge in research activities: (1) to determine the exact nature of the reaction (which is not yet fully understood); (2) to define the acceptable limits of alkali content, moisture content, and reactive aggregate content; and (3) to determine methods to reduce the degree of destructive expansion. Extensive research has been made in two directions. One is the development of testing methods, which are designed to reveal whether an aggregate is potentially reactive and can cause abnormal expansion and cracking of concrete. The other is the development of effective methods to prevent damage induced by AAR.

Although it is possible to determine what types of aggregates have a trend toward AAR, it is impossible to predict whether their use will result in excessive expansion or not. It has been found that a critical amount of reactive minerals exists for each type of aggregate, which can result in serious expansion and an amount smaller or larger than this value will not cause significant swelling. Measuring the expansion of test specimens has been considered to be the most dependable way to evaluate aggregate reactivity, and a number of test procedures have been devised. The testing methods that have been established include standard test methods, such as the mortar bar test for potential alkali reactivity of "cement-aggregate combinations" (ASTM C227, withdrawn in 2018 due to its limited use by industry), the chemical method for potential alkali-silica reactivity of aggregates (ASTM C289, withdrawn in 2016), the rock cylinder method for potential alkali reactivity of carbonate rocks as concrete aggregate (ASTM C586), and a rapid test method, a mortar-bar method for potential alkali reactivity of aggregates (ASTM C1260). One of the disadvantages of the older standard testing methods is that most of them are very time-consuming, which is incompatible with the demands of the construction industry. The ASTM C1260 method uses an NaOH solution to accelerate the AAR so the alkali content of the cement is not a significant factor in affecting expansions. It is especially useful for aggregates that react very slowly in the service environment. It is worth noting that ASTM C1260 only gives the potential reactivity of the aggregate, but it does not predict expansion in the real situation. Another rapid test method is the concrete prism testing (CPT) which uses concrete prisms of $75 \times 75 \times 300–400$ mm as specimens. The prisms should be prepared with a cement content of 420 kg/m$^3$ and the alkali content is boosted to 1.25% $Na_2O_{eq}$ by cement mass (total alkali content of 5 kg/m$^3$). The maximum aggregate size can be up to 19 mm. The test setup of CPT is shown in Figure 5-73. This method has been documented in ASTM C1293.

To overcome the limitations of the above-mentioned test methods, other rapid testing methods have been developed, such as the dynamic modulus test and the gel fluorescence test. The dynamic modulus can be obtained by measuring the resonant frequency and pulse velocity. It has been shown that it can provide a good indication of deterioration due to AAR. The measurement can even detect deterioration before any expansion and visible cracking occur. It is also sensitive to the changes in environmental conditions, which activate or suppress the AAR. In the gel fluorescence test method, a 5% solution of uranyl acetate is applied to the surface of the specimen, then the specimen is viewed under an ultraviolet (UV) light. A yellowish-green fluorescent glow means that AAR is present.

Since AAR can cause significant deterioration and damage to concrete structures, much research has been conducted to decrease the effect of the reaction. There are numerous recommendations for minimizing the risk caused by AAR, including the following:

**(a)** Use non-deleterious aggregates and/or nonreactive aggregates when the alkali content of the cement is high (more than 3 kg/m$^3$).

**(b)** Use low-alkali Portland cement or blended cements with sufficient amounts of fly ash or slag when the active silica content of aggregate is high.

**(c)** Keep the concrete dry (relative humidity of the concrete less than 80%). The choices of the types of cements and aggregates at a construction site are usually very limited, and the environment surrounding the concrete is obviously unchangeable. In many cases, the content of silica in aggregates and/or alkalis in cement paste cannot be reduced effectively, either. The

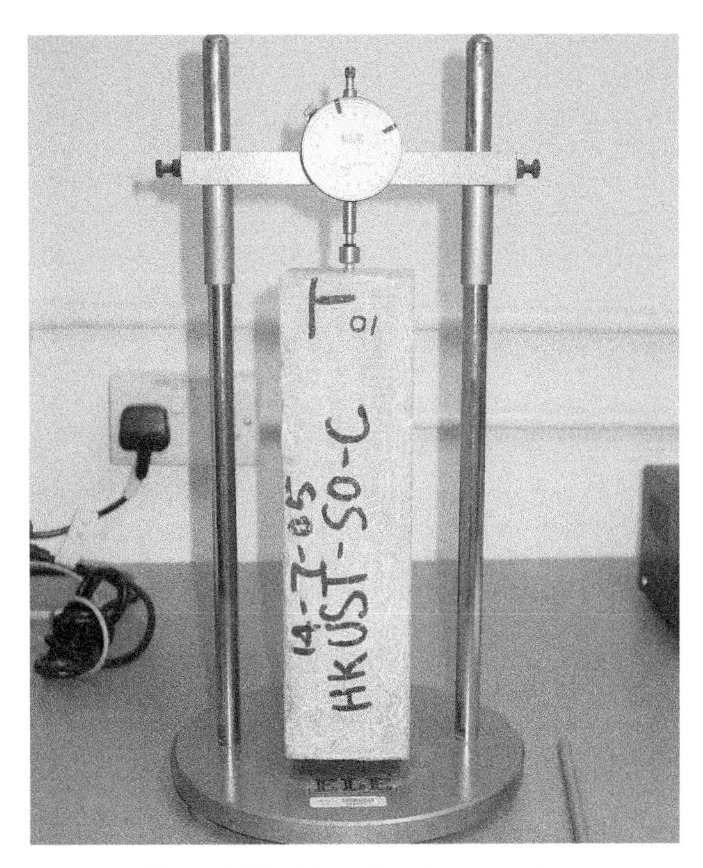


**Figure 5-73** Concrete prism test setup

only effective way to reduce the risk of AAR is to control moisture migration in the concrete, since no AAR will occur in dry concrete even if there are alkalis and reactive silica present. Control of moisture diffusion in concrete can be implemented at two different levels. One is to control the local moisture diffusion around the boundary of each aggregate, that is, to control the moisture exchange between the aggregate and the surrounding cement paste (see item d below). The other is to control the diffusion of moisture into and out of the surface of concrete members (e.g., by using a polymeric coating (see item e below)).

**(d)** Use a local diffusion-control surface coating. The control of local moisture diffusion could be very effective because AAR occurs right on the boundary of the aggregate. Distributions of chemicals, $Na_2O$, $SiO_2$, $K_2O$, and CaO, around the aggregate boundary have shown quantitatively that the reaction rim is in the range of 300 μm. Local diffusion control coating has been developed based on the crystallization technique. The coating product consists of powders of finely ground rapid-setting Portland cement, treated silica sand (de-alkaline silicate), and proprietary chemical additives. They are mixed with water to form a slurry. The slurry is applied to the reactive aggregate before the mixing of the concrete. A coating is then formed around the aggregates. Application of this technique aims at reducing and slowing down the AAR rate rather than completely eliminating AAR. As a result of a slow reaction, the product of the reaction can be accommodated and deposited in the large capillary pores. Thus, the detrimental damage due to the expansion will be minimized.

**(e)** Use a global diffusion control coating. This method applies hydrophobic weather-resistant coatings which can set on the surface of concrete members to reduce moisture diffusion.

**(f)** The use of lithium nitrate ($LiNO_3$), calcium nitrate, or other types of nitrates, as well as other lithium compounds can control AAR-induced damage. The work of Li, Mu et al. (1999) has demonstrated that calcium nitrate can effectively reduce AAR.

Summarizing these approaches, ASTM C1778 has been developed to provide recommendations for identifying the potential for deleterious AAR and selecting appropriate preventive measures based on a prescriptive-based or performance-based approach to minimize the risk of AAR.

#### 5.4.3.3 Deterioration Caused by Freeze-Thaw Cycling

Concrete is a porous material. As the cement and water in fresh concrete react to form a hardened paste binding the coarse and the fine aggregates together, the capillary pores are left in the originally water-filled space among the cement grains. The size of capillary pores can range from approximately 5 nm to 1 μm and sometimes are even larger. In addition to the capillary pores, cement paste also contains a significant volume of gel pores. Water content and capillary forces in these pores are important to the durability of hardened concrete, especially for those subjected to repeated cycles of freezing and thawing, which can cause disintegration of the concrete surface layers.

Concrete deterioration caused by freeze-thaw cycling is linked to the presence of water in concrete but cannot be explained simply by the expansion of water on freezing. While pure water in the open space freezes at 0°C, in concrete the water is really a solution of various salts, so its freezing point is lower. Moreover, the temperature at which water freezes is a function of the size of the pores. The freezing temperature of water in concrete pores decreases with decreasing pore size. In concrete, pore sizes cover a wide range, so there is no single freezing point. Indicatively, for saturated Portland cement paste, free water in pores larger than 0.1 mm freezes between 0 and −10°C; water in pores sized between 0.1 and 0.01 mm freezes between −20 and −30°C; and water in gel pores (pores less than 10 nm) freezes below −35°C (Beddoe and Setzer, 1990). Freezing begins in the outer layers and in the largest pores and extends to the inner parts and to smaller pores only if the temperature drops further. Specifically, the gel pores are too small to permit the formation of ice, and the greater part of freezing takes place in the capillary pores. It is also noted that larger voids, arising from incomplete compaction, are usually air-filled and are not appreciably subjected to the initial action of frost.

When water freezes, there is an increase in volume of approximately 9%. As the temperature of concrete drops, freezing occurs gradually inward and exerts hydraulic pressure on the unfrozen water in the capillary pores due to the volume expansion of ice. Such pressure, if not relieved, can result in internal tensile stresses that may cause local failure of the concrete. On subsequent thawing, the expansion caused by ice is maintained so that there is now new space available for additional water, which may be subsequently imbibed. During refreezing, further expansion occurs. Thus, repeated cycles of freezing and thawing have a cumulative effect. In most cases, it is the repeated freezing and thawing, rather than a single occurrence of frost that cause damage. Frost action is an important factor causing concrete degradation in cold regions.

There are two other processes that could increase the hydraulic pressure of the unfrozen water in the capillaries. First, since there is a thermodynamic imbalance between the gel water and the ice, diffusion of gel water into capillaries can lead to a growth in the ice body and thus to an increase of hydraulic pressure. Second, the hydraulic pressure is increased by the pressure of osmosis caused by local increases in solute concentration due to the removal of frozen (pure) water from the original solution. The extent of damage caused by repeated cycles of freezing and thawing varies from surface scaling to complete disintegration as layers of ice are formed, starting at the exposed surface of the concrete and progressing through its depth. In general, concrete members that remain wet for long periods are more vulnerable to frost than any other concrete. It is clear that the hydraulic-pressure mechanism of frost damage has more severe consequences in a system of fully saturated pores, because in that case the pressure can be released only if the microstructure expands, which may quickly result in cracking.

From the above discussion, it is known that the frost damage of concrete is tied to a degree of saturation (DOS), i.e., the ratio of the fluid-filled pores to the total volume of pores inside the cement matrix. Many studies have proven the existence of a critical DOS for a specific concrete (Li et al., 2012; Smith et al., 2019). If the concrete has a DOS lower than the critical DOS, frost damage is not observed, while the damage occurs (or is accelerated) when DOS is higher than the critical DOS. The critical DOS has been found to be 86–88% for normal concrete (Li et al., 2012; Farnam et al., 2014; Smith et al., 2019) and it depends on body size, homogeneity, the rate of freezing, and air-entrainment.

During freeze-thaw cycling attacks on concrete, the presence of de-icing salts, like calcium chloride and sodium chloride, in contact with concrete is a detrimental factor. When de-icing salts are present in the pore solution, they can lower the temperature of ice formation, which may be viewed as a positive effect. However, they may also bring the following negative effects: (1) an increase in the DOS of concrete due to the hygroscopic character of the salts; (2) an increase in the disruptive effect; (3) the development of differential stresses as a result of layer-by-layer freezing of concrete due to salt concentration gradients; (4) temperature shocks; and (5) salt crystallization in supersaturated solutions in pores (Mehta and Monteiro, 2006). Overall, the negative effects far outweigh the positive effects. As a result, these chemicals generally lead to the early appearance of scaling and detachment of the cement paste that covers the aggregate, as shown in Figure 5-74. In addition, it has also been proven that the chloride de-icing chemicals (especially magnesium chloride) tend to dissolve the hydration products of cement, leading to the formation of calcium oxychloride which keeps inflicting damage even during the thawing period such as in summer (Smith et al., 2019).

In general, the loss of mass or the decrease of dynamic modulus are used as indexes of degradation induced by freeze-thaw cycling. The resistance of a particular concrete to freeze-thaw damage is determined by the number of freeze–thaw cycles that the concrete can withstand before

**Figure 5-74** Scaling deterioration caused by freezing-thawing

reaching a given level of degradation. Such resistance can normally be evaluated using ASTM C666. To prevent/mitigate the damage caused by repeated cycles of freezing and thawing, an air-entraining agent can be used. An air-entraining agent is a chemical admixture that can deliberately entrain the air into cement paste in an enclosed space (<0.3 mm). The main types of air-entraining agents include: (1) animal and vegetable fats and oils and their fatty acids; (2) natural wood resins, which react with lime in the cement to form a soluble resonate; the resin may be pre-neutralized with NaOH so that a water-soluble soap of a resin acid is obtained; and (3) wetting agents, such as alkali salts of sulfated and sulfonated organic compounds.

The performance of an air-entraining admixture can be checked by trial mixes in terms of the requirements of ASTM C260. The essential requirement of an air-entraining admixture is that it rapidly produces a system of finely divided and stable foam, the individual bubbles of which resist coalescence; also, the foam must have no harmful chemical effect on the cement and cement paste. The major beneficial effect of air entrainment on concrete subjected to freezing and thawing cycles is to create space for the movement of water under hydraulic pressure. However, there can be some side effects on the properties of concrete, some beneficial (e.g., the water-reducing effect of the air entrainers), while others are not (e.g., the adverse influence of the air bubbles on the strength of concrete at all ages).

#### 5.4.3.4 Deterioration Caused by Sulfate Attack

Sulfate attack is one of main factors causing deterioration of concrete durability. It is generally regarded as an expansion due to the reaction of sulfate with some hydration products in cement paste. Portland cement itself contains gypsum which is a sulfate. In most cases, however, the sulfates inducing sulfate attack are from external sources. The sulfates to which concrete can be exposed may come from contaminated aggregates, groundwater (clay soil), and seawater. Solutions of sulfates of sodium, potassium, magnesium, and calcium are common salts that may cause severe deterioration to concrete. The total sulfate content in a concrete mixture serves as a controlling factor of the degree of sulfate attack. The acceptable concentration of sulfate in concrete is about 4% by weight of cement. It is generally accepted that a sulfate attack of hydrated cement takes place by the reaction of sulfate ions with calcium hydroxide and hydrated calcium aluminates to form gypsum and/or ettringite. The sulfate also reacts with the tricalcium aluminate ($C_3A$) in the cement to form the compound ettringite. The deterioration of Portland cement concretes exposed to sulfates may be ascribed to the following reactions (Campbell-Allen and Roper, 1991):

1. The conversion of calcium hydroxide derived from cement hydration reactions to calcium sulfate, and the crystallization of this compound with resulting expansion and disruption.

$$Ca(OH)_2 + SO_4^{2-} + 2H_2O \rightarrow CaSO_4 \cdot 2H_2O + 2OH^- \qquad (5\text{-}118)$$

2. The conversion of hydrated calcium aluminates and ferrites to calcium-sulfo-aluminates and sulfo-ferrites or the sulfate enrichment of the latter minerals. The products of these reactions occupy a greater volume than the original hydrates, and their formation tends to result in expansion and disruption.

$$C_3A + 3CaSO_4 + 32H_2O \rightarrow 3CaO \cdot Al_2O_3 \cdot 3CaSO_4 \cdot 32H_2O \qquad (5\text{-}119)$$

3. The decomposition of hydrated calcium silicates. In the presence of calcium sulfate, only reaction (2) can occur, but with sodium sulfate, both reactions (1) and (2) may proceed. The

sodium sulfate reacts with calcium hydroxide to form gypsum first, then with hydrated calcium aluminates or unhydrated $C_3A$ to form ettringite, as shown in the following equations:

$$CH + N\bar{S} + 2H \longrightarrow C\bar{S}H_2 + NH \tag{5-120}$$

$$3C\bar{S}H_2 + C_3A + 26H \longrightarrow C_6A\bar{S}_3H_{32} \tag{5-121}$$

$$3C\bar{S}H_2 + C_4AH_{13} + 14H \longrightarrow C_6A\bar{S}_3H_{32} + CH \tag{5-122}$$

With magnesium sulfate, all the three reactions may occur. It means that in additional to the reaction with CH to form gypsum first and then with $C_3A$ to form ettringite, magnesium sulfate can react with C–S–H directly as shown in the following equation:

$$C_xS_yH_z + xM\bar{S} + (3x + 0.5y - z)\,H \longrightarrow xC\bar{S}H_2 + xMH + 0.5yS_2H \tag{5-123}$$

From the different reaction mechanisms of different sulfates, it can be deduced that the severity of attack depends on the type of sulfate. Calcium sulfate undergoes an expansion reaction with ettringite (coming from calcium aluminate in cement), which gives rise to greater expansive effects than gypsum. Part of the formed ettringite is commonly located in the interface between the paste and aggregate, resulting in loss of bond. Sodium sulfate ($Na_2SO_4$) also reacts with calcium hydroxide to form gypsum, which reduces the paste strength and stiffness. Magnesium sulfate ($MgSO_4$) reacts to form gypsum and ettringite, and destabilizes C–S–H, the strength-governing phase in cement paste. This is because $Mg^{2+}$ and $Ca^{2+}$ ions associate well, since they have equal valence and similar ionic radii, which can lead to a reaction between magnesium sulfate and the C–S–H gel. Severity of attack therefore increases from calcium sulfate to sodium sulfate to magnesium sulfate. In other words, magnesium sulfate attacks have the most damaging effect on concrete.

Generally speaking, the formation of gypsum and ettringite is responsible for the expansion of concrete. The expansion caused by a sulfate attack can lead to the development of cracks in concrete. When concrete cracks, its permeability increases. Aggressive water penetrates more easily into the interior of the concrete through the cracks, thus accelerating the process of deterioration. The deterioration of concrete due to the sulfate attack is a complicated phenomenon of physical and chemical processes. Sulfate attacks can also be manifest as a progressive loss of strength of the cement paste due to loss of cohesion between the hydration products, and loss of adhesion between the hydration products and the aggregate particles in concrete.

The concrete attacked by sulfate could have a whitish appearance. Usually, damage starts at the edges/corners, followed by progressive cracking/spalling. The rate of sulfate attack also increases with the concentration of sulfate and the replenishment rate (e.g., sulfate attacks on concrete can be faster in flowing groundwater due to a faster replenishment rate). Tests on sulfate resistance are normally conducted through storing specimens in a solution of sodium or magnesium sulfate, or a mixture of the two, see Figure 5-75. The tests may be accelerated with wetting/drying cycles that will induce salt crystallization in the pores. Effect of exposure can be estimated from (1) a change in dimensions; (2) the loss of strength; (3) the change in the dynamic modulus of elasticity; and (4) the loss of weight. The test method of ASTM C1012 immerses well-hydrated mortar in a sulfate solution and considers excessive expansion a criterion of failure under sulfate attack. However, this method is used only for mortar, not for concrete. Besides, the method is slow—it normally takes several months to finish. As an alternative, ASTM C452 prescribes a method in which a certain amount of gypsum is included in the original mortar mix. This could speed up the reaction of $C_3A$ with sulfate, but this method is not appropriate for use with blended cements. The criterion of sulfate resistance in this method is specified in terms of the expansion at the age of 14 days.

**Figure 5-75** Storage pond for sulfate attack test

The resistance of concrete to sulfate attack depends primarily on the permeability and diffusivity of the concrete, the type and amount of cement in the concrete, and the type and amount of mineral additives in the concrete. Low permeability and diffusivity provide the best defense against sulfate attack by reducing sulfate penetration. To achieve a low permeability and diffusivity, the water/cement ratio should be decreased to a value as small as possible and pozzolanic additives (e.g., fly ash and GGBS) that can reduce the calcium hydroxide content and refine the pore structure of the matrix through pozzolanic reaction should be used. Reducing the water/cement ratio is generally more effective than using mineral additives. Given the comparable permeability and diffusivity, the severity of a sulfate attack depends on the content of $C_3A$ and, to a lesser extent, the content of $C_4AF$ in the cement. Therefore, reducing the calcium aluminate content of cement is effective in improving the sulfate resistance of concrete. In practice, sulfate-resisting cements (e.g., Types II, IIA, and V Portland cements, as specified in ASTM C150/C150M-21) normally have a low tricalcium aluminate content and hence less potential for expansive reaction and better resistance to sulfate attack. Reducing the amount of calcium hydroxide and hydrated calcium aluminate is also helpful in improving the sulfate resistance of concrete. Blended cements with pozzolanic materials or GGBS show enhanced resistance to sulfate attack (Mehta, 1988b). For example, low-calcium fly ash is an effective blending material for combating a sulfate attack on concrete; and the incorporation of silica fume into a concrete mixture also greatly improves the sulfate resistance, owing to the reduced amount of gypsum formation (compared with mixtures that do not contain silica fume) and improved impermeability. In a sulfate-rich service environment, the mixed utilization of silica fume and fly ash should be a much better choice for the production of sulfate-resisting, high-performance concrete.

On a related note, Mehta (1988a) has pointed out that it is the mineralogy rather than the chemical composition that determines the resistance to attack, and empirical guidelines based on chemical compositions of a mineral additive do not prove to be reliable. Nevertheless, Mehta (1988b) has concluded that, as a first approximation, a blended cement will resist sulfate attack when made with a highly siliceous natural pozzolanic or low-alumina fly ash or slag, provided that the proportion of the blending material is such that most of the free calcium hydroxide can be used up during the course of cement hydration. Under severe conditions, alternative hydraulic cements other than Portland cement should be used, for example, supersulfated cement and high-alumina cement. Supersulfated cement is mainly comprised of GGBS (80–85%), $CaSO_4$ (10–15%), and a small amount of alkali-activator (i.e., Portland cement or lime). It offers very high resistance to sulfates, especially if its Portland cement component is of the sulfate-resisting variety. It should be noted here that high-alumina cement should not be used in continuously warm and damp conditions, or in mass construction from which the heat of hydration cannot easily be dissipated. In the prefabrication industry, high-pressure steam curing can improve the resistance of concrete to sulfate attack due to the transition of $C_3AH_6$ into a less reactive phase, and also to the removal of $Ca(OH)_2$ by accelerated pozzolanic reactions if relevant mineral additives are used.

### 5.4.4 Durability in a Marine Environment

Many infrastructures are built in marine environments, such as sea bridges, tunnels, oil platforms, and piers. The marine environment can be distinguished as being under seawater, above seawater (i.e., in the tidal zone), and in the atmospheric zone. Very often, a splash zone is also considered between the tidal zone and the atmospheric zone. The deterioration of concrete structures in the marine environment in a tidal zone could be the most serious because it is facing a combination of mechanisms, both chemical and physical in nature, as shown in Figure 5-76. Attacks on concrete in the tidal zone can be of various types: (1) physical erosion; (2) salt crystallization pressure; (3) leaching of sulfate attack products; (4) freezing/thawing in cold regions; and (5) chloride penetration. The tidal zone refers to the range between the mean high and the mean low water levels; and, thus, it is periodically immersed and dried, generally on a daily basis. Such a wet-dry cycling could severely accelerate the ingress of the seawater-borne chloride, which when it reaches steel reinforcement and goes beyond the threshold chloride concentration would induce and accelerate steel corrosion (the commonest and most severe mechanism of concrete deterioration). The cumulation of corrosion products of steel can result in expansion of the reinforcing steel surface, ultimately spalling the concrete surface. This main degradation mechanism, when coupled with the other physicochemical mechanisms listed above, will facilitate the degradation of marine concrete structures, especially those located in the tidal zone.

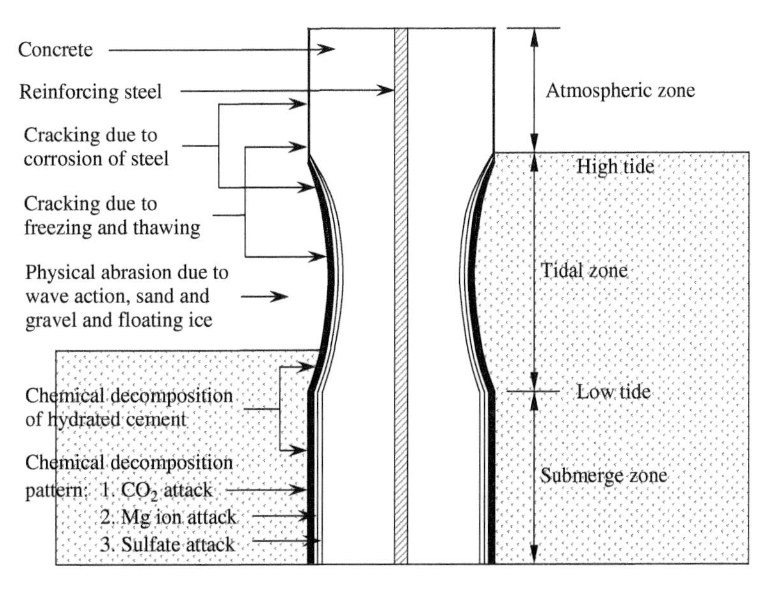


**Figure 5-76** Deterioration mechanisms of concrete structure in marine environment

**Figure 5-77** Seawater tank for marine environmental simulation test

Abrasion of structural elements from ice and debris occur within the tidal zone. The alternating motion of waves and tides also contributes to deterioration of the concrete. A high-velocity wave and depressions and irregularities of the surface layer can cause cavitation to form in marine concrete structures. If the absolute pressure at points of surface irregularities approaches the vapor pressure of the water, minute bubbles will form and quickly collapse. The collapse of these bubble can produce minute water jets having extremely high velocities and create an intense impact wave splashing to the concrete surface. This effect is very destructive to high-strength concrete. The deterioration caused by cavitation can take the form of tearing out of large pieces of concrete. Cyclic drying and capillary suction occur in the concrete just above sea level, and water carries the dissolved salts into the concrete. Subsequent evaporation causes these salts to crystallize in the pores, producing stresses that can cause microcracking. Below the low-tide line is an area where continual immersion occurs. In general, this is an area of less severe attack to the steel and concrete structures, compared with those of exposed at tidal or splash zones.

The reaction between sulfate ions and both $C_3A$ and C–S–H takes place in concrete exposed in a marine environment, resulting in the formation of ettringite, which, as well as gypsum, fortunately, is soluble in the presence of chlorides and can be leached out of the concrete by the seawater (Lea, 1970). This happens with no expansion but gradual material loss, which is quite different from sulfate attack in other environments. To investigate the deterioration mechanism of a concrete structure in a marine environment, simulation experiments can be conducted in the laboratory with an environmental tank as shown in Figure 5-77. However, to obtain more realistic information on marine structure deterioration, exposure testing to real ocean environments should be conducted. Figure 5-78 shows the marine exposure test site in Qingdao, China.

Keys to improve concrete durability against deteriorations caused by the marine environment include: (1) using concrete with low permeability, and (2) limiting the $C_3A$ content of the cement or using pozzolans for partial cement replacement. Mehta (1980) has indicated that the permeability of concrete is the most important factor influencing concrete performance exposed to marine environments. Low permeability can reduce the penetration of salt, sulfate, and water; besides, low-permeability concrete normally has high strength and good erosion resistance to the marine

**Figure 5-78** The marine exposure test site in Qingdao, China

environment. The rate of deterioration caused by seawater depends on the quantity of seawater absorbed by the concrete, so that all the factors that contribute to obtaining a lower permeability will improve concrete structures' resistance to attacks by seawater and the marine environments. Mehta (1980) has also noted that concretes containing even high tricalcium aluminate cements have excellent service lives in marine environments if the permeability is sufficiently low.

Low-permeability concrete can be achieved by the use of a low water/cement ratio, an appropriate choice of cementitious materials, good compaction, absence of cracking, and good curing. Since salt in water can contribute to the corrosion of reinforcing steel, normally 50–75 mm of dense concrete cover is required for reinforced concrete structures in the marine environment. To avoid the alkali-silica reaction, Mehta (1988b) has recommended the use of a cement content that is no lower than 400 kg/m$^3$, a tricalcium aluminate content of cement below 12% and preferably between 6 and 10%, and good-quality aggregate. For ocean structures, it is preferable to use blast-furnace slag cement, fly ash cement, or pozzolanic cement. Due to pozzolanic reaction of slag, fly ash, and pozzolans, a considerable amount of calcium hydroxide will be consumed to form secondary C–S–H. As a result, concrete produced by these cements has a finer pore structure, which largely reduces the transport rate of both sulfate and chloride ions.

### 5.4.5 General Methods to Enhance the Durability of Concrete

In common practice, concrete structures are exposed to different environmental conditions and deteriorate by the coupling effect of various factors. The marine structure mentioned above under erosion, chloride diffusion, and sulfate attack simultaneously is a good example of multi-factor deterioration. Moreover, concrete structures have to carry the mechanical loads while being exposed to different environments. Hence, it is more realistic to study the deterioration mechanism of a concrete structure under the coupling effects of mechanical loading and a combination of environmental factors. Many researchers have undertaken studies in this direction. For example, Sun et al. (1999, 2002) have conducted experiments on concrete with a combination of loading, chloride diffusion, and freeze–thaw cycling to investigate the coupling effect on the deterioration process. However, this area of research is still in its initial stage, and more efforts have to be made to better understand the deterioration mechanism under a combination of loading and environmental factors. In spite of the lack of understanding of mechanisms of multi-factor degradation, some general methods are known to enhance the durability of concrete, which are discussed as follows. More specific methods to mitigate single-factor degradations have been discussed in Section 5.4.3.

The chemical mechanisms of concrete deterioration (e.g., sulfate attack, steel corrosion, and AAR) are all associated with migrations of substances, such as those penetrating into concrete from the surroundings (e.g., chloride, sulfate, oxygen, carbon dioxide, and moisture) and ones moving in the interior of concrete (e.g., alkalis). Physical mechanisms of concrete deterioration (e.g., frost damage) could also be tied to ingress of water. Therefore, reducing the transport properties of concrete is an essential way of enhancing durability. This can be achieved by improving the compaction of concrete, reducing porosity and pore connectivity, and mitigating cracking. Compaction can be improved through better manipulation of fresh concrete. The pore structure of concrete can be refined by using a lower water-to-cement ratio or water-to-binder ratio, replacing part of the cement with quality supplementary cementitious materials (e.g., silica fume, ground granulated blast-furnace slag, and on-specification coal fly ash), and modifying the pore structure with polymers. Cracking mitigation, depending on its origins of cracks (i.e., induced by expansion, shrinkage, or loading), may be achieved by fiber reinforcement, enhanced curing, shrinkage-reducing admixtures, expansion agents, internal curing, or specific expansion controlling methodologies.

Since most of the chemical deteriorations of concrete involve reactions with specific cement components (e.g., alkalis and aluminates) or hydration products (e.g., calcium hydroxide), reducing the availability of these substances could be a general method to enhance durability. A mature method of mitigating sulfate attack following this strategy is to use sulfate-resistant cements instead of general-purpose cement. In ASTM C150, both types of sulfate-resistant cements (i.e., type II and type V) contain less aluminate phases than type I cement. A more general method that could improve resistance against multiple deterioration mechanisms (e.g., carbonation, sulfate attack, AAR, and leaching) is to reduce the alkalinity, especially the amount of calcium hydroxide, by replacing part of cement using the above-listed quality supplementary cementitious materials.

Almost all the above-discussed deterioration mechanisms need water to take action. So, theoretically, if water/moisture is not available in the service environment or in the microstructure of concrete, the deterioration can be prevented. Practically, this can be done by using film-forming admixtures (in small dosage) to block the pore network (i.e., pathways of moisture ingress), coating the surface of concrete using a waterproof (e.g., polymeric) or super-hydrophobic (e.g., silane) coating, or enhancing drainage in the service environment.

## DISCUSSION TOPICS

What are the strengths of concrete? Why is compressive strength an important index?

Does concrete have equal values of compressive and tensile strength? Why?

How is the indirect tension test conducted?

How is the bending test conducted?

How can shrinkage influence the quality and serviceability of a concrete structure?

How can creep influence the quality and serviceability of a concrete structure?

What are the main factors affecting the durability of concrete?

How is a permeability test conducted?

How is a chloride diffusion test conducted?

Why is durability important to a concrete structure?

How can the corrosion of reinforcing steel cause damage of concrete structure?

How can the corrosion of reinforcing steel be detected?

How can the corrosion of reinforcing steel be prevented?

Why is AAR harmful to concrete structure?

What are the methods to prevent AAR in concrete? What are their suitability and limitation?

What are the relationships between tensile, flexural, and compressive strength?

What is strain softening?

Why is Poisson's ratio of concrete not a constant in the stress range up to peak?

What is the main reason responsible for drying shrinkage?

What causes creep in concrete?

Discuss the factors affecting shrinkage and creep.

Define the durability of concrete.

Discuss the main factors affecting the durability of concrete.

Do you know any incidents caused by reinforcing steel corrosion?

Can you prevent or minimize corrosion if you work as an engineer on a construction site?

## PROBLEMS

1. Take the transition zone into consideration and derive the Young's modulus expression using parallel and series model. Suppose that volume fraction ratio of transition zone is 1% of the volume of concrete. If $E_c$ for a parallel model is 80 GPa and for a series model is 58 GPa, calculate the volume fraction of aggregate and elastic modulus of the transition zone. Assume that $E_a = 100$ GPa, and $E_m = 30$ GPa.
2. A specimen of shrinkage compensating concrete of 0.25 × 0.25 × 2.5 m is made with 12 steel rebars (12 mm in diameter) inside of the same length. The specimen expands during the wet curing period. If $Ec = 15$ GPa, $Es = 200$ GPa, and free expansion strain = 0.0003 at the time, how much will be the change in the specimen length? What is the force generated in the concrete and the steel?
3. Assume the behavior of a creeping material to be represented by the Burger's body given below. The material parameters are: $E_1 = 20$ GPa, $\eta_1 = 10{,}000$ GPa·days, $E_2 = 50$ GPa, $\eta_2 = 5{,}000$ GPa·days.

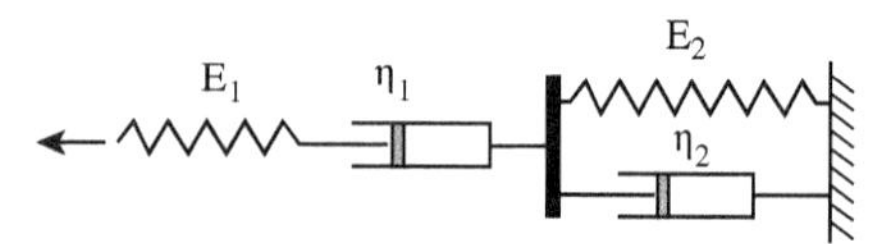


(a) Find a mathematical expression for the Creep Compliance J in terms of time in days (Note: the unit should be /MPa.)

(b) The material is placed under the following stress history (stress unit is MPa):

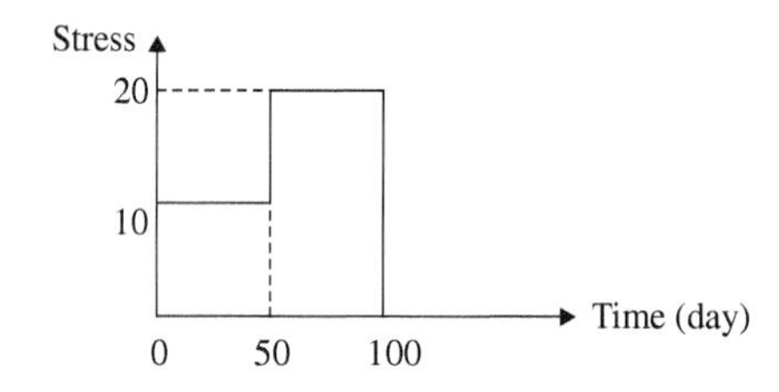


Derive expressions for the strain for:
   (i) $0 < \text{time} < 50$ days
   (ii) $50 \leq \text{time} < 100$ days
   (iii) $\text{time} \geq 100$ days

Specifically, what are the strain values?
   (i) at time = 0
   (ii) at 50 days, before and after loading increase
   (iii) at 100 days, before and after unloading

Draw, schematically, the variation of strain with time up to 150 days.

4. A specimen of concrete column (200 × 200 mm) of 1.5 m long has 4 steel rebars (20 mm in diameter) inside with same length. The specimen was loaded with a constant load of 600 kN under 100% relative humidity. Calculate the initial stress in steel and concrete. If the basic creep for plain concrete at 100 days is $600 \times 10^{-6}$, and the measured stress in steel at 100 days for this column is 160 MPa, estimate the percentage of creep restrained by steel. (Young's modulus of steel is 210 GPa, the modulus of elasticity of concrete is 25 GPa.) Also, calculate the length change of the specimen.

5. Suppose that the self-balanced strain due to shrinkage for a circular-shaped beam (150 × 500 mm) along the diameter direction follows the distribution of $\varepsilon = 0.00016\left(\frac{x}{R}\right)^2 - 0.0001$, where R is the radius of the circular-shape. The $Ec = 20$ GPa. If the peak load, P, for this beam under pure bending is 6,600 N, calculate the tensile strength for the concrete by eliminating the influence of shrinkage. The self-weight of the beam can be ignored in this problem.
6. A reinforced concrete column with a cross-section of 250 × 250 mm was repaired with a polymer concrete. The repaired portion is 250 × 100 mm at the bottom of the column. The length of the column is 3 m and carrying an axial load of 150 tons. The reinforcing ratio of the column is 1%. The Young's modulus of rebar, old concrete is 200 GPa and 28 GPa. The density of reinforced concrete is 2.4. What is the minimum E requirement of the polymer concrete if the stress in old concrete has to be less than 25 MPa?

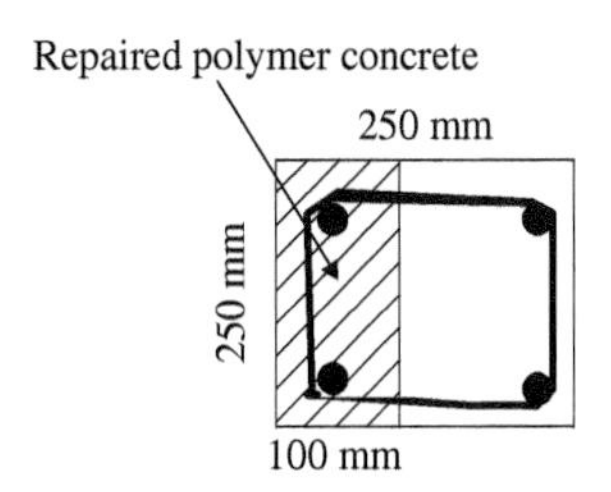


## REFERENCES


Agarwala, S., Reed, P., and Ahmad, S. (2000) "Corrosion detection and monitoring: A review," *in Proceedings of Corrosion 2000* (paper number: NACE-00271), Orlando, Florida, March 2000.

Allen, R.T.L., Edwards, S.C., and Shaw, J.D.N. (1993) *The repair of concrete structures*, London: Blackie Academic & Professional.

Alonso, C. and Andrade, C. (1994) "Lifetime of rebars in carbonated concrete," In: Costa, J.M. and Mercer, A.D., eds., *Progress in understanding and prevention of corrosion*, London: Institute of Materials.

Alonso, C., Andrade, C., and González, J.A. (1988) "Relation between resistivity and corrosion rate of reinforcements in carbonated mortar made with several cement types," *Cement and Concrete Research*, 18(5), 687–698.

Angst, U., Elsener, B., Larsen, C., and Vennesland, Ø. (2009) "Critical chloride content in reinforced concrete A review," *Cement and Concrete Research*, 39(12), 1122–1138.

Anonymous (1988) "Concrete durability—a multibillion-dollar opportunity," *Concrete International*, 10(1), 33–35.

ASTM C1202-09 *Standard test method for electrical indication of concrete's ability to resist chloride ion penetration*. ASTM.

Beddoe, R. and Setzer, M.J. (1990) "Phase transformations of water in hardened cement paste, a low temperature DCS investigation," *Cement and Concrete Research*, 20, 236–242.

Bentur, A., Diamond, S., and Berke, N.S. (1997) *Steel corrosion in concrete: fundamentals and civil engineering practice*, London: E & FN Spon.

Berke, N.S., Dallaire, M.P., Hicks, M.C., and Kerkar, A. (1999) "New development in shrinkage-reducing admixtures," ACI SP 173-48, Farmington Hills, MI: American Concrete Institute, pp. 973–998.

Bertolini, L., Elsener, B., Pedeferri, P., and Polder, R. (2004) *Corrosion of steel in concrete: prevention, diagnosis and repair*, Weinheim, Germany; Wiley-VCH.

Birnin-Yauri, U., and Glasser, F. (1998) "Friedel's salt, $Ca_2Al(OH)_6(Cl,OH)\cdot 2H_2O$: its solid solutions and their role in chloride binding," *Cement and Concrete Research*, 28(12), 1713–1723.

Blechman, I. (1992) "Differential equation of concrete behaviour under uniaxial short-term compression in terms of atrophy (degeneration) and its solution," *Magazine of Concrete Research*, 44(159), 107–115.

Bloom, R. and Bentur, A. (1995) "Free and restrained shrinkage of normal and high-strength concretes," *ACI Materials Journal*, 92(2), 211–217.

Broomfield, J.P. (1997) *Corrosion of steel in concrete: understanding, investigation and repair*, New York: E & FN Spon.

Burrows, R.W. (1998) *The visible and invisible cracking of concrete*, ACI Monograph No.11, Farmington Hills, MI: American Concrete Institute.

Campbell-Allen, D. and Roper H. (1991) *Concrete structures: materials, maintenance and repair*, New York: Longman Scientific & Technical.

CCES (China Civil Engineering Society Standard) (2004). *Guidelines of design and construction of durable concrete structure*, CCES01-2004.

Chandler, K.A. and Bayliss, D.A. (1985) *Corrosion protection of steel structures*, Oxford: Elsevier Science.

Chen, W.F. (1981) *Plasticity in reinforced concrete*, New York: McGraw-Hill.

Clear, K.C. (1992) "Effectiveness of epoxy-coated reinforcing steel," *Concrete International*, 14(5), 58–64.

Clifton, J.R., Beeghley, H.F., and Mathey, R.G. (1975) "Nonmetallic coatings for concrete reinforcing bars," *Building Science Series* 65, Washington, DC: U.S. Department of Commerce, National Bureau of Standards.

Damidot, D., Nonat, A., and Barret, P. (1990) "Kinetics of tricalcium silicate hydration in diluted suspension by microcalorimetric measurements," *Journal of the American Ceramic Society*, 73(11), 3319–3322.

Du, Y., Chen, Y., Zhuang, Y., Zhu, C., Tang, F., and Huang, J. (2017) "Probing nanostrain via a mechanically designed optical fiber interferometer," *IEEE Photonics Technology Letters*, 29(16), 1348–1351.

DuraCrete (2000) "DuraCrete—probabilistic performance based durability design of concrete structures," Final Technical Report (The European Union–Brite EuRam III), document BE95-1347/R17.

Farnam, Y., Bentz, D., Sakulich, A., Flynn, D., and Weiss, J. (2014) "Measuring freeze and thaw damage in mortars containing deicing salt using a low-temperature longitudinal guarded comparative calorimeter and acoustic emission," *Advances in Civil Engineering Materials*, 3(1), 316–337.

Glass, G.K., Page, C.L., and Short, N.R. (1991) "Factors affecting the corrosion rate of steel in carbonated mortars," *Corrosion Science*, 32, 1283–1294.

Goto, T., Sato, T., Sakai, K., and Motohiko I. (1985) "Cement-shrinkage-reducing agent and cement composition," United States Patent Number 4,547,223.

Gouda, V.K. (1970) "Corrosion and corrosion inhibition of reinforcing steel, I: immersed in alkaline solutions," *British Corrosion Journal*, 5(9), 198–203.

Hartt, W.H. and Rosenberg, A.M. (1989) "Influence of $Ca(NO_2)_2$ on seawater corrosion of reinforcing steel in concrete," *American Concrete Institute, Detroit, SP* 65–33, 609–622.

Hausman, D.A. (1967) "Steel corrosion in concrete," *Materials Protection*, 6(11), 19–22.

He, Z., and Li, Z. (2005) "Influence of alkali on restrained shrinkage behavior of cement-based materials," *Cement and Concrete Research*, 35, 457–463.

He, Z., Zhou, X., and Li, Z. (2004) "New experimental method for studying early-age cracking of cement-based materials," *ACI Materials Journal*, 101(1), 50–56.

Huang, J., Wei, T., Lan, X., Fan, J., and Xiao, H. (2012) "Coaxial cable Bragg grating sensors for large strain measurement with high accuracy," In *Sensors and Smart Structures Technologies for Civil, Mechanical, and Aerospace Systems 2012*, 8345, 83452Z.

Kuhl, D., Bangert, F., and Meschke, G. (2004) "Coupled chemo-mechanical deterioration of cementitious materials, Part I: modeling; and Part II: numerical methods and simulations," *International Journal of Solids and Structures*, 41(1), 15–67.

Lea, F.M. (1970) *The chemistry of cement and concrete*, London: Arnold.

Le Bellégo, C., Pijaudier-Cabot, G., Gérard, B., Dubé, J.F. and Molez, L. (2003) "Coupled mechanical and chemical damage in calcium leached cementitious structures," *ASCE Journal of Engineering Mechanics*, 129(3), 333–341.

Le Chatelier, H. (1900) "Sur les changements de volume qui accompagnent le durécissement des ciments," *Bulletin de Société pour l'Encouragement de l'Industrie Nationale*, 5(5), 54–57.

L'Hermite, R.G. (1960) "Volume changes of concrete,"in *Proceedings of the 4th international symposium on the chemistry of cement*, Washington, DC, Vol. 2, pp. 659–694.

Li, W., Pour-Ghaz, M., Castro, J., and Weiss, J. (2012) "Water absorption and critical degree of saturation relating to freeze-thaw damage in concrete pavement joints," *Journal of Materials in Civil Engineering*, 24(3), 299–307.

Li, Z. and Chau, C.K. (2000) "A new water permeability test scheme for concrete," *ACI Materials Journal*, 97(1), 84–90.

Li, Z., Li, F., Zdunek, A., Landis, E., and Shah, S.P. (1998) "Application of acoustic emission technique to detection of reinforcing steel corrosion in concrete," *ACI Materials Journal*, 95(1), 68–76.

Li, Z., Mu, B., and Peng, J. (1999) "The combined influence of chemical and mineral admixtures upon the alkali–silica reaction," *Magazine of Concrete Research*, 51(3), 163–169.

Li, Z., Mu, B., and Peng, J. (2000) "Alkali–silica reaction of concrete with admixtures—experiment and prediction," *Engineering Mechanics, ASCE*, 126(3), 243–249.

Li, Z., Peng, J., and Ma, B. (1999) "Investigation of chloride diffusion for high-performance concrete containing fly ash, microsilica and chemical admixtures," *ACI Materials Journal*, 96(3), 391–396.

Liao, W., Zhuang, Y., Zeng, C., Deng, W., Huang, J., and Ma, H. (2020) "Fiber optic sensors enabled monitoring of thermal curling of concrete pavement slab: temperature, strain and inclination," *Measurement*, 165, 108203.

Lyman, C.G. (1934) *Growth and movement in Portland cement concrete*, London: Oxford University Press.

Ma, H. (2013) "Multi-scale modeling of the microstructure and transport properties of contemporary concrete," PhD thesis. Hong Kong: The Hong Kong University of Science and Technology.

Ma, H. (2014) "Mercury intrusion porosimetry in concrete technology: tips in measurement, pore structure parameter acquisition and application," *Journal of Porous Materials*, 21 (2), 207–215.

Ma, H., and Li, Z. (2014) "Multi-aggregate approach for modeling interfacial transition zone in concrete," *ACI Materials Journal*, 111 (2), 189–200.

Ma, H., Xu, B., Liu, J., Pei, H., and Li, Z. (2014) "Effects of water content, magnesia-to-phosphate molar ratio and age on pore structure, strength and permeability of magnesium potassium phosphate cement paste," *Materials & Design*, 64, 497–502.

Mailvaganam, N.P. (1992) *Repair and protection of concrete structures*, Boca Raton, FL: CRC Press.

Masoero, E., Jennings, H.M., Ulm, F.-J., Del Gado, E., Manzano, H., Pellenq, R.J.-M, and Yip, S. (2014) "Modelling cement at fundamental scales: From atoms to engineering strength and durability." In Bicanic et al., ed., *Computational modelling of concrete structures*, London: Taylor & Francis.

Mehta, P.K. (1980) "Durability of concrete in marine environment—a review," *CI* SP-65, 1–20.

Mehta, P.K. (1988a) "Durability of concrete exposed to marine environment—a fresh look." In Malhotra, V.M., ed., *Concrete in marine environment*, ACI SP-109, 1–29.

Mehta, P.K. (1988b) "Sulfate resistance of blended cements." In: Ryan, W.G., ed. *Concrete 88 workshop*, Concrete Institute of Australia, pp. 337–351.

Mehta, P.K., and Monteiro, P.J.M. (2006) *Concrete: microstructure, properties and materials*, 3rd ed., New York: McGraw-Hill.

Mindess, S., Young, J.F., and Darwin, D. (2003) *Concrete*, New York: Prentice Hall.

Ministry of Development of the People's Republic of China (2007) "Chinese design code of concrete structure durability," Beijing.

Mitsui, K., Li, Z., Lang, D., and Shah, S.P. (1994) "Relationship between microstructure and mechanical properties of the paste–aggregate interface," *ACI Materials Journal*, 91(1), 30–39.

Montemor, M., Simoes, A., and Ferreira, M. (2003) "Chloride-induced corrosion on reinforcing steel: from the fundamentals to the monitoring techniques," *Cement & Concrete Composites*, 25, 491–502.

Nawa, T. and Horita, T. (2004) "Autogenous shrinkage of high-performance concrete," *Proceedings of the International Workshop on Microstructure and Durability to Predict Service Life of Concrete Structures*, Sapporo, Japan.

Nepomuceno, E., Sena-Cruz, J., Correia, L., and D'Antino, T. (2021) "Review on the bond behavior and durability of FRP bars to concrete," *Construction and Building Materials*, 287, 123042.

Nguyen, V.H., Colina, H., Torrenti, J.M., Boulay, C., and Nedjar, B. (2007) "Chemo-mechanical coupling behavior of leached concrete, part I: experimental result; and part II: modeling," *Nuclear Engineering and Design*, 237(20/21), 2083–2097.

Page, C.L. (1992) "Nature and properties of concrete in relation to reinforcement corrosion," paper presented at Corrosion of Steel in Concrete conference, Aachen, Feb. 17–19.

Pei, H., Li, Z., Zhang, B., and Ma, H. (2014) "Multipoint measurement of early age shrinkage in low w/c ratio mortars by using fiber Bragg gratings," *Materials Letters*, 131, 370–372.

Popovics, S. (1973) "A numerical approach to the complete stress–strain curve of concrete," *Cement and Concrete Research*, 3(5), 583–599.

Powers, T.C. (1965) "Mechanism of shrinkage and reversible creep of hardened cement paste," *Proceedings of the International Conference on Structural Concrete*, London, pp. 319–334.

Price, W.H. (1951) "Factors influencing concrete strength," *ACI Journal Proceedings*, 47(2), 417–432.

Qian, Z. (2012) "Multiscale modeling of fracture processes in cementitious materials," doctoral thesis, Delft University of Technology. Alblasserdam, the Netherlands: Haveka Holding B.V.

Rodrigues, R., Gaboreau, S., Gance, J., Ignatiadis, I., and Betelu, S. (2021) "Reinforced concrete structures: A review of corrosion mechanisms and advances in electrical methods for corrosion monitoring," *Construction and Building Materials*, 269, 121240.

Satake, J., Kamakura, M., Shirakawa, K., Mikami, N., and Swamy, R.N. (1983) "Long term resistance of epoxy-coated reinforcing bars." In: Crane, A.P., ed., *Corrosion of reinforcement in concrete construction*, Chichester: The Society of Chemical Industry/London: Ellis Horwood, pp. 357–377.

Smith, L.L., Kessler, R.J., and Powers, R.G. (1993) "Corrosion of epoxy coated rebar in a marine environment," *Transportation Research Circular*, 403, Transportation Research Board, National Research Council, pp. 36–45.

Smith, S.H., Qiao, C., Suraneni, P., Kurtis, K.E., and Weiss, W.J. (2019) "Service-life of concrete in freeze-thaw environments: Critical degree of saturation and calcium oxychloride formation," *Cement and Concrete Research*, 122, 93–106.

Soliman, A. and Nehdi, M. (2014) "Effects of shrinkage reducing admixture and wollastonite microfiber on early-age behavior of ultra-high performance concrete," *Cement and Concrete Composites*, 46, 81–89.

Sun, W., Mu, R., Luo, X., and Miao, C. (2002) "Effect of chloride salt, freeze–thaw cycling and externally applied load on the performance of concrete," *Cement and Concrete Research*, 32(12), 1859–1864.

Sun, W., Zhang, Y.M., Yan, H.D., and Mu, R. (1999) "Damage and damage resistance of high strength concrete under the action of load and freeze–thaw cycles," *Cement and Concrete Research*, 29(9), 1519–1523.

Sun, X., Du, Y., Liao, W., Ma, H., and Huang, J. (2019) "Measuring the heterogeneity of cement paste by truly distributed optical fiber sensors," *Construction and Building Materials*, 225, 765–771.

Suryavanshi, A., and Swamy, R. (1996) "Stability of Friedel's salt in carbonated concrete structural elements," *Cement and Concrete Research*, 26(5), 729–741.

Tang, F., Chen, G., Brow, R., Volz, J., and Koenigstein, M. (2012) "Corrosion resistance and mechanism of steel rebar coated with three types of enamel," *Corrosion Science*, 59, 157–168

Tang, L. (1996) "Electrically accelerated methods for determining chloride diffusivity in concrete: Current development," *Magazine of Concrete Research*, 48(176), 173–179.

Tang, L. (1999) "Concentration dependence of diffusion and migration of chloride ions: Part 1. Theoretical considerations," *Cement and Concrete Research*, 29(9), 1463–1468.

Tang, L. and Nilsson, L.-O. (1992) "Rapid-determination of the chloride diffusivity in concrete by applying an electrical-field," *ACI Materials Journal*, 89(1), 49–53.

Tang, L. and Nilsson, L.-O. (1999) "Ionic migration and its relation to diffusion," In *Proceedings of the International Conference on Ion and Mass Transport in Cement-Based Materials*, Toronto: University of Toronto, pp. 81–96.

Tang, M. (1987) "Studies on the effect of alkali in cement and concrete in China—a review," *Durability of Building Materials*, 4(4), 371–376.

Thomas, M.D.A. and Folliard, K.J. (2007) "Concrete aggregates and the durability of concrete," in *Durability of Concrete and Cement Composites*, Cambridge: Woodhead.

Trabanelli, G. (1986) "Corrosion inhibitors." In: Mansfield, F., ed., *Corrosion mechanism*, New York: Marcel Dekker.

Troxell, G.E., Davis, H.E., and Kelly, G.W. (1968) *Composition and properties of concrete*, 2nd ed. New York: McGraw-Hill.

Ulm, F.J., Coussy, O., Li, K.F., and Larive, C. (2000) "Thermo-chemo-mechanics of ASR expansion in concrete structure," *ASCE Journal of Engineering Mechanics*, 126(3), 233–242.

Valenta, O. (1969) "Kinetics of water penetration into concrete as an important factor of its deterioration and of reinforced corrosion," *RILEM International Symposium on the Durability of Concrete*, Prague, part I, pp. 177–193.

Valipour, M. and Khayat, K. (2018) "Coupled effect of shrinkage-mitigating admixtures and saturated lightweight sand on shrinkage of UHPC for overlay applications," *Construction and Building Materials*, 184, 320–329.

Vandamme, M. and Ulm, F.-J. (2009) "Nanogranular origin of concrete creep," *Proceedings of the National Academy of Sciences*, 106 (26), 10552–10557.

Wagh, A.S. and Drozd, V. (2014) "Inorganic phosphate corrosion resistant coatings," US Patent US2014/0044877 A1.

Wang, D., Yue, Y., Mi, T., Yang, S., McCague, C., Qian, J., and Bai, Y. (2020) "Effect of magnesia-to-phosphate ratio on the passivation of mild steel in magnesium potassium phosphate cement," *Corrosion Science*, 174, 108848.

Wang, P.T., Shah, S.P., and Naaman, A.E. (1978) "Stress–strain curves of normal and lightweight concrete in compression," *ACI Journal Proceedings*, 75(11), 603–611.

Yip, W.K. (1998) "Generic form of stress–strain equations for concrete," *Cement and Concrete Research*, 28(1), 33–39.

Zaki, A., Chai, H., Aggelis, D., and Alver, N. (2015) "Non-destructive evaluation for corrosion monitoring in concrete: A review and capability of Acoustic Emission Techniques," *Sensors*, 15(8), 19069–19101.

Zhang, J., Ma, H., Pei, H., and Li, Z. (2017) "Steel corrosion in magnesia–phosphate cement concrete beams," *Magazine of Concrete Research*, 69(1), 35–45.

CHAPTER 6

# ADVANCED CEMENTITIOUS COMPOSITES

In this chapter, the advanced cement-based composites, including fiber-reinforced concrete, (ultra-) high-strength and ultra-high-performance concretes, polymer-modified concrete, shrinkage compensating concrete, self-compacting concrete, engineered cementitious composites, confined concrete, high-volume fly ash concrete, structural lightweight concrete, and sea water and sea sand concrete, are introduced. These materials have been developed to achieve unique advantages regarding performance or sustainability.

## 6.1 FIBER-REINFORCED CEMENTITIOUS COMPOSITES

### 6.1.1 Introduction

Fiber-reinforced cementitious composites (FRC) are cement-based composites incorporated with fibers, mainly short and discontinuous fibers. Although fiber-reinforced concrete has a long history, a steady increase in the use of FRC began in the 1960s. The development of FRC mainly attempts to overcome the two major deficiencies of cement-based composites: a relatively low tensile strength and a rather low energy consumption capacity or toughness. The functions of the fibers in cement-based composites can be classified into two categories: shrinkage crack control and mechanical property enhancement. For shrinkage crack control, usually small amounts of low-modulus and low-strength fibers are added to restrain the early-age shrinkage and to suppress shrinkage cracking (Shah, 1991). For mechanical property enhancement, fiber reinforcement has been employed in various concrete structures to improve the flexural performance (Xu and Hannant, 1992; Maalej and Li, 1994), to increase impact resistance (Mindess et al., 1987), and/or to change the failure mode (Li and Leung, 1992).

The amount of fiber added has a significant influence on the mechanical properties and failure mode of FRC. Based on how much fiber is added, fiber-reinforced cementitious composites can be classified into three groups. FRCs employing low fiber volume fractions (<1%) utilize the fibers for reducing shrinkage cracking (Balaguru and Shah, 1992). FRCs with moderate fiber volume fractions (between 1 and 2%) exhibit improved mechanical properties, including modulus of rupture (MOR), fracture toughness, and impact resistance. The fibers in this class of FRC can be used as secondary reinforcement in structural members, such as in partial replacement of shear steel stirrups (Batson et al., 1972; Sharma, 1986), or for crack width control in structures (Stang and Aarre, 1992; Stang et al., 1995). In the last several decades, a new class of FRC, generally labeled high-performance FRC, or simply HPFRC, has been introduced. HPFRCs exhibit apparent strain-hardening behavior by employing high fiber content. These HPFRCs include: SIFCON, slurry infiltrated fiber (e.g., 5–20 vol.% of steel fibers) concrete (Naaman and Homrich, 1989); SIMCON, slurry infiltrated mats concrete, which is a special type of SIFCON wherein 6% steel fiber mat is used instead of discrete fibers (Krstulovic-Opara and Toutanji, 1996); and COMPACT

REINFORCED COMPOSITE (CRC), using 5–10% fine, strong and stiff steel fibers to enable a strain-hardening matrix, coupled with densely arranged reinforcing bars (Bache, 1987). The tensile strain capacity of HPFRC is typically about 1.5%.

In conventional applications of fiber-reinforced cementitious composites, usually with a low volume fraction ($V_f$) of fibers (e.g., 0.5% steel fibers or 0.05% polypropylene fiber), the function of the fibers is apparent only after a major crack has formed in the composite. Although there is still only one major crack and the overall behavior of the composites is still characterized by strain softening after the peak load is reached, the incorporation of fibers leads to a significant increase in the total energy consumption and overall toughness of the composites, represented by the area under a stress–strain or load–displacement curve, as shown in Figure 6-1. In such cases, as long as there is no fiber fracture, the fiber debonding and pullout process can consume a great amount of energy. On the other hand, with an increase in fiber volume fraction, it is possible that microcracks formed in the matrix will be stabilized due to the interaction between the matrix and the fibers through bonding, hence postponing the formation of the first major crack in the matrix. Thus, the apparent tensile strength of matrix can be increased.

Moreover, when a sufficient volume fraction of small-diameter steel, glass, or synthetic fibers is incorporated into the cement-based matrix, the fiber/matrix interaction can lead to strain-hardening and multiple cracking behaviors, changing the failure mode from quasi-brittle to pseudo-ductile. As a result, not only the composites' toughness, but also the deformation capacity and tensile strength can be significantly improved. One of the mechanisms in slowing down growth of a transverse crack in unidirectional fiber composites can be attributed to the development of longitudinal cylindrical shear microcracks located at the boundary between the fiber and the bulk matrix, allowing the fibers to debond while transferring the force across the faces of the main crack. In addition to enhancing the toughness and tensile strength, the addition of fibers can also improve the bending resistance of cement-based composites. However, adding fibers has only a minor influence on the compressive strength of cement-based composites. At a small fiber volume fraction, there is almost no effect. Hence, it is not worthwhile incorporating fibers for the purpose of enhancing compressive strength. The order of enhancement of fibers on the mechanical properties of cement-based composites is toughness > flexural strength > tensile strength > compressive strength.

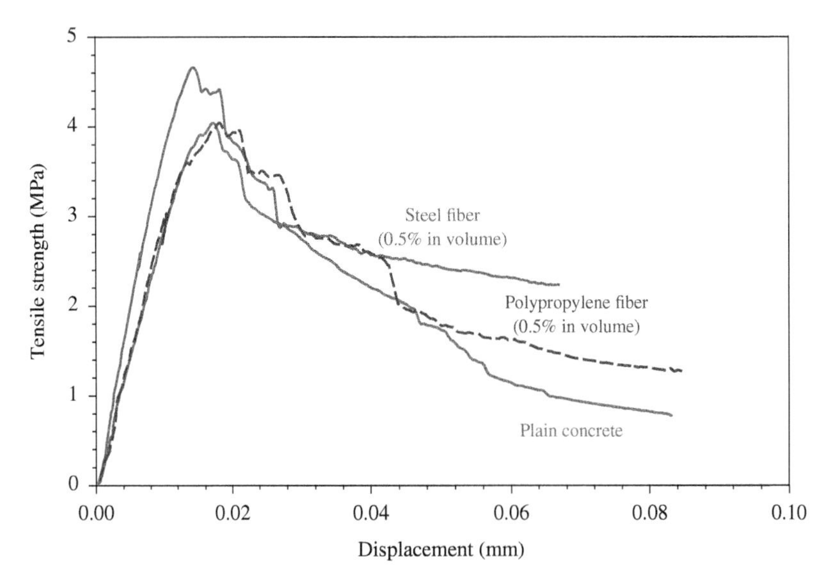


**Figure 6-1** Load-displacement curves of plain and fiber-reinforced concretes

Although fibers can improve the mechanical properties of cement-based materials, it is important to recognize that, in general, fiber reinforcement is not a substitute for conventional reinforcement or rebars. Fibers and rebars have different roles to play in contemporary concrete technology, and there are many applications in which both are used, such as CRC and some critical structural members or joints.

### 6.1.2 Factors Influencing the Properties

The properties of fiber-reinforced cement-based composites can be influenced by many parameters, such as fiber type, fiber amount, matrix variation, and manufacturing methods. In this section, these parameters are discussed in detail.

**(a)** *Fiber type*: The fiber type can be viewed with different criteria. From the size point of view, fibers can be classified into macro- and microfibers. The diameter of macrofibers is in the range of 0.2 to 1 mm and for microfibers is in a range of a few to tens of micrometers. Basically, microfibers are efficient in restraining microcracks and macrofibers in restraining macroscopic cracks. From the materials point of view, the fibers that are commonly used in fiber-reinforced cement-based composites are carbon, glass (borosilicate and alkali resistant), polymeric (acrylic, aramid, nylon, polyester, polyethylene, polypropylene, and poly vinyl alcohol), natural (wood cellulose, sisal, coir or coconut, bamboo, jute, akwara, and elephant grass), and steel (high tensile strength and stainless). Different types of fibers have different values of Young's modulus, different tensile strength, different surface texture, and different elongation ability, as can be seen in Table 6-1. These characteristics influence the bond between the fibers and the matrix, the crack-restraining ability, the matrix property enhancement, and, hence, the overall behavior of fiber-reinforced cement-based composites.

Steel fiber, having a high modulus, high fractural strain, and high tensile strength, is the most widely used fiber in cement-based composites. Steel fibers have a large distribution of diameters, and range from microfiber to macrofiber. Steel has various shapes and surface textures, as shown in Figure 6-2. Steel fibers with deformed shapes of ribs or hooks and rough surfaces are good for bond strength. One disadvantage of the steel fibers is its high specific gravity, which can increase the dead load (i.e., self-weight) of a composite.

Glass fiber is also commonly used in FRC, in the form of filaments. Each glass strand bundle has 204 filaments. Glass fibers are supplied in a continuous roving and can be chopped into short fibers, using a commercially available chopper. Glass fibers have high tensile strength (i.e., 2.1–2.5 GPa) and high fracture strain, but relatively low modulus (~70 GPa). Moreover, ordinary borosilicate glass fibers (E-glass) and soda-lime glass fibers (A-glass) can be easily attacked by alkali solution in cement-based composites and are thus less durable and should be used with caution. Alkali-resistant (AR) glass fibers contain about 16–20%

**Table 6-1** Properties of different types of fibers

| Fiber | Diameter (μm) | Specific gravity | Tensile strength (GPa) | Elastic modulus (GPa) | Fracture strain (%) |
|---|---|---|---|---|---|
| Steel | 5–500 | 7.84 | 0.5–2.0 | 210 | 0.5–3.5 |
| Glass | 9–15 | 2.6 | 2.0–4.0 | 70–80 | 2.0–3.5 |
| Fibrillated polypropylene | 20–200 | 0.9 | 0.5–0.75 | 5–77 | 8.0 |
| Cellulose | — | 1.2 | 0.3–0.5 | 10 | — |
| Carbon (high strength) | 9 | 1.9 | 2.6 | 230 | 1.0 |
| Cement matrix for comparison | — | 2.5 | $3.7 \times 10^{-3}$ | 10–45 | 0.02 |

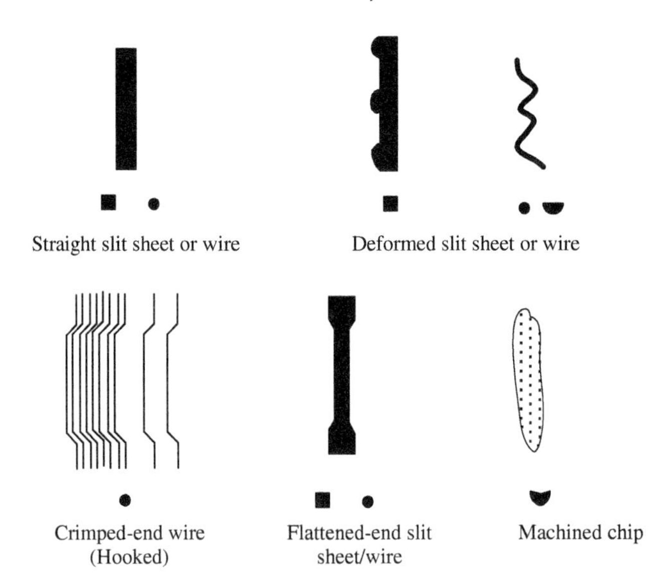


**Figure 6-2** Different shapes and surface textures of steel fibers

of zirconia ($ZrO_2$), which protects the fibers from high-alkalinity attacks. AR glass fiber is the most popular glass fiber used in cement-based composites. Disadvantages of glass fibers include low resistance to moisture, sustained loads, and cyclic loads.

Carbon fibers possess much higher strength and stiffness than glass fibers but are far more expensive. The strength of carbon fibers ranges from 2.2 to 5 GPa, modulus from 800 to 250 GPa, and the ultimate strain from 0.3 to 1.8%. The specific gravity ranges from 2.2 for a high-modulus fiber to 1.8 for a low-modulus fiber. The stronger fibers are associated with lower moduli and higher ultimate strain. Besides high strength and stiffness, carbon fibers also possess excellent resistance to moisture and chemicals, and are insensitive to fatigue. The weaknesses of carbon fibers are their low impact resistance, low ultimate strain, and high price.

Basalt fiber—made from extremely fine fibers of basalt (composed of minerals such as plagioclase, olivine, and pyroxene)—has been attracting increasing attention in recent years. Having a tensile strength of 2.8–3.1 GPa, a modulus of ~85 GPa, and an ultimate strain of 3.1%, the physicochemical properties of basalt fiber are between those of glass fibers and carbon fibers. One of the advantages of basalt fiber is that it is cheap, because of its easy and energy-efficient processing (crushing and washing, one-time heating/melting, followed by extrusion).

In all types of fibers, polypropylene fiber has the lowest density, the highest fracture strain, and reasonable tensile strength, but its modulus is quite low. Polypropylene fiber has both microfiber and macrofiber sizes which are commercially available.

To get multiple-cracking and strain-hardening responses, two fundamental requirements have to be satisfied when a fiber is selected. First, the fibers should be strong enough to carry the total load at the position of the first matrix transverse crack. Second, the bond at the fiber–cement interface should be strong enough to transfer the forces from the fiber to the matrix and thus build up the tensile stress in the matrix. According to previous studies, polyvinyl acetate (PVA) fiber shows a very promising potential in improving the interfacial bond and achieving multiple cracking.

**(b)** *Fiber volume fraction*: Another important factor that greatly influences FRC properties is the fiber volume fraction ratio, which is defined as the ratio of the fiber volume to the total volume of FRC. As mentioned earlier, at a low fiber volume ratio, the addition of fibers mainly contributes to the energy-consuming property. At a higher fiber volume fraction ratio, the tensile strength of the matrix can be enhanced and the failure mode can be changed.

For a low fiber volume fraction case, the ultimate stress of FRC is usually reached when the matrix reaches its ultimate stress. According to the rule of mixtures, the ultimate stress of the composites can be written as

$$\begin{aligned} \sigma_{\mathrm{cu}} &= V_{\mathrm{m}}\sigma_{\mathrm{mu}} + V_{\mathrm{f}}\sigma_{\mathrm{f}} \\ &= V_{\mathrm{m}}E_{\mathrm{m}}\varepsilon_{\mathrm{mu}} + V_{\mathrm{f}}E_{\mathrm{f}}\varepsilon_{\mathrm{mu}} \\ &= V_{\mathrm{m}}E_{\mathrm{m}}\varepsilon_{\mathrm{mu}} + V_{\mathrm{f}}\frac{E_{\mathrm{f}}}{E_{\mathrm{m}}}\sigma_{\mathrm{mu}} \end{aligned} \tag{6-1}$$

where

$\sigma_{\mathrm{cu}}$ = ultimate stress in fiber-reinforced concrete
$\sigma_{\mathrm{mu}}$ = ultimate stress in the matrix
$\sigma_{\mathrm{f}}$ = stress in the fiber
$E_{\mathrm{m}}$ = Young's modulus of matrix
$\varepsilon_{\mathrm{mu}}$ = ultimate strain
$V_{\mathrm{m}}$ = matrix volume fraction ratio
$V_{\mathrm{f}}$ = fiber volume fraction ratio

For a high fiber volume fraction, the ultimate strength is essentially determined by the ultimate strength of the fibers, $\sigma_{\mathrm{fu}}$, as the matrix's contribution can be ignored due to cracking. Hence,

$$\sigma_{\mathrm{cu}} = V_{\mathrm{f}}\sigma_{\mathrm{fu}} \tag{6-2}$$

where $\sigma_{\mathrm{fu}}$ = ultimate stress of fiber

If we plot Equations 6-1 and 6-2 as a function of $V_{\mathrm{f}}$, Figure 6-3 can be obtained. The intersection of the two lines separates the failure modes of the single crack and multiple cracks. Thus, by equating the above two equations, the minimum fiber volume fraction for obtaining multiple cracks failure mode can be obtained as

$$V_{\mathrm{f}}^{\mathrm{minimum}} = \frac{\sigma_{\mathrm{mu}}}{\sigma_{\mathrm{fu}} + \left(1 - \frac{E_{\mathrm{f}}}{E_{\mathrm{m}}}\right)\sigma_{\mathrm{mu}}} \tag{6-3}$$

It can be seen from Equation 6-3 that the critical fiber volume fraction depends only on the fiber and matrix ultimate strength, according to the theory applied here. However, in a real situation, many other factors, such as the bond between the fiber and the matrix and the ultimate strain of the fiber, can influence the minimum fiber volume ratio that is required to obtain a multi-crack failure mode.

**(c)** *Matrix variation*: The properties of the matrix influence the bond with the fibers and the mechanical properties of FRC, such as ultimate tensile strength. The FRC matrix can be modified using mineral admixtures, such as fly ash, slag, silica fume, and metakaolin. It can also be modified by adding some water-soluble polymers (e.g., PVA). Changing the matrix composition can increase the bond properties with the fibers, improve the matrix toughness, and enhance the matrix tensile strength and, hence, the mechanical properties of FRC (Sun et al., 2016).

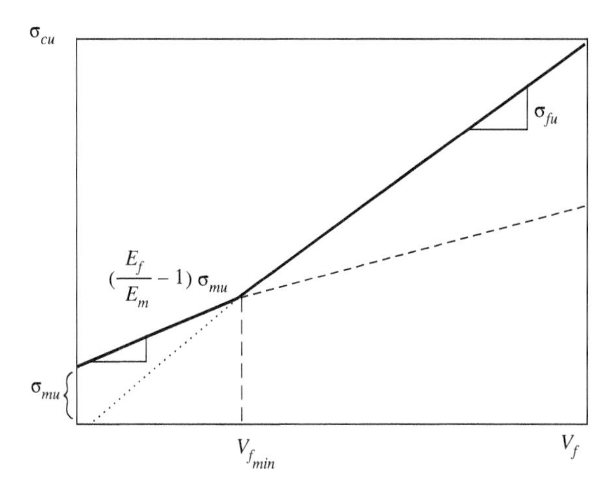


**Figure 6-3** Relationship between stress and fiber volume fraction

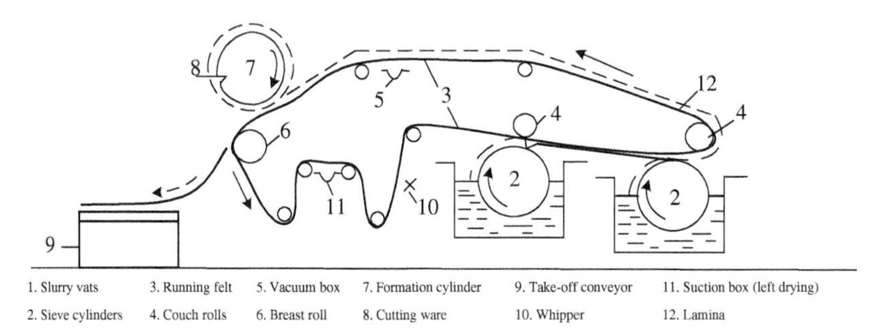


**Figure 6-4** The Hatschek process

**(d)** *Processing methods*: Recent studies have shown that the response of fiber-reinforced composites also depends on the methods of processing. The commonly used industrial processing methods for producing FRC products include the Hatschek process, normal casting, the pultrusion process, the Reticem process, and the extrusion process (Aguado et al., 1994). Different processing methods produce FRC products with different densities, and bonds and, hence, different mechanical properties. The Hatschek process was originally developed for producing FRC products with asbestos fibers using a dewater process, as shown in Figure 6-4. However, as more and more countries have restricted the use of asbestos fiber due to health considerations, the process is now mainly used to produce thin-wall products, such as sidings, using cellulose fibers. The normal casting method is the same as for normal concrete casting. The Hatschek process and casting are suitable only for short and discrete fibers. Both the pultrusion process and Reticem process are used for FRC products with continuous fibers. The pultrusion setup is shown in Figure 6-5. It can be seen that after passing through a slurry bath, the fibers are enforced by passing through a die or by forming a designed shape by the filament method. Here pultrusion was employed to incorporate continuous fibers into the cement matrix. Using the pultrusion method, composites with a fiber volume ratio of more than 10% can be produced. The strain-hardening type of response was a direct result, with a tremendous enhancement in tensile strength.

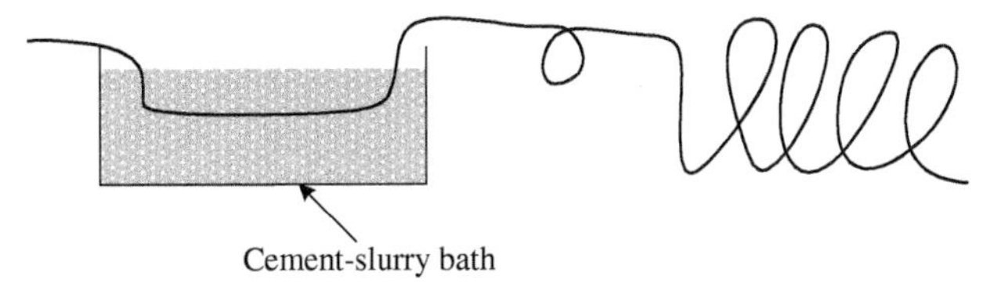


**Figure 6-5** The pultrusion process

**Figure 6-6** The extrusion process

Extrusion technology is an economical mass-production method. A typical extrusion process is shown in Figure 6-6. During the extrusion process, the dough-like material is pushed out through a die by an auger. The advantages of extrusion in producing FRC products are that the products formed under high shear and high compressive forces are denser and stronger, fiber alignment is controllable, and the product shape is flexible and good for mass production. With properly designed dies and properly controlled material mixes and viscosities, the fibers can be aligned in a load-bearing direction (Qian et al., 2003). Thus, extrusion can lead to a strain-hardening type of response, even for structural-functional-integrated composite incorporating phase change materials loaded lightweight aggregates (e.g., expanded perlite) (Lu et al., 2015). It is obvious that special processing compacted matrices with fibers of a low porosity controls fiber direction and distribution and improves the interfacial bond between the fibers and the matrix, which, in turn, leads to a class of high-performance fiber-reinforced composites with strain hardening. Extruded specimens can achieve a better performance than cast specimens, provided other conditions are similar.

Nowadays, except for the traditional products, such as bricks, tiles, and pipes, finely structured ceramic honeycomb units and completely shaped plastic doors and windows can also be produced using the extrusion technique. The advantages of the extrusion technique lie in its mass production capability, manufacturing flexibility, ability to manufacture products with complicated shapes, and capability of improving the material properties under high shear and high compressive force. Great research attention has been devoted to the feasibility of applying the extrusion technique in the manufacture of cement-based materials. Shao et al. (1995) at Northwestern University experimented with a screw-type extruder and successfully extruded various types of products, including sheets and tubes. They proved that extrusion is a promising alternative for fabricating cement composites with short fiber reinforcement. Stang and Pedersen (1996) applied the extrusion technique for manufacturing of pipes. They developed a novel type of extrusion that combines the ease of material mixing and only a few requirements for material preprocessing, with a high degree of accuracy and stability of the newly extruded material. Li and Mu (1998) extruded fiber-reinforced cement-based thin plates with a width to thickness ratio of 50:1. Li et al. (1996) studied the influence of mineral admixtures such as silica fume and metakaolin on cement-based extrudates, and showed beneficial effects and formulation hits. Li et al. (2005) have successfully extruded short-fiber-reinforced geopolymer thin plates (6 mm thick) and made a significant improvement in extrudate formulas. Generally speaking, the matrix formed through the extrusion process is dense and good in flexural stress withstanding. It has also been proven that such a matrix can effectively prevent phase change materials from leakage in energy-efficient building materials (Lu et al., 2015).

### 6.1.3 Fiber-Cement Bond Properties

Since the bond between the fiber and cement matrix plays a very important role in FRC, significant research activity has been conducted to characterize the interfacial bond properties and fiber debonding/pullout behaviors. Theoretically, two analytical approaches have been developed to interpret the material properties for fiber debonding and pullout problems (Stang et al., 1990): stress-based and energy release rate-based approaches. According to the stress-based criterion, debonding of the fiber from a matrix takes place when the maximum shear stress at the interface reaches a critical value. According to the energy release rate criterion, debonding propagates only when the energy flowing into the interface exceeds the value of the specific resistance energy.

The fiber-cement matrix bonding properties can be measured by a pullout test. Fabrication of the specimens for pullout testing is complicated and can be achieved using different methods. Li et al. (1991) have utilized a specially made brass mold to prepare specimens. A brass-guided plate with 16 holes was used to provide alignment of fibers during construction. Small springs attached to the frame of the mold were used to keep the fibers straight. During testing, this guide plate was used to separate the pullout section of the specimen from the anchored end and to provide for the transfer of the load. Since measurement of debonding on only one side (bottom portion) was desired, epoxy droplets were placed on the other side (top portion) of the specimens, thus providing anchorage of the fibers in the matrix. Hence, it could be reasonably assumed that the slip of the fiber during the tests would occur only on the pullout side. Furthermore, one specimen was prepared with epoxy-resin anchorage provided on both sides. This specimen was used to measure the deformation of the fiber inside the guide plate, which was subsequently subtracted from the slip displacement of pullout test specimens, as described in the following sections.

The specimen being held in the mechanical fixture is shown in Figure 6-7. The specimen is connected to the U-shaped loading fixture by means of a stainless steel rod (loading rod) 6.3 mm (0.25 in.) in diameter. The loading fixture is connected to the servo hydraulic actuator and the entire specimen fixture moves with the actuator. A restraining frame that can make contact with the specimen's brass guide plates is connected to a load cell and is used to resist the upward movement of the guide plate. Subsequently, the load is transferred from the brass guide plate to the matrix. This load is being reacted against by the fibers, which are the only means of connecting the top and bottom portion of the specimen. The load is transferred to the top of the specimen and thereafter to the loading fixture through the bending of the loading rod.

Two extensometers (1.905 mm or 0.075 in. range) mounted across the guide plate with a 12.5-mm (0.5-in.) gauge length were used to measure the slip displacement of the fiber debonding and pullout. The average output of the extensometers was used as the feedback signal in the control of the servo hydraulic system. The pullout tests were performed at a rate of 0.0254-mm opening of extensometer transducers per minute. Once the magnitude of slip exceeded the transducers

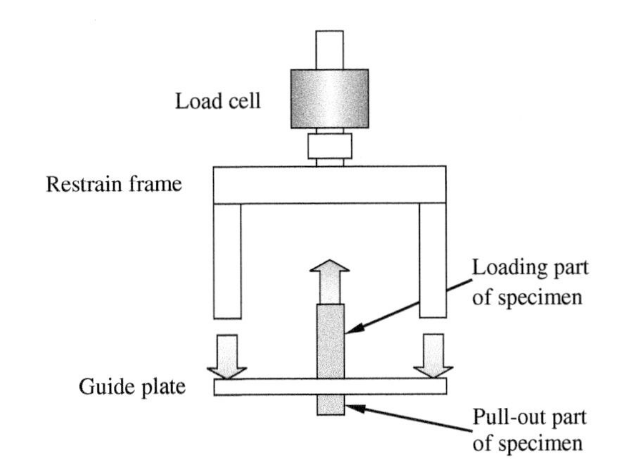


**Figure 6-7** Specimen being held in the mechanical fixture

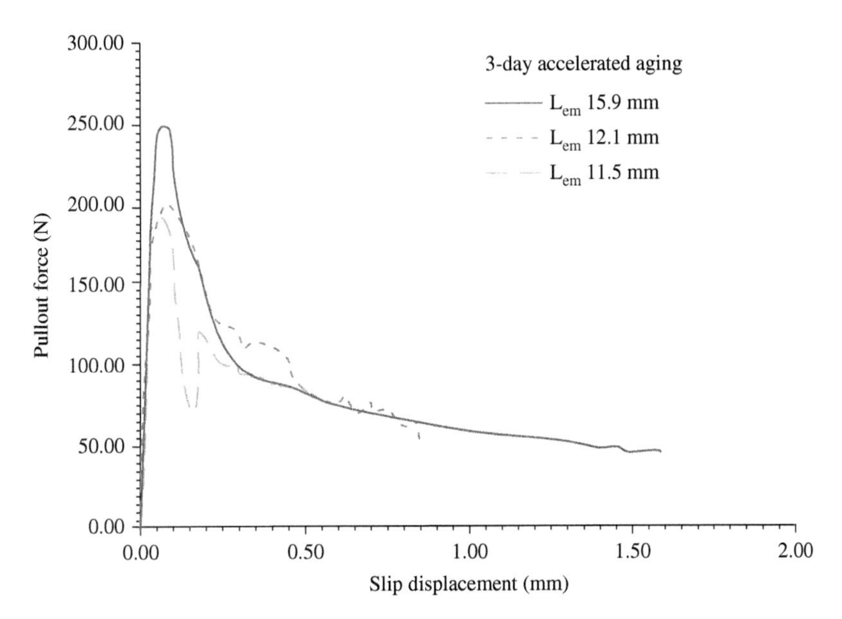


**Figure 6-8** Experimental results of fiber pullout test

range, the mode of the control was switched to stroke control and the entire fiber length was pulled out of the matrix.

A typical pullout load–slip displacement plot is shown in Figure 6-8. It can be seen from Figure 6-8 that, originally, the load and displacement had a linear relationship. Then, nonlinear behavior appears until the leak load is reached. After that, a sudden drop of the load is experienced and is followed by a gradual decrease of the load with displacement, representing a fiber pullout from the matrix. By utilizing the characteristic points on the curve, the interface properties, such as the interface stiffness parameter, shear bond strength, frictional bond strength, and surface energy release rate can be interpreted, as demonstrated by Li et al. (1991).

Typical pullout fibers from a three-point bending test are demonstrated in Figure 6-9. It can be seen that the steel fibers are kept intact after being pulled out, and the two sides have similar amounts of fibers left. This proves that the pullout test can reflect the true situation in fiber-reinforced, cement-based composites under mechanical loading.

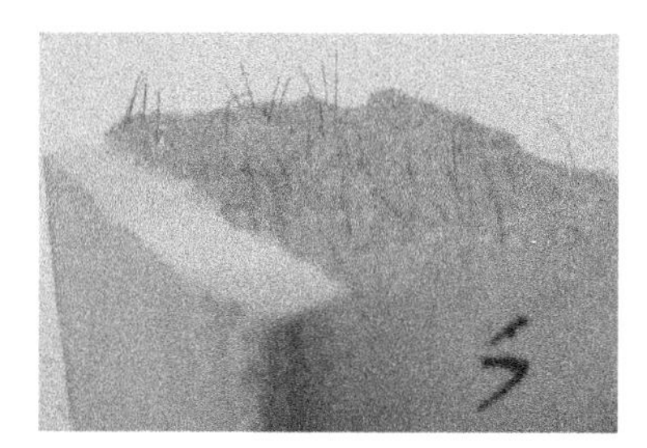
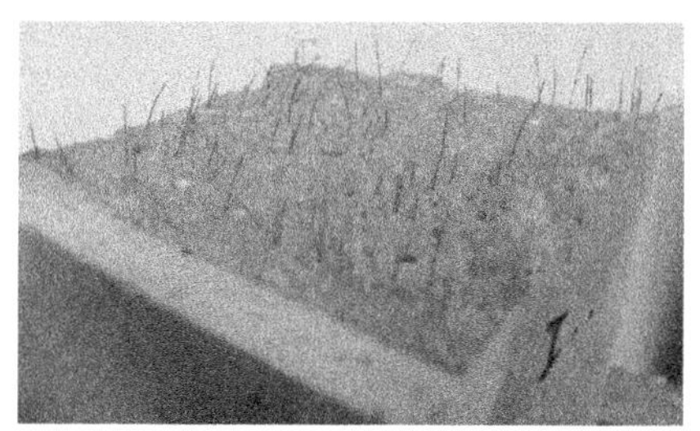

**Figure 6-9** Typical pullout fibers from a three-point bending test

### 6.1.4 Mechanical Properties

As mentioned earlier, the incorporation of fibers into cement-based composites mainly improves their toughness, bending, and tensile properties. In this section, we discuss the effect of fibers on these properties.

**(a)** *Tension*: Incorporation of fibers into cement-based composites can largely improve the tensile behavior of the composites, including toughness, tensile strength, and failure mode. Usually, fiber-reinforced cement-based composites can be classified into two categories according to their global tensile responses, i.e., strain softening and strain hardening. The strain-softening type of fiber-reinforced cement-based composites is usually reinforced with a low volume of short fibers. This kind of composite, containing about 1% fiber, is typically used for bulk field applications involving massive volumes of concrete. Figure 6-1 illustrates the tensile responses of such composites, with the incorporation of low-volume fiber fraction. For comparison, it also shows the stress–deformation curves for a plain concrete specimen. For steel fiber-reinforced concrete specimens, the length, diameter, and volume fraction of the steel fiber are 48 mm, 0.5 mm, and 0.5%. For polypropylene fiber-reinforced concrete specimens, the length and volume fraction of the fibrillated polypropylene fiber are 50 mm and 0.5%. It can be seen from Figure 6-1 that although all the specimens display a softening behavior after the peak load is reached, the areas under the load–deformation curve are different and are larger for fiber-reinforced specimens than for plain concrete specimens. This implies that the fiber-reinforced specimens require more energy for fracture than do the plain concrete specimens. For a strain-softening type of failure, usually only one major crack is formed in the specimen. The fracture process of the material can be divided into four stages: the linear elastic stage (0–35% of peak load), the randomly distributed damage stage (35 to about 80% of the peak load in the pre-peak stage), the microcrack localization stage (during the loading period between 80% of pre-peak load and 80% of post-peak load), and the major crack propagation stage (the period after 80% of post-peak load).

A failure mode with multiple matrix cracking can be obtained when the cement-based composite is reinforced with either a high-volume fraction of short fibers or with aligned continuous fibers. Such high-volume fiber composites have been used in thin sheets (e.g., curtain walls), slurry-filled cementitious composites, and concrete. A typical strain-hardening curve for steel fiber-reinforced cementitious composites is shown in Figure 6-10. There is a special point (marked B on the curve) called the *bend over point* (BOP) at which the matrix contribution to the tension capacity reaches a maximum. The stress–strain curve shown in Figure 6-10 can be roughly divided into four stages. Stage 1 is from O to A in the figure, which is characterized by the elastic behavior of the composite until a few microcracks are

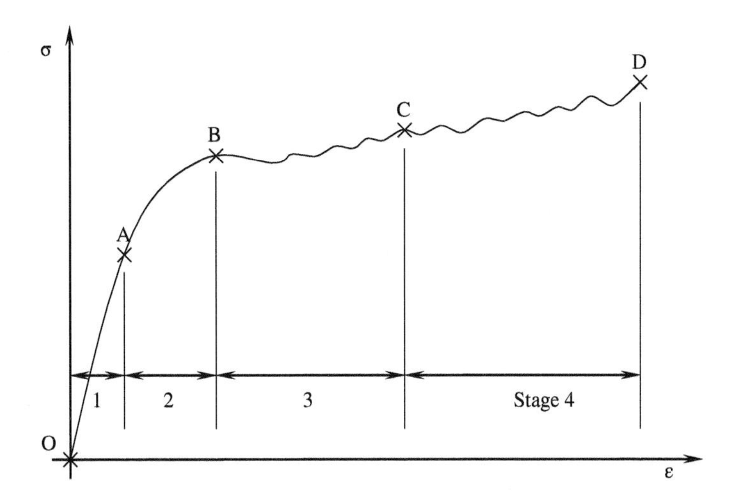


**Figure 6-10** Strain hardening behavior of steel fiber-reinforced cementitious composite

initiated and are randomly dispersed at end of this stage. Stage 2 is from A to B. In this stage, randomly distributed microcracks start to localize and form the first major crack across the specimen's cross section at point B. The BOP corresponds to the end of stage 2. The third stage is from B to C. In this stage, the incremental loading carried purely by the fibers at the position of the first crack is transferred back to the matrix through the bond and this process builds up the tensile stresses in the matrix again. When the tensile stress exceeds the tensile strength of the matrix, the next matrix crack will form. The multiple cracking process continues until the minimum crack spacing is reached, at which the tensile stress is transferred from the fibers back to the matrix cannot reach the tensile strength of the matrix. The process leads to an almost uniform distribution of fine matrix cracks. The fourth stage is from C to D. In this stage, no further matrix crack is expected and the additional load is sustained only by the fibers until failure of the fibers is reached. The phenomenon of strain hardening with multiple cracks is observed not only for continuous fiber-reinforced cement-based composites, but also for short fiber-reinforced concrete. Figure 6-11 shows the test results of short steel fiber-reinforced concrete under uniaxial tension test (Li and Li, 1998). The steel fiber used has a length of 32 mm, and an equivalent diameter of 0.8 mm with a tensile strength of 810 MPa and Young's modulus of 200 GPa.

It can be seen from Figure 6-11 that with the increase of the fiber volume fraction ratio, not only is the area under the stress–strain curves increased, but also the tensile strength is increased. Moreover, the failure mode changes from strain-softening to strain hardening. To study the multiple cracking and associated interfacial bond behavior, experiments have been conducted on the cementitious composites reinforced with continuous glass, polypropylene, and steel fibers. Three recently developed techniques—laser holographic interferometry, quantitative optical microscopy, and MRO interferometry—were employed in the investigation.

So far, the applications of FRCs in structural components are limited for inhibiting cracking, improving resistance to impact or dynamic loading, and resisting material disintegration, such as for airport runways and pavements. The limitation of the applications of FRC is partly due to the high cost of the fibers. For example, for steel fiber-reinforced concrete (SFRC), 1% by volume will increase the cost by more than US$55 per cubic meter of concrete.

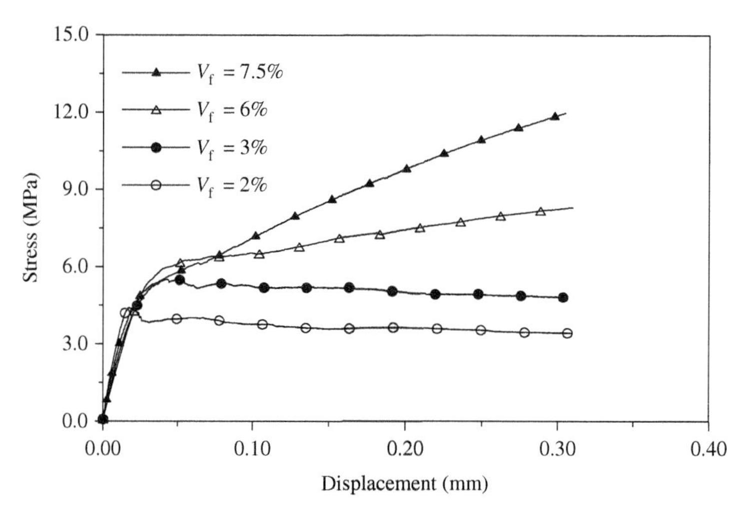


**Figure 6-11** Tensile response of short steel fiber-reinforced concrete

**Figure 6-12** Large deflection of an extruded thin plate

**(b)** *Bending*: Fiber-reinforced cement-based composites can achieve very good flexural strength and ductility. Figure 6-12 shows that an extruded fiber-reinforced cement-based plate exhibits a large, seeable deflection under four-point bending. Figure 6-13 shows the load–deflection curves for extruded thin FRC plates with and without fly ash. These curves demonstrated large deformations with reasonably high strengths (maximum stresses), hence the areas under the stress–strain curves are large (implying high toughness). Increased toughness also means improved performance in resisting fatigue, impact, and impulse loading. Moreover, according to the comparison shown in Figure 6-13, the discussed mechanical behaviors can further benefit from the incorporation of fly ash in the FRC.

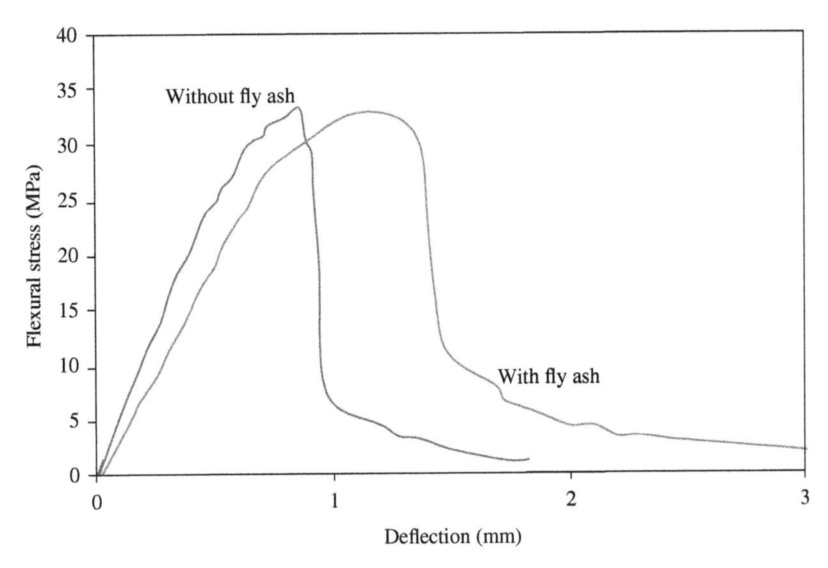


**Figure 6-13** Stress-deflection curves of extruded FRC thin plates with and without fly ash

To characterize the toughness of FRC beam specimens under bending, the concept of toughness index has been proposed. The toughness index utilizes a ratio of the area under the load–deflection curve of an FRC beam up to a specified deflection value to the area up to the first crack, or simply an area up to a specified deflection value. For instance, the ACI Committee 544 defines the toughness index as the ratio of the area of the load–midspan deflection curve to 1.9 mm to the area of the curve up to the first cracking, as shown in Figure 6-14. ASTM C1018 defines three toughness indexes, $I_5$, $I_{10}$ and $I_{30}$, as follows:

$$I_5 = \text{area of the load} - \text{midspan deflection curve to } 3\delta/\text{area of the curve up to } \delta$$

$$I_{10} = \text{area of the load} - \text{midspan deflection curve to } 5.5\delta/\text{area of the curve up to } \delta$$

$$I_{30} = \text{area of the load} - \text{midspan deflection curve to } 15.5\delta/\text{area of the curve up to } \delta$$

where $\delta$ is the deflection of the beam corresponding to the first crack of the beam under bending. As shown in Figure 6-15, $I_5$, $I_{10}$, and $I_{30}$ roughly represent the areas of 5, 10, and 30 times the area up to the first crack. The Japan Concrete Institute (JCI) defines the toughness index as the area of the load–midspan deflection curve up to the value of 1/150 span, as shown in Figure 6-16. It is noted herein that ASTM C1018 was withdrawn in 2006 due to limited use, and the currently active ASTM C1609 has adopted the JCI's method to assess the toughness of FRCs.

### 6.1.5 Hybrid FRC

As mentioned earlier, fibers can be distinguished as microfibers and macrofibers, according to their normal diameter and length. Fibers are also made from different materials that leads to different properties. A hybrid FRC uses more than two different fibers simultaneously to optimize their advantages. For example, by incorporating micro- and macrofibers into a cement-based composite simultaneously, the microfibers can restrain the development of the microcrack, while the macrofibers can control the propagation of macrocracks, as shown in Figure 6-17. Hence, the benefits of different size scales in restraining cracks of different sizes can be fully leveraged.

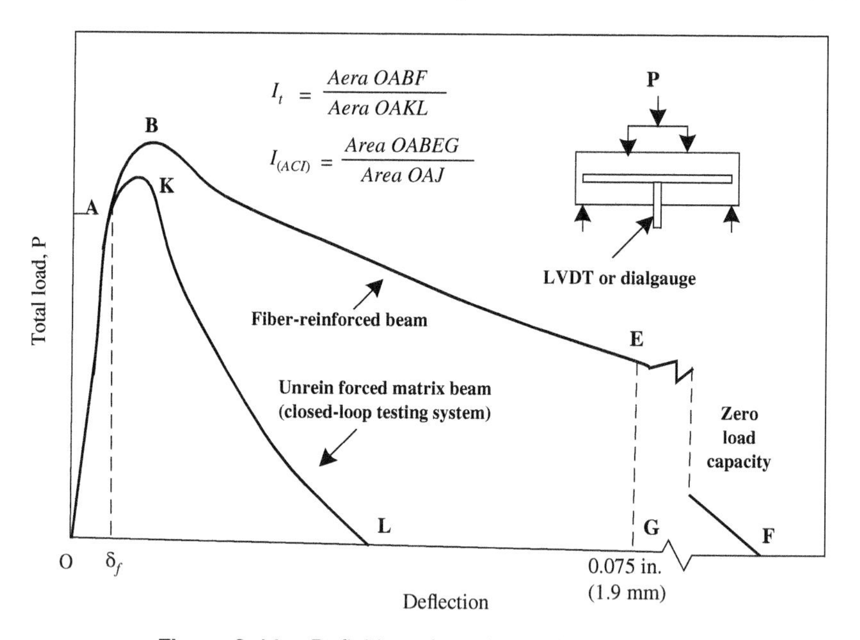


**Figure 6-14** Definition of toughness index by ACI

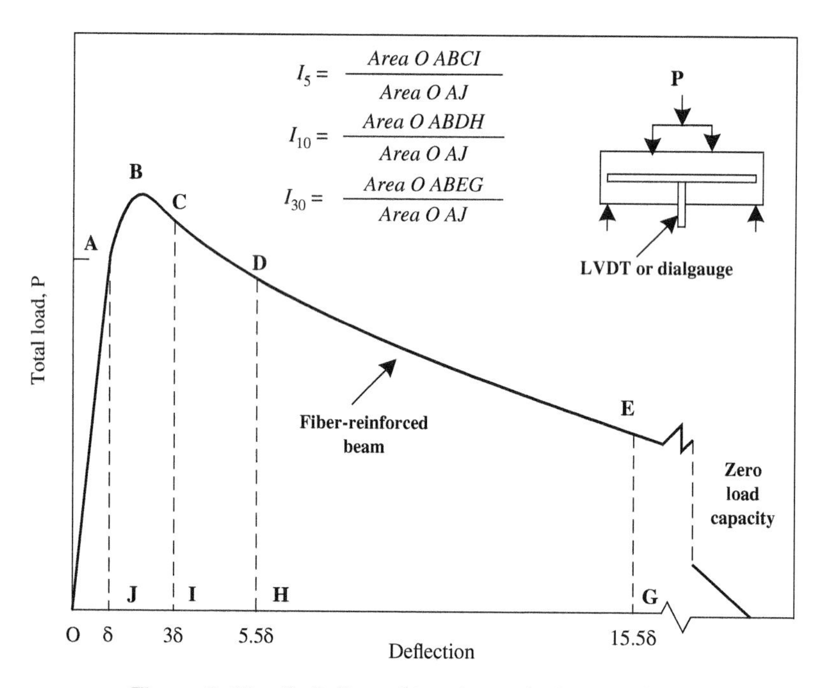


**Figure 6-15** Definition of toughness index by ASTM

Chen and Qiao (2011) demonstrated that a proper hybrid combination of steel fibers and polyvinyl alcohol (PVA) microfibers enhances the resistance to both crack nucleation and crack growth. On the other hand, combining different types and lengths of fibers can optimize the performance of FRC for specific benefits, such as improved processing, improved mechanical

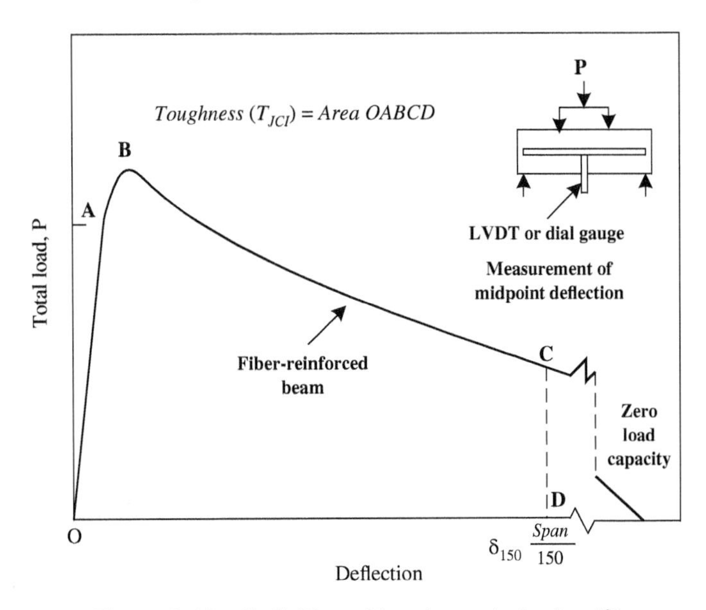


**Figure 6-16** Definition of toughness index by JCI

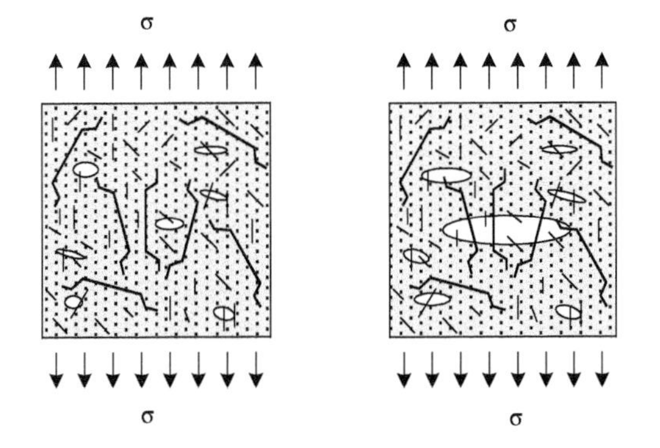


**Figure 6-17** Hybrid macro-fiber and micro-fiber reinforced cementitious composites

performance, improved durability, and reduced cost. Yao et al. (2003) have constructed three types of hybrid composites using fiber combinations of polypropylene (PP) and carbon, carbon and steel, and steel and PP fibers. Their test results showed that the fibers, when used in a hybrid form, could result in superior composite performance compared to their individual fiber-reinforced concretes. Among the three types of hybrids, the carbon–steel combination showed the highest strength and flexural toughness, because of the similar modulus and the synergistic interaction between the two reinforcing fibers. Peled et al. (2000) studied the performance of extruded thin sheet with hybrid fibers and found that the bending properties can be significantly improved, as shown in Figure 6-18. Recently, it was also reported that adding 0.3% carbon nano-fiber to a concrete reinforced with 0.5% steel fiber resulted in a 55% increase in direct tensile strength and a 110% increase in toughness (Meng and Khayat, 2018). A preliminary study on the hybrid FRC has demonstrated the good potential of such a composite, but how to attain uniform dispersion of the different fibers remains a big issue in this type of composites.

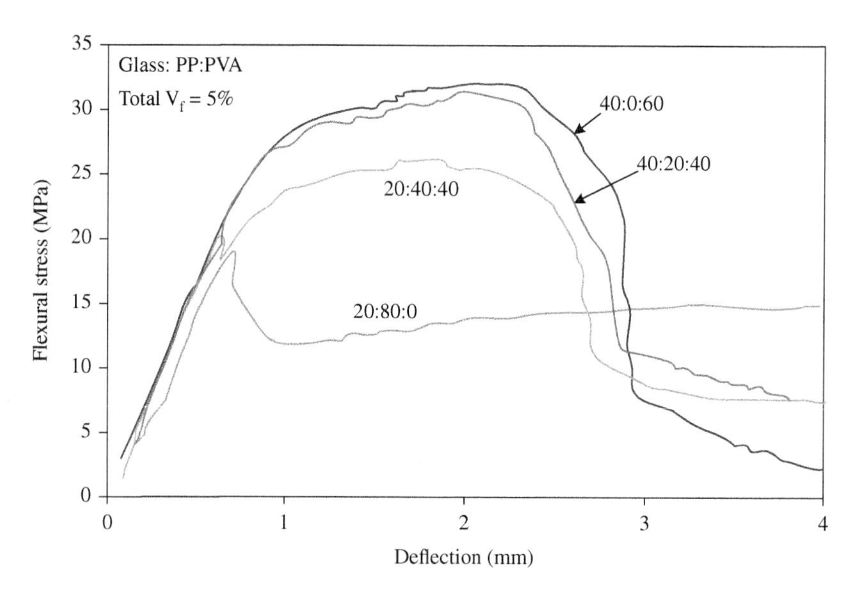


**Figure 6-18** Flexural response of hybrid composites combing glass/ PP/PVA fibers (Peled, et al., 2000)

### 6.1.6 FRC Products

Based on the matrix composition, fiber volume fraction, type of fibers, and manufacturing process, many fiber-reinforced cement-based products can be manufactured. The following products are introduced due to their common applications in practice or potential utilization in the future.

*Fiber-reinforced concrete*: Concrete contains coarse aggregates and the existence of coarse aggregates in the matrix brings some difficulties for fiber distribution or dispersion. Thus, the volume fraction of fibers is highly limited. For fiber-reinforced concrete used in practice, the applicable fiber volume fraction ranges from 0.4 to 2% for steel fibers and 0.06 to 0.5% for polypropylene fibers. The matrix is usually proportioned following the procedures used for plain concrete.

*Fiber-reinforced mortar*: Since mortar contains only fine aggregates, it is relatively easy to incorporate a larger amount of fibers into the matrix. The fiber volume fraction ranges from 1 to 5%. It applies to a wide range of manufactured products, such as glass fiber-reinforced sheets, polymeric fiber-reinforced panels, and tiles.

*Fiber-reinforced cement paste*: There are no aggregates in cement paste, so the incorporation of fiber is much easier. Larger amounts of fibers can be incorporated, with volume fractions ranging from 4 to 10% or even higher.

*Slurry-infiltrated fiber concrete (SIFCON)*: SIFCON is a special type of fiber-reinforced cement paste or mortar, which is produced by infiltrating a bed of fibers with cement or mortar slurry (Naaman, 1992). Since the fibers are placed into the mold (bed) by hand, large amounts of fiber can be added, with volume fractions from 4 to 22%. This type of high-volume fiber-reinforced cement composite has demonstrated a unique high performance and hence attracted great attention. Table 6-2 compares the mechanical properties of SIFCON with normal FRC. It can be seen that SIFCON can largely increase the compressive strength of the composites due to a large amount of steel fibers added. At such a high percentage, the fibers contribute not only directly to the compressive resistance of the composite through

**Table 6-2** Comparison of properties of SIFCON with normal FRC

| Property | SIFCON | FRC |
|---|---|---|
| Compressive strength (psi) | 12,000–22,000 | 5000–8000 |
| Flexural strength (psi) | 4000–8000 | 1000–2000 |
| Compressive/flexural | About 3:1 | 4–5:1 |
| Strain at failure | 0.02–0.08 | 0.005–0.01 |

parallel support, but also indirectly to the resistance through the confining effect on the matrix. The flexural strength of SIFCON can reach 60 MPa, which is about 4–10 times that of normal FRC. Moreover, SIFCON demonstrates much better ductility than normal FRC. The ratio of compressive strength to flexural strength is 3 and the fracture strain can reach 8%. Hence, SIFCON can be advocated as an ultra-high-performance, cement-based composite. However, it requires a hand layup to make a fiber bed and thus is time-consuming and labor-intensive.

*Compact reinforced composite*: The compact reinforced composite is made of cement, silica fume, steel microfibers, fine aggregate, and rebars (Bache, 1987). The fiber volume ratio can reach 15% or even higher. Fibers in such large quantities fundamentally alter the nature of cementitious matrices. The responses clearly indicate the large increase in both strength and ductility. Microcracking is stabilized, and homogeneous distribution of microcracks can be found even at very high strain levels of 1%, which is 100 times the strain of plain concrete. Thus the bending capacity of a beam made up of this kind of material approximates to that of structural steel.

*Extruded fiber-reinforced products*: As discussed earlier, extrusion is a good method for producing high-quality fiber-reinforced, cement-based composites. In the last few decades, the extrusion technique has been utilized in laboratories for new product development, and in industry for mass production. Along this line, researchers at Northwestern University and the Hong Kong University of Science and Technology have studied the feasibility of applying the extrusion technique in the manufacture of short-fiber-reinforced, cement-based materials. Shao et al. (1995) have successfully extruded various types of products, including sheets (6 mm thick and 75 mm wide) and tubes. Stang and Pedersen (1996) applied the extrusion technique specifically in pipe manufacture. They developed a novel type of extrusion combining ease of material mixing and few requirements for material preprocessing, with a high degree of the accuracy and stability of the newly extruded material. Mori and Baba (1994) produced molded cementitious products by screw extruders. Their work mainly focused on material aspects of the extrusion process control. A series of experiments were performed by them to show the relationship between particle size (silica sand), extrusion velocity and pressure, water/cement ratio, and flexural strength for plates 12 mm in depth and 60 mm in width. Li and Mu (1998) have extruded a wide and thin sheet, with a cross-section of 6 mm × 300 mm, using a single screw and a two-section vacuum extruder. The barrel of the above extruder is 100 mm in diameter. In the extrusion process, elastic and viscous properties were studied by a coaxial cylinder rheometer (Rheocord 9000, HAAKE) to obtain smooth and homogeneous products under continuous operation. HKUST also manufactured products with the incorporation of foam plastic and phase-changing material for heat insulation purposes, as shown in Figure 6-19. Such products are very light with a unit weight less than 800 kg/m$^3$, and can float on water. Li et al. (2004) studied the heat resistance of the extruded product with the incorporation of perlite. In addition, imitation wood-products using wood

**Figure 6-19** Extruded heat insulation products

**Figure 6-20** Extruded wood-imitation products

particles or dust have also been developed, as shown in Figure 6-20. These products can be sawed and nailed, just like wood.

The extrusion technique has been transferred into industry-scale production. Figure 6-21 shows a wall panel with a hollow cross-section being extruded out of the die and on the conveyor belt. The panel has a cross-section of 60×600 mm, with a hollow area ratio of 50%. Several other types of products have also been manufactured. One is a panel of solid cross section, with the incorporation of expanded perlite or foam plastics. Such a product is very light and has very good heat and acoustic insulation properties. Moreover, the panel is very easily machined, as shown in Figure 6-22. Other industrially extruded products include functional blocks with cable channels and a heat insulation core, as shown in Figure 6-23, and a structural wall panel with a heat insulation core, as shown in Figure 6-24. Such a wall panel has a compressive strength of around 15 MPa and thus can carry the loads transferred from other components. Meanwhile, the heat insulation core provides excellent performance in reducing the heat transfer from the outside to the inside of a room.

Other popular or well-known products of fiber-reinforced cement-based composites, called ultra-high-strength concrete (UHSC), ultra-high-performance concrete (UHPC), and the engineered cementitious composite (ECC), will be introduced in detail in later sections.

**Figure 6-21** Extrusion process in a plant

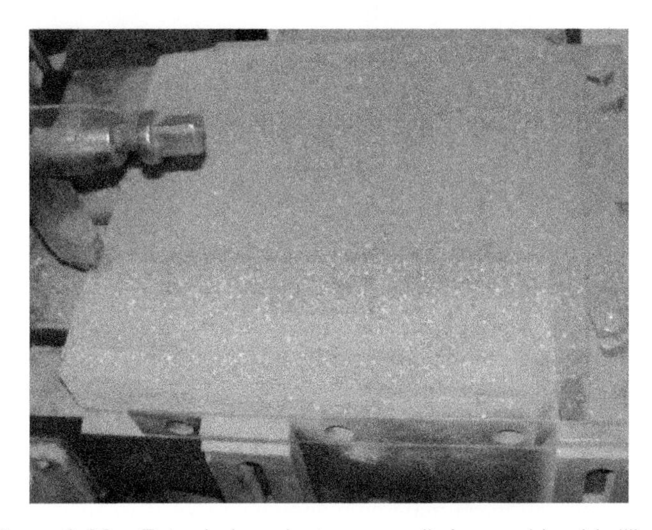

**Figure 6-22** Extruded products can easily be machined (milling)

**Figure 6-23** Industry-extruded products with several different functions

**Figure 6-24** Industry-extruded wall panel with heat insulation core

## 6.2 HIGH-STRENGTH CEMENTITIOUS COMPOSITES

### 6.2.1 High-Strength Concrete

High-strength concrete is defined as concrete made with normal-weight aggregates with a compressive cylindrical strength higher than 50 MPa. Here high strength means high compressive strength only. Although there are no differences in the raw materials used to make normal-strength concrete and those used to make high-strength concrete, more stringent quality control and more care in the section of materials are needed.

Usually, for the aggregate to be used in making high-strength concrete, it is better to choose one with a high crushing strength, if possible. The maximum size of aggregate is usually limited to 20 mm. The limitation on maximum aggregate size is intended to reduce the influence of the transition zone and to get a more homogeneous material. The moisture content in aggregates has to be carefully calculated to make sure the right *w/c* or *w/b* ratio is secured. The materials proportions of high-strength concrete are different from normal-strength concrete. The cement content is usually high, in the range of 400–600 kg/m$^3$. The higher cement content is the result of limiting the maximum aggregate size and the need for workability under the smaller *w/c* ratio condition, as will be discussed later. Moreover, the higher cement content also leads to a more homogenous concrete structure. Frequently, water-reducing admixtures and mineral admixtures such as fly ash, slag, and silica fume are incorporated into the mixtures for high-strength concrete.

The *w/c* or *w/b* ratio is a key parameter in making high-strength concrete. From Abrams' law and the gel/space ratio introduced earlier, it is easily seen that a low *w/c* or *w/b* ratio can lead to high compressive strength. Thus, the basic measure usually taken in making high-strength concrete is to reduce the *w/c* or *w/b* ratio from the values for normal-strength concrete (i.e., about 0.5 or higher) to about 0.3 or lower. However, lowering the *w/c* or *w/b* ratio will lead to the loss of workability, especially fluidity. This disadvantage is overcome in concrete proportions by adding superplasticizers and increasing the cement content. As discussed earlier, a superplasticizer can separate cement particles from flocculation and thus release the entrapped water by cement particle clusters, either by electrostatic repulsion or steric hindrance. Hence, the workability of concrete can be improved. Increasing the cement content can increase the total amount of water used in the concrete mix and provide more paste for the lubrication effect, which leads to an enhanced workability.

Incorporating mineral admixtures, especially silica fume, is another key factor in producing high-strength concrete. As discussed in Chapter 3, silica fume is a by-product of the ferro-silicon manufacturing process, with very small particle size and high reactivity. Due to the small size of around 100 nanometers, silica fume can easily pack into the gaps among cement particles of 20 microns on average, to form a much denser microstructure. Due to its highly reactive amorphous nature, silica fume can react with calcium hydroxide (CH, generated from the hydration of cement) and water to produce secondary C–S–H, which is very efficient in filling up large capillary spaces, thus further improving the density of the microstructure and reducing the porosity of concrete. Subsequently, the strength and permeability of the concrete are enhanced. Moreover, due the reduction of CH and permeability, the penetration rate and amount of carbon dioxide, oxygen, chloride, and moisture in the concrete are reduced. The possibilities for leaching, alkali aggregate reaction (AAR), carbonation, corrosion, and sulfate attack are all reduced and hence the overall durability of high-strength concrete is enhanced. Table 6-3 provides typical mix proportions for a normal-strength and a high-strength concrete. It can be seen that the high-strength concrete has incorporated both silica fume and a superplasticizer. Figure 6-25 shows the responses of the two concretes under compressive load. It can clearly be seen that the stress development of high-strength concrete is much faster than normal-strength concrete. It implies that high-strength concrete also has high early-age strength.

**Table 6-3** Mix proportions of normal concrete and high-strength concrete

| | **Binder** | | | | |
|---|---|---|---|---|---|
| | **Cement** | **Silica Fume** | **Water** | **Sand** | **Gravel** |
| NSC | 1 | 0 | 0.5 | 1.5 | 2.4 |
| HSC | 0.9 | 0.1 | 0.3 | 1 | 1.8 |

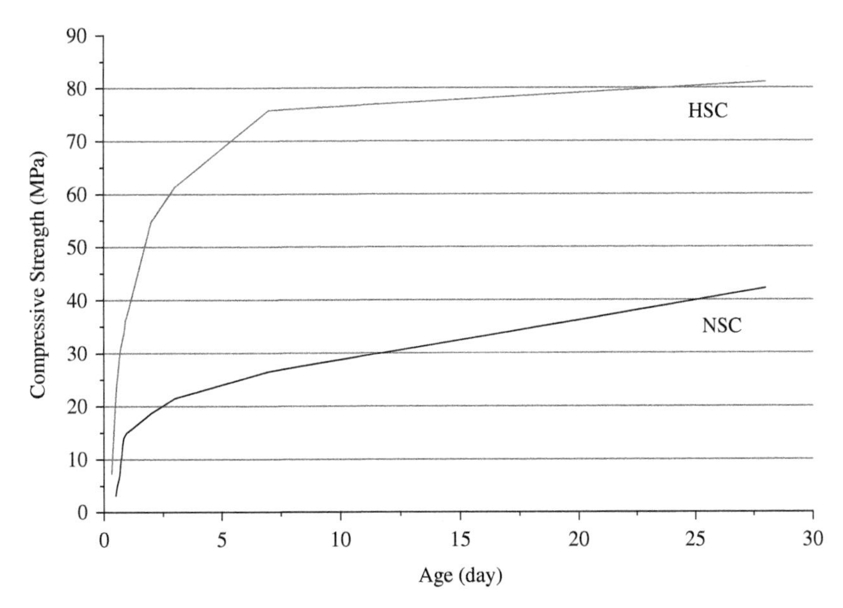


**Figure 6-25** Compressive stress development of normal-strength concrete (NSC) and high-strength concrete (HSC)

In addition to compressive strength, the microstructure and failure modes under compression of high-strength concrete are quite different from those of normal-strength concrete. From a microstructure point of view, high-strength concrete looks more homogeneous than normal-strength concrete due to the limitation of the maximum aggregate size and the increase of cement content. The extent of porosity in the transition zone is greatly reduced. Sometimes, the existence of the transition zone is almost eliminated. Also, the number of microcracks in high-strength concrete associated with shrinkage, short-term loading, and sustained loading is significantly less than that in normal-strength concrete. As for the failure mode, high-strength concrete usually has a vertical crack going through the aggregate and shows a more brittle mode of fracture and less volumetric dilation, while for normal-strength concrete the major crack usually passes around coarse aggregate, as shown in Figure 6-26.

### 6.2.2 MS Concrete

MS stands for *microsilica*, and an MS concrete contains a large amount of microsilica, usually up to 15% by weight of cementitious materials. There are some commercially available MS concretes in the United States and Europe. The microstructure and properties of MS concrete have the following characteristics.

In MS concretes, the amount of CH is greatly reduced due to the pozzolanic reaction of the silica. The C–S–H formed in MS concrete in the presence of the reactive pozzolanic material (i.e., the amorphous microsilica) is slightly different from that forms due to normal hydration of cement (Liao et al., 2019). In particular, the C/S ratio is reduced from ~1.7 to ~1.1, but the silicate structure is not significantly different. It is likely that both types, "normal" (C/S = ~1.7) and "pozzolanic" (C/S = ~1.1), co-exist, giving rise to a mean value of around 1.5. The addition of microsilica can reduce or eliminate the macroporosity (>100 nm diameter) that dominates permeability. Hence, MS concrete can be very impermeable and durable.

With the incorporation of large amounts of microsilica, the interfacial zone of MS concrete has a denser and more uniform structure that is similar to that of the bulk paste. The small particle size of the microsilica can inhibit the development of segregated water film surrounding aggregate particles and mitigate the wall-effect of binder particle packing against the aggregate surface, which probably are the causes of the porous interfacial transition zones (Ma and Li, 2014). Similarly, the bond between the paste and other embedded materials with hydrophobic surfaces, such as steel and glass fibers, will also be improved.

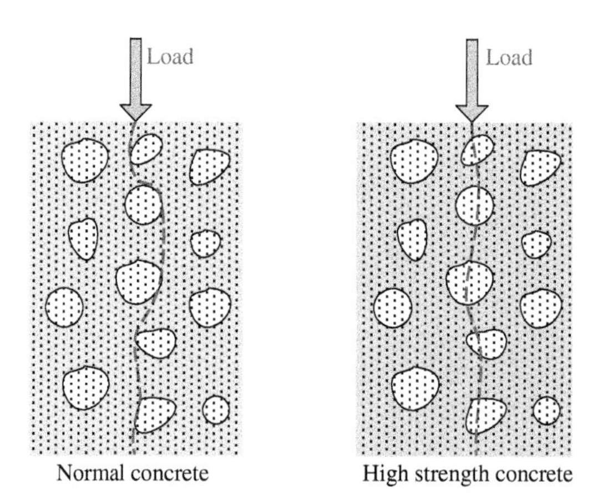


**Figure 6-26** Failure modes of normal-strength concrete (NSC) and high-strength concrete (HSC) under compression

The effectiveness of microsilica in reducing the permeability of concrete was demonstrated a long time ago. Typical figures are given in Table 6-4. These data show that even with low cement content and high *w/c* ratios, low permeability coefficients can be attained due to the incorporation of microsilica. The efficiency factor is about 10 for a 10% incorporation of silica, i.e., replacement of 1 part cement for 1 part microsilica is equivalent to adding 10 parts cement. In other words, the reduction in permeability with 10% of microsilica is approximately equivalent to doubling the cement content. This assessment is based on well-cured concretes. Inadequate curing will have a greater detrimental effect on MS concrete than on conventional concrete, because the pozzolanic reaction can be suppressed without proper curing.

The refined pore structure increases the chemical resistance of MS concrete and reduces the access of oxygen and other deleterious substances. Alkalinity reduction caused by the pozzolanic reaction is not significant. Diamond (1985) measured a pH > 12 after 145 days hydration in a concrete with the presence of 30% microsilica. MS concrete has been exposed to salt spray in Kristiansand, Norway, for over 20 years without any outward sign of corrosion.

When chlorides are added to MS concrete, the chloride ions do not appear to be bound by the hydration products to the same extent as in conventional concretes, due to the reduction in effective amount of aluminates which react with chloride ions to form Friedel's salt. Thus, de-passivation of steel can occur at lower dosages of added chlorides. However, ingress of external chloride ions is significantly reduced, by at least a factor of 10. The rate of corrosion in the presence of equal amounts of chloride ions is the same regardless of the microsilica content of the concrete; i.e., there is no negative influence of microsilica to nullify or outweigh the beneficial effects of a refined pore structure.

The effectiveness of microsilica in controlling AAR has been well established in several independent studies. In Iceland, 7.5% microsilica is routinely added to all cement for this purpose. Diamond et al. (2004) have shown that the addition of microsilica does reduce the concentration of alkalis ($K^+$ and $OH^-$) in the pore solution. The low permeability and reduction in CH

**Table 6-4** Effectiveness of microsilica in reducing permeability of concrete

| Cement content (lb/yd$^3$) | WDRA[a] (wt.%) | Microsilica (wt.%) | Water-to-powder ratio | Permeability coefficient, $K$ (m/s) |
|---|---|---|---|---|
| 170 | 0 | 0 | 2.38 | $120 \times 10^{-10}$ |
| 170 | 1 | 0 | 2.09 | $1160 \times 10^{-10}$ |
| 170 | 0 | 10 | 2.32 | $10 \times 10^{-10}$ |
| 170 | 1 | 10 | 2.10 | $4 \times 10^{-10}$ |
| 170 | 2 | 20 | 2.02 | $0.6 \times 10^{-10}$ |
| 420 | 0 | 0 | 0.89 | $0.5 \times 10^{-10}$ |
| 420 | 1 | 0 | 0.81 | $620 \times 10^{-15}$ |
| 420 | 0 | 10 | 0.97 | $95 \times 10^{-15}$ |
| 420 | 1 | 10 | 0.82 | $18 \times 10^{-15}$ |
| 420 | 2 | 20 | 0.79 | $21 \times 10^{-15}$ |
| 675 | 0 | 0 | 0.52 | $7 \times 10^{-15}$ |
| 675 | 0 | 10 | 0.56 | $136 \times 10^{-15}$ |
| 675 | 1 | 10 | 0.47 | $40 \times 10^{-15}$ |
| 675 | 1[b] | 10 | 0.44 | $8 \times 10^{-15}$ |
| 845 | 0 | 0 | 0.43 | $14 \times 10^{-15}$ |
| 845 | 1 | 0 | 0.49 | $41 \times 10^{-15}$ |

[a] Lignosulfonate-based water-reducing agent.

[b] Naphthalene-based superplasticizer.

content should also improve the resistance to sulfate and acid attacks. It has been demonstrated that microsilica performs better than other pozzolans at equivalent additions. Field trials under severe conditions in Oslo, Norway, have shown good performance over several decades (Herfurth and Nilsen, 1993).

### 6.2.3 DSP Materials

DSP stands for *densified with small particles*. DSP concrete is very similar in composition to MS concrete. It is reasonable to expect that the behavior of DSP materials can be estimated from MS concrete properties. Although there is no clear gap between MS concrete and DSP materials, DSP materials have the following unique characteristics: (1) DSP concretes usually have much lower *w/b* ratio (~0.20); (2) DSP concretes incorporate much higher microsilica content (>20 wt.%); and (3) DSP concrete's requirement on particle packing is more stringent.

It is critical to achieve an optimum particle packing for DSP concrete and hence the grading of aggregates and size distribution of binder particles have to be carefully selected. The aggregates used for DSP concrete are usually very special. For instance, convoluted stainless steel, granite, diabase and calcine bauxite aggregate are frequently used.

With such compacted particle packing and low *w/b*, a large quantity of dispersing agent (e.g., superplasticizer) has to be used. Under such low *w/b* ratios, only about 50% of the cement is able to hydrate. Thus, roughly, DSP concrete has 50% unhydrated cement particles. These particles act like a hard core to help the hydrates carry the compressive load. Due to the pozzolanic reaction of a large amount of silica fume, the CH formed as a hydration product is reduced to very low levels. Large massive crystals of CH cannot be observed, and the microstructure is very uniform in contrast to that of regular cement paste. The porosity of DSP concrete is very low and the pore structure is quite different than normal Portland cement concrete, as demonstrated in Figure 6-27. It can be seen that no capillary pores larger than 100 nm are present in DSP materials.

The strength development of DSP concrete is very rapid, after a delayed setting time of 15–20 h. In about 2 days, the compressive strength of DSP concrete can reach 100 MPa. The strengths can continue to increase with extended moist curing. A compressive strength of 345 MPa is the highest reported value for DSP materials. This should be the result of eliminated microporosity and densified microstructure. It is important that a good dispersion of the microsilica can

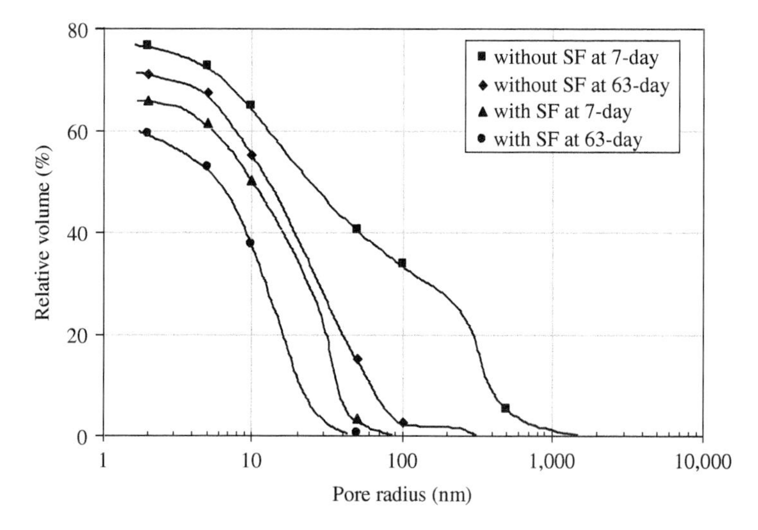


**Figure 6-27** Effects of the addition of silica fume on pore size distribution of DSP materials

be achieved. The use of a chemical dispersion agent is insufficient if low shear mixing is used. The application of high-shear mixing will allow better dispersion of microsilica and, thus, significantly increased compressive strength. By optimizing processing, it is possible to achieve DSP paste and mortar strengths over 200 MPa (see Figure 6-28).

Like high-strength concrete, DSP materials are extremely brittle. The ratio of compressive to flexural strength is in the range of 10–12:1 for paste, and 7–12:1 for mortar. Thus, reinforcement will be needed to improve ductility. Regarding other physical properties, the abrasion resistance of DSP materials is excellent. A screw feeder for fly ash made from DSP materials has a reported lifetime of 1250 h compared with the 250 h of one made of steel with a wear-resistant carbide coating. The permeability of DSP materials would be expected to be extremely low because of the optimized particle packing, low *w/b* ratio, and the pozzolanic reaction products of the high-dose microsilica. The DSP commercial product DASH 47 has shown an air permeability even lower than that of cast aluminum.

DSP materials are usually used to make molds and tools. DSP has an excellent ability to withstand temperature cycling without losing its properties up to 200–250°C. When the temperature is higher than this limit, large degradation in mechanical properties occurs. To overcome this limit, impregnation with high-temperature polymers (Novolac epoxy resin) has been used. It is verified by experiments that, as shown in Table 6-5, vacuum impregnation of DSP material using polymer can significantly improve the flexural strength and reduce the leaking rate.

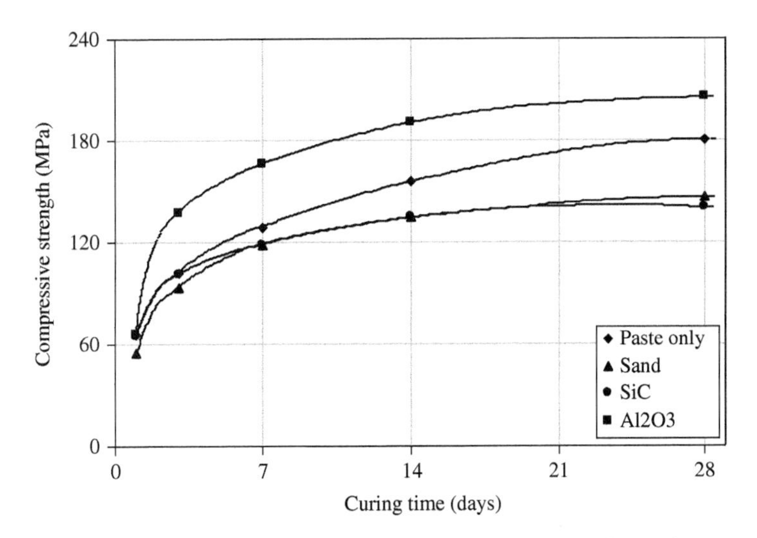


**Figure 6-28** Strength development of DSP paste and mortars with various aggregates

**Table 6-5** Temperature resistance of DSP materials

| Temp. (°C) | Wt. loss (%) | Flexural strength (psi) before PI | Flexural strength (psi) after PI | Vac. leak rate (torr/min) before PI | Vac. leak rate (torr/min) after PI |
|---|---|---|---|---|---|
| 200 | 4.5 | 4,650 | 4,850 | 0.1 | n/a |
| 300 | 5.0 | 4,200 | 4,900 | 6.0 | 0 |
| 400 | 6.2 | 3,850 | 6,550 | 11.0 | 0 |
| 550 | 6.6 | 3,500 | 6,250 | 18.0 | 0.3 |
| 700 | 7.0 | 2,450 | 12,850 | 100.0 | n/a |

Note: PI: polymer impregnation.

### 6.2.4 MDF Materials

MDF stands for *macro-defect-free*. MDF cements/materials were invented by the British chemical company ICI in 1982 (Birchall et al., 1982). The term macro-defect alludes to the relatively large internal voids that are present in conventionally mixed pastes, due to entrapped air and inadequate dispersion. These internal flaws act as crack initiators and reduce the mechanical properties. MDF materials are composites of cement and polymer. The raw materials used to produce MDF are cement, either Portland cement or calcium aluminate cement, water-soluble polymers, such as polyvinyl alcohol or polyvinyl acetal, and water. Water-soluble polymer is added to the cement as a rheology-management aid to make a dough-like material.

The processing method for MDF materials is called "twin roll mixing", as shown in Figure 6-29. The dough-like MDF material is processed under high shear and then stretched into a thin sheet. The polymer is indispensable for this kind of mixing, providing necessary cohesion and allowing a close packing of cement grains to be achieved. Past studies have shown that the polymer is more than a rheology aid. It reacts with the cement to create a viscoelastic material that is capable of forming a cohesive dough, which can easily be shaped. The extracted thin sheet is then put into a press under a modest pressure (typically around 6 MPa) (Donatello et al., 2009). The pressing process removes the entrapped air and makes the paste free of large defects, giving it superior mechanical properties.

The typical formulation of MDF cement is given in Table 6-6, using calcium aluminate cement and polyvinyl alcohol. This cement–polymer combination gives the highest flexural strength of 250–300 MPa. When Portland cement is used, the strength can also exceed 100 MPa. The microstructure of MDF is very dense and the cement grains are close-packed and bound together by the polymer matrix. Surrounding each cement grain is a rim of hydrated material, about 0.25 μm wide, which also contains polymer. This interface region also contributes to bonding since it forms a percolating network. There is no capillary porosity present in the matrix of MDF materials.

When the polymer is removed by heating at high temperature, a capillary pore network is formed, with mean pore diameters around 0.1 to 0.0075 μm. The residual flexural strength is about 20 MPa. The pore system can be filled by further hydration or by impregnation with another

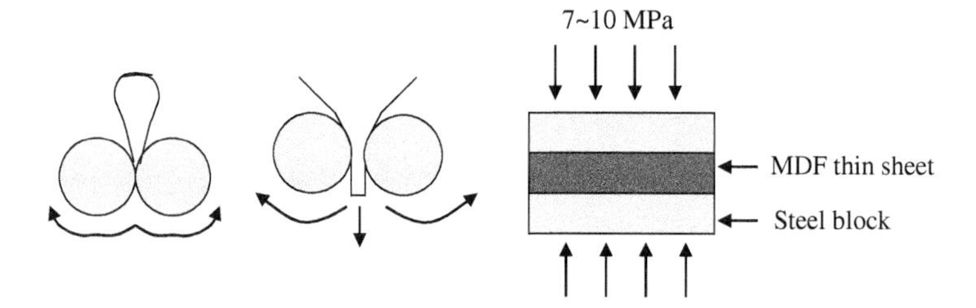


**Figure 6-29** Processing method of MDF materials

**Table 6-6** Compositions of representative MDF materials

| Constituent | Parts of weight | Weight (%) | Volume (%) |
|---|---|---|---|
| Calcium aluminate cement | 100 | 84.3 | 65.2 |
| Polyvinyl alcohol/acetate | 7 | 5.9 | 12.3 |
| Glycerin | 0.7 | 0.6 | 1.4 |
| Water | 11 | 9.3 | 21.1 |

**Table 6-7** Comparisons of the properties of sintered alumina, MDF cement, and conventional cement paste

| Properties | Sintered alumina | MDF cement | Portland cement |
|---|---|---|---|
| Compressive strength (MPa) | 2,100 | 300 | 70 |
| Flexural strength (MPa) | 350 | >150 | <10 |
| Modulus of elasticity (GPa) | 40 | 45 | 35 |
| Fracture toughness ($MPa \cdot m^{1/2}$) | 5 | 3 | 0.3 |
| Density ($kg/m^3$) | 3700 | 2400 | 2700 |
| Thermal expansion ($10^{-6}$/°C) | 7 | 10 | 15 |

polymer. A polymer-free MDF cement (formed under pressure) has a flexural strength of about 75 MPa. A major drawback to the commercialization of MDF cement is its sensitivity to water. Nearly two-thirds of the dry strength can be lost after 3 weeks of immersion in water. The exact mechanism of strength loss is not yet known with certainty, but the following explanation is proposed. MDF develops its high strength only after drying at 80–100°C. Removal of water creates a strong but brittle PVA matrix. Water can be reabsorbed to form a soft, weak, rubbery polymer. The water eventually reaches the cement grains, which continue to hydrate and destroy the interface region. Eventually a hydrated matrix will become the sole binder and the limiting wet strength will approach 20 MPa. The addition of a commercial organo-titanate can effectively prevent this loss of strength. It appears that the titanate reacts chemically with PVA, thereby reducing the rate at which it can absorb water. It may also modify the interphase region.

The properties of MDF lie between conventional cement paste and a sintered ceramic, as listed in Table 6-7. Its fairly high fracture toughness makes it a good candidate for a ballistic protection. Frost resistance is not a problem even when the strength is reduced by exposure to moisture, because there should be no freezable water in the matrix of MDF materials.

## 6.3 ULTRA-HIGH-STRENGTH CONCRETE

Ultra-high-strength concrete (UHSC) is defined as concrete that has a compressive strength greater than 150 MPa. When this type of composite was developed in France, it was called reactive powder concrete (RPC) because it contained a larger amount of active silica fume, as discussed in Chapter 1. It is different from high-strength concrete not only because it does not have coarse aggregates, but also because of its large amount of powders and incorporation of fibers. A comparison of normal-strength, high-strength, and ultra-high-strength concrete is given in Table 6-8.

### 6.3.1 Composition of Ultra-High-Strength Concrete

To illustrate the composition of UHSC, two examples are given in Table 6-9. The key characteristics of the ultra-high-strength concrete can be summarized as follows: (1) UHSC has very low *w/b* ratio. The *w/b* ratio of UHSC can be less than 0.15, very close to that of MDF. However, since there is no polymer, UHSC can have very good fluidity just like normal concrete. (2) UHSC has a large quantity of silica fume (and/or other fine mineral powder). The incorporation of silica fume in UHSC can reach 25% by weight of cementitious materials, similar to that of DSP. (3) UHSC aggregates contain only fine sand. Since there are no coarse aggregates, UHSC looks more like a homogenous material. Also, it is easy to incorporate large amounts of fibers into UHSC due to the lack of coarse aggregates. (4) UHSC needs a high dosage of superplasticizers, because of the high percentage of silica fume and very low *w/b* ratio. With such high dosage of superplasticizer, UHSC can achieve a very good flowability.

**Table 6-8** Characteristics of conventional concrete, high-strength concrete, and ultra-high-strength concrete

| | **Conventional concrete** | **High-strength concrete** | **Ultra-high-strength concrete** |
|---|---|---|---|
| Compressive strength (MPa) | <50 | ~100 | >150 |
| Water–binder ratio | >0.5 | ~0.3 | <0.2 |
| Chemical admixture | Not necessary | WRA/HRWRA necessary | HRWRA essential |
| Mineral admixture | Not necessary | Fly ash (and/or) silica fume commonly used | Silica fume, (and/or) fine powder essential |
| Fibers | Beneficial | Beneficial | Essential |
| Air entrainment | Necessary | Necessary | Not necessary |
| Processing | Conventional | Conventional | Heat treatment and pressure |
| Steady-state chloride diffusion ($\times 10^{-12}$ m$^2$/s) | 1.00 | 0.60 | 0.02 |

**Table 6-9** Examples of compositions of ultra-high-strength concrete

| | **Cement** | **Water** | **Superplasticizers** | **Silica fume** | **Fine sand** | **Quartz flour** |
|---|---|---|---|---|---|---|
| No. 1 | 1 | 0.28 | 0.060 | 0.33 | 1.43 | 0.3 |
| No. 2 | 1 | 0.15 | 0.044 | 0.25 | 1.10 | — |

Depending on the level of desired compressive strength, post-setting heat treatment and application of pressure before or during setting may be necessary to boost hydration and achieve a densified microstructure.

### 6.3.2 Microstructure of Ultra-High-Strength Concrete

Based on the observations made by optical microscopy, SEM, and TEM, the microstructure of ultra-high-strength concrete can be described as follows: (1) On a millimeter scale, the material is more homogeneous as there is no coarse aggregate, and a large amount of binders is used. (2) There is an absence of a pronounced transition zone between the sand particles and the paste. This is the result of a low *w/b* ratio and the dispersion effect of the superplasticizer. (3) There is a low or negligible presence of Portlandite because of the pozzolanic reaction of a large amount of silica fume. By using mercury intrusion porosimetry, it was also found that UHSC has (4) a low total porosity (1–3%) as a result of the low *w/b* ratio and pozzolanic reaction, (5) a proportionally lower volume of capillary porosity (10% of total porosity), and (6) a strong component of pores with an average diameter of 2.5 nm. The above-mentioned microstructural features lead to low water absorption, low gas permeability, and low chloride diffusivity, which imply superior durability of UHSC.

### 6.3.3 Brittleness

UHSC is considerably more brittle than conventional concrete. This can be seen by comparing the uniaxial compressive stress–strain curve of conventional concrete with that of an ultra-high-strength concrete, as shown in Figure 6-30. These curves were obtained with a digitally controlled, closed-loop test system. The post-peak response of UHSC is considerably steeper than that of conventional concrete. In practice, fibers are essential for UHSC to overcome the brittleness. With the incorporation of fibers, the flexural strength can reach 50 MPa, as shown in Figure 6-31.

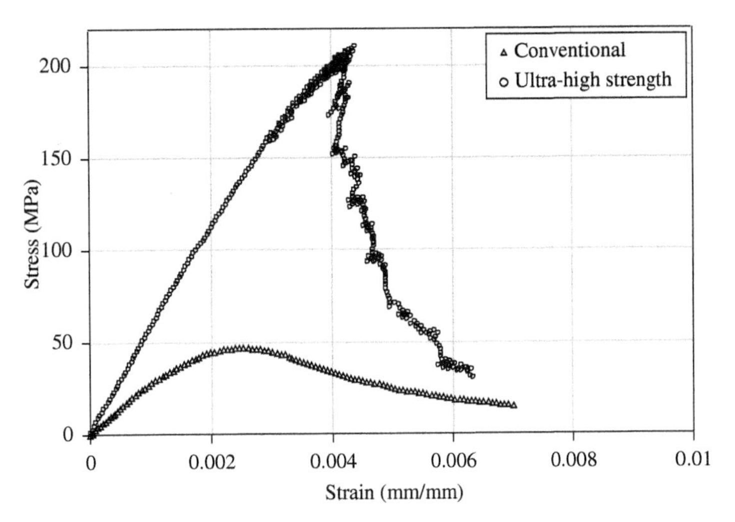


**Figure 6-30** Comparison of stress–strain curves of representative conventional and ultra-high-strength concretes

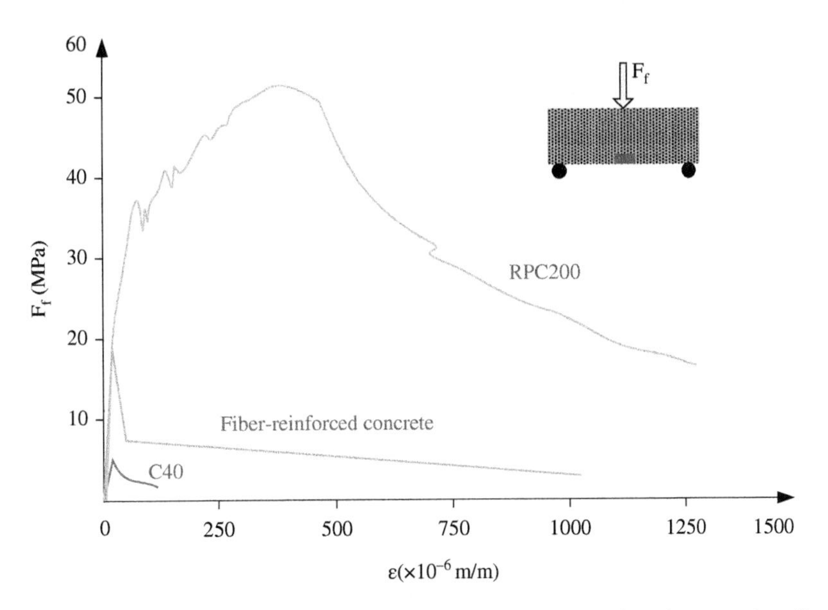


**Figure 6-31** Comparison of flexural behaviors of conventional concrete, fiber-reinforced concrete, and ultra-high-strength concrete (denoted as RPC 200)

### 6.3.4 Ultra-High-Performance Concrete

Ultra-high-performance concrete (UHPC) has been a popular term in recent years. However, it is herein classified as an UHSC, instead of a separate type of advanced cementitious composite, because its design philosophy followed the route of DSP, SIFCON, etc. (e.g., maximizing particle packing, enhancing homogeneity by limiting aggregate size, and using fiber to provide overall toughness), and the first commercial "UHPC" was developed based on the recipe of RPC/UHSC (Du, Meng et al., 2021; Richard and Cheyrezy, 1995). Rooted in RPC, the mixtures of UHPC are generally featured by low *w/b* ratio (0.15–0.24), high solid particle packing density (0.825–0.855), limited aggregate size (<0.6 mm), high contents of cement (800–1100 $kg/m^3$) and silica fume (150–300 $kg/m^3$), and high steel fiber volume fraction (2–5%) (Chen et al., 2019; Du, Meng et al., 2021). There are no explicitly defined criteria to judge if a mixture can be called UHPC, but generally UHPC mixtures should meet high flowability (e.g., mini-slump flow is no less than 160 mm), high compressive strength (no less than 120 MPa at an age of 28 days), and high tensile strength (no less than 5 MPa) (ACI, 2018; Russell and Graybeal, 2013). If steam curing is used, higher-level mechanical properties should be expected. Further, because of the incorporation of steel fibers, most of UHPC materials exhibit strain-hardening behaviors when subjected to flexure or tension (Graybeal, 2008; Meng and Khayat, 2017). The high flowability normally needs to be ensured by a relatively high dosage of superplasticizers. The good flowability also ensures compaction so the designed dense microstructure can be achieved, which implies resistance to deleterious substances (e.g., chloride, sulfate, oxygen, moisture, etc.) and, thus, superior durability.

Due to the rigorous requirements on raw materials and processing, as well as high contents of cementitious materials and steel fiber, both the cost and the $CO_2$-embodiment of UHPC are high. This is one of the major limitations of UHPC in engineering practice. To overcome this limitation, many attempts have been made in the past two decades. These efforts include reducing cementitious binder content and replacing cement/silica fume by cheaper and more eco-efficient supplementary cementitious materials (e.g., coal fly ash) and fillers (e.g., ground limestone powder); replacing finely-ground quartz sand by conventional sand; reducing steel fiber content by using hybrid fiber systems; using standard curing instead of energy-intensive steam-/heat-curing (Huang et al., 2017; Li et al., 2018; Meng et al., 2017; Soliman and Tagnit-Hamou, 2016; Van et al., 2014; Wang et al., 2012). Some recent developments have also enabled using coarse aggregate to make UHPC (Wang et al., 2021; Zhang et al., 2018). In addition, the wide application of UHPC is also challenged by the difficulty of mixing in the field, which limits the batch size (Berry et al., 2020); and the large autogenous shrinkage, which is typically larger than 800 $\mu\varepsilon$ (Yoo et al., 2014). The large autogenous shrinkage of UHPC is attributed to the low *w/b* and high content of cementitious materials. If it is not controlled well, cracking or debonding will be induced, which counteracts the advantages of UHPC, such as superior durability. Apart from leveraging the shrinkage restraining effect of fibers, the shrinkage of UHPC can also be mitigated effectively by regulating hydration of cementitious materials (e.g., replacing part of cement using fly ash and/or slag), maintaining internal humidity (e.g., via internal curing using saturated lightweight sand or superabsorbent polymer), and adding functional admixtures (e.g., shrinkage reducing admixture and expansive agent) (Du, Meng et al., 2021).

### 6.3.5 Applications

In spite of some existing challenges in cost and eco-efficiency, UHSC and UHPC have been attracting increasing interest from the industry because of their technical merits (especially, the superior strengths allowing small cross-sections of structural members). Both UHSC and UHPC can be used for cast-in-place and prefabricated (e.g., prestressed) concrete structures. The first application of UHSC was a pedestrian bridge constructed in Sherbrook, Canada, with an ultra-high compressive strength of 200 MPa. The footbridge was built with post-tensioned, precast elements made with UHSC, having a fiber amount of 200 $kg/m^3$. The thickness of the bridge deck was only 35 mm. UHSC has also been used to build the Shawnessy Light Rail Train Station in Calgary, Canada. The thin-shelled canopies, 5.1 by 6 m, and just 20 mm thick, supported on single columns, protect commuters from the elements, as shown in Figure 6-32a. Similarly, UHPC has been used to build the cladding for the Qatar National Museum, as shown in Figure 6-32b (Menétrey, 2013). In China, UHSC has been applied as a pavement cover plate for high-speed railways as shown in Figure 6-33. The UHSC has a compressive strength ≥130 MPa, a flexural strength of 18 MPa, a modulus of elasticity of 48 GPa, a penetration quantity of chloride ions <40 Coulombs, and a frost-resistance grade >500 cycles. Investigations have also been conducted on the resistance to impact of UHSC. As shown in Figure 6-34, UHSC is extremely brittle without the addition of fibers. However, when 3% or 4% of fiber is added, its resistance to impact is significantly improved. UHSC has been used to cast reinforced beams. The cross-section of a beam made of UHSC is very close to that of steel and so is the self-weight. UHSC and/or UHPC have been commercialized by different entities under various brand names, such as the Ductal® by LaFargeHolcim, the BSI® by Eiffage, and BCV® by Vicat.

The fibers used in UHSC/UHPC are usually special steel fibers 12 mm long and 0.2 mm in diameter, originally used in rubber tire production. The cost of UHSC/UHPC with fibers is up to

(a)

(b)

**Figure 6-32** Applications of UHSC/UHPC in structures: (a) the thin-shelled canopies, Shawnessy Light Rail Train Station in Calgary, Canada; and (b) the cladding, Qatar National Museum.7 (Source: Menétrey, P. (2013), Vol. 360 / with permission of RILEM Publications SARL.)

**Figure 6-33** Application of UHSC in pavement cover plate in high-speed railway in China

**Figure 6-34** Influence of fiber on UHSC impact resistance

US$1242 per cubic meter, whereas without fibers, the cost is US$332 per cubic meter. Lowering the cost will be the key to promote applications of these materials at scale, provided that other technical issues (e.g., batch production) can be solved.

## 6.4 POLYMERS IN CONCRETE

### 6.4.1 Introduction

Polymers are used in concrete production in three major forms. First, polymer is used directly as the binder to replace Portland cement in gluing aggregates together, which is called polymer concrete. Second, polymer is applied as an impregnating agent to penetrate a Portland cement concrete member to enhance its properties, which is called polymer-impregnated concrete. And, lastly, polymer is used as an admixture added into concrete, known as polymer-modified concrete or latex-modified concrete, as the polymers used in this category are mainly latexes.

### 6.4.2 Polymer Concrete

Polymer concrete (PC) is produced by premixing a two-part polymer system, composed of monomers or prepolymers together with hardeners (a cross-linking agent), which is then added to aggregates to produce a hardened plastic material with aggregate as filler. Other components used may include catalyst, plasticizer, fire retardant, and fibers. The function of the plasticizer is to improve the workability of polymer concrete. The reason for adding the fire retardant is to improve the resistance of polymer concrete to high temperature or fire hazard. The polymer is a homopolymer if it is made by the polymerization of one monomer, and a copolymer when two or more monomers are polymerized. Polymer concrete has been made with a variety of resins and monomers. The commonly used monomers or prepolymers for PC include: (1) methyl methacrylate (MMA); (2) polyester prepolymer-styrene; and (3) epoxide prepolymers. Polyester resins are attractive because of moderate cost, the availability of a great variety of formulations, and moderately good properties. Epoxy resins are generally more expensive, but may offer advantages such as adhesion to wet surfaces (ACI, 1998). The properties of polymer concrete are largely dependent on the properties and the amount of the polymer used, but can be modified somewhat by the effects of the aggregate and selected filler materials. The composites do not contain a hydrated cement phase, although Portland cement can be used as a filler.

The term "polymer concrete" refers to a family of products, including mixtures of concrete and mortar. The improvement of properties of hardened concrete by the addition of polymers is well documented (ACI, 1997). The advantages of PC include high tensile and flexural strengths, excellent adhesion, good resistance to attack by chemical, very low water sorption, high water resistance, good resistance to abrasion, and good freeze–thaw stability. The main disadvantages of PC include low Young's modulus, high creep values, shrinkage varying with the polymer used, sensitivity to high temperature, and high price. The main application areas for polymer concrete is in making façade plates, sanitary products, panels, floor tiles, pipes, and industrial flooring. It is used in various precast and cast-in-place applications in construction work, skid-resistant overlays in highways, plaster for exterior walls, and resurfacing of deteriorated structures. In addition, PC is also widely used as repair material, in particular for concrete carriageways, around the world nowadays. In Japan, PC mortars are used as grouts for repairing cracks and delaminations of concrete structures, patching materials for damaged concrete structures, and rustproof coatings for corroded reinforcing bars. In the United States, polymer mortar and concrete are employed as the main patching materials for repair work and overlays for bridge decks in cast-in-place applications, and in precast applications (Chandra and Ohama, 1994).

Only limited numbers of polymer systems are appropriate for the repair of wet concrete surfaces. In general, the aggregates used in polymer concrete should be dry in order to obtain the highest strengths. High temperatures can adversely affect the physical properties of certain polymer concrete, causing softening. Service temperatures should be evaluated prior to selecting polymer concrete systems for such uses. Epoxy systems may burn out in fires where the temperature exceeds 230°C and can significantly soften at lower temperatures. Users of polymer concrete must consider its poor fire and high-temperature resistance. Conventional concrete generally cannot bond to cured polymer concrete, and compatibility of the systems should be considered.

Many polymer concrete patching materials are primarily designed for the repair of highway structures where traffic conditions allow closing of a repair area for only a few hours. However, polymer concrete is not limited to that usage and can be formulated for a wide variety of applications. Polymer concrete is used in several types of applications: (1) fast-curing, high-strength patching of structures, and (2) thin (5–9 mm thick) overlays for floors and bridge decks. Polymer mortars have been used in a variety of repairs where only thin sections (patches and overlays) are required. Polymers with high elongation and low modulus of elasticity are particularly suited for bridge overlays. Polymer concrete is especially suitable for areas subject to chemical attack.

Polymer concrete is mixed, placed, and consolidated in a manner similar to conventional concrete. With some harsh mixtures, external vibration is required. A wide variety of prepackaged polymer mortars is available, which can be used as mortars or added to selected blends of aggregates. Depending on the specific use, mortars may contain variable aggregate gradations intended to impart unique surface properties or aesthetic effects to the structure being repaired. Polymer mortars are trowelable and are specifically intended for overhead or vertical applications. Epoxy mortars generally shrink less than polyester or acrylic mortars. Shrinkage of polyester and acrylic mortars can be reduced by using an optimum aggregate grading.

Rapid curing generally means less time for placing and finishing operations. Working times for these materials are variable and depending on ambient temperatures, may range from less than 15 minutes to more than one hour. Also, high or low ambient and concrete temperatures may significantly affect polymer cure time or performance. The coefficients of thermal expansion of polymer materials are variable from one product to another, and are significantly higher than conventional concrete. Shrinkage characteristics of polymer concrete must be evaluated so that unnecessary shrinkage cracking is avoided. The modulus of elasticity of polymer concrete may be significantly lower than that of conventional concrete, especially at higher temperatures. Its application in load-carrying members must be carefully considered.

Organic solvents may be needed to clean equipment when using polyesters and epoxies as binders to produce polymer concrete. Volatile systems such as methyl methacrylate evaporate quickly and present no cleaning problems. However, such systems are potentially explosive and require nonsparking and explosion-proof equipment.

### 6.4.3 Polymer-Impregnated Concrete

Polymer-impregnated concrete (PIC) is produced by impregnating a monomer and catalyst into a hardened concrete and stimulating *in situ* polymerization using steam or infrared heater. The procedures for making PIC are as follows: (1) precasting hardened Portland cement concrete; (2) drying the precast conventional concrete; (3) displacing the air from the open pores (e.g., via vacuuming); (4) saturating the open pore structure by diffusion of low-viscosity monomers or a prepolymer–monomer mixture; and (5) *in situ* polymerization with heating.

The compressive strength and tensile strength of PIC can reach 140 MPa and 15 MPa, respectively. PIC usually has very low permeability and diffusivity. Subsequently, PIC has excellent chemical resistance and superior durability. Thus, PIC is commonly used to produce structural industry floors, sewage pipes, storage tanks for sea water, structural members for desalination plants and distilled water plants, and tunnel liners. PIC is also very expensive due to the complicated processing procedures and the cost of the polymer; thus, usage of PIC in practice has been limited.

### 6.4.4 Polymer (Latex)-Modified Concrete

Polymer (usually latex)-modified concrete is produced by adding polymer (latex) to a Portland cement concrete mixture during the mixing process. The processing procedures are the same as normal Portland cement concrete. Since the early 1950s, it has been known that certain polymers can be added to cementitious mortars, to overcome some of the problems of pure cement-based repair materials (e.g., poor interfacial bonding). The polymers used as admixtures for cementitious systems are normally supplied in the form of latex and are used to gauge the cementitious mortar as a whole or as partial replacement of the mixing water. Such mortars afford similar alkaline passivation protection to the steel as conventional cementitious materials do, and can be readily placed in a single application with a 12- to 15-mm thickness that gives adequate protective cover.

Latex can be either natural or man-made. Natural latex can come from rubber trees and man-made latex is created by synthetic processes. The most widely used latexes with Portland cement are styrene–butadiene rubber (SBR), 100% acrylic copolymers (PAE), styrene acrylic (SA), vinyl acetate ethylene (VAE), and polyvinyl acetate (PVA). As the names indicate, these latexes are composed of organic polymers that are combinations of various monomers, such as styrene, acrylic, butadiene, and vinyl acetate. The general properties of the above-mentioned copolymers are given in Table 6-10. The form of latex is usually a dispersion of very small particles of an organic polymer

**Table 6-10** Characteristics of different latexes

| Latex type | Acronym | Solids (%) | Viscosity (cps) | MFFT[a] (°C) | pH |
|---|---|---|---|---|---|
| Styrene–butadiene rubber | SBR | 47 | 20–50 | 12 | 10 |
| Acrylic copolymers | PAE | 47 | 20–100 | 10–12 | 9–10 |
| Styrene acrylic copolymers | SA | 48 | 75–5000 | 10–18 | 6–9 |
| Polyvinyl acetate | PVA | 55 | 1000–2500 | 15 | 4–5 |

[a]MFFT: Minimum film forming temperature.

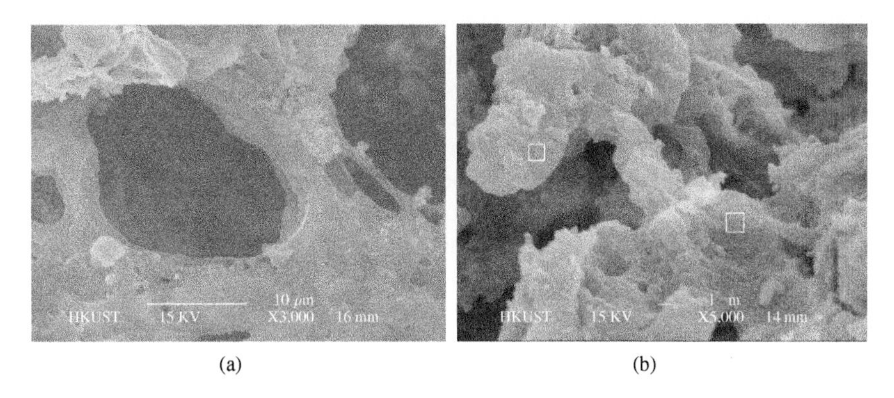


(a) (b)

**Figure 6-35** Microscopic morphology of mortars modified with different polymer latexes: (a) inert latex; (b) reactive latex
Note: EDX analysis in the marked areas shows both carbon and hydration products of cement.

in water. These particles are typically in the range of 1500–2500 angstroms (i.e., 0.15 to 0.25 μm) in diameter. The solid content in a latex is about 50%. The application of latex in Portland cement concrete requires about 15% of solid content by weight of cement. In this case, about 30% of latex by weight of cement has to be added. Thus, 15% of water by weight of cement is brought into concrete. The water has to be reduced from the free water for concrete mixing.

The reaction of latex-modified Portland cement involves two processes: the hydration of the cement and the coalescence of the latex. The chemistry and reaction processes of cement hydration occur in the same way as in conventional mortar and concrete. Meanwhile, with water being removed from the latex, the polymer particles get closer. With continual water removal, the latex particles can eventually coalesce into a film, coating the hydrates and aggregate surface with a semi-continuous plastic film. It was shown (Isenberg and Vanderhoff, 1974) that this process could form a network of polymer strands, interpenetrating with the cement hydration products. It has also been reported that the polymer film formation may depend on the nature of polymer: while inert latexes (e.g., SBR) could result in an organic-inorganic interpenetrated structure, reactive latexes (e.g., PAE and SA) can interact with the hydration products to form an organic-inorganic hybrid matrix which yields similar properties to those modified by inert latexes (Ma et al., 2011; Ma and Li, 2013). The differences in microstructure are shown in Figure 6-35.

As the latex particles coalesce or interact with the cement hydrates and form a semi-continuous film or hybrid matrix, moisture can be maintained around the cement particles, permitting the cement hydration process to continue and reduce the need for an external wet curing. This process occurs after the material reaches initial set and is common in both mortar and concrete applications. For many mortar applications, the materials are formulated with some latex so that no external curing is needed, even right after placement, whereas for most concrete applications, the normal curing procedure is 1 day of moist cure followed by air drying for the remainder of the curing time. For the moist curing of concrete, damp burlap and polyethylene films are typically used.

Usually, surfactants are added to the latex formulation to prevent polymer particles from coagulating due to the influence of severe mechanical or chemical conditions, such as high shear or calcium ions in Portland cement. However, these surfactants can introduce a large number of air voids into latex-modified mortar or concrete due to the nature of the surfactant (which generates foams). Thus, an antifoaming agent may need to be added to the latex-modified system to control the air content in the Portland cement mixtures.

Latexes can influence the properties of mortar or concrete in both the plastic stage and the hardened stage. In the plastic stage, the combined characteristics of latex and antifoam will cause some air to be entrained in the mix. Values are typically in the range of 5–9% for mortar and 4–6% for concrete. The entrained air voids act as a lubricant and hence improve the workability of the modified concrete. One more factor contributing to the improved workability is the dispersing effect of surfactants when the latex is combined with the mixing water. The improvement on the slump value is evident from the experimental results, as shown in Figure 6-36.

The setting time of latex-modified mixes is usually controlled by the hydration of the Portland cement. The available working time for finishing, however, can be considerably shorter than an unmodified composite due to the latex drying and film formation initiated by evaporation of water from the surface of the mixture. This causes the formation of a crust on the surface, which can tear if over-finished and is the major difference in finishing latex-modified Portland cement mixes and mixes without latex. In the hardened stage, latexes can improve adhesion properties, tension and flexural properties, impermeability, and the chemical resistance of cement-based materials.

The bond between the latex-modified mixes and existing concrete is an important property when latex-modified concrete (LMC) is used as a repair material. The bond as a repair material is desired so that the new material remains in place, preferably for the rest of the life of the parent concrete. The ideal bond would be one that exceeds the tensile and shear strength of the parent concrete so that if failure occurs, it does so in the parent concrete. Different bond strengths can be measured by different methods. The tension-bond strength can be measured by a pull-off test.

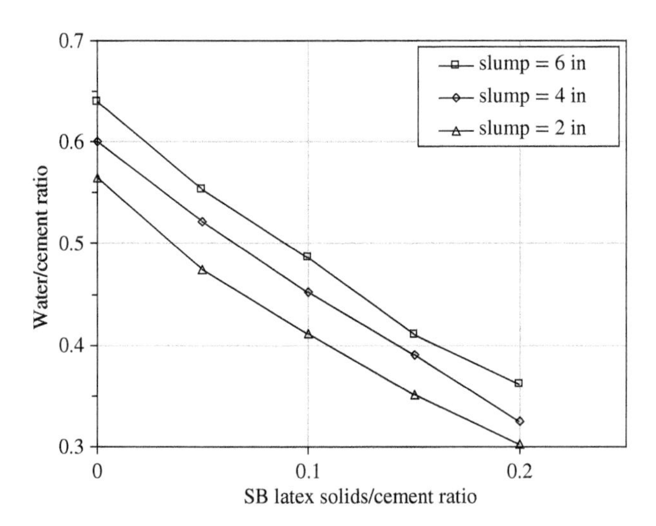


**Figure 6-36** Workability of SBR latex-improved concrete

It is done on a parent concrete slab with the disk-like specimens of LMC material adhering to it (Figure 6-37a). Then dollies are attached on the top surface of the disk specimen, which is followed by a pull-off process to measure the bond strength (Figure 6-37b). The shear-bond strength can be measured with a prismatic specimen of size $100 \times 100 \times 300\,mm^3$. The specimen is cast using normal concrete into a slanted shape. After the concrete is hardened, another half of the beam specimen is cast with LMC material on top of the parent concrete to form a test prism, as shown Figure 6-38. The specimen is then tested under bending to determine the shear bond strength.

The bond of latex-modified mortar to concrete is very strong, thus creating a homogeneous combination of the two materials. Bond tests have demonstrated that ultimate debonding failure

**Figure 6-37** Measurement of pull-off strength of LMC repair materials: (a) casting of disk-like specimens; and (b) pull-off tester

**Figure 6-38** Casting of prisms for slant shear measurement

**Table 6-11** Permeability of different types of concrete

| Permeability rating | Charge passed (coulombs) | Type of concrete |
|---|---|---|
| High | >4000 | High water/cement, conventional (>0.6) PCC |
| Moderate | 2000–4000 | Moderate water/cement, conventional (0.4–0.5) PCC |
| Low | 1000–2000 | Low water/cement, conventional (<0.4) PCC |
| Very low | 100–1000 | Latex-modified concrete, internally sealed concrete |
| Negligible | <100 | Polymer-impregnated concrete, polymer concrete |

can occur in the parent concrete. It is reported that mortars modified with SB, PAE, and PVA latexes exceed the bond strength of the unmodified control by factors of 2–3.

The pore-sealing or blocking effects of latex in a concrete mix result in a major reduction of its permeability to gases, liquids, and ions (e.g., chloride ions), as listed in Table 6-11. Carbonation studies have shown that the addition of latex to concrete significantly reduces the depth of carbonation of the concrete.

The compressive strength of LMC can be measured by a cube specimen. If the LMC is very sticky, the casting of cubic specimens for determination of compressive strength should use plastic molds with a size of $50 \times 50 \times 50\,mm^3$, as shown in Figure 6-39. In addition, since shrinkage is an important parameter for LMC, drying shrinkage measurement should be made. The specimen size for the determination of drying shrinkage is $25 \times 25 \times 285\,mm^3$. Moreover, the modulus of elasticity of LMC can be measured with a cylinder specimen of size $\varphi 100 \times 200\,mm^3$. The typical mechanical properties of concretes containing polymers are given in Table 6-12. It can be seen that both PC and PIC have much better mechanical properties than LMC.

In the early 1980s, polymer-modified repair mortars were blended on site using sand, cement, latex, and water. This resulted in some problems of unsatisfactory mortars due to the lack of

**Figure 6-39** Casting of cubic specimens for determination of compressive strength

**Table 6-12** Typical mechanical properties of concretes containing polymers

| | PC | | LMC | | | PIC | |
|---|---|---|---|---|---|---|---|
| | Polyester | Polymerized NMA | Control | | LMC containing styrene butadiene, air cured | Control unimpregnated | NMA impregnated thermal-catalytical polymerization |
| | Polymer/agg. ratio of 1:10 | 1:15 | Moist cured | Air cured | | | |
| Compressive strength | 18,000 | 20,000 | 5,800 | 4,500 | 4,800 | 5,300 | 18,000 |
| Tensile strength | 2,000 | 1,500 | 535 | 310 | 620 | 420 | 1,500 |
| Flexural strength | 5,000 | 3,000 | 1,070 | 610 | 1,430 | 740 | 2,300 |
| Elastic modulus $\times 10^6$ | 5 | 5.5 | 3.4 | - | 1.56 | 3.5 | 6.2 |

adequate quality control (poor proportion, inadequate labor, unsatisfactory mixes, etc.). To overcome this problem, complete "bag and bottle" mixes of latex and preblended sand cement were developed. These packs were ready to be used on site without any further addition. A further development was packing redispersible spray-dried polymer powders blended with graded sand, cement, and other additives into a bag. The site application was carried out only by adding appropriate amount of water (Mirza et al., 2002). Polymer-modified cementitious mortars are mainly used for the repair of reinforced concrete where the cover to be replaced is less than 30 mm in thickness. In some instances, they are used in conjunction with a protective coating in thin cover situations when the cover is less than 12 mm (Shaw, 1984).

LMC can be used for emergency concreting jobs in mines, tunnels, ceramic tile adhesives and grouting, swimming pool finishes, and industrial floor toppings, as well as the production of high-strength precast products. As a coating, it has been used for basement and exterior walls, skid-resistant surfaces on concrete pavement and ship decks, and maintenance coatings on steel. LMC is the most commonly used repair material, especially for overlays on bridge decks or concrete highways.

Concrete roadways are normally reinforced with steel mesh, and, in general, most highway deterioration is associated with corrosion of the reinforcement (Batis et al., 2003). Under heavy traffic loading and exposure to harsh environments, deterioration of concrete roadways frequently occurs. Such problems are faced by most modern cities, for example, Hong Kong, one of the busiest cities in Asia. The roads in Hong Kong are heavily used with over 580,000 vehicles running daily on 1900 km of roads, including over 900 highway bridges within a small territory of size about 1100 $km^2$. To maintain these roads in good condition is indeed a challenging job as well as a multimillion-dollar burden. In the financial year 1998/1999, the Government of the Hong Kong Special Administrative Region (HKSAR) spent a total of around HK$1.22 billion on road maintenance and minor improvement, in which about 51%, or HK$624 million, was used on the maintenance of concrete roadways (HKSAR Highways Department, 2006).

Roadway maintenance works inevitably occupy road space and disrupt normal traffic. Hence, one basic requirement is that the repaired portion should set and harden fast to minimize the disturbance to traffic. Concrete roadway maintenance works can be broadly classified into corrective repairs and programmed works. With a background of high traffic demand, acute land constraints, and growing community expectation, all roadway repair works are carried out under an extremely tight working program and within limited hours.

If the thickness of an overlay is less than 30 mm, repair mortar for a concrete roadway is demanded (Hassan et al., 2001) because latex-modified mortar has no minimum thickness requirement and has the following properties: (1) consistent physical properties with concrete; (2) fast hardening, within hours; (3) superior bonding strength to the substrates; (4) good volume stability and matching color; and (5) appropriate workability, surface finishing, and durability. If the thickness of an overlay is greater than 30 mm, latex-modified concrete is preferred. The majority of concrete applications incorporating a latex modifier have been used in bridge and carpark deck overlays for over thirty years. Thousands of bridges in the United States have been protected with latex-modified concrete. The overlay can last over 20 years.

For most applications, type I/II Portland cements are suitable for latex-modified mixes. Most sands that are suitable for quality mortar and concrete are suitable for latex-modified mixes. The *w/c* ratio is usually less than 0.4, and the latex solids–cement ratio for many applications is around 15%. The two typical mixture compositions of mortar and concrete modified with latex are given in Table 6-13. It should be pointed out that LMC is very sensitive to high ambient temperature. At high temperature (i.e., 29.4°C or above) and under conditions of rapid drying, latex modification of fresh cement-based materials will cause a skin or crust to form.

The property modifications of LMC depend on the composition of the polymer. For instance, concrete modified with SBR and PAE latexes is superior in water resistance to concrete modified with PVA latex. In addition, the monomer ratio can affect concrete properties. An SBR latex with a styrene-to-butadiene ratio of 35/65 can produce concrete with a lower compressive strength than latex with a styrene-to-butadiene ratio of 50/50.

### 6.4.5 Selection of LMC as Repair Materials

The selection of LMC as a repair material is based on the following factors: (1) job nature, including the total repair area (volume), thickness, and allowed working time before reopening to traffic; (2) compatibility between the repair materials and the original concrete; (3) the total budget for the repair work; and (4) service quality of suppliers. Patch repairs are discrete repairs carried out

**Table 6-13** Typical mixture proportioning of mortar and concrete modified with latex

| | Mortar parts by weight | Concrete parts by weight |
|---|---|---|
| Portland cement | 1.00 | 1.00 |
| Sand | 3.50 | 2.50 |
| Stone | — | 2.00 |
| Latex, 48% solids | 0.31 | 0.31 |
| Water | 0.24 | 0.24 |
| Flow, ASTM C230, cm | 12 | 0 |
| Slump, ASTM C143, in. | — | 6–8 |

**Table 6-14** Physical properties of typical products used in concrete roadway repairs

| | Epoxy resin grouts, mortars, and concretes | Polyester resin grouts, mortars, and concretes | Cementitious grouts, mortars, and concretes | Polymer-modified cementitious systems |
|---|---|---|---|---|
| Compressive strength (MPa) | 55–110 | 55–110 | 20–70 | 10–80 |
| Modulus of elasticity (GPa) | 0.5–20 | 2–10 | 20–30 | 1–30 |
| Flexural strength (MPa) | 25–50 | 25–30 | 2–5 | 6–15 |
| Tensile strength (MPa) | 9–20 | 8–17 | 1.5–3.5 | 2–8 |
| Elongation at break (%) | 0–15 | 0–2 | 0 | 0–5 |
| Linear coefficient of thermal expansion per °C | $25–30 \times 10^{-6}$ | $25–35 \times 10^{-6}$ | $7–12 \times 10^{-6}$ | $8–20 \times 10^{-6}$ |
| Water absorption, 7 days at 25°C (%) | 0–1 | 0.2–0.5 | 5–15 | 0.1–0.5 |
| Maximum service temperature under load (°C) | 40–80 | 50–80 | >300°C (dependent on mix design) | 100–300 |
| Rate of development of strength at 20°C | 6–48 h | 2–6 h | 1–4 weeks | 1–7 days |

in small areas on a structure or concrete roadway. They are generally less than half a square meter in area and are applied by hand (Kay, 1992). The repair material may be cementitious, polymer-modified, or a straight polymer mortar with an aggregate of fine sand or other fillers. The repair system may include bonding aid, reinforcement primer, repair mortar, pore filler, leveling mortar, and protective coatings. In massive repairs it may not be appropriate or economical to use hand-applied mortars.

Although there is a wide variety of materials available for repair of concrete highways, in most repair work, polymer-modified and resin-based repair materials are bonded directly to concrete or other cementitious materials. It is, therefore, important to understand the similarities and differences in the mechanical and physical properties of repair materials and the original concrete. Typical properties of the different repairing systems are given in Table 6-14. Bond and compressive strengths are obviously important properties in any repair case, and they have to be satisfied first. ASTM C881 covers two-component, epoxy resin bonding systems for application to Portland cement concrete, which are able to be cured under humid conditions and bond to damp surfaces. ASTM C881 provides the compressive strength requirement of an epoxy resin bonding system for use in load-bearing applications for bonding concrete to hardened concrete and other materials, and as a binder for epoxy mortars and concretes, showing that the strength at 24 and 48 h were 14 and 40 MPa, respectively. Generally, a minimum early-age compressive strength development of the repair material for a concrete roadway of 20 MPa would be considered strong enough for rubber-tired traffic (Crovetti, 2005). At such a strength level, the corresponding flexural strength of the repair material is normally within 10% of its compressive strength, that is, around 2 MPa, which is another important criterion for allowing early opening to traffic (FHWA, 2006).

There are some other material properties that can be of equal or of greater importance than the bond and compressive strengths of the repair materials (Warner, 1984). These properties include:

**(a)** *Coefficient of thermal expansion*: It is important to use a repair material with a coefficient of thermal expansion similar to that of the parent concrete. Thermal compatibility is particularly important when patches have a large volume. If there is a large difference in the thermal properties of the two materials, significant changes in temperature could lead to failure either at the interface or within the material of lower strength. This should be considered in all situations.

**(b)** *Shrinkage*: Since most repairs are made on older Portland cement concrete in which major shrinkage has been completed, the repair material should be essentially low shrinkage in nature to avoid shear at the interface. Shrinkage of cementitious repair materials can be reduced in a number of ways, including modification using polymer latex.

**(c)** *Modulus of elasticity*: The modulus of elasticity of repair materials should be as close as possible to the old concrete being repaired. If the moduli of the two materials differ too much, large differences in stress can be generated under parallel loading or in deformations under a series loading in the two materials. Either case may result in failure of the repaired structures.

**(d)** *Color consistency*: For repair of architectural concrete surfaces and concrete highways, the color of the repair material should not differ appreciably from the adjacent surface. Trials should be made on the jobsite mockup prior to start of the actual production and repairing work.

**(e)** *Cost*: The total cost of the materials is another important consideration in material selection for repairs of concrete architectures and pavements.

**(f)** *Compatibility*: Compatibility is the most important property for a repair material. Compatibility means the properties of a repair material match well the properties of the materials to be repaired, and hence two materials can work together to carry external loadings and withstand the impact of the environment.

The widely used methods for evaluating the performance and properties of polymeric repair mortars are listed in Table 6-15.

**Table 6-15** Testing methods for evaluating the performance of polymeric repair mortars

| Test | Method of testing |
|---|---|
| Determination of compressive strength | BS 6319-2, ASTM C579 |
| Measurement of pull-off strength | BS 1881-207, ASTM C1583 |
| Determination of shear strength | BS EN 12615, ASTM C882 |
| Measurement of shrinkage and coefficient of thermal expansion | ASTM C531 |
| Determination of modulus of elasticity | BS 1881-121, ASTM C469 |
| Compressive strength of parent concrete | BS EN 12390-3, ASTM C39 |

### 6.4.6 General Application Guidelines

Different types of repair mortars require different kinds of treatment and are suitable for specific application conditions. Therefore, the application procedures for repair materials recommended by the corresponding manufacturers should be followed. The general steps in the repair of concrete structures and pavement using different repair mortars are as follows:

**Step 1:** Cut (saw cut is recommended) the boundary of the repair location to avoid feather-edging and to provide a square edge.

**Step 2:** Clean the surface and remove all faulty materials, including dust, unsound or contaminated material, oil, paint, grease, and corrosion deposits.

**Step 3:** The repair surface should be water saturated or primed with a primer in accordance with the manufacturer's instructions.

**Step 4:** A fresh mixture of the repair materials should be mixed in the recommended proportions, and then placed in a position with proper compaction, according to the manufacturer's instructions;

**Step 5:** The repair portion should be protected from damage and for a suitable curing period before opening to regular use.

## 6.5 SHRINKAGE-COMPENSATING CONCRETE

Shrinkage-compensating concrete is an expansive concrete. Its expansion occurs at an early stage before drying shrinkage occurs and is compensated for by the drying shrinkage occurring later on. In Figure 6-40, the early-age volume changes of a normal Portland cement concrete specimen and an expansive concrete are schematically drawn to compare the behaviors of the two types of concretes and to illustrate the mechanism shrinkage-compensating concrete.

### 6.5.1 Expansive Materials and Mechanisms

Shrinkage-compensating concrete can be made in two ways. One is to use expansive cement, such as types K, M, S, and O; and the other way is to add suitable expansive admixtures to normal Portland cement concrete.

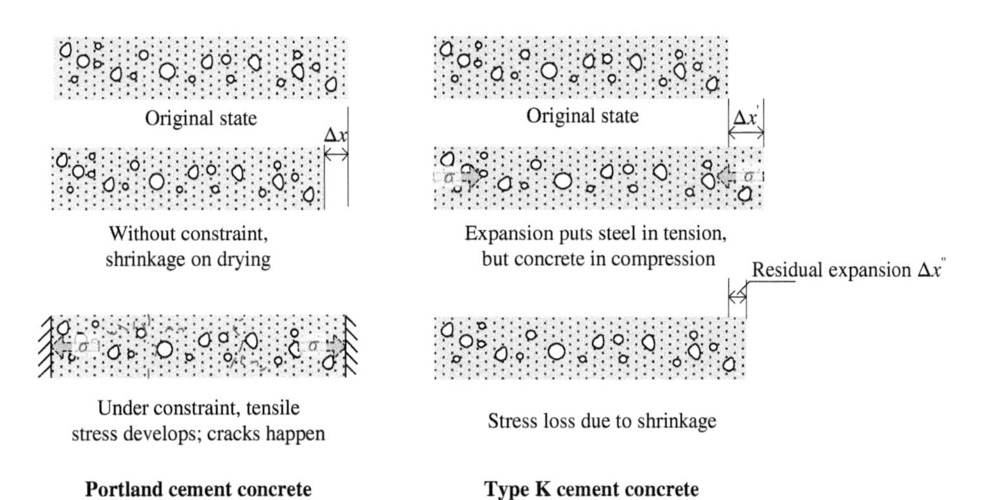


**Figure 6-40** Performance of shrinkage-compensating (Type-K cement) concretes

Expansive cements are usually special Portland cements with some expansive chemical compounds added. For instance, type K cement is Portland cement plus anhydrous tetracalcium trialuminosulfate, $C_4A_3S$, additional calcium sulfate, and uncombined calcium oxide (lime). Type M cement is a blend of Portland cement and high alumina cement plus additional gypsum. Type S cement is Portland cement with a large amount of $C_3A$ and gypsum. In China, in special Portland expansion cements, aluminum sulfate expansive cement (or sulfoaluminate expansive cement) is normally used for shrinkage compensation, which is produced with (Ye'elimite + belite) clinker and gypsum. The expansive admixtures used include various expansive agents made of Ye'elimite, alumite, and anhydrous gypsum or MgO.

The main mechanism of expansive concrete is the growth of the expansive hydration products right after setting under moist conditions, such as ettringite (AFt), monosulfoaluminate (AFm), and magnesium hydroxide (MH). These hydration products have a good crystal structure and large volume expansion with a high crystal growth pressure. They can increase the volume of concrete through the growth of the crystals (Chatterji and Jeffery, 1966; Okushima et al., 1968). It should be pointed out that, even though these hydration products have a large volume expansion compared to the original raw compounds, the volume is still less than the total volume of the compounds and water. Hence, the expansion is only realized under moist conditions. Moreover, the growth of such crystals should occur before the concrete matures. Otherwise, as discussed in Chapter 2, unsoundness can occur and the hardened concrete can be damaged by the cracks formed due to the aggressive movement of the later-formed crystals. An air entraining agent can also act as an expansion admixture because it causes an increase of the concrete volume due to the entrained air bubbles.

In practice, it is more convenient to produce an expansive concrete using an expansive agent plus normal Portland cement than to produce it with expansive cement. In China, the production of expansion agents has been rapidly increased. A common dosage for expansive agents is from 5 to 15% by weight of cement, for replacement. The main characteristic of shrinkage-compensating concrete is its expansion, usually expressed as the restrained expansion rate. The expansion rate can be measured using a restrained frame (Wang, 2006), as shown in Figure 6-41. To measure the restrained expansion, an expansion concrete or mortar is cast around the frame and the restrained strain is measured after demolding. The experiment is designed in this way so as to reflect the true situation of the concrete expansion by considering the minimum reinforcement required. It has been found from experiments that the restrained expansion rate can reach $(11-18) \times 10^{-4}$ if the expansive agent replacement percentage is in the range of 12 to 15% (Wang, 2006).

The mix proportion of expansive cement concrete is slightly different than that of normal Portland cement concrete. When type K cement is used in concrete, a minimum cement content

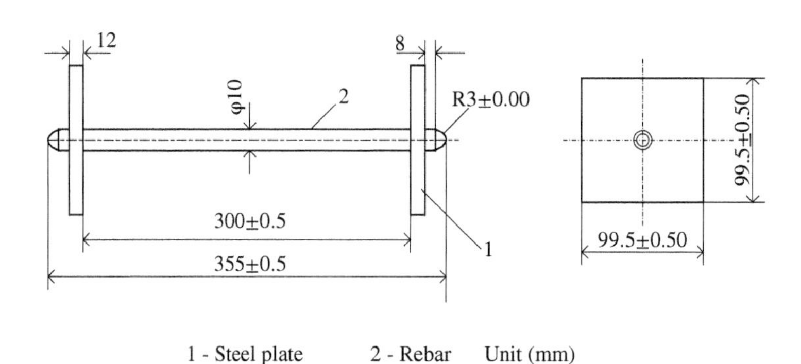


**Figure 6-41** A restrained frame for expansion rate measurement on concrete. (Source: Wang, 2006)

of 305 kg/m$^3$ is required to ensure adequate expansion. Moreover, a minimum of 0.15% reinforcement has to be added. The incorporation of reinforcement is aimed at providing a suitable restraint before compensating the dry shrinkage to avoid even earlier cracking due to expansion. When an expansive agent is used to produce shrinkage-compensating concrete, the mix proportion procedures are similar to those for normal Portland cement concrete. As discussed above, the amount of expansive agent used is expressed as the percentage of binder to be replaced. Usually, other mineral admixtures, such as slag or fly ash and water-reducing admixtures are used together with the expansive agent. In this case, it should be noted that pozzolans and water-reducing admixtures have a trend in reducing expansion. Typical mix proportions of shrinkage-compensating concretes using either type K cement or expansion agents are given in Table 6-16.

### 6.5.2 Properties

The properties of shrinkage-compensating concrete, both in fresh and hardened stages, have their own characteristics different from those of normal Portland cement concrete, which need to be paid attention to in order to fully utilize the advantages and avoid the limitations.

(a) *Workability*: In the fresh stage, shrinkage-compensating concrete tends to be stiff but highly cohesive. The slump loss of the expansive concrete at a temperature of 32°C or higher and in dry conditions is more serious than that of normal concrete. This will cause a reduction of both strength and expansion. The reason for this is the large amount of AFt formed in the early stage. To solve this problem, cool water with ice added should be used to lower the temperature of the concrete mixture to below 29°C. Because of the quicker stiffening and setting of expansive concrete under hot and windy conditions, plastic shrinkage cracks easily occur and create more serious problems than normal Portland cement concrete. Hence, a careful early-age curing is a must for expansive concrete.

(b) *Strength*: Shrinkage-compensating concrete made of type K cement is usually prepared with a *w/c* ratio from 0.4 to 0.65. Under such conditions, the paste matrix in shrinkage-compensating concrete is denser, and the interface between the aggregates and paste is stronger. Subsequently, the compressive strength of shrinkage-compensating concrete is much higher than that of normal Portland cement concrete. For shrinkage-compensating concrete with an expansive agent, an enhanced compressive strength is also normally observed. However, for different grades of concrete, the improvement levels are different. For normal-strength concrete with a projected 28-day compressive strength of 30 MPa, the strength improvement due to the incorporation of an expansive agent is not obvious. However, for high-strength concrete with a projected 28-day compressive strength of 60 MPa, the improvement of compressive strength can reach 15 to 20% (Wang, 2006).

(c) *Microstructure*: The hydration rate of shrinkage-compensating concrete is usually higher than that of normal Portland cement concrete. The microstructure of shrinkage-compensating

**Table 6-16** Typical mix proportions of shrinkage-compensating concrete (kg/m$^3$)

| Series | Binder | Coarse aggregate | Fine aggregate | Water | Water reducer | w/c |
|---|---|---|---|---|---|---|
| 1[a] | 347 | 1056 | 754 | 184.1 | 0.00 | 0.53 |
| 2[b] | 338 + 22 | 1030 | 810 | 187.2 | 2.16 | 0.55 |
| 3[b] | 470 + 30 | 1038 | 692 | 172.5 | 5.00 | 0.37 |

[a]M. W. Hoffman and E. G. Opbroek (1997) ACI, *Concrete International*, **1**(3), 19–25. Binder is type K cement.
[b]D. M. Wang (2006), *High performance expansive concrete*, Publisher of Hydraulic and Hydropower of China, Beijing. Binder is cement plus expansive agent (2: fc = 30 MPa and 3: fc = 60 MPa).

concrete at an early age contains a large number of AFt crystals. With the process of hydration, C–S–H can fill in the spaces occupied by water and surround the AFt crystals. Hence, a much denser microstructure can be produced.

**(d)** *Durability*: As shrinkage-compensating concrete has a denser microstructure and a higher gel-to-space ratio, its resistance to water penetration and ion diffusion is greatly improved. As a result, the migration rate of water, and the diffusion rate of chloride ions, oxygen, and carbon dioxide into the concrete are reduced. Generally speaking, shrinkage-compensating concrete has better durability than normal concrete in respect of carbonation, corrosion of steel bar, freezing–thawing, chemical resistance, and leaching.

### 6.5.3 Applications

Although studies on expansion cement and expansive concrete started in the 1950s, the application of expansive concrete only became somewhat popular in the 1980s. It has been reported that shrinkage-compensating concrete has been mostly used in structural elements, such as slabs, pavement, runways, prestressed beams, and water- and sewage-handling structures (Mehta and Monteiro, 2006). In China, many buildings and infrastructures have been built with shrinkage-compensating concrete. The following are good examples (You and Li, 2005):

**(a)** *Bank of China in Qingdao*: The underground structure of the building with an area of 6500 $m^2$ has been built partially with shrinkage-compensating concrete. The total amount used was 1200 $m^3$ with 12 to 13.5% of expansion agent added.

**(b)** *Car park at Beijing Airport*: This structure has four stories underground and two above ground. To solve the leakage problem due to shrinkage cracking, shrinkage-compensating concrete was used. The binder composition for the concrete was 340 kg/$m^3$ Portland cement, 110 kg/$m^3$ slag, and 55 kg/$m^3$ of expansion agent. The slump flow was greater than 550 mm and the 28-day compressive strength was 80 MPa.

**(c)** *Beijing National Stadium* (the Bird's Nest): The ground-floor slab of the stadium has a thickness of 650 mm and a large area. It adopted a shrinkage-compensating concrete for construction.

Other examples include Terminal 3 Building of Beijing National Airport, the Beijing Tongchan building, Gongbei Port Square in Zhuhai, and the East Square building in Beijing.

## 6.6 SELF-COMPACTING CONCRETE

Self-compacting concrete (SCC) or self-consolidation concrete is a high-fluidity concrete (HFC). Self-compacting means it can easily be placed and consolidated by its own gravity in a formwork, even with highly congested reinforcements, without external consolidation by vibration. It is characterized by its high filling capacity caused by high visco-plastic deformability, resistance to segregation, and an ability to maintain a stable composition throughout transportation and placing. SCC has the advantages of fast construction, noise reduction, good formability, and energy effectiveness. SCC enables concrete to be placed in structures that would otherwise have been impossible to construct with concrete, and has undergone dramatic improvements in construction efficiency of extralarge-scale structures, such as bridge tower pile cap (Figure 6-42).

Self-compacting concrete was first systematized by Hajime Okamura in 1988 in Japan (Okamura, 1988). In the beginning, he named it self-compacting "high-performance concrete" (HPC). SCC/HPC was initially developed to meet the need of construction in Japan. The postwar reconstruction of Japan in the 1960s led to a boom in building and infrastructure construction. As Japan is a major earthquake region, most structures have to be heavily reinforced to be earthquake-proof,

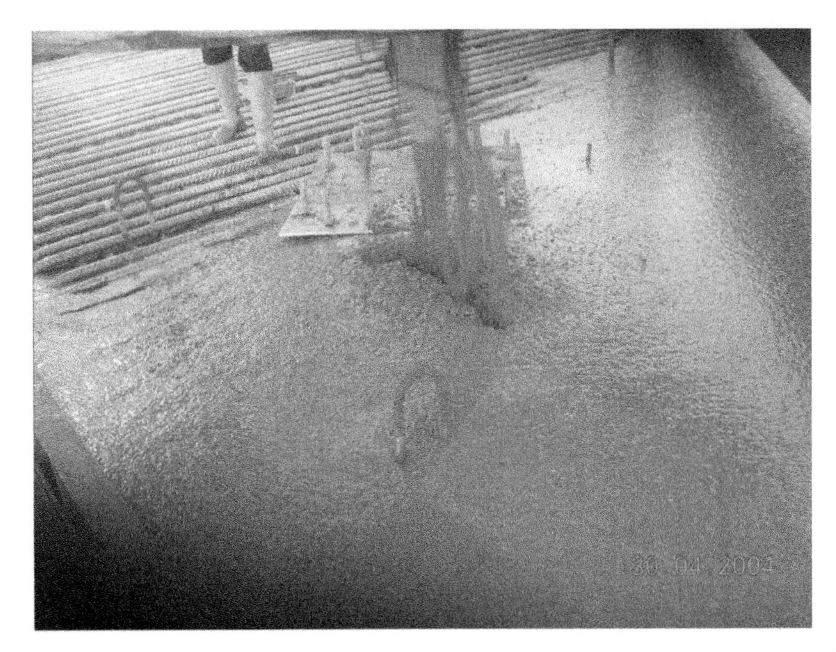

**Figure 6-42** Application of self-compacting concrete for *in situ* construction of a pile cap (Photo provided by Ove Arup, HK)

especially at beam–column joints. Because of these situations and the need for speedy project delivery, many structures could not be built with sufficiently compacted concrete. Within a decade or two, many reinforced concrete structures deteriorated. To solve the durability issue of new concrete construction, a project was started by H. Okamura of the University of Tokyo. His investigation found that insufficient compaction was the most common cause of deterioration of concrete structures and suggested an increase in the flowability of the fresh mixture to eliminate the need of compaction, i.e., by creating "self-compacting" concrete mixtures (Kuroiwa et al., 1993).

In Europe, a large, industry-led (NCC Sweden, GTM-Vinci France, and six other partners) research project was funded by the European Commission in 1997 (Grauers, 1997). The European SCC project proved that SCC was practical for applications using a variety of local materials, and that the expected benefits were obtainable in real construction practice. The working group on SCC set up under RILEM TCI45-WSM was converted to a new RILEM workgroup (TCI74-SCC), and its guidelines on SCC were published in 2000 (Petersson and Skarendahl, 2000).

The fluidity of self-compacting concrete was realized with the invention of high range of water-reducing admixtures or superplasticizers. Supplementary approaches to raise consistency were to increase the cement paste content and sand ratio of the concrete mixtures.

### 6.6.1 Advantages of Self-Compacting Concrete

SCC is considered an environmentally friendly material. First, the application of SCC on a construction site can eliminate the need for vibration to compact the concrete. The compaction of fresh concrete by vibration is generally recognized as a heavy physical job and an unpleasant activity in the concrete construction process. The vibration can cause high noise levels, up to 90 dB, that are not good for public health, especially the health of the operators. Second, the acceleration generated by vibrators can reach 0.70 to 4 $m/s^2$ and has potential to injure the vibrator operator. Hence, eliminating the vibration significantly improves health and the environment on a concrete construction site (Bartos and Cechura, 2001). Third, some skill and experience are needed for vibrator operators in compacting fresh concrete into a satisfactory concrete. Lack of experienced workers in many regions may lead to low quality of concrete. The application of SCC can solve this problem and ensure consistent high quality for concrete structures. The application of SCC also makes automation of construction possible, leading to higher productivity. This is envisaged not only in precast concrete production, but also for *in situ* concreting operations.

The introduction of SCC benefits architects and structural engineers, and ultimately the users of buildings. New types of structural elements, which it was not possible to make with traditional concrete, can be produced using SCC. Such elements include different types of steel–concrete structural elements with more complex shapes, that are thinner with a much heavier reinforced cross-section.

### 6.6.2 Property Evaluation of Fresh Self-Compacting Concrete

The major difference between traditional concrete and SCC is the flowability in the fresh stage. Due to its water-like fluidity, the methods used to evaluate the workability of traditional concrete in the fresh stage are not suitable for SCC. Thus, starting from the invention of SCC, much research has focused on the evaluation of the properties of SCC in the fresh stage. The test methods for evaluation of properties of fresh SCC, standardized or not, are presented in this section.

#### 6.6.2.1 Slump Flow Test

The slump flow test is the most common test method for measuring the flowability of SCC. The apparatus employed and the procedures followed are very similar to the slump test for conventional concrete stipulated in BS EN 12350-2 or ASTM C143. The only differences between slump and slump flow tests are the following: (1) no tamping is allowed when filling the concrete into the slump cone for the slump flow test; (2) the slump flow test measures the time that is needed for SCC to spread into a circular-shaped configuration with a nominal diameter of 500 mm, named T500, as the index of flow velocity; Usually, it requires SCC to have a T500 less than 8 seconds (Felekoglu et al., 2007; Gesoglu and Ozbay, 2007; Khayat et al., 2004; Shindoh and Matsuoka, 2003; Sonebi et al., 2007); (3) the slump flow test also measures the maximum nominal diameter that the SCC can reach after lifting the slump cone without limiting the time; usually, the SCC should be able to reach a nominal diameter of 600 mm; and (4) the flow test can also provide some information on the cohesiveness of SCC through visual inspection. If the SCC has insufficient cohesiveness, the coarse aggregate in SCC cannot flow with the mortar, and the sand particles cannot flow with the cement paste. Thus, the coarse aggregates will be left in the central region of the slump patty, and a layer of thin mortar will appear on the outside ring of the slump patty, as shown in Figure 6-43. To evaluate this type of phenomenon, Daczko and Kurtz (2001) proposed a visual stability index (VSI), and the assessment criteria adopted by ASTM C1611 are summarized in Table 6-17.

#### 6.6.2.2 V-Funnel Test

The V-funnel test was first proposed by Ozawa et al. (1995), and can be used to evaluate the ability of SCC to pass through a small space. As shown in Figure 6-44, the apparatus for a V-funnel test consists of two parts: a V-shaped top hopper connected to a rectangular bottom channel. There is a gate at the bottom of the apparatus. To perform the V-funnel test, the bottom gate is first closed to shut off the orifice and then SCC is gently poured into the V-funnel until it reaches the top edge. After leveling the top surface of the SCC, the bottom gate is opened to discharge the SCC.

**Figure 6-43** Slump flow of SCC that suffers from lack of cohesiveness

**Table 6-17** Visual stability index (VSI) values

| VSI value | State | Description and criteria |
|---|---|---|
| 0 | Highly stable | No evidence of segregation or bleeding |
| 1 | Stable | No evidence of segregation; slight bleeding observed as a sheen on the concrete mass |
| 2 | Unstable | A slight mortar halo $\leq$ 10 mm and/or aggregate pile in the center of the concrete mass |
| 3 | Highly unstable | Clear segregation evidenced by a large mortar halo $\geq$10 mm and/or a large aggregate pile in the center of the concrete mass |

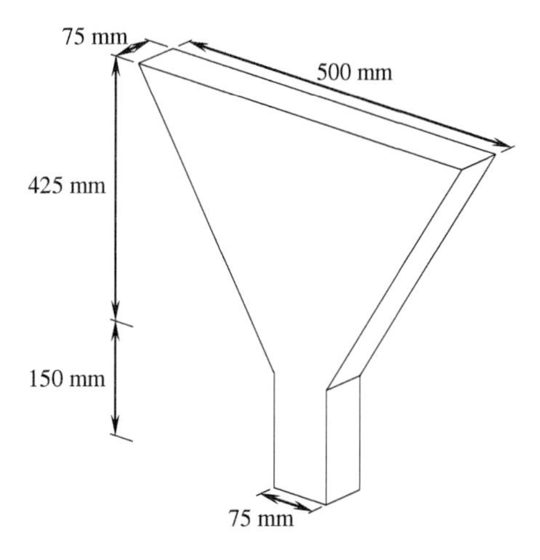


**Figure 6-44** Schematic diagram of a V-funnel

The period from the time the bottom gate is opened to the time when the concrete has been discharged to such extent that light can be seen through the orifice is recorded as the V-funnel time. The V-funnel time is a measure of the flowability of SCC. A shorter V-funnel time means a higher flowability, while a longer V-funnel time means a lower flowability. The recommended ranges of V-funnel time, according to various researchers and guidelines, are summarized in Table 6-18 for comparison. It can be observed that the recommended values vary in a rather wide range.

### 6.6.2.3 J-Ring Test

The J-ring test was developed in Japan (J-ring means Japanese ring). The J-ring test measures the ability of SCC to pass through reinforcing steel, which has been standardized in ASTM C1621. A typical J-ring is shown in Figure 6-45. A number of 12-mm-diameter, plain, round steel bars are fixed uniformly along the circumference of a circular steel ring with an external diameter of 360 mm, an internal diameter of 300 mm, and a thickness of 30 mm. They are used to simulate the obstructive effects of the steel reinforcing bars in the formwork. The J-rings used by different researchers are basically similar except for the spacing between the steel bars, which varies from 30 to 120 mm (Daczko, 2003; Grunewald and Walraven, 2001; Khayat et al., 2004). Ideally, the clear spacing between the steel bars of the J-ring should be the same as the minimum clearance of the steel reinforcing bars in the formwork of a target project to truly reflect the actual situation.

**Table 6-18** Recommended values of V-funnel time

| Source | Range (sec) |
|---|---|
| European Federation of National Trade Associations 2002 | 6–12 |
| Precast/Prestressed Concrete Institute 2003 | 6–10 |
| Self-compacting Concrete European Project Group 2005 | ≤25 |
| Hwang et al. (2006) | <8 |

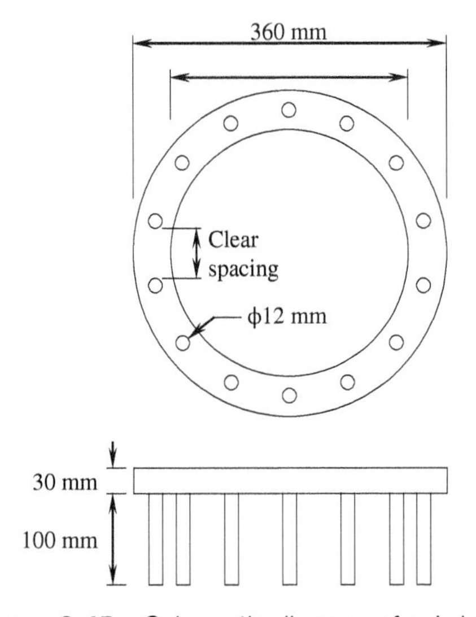


**Figure 6-45** Schematic diagram of a J-ring

The J-ring is usually used together with a consistency test setup by putting the ring outside the slump cone to obstruct the flow of SCC. The flowability of SCC is measured in terms of the reduction in flow of SCC after passing the J-ring as compared to the free slump flow without a J-ring (Brameshuber and Uebachs, 2001; Daczko, 2003; Khayat et al., 2003). The passability of SCC can also be indexed in terms of the difference in the height of concrete between the inside and outside of the J-ring (RILEM Technical Committee 145, 2002; European Federation of National Trade Associations, 2002; Precast/Prestressed Concrete Institute, 2003). Similar to the slump flow test, visual inspection after the J-ring test should be carried out to detect any signs of segregation. In general, segregation tends to be more serious with the J-ring in place due to the blockage effect of the reinforcing bars.

Table 6-19 shows the acceptance requirements of a J-ring test recommended by different researchers, and the guidelines either in the flow reduction or in the height difference as measures of the passability of SCC. Regarding the flow reduction, 50 mm is an index accepted by all researchers and guidelines.

#### 6.6.2.4 U-Box Test

The U-box test was originally developed in Japan (Hayakawa et al., 1993). In this test, a specially designed U-shaped tube, comprised by a storing compartment, a filling compartment, and an opening between the two compartments, is employed (Okamura et al., 2000).

The procedure for the U-box test is as follow. First, the sliding gate separating the two compartments is closed and the concrete is poured gently into the storing compartment until the compartment is full; the top surface of the concrete is then troweled flat. Second, the sliding gate is opened quickly. Due to the hydrostatic pressure, the concrete in the storage compartment is pushed to fill the filling compartment until it completely stops. Ideally, the height in the storage and filling compartment should become the same. However, a level difference between the two compartments may exist due to the friction of the wall of the compartments and the viscosity of the concrete. After the concrete stops flowing, the height of the concrete in the filling compartment is measured. The filling height is an integrated measure of the filling ability and passing ability of the concrete.

#### 6.6.2.5 L-Box Test

The L-box test was another method developed to assess the flowability and passing ability of SCC (Yonezawa et al., 1992). A typical apparatus for an L-box test is illustrated in Figure 6-46. It consists of a vertical compartment to store the concrete and a horizontal one to be filled by the concrete flowing out from the vertical compartment. The dimensions of the vertical compartment can range from 60 to 100 mm in breadth, 150–300 mm in width, and 400–600 mm in height. The length of the horizontal trough varies among different L-box versions, ranging from 600 to 710 mm (RILEM Technical Committee 145, 2002; Self-Compacting Concrete European Project Group, 2005; Sonebi, 2004; Testing-SCC Project Group, 2005a), but the most commonly used version

**Table 6-19** Recommended acceptance requirement for J-ring test

| Source | Measurement | Range (mm) |
|---|---|---|
| Brameshuber and Uebachs (2001) | Flow reduction | ≤50 |
| European Federation of National Trade Associations (2002) | Height difference | <10 |
| RILEM Technical Committee 145 (2002) | Height difference | <10 |
| Precast/Prestressed Concrete Institute (2003) | Height difference | <15 |
| ASTM C1621 | Flow reduction | ≤50 |

**Figure 6-46** An L-box apparatus

has a length of 700 mm. A sliding gate is inserted at the bottom of the vertical compartment. A number of plain, round steel bars are also inserted as obstacles to simulate the reinforcement in the formwork. The number of reinforcing bars to be inserted should be decided according to the net spacing of reinforcing bars in the formwork of a specific project. If no information on the real situation is available, an obstacle arrangement of three 12-mm, plain, round steel bars with center-to-center spacing of 53 mm can be used.

The procedure for the L-box test is as follows. First, the sliding gate separating the vertical compartment and the horizontal trough is closed and the concrete is poured gently into the vertical compartment. Then, after the top surface of the concrete is troweled flat, the sliding gate is opened quickly. Next, the concrete in the vertical compartment will be pushed by the hydrostatic pressure to flow along the horizontal trough, passing the obstacle. If the concrete has sufficient fluidity and passability, it will reach the end wall of the horizontal trough with a certain height.

After the flow of the concrete ceases, the height of the concrete at the end wall, $h_1$, and the height of the concrete at the vertical compartment, $h_2$, are measured. The blocking ratio defined by $h_1/h_2$ can be estimated as a measure of the passability of the concrete. In addition, the time difference between two given points in the horizontal channel can be used as a measure of the viscosity of the concrete (Gesoglu and Ozbay, 2007; RILEM Technical Committee 145, 2002). The recommended values of blocking ratio are summarized in Table 6-20. A blocking ratio higher than 0.80 is usually suggested.

#### 6.6.2.6 Sieve Segregation Test

The sieve segregation test was developed in the 1990s by Fujiwara (1992) to measure the segregation stability of SCC. In this method, a 2-liter (about 4.8 kg) concrete sample is taken from a concrete batch and placed gently over a 5-mm aperture sieve. Because the sieve is porous, the fine portion of the concrete that is incapable of adhering to the aggregate particles on the sieve will drip through the apertures of the sieve to the base receiver placed underneath. After 5 min, the sieve is removed without agitation and the weight of the material collected in the base receiver

**Table 6-20** Recommended values of blocking ratio

| Source | Range |
|---|---|
| European Federation of National Trade Associations (2002) | >0.8 |
| Swedish Concrete Association (2002) | >0.8 |
| RILEM (2002) | 0.80–0.85 |
| Precast/Prestressed Concrete Institute 92003) | >0.75 |
| Self-compacting Concrete European Project Group (2005) | ≥0.8 |

**Table 6-21** Recommended values of segregation index

| Source | Pouring method | Range (5) |
|---|---|---|
| Fujiwara (1992) | Direct | <5 |
| Khayat et al. (1998) | Direct | <10 |
| European Federation of National Trade Associations (2002) | Indirect | ≤15 |
| Lachemi et al. (2003) | Direct | <10 |
| Testing-SCC Project Group (2005b) | Indirect | 5–15 |
| Self-Compacting Concrete European Project Group (2005) | Indirect | ≤20 |

is measured. Since this method pours SCC directly from the batch to the sieve, it is called a direct pouring sieve method.

A modified sieve segregation test was developed in France (Association Française de Génie Civil, 2000) and later adopted in many countries in Europe (RILEM Technical Committee 145, 2002; Self-Compacting Concrete European Project Group, 2005; Testing-SCC Project Group, 2005b). In this method, 10 liters of a concrete sample taken from a batch is placed gently in a 300-mm-diameter container. The sample is then covered by a lid, put in a level position, and left undisturbed to allow any sedimentation of the coarse aggregate particles and bleeding of water to take place. After 15 minutes, 2 liters (about 4.8 kg) of the top portion of the concrete sample is poured gently onto the sieve at a height of 500 mm. The fine portion of the concrete without sufficient adhesion to the aggregate particles will drip through the openings of the sieve to the base receiver underneath. After 2 min, the sieve is removed without agitation and the weight of the material collected in the base receiver is measured. Since this method pours SCC on the sieve after placing it in a container for a while, it is called an indirect pouring method. The segregation index, SI, for both methods, is calculated using the formula

$$SI = \frac{M_p}{M_c} \times 100\% \tag{6-4}$$

where SI is the segregation index of SCC; $M_p$ is the mass of the material collected from the base receiver; and $M_c$ is the mass of the concrete poured onto the sieve. The segregation index measures the adhesive ability of the SCC matrix to coarse aggregates. A low segregation index represents a high adhesion, whereas a high segregation index represents a low adhesion of SCC.

As revealed in Table 6-21, all the recommendations impose an upper limit on the acceptable segregation index for SCC, no matter which method is used. In general, the recommended upper limit of the segregation index for the direct pouring method is 10%, while the recommended upper limit of the segregation index for the indirect pouring method is 20%.

#### 6.6.2.7 Noncontact Resistivity Measurement

Li and Xiao (2009) characterized fresh SCC properties with electrical resistivity measurements. The study adopted a noncontact resistivity measurement method that can provide accurate and reliable information for fresh SCC. The influence of various minerals, chemical admixtures, and rheological modifying agents on fresh SCC's properties can be identified by resistivity measurement. Figure 6-47 shows the electrical resistivity development with time for samples SCC-0 (without viscosity modifying agent) and NC (a normal concrete). The proportions of the two concretes are shown in Table 6-22. It can be seen from the figure that the curve shapes of samples SCC-0 and NC are different. First, the resistivity of NC develops faster than that of SCC-0. Second, there is an extreme point (*P*), as shown in Figure 6-47, followed by a temporary decreasing range, on the curve of sample SCC-0, which is not experimentally observed in normal concrete. From Table 6-22, it can be seen that a superplasticizer was added in SCC-0 while no superplasticizer was used in NC, which caused the delay of the C–S–H formation. The transformation of hydration products AFt to AFm occurred when sulfate ions were not sufficient for the hydration system, releasing ions ($Ca^{2+}$ and $SO_4^{2-}$) and increasing the solid volume (AFt has lower density than AFm). The consequence is an electrical resistivity decrease until enough C–S–H formation offsets such a trend and the electrical resistivity increases again. In sample NC, the C–S–H formation is not delayed and its content increases with the hydration process covering the possible decreasing trend of the electrical resistivity caused by the phase transformation.

A superplasticizer is necessary in SCC to improve fluidity, which often has a retarding effect on the hydration process. Therefore, the extremum point becomes an obvious characteristic of SCC on the electrical resistivity curve. Figure 6-48 shows the electrical resistivity development with time for four SCC samples: SCC-0 without a viscosity-modifying agent (VMA) and SCC-1

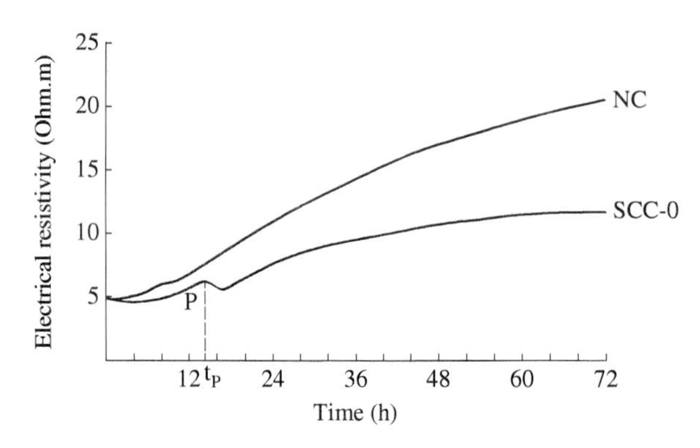


**Figure 6-47** The electrical resistivity curves of SCC and normal concrete

**Table 6-22** Mix proportions of normal concrete and SCCs with different VMAs (kg/m$^3$)

| | | | | | | | Viscosity modifying agent | |
|---|---|---|---|---|---|---|---|---|
| **Name** | **Water** | **Cement** | **Fly Ash** | **Sand** | **Coarse Aggregate** | **SP** | **Weight** | **Type** |
| NC | 176 | 352 | 235 | 744 | 819 | — | — | — |
| SCC-0 | 176 | 352 | 235 | 744 | 819 | 1.76 | — | — |
| SCC-1 | 176 | 352 | 235 | 744 | 819 | 1.76 | 0.29 | MC |
| SCC-2 | 176 | 352 | 235 | 744 | 819 | 1.76 | 0.29 | PAM |
| SCC-3 | 176 | 352 | 235 | 744 | 819 | 1.76 | 0.29 | CMC |

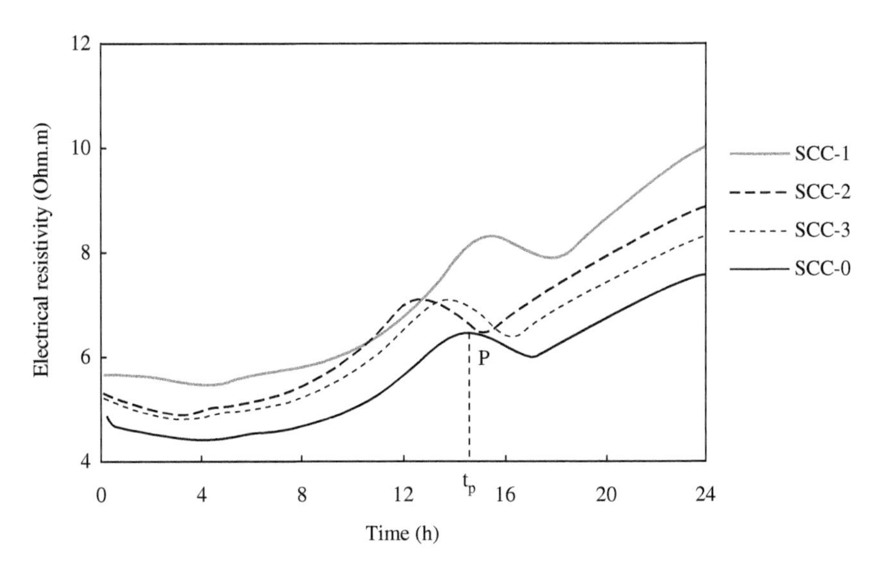


**Figure 6-48** The electrical resistivity curves of SCC with various VMAs up to 24 hours

through SCC-3 containing one specific type of VMA, as shown in Table 6-22. All the curves follow a similar pattern to that presented earlier. There is an extremum point (*P*), followed by a temporary decreasing range on each curve. It can be seen that the addition of each of the VMAs leads to an electrical resistivity increase during the measurement period and the extremum point shifts to the left or the right.

The electrical resistivity increase in the SCC mixes with a VMA is attributed to the air entrapped by the VMA addition. The air bubbles act as insulators for ion conduction and cause the electrical resistivity to become higher than that of mix SCC-0 without VMA, when other factors are kept the same.

### 6.6.3 Characteristics of the Mix Proportion of SCC

The SCC mix proportion has some unique characteristics compared to that of conventional concrete. For example, the cement-based paste content in SCC is much higher due to the need to keep a good workability under low water–powder ratios. The increased paste content is used to supplement the decreased volume of the coarse aggregate, and to assure the required viscosity of the paste at lower water–powder ratios. Usually, the powder content is increased to 600 kg/m$^3$ or more, and the water content is restrained to an adequately low level. Various mineral admixtures are utilized to increase the powder content. High fluidity is produced with the help of high-performance superplasticizers, and suitable cohesiveness is reached with the addition of a viscosity-modifying agent or powder.

The characteristics of the mix proportions of SCC are somewhat near to those of high-strength concrete. Hence, high-strength concrete can be easily transformed to SCC, and vice versa. However, to make SCC, the rheology of the matrix must be measured first to ensure that the matrix can maintain a uniform distribution of coarse aggregates. From the rheological point of view, a fresh SCC mixture should have low yield stress and moderate viscosity (Bonen and Shah, 2005; Khayat et al., 1999). The low yield stress is necessary to achieve high flowability and high filling ability, while a moderate viscosity is essential to maintain good cohesiveness. Based on the mix design methodology and the methods to achieve the rheological requirements, currently, two major categories of SCC can be identified: the powder-type SCC and the VMA-type SCC

(Bonen and Shah, 2005). The characteristics of the mix proportions of the two categories of SCCs are discussed separately in the following sections.

### 6.6.3.1 Powder-Type SCC Mixtures

In general, powder-based SCC mixtures are developed from traditional concrete mixtures through necessary modifications. However, the relative proportions of SCC mixtures are quite different from those of conventional concrete. As shown in Figure 6-49, for a given total volume, the coarse aggregate volume of SCC mixes is much smaller than that of conventional concrete; the sand volume of SCC mixes is larger than that of conventional concrete; the cementitious material content of SCC mixes is much higher than that of conventional concrete; while the water volume of SCC mixes is similar to that of conventional concrete. Compared with relative mix proportions of a conventional concrete mixture with that of a SCC mixture, it can be observed that a SCC mixture generally consists of a lower coarse aggregate content, higher paste volume, and lower water/cement (*w*/*c*) ratio. Actually, it is these characteristics in SCC mixtures that are responsible for the excellent workability of SCC.

A low aggregate content in the concrete mixture is essential for the high performance of SCC mixtures because it can reduce the frequency of collision and contact between aggregate particles and thus reduce the solid-to-solid friction resulting from the particle interactions and shearing actions of the aggregate particles when they are moving relative to each other. Hence, the deformability and the flow speed of fresh SCC can be enhanced.

With a reduced coarse aggregate content but slightly increased fine aggregate content in SCC mixtures, the fine aggregate to total aggregate (*F*/*T*) ratio is increased. Typical values are increased from 0.35-0.40 for conventional concrete mixtures to higher than 0.50 for SCC mixtures (Bonen and Shah, 2005; Saak et al., 2001; Yurugi et al., 1993). Such an increase in the *F*/*T* ratio helps maintain a moderate viscosity of the concrete to avoid localized increase in the shear stress and improve the deformability (Okamura and Ozawa, 1994).

Apart from the *F*/*T* ratio, the volume of paste (i.e., water + powder) of an SCC mixture is increased as compared to that of a conventional concrete mixture. In conventional concrete mixtures, the function of the paste when the concrete is still plastic is mainly to fill the voids among the aggregate particles. However, in SCC mixes, the paste functions as a medium for lubricating the aggregate particles suspended therein, to provide the required workability and deformability (Petrou et al., 2000; Roussel, 2006; Toutou and Roussel, 2006). It should be pointed out that the increase of the paste volume of SCC mixtures is achieved by an increase in the powder content with a lower water/binder (*w*/*b*) ratio. The total water content remains almost unchanged as compared

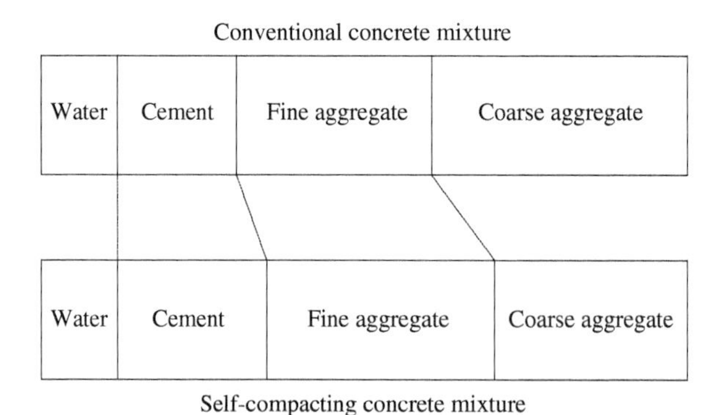


**Figure 6-49** Mix proportions of conventional and self-compacting concrete mixtures

with typical conventional concrete mixes. The *w/b* ratio of SCC mixtures should be set in a range in which both deformability and cohesiveness of the SCC are high. It has been found, from practice, that a *w/b* ratio ranging from 0.30 to 0.45 gives the SCC mixtures satisfactory performance.

The powders used in powder-type SCCs include silica fume, fly ash, slag, and ground limestone with different size distribution. In Europe and Japan, all of these powders have been used. In the United States, the powders used are mostly fly ash and silica fume.

According to Khayat (2000), SCC mixtures having a powder content (powder here means particles finer than 80 μm) of 500–600 kg/m$^3$ can effectively improve the cohesiveness and reduce the superplasticizer demand. Likewise, the ACI Committee 237 (2007) recommended that, for initial trial mixing, a relatively high powder content (powder here means particles finer than 125 μm) of 386 to 475 kg/m$^3$ should be adopted.

#### 6.6.3.2 VMA-Type SCC Mixtures

In a viscosity-modifying agent (VMA) type SCC, to satisfy the rheological requirements of SCC, superplasticizer and VMAs are incorporated into the mixture. The incorporation of superplasticizer can largely decrease the SCC yield stress with a limited influence on the viscosity (Khayat, 1999; Wallevik, 2003). The incorporation of VMA can significantly improve the cohesiveness of the SCC. VMAs are usually water-soluble polymers, and can dissolve in the mixing water and adsorb cement particles. The polymer can hold water tightly and provide sufficient adhesion to the aggregate. As reported by Khayat and Guizani (1997), incorporation of 0.07% by mass of cementitious materials of a welan gum VMA in a highly fluid concrete significantly reduces the bleeding by more than 85% and the segregation by 45%. Rols et al. (1999) also reported that incorporation of 0.03% of starch-based VMA significantly reduced the aggregate sedimentation of concrete. Sometimes, VMA can be used to overcome the negative impacts resulting from the poor aggregate grading used in SCC mixtures (Lachemi et al., 2003). In summary, the VMA-type SCC uses polymers to improve the viscosity of SCC and is proportioned based on the use of both superplasticizer and VMA. In these kinds of concrete mixtures, the required workability or flowability is controlled by the superplasticizer dosage, while the required segregation stability (viscosity/cohesiveness) of the SCC mixture is adjusted by the VMA usage.

The differences in mix proportions of traditional concrete, powder-type SCC, and VMA-type SCC are schematically shown in Figure 6-50. It can be seen from the figure that among the three types, traditional concrete contains the highest amount of coarse aggregates, the smallest amount of sand, the lowest cementitious materials, and the least water. The VMA type of SCC contains

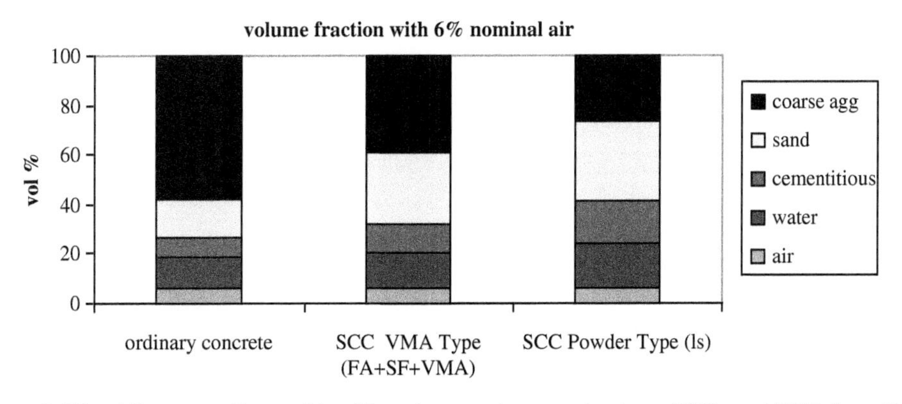


**Figure 6-50** Mix proportions of traditional concrete, powder-type SCC, and VMA-type SCC

a moderate to high content of cementitious material, moderate water content, almost no or very small amounts of mineral admixture, moderate sand content, moderate coarse aggregate content, high superplasticizer content, and a certain amount of VMA. The powder type of SCC contains the highest content of cementitious materials, the highest sand content, the lowest coarse aggregate content, high mineral powder content, relatively high water content, and relatively high superplasticizer content.

### 6.6.4 SCC Pressure on Formwork

The applications of SCC in precast plants are more extensive than those in cast-in-place concrete work on construction sites. One technical issue limiting SCC's application on a construction site is a lack of knowledge about the lateral pressure generated by SCC on the formwork (Shah et al., 2009).

Formwork pressure depends on the fluidity and cohesion of the SCC, rate of vertical rise, and the method of placing (from the top or from the bottom). Some design guidelines, such as "The European Guidelines for Self-Compacting Concrete (SCC) Specification, Production and Use," suggest that in formwork design, including supports and fixing systems, it should normally be assumed that the full hydrostatic concrete pressure is applied to the formwork. Such type of design, of course, is conservative and safe. However, it may overestimate the pressure that leads to unnecessary increased cost. It has been reported and confirmed that the formwork pressure of SCC can be smaller than the theoretical hydrostatic pressure (Fedroff and Frosch, 2004). This is due to the buildup of a 3D structure when the concrete is left to rest. A similar phenomenon has been observed by Khayat and Omran (2009) and Lange et al. (2009). In Khayat and Omran's (2009) work, a portable device was developed to monitor lateral pressure exerted by SCC. Their study shows that the formwork pressure exerted by SCC on site can be predicted by the simulation of the device through the analytical model that they built up. In the work of Lange et al. (2009), a formwork pressure model was built based on a threefold SCC performance. The model considers that SCC exhibits a pressure decay. The model is relatively simple and easy-to-use, and has been successfully used in tall wall constructions.

Formwork pressure and 3D structure rebuilding of SCC is highly influenced by the mixture proportion and raw materials composition, in which the powder materials play an important role. It has been found that cement with lower alkali/$C_3A$ content produces lower formwork pressure (Shah et al., 2009). Moreover, shows the same study that replacement of cement with metakaolin clay is very effective in reducing the formwork pressure of SCC.

If the SCC is pumped from the bottom, then, locally, pressure can be increased close to the pump entry point, especially on restart, if there is an interruption in pumping. This pressure increase has to be taken into account in formwork design. If not, the pressure is underestimated, which may cause large deformation and even failure of the structural formwork.

### 6.6.5 Applications of SCC

Due to its attractive advantages, SCC has been widely used throughout the world. Full-scale trials and demonstrations were carried out in the early 1990s, and self-compacting concrete was first used in significant practical applications in Japan (Hayakawa et al. 1993). The gradual increase in volume consumption in Japan reached its peak in the construction of the two anchorages of the Akashi Kaikyo Bridge in 1997. A total of 600,000 $m^3$ of different varieties of SCC were placed, which remains the most prominent landmark application of SCC and the largest single application to date. In India, a large application of SCC was commenced during the Delhi Metro Project in 2003 (Vachhani et al., 2004). However, the annual consumption of SCC in India ranges only from 50,000 to 100,000 $m^3$, about 0.1% of the total concrete production (Asmus and

Christensen, 2009). In China, due to the booming infrastructure construction in the last few decades, the applications of SCC have been speeded up. A steel fiber-reinforced SCC with compressive strength of 60 MPa has been used in the construction of the new China Central TV headquarters building. The project comprises two towers of 234 m tall, inclined at 6 degrees toward each other. Columns of 1900 × 1500 mm are used to support the towers. C60 steel fiber-reinforced SCC was used to build up the columns to meet the requirements of crack control and to fit the complex shape (Yan and Yu, 2009). In Beijing, a C60 SCC was also used in Tower A of the International Trade Center, Phase 3, which is 330 m in height. The columns in the core wall of this tower are huge, with a cross-sectional area of 3.6 $m^2$. The mix proportion of the concrete for this project is given in Table 6-23 and its properties are given in Table 6-24. Moreover, a C50 SCC has been applied in a famous project in Beijing—the Birdsnest, which is the National Olympics stadium. The outer shell of the stadium is composed of irregular steel elements. Many concrete columns have been built up between the outer steel structure and the inner concrete stand to support the stand system. There are a total of 124 steel tubular columns, 228 inclined beams, 600 inclined poles, and 112 Y-shaped poles. They had to be interwoven with 3D contorted annular beams. It is very difficult, if not impossible, to use traditional concrete due to the limited space for compacting. Hence, SCC was selected for construction of these columns and poles (Yan and Yu, 2009). In Hong Kong, a multipurpose commercial building (International Commerce Center) has been designed and constructed, that has 108 stories and is 484 m in height. High-strength SCC of grade 90 was used to construct the concrete core of the building to reduce the dead load and column size (Zhang et al., 2009). The main mix parameters for the SCC were cementitious material content of 600 kg/$m^3$, water to binder ratio of 0.26, and a liquid PC-based superplasticizer at a dosage of 3 L per 100 kg of cementitious material. The main requirements for the SCC are (1) mean compressive strength of 90 MPa and a target strength at 28 days of 104 MPa; (2) mean modulus of elasticity of 39 GPa at 28 days; (3) T500 $<$ 15 sec and slump flow of 700 mm in 120 sec; (4) pumpable to a height of 320 m; and (5) casting temperature of concrete lower than 25°C on site. In the United States, the 15-km-long Los Angeles Sewer Tunnel renovation project has used 37,000 $m^3$ SCC, which made it the largest SCC project in North America.

The application of SCC has also been extended into hydraulic engineering, such as dam construction (An et al., 2009). Rock-filled concrete (RFC) is a new type of mass concrete structure developed in China. It uses large rock blocks, up to 300 mm, as the filler and SCC as the "binder" or "grout" to build up the dam and other massive concrete structures. The first application of RFC in China was a gravity dam in a reservoir project in Beijing. The dam, 13.5 m in height and 2000 $m^3$

**Table 6-23** The mix proportion of a c60 self-compacting concrete (kg/$m^3$)

| Cement | Fly ash | Sand | Crushed limestone | Water | Superplasticizer |
|---|---|---|---|---|---|
| 360 | 240 | 800 | 840 | 175 | 7.1 |

**Table 6-24** Properties of fresh and hardened C60 concrete

| Fresh concrete after 1 hour of casting | | | | Compressive strength (MPa) | | |
|---|---|---|---|---|---|---|
| Air content (%) | Slump (mm) | Slump flow (mm) | Flowing time[a] (sec) | 3-day | 7-day | 28-day |
| 2.4 | 250 | 665 | 21 | 43 | 60 | 84 |

[a]Flowing time through V-funnel.

in volume, was constructed in 2005. Following the success of this project, RFC has also been applied in the construction of an auxiliary dam in the Baoquan water-storage project. The dam is 42 m in height, in which the top 3 m are constructed with RFC. In addition, RFC has also been applied in the caisson backfill construction of the Xiangjiaba hydropower project, the third-largest hydropower station in China. The greatest depth of the caisson is 54.7 m and the total volume is 80,000 $m^3$. The consumption of RFC for this project was about 70,000 $m^3$.

## 6.7 ENGINEERED CEMENTITIOUS COMPOSITE

Engineered cementitious composites (ECC) are special types of random short-fiber-reinforced cement-based composites. The representative characteristic of ECC is its excellent pseudo-ductility and toughness. The early work at the National Science Foundation Center of Advanced–Cement-Based-Materials at Northwestern University in the United States has demonstrated that cement-based composite manufactured by the extrusion technique with a few percent of polyvinyl alcohol fibers does show great improvement in ductility (Shao et al., 1995). The work continued at HKUST showed that extruded plates can reach a large deflection under four-point bending (Li and Mu, 1998). The concept of ECC was developed by Li (1998) and Li, Wang, and Wu (2001).

The theoretical mechanism in the development of ECC was an extension from the pioneering research by the IPC group (Aveston et al., 1971), which applied fracture mechanics concepts to analyzing fiber-reinforced cementitious composite systems. In ECC, micromechanics was employed to optimize the microstructure of the composite, in which the mechanical interactions between the fiber, matrix and interface were taken into account. Marshall and Cox (1988) showed that steady-state crack propagation can prevail over the typical Griffith-type crack when the following equation is satisfied:

$$J_{\mathrm{tip}} = \sigma_{\mathrm{ss}}\delta_{\mathrm{ss}} - \int_0^{\delta_{ss}} \sigma(\delta)\, d\delta \tag{6-5}$$

where $J_{\mathrm{tip}}$ is the energy density at the crack tip; $\sigma_{\mathrm{ss}}$, the constant ambient tensile stress; $\delta_{\mathrm{ss}}$, the constant crack opening; and $\sigma(\delta)$, the closing pressure as a function of the crack opening, $\delta$. $J_{\mathrm{tip}}$ approaches the matrix toughness $K_m^2/E_m$at small fiber content. The matrix fracture toughness $K_{\mathrm{m}}$ and Young's modulus $\mathrm{E_m}$ are sensitive to the mix design, such as *w*/*c* ratio and sand content. The right-hand side of Equation 6-5 can be interpreted as the energy supplied by external work minus the energy dissipated by the deformation of the inelastic springs at the fracture process zone. The inelastic spring concept is a convenient means of capturing the inelastic processes of fiber deformation/breakage and interface debonding/slippage of those fibers bridging across the crack faces in the process zone. Hence, Equation 6-5 expresses the energy balance during steady-state crack propagation.

Figure 6-51 schematically illustrates this energy balance concept on a fiber bridging the stress–crack opening relationship. The right-hand side of Equation 6-5 is shown as the area A and is referred to as the complementary energy. Since the maximum value (area B) of this complementary energy $J_{\mathrm{b}}'$ occurs when $\sigma$ increases to the peak stress $\sigma_0$, and $\delta$ to the crack opening $\delta_0$, it implies an upper limit on the matrix toughness as a condition for strain hardening:

$$\frac{K_{\mathrm{m}}^2}{E_{\mathrm{m}}} \leq \sigma_0\delta_0 - \int_0^{\delta_0} \sigma(\delta)\, d\delta \equiv J_{\mathrm{b}}' \tag{6-6}$$

It is clear from Equation (6-6) that the successful design of an ECC requires the tailoring of fiber, matrix, and interface properties. Specifically, the fiber and interface properties control the shape of the $\sigma$–$\delta$ curve and are therefore the dominant factors governing $J_{\mathrm{b}}'$. Composite design

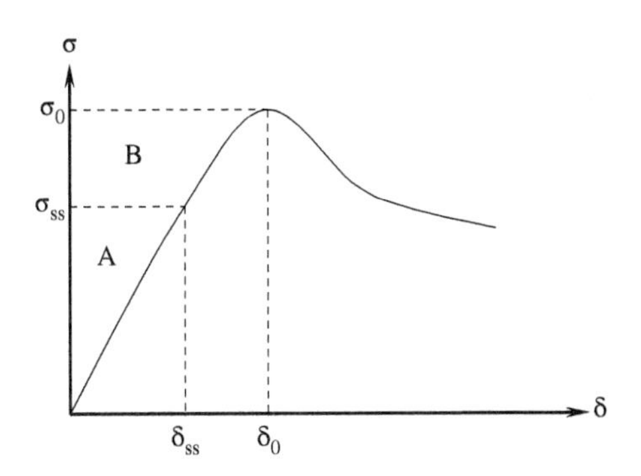


**Figure 6-51** Schematic illustration of energy balance concept on a fiber bridging stress-crack opening stress-strain curve

for strain hardening requires the tailoring of the fiber/matrix interface to maximize the value of $J_b'$. Similarly, the matrix composition must be designed so that the value of $J = K_m^2/E_m$ is not excessive.

The shape of the curve and especially the rising branch associated with $J_b'$, shown in Figure 6-51, are related to a number of fiber/matrix interaction mechanisms, including debonding and frictional pullout. It also involves many parameters such as fiber content $V_f$, diameter $d_f$, length $L_f$, and stiffness $E_f$, chemical bond $G_d$, and the interface frictional bond $\tau_0$. The peak value of the $\sigma-\delta$ curve is mainly governed by $V_f$, $d_f$, $L_f$, $\tau_0$ in the case of simple friction pull-out. An analytic expression of $\sigma-\delta$ and $\sigma_0$ obtained by Li (1992) can be used as a base for theoretical analysis of ECC. In the case when both the interface chemical bond and slip hardening are present, an analytic expression of $\sigma-\delta$ found by Lin et al. (1999) can be employed.

ECC can be made with the same ingredients as regular concrete but without the use of a coarse aggregate. ECC can be produced by mixing, in a similar way to produce traditional mortar. ECC is suitable for mass production. It can easily be molded and shaped into different products reinforced with random short fibers, usually polymeric fibers (e.g., PVA fibers). Since the introduction of the concept in the 1990s, ECC has undergone major evolution in materials development. Li and co-workers (1998; Li, Kong, and Bike, 2000; 2001) developed self-compacting ECC via a constitutive rheological approach. The high flowability was achieved with the use of a polyelectrolyte-type superplasticizer. In addition, extrusion of ECC has also been demonstrated (Stang and Li, 1999). Spray ECC, equivalent to shotcreting but replacing the concrete with ECC, has also been developed at the University of Michigan.

In mechanical behaviors, ECC exhibits tensile strain-hardening with a strain capacity in the range of 3–7%. The fiber content of ECC is typically 2% by volume or less, which allows ECC to have very good bending ductility, like a piece of metal sheet. The fact that ECC can be bent to such a large deflection is due to the distinctively coated fibers embedded in the cementitious materials that are allowed to slide within the cementitious matrix.

Because ECC has excellent pseudo-ductility, investigations have been conducted on applications of ECC in practice by leveraging this advantage. These include the repair and retrofitting of pavements or bridge decks (Kamada and Li, 2000; Lim and Li, 1997; Zhang and Li, 2002), the retrofitting of building walls to withstand strong seismic loading (Kanda et al., 1998; Kesner and Billington, 2002), and the design of new frame systems. These studies reveal the unique characteristics of ECC or R/ECC (steel-reinforced ECC) in a structural context, such as high damage tolerance in beam–column joints (Parra-Montesinos and Wight, 2000), super resistance to shear load (Kanda et al., 1998; Fukuyama et al., 1999), higher energy absorption (Fischer and Li, 2003), excellent delamination and spall resistance (Kamada and Li, 2000; Lim and Li, 1997; Zhang and Li, 2002), high deformability, and tight crack width control for durability (Maalej and Li, 1995). ECC has been used all over the world, including projects in Japan, Korea, Switzerland, Australia, China, and the United States. Many unsolved problems of traditional concrete—lack of durability, failure under severe strain, and the resulting expenses of repair—have been a driving force in the application of ECC.

## 6.8 CONFINED CONCRETE

Steel has the advantages of high strength and excellent ductility. When steel is used as a structural member, especially a compressive member, the cross-section is usually small and hence stiffness is limited. Concrete has advantages in compressive strength and stiffness as a compressive member, due to a large cross-section. Hence, by using a steel tube and concrete to form a steel tube-reinforced concrete column, the compressive strength, ductility, and stiffness of the column can easily be satisfied simultaneously. Moreover, as discussed in Chapter 5, due to the confinement of the steel tube, the performance of a concrete-filled steel tube (CFST) column is greater than that of just a steel tube plus a concrete core. The enhancement in construction speed, load-carrying capability, and ductility are superior to that of reinforced concrete columns.

The tubes used for CFSTs can be circular, square, or even rectangular in shape, as shown in Figure 6-52. The behavior of a circular CFST is much better than that of a square or rectangular CFST. Moreover, the processing method for circular tubes is usually relatively easier and cheaper. Spiral welding is always used for circular steel tube when the thickness of the plate is less than 20 mm. Under such a welding method, the quality of the weld can be guaranteed. When the thickness of the plate is greater than 20 mm, longitudinal butt welds are adopted, with only one weld necessary for a circular tube. However, for a square tube, at least two or even four welds are necessary to form a box cross-section. It is obvious that the cost of circular tube processing is cheaper than that for square tubes. More importantly, the butt weld in a circular tube carries the tensile force only during the loading period, while the butt welds in square tubes are under complex stress conditions.

CFST has been used for many construction and structural purposes, such as saving the formwork and speeding up the construction work, compressive strength enhancement for short axially loaded columns, stiffness enhancement of steel structures, and ductility enhancement for

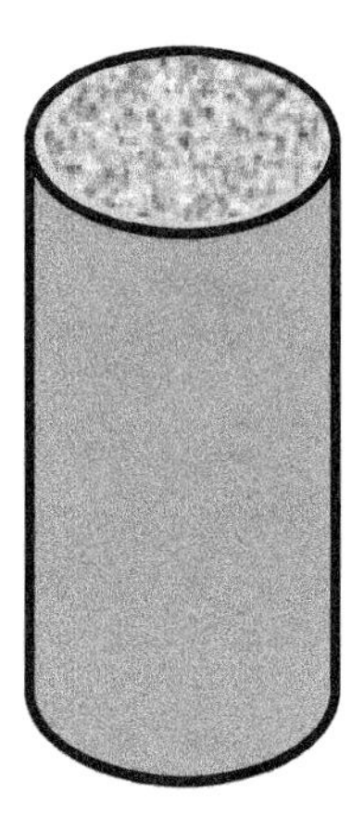
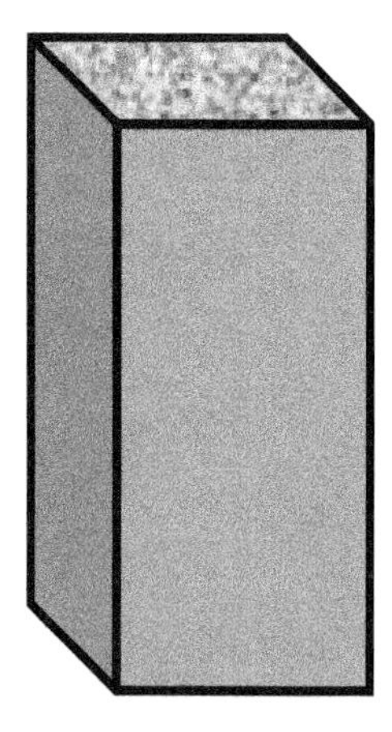
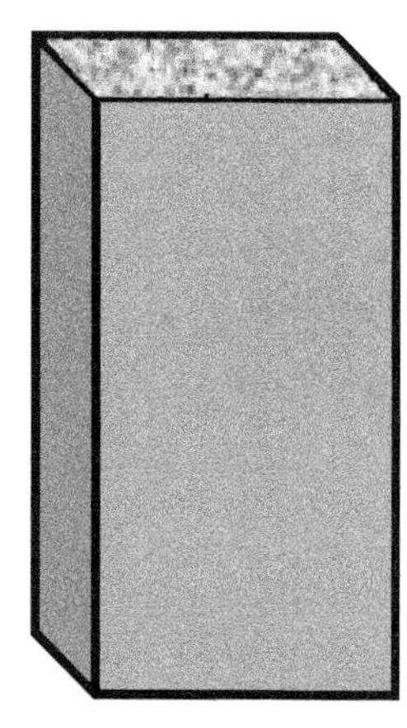

**Figure 6-52** Tubes used for CFST

seismic-resistant structures. In a CFST structural member, the concrete fill provides compressive strength and stiffness to the steel and reduces the potential for local buckling. The steel tube provides formwork to minimize the cost of the concrete placement, constrains the concrete under uniaxial compression, and reinforces the concrete for axial tension, bending, and shear, as well as providing sufficient ductility for the structure. Both circular and square CFSTs can be designed according to the "unified theory" suggested by Zhong (1995). CFST columns are particularly well suited for short axially loaded columns due to the significant increase in compressive strength of concrete because of the confinement effect of the steel tube. A good example that demonstrates the application of the short CFST in Tianjin Station, Tianjin, China, is shown in Figure 6-53.

CFST columns are particularly well suited for braced frames, where great axial strength and stiffness are needed, so braced frames have been adopted for the majority of the buildings constructed with CFST columns. Braced-frame–CFST column connections are fundamentally different from moment frame connections, since they transfer relatively large axial forces to the column, which must be distributed between the steel and concrete. The ductility demands are different in braced frames because dissipation is achieved through brace buckling rather than flexural yielding of the beams. In 2009, the Guangzhou TV Astronomical and Sightseeing Tower (aka, the Canton Tower) was topped out with a braced frame–CFST system, see Figure 6-54. The Tower is 604 m in height, and the 604-m-tall twisted, tapering tube is formed by the rotation between two ellipses, which form a "slim waist" in the center of the building. The tower was completed and became operational in September 2010.

The CFST structure is also suitable for arch structures due mainly to compressive forces that are generated in the structure. In recent years, CFST arch bridges have been widely applied in China. More than 30 arch bridges, with spans longer than 200 m, have been built. The CFST arch bridge with the longest span of 460 m (i.e., the Wushan Bridge), shown in Figure 6-55, was built in Chongqing, China.

**Figure 6-53** Application of CFST as short axially load column in Tianjin Station, China (Photo provided by Linhai Han)

**Figure 6-54** Guangzhou TV Astronomical and Sightseeing Tower: (a) under construction; and (b) site view of finished tower (Photo provided by Linhai Han)

**Figure 6-55** Wushan CFST arch bridge, Chongqing, China (Photo provided by Linhai Han)

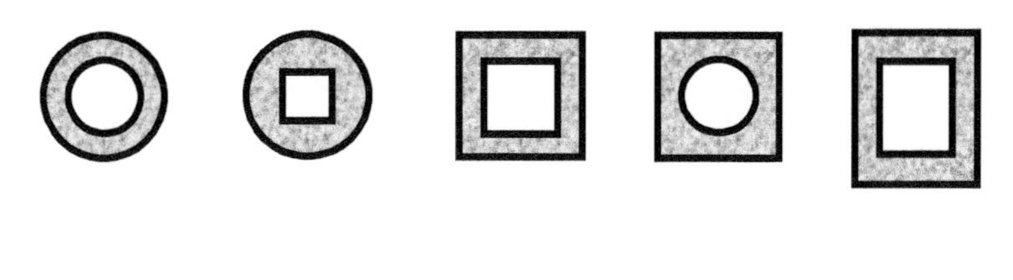

(a) Cross-sections of hollow CFST

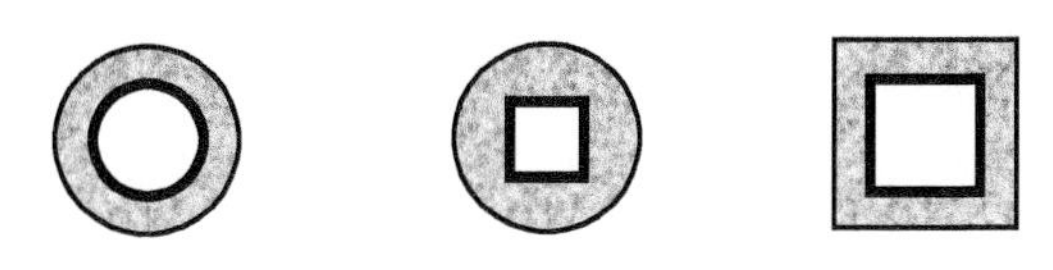

(b) Cross-sections of fibrous-composite-tube (Outside) reinforced concrete with steel inside

**Figure 6-56** Typical cross-sections of hollow CFST or composite concrete-filled tubular members

Recently, hollow concrete-filled tubular structures or composite concrete-filled tubular structures have attracted great attention in both research and practice. The advantages of hollow CFST over solid CFST are: (1) saving materials and reducing self-weight; (2) providing room for utilities, such as cables or pipe installations; (3) increasing the effectiveness of the cross-section (such as a higher moment of inertia); and (4) easier and more economic assembly. Typical cross-sections of a hollow CFST or composite concrete-filled tubular member are shown in Figure 6-56. Past studies have shown that the structural behavior of hollow CFST members is highly influenced by the coefficient of the confining effect of the steel tube on the concrete circular core, and the ratio of the empty area to the solid area (Huang et al., 2010). Typical failure modes of hollow CFST members are shown in Figure 6-57.

Due to the advantages of hollow CFST, it has been rapidly used in practice, especially for tall bridge piers. In China, more than 10 bridge piers higher than 100 m have used composite-type hollow CFST piers shown in Figure 6-58. This CFST structure includes a steel tube having an inner surface and a concrete core cast within the steel tube. Under normal conditions, the concrete core is bonded with the inner surface of steel tube to form a unique structure. However, a separating layer interposed between the inner surface of the steel tube and the surface of the concrete core may form so that the steel tube may not be bonded to the concrete core. Such a layer may be caused by shrinkage or bleeding. To eliminate the separating layer, expansive concrete is preferred in such a tube-confined concrete structure.

## 6.9 HIGH-VOLUME FLY ASH CONCRETE

Fly ash is a by-product of an electricity generating plant using coal as fuel. It used to be disposed of as industrial waste. Because a large amount of fly ash can be produced by power plants every year, fly ash disposal became a major environmental issue in many countries. To solve the disposal problem, since the early 1960s, many countries have started to incorporate fly ash into concrete. It was found that the incorporation of fly ash has certain advantages and some disadvantages as a result of research carried out for the past 50 years or so. The major advantages of fly ash concrete are low cost, low $CO_2$-embodiment and energy consumption, low hydration heat, and improved long-term strength and durability. The disadvantages of fly ash concrete are longer initial setting time and low early-age strength. Depending on the amount of fly ash incorporated in the concrete,

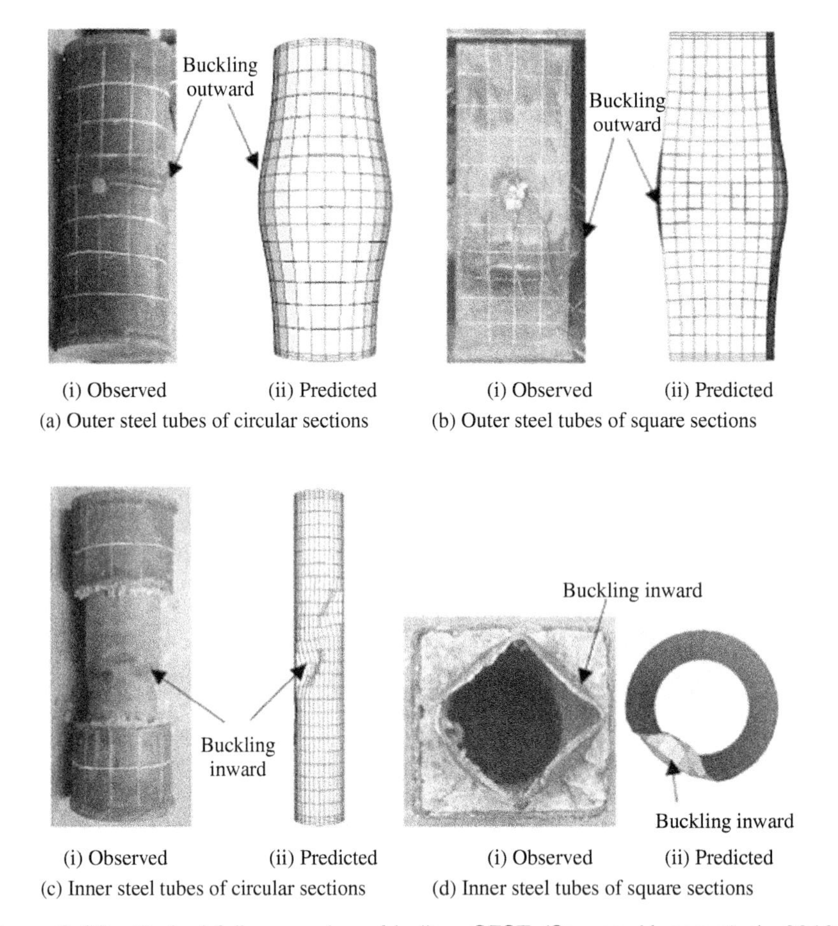


**Figure 6-57** Typical failure modes of hollow CFST. (Source: Huang et al., 2010 / With permission of Elsevier)

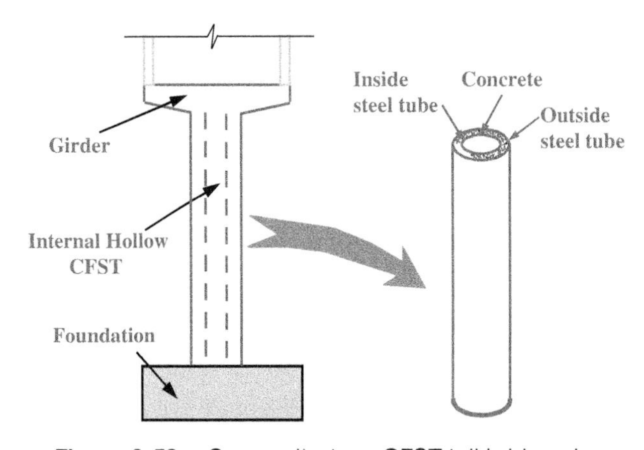


**Figure 6-58** Composite-type CFST tall bridge pier

there are two categories of fly ash concrete: regular fly ash concrete and high-volume fly ash concrete. The regular fly ash concrete utilizes about 25–30% of fly ash by weight of cementitious binder and is widely used for massive structures, such as gravity dams. Attempts have also been made in many countries to use a large amount of fly ash in concrete without losing its early-age compressive strength. The high-volume fly ash (HVFA) concrete can use fly ash to replace 60% or more by weight of cement. Desirable mechanical and durability properties of HVFA concrete have been achieved by careful selection of mix proportions and utilization of chemical admixtures such as superplasticizers (Langley et al., 1989). The useful features of HVFA concrete are low Portland cement content (about 150 kg or less per cubic meter), which means fly ash concrete can be produced at less cost and less carbon footprint than conventional concrete; high early-age and ultimate strengths, so that some HVFA concrete's compressive strength is greater than 45 MPa at 28 days, with early-age strength in the range of 30–35 MPa at 7 days; and high durability in chemically aggressive environments. Arezoumandi et al. (2013) have also shown that the normalized shear capacity of the HVFA concrete—in which fly ash replaced 70% cement—can be higher than the fly ash-free reference concrete.

## 6.10 STRUCTURAL LIGHTWEIGHT AND HEAVYWEIGHT CONCRETE

Structural lightweight concrete is defined as a concrete having a compressive strength in excess of 17 MPa and a bulk density less than 1950 kg/m$^3$. To make lightweight concrete, lightweight aggregate (expanded clay, expanded shale, expanded perlite, zeolite, cenosphere, etc.) has to be used. Nowadays, structural lightweight concrete can be more than 25% lighter than normal-weight concrete while yielding a compressive strength of 60 MPa or higher (Guo et al., 2000; Huang et al., 2018). The coefficient of thermal expansion of structural lightweight concrete can be much lower than that of normal-weight concrete, with a typical value of $7 \times 10^{-6}$/°C; and the insulating thermal conductivity of lightweight concrete can be over 25% lower than that of normal concrete. The improved fire resistance of structural lightweight concrete, as widely reported, can be attributed to a lower coefficient of thermal expansion and a smaller reduction of strength at elevated temperatures. Structural lightweight concrete has been applied in building and infrastructure construction since the 1920s. In the United States, Europe, and China, many bridges and tall buildings have used high-performance lightweight concrete. Surprisingly, some structural lightweight concretes have shown enhanced durability in chemical resistance, frost resistance, and permeability reduction (Haque et al., 2004; Vijayalakshmi and Ramanagopa, 2018).

Heavyweight concrete is defined as a concrete having bulk density in the range of 3360–3840 kg/m$^3$, and is used for special purposes. Heavyweight concrete has been used for the prevention of seepage from radioactive structures due to the harmful effect of radioactive rays (x-rays and gamma rays as well as nuclear radiation) on living organisms. It is mainly used in laboratories, hospitals, and nuclear power plants. The most important point about heavyweight concrete is the determination of the *w/c* ratio. The favored *w/c* ratio for heavyweight concrete is 0.40. Another important point for heavyweight concrete is that the cement dosage should be both high enough to allow for radioactive impermeability and low enough to prevent splits originating from shrinkage. The recommended cement dosage should be greater than 350 kg/m$^3$.

Specially designed heavyweight concrete can also be used to provide soundproofing (Li et al., 2003). In this type of concrete, the aggregates may be made of heavyweight metal balls coated with a layer of soft rubber as shown in Figure 6-59. In fact, such aggregates can play the role of a local resonance unit, and with this effect, sound at certain frequencies can be effectively blocked. Figure 6-60 shows experimental results of sound shielding with different construction materials, including gypsum board, concrete plate, and plastics. It can be seen that at around 180 Hz, the sound transmission loss for heavyweight concrete reaches 45 dB while for all other materials is

Figure 6-59 Heavyweight concrete for soundproofing

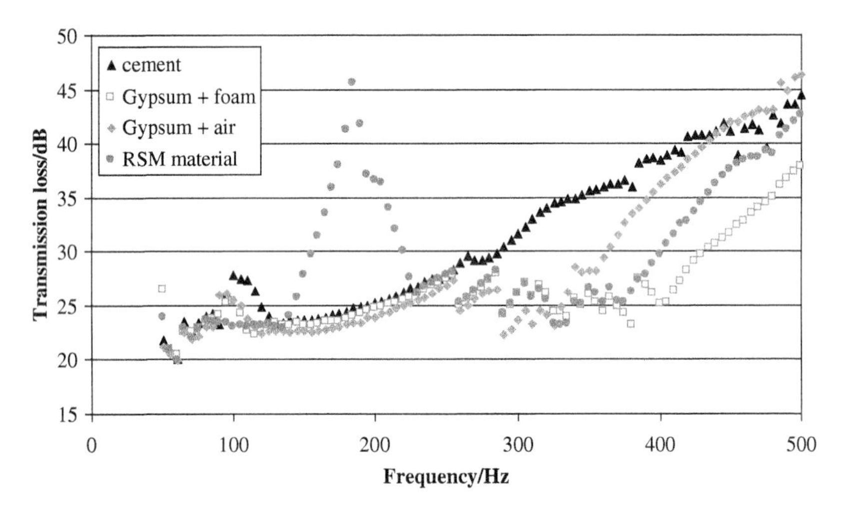


Figure 6-60 Experimental results of sound shielding with different construction materials

25 dB. This implies that the ability of the heavyweight concrete in sound shielding at a frequency of 180 Hz is 100 times higher than that for other building materials.

## 6.11 SEA SAND AND SEA WATER CONCRETE

Following the continuous increase in infrastructure construction, the whole world is facing a shortage of river sand for concrete production (Bendixen et al., 2019). Scarcity of fresh water as mixing water is also becoming a supply chain issue in some regions. In specific areas, especially coastal regions, using sea sand and sea water for concrete making is a potential solution for these problems.

However, mainly because of the chloride salts in sea sand and sea water, they are conventionally prohibited as concrete raw materials in most countries and regions. Actually, many premature deteriorations of reinforced structures and earthquake-induced collapse or damage of buildings (e.g., in Taiwan, Shenzhen, Japan, Turkey, and Ecuador) have been attributed to the use of sea sand and/or sea water (Çağatay, 2005; Yépez and Yépez, 2017). In recent years, there has been burgeoning interest in exploring the feasibility of using sea sand (especially washed sea sand) and sea water to make concrete, and the core of these efforts is to elucidate the potential effects of sea sand and sea water on the hydration process, fresh properties, the microstructure, the mechanical properties, and the durability, as well as to figure out how to effectively manage the potential adverse effects.

Details of sea sand and sea water were introduced in Chapter 2. In brief, both contain certain levels of deleterious materials with respect to concrete performance. Major ions in sea water that are potentially deleterious to concrete include chloride, sulfate, and magnesium. While chloride is well known to accelerate corrosion of steel reinforcement and sulfate induces expansive cracking of concrete by forming excess gypsum and ettringite, magnesium ions may tend to destabilize C–S–H gel (Zhao et al., 2021). In addition, sulfate and chloride salts associated with magnesium cation are known to be more harmful than calcium and sodium salt (Smith et al., 2019). The salts in sea water can be adsorbed on the surface of sea sand, and more or less remain even after washing. These salts can be brought into the concrete and cause damage if sea sand is used as the fine aggregate. The physical properties of sea sand may vary significantly from region to region, but the chloride content of sea sand (unwashed) mainly depends on fineness: the finer the sand, the higher the chloride content (Limeira et al., 2010). The chloride content in wet, fine dredged sea sand from off-shore zones can be higher than 7%; however, after washing in fresh water, the chloride content can be reduced to below 0.015% (Dhondy et al., 2020; Sun et al., 2016). In Ningbo and Shenzhen, China, the sea sand is dredged from off-shore or estuary zones, transported upstream in vessels, washed using river water, and stockpiled along the river. This practice makes the production of sea sand economical. Apart from salts, sea sand can contain other deleterious materials, such as seashells, organic matter, and soft materials (e.g., clay). Seashells, while sometimes reported to benefit the workability of concrete, can compromise the mechanical properties of concrete (especially in high contents) (Zhao et al., 2021).

Using sea sand and/or sea water in concrete does not significantly change the process of cement hydration, which is normally divided into dissolution, induction (dormant), acceleration, deceleration, and stable stages. It has been found that using sea water as mixing water may lead to slightly faster dissolution of $C_3A$ and gypsum, thus accelerating the dissolution stage; shortening the induction stage, thus leading to shorter setting times; and resulting in a higher heat release peak in the acceleration stage (Wang et al., 2018). As a result, the early-age hydration heat of sea water concrete is normally higher than its fresh water counterpart; however, this "advantage" tends to vanish after 7 days. How sea water affects the total hydration heat of cement may depend on the *w/c*: at a low *w/c* (e.g., 0.2), using sea water tends to generate lower total hydration heat; and at a high *w/c* (e.g., >0.4), using sea water may lead to more total hydration heat than fresh water (Li et al., 2018; Wang et al., 2018). Sea sand affects cement hydration in a similar way, but to a smaller extent, as the salt content is relatively small (Ma et al., 2007b). Due to the influence of sea sand and sea water on the hydration process, one can observe differences in microstructure, such as volume fractions of ettringite and calcium hydroxide at certain ages (Li et al., 2020). However, the biggest characteristic of sea sand/sea water concrete should be the occurrence of flake-like Friedel's salt produced by the reaction between chlorides and the aluminate phases in cement (Cheng et al., 2018).

Regarding the influence of sea sand and sea water on the compressive strength of concrete, many contradictory results can be found in the literature. In general, sea water tends to improve early-age (up to 28 days) compressive strength, and the effect of sea sand on early-age compressive strength of concrete is distributed between −20% and +20%. The long-term compressive strength of sea sand and sea water concrete can be comparable to that of the reference concrete made of river sand and fresh water (Zhao et al., 2021). It was reported recently that sea sand and sea water have positive effects on the flexural strength and dynamic modulus of concrete, but the effect on the static modulus of elasticity is negligible (Du, Yaseen, et al., 2021).

Because of the salts that are introduced into concrete, sea water could induce larger autogenous shrinkage and drying shrinkage than fresh water concrete (Vafaei et al., 2021; Younis et al., 2018). Microscopically, this could be attributed to the larger volume fractions of gels in sea water concrete (Du, Yseeen, et al., 2021). The effect of sea sand on the shrinkage of concrete is insignificant.

There are many contradictory results in the literature regarding the effects of sea sand and sea water on durability (represented by carbonation depth, resistance to freeze-thaw cycles, sulfate attack, permeability, etc.) of concrete. However, most of the identified effects are insignificant (Du, Yaseen, et al., 2021; Vafaei et al., 2021; Xiao et al., 2017). The major concern of using sea sand and sea water regarding durability is that the introduced chloride ions can accelerate the corrosion of steel reinforcement. It has been found that, given the same total amount of chloride, the harmfulness of sea sand is less than sea water, since the chloride ions adsorbed on sea sand particles cannot be completely dissolved into the pore solution (Ma et al., 2007a). If the introduction of excess chloride ions cannot be avoided, the use of fly ash, slag, and metakaolin as supplementary cementitious materials can effectively improve the chloride binding capacity of the concrete and mitigate the adverse effects on steel reinforcement. Furthermore, led by the University of Miami, an international project entitled SEACON has demonstrated the safe use of sea sand and sea water as concrete raw materials, where bars made from noncorrosive materials (e.g., fiber-reinforced polymers and stainless steel) are used as reinforcement (Morales et al., 2021; Xiao et al., 2017). In addition, using anti-corrosion cementitious binders (e.g., magnesium phosphate cement, which has been proven to passivate steel bars) to replace Portland cement may be another possible solution to the corrosion of steel rebar induced by sea sand and sea water.

## 6.12 THE 3D PRINTED CONCRETE

Additive manufacturing (AM) is a disruptive technology that can integrate digitalization and automation in construction. Its application greatly compensates for the shortage of skilled laborers, resources, and construction efficiency. The application of large-scale AM with cement-based "inks" in the field of civil engineering is also known as 3D printing of concrete (3DPC). A concise definition of 3D printing technology in construction is "transforming an imagination of a facility, in whole or in part, depicted through a computer model, into a real facility (bridges, highways, buildings, etc.), with least human involvement and most conservation of natural resources" (Khan et al., 2020). Using 3DPC, innovators all over the world have been creating buildings, structures, components, and artworks. Figure 6-61 shows two 3D printed concrete components. The state of the art of 3D printed concrete, covering materials and fresh properties, the construction process (including reinforcement installation), and hardened-state properties, is briefly reviewed in this section.

### 6.12.1 Materials and Key Fresh-State Performance

The 3DPC is a highly adaptive process. Apart from regular Portland cement-based materials, alternative cements (Cao et al., 2016; Huang et al., 2019), geopolymer-based materials (Guo et al., 2020; Panda et al., 2018), and polymer-based/fiber-reinforced composites (Kabir et al., 2020; Wang et al., 2017) can all be processed through 3D printing. In the literature about "cement"-based 3D

**Figure 6-61** Concrete components fabricated using 3D printing

printing, nearly 80% of the published studies used Portland cement as the major printing material; and geopolymer/alkali-activated materials and alternative cements (e.g., magnesia cements and sulfoaluminate cements) each took a percentage of around 10%. In most cases, 3D printing is a process where an "ink" material is deposited, joined, or solidified (hardened) layer-by-layer under computer control to create a 3D object. The 3DPC is usually an easy production process that is not limited by formwork and vibration processes. Also, construction wastes, costs, and time can be greatly reduced with 3DPC (Zhang et al., 2019). However, it should also be noted that these methods put stricter demands on cement-based materials' workability. Designers of 3D printed concrete often face contradictory requirements: the fresh mixture needs to be fluid enough to be mixed, pumped, and extruded, while still be consistent enough (after being printed) to maintain the intended shape without external support. Apart from the slump, slump flow, and rheological properties well discussed in Chapter 3, Section 3.2, the setting and hardening behaviors (e.g., setting on demand), pumpability, extrudability, and buildability also need to be taken into consideration (Khan et al., 2020; Ma and Wang, 2018; Reiter et al., 2020). The desired workability of 3D printed concrete can normally be tailored and/or optimized by using mineral additives (e.g., micro-silica), chemical admixtures (e.g., superplasticizer, viscosity modifying agents, and accelerator/retarder), and fibers. Herein, the evaluations of pumpability, extrudability, and buildability of 3D printed concrete are summarized.

**(a)** *Pumpability*: Pumpability measures how easy the fresh concrete can be transported in the printing device (e.g., in pipelines) to deliver the desired extrusion performance in a later phase. Although this property is known to depend on the composition of the concrete, it is difficult—if at all possible—to predict the pumpability from the composition. More practically, the pumpability of specific concrete mixtures needs to be measured (Nerella and Mechtcherine, 2019). For the evaluation of pumpability, a dedicated test device, i.e., the sliding pipe rheometer (aka, "sliper", shown in Figure 6-62), has been used (Nerella and Mechtcherine, 2019). Based on this method, the pumpability is measured by filling the top position of the upper pipe with fresh concrete, and the pipe slides down under the self-weight of both the concrete and the pipe. If flow speed needs to be considered as a factor, extra weights can be added on the top to accelerate the sliding. Displacement sensors are installed to measure the sliding speed of the pipe, which is used to reflect the flow speed of the concrete ($Q$); and the pressure at the piston head, which can be measured by a pressure sensor, reflects the pumping pressure ($P$) of the concrete. Eventually, a graph can be plotted with the measured parameters ($P$-$Q$). According to the slope and $P$-intercept of the $P$-$Q$ diagram and in combination with the specifications of the pumping circuit, the pumpability of the

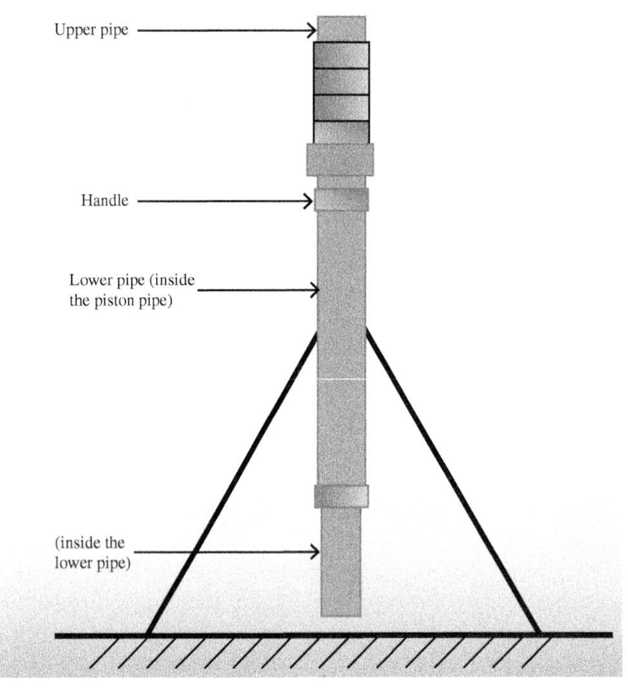


**Figure 6-62** Sliding pipe rheometer (sliper). Source: Based on Nerella and Mechtcherine, 2019.

concrete can be evaluated. The required discharge pressure $P$ for a specific pumping circuit can be estimated as (Nerella and Mechtcherine, 2019)

$$P = \frac{4L}{D}a + \frac{16LQ}{\pi D^3}b + \rho g H \tag{6-7}$$

where $D$ is the diameter of the pipeline, $H$ is the pumping height, $L$ is the length, $\rho$ is the density of concrete, and $a$ and $b$ are constants obtained by fitting.

**(b)** *Extrudability*: The extrudability measures the ability of fresh concrete, when being delivered to the printhead through pipelines, to pass through the nozzle with minimal energy input and without blockage. This property appears to be determined mainly by the viscosity of the cement-based material and the size of aggregate. The extrudability of visco-plastic materials like concrete can be quantified by ram extruders, the squeeze-flow method (Toutou et al., 2005), or the penetration resistance method (Chen et al., 2006). However, the true effectiveness of these methods has not been clear, especially these methods may tend to oversimplify the flow field and shear history in a realistic printer, leading to systematic errors. Recently, a novel extrudability test method has been developed with the help of a 3D printing test device at TU Dresden. This method intends to determine the energy required for the concrete to pass through the nozzle on the 3D printing device; and the energy is calculated from the electric power consumption that is associated with the concrete flow rate at various rotational speeds (Nerella and Mechtcherine, 2019).

**(c)** *Buildability*: The buildability of 3D printed concrete is the resistance of a printed layer to deformation caused by its self-weight and the weight of subsequent layers. Buildability is significantly affected by the time-dependent rheological properties of the concrete. In the layer-by-layer construction process, the pressure on the bottom concrete can be regarded as the vertical stress of $\rho gh_{max}$ (Perrot et al., 2016) or shear stress of $\rho gh_{max}/\sqrt{3}$ (Wangler et al., 2016), when the structure's height reaches a highest point of $h_{max}$. Based on these theories, by comparing time-resolved vertical pressure with the corresponding static yield stresses (vertical and shear) of the printed concrete, rheology-based failure criteria can be formulated. This idea can be expressed by Equation (6-8) for vertical stress and Equation (6-9) for shear stress (Nerella and Mechtcherine, 2019):

$$\rho gh(t) \leq \alpha_g \left( \tau_{0,0} + A_{thix} t_c \left( e^{\frac{t}{t_c}} - 1 \right) \right) \tag{6-8}$$

$$\rho gh(t) \leq \sqrt{3} \left( \alpha_g \left( \tau_{0,0} + A_{thix} t_c \left( e^{\frac{t}{t_c}} - 1 \right) \right) \right) \tag{6-9}$$

where $t$ is time; $\alpha_g$ is the geometry factor; $\tau_{0,0}$ is yield stress at the beginning of rest-time; $A_{thix}$ is the yield stress increasing rate; $t_c$ is a characteristic rest-time, after which static yield stress grows exponentially.

In addition to pumpability, extrudability and buildability, the term "printability" has also been used widely to describe the fresh properties of 3D printed concrete. However, this concept has not been defined explicitly so it cannot be quantified. In most cases, printability seems to be used to describe the ability of the material/device combo to produce a well-controlled filament of the extruded concrete (Wangler et al., 2019).

### 6.12.2 Construction Processes

The process of 3D printing in the construction industry can generally be divided into two categories based on how the solid and liquid materials are fed into the printing process: (1) extrusion-based 3D printing; and (2) powder-based 3D printing. For the former category, all solid and liquid raw materials are mixed before being poured into the printing device, and then extruded out of the printing nozzle to the printing platform. Typical industrialized techniques in this category include Contour Crafting (CC) and Concrete Printing (CP). Techniques in the latter category manipulate the solid and liquid raw materials in a different way: the printing nozzle injects a liquified phase (e.g., water or slurry) into a "powder bed" of the majority of the solid materials where chemical reactions are forced to occur, leading to the formation of the desired solid skeleton. D-Shape (DS) is a representative industrialized technique in the latter category. The three types of techniques (i.e., CC, CP, and DS) are introduced below, and their similarities are shown in Figure 6-63.

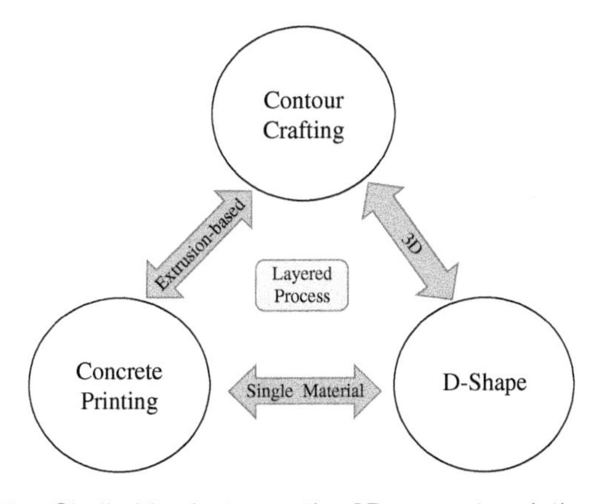


**Figure 6-63** Similarities between the 3D concrete printing processes

**(a)** *Contour Crafting*: CC is a free-form construction method that constructs concrete buildings through the extrusion of cementitious material. This technology has been in development for more than 20 years (Khoshnevis and Dutton, 1998). It extrudes a cement-based material layer-by-layer, and a trowel is attached to the nozzle to achieve a smooth lateral surface finish following the build-up of subsequent layers. The CC technology is tolerant to a variety of materials (such as mortar, concrete, cement pastes, and fiber-reinforced concrete), and it is also adaptive to various types of printers (e.g., gantry, frame, robotic arm, polar, and delta, as shown in Figure 6-64). Because of the advantages of this technology, it was selected by NASA for its Innovative Advanced Concepts (NIAC) to explore the feasibility of building a Lunar Settlement infrastructure (Khoshnevis et al., 2012). Many projects have been constructed using contour crafting. In 2014, the Chinese company, WinSun, built 10 basic houses in less than 24 hours, with the area and cost of each being about 195 $m^2$ and US$4800, respectively. The company used a large extrusion-based 3D printer to manufacture the basic house components separately off-site before they were transported and assembled on site (Hock, 2014). In 2015, the company built a five-story apartment building with a total area of about 1100 $m^2$, which is so far the tallest 3D printed structure in China. The researchers at the University of Naples Federico II, Italy, used a 4-m high BIGDELTA WASP (World's Advanced Saving Project) printer to build the first modular, reinforced-concrete beam of about 3 m long. With this WASP printer, the researchers have developed a system to produce concrete elements that can be assembled with steel bars (WASP, 2017). Using the same type of printer, Supermachine Studio and the Siam Cement Group collaborated to build a 3 m tall cave structure called the "Y-Box Pavilion, 21st-century Cave" in Thailand.

**(b)** *Concrete printing*: Lim et al. (2009) pointed out some limitations of CC regarding its excessive steps, and then proposed the CP technology to overcome these limitations to some extent (Lim et al., 2012). Similar to CC, the CP is also based on the extrusion of concrete, but a support material is used to improve the 3D freedom and enable better control of internal and external geometries. The CP technology has been used to manufacture Bloom, which was a 2.74-m-tall, freestanding tempietto composed of 840 customized 3D printed blocks. Each block was printed using a farm of 11 powder 3D printers with an optimized cement composite formulation. The blocks were held in place using stainless steel hardware and assembled into 16 large, lightweight, prefabricated panels that could be assembled in just a few hours (Rael et al., 2015). The technology was also used to manufacture Shed, which was a small prototype building constructed with modular 3D printed building blocks (Rael and Fratello, 2017).

**(c)** *D-shape*: Dini (2007) invented a large-scale powder bed-based 3D printing technology named D-shape, which is structurally similar to stereolithography. This process uses a powder deposition process by which the powder is selectively hardened using an activator

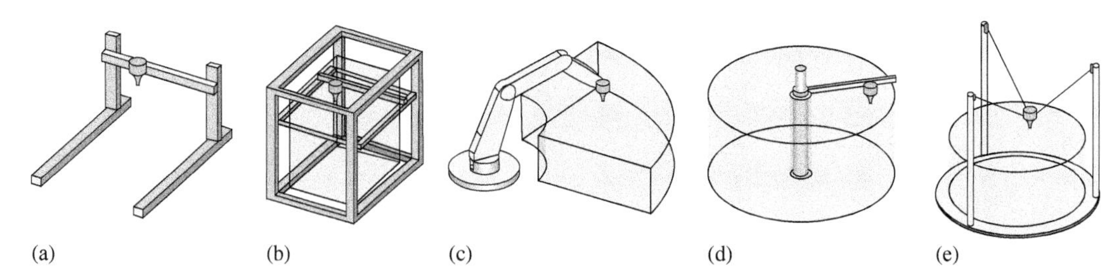


**Figure 6-64** Existing motion systems for extrusion-based 3D printing of concrete: (a) gantry; (b) frame; (c) robotic arm; (d) polar; and (e) Delta

(e.g., an alkali solution), reactant (e.g., water), or a binder (e.g., cement paste). Once a layer of the powder material is laid to the desired thickness, it is compacted properly, and then the automatically controlled nozzle applies the activator, reactant, or binder. The activator/reactant/binder is applied only in positions where the building material should become solid and the rest is kept loose so it can be removed at a later stage. Once the designed structure/object is completed, it can be dug out of the loose powder bed. The process has been used to create "Radiolaria," the tallest printed sculpture, in 2009.

The 3D printing processes cannot improve the inherent brittleness of cement-based materials (represented by low tensile/flexural-to-compressive strength ratios). To build safe structures, reinforcements (micro or macro) are indispensable to ensure sufficient structural toughness. Many researchers have paid attention to toughening 3D printed concrete using fibers (e.g., steel, glass, basalt, and synthetic fibers), and it has been widely reported that the fibers can be oriented in the flow direction to effectively improve the tensile strength along the direction of the fibers/filament. Well-designed fiber-reinforced cement-based materials can definitely achieve strain-hardening, as demonstrated in earlier sections of this chapter. Such materials may be promising if challenges such as performance sensitivity to mixing procedures can be overcome. However, these materials cannot serve as structural materials on their own without macro-reinforcement since that does not meet current structural design codes.

Realizing that 3D printed concrete (even reinforced with fibers) can hardly serve as the primary structural material, many 3D printing projects actually used their technology to print concrete "permanent formwork." Conventional steel bars/cages were inserted into the 3D printed formwork, followed by conventional concrete casting (Wu et al., 2016). One example of this practice is WinSun's five-story apartment building, printed in 2015. An alternative method of adding macro-reinforcement into 3D printed concrete structure is to place—manually or automatically (e.g., using a computer-controlled welding robot)—reinforcement (e.g., rebar, cage, mesh, or textile), first, and then extrude concrete on or around the reinforcement (Hack and Lauer, 2014; Pfändler et al., 2019). The concrete printing process is adaptive to this type of technology. Some recently developed integrated printing technologies have attempted to entrain steel cables or spiral reinforcement into the extruded concrete filament (Bos et al., 2017; Lim et al., 2018). This could mimic conventional reinforced concrete to some extent, but it is hard to introduce stirrups. In the future, more innovations regarding reinforcement of 3D printed concrete are desired to improve the construction efficiency and structural toughness. Whether reinforcement, especially macro-reinforcement, can be effectively and efficiently integrated into 3D printed concrete will determine the scale of application of 3DPC to a large extent.

### 6.12.3 Hardened-State Properties

Theoretically, the hardened-state properties (e.g., strengths and durability properties) of 3D printed concrete are mainly determined by the binder material and mixture proportions. Therefore, these properties can be tailored through material design. However, these properties may be affected by the 3D printing processes. Lim et al. (2012) suspected that the compressive strength and flexural strength of extruded material are lower than the cast counterpart. Nevertheless, more recent test results have shown that the 3D printed (extrusion-based) concrete could have higher compressive strength (>80 MPa vs 73 MPa) and flexural strength (5.8 MPa vs 5.1 MPa) than the corresponding cast specimen at the same age (i.e., 21 days) (Nerella and Mechtcherine, 2019). This improvement in strength could be attributed to the pressure applied to extrude the concrete which densifies the microstructure and excludes some macro-defects.

It is a general assumption that the inter-layer bonding is the weakest link in the structure of 3D printed concrete, so the lateral load-carrying capacity of 3D printed concrete may be lower than the vertical load-carrying capacity. However, this may only be an issue when direct shear is applied in parallel to the interface plane. According to the test results of Nerella and Mechtcherine (2019), no significant differences of compressive and flexural strengths could be measured along the vertical and lateral directions. This confirms that, to a large extent, satisfactory inter-layer bonds could lead to isotropic material behaviors (in spite of structural anisotropy). The inter-layer bond strength can be related to several factors, including major ones (e.g., interlayer time interval and surface moisture content) and minor ones (e.g., print head speed and nozzle height). The interlayer time interval measures the time gap between the printings of two sequential layers. A longer time interval tends to reduce the inter-layer bond strength (Wangler et al., 2019). Panda et al. (2018) and Kim et al. (2017) found a critical time window, within which the influence of time interval on inter-layer bond strength is negligible and beyond which the influence can be significant. Surface moisture content can also potentially affect the inter-layer bond significantly (by up to 50%) (Keita et al., 2019; Wolfs et al. 2019). Normally, protecting the printed concrete from drying helps achieve a better interfacial bond. The minor factors do not have a significant influence on the inter-layer bond, and findings on these factors in the literature are often inconsistent. If one intends to improve the inter-layer bond in addition to manipulation of the above-mentioned factors, a primer, an additional layer of cement mortar, or an accentuated top surface of the filament could be applied (Wangler et al., 2019).

The long-term stability and durability of 3D printed concrete—though essential for large-scale deployment of this technology—have not been adequately addressed. Very limited studies can be found in the literature. One may hypothesize that the incorporation of large amounts of chemical admixtures, especially aggressive admixtures (e.g., accelerators), may lead to detrimental effects not only to the long-term stability of concrete but also to the durability of the steel reinforcement. It could also be hypothesized that 3D printed concrete itself has superior durability to its cast counterpart, since it has been shown that the extrusion can increase compressive strength (Nerella and Mechtcherine, 2019), which normally implies a denser microstructure and lower permeability. These hypotheses will need to be tested in future studies. When it comes to durability of 3D printed concrete structure, the inter-layer bond seems to play a key role. Past studies have shown that poorer inter-layer bonding (e.g., due to prolonged interlayer time interval) can lead to increased ingress of water and chloride (Schröfl et al., 2019; Wangler et al., 2019). Therefore, manipulation of the above-discussed factors affecting the interlayer bond is key to achieve superior durability for 3D printed concrete structures.

To sum up, the mechanical and durability properties of 3D printed concrete have not been adequately addressed, compared to studies on mix proportion optimizations and printing processes. To further promote and deploy this technology at scale, in the near future, holistic evaluations of the hardened-state properties as well as development of methodologies to enhance the mechanical and durability properties of 3D printed concrete materials and structures are required.

## DISCUSSION TOPICS

Which factors influence the properties of fiber-reinforced concrete?

What is the most beneficial effect of incorporating fibers into concrete?

What are two types of responses in tension of fiber-reinforced concrete?

Define the toughness index in bending for fiber-reinforced concrete.

Define ultra-high-strength and high-strength concrete.

What are the major differences between normal-strength and high-strength concrete?

What is MS concrete?

What does DSP stand for?

What are the similarities and differences between MS and DSP?
What does MDF stand for?
Why has MDF not yet become popular?
How many common ways are there to incorporate polymer into concrete?
What are the main applications for polymer-modified concrete?
How can shrinkage be compensated for?
Can you think of a case suitable for the application of fiber-reinforced concrete?
Can you design a fiber-reinforced concrete with strain-hardening behavior?
How can you improve concrete strength using micro engineering?
Suppose that you are a civil engineer. Which materials would you recommend for the emergency repair of a concrete road?
What aspects should you pay attention when you apply shrinkage-compensating concrete?
What is the mechanical background in the design of ECC?
Why can ECC sustain a large deformation without collapse?
Define SCC.
Define UHSC.
How can high flowability be achieved for SCC?
How can the viscosity or cohesiveness of SCC be improved?
List two common types of SCC and discuss the differences between them.
List a few examples in which ECC has obvious advantages over common concrete.
Discuss the advantages of UHSC. What will be the potential of the applications of UHSC?
What is CFST?
Discuss the advantages of CFST.
What are the main functions of a tube and concrete in a CFST?
Discuss the differences between a short, axially loaded CFST and a slender CFST column with bending.
What are the advantages and disadvantages of high-volume fly ash concrete?
List some applications of lightweight concrete.
List some applications of heavyweight concrete.

## PROBLEMS

1. A continuous fiber reinforced cement specimen has the following properties. The Young's modulus of the cement paste is 20 GPa and the tension strength is 6 MPa. The Young's modulus of the fiber is 210 GPa. Vf is 5% and a total of 60 fibers with a diameter of 0.4 mm are used. The specimen started to carry a uniaxial load at its age of 60 days (the free drying shrinkage strain of cement paste at 60 days is 0.00038).
   **(a)** At which load, will the cement paste get its first crack?
   **(b)** Draw the stress distribution in a fiber after cement paste cracks.
2. A continuous fiber-reinforced cement specimen is loaded in uniaxial tension. The Young's modulus of the cement paste is 20 GPa and the tension strength is 5 MPa. The Young's modulus of the fiber is 210 GPa.

**(a)** If Vf is 1% and a total of 50 fibers with a diameter of 0.2 mm are used, when the stress in FRC reaches 5 MPa, what is the stress in cement paste and fiber?

**(b)** At which stress in FRC will a crack appear in the cement paste? Draw the stress distribution in the fiber and in the matrix along the fiber length, starting from the matrix crack position right after the cement paste cracks (suppose that the force incremental is negligible during the crack development).

**3.** A continuous fiber-reinforced cement specimen is loaded in uniaxial tension. The Young's modulus of cement paste is 20 GPa and the tension strength is 5 MPa. The Young's modulus of the fiber is 210 GPa.

**(a)** When the cement paste reaches its tensile strength, what is the stress in the fiber?

**(b)** If $V_f$ is 0.3% and a total of 25 fibers with a diameter of 0.2 mm are used, what is the load at that moment?

## REFERENCES


ACI (1997) "503.4-92 Standard specification for repairing concrete with epoxy mortars," American Concrete Institute.

ACI (1998) "503R-93 Use of epoxy compounds with concrete," (Reapproved 1998) American Concrete Institute.

ACI (2018) "239R-18 Ultra-high-performance concrete: an emerging technology report," American Concrete Institute.

ACI Committee 237 (2007) "ACI 237R-07, Self-consolidating concrete," American Concrete Institute.

Aguado, A., Gettu, R., and Shah, S.P. (1994) *Concrete technology: New trends, industrial applications*, RILEM, London: E & FN Spon.

An, X.H., Huang M.S., Zhou, H., and Jin, F. (2009) "Rock-filled concrete in China—self compacting concrete for massive concrete." In: Shi, Yu, Khayat, and Yan, eds. RILEM proceedings, *PRO 65, Design, performance and use of self-consolidating concrete, SCC'2009,* Beijing, pp. 615–627.

Arezoumandi, M., Volz, J. S., and Myers, J. (2013) "Shear behavior of high-volume fly ash concrete versus conventional concrete," *Journal of Materials in Civil Engineering*, 25(10), 1506–1513.

Asmus, S.M.F. and Christensen, B. J. (2009) "Status of self consolidating concrete (SCC) in Asia Pacific," in: Shi, C., Yu, Z., Khayat, K.H., and Yan, P., eds. *RILEM proceedings, PRO 65, Design, performance and use of self-consolidating concrete, SCC'2009*, pp. 35–42.

Association Française de Génie Civil (2000) "*Betons auto-placants—recommendations provisoires*," Documents *Sciéntifiques et Techniques*.

Aveston, J., Cooper, G.A. and Kelly, A. (1971) "Single and multiple fracture." In: *The properties of fiber composites*. Guildford, UK: IPC Science and Technology Press, pp. 15–26.

Bache, H.H. (1987) "Introduction to compact reinforced composites," *Nordic Concrete Research*, 6, 19–33.

Balaguru, P. and Shah, S. (1992) *Fiber reinforced cement composites*. New York: McGraw-Hill.

Bartos, P.J.M. and Cechura, J. (2001) "Improvement of working environment in concrete construction by the use of self-compacting concrete," *Structural Concrete*, 2(3), 127–131.

Batis, G., Routoulas, A., and Rakanta, E. (2003) "Effects of migrating inhibitors on corrosion of reinforcing steel covered with repair mortar," *Cement & Concrete Composites*, 25, 109–115.

Batson, G., Jenkins, E., and Spatney, R. (1972) "Steel fibers as shear reinforcement in beams," *ACI Journal*, 69(10), 640–644.

Bendixen, M., Best, J., Hackney, J., and Iverson, L.L. (2019) "Time is running out for sand," *Nature*, 571, 29–31.

Berry, M., Scherr, R., and Matteson, K. (2020) "Feasibility of non-proprietary ultra-high performance concrete (UHPC) for use in highway bridges in Montana: phase II field application," Montana Department of Transportation.

Birchall, J.D., Kendall, K., and Howard, A.J. (1982) "Hydraulic cement composition," US patent 4,353,74.

Bonen, D. and Shah, S.P. (2005) "Fresh and hardened properties of self-consolidating concrete," *Progress in Structural Engineering and Materials*, 7(1) 14–26.

Bos, F.P., Ahmed, Z.Y., Jutinov, E.R., and Salet, T.A.M. (2017) "Experimental exploration of metal cable as reinforcement in 3D printed concrete," *Materials*, 10, 1314.

Brameshuber, W. and Uebachs, S. (2001) "Practical experience with the application of self-compacting concrete in Germany," in *Proceedings of the 2nd international symposium on self-compacting concrete*, Tokyo, Japan, pp. 687–696.

Çağatay, I. H. (2005) "Experimental evaluation of buildings damaged in recent earthquakes in Turkey," *Engineering Failure Analysis*, 12(3), 440–452.

Cao, X., Li, Z., Liang, R., and Sun, G. (2016) "Factors influencing the mechanical properties of the 3D printed product out of magnesium potassium phosphate cement material," in *Proceedings of the 11th FIB International PhD Symposium in Civil Engineering*, pp. 411–418.

Chandra, S. and Ohama, Y. (1994) *Polymers in concrete*, Boca Raton, FL: CRC Press.

Chatterji, S. and Jeffery, J.W. (1966) "The volume expansion of hardened cement paste due to the presence of 'dead burned' CaO," *Magazine of Concrete Research*, 19(55), 65–68.

Chen, Y., Matalkah, F., Soroushian, P., Weerasiri, R., and Balachandra, A. (2019) "Optimization of ultra-high performance concrete, quantification of characteristic features," *Cogent Engineering*, 6(1), 1558696.

Chen, Y. and Qiao, P.Z. (2011) "Crack growth resistance of hybrid fiber reinforced cement matrix composites," *Journal of Aerospace Engineering*, 24(2), 154–161.

Chen, Y., Struble, L.J., and Paulino, G.H. (2006) "Extrudability of cement-based materials," *American Ceramic Society Bulletin*, 85, 9101–9105.

Cheng, S., Shui, Z., Sun, T., Huang, Y., and Liu, K. (2018) "Effects of seawater and supplementary cementitious materials on the durability and microstructure of lightweight aggregate concrete," *Construction and Building Materials*, 190, 1081–1090.

Crovetti, J. (2005) "Early opening of Portland cement concrete pavements to traffic," report submitted to Wisconsin Department of Transportation.

Daczko, J.A. (2003) "A comparison of passing ability test methods for self-consolidating concrete," in *Proceedings of the 3rd international RILEM symposium on self-compacting concrete*, RILEM Publication SARL, pp. 335–344.

Daczko, J. and Kurtz, M. (2001) "Development of high volume coarse aggregate self-consolidating concrete," in *Proceedings of the 2nd international symposium on self-compacting concrete*, pp. 403–412.

Dhondy, T., Remennikov, A., and Sheikh, M. (2020) "Properties and application of sea sand in sea sand-seawater concrete," *Journal of Materials in Civil Engineering*, 32(12), 04020392.

Diamond, S. (1985) "Very high strength cement-based materials," *Materials Research Society Symposia Proceedings*, 42, 233–243.

Diamond, S., Sahu, S., and Thaulow, N. (2004) "Reaction products of densified silica fume agglomerates in concrete," *Cement and Concrete Research*, 34(9), 1625–1632.

Dini, E. (2007) "D-shape printers," [Online], Available: http://www.d-shape.com/d-shape-printers.

Donatello, S., Tyrer, M., and Cheeseman, C. R. (2009) "Recent developments in macro-defect-free (MDF) cements," *Construction and Building Materials*, 23(5), 1761–1767.

Du, J., Meng, W., Khayat, K., Bao, Y., Guo, P., Lyu, Z., Abu-obeidah, A., Nassif, H., and Wang, H. (2021a) "New development of ultra-high-performance concrete (UHPC)," *Composites Part B: Engineering*, 224, 109220.

Du, P., Yaseen, S., Chen, K., Niu, D., Leung, C., and Li, Z. (2021b) "Study of the influence of seawater and sea sand on the mechanical and microstructural properties of concrete," *Journal of Building Engineering*, 42, 103006.

European Federation of National Trade Associations (2002) "Specification and guidelines for self-compacting concrete."

Fedroff, D. and Frosch, R. (2004) "Formwork for self-consolidating concrete," *Concrete International*, 26(10), 32–37.

Felekoglu, B., Turkel, S., and Baradan, B. (2007) "Effect of water/cement ratio on the fresh and hardened properties of self-compacting concrete," *Building and Environment*, 42(4), 1795–1802.

FHWA (Federal Highway Administration) (2006) "Highway concrete technology development and testing," IV, FHWA-RD-02-085, Washington, DC: U.S. Department of Transportation.

Fischer, G. and Li, V.C. (2003) "Deformation behavior of fiber-reinforced polymer reinforced engineered cementitious composite (ECC) flexural members under reversed cyclic loading conditions," *ACI Structures Journal*, 100(1), 25–35.

Fujiwara, H. (1992) "Fundamental study on the self-compacting property of high-fluidity concrete," *Proceedings of Japan Concrete Institute*, 14(1), 27–32.

Fukuyama, H., Matsuzaki, Y., Nakano, K., and Sato, Y. (1999) "Structural performance of beam elements with PVA-ECC," in H. Reinhardt, and A. Naaman, eds., *Proceedings of High performance fiber reinforced cement composites 3 (HPFRCC 3),* Mainz, Germany: Chapman & Hall, pp. 531–542.

Gesoglu, M. and Ozbay, E. (2007) "Effects of mineral admixtures on fresh and hardened properties of self-compacting concretes: binary, ternary and quaternary systems," *Materials and Structures*, 40(9), 923–937.

Grauers, M. (1997) "Rational production and improved working environment through using self-compacting concrete," EC Brite-EuRam Contract No. BRPR-CT96-0366.

Graybeal, B.A. (2008) "Flexural behavior of an ultrahigh-performance concrete I-girder," *Journal of Bridge Engineering*, 13(6), 602–610.

Grunewald, S. and Walraven, J.C. (2001) "Parameter-study on the influence of steel fibers and coarse aggregate content on the fresh properties of self-compacting concrete," *Cement and Concrete Research*, 31(12), 1793–1798.

Guo, X., Yang, J., and Xiong, G. (2020) "Influence of supplementary cementitious materials on rheological properties of 3D printed fly ash based geopolymer," *Cement and Concrete Composites*, 114, 103820.

Guo, Y.S., Kimura, K., Li, M.W., Ding, J.T., and Huang, M.J. (2000) "Properties of high performance lightweight aggregate concrete." In: Helland, S., Holand, I., and Smeplass, S., eds. *Proceedings of second international symposium on structural lightweight aggregate concrete*, Norway, pp. 548–561.

Hack, N., and Lauer, W.V. (2014) "Mesh-mould: robotically fabricated spatial meshes as reinforced concrete formwork," *Architectural Design*, 84, 44–53.

Haque, M., Al-Khaiat, H., and Kayali, O. (2004) "Strength and durability of lightweight concrete," *Cement and Concrete Composites*, 26(4), 307–314.

Hassan, K.E., Brooks, J.J., and Al-Alawi, L. (2001) "Compatibility of repair mortars with concrete in a hot-dry environment," *Cement and Concrete Composites*, 23, 93–101.

Hayakawa, Y., Matsuoka, Y., and Shindoh, T. (1993) "Development and application of super-workable concrete." In: Bartos, P.J.M., ed., *Special concrete: workability and mixing,* London: E & FN Spon, pp. 183–190.

Herfurth, E.A. and Nilsen, T. (1993) "Advances in dry shotcrete technology by means of microsilica," in: *Proceedings of the 5th engineering foundation conference*, Uppsala, Sweden, Code 18378.

HKSAR Highways Department (2006) www.hyd.gov.hk/eng/home/index.htm.

Hock, L. (2014) "3D printing builds up architecture," *Product Design & Development*, Highlands Ranch.

Huang, H., Han, L.H, Tao, Z., and Zhao, X.L. (2010) "Analytical behaviour of concrete-filled double skin steel tubular (CFDST) stub columns," *Journal of Constructional Steel Research*, 66(4), 542–555.

Huang, T., Li, B., Yuan, Q., Shi, Z., Xie, Y., and Shi, C. (2019) "Rheological behavior of Portland clinker-calcium sulphoaluminate clinker-anhydrite ternary blend," *Cement and Concrete Composites*, 104, 103403.

Huang, W., Kazemi-Kamyab, H., Sun, W., and Scrivener, K. (2017) "Effect of cement substitution by limestone on the hydration and microstructural development of ultra-high performance concrete (UHPC)," *Cement and Concrete Composites*, 77, 86–101.

Huang, Z., Padmaja, K., Li, S., and Liew, J. (2018) "Mechanical properties and microstructure of ultra-lightweight cement composites with fly ash cenospheres after exposure to high temperatures," *Construction and Building Materials*, 164, 760–774.

Hwang, S.D., Khayat, K.H., and Bonneau, O. (2006) "Performance-based specifications of self-consolidating concrete used in structural applications," *ACI Materials Journal*, 103(2), 121–129.

Isenberg, J.E. and Vanderhoff, J.W. (1974) "Hypothesis for reinforcement of Portland cement by polymer latexes," *Journal of American Ceramic Society*, 57(6), 242–245.

Kabir, S., Mathur, K., Seyam, A. (2020) "A critical review on 3D printed continuous fiber-reinforced composites: History, mechanism, materials and properties," *Composite Structures*, 232, 111476.

Kamada, T. and Li, V.C. (2000) "Effects of surface preparation on the fracture behavior of ECC/concrete repair system," *Cement and Concrete Composites*, 22(6), 423–431.

Kanda, T., Watanabe, S., and Li, V. C. (1998) "Application of pseudo strain hardening cementitious composites to shear resistant structural elements." In: *Fracture mechanics of concrete structures, proceedings FRAMCOS-3, Freiburg*, Germany: AEDIFICATIO Publishers, pp. 1477–1490.

Kay, T. (1992) *Assessment and renovation of concrete structures,* Harlow: Longman Scientific & Technical.

Keita, E., Bessaies-Bey, H., Zuo, W., Belin, P., and Roussel, N. (2019) "Weak bond strength between successive layers in extrusion-based additive manufacturing: measurement and physical origin," *Cement and Concrete Research*, 123, 105787.

Kesner, K. and Billington, S.L. (2002) "Experimental response of precast infill panels made with DFRCC," in *Proceedings of DFRCC international workshop*, Takayama, Japan, pp. 289–298.

Khan, M.S., Sanchez, F., and Zhou, H. (2020) "3-D printing of concrete: beyond horizons," *Cement and Concrete Research*, 133, 106070.

Khayat, K.H. (1999) "Workability, testing, and performance of self-consolidating concrete," *ACI Materials Journal*, 96(3), 346–353.

Khayat, K.H. (2000) "Optimization and performance of air-entrained, self-consolidating concrete," *ACI Structural Journal*, 97(5), 526–535.

Khayat, K.H., Assaad, J., and Daczko, J., (2004) "Comparison of field-oriented test methods to assess dynamic stability of self-consolidating concrete," *ACI Materials Journal*, 101(2), 168–176.

Khayat, K.H., Ghezal, A., and Hadriche, M.S. (1998) "Development of factorial design method models for proportioning self-consolidating concrete." In: Malhotra, V.M., ed., *Nagataki symposium on vision of concrete: 21st century*, pp. 173–197.

Khayat, K.H., Ghezal, A., and Hadriche, M.S. (1999) "Factorial design models for proportioning self-consolidating concrete," *Materials and Structures*, 32(223), 679–686.

Khayat, K.H. and Guizani, Z. (1997) "Use of viscosity-modifying admixture to enhance stability of fluid concrete," *ACI Materials Journal*, 94(4), 332–340.

Khayat, K.H., Hu, C. and Laye, J.M. (2003) "Importance of aggregate packing density on workability of self-consolidating concrete," in *Proceedings of the 1st North American conference on the design and use of self-consolidating concrete*, Addison, MA: Hanley-Wood Publication, pp. 55–62.

Khayat, K.H. and Omran, A.F. (2009) "Evaluation of SCC formwork pressure," in: Shi, C., Yu, Z., Khayat, K.H., and Yan, P., eds., *RILEM Proceedings, PRO 65, Design, performance and use of self-consolidating concrete, SCC'2009*, pp. 43–55.

Khoshnevis, B., Carlson, A., Leach, N., and Thanavelu, M. (2012) "Contour crafting simulation plan for lunar settlement infrastructure build-up," in *Earth and Space 2012@ Engineering, Science, Construction, and Operations in Challenging Environments*, ASCE, Pasadena CA, pp. 1458–1467.

Khoshnevis, B., and Dutton, R. (1998) "Innovative rapid prototyping process makes large sized, smooth surfaced complex shapes in a wide variety of materials," *Materials Technology*, 13(2), 53–56.

Kim, K., Park, S., Kim, W., Jeong, Y., and Lee, J. (2017) "Evaluation of shear strength of RC beams with multiple interfaces formed before initial setting using 3D printing technology," *Materials*, 10, 1349.

Krstulovic-Opara, N. and Toutanji, H. (1996) "Infrastructural repair and retrofit with HPFRCCs." In: Naaman, A.E. and Reinhardt, H.W., eds., *High performance fiber reinforced cement composites 2*, London: E & FN Spon, pp. 423–439.

Kuroiwa, S., Matsuoka, Y., Hayakawa, M., and Shindoh, T. (1993) "Application of super workable concrete to construction of a 20-story building." In: Zia, P., ed., *Proceedings of symposium on high performance concrete in severe environment*, American Concrete Institute, pp. 147–161.

Lachemi, M., Hossain, K.M.A., Lambros, V., and Bouzoubaa, N. (2003) "Development of cost-effective self-consolidating concrete incorporating fly ash, slag cement, or viscosity-modifying admixtures," *ACI Materials Journal*, 100(5), 419–425.

Lange, D.A., Lin, Y.S., and Henschen, J. (2009) "Modeling formwork pressure of SCC." In: Shi, C., Yu, Z., Khayat, K.H., and Yan, P., eds., *RILEM Proceedings, PRO 65, Design, performance and use of self-consolidating concrete, SCC'2009*, Beijing, pp. 56–63.

Langley, W., Carette, G., and Malhotra, V. (1989) "Structural concrete incorporating high volume of class-F fly ash," *ACI Materials Journal*, 86(5), 507–514.

Li, F. and Li, Z. (1998) "Tensile strain hardening behavior of cementitious composites reinforced with short steel fibers," *Key Engineering Materials*, 145–149, 965–970.

Li, H., Farzadnia, N., and Shi, C. (2018a) "The role of seawater in interaction of slag and silica fume with cement in low water-to-binder ratio pastes at the early age of hydration," *Construction and Building Materials*, 185, 508–518.

Li, P., Li, W., Yu, T., Qu, F., and Tam, V. (2020) "Investigation on early-age hydration, mechanical properties and microstructure of seawater sea sand cement mortar," *Construction and Building Materials*, 249, 118776.

Li, P., Yu, Q., and Brouwers, H. (2018b) "Effect of coarse basalt aggregates on the properties of Ultra-high Performance Concrete (UHPC)," *Construction and Building Materials*, 170, 649–659.

Li, V.C. (1992) "Post-crack scaling relations for fiber reinforced cementitious composites," *ASCE Journal of Materials in Civil Engineering*, 4(1), 41–57.

Li, V.C. (1998) "Engineered cementitious composites–tailored composites through micromechanical modeling," In: Banthia, N. et al. eds., *Fiber reinforced concrete: present and the future,* CSCE, Montreal, pp. 64–97.

Li, V.C., Kong, J. and Bike, S. (2000) "High performance fiber reinforced concrete materials." In: Leung, C.K.Y., et al., eds., *Proceedings of high performance concrete—workability, strength, and durability*, China, pp. 71–86.

Li, V.C., Kong, J., and Bike, S. (2001a) "Constitutive rheological design for development of self-compacting engineered cementitious composites," in *Proceedings of the 2nd international workshop on self-compacting concrete*, Tokyo, Japan.

Li, V.C., Kong, H.J., and Chan, Y.W. (1998) "Development of self-compacting engineered cementitious composites," in *Proceedings, international workshop on self-compacting concrete*, Kochi, Japan, pp. 46–59.

Li, V.C. and Leung, C.K.Y. (1992) "Steady state and multiple cracking of short random fiber composites," *ASCE Journal of Engineering Mechanics*, 118(11), 2246–2264.

Li, V.C., Wang, S. and Wu, C. (2001b) "Tensile strain-hardening behavior of PVA-ECC," *ACI Materials Journal*, 98(6), 483–492.

Li, Z., Mobasher, B., and Shah, S.P. (1991) "Characterization of interfacial properties in fiber-reinforced cementitious composites," *Journal of the American Ceramic Society*, 74(9), 2156–2164.

Li, Z. and Mu, B. (1998) "Application of extrusion for manufacture of short fiber reinforced cementitious composite," *Journal of Materials in Civil Engineering, ASCE*, 10, 2–4.

Li, Z., Mu, B., and Chui, S. (1996) "The systematic study of the properties of extrudates with incorporated metakaolin or silica fume," *ACI Materials Journal*, 96(5), 574–579.

Li, Z., Shen, P., and Siu, A. (2003) "Soundproof concrete," *Magazine of Concrete Research*, 55(2), 177–181.

Li, Z. and Xiao, L.Z. (2009) "Characterization of fresh self consolidating concrete properties with electrical resistivity," Second international symposium on design, performance and use of self-consolidating concrete, Beijing, China.

Li, Z., Zhang, Y., and Zhou, X. (2005) "Short fiber reinforced geopolymer composites manufactured by extrusion," *Journal of Materials in Civil Engineering*, 17(6), 624–631.

Li, Z. Zhou, X., and Shen, B. (2004) "Systematic study of fiber-reinforced cement extrudates with perlite subjected to high temperatures," *Journal of Materials in Civil Engineering, ASCE*, 16(3) 221–229.

Liao, W., Sun, X., Kumar, A., Sun, H., and Ma, H. (2019) "Hydration of binary Portland cement blends containing silica fume: a decoupling method to estimate degrees of hydration and pozzolanic reaction," *Frontiers in Materials*, 6, 78 (13 pages).

Lim, J.H., Panda, B., and Pham, Q.-C. (2018) "Improving flexural characteristics of 3D printed geopolymer composites with in-process steel cable reinforcement," *Construction and Building Materials*, 178, 32–41.

Lim, S., Buswell, R.A., Le, T.T., Austin, S.A., Gibb, A.G., and Thorpe, T. (2012) "Developments in construction-scale additive manufacturing processes," *Automation in Construction*, 21, 262–268.

Lim, S., Le, T., Webster, J., Buswell, R., Austin, S., and Gibb, A. (2009) "Fabricating construction components using layer manufacturing technology, Global Innovation in Construction Conference," Loughborough University, Loughborough, UK.

Lim, Y.M. and Li, V.C. (1997) "Durable repair of aged infrastructures using trapping mechanism of engineered cementitious composites," *Journal of Cement and Concrete Composites*, 19(4), 373–385.

Limeira, J., Agullo, L., and Etxeberria, M. (2010) "Dredged marine sand in concrete: an experimental section of a harbor pavement," *Construction and Building Materials*, 24(6), 863–870.

Lin, Z., Kanda, T. and Li, V.C. (1999) "On interface property characterization and performance of fiber reinforced cementitious composites," *Journal of Concrete Science and Engineering*, 1, 173–184.

Lu, Z., Zhang, J., Sun, G., Xu, B., Li, Z., and Gong, C. (2015) "Effects of the form-stable expanded perlite/paraffin composite on cement manufactured by extrusion technique," *Energy*, 82, 43–53.

Ma, H., and Li, Z. (2013) "Microstructures and mechanical properties of polymer modified mortars under distinct mechanisms," *Construction and Building Materials*, 47, 579–587.

Ma, H., and Li, Z. (2014) "Multi-aggregate approach for modeling interfacial transition zone in concrete," *ACI Materials Journal*, 111(2), 189–200.

Ma, H., Tian, Y., and Li, Z. (2011) "Interactions between organic and inorganic phases in PA-and PU/PA-modified-cement-based materials," *Journal of Materials in Civil Engineering*, 23(1), 1412–1421.

Ma, H., Xing, F., Dong, B., Liu, W., and Huo, Y. (2007a) "Study of electrochemical characteristics for steel corrosion in sea sand concrete," *Concrete*, 2007(6), 20–23.

Ma, H., Xing, F., Dong, B., Liu, W., and Huo, Y. (2007b) "Study on chloride bonding regulation in sea sand concrete," *Low Temperature Architecture Technology*, 2007(6), 1–3.

Ma, G., and Wang, L. (2018) "A critical review of preparation design and workability measurement of concrete material for largescale 3D printing," *Frontiers of Structural and Civil Engineering*, 12(3), 382–400.

Maalej, M. and Li, V.C. (1994) "Flexural/tensile strength ratio in engineered cementitious composites," *Journal of Materials in Civil Engineering, ASCE*, 6(4), 513–528.

Maalej, M. and Li, V.C. (1995) "Introduction of strain hardening engineered cementitious composites in the design of reinforced concrete flexural members for improved durability," *ACI Structural Journal*, 92(2), 167–176.

Marshall, D. and Cox, B.N. (1988) "A J-integral method for calculating steady-state matrix cracking stress in composites," *Mechanics of Materials*, 7, 127–133.

Mehta, P.K. and Monteiro, P.J.M. (2006) *Concrete: microstructure, properties, and materials*, 3rd ed. New York: McGraw-Hill.

Menétrey, P. (2013) "UHPFRC cladding for the Qatar National Museum," In: *Proceedings of international symposium on ultra-high performance fiber-reinforced concrete*, Marseille, France.

Meng, W., and Khayat, K. (2017) "Improving flexural performance of ultra-high-performance concrete by rheology control of suspending mortar," *Composites Part B: Engineering*, 117, 26–34.

Meng, W., and Khayat, K. (2018) "Effect of graphite nanoplatelets and carbon nanofibers on rheology, hydration, shrinkage, mechanical properties, and microstructure of UHPC," *Cement and Concrete Research*, 105, 64–71.

Meng, W., Valipour, M., and Khayat, K.H. (2017) "Optimization and performance of cost-effective ultra-high performance concrete," *Materials and Structures*, 50(1), 29.

Mindess, S., Banthia, N., and Yan, C. (1987) "The fracture toughness of concrete under impact loading," *Cement and Concrete Research*, 17(2), 231–241.

Mirza, J., Mirza, M.S., and Lapointe, R. (2002) "Laboratory and field performance of polymer-modified cement-based repair mortars in cold climates," *Construction and Building Materials*, 16, 365–374.

Morales, C., Guillermo, C., Emparanza, A., and Nanni, A. (2021) "Durability of GFRP reinforcing bars in seawater concrete," *Construction and Building Materials*, 270, 121492.

Mori, A. and Baba, A. (1994) "A method for predicting the operating characteristics curing extrusion molding process for cementitious materials." In: Brandt, A.M., Li, V.C., and Marshall, I.H., eds., *Proceedings of the international symposium on brittle matrix composites* 4, Warsaw, pp. 492–501.

Naaman A.E. (1992) "SIFCON: Tailored properties for structural purpose," High performance fiber reinforced cement composites, *RILEM proceedings*, 15. London: E & FN SPON, pp. 18–38.

Naaman, A.E., and Homrich, J.R. (1989) "Tensile stress-strain properties of SIFCON," *ACI Materials Journal* 86(3), 244–251.

Nerella, V., and Mechtcherine, V. (2019) "Studying the printability of fresh concrete for formwork-free concrete onsite 3D printing technology (CONPrint3D)," in *3D Concrete Printing Technology*, Oxford: Butterworth-Heinemann, pp. 333–347.

Okamura, H. (1988) "Self-compacting high performance concrete," *Progress in Structural Engineering and Materials*, 1(4), 378–383.

Okamura, H. and Ozawa, K. (1994) "Self-compactable high-performance concrete in Japan." In: Zia, P., ed. *Proceedings of the international workshop on high performance concrete*, Detroit, MI: American Concrete Institute, pp. 31–44.

Okamura, H., Ozawa, K., and Ouchi, M. (2000) "Self-compacting concrete," *Structural Concrete*, 1(1), 3–17.

Okushima, M., Kondo, R., Muguruma, H. and Ono, Y. (1968) "Development of expansive cement with sulphoaluminous cement clinker," *5th international symposium on the chemistry of cement*, Tokyo, 4, pp. 419–430.

Ozawa, K., Sakata, N., and Okamura, H. (1995) "Evaluation of self-compactability of fresh concrete using the funnel test," *Concrete Library of JSCE*, 25, 59–75.

Panda, B., Unluer, C., and Tan, M. (2018) "Investigation of the rheology and strength of geopolymer mixtures for extrusion-based 3D printing," *Cement and Concrete Composites*, 94, 307–314.

Parra-Montesinos, G. and Wight, J.K. (2000) "Seismic response of exterior RC column-to-steel beam connections," *ASCE Journal of Structural Engineering*, 126(10), 1113–1121.

Peled, A., Cyr, M., and Shah, S.P. (2000) "Hybrid fibers in high performance extruded cement composites," *Proceedings of fifth RILEM symposium on fiber reinforced concretes*, Lyon, France, pp. 139–147.

Perrot, A., Rangeard, D., and Pierre, A. (2016) "Structural built-up of cement-based materials used for 3D-printing extrusion techniques," *Materials and Structures*, 49, 1213–1220.

Petersson, O. and Skarendahl, A. (2000) "Self-compacting concrete," in *Proceedings of the 1st international symposium*, Stockholm 1999. Cachan, France: RILEM Publications.

Petrou, M.F., Harries, K.A., Gadala-Maria, F., and Kolli, V.G. (2000) "A unique experimental method for monitoring aggregate settlement in concrete," *Cement and Concrete Research*, 30(5), 809–816.

Pfändler, P., Wangler, T., Mata-Falcón, J., Flatt, R.J., and Kaufmann, W. (2019) "Potentials of steel fibres for mesh mould elements," in *Proceedings of the First RILEM International Conference on Concrete and Digital Fabrication – Digital Concrete 2018*, pp. 207–216, Springer International Publishing.

Precast/Prestressed Concrete Institute (2003) "PCI interim SCC guidelines TR-6-03: interim guidelines for the use of self-consolidating concrete in precast/prestressed concrete institute member plants."

Qian, X., Zhou, X., Mu, B., and Li, Z. (2003) "Fiber alignment and property direction dependency of FRC extrudate," *Cement and Concrete Research*, 33(10), 1575–1581.

Rael, R., and Fratello, V. (2017) "Shed," [Online], Available: http://emergingobjects.com/project/shed/.

Rael, R., Fratello, V., Wilson, K., Schofield, A., Anastassiou, S., Dong, Y., et al. (2015) "Bloom," [Online], Available: http://emergingobjects.com/project/bloom-2/.

Reiter, L., Wangler, T., Anton, A., and Flatt, R. (2020) "Setting on demand for digital concrete – principles, measurements, chemistry, validation," *Cement and Concrete Research*, 132, 106047.

Richard, P., and Cheyrezy, M. (1995) "Composition of reactive powder concretes," *Cement and Concrete Research*, 25(7), 1501–1511.

RILEM Technical Committee 145 (2002) "Workability and rheology of fresh concrete: compendium of tests," RILEM Publications SARL.

Rols, S., Ambroise, J., and Pera, J. (1999) "Effects of different viscosity agents on the properties of self-leveling concrete," *Cement and Concrete Research*, 29(2), 261–266.

Roussel, N. (2006) "A thixotropy model for fresh fluid concretes: theory, validation and applications," *Cement and Concrete Research*, 36(10), 1797–1806.

Russell, H.G., and Graybeal, B.A. (2013) "Ultra-high performance concrete: a state-of-the-art report for the bridge community," United States Federal Highway Administration, Office of Infrastructure Research and Development.

Saak, A.W., Jennings, H.M., and Shah, S.P. (2001) "New methodology for designing self-compacting concrete," *ACI Materials Journal*, 98(6), 429–439.

Schröfl, C., Nerella, V. N., and Mechtcherine, V. (2019) "Capillary water intake by 3D-printed concrete visualised and quantified by neutron radiography," in *Proceedings of the First RILEM International Conference on Concrete and Digital Fabrication – Digital Concrete 2018*, Springer International Publishing, pp. 217–224.

Self-Compacting Concrete European Project Group (2005) *The European guidelines for self-compacting concrete*, BIBM, CEMBUREAU, EFCA, EFNARC and ERMCO.

Shah, S.P. (1991) "Toughening of cement-based materials with fiber reinforcement," *Materials Research Society Symposium Proceedings*, 211, 3–13.

Shah, S.P., Ferron, R.F., Tregger, N.A., Ferrra, L., and Beacraft, M.W. (2009) "Self-consolidating concrete: now and future." In: Shi, C., Yu, Z., Khayat, K.H., and Yan, P., eds., *RILEM Procedings, PRO 65, Design, Performance and use of self-consolidating concrete, SCC'2009*, Beijing, pp. 3–15.

Shao, Y., Marikunte, S., and Shah, S.P. (1995) "Extruded fiber-reinforced composites," *Concrete International*, 17(4), 48–52.

Sharma, A.K. (1986) "Shear strength of steel fiber reinforced concrete beams," *ACI Proceedings*, 83(4), 624–628.

Shaw, J.D.N. (1984) "Concrete repair—materials selection," *Civil Engineering*, August, pp. 53–58.

Shindoh, T. and Matsuoka, Y. (2003) "Development of combination-type self-compacting concrete and evaluation test methods," *Journal of Advanced Concrete Technology*, 1(1), 26–36.

Smith, S.H., Qiao, C., Suraneni, P., Kurtis, K.E., and Weiss, W. J. (2019) "Service-life of concrete in freeze-thaw environments: Critical degree of saturation and calcium oxychloride formation," *Cement and Concrete Research*, 122, 93–106.

Soliman, N., and Tagnit-Hamou, A. (2016) "Development of ultra-high-performance concrete using glass powder–Towards ecofriendly concrete," *Construction and Building Materials*, 125, 600–612.

Sonebi, M. (2004) "Applications of statistical models in proportioning medium-strength self-consolidating concrete," *ACI Materials Journal*, 101(5), 339–346.

Sonebi, M., Grunewald, S., and Walraven, J. (2007) "Filling ability and passing ability of self-consolidating concrete," *ACI Materials Journal*, 104(2), 162–170.

Stang, H. and Aarre, T. (1992) "Evaluation of crack width in FRC with conventional reinforcement," *Cement and Concrete Composite*, 14(2), 143–154.

Stang, H. and Li, V.C. (1999) "Extrusion of ECC-Material." In: Reinhardt, M., and Naaman, A., eds., in *Proceedings of high performance fiber reinforced cement composites* 3 (HPFRCC 3), London: Chapman & Hall, pp. 203–212.

Stang, H., Li, V.C., and Krenchel, H. (1995) "Design and structural applications of stress-crack width relations in FRC," *Materials and Structures*, 28(4), 210–219.

Stang, H., Li, Z., and Shah, S.P. (1990) "Pull-out problem: stress versus fracture mechanical approach," *Engineering Mechanics, ASCE,* 116(10), 2136–2150.

Stang, H. and Pedersen, C. (1996) "HPFRCC-extruded pipes," In: Chong, K.P., ed., *Materials for the new millennium*, New York: American Society of Civil Engineers 2, pp. 261–270.

Sun, G., Liang, R., Lu, Z., Zhang, J., and Li, Z. (2016a) "Mechanism of cement/carbon nanotube composites with enhanced mechanical properties achieved by interfacial strengthening," *Construction and Building Materials*, 115, 87–92.

Sun, W., Liu, J., Dai, Y., and Yan, J. (2016b) "Study on the influence of chloride ions content on the sea sand concrete performance," *American Journal of Civil Engineering*, 4(2), 50–54.

Swedish Concrete Association (2002), "Self-compacting concrete, recommendations for use," *Concrete Report No.* 10(E) 84 pp.

Testing-SCC Project Group (2005a) "Guidelines for testing fresh self-compacting concrete."

Testing-SCC Project Group (2005b) "Measurement of properties of fresh self-compacting concrete."

Toutou, Z. and Roussel, N. (2006) "Multiscale experimental study of concrete rheology: from water scale to gravel scale," *Materials and Structures*, 39(286), 189–199.

Toutou, Z., Roussel, N., and Lanos, C. (2005), "The squeezing test: a tool to identify firm cement-based material's rheological behavior and evaluate their extrusion ability," *Cement and Concrete Research*, 35, 1891–1899.

Vachhani, S.R., Chaudary, R., and Jha, S.M. (2004) "Innovative use of SCC in Metro Construction," *ICI Journal*, 4, 27–32.

Vafaei, D., Ma, X., Hassanli, R., and Zhuge, Y. (2021) "Effects of seawater and sea-sand on concrete properties: A review paper," in *Proceedings of the 16th East Asian-Pacific Conference on Structural Engineering and Construction, 2019*, pp. 2037–2049.

Van, V., Rößler, C., Bui D., and Ludwig H. (2014) "Rice husk ash as both pozzolanic admixture and internal curing agent in ultra-high performance concrete," *Cement and Concrete Composites*, 53, 270–278.

Vijayalakshmi, R., and Ramanagopa, S. (2018) "Structural concrete using expanded clay aggregate: A review," *Indian Journal of Science and Technology*, 11(16), 121888.

Wallevik, O.H. (2003) "Role in rheology in developing new breeds of concrete and construction techniques," in *Proceedings of the international symposium—celebrating concrete: people and practice*, pp. 441–450.

Wang, C., Yang, C., Liu, F., Wan, C., and Pu, X. (2012) "Preparation of ultra-high performance concrete with common technology and materials," *Cement and Concrete Composites*, 34(4), 538–544.

Wang, D.M. (2006) *High performance expansive concrete*. Beijing: Publisher of hydraulic and hydropower of China [in Chinese].

Wang, J., Liu, E., and Li, L. (2018) "Multiscale investigations on hydration mechanisms in seawater OPC paste," *Construction and Building Materials*, 191, 891–903.

Wang, J., Liu, J., Wang, Z., Liu, T., Liu, J., and Zhang, J. (2021) "Cost-effective UHPC for accelerated bridge construction: material properties, structural elements, and structural applications," *Journal of Bridge Engineering*, 26(2), 04020117.

Wang, X., Jiang, M., Zhou, Z., Gou, J., Hui, D. (2017) "3D printing of polymer matrix composites: a review and prospective," *Composites Part B: Engineering*, 110, 442–458.

Wangler, T., Lloret, E., Reiter, L., Hack, N., Gramazio, F., and Kohler, M. (2016) "Digital concrete: opportunities and challenges," *RILEM Technical Letters*, 1, 67–75.

Wangler, T., Roussel, N., Bos, F., Salet, T., and Flatt, R. (2019) "Digital concrete: A review," *Cement and Concrete Research*, 123, 105780.

Warner, J. (1984) "Selecting repair materials," *Concrete Construction*, 29(10), 865–873.

WASP (2017) "Concrete beam created with 3D printing," [Online] Available: http://www.waspro-ject.it/w/en/concrete-beam-created-with-3d-printing/.

Wolfs, R.J.M., Bos, F.P., and Salet, T.A.M. (2019) "Hardened properties of 3D printed concrete: the influence of process parameters on interlayer adhesion," *Cement and Concrete Research*, 119, 132–140.

Wu, P., Wang, J., and Wang, X. (2016) "A critical review of the use of 3-D printing in the construction industry," *Automation in Construction*, 68, 21–31,

Xiao, J., Qiang, C., Nanni, A., and Zhang, K. (2017) "Use of sea-sand and seawater in concrete construction: Current status and future opportunities," *Construction and Building Materials*, 155, 1101–1111.

Xu, G. and Hannant D.J. (1992) "Flexural behavior of combined polypropylene network and glass fiber reinforced cement," *Cement and Concrete Composites*, 14, 51–61.

Yan, P.Y. and Yu, C.H. (2009) "Application of self-consolidating concrete in Beijing,." In: Shi, C., Yu, Z., Khayat, K.H., and Yan, P., eds., *RILEM Proceedings, PRO 65, Design, performance and use of self-consolidating concrete, SCC'2009*, Beijing, pp. 817–822.

Yao, W., Li, J., and Wu, K.R. (2003) "Mechanical properties of hybrid fiber-reinforced concrete at low fiber volume fraction," *Cement and Concrete Research*, 33(1), 27–30.

Yépez, F., and Yépez, O. (2017) "Role of construction materials in the collapse of R/C buildings after Mw 7.8 Pedernales – Ecuador earthquake, April 2016," *Case Studies in Structural Engineering*, 7, 24–31.

Yonezawa, T., Izumi, I., Okuno, T., Sugimoto, M., Shimuno, T., and Azakuna, T. (1992) "Reducing viscosity of high strength concrete using silica fume," in *Proceedings of the 4th CANMET/ACI international conference on fly ash, silica fume, slag and natural pozzolans in concrete*, pp. 665–680.

Yoo, D., Park J., Kim S., and Yoon, Y. (2014) "Influence of reinforcing bar type on autogenous shrinkage stress and bond behavior of ultra-high performance fiber reinforced concrete," *Cement and Concrete Composites*, 48, 150–161.

You, B.K. and Li, N.Z. (2005) "Expansive agent and shrinkage compensating concrete," Beijing: Publisher of China Building Materials Industry, [in Chinese].

Younis, A., Ebead, U., Suraneni, P., and Nanni, A. (2018) "Fresh and hardened properties of seawater-mixed concrete," *Construction and Building Materials*, 190, 276–286.

Yurugi, M., Sakata, N., Iwai, M., and Sakai, G. (1993) "Mix proportion for highly workable concrete," in *Proceedings of the international conference of concrete 2000*, Dundee, UK.

Zhang, J. and Li, V.C. (2002) "Monotonic and fatigue performance in bending of fiber-reinforced engineered cementitious composite in overlay system," *Cement and Concrete Research*, 32(3), 415–423.

Zhang, J., Wang, J., Dong, S., Yu, X., and Han, B. (2019) "A review of the current progress and application of 3D printed concrete," *Composites Part A: Applied Science and Manufacturing*, 125, 105533.

Zhang, L., Liu, J., Liu, J., Zhang, Q., and Han, F. (2018) "Effect of steel fiber on flexural toughness and fracture mechanics behavior of ultrahigh-performance concrete with coarse aggregate," *Journal of Materials in Civil Engineering*, 30(12), 04018323.

Zhang, S.Q., Hughes, D., Jeknavorlan, A.A., Nishimura, T., and Yang, K. (2009) "Self-compacting concrete, worldwide experience," In: Shi, C., Yu, Z., Khayat, K.H., and Yan, P., eds., *RILEM Proceedings, PRO 65, Design, performance and use of self-consolidating concrete, SCC'2009*, Beijing, pp. 831–840.

Zhao, Y., Hu, X., Shi, C., Zhang, Z., and Zhu, D. (2021) "A review on seawater sea-sand concrete: mixture proportion, hydration, microstructure and properties," *Construction and Building Materials*, 295, 123602.

Zhong, S.T. (1995) *Concrete filled steel tubular structures*, Harbin: Heilongjiang Science-Technology Publisher.

CHAPTER 7

# CONCRETE FRACTURE MECHANICS

## 7.1 INTRODUCTION

According to the features of a tensile stress-strain curve, materials can be categorized as brittle materials, quasi-brittle materials and ductile materials (see Figure 1-12). As shown in Figure 1-12a, brittle materials break suddenly when the stress reaches maximum. As shown in Figure 1-12b, quasi-brittle materials can show a strain-softening behavior, i.e., stress decreases with strain increase, if a proper loading control is provided. As shown in Figure 1-12c, ductile materials show a long plastic plateau before failure. Glass is a typical brittle material, concrete is a typical quasi-brittle material, while low-carbon steel at normal temperatures is a typical ductile material. Fracture occurs as a brittle failure mode. Fracture mechanics is a subject in the study of stress and displacement fields in a material in the region surrounding a crack tip.

It has been long known that there are a lot of sources of microcracks inside a cement-based material such as concrete. It has been shown that both stress-strain behavior and the failure mode of cement paste and concrete are governed by a process of microcrack propagation. Thus, the failure of concrete under loading is closely related to crack formation and propagation. Hence, fracture mechanics fits concrete naturally. To understand the nature of concrete behavior, it is essential to understand its fracture mechanism. Before discussing the concrete fracture mechanism, it is necessary to briefly review the basic concepts of fracture mechanics first.

### 7.1.1 History of Fracture Mechanics

The industrial revolution in Europe in the nineteenth century broadened the application of materials, especially metals. The vastly increasing use of metals, moreover, caused a number of accidents, including broken axles or wheels of trains, explosion of boilers, and chains giving way on suspension bridges. Most of the accidents were a result of the fracture of metal components. It was found that the failures often occurred under conditions of low stress which made them seemingly unexpected. These accidents attracted great attention from the science and engineering community and initiated extensive scientific investigations. The investigations revealed that the pre-existing flaws in materials subsequently caused stress concentrations which were mainly responsible for such accidents. To explain this, let us look at the theoretical cohesive stress.

In a metal, the structure is formed by metallic lattices. The strength of a metal is sourced from the metallic bonds. With metallic bonding, the valence electrons become detached from individual parent atoms and move freely within the solid as an electron gas. When the atoms give up their electrons, they become positive ions. It is easy to see that there is an equilibrium condition for two ions at which the forces between them are balanced.

Ideally, the strength of a solid depends on the strength of its atomic bonds. Thus, to obtain an approximation to at least the order of magnitude of the fracture strength, we can consider the interaction between ions. Let us consider the cubic lattice with a balanced spacing, $b_0$, as shown in Figure 7-1. Furthermore, if we assume that only adjacent neighbors interact, we reach the scenario

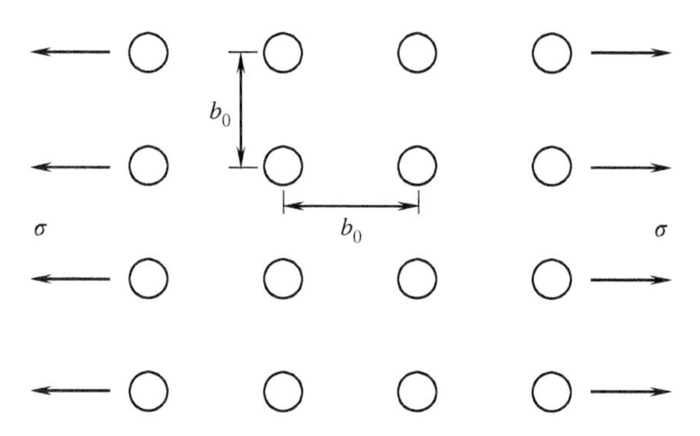


**Figure 7-1** Lattice structure under stress

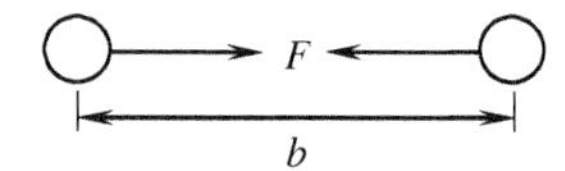


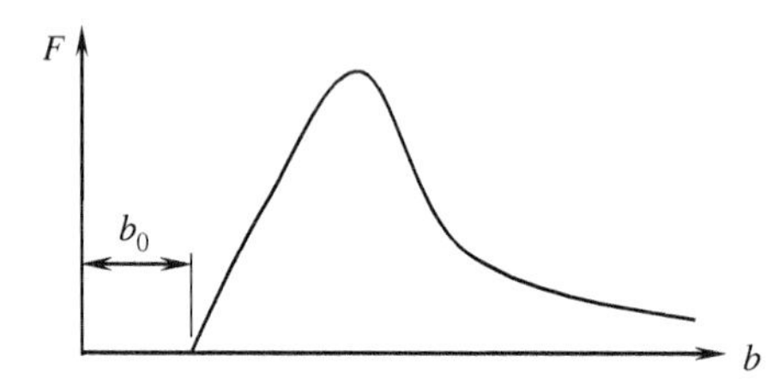


**Figure 7-2** Force acting on two neighboring ions

shown in Figure 7-2. Two ions will attract each other when $b > b_0$ and repel each other when $b < b_0$, where $b_0$ is the balanced distance between the two ions and $b$ is the distance of the two ions after the force $F$ acting on the ions is applied. Let $\Delta l$ be the distance change between two ions (i.e., $\Delta l = b - b_0$), and we can write the stress and strain expressions as:

$$\sigma = \frac{F}{b_0^2} \tag{7-1}$$

and

$$\varepsilon = \frac{\Delta l}{b_0} \tag{7-2}$$

Subsequently, the $F$-$b$ form can be transferred into a $\sigma$-ε curve that can be further simplified as a sine curve

$$\sigma = \sigma_{\max} \sin(2\pi\varepsilon) \tag{7-3}$$

The differential of the above equation yields

$$\left.\frac{d\sigma}{d\varepsilon}\right|_{\varepsilon=0} = \sigma_{\max} 2\pi = E \tag{7-4}$$

Thus, we can estimate the strength for ideal materials as

$$\sigma_{\max} = \frac{E}{2\pi} \approx \frac{E}{6} \tag{7-5}$$

However, in the real world, the material strength is usually much lower than the above theoretical value. For instance, E for steel is 210 GPa, thus according to the above equation, we have

$$\sigma_{max} \approx \frac{E}{6} = 35 \text{ GPa} \tag{7-6}$$

The tensile strength measured for reinforcing steel is only 0.3–0.6 GPa. There is a big difference between the real value and the theoretical value. Why? This is because of the pre-existing flaws inside a material and they cause stress concentrations and uneven distributions. The stress concentration may lead to very high stress that causes a local failure, which results in low apparent strength, indexed with normalized stress. These phenomena initiated the study of fractures.

Linear elastic fracture mechanics (LEFM) theory was developed in 1920 with Griffith as the founder. Griffith's original interest was in the effect of surface treatment on the strength of solids. It had been observed, experimentally, that small imperfections had a much less damaging effect on the material properties than large imperfections. This was theoretically puzzling because at that time people believed that the stress concentrations caused by imperfections should be the same for imperfections of different sizes. Griffith tackled this dilemma by developing a new criterion for fracture prediction by considering the crack size. LEFM had essentially two criteria: stress-based and energy-based criteria. What Griffith developed was an energy-based criterion. He used a compliance concept and dealt with both fixed-load and fixed-displacement cases. Griffith showed that an instability criterion could be derived for cracks in brittle materials based on the variation of potential energy of the structure, as the crack grew. The Griffith approach was global but could not easily be extended to accommodate structures with finite geometries subjected to various types of loads. The theory was considered to apply only to a limited class of extremely brittle materials, such as glasses or ceramics.

The stress-based LEFM criterion was first developed by Inglis (1913), who provided the solution of maximum stress for a solid weakened by an elliptical cavity subjected to uniform stress normal to the semi-axis of the ellipse. Later on, Irwin defined the fundamental concept of a *Stress Intensity Factor* (SIF) and the *critical stress intensity factor* ($K_{IC}$) which is a material property. The fundamental postulate of LEFM is that crack behavior is determined solely by the value of the stress intensity factor at the crack tip, which is a function of the applied load and the geometry of the cracked structure. The SIF, thus, plays a very important role in LEFM applications, for example, in fatigue crack growth analyses. The stress values in the vicinity of a crack tip are usually expressed as a function of SIF. According to the stress expressions, it can reach infinity, termed a singularity. However, in the real material, this is impossible. Hence, researchers inferred that an inelastic zone must exist in front of a crack tip. This assumption trigged the study of nonlinear fracture mechanics. Nonlinear fracture mechanics mainly focused on the determination of the size of the plastic zone in front of a crack tip. The representative work in determining the size of the plastic zone included Irwin's plastic zone correction, (Irwin, 1958, 1960), Dugdale's cohesive zone model (1960) and Rice's J-integral (1968). These works confirmed that there is a small plastic zone in front of a crack tip and provided equations to calculate the size of the plastic zone. It was also proved that for a material with a small plastic zone, LEFM could still be applied.

A broad range of disciplines, from nano-scaled materials science to meter-scaled engineering applications, was involved in the development of fracture mechanics. Small-scale material science concerned itself with the fracture processes on the scale of atoms and dislocations to that of impurities and grains and provided an understanding of the origin and driving force of microcracks. The large-scale engineering mechanics provided analysis methods for stress and deformation determination around the crack tip as well as design criteria based on crack analysis for structures. The prediction on fracture strength could be checked experimentally.

With the study of fracture mechanics, the following questions should be able to be answered:

**(a)** How is a crack formed? What is the driving force of a crack?
**(b)** How to determine the maximum tolerated crack size corresponding to the service load?
**(c)** How to determine the maximum stress a structure can carry with an existing crack size?
**(d)** How to decide the fatigue life of a structure with a certain initial crack size?
**(e)** What are the essential conditions for a crack to grow?
**(f)** How often should a structure be inspected for cracks?

### 7.1.2 Development of Concrete Fracture

Concrete is a multiple scale composite material, mainly comprised of hardened cement paste and aggregates. Due to the nature of the cement hydration, shrinkage and creep occurs. Microcracks exist, and sometimes even macrocracks, in concrete, from the very beginning. Hence, the failure of concrete structures is usually accompanied by the propagation of cracks in the concrete matrix. Understanding and modeling of how and when concrete fails are not only critical for designing concrete structures, but also important for developing new cement-based materials. Except for the physical formation of a crack, the failure of concrete is softening without a plateau. The plasticity does not apply to concrete. On the other hand, it is natural to apply fracture mechanics to concrete due to the pre-existence of cracks inside.

Fracture mechanics was first applied to concrete in 1961 by M. F. Kaplan (1961). Based on the observations that microcracks occurred in concrete, Kaplan made an effort to ascertain whether the Griffith crack theory was a necessary condition for rapid crack propagation and consequent fracture of concrete or not, by conducting three-point bending tests. Since then, a large number of tests have been conducted to examine the applicability of fracture mechanics to concrete. For instance, Shah and McGarry (Shah and McGarry, 1971) concluded that the critical crack length was likely to depend on the volume, type, and size of aggregate particles from the experimental observations. One characteristic of concrete fracture was that it had a large fracture process zone. Castro-Montero et al. (1990) conducted a laser holographic interferometry test on a concrete plate loaded vertically at a central circular hole with two notches along the horizontal diameter from the edge of the hole. The fracture process zone was identified by plotting the difference between a strain field experimentally measured by holographic and a strain field from a linear elastic solution. The results showed that a large fracture process zone existed both behind and in front of the crack tip. Ouyang et al. (1991) applied the acoustic emission technique to determine internal microcrack positions for three-point bending testing of a concrete beam. They found that most of the acoustic emission events were located at both sides of the plane containing the notch, which indicated that the fracture process zone had a substantial width. From the observation of the large fracture process zone in concrete, it was concluded that the criterion of LEFM could not be directly applied to predict concrete fracture behavior. Thus, researchers had to look for a modified method of fracture mechanics to solve concrete fracture problem.

Since the 1970s, models with more than one fracture parameter have been proposed to explain the fracture processes in concrete. Hillerborg et al. (1976) proposed a fictitious crack model, which is somewhat similar to Dugdale's model used in metals. Bažant and Oh (1983) have proposed a crack band model based on the concept of strain-softening to explain the fracture process of concrete. Wecharatana and Shah (1982, 1983) used a compliance-based model to calculate the length of the fracture process-zone, which was represented by traction forces resulting from aggregate or fiber interlock. They observed that R-curves could be considered a material property provided the length of the process zone and the inelastic energy absorbed in the process zone were included in the analysis. Moreover, Jenq and Shah (1985) introduced a two-parameter fracture model. It was demonstrated that these two parameters, the critical stress intensity factor at the tip of the effective crack and the elastic critical crack opening displacement, were size independent. In 1984, Bažant developed a size effect law to model concrete fracture.

In the later 1980s and 1990s, intensive research was undertaken and trial applications of fracture mechanics in the design of beams, anchorage, and large dams are becoming more common. It has finally been recognized by the American Concrete Institute, whose 2019 design code is the first based on fracture mechanics, underlying its 2019 design specifications for the size effect in shear strength of beams and slabs.

## 7.2 LINEAR ELASTIC FRACTURE MECHANICS

Linear elastic fracture mechanics studies the stress and deformation distribution in the region in front of a crack tip of brittle materials. It uses a single parameter as an index of the fracture criterion. Two criteria are used in LEFM: one is stress-based and the other energy-based. The fracture index in the stress-based criterion is stress intensity factor and, in the energy-based case, it is the surface energy release rate.

Originally, fracture mechanics was developed from elasticity problems dealing with inside defects, such as circular or elliptical holes.

### 7.2.1 Stress Concentration Factor and Intensity Factor at a Crack Tip

**(a)** *Stress concentration factor*. The presence of defects in materials influences the stress distribution in the materials and subsequently the mechanical properties of the materials. This can be illustrated by a plate with an elliptical hole as shown in Figure 7-3, where the elliptical hole represents the defect. The existence of the hole in the plate alters the stress distribution in the surrounding area and leads to the maximum stress, $\sigma_{\mathrm{max}}$, being formed at the edge of the hole. The $\sigma_{\mathrm{max}}$ is much larger than the normal stress, $\sigma_{\mathrm{N}}$. This phenomenon is called a stress concentration. The stress concentration factor, $K_{\mathrm{t}}$, for the given loading condition in Figure 7-3, can be expressed as:

$$K_{\mathrm{t}} = \frac{\sigma_{\mathrm{max}}}{\sigma_{\mathrm{N}}} = 1 + \frac{2a}{c} \tag{7-7}$$

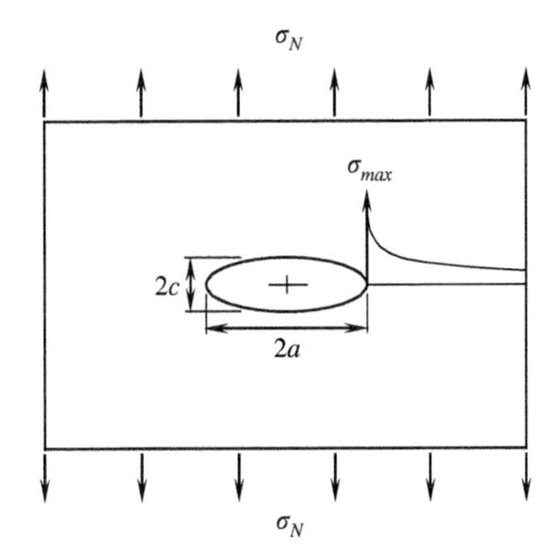


**Figure 7-3** A specimen with an elliptical hole under tensile loading

where $a$ and $c$ are the long and short radii of the ellipse, respectively. It can be seen from Equation (7-7) that when $a = c$, i.e., for a circular hole, $K_t = 3$. It should be noted that $K_t$ depends not only on the shape of the hole, but also the loading pattern on the specimen. For instance, for a plate with a circular hole under pure shear, as shown in Figure 7-4, the value of $K_t$ is 4.

It can also be seen from Equation (7-7) that the value of $K_t$ largely depends on the ratio of $a/c$. With the increase of the $a/c$ ratio, the value of $K_t$ increases. The value of $K_t$ tends to approach infinity for a very narrow ellipse or sharp crack. Hence, $K_t$ cannot be a material parameter under such a condition. This indicates that the conventional analysis based on the concentration factor is not valid for a structure with a sharp crack. Therefore, fracture mechanics should be introduced to solve such a problem.

**(b)** *Stress intensity factor*. In fracture mechanics, cracks are classified into mode I (opening mode), mode II (shearing mode), mode III (tearing mode) and mixed mode (I+II or I+III). The stress expressions in the vicinity of a crack tip for the three basic modes can be derived from the elasticity theory (Irwin, 1958). Similar to the case of stress concentrations, the stress expressions not only depend on the specimen geometry, but also the loading pattern acting on the specimen. Figure 7-5 shows a thin plate with a central crack subjected to two-dimensional

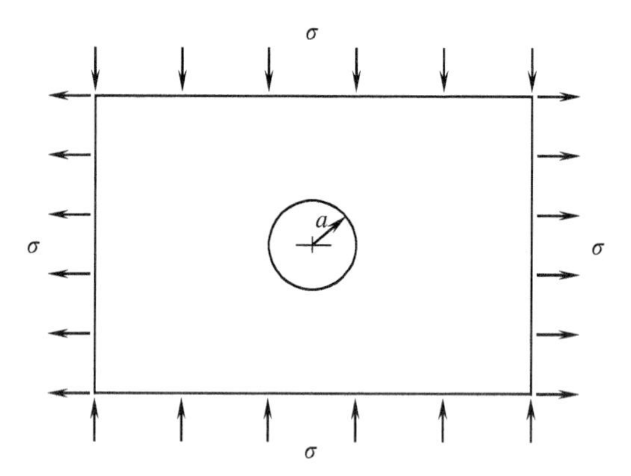


**Figure 7-4** A thin plate with a circular hole subjected to pure shear

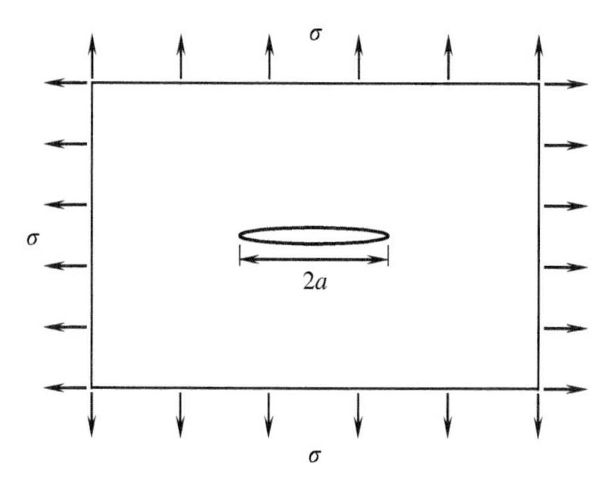


**Figure 7-5** A specimen with a mode I crack

axial tension. The crack length is usually expressed as '$a$'. It corresponds to a crack tip. If a crack has two crack tips such as central crack, it has a crack length of '$2a$'. If a crack has only one crack tip, such as an edge crack, it has a crack length of '$a$'.

Clearly, Figure 7-5 is a mode I case and the elastic stress field in the vicinity of a crack tip has the form:

$$\begin{aligned}
\sigma_{xx} &= \frac{K_{\mathrm{I}}}{\sqrt{2\pi r}} \cos\frac{\theta}{2}\left(1 - \sin\frac{\theta}{2}\sin\frac{3\theta}{2}\right) \\
\sigma_{yy} &= \frac{K_{\mathrm{I}}}{\sqrt{2\pi r}} \cos\frac{\theta}{2}\left(1 + \sin\frac{\theta}{2}\sin\frac{3\theta}{2}\right) \\
\sigma_{xy} &= \frac{K_{\mathrm{I}}}{\sqrt{2\pi r}} \cos\frac{\theta}{2}\sin\frac{\theta}{2}\cos\frac{3\theta}{2}
\end{aligned} \tag{7-8}$$

where $r$ is the distance from crack tip to the point where the stress is calculated, $\theta$ the angle between $r$ and the $x$-axis, and $K_{\mathrm{I}}$ is called the stress intensity factor and defined as:

$$K_{\mathrm{I}} = \sigma\sqrt{\pi a} f(a/b) \tag{7-9}$$

where $\sigma$ is the applied stress (or normal stress) on the structure, $a$ is the crack length, $b$ the size of the structure, and $f(a/b)$ the geometry factor.

For a uniaxial tensile plate with a central crack, we have:

$$f\left(\frac{a}{b}\right) = \sqrt{\sec\left(\frac{\pi a}{2b}\right)} \tag{7-10}$$

For a uniaxial tensile plate with a single edge crack, we have

$$f\left(\frac{a}{b}\right) = 1.12 - 0.231\frac{a}{b} + 10.55\left(\frac{a}{b}\right)^2 - 21.72\left(\frac{a}{b}\right)^3 + 30.39\left(\frac{a}{b}\right)^4 \tag{7-11}$$

For a three-point bending beam with span equal 4 times of beam height,

$$f\left(\frac{a}{b}\right) = \frac{1.99 - \frac{a}{b}\left(1 - \frac{a}{b}\right)\left[2.15 - 3.93\left(\frac{a}{b}\right) - 2.70\left(\frac{a}{b}\right)^2\right]}{\sqrt{\pi}\left(1 + 2\frac{a}{b}\right)\left(1 - \frac{a}{b}\right)^{\frac{3}{2}}} \tag{7-12}$$

The expressions of $f(a/b)$ for other geometries can be obtained from the tables in the fracture analysis handbook (Tada et al., 2000). The value of $f(a/b)$ approaches unity when the ratio $a/b$ approaches zero, for most loading cases and geometries.

Both the stress intensity factor $K_{\mathrm{I}}$ and the stress concentration factor $K_{\mathrm{t}}$ are used to account for the increase of stresses due to a defect. When the defect is a sharp crack, $K_{\mathrm{I}}$ has a limiting value but $K_{\mathrm{t}}$ approaches infinity. The value of $K_{\mathrm{I}}$ accounts for the singularity of the stress field in the crack tip, and is a function of load, specimen geometry and size, boundary condition, and crack length.

The fundamental postulate of linear elastic fracture mechanics (LEFM) is that crack behavior is determined solely by the value of the stress intensity factor (SIF) at the crack tip which is a function of the applied load and the geometry of the cracked structure. The SIF hence plays a very important role in LEFM applications, for example, in fatigue crack growth analyses. SIF can be obtained using the Airy stress function with complex variables.

$$\Phi(Z) = \mathrm{Re}\, Z + iIMZ \tag{7-13}$$

where $Z = x + iy$.

However, numerical methods are usually used for the evaluation of the SIFs in engineering structures, because of their complex configurations and loading status. A 1–5% accuracy of SIFs can be obtained by the most used numerical methods. It is good enough for general engineering problems except for numerical simulation of fatigue crack growth. The finite element method (FEM) has a long and well-documented history in fracture mechanics applications. The boundary element method (BEM) is also a well-established numerical technique in fracture mechanics. The procedures of the general displacement-based SIF computation method in fracture mechanics problems can be expressed as follows.

The displacement fields of the crack problems are first obtained through the application of numerical methods, such as FEM or BEM. In order to achieve better results for the displacement fields, specially designed crack tip elements, such as quarter-point elements, have often been used. Then, SIFs are extracted from the obtained displacement fields by the techniques deduced from LEFM. A lot of techniques have been developed to extract SIFs from the displacement fields. These techniques can be divided into two classes of procedures: local displacement field procedures and global displacement field procedures. Local displacement field procedures are based on the asymptotic analysis of the displacements on the crack line near a crack tip with the following expressions:

$$v = \frac{8}{E}\sqrt{\frac{r}{2\pi}}K_{\mathrm{I}} + O\left(r^{\frac{3}{2}}\right) \tag{7-14}$$

$$u = \frac{8}{E'}\sqrt{\frac{r}{2\pi}}K_{\mathrm{II}} + O\left(r^{\frac{3}{2}}\right) \tag{7-15}$$

where $v$ and $u$ are the crack opening displacements (COD) normal and tangential to the crack line, respectively; $r$ the distance from the crack tip along the crack line; $K_{\mathrm{I}}$ and $K_{\mathrm{II}}$ the mode I and mode II SIFs at the crack tip, respectively. $E' = E$ is for plane stress, and $E' = E/\left(1 - \upsilon^2\right)$ for plain strain, where $E$ and $v$ are the Young's modulus and Poisson's ratio, respectively, of the material.

Apparently, by using Equations (7-14) and (7-15), only the displacements near a crack tip are used to extract the SIFs. An advantage of using the local displacement extraction techniques is that they can readily be programmed with a simple post-processing algorithm. Note that Equations (7-14) and (7-15) are exactly valid only for $r \to 0$, so in numerical calculations the nodal positions, which will be used to extract SIFs, must be chosen to be very small. On the other hand, since the gradient of the displacement field is singular as $r \to 0$, the errors of the displacements near the crack tip, even with the use of specially designed crack tip elements, are much bigger than those far away from the crack tip. Unfortunately, when using Equations (7-14) and (7-15) to extract SIFs, it is not clear what value of $r$ should be chosen to obtain the best results. Lim et al. (1992) and Pang (1993) systematically studied these techniques, and they indicate that the displacement extrapolation techniques exhibited some erratic characteristics and are highly sensitive to the nodal displacement distribution. In other words, fortuitously, more accurate displacements do not always give more accurate SIFs because of the interaction between the error in the approximate displacements and the error in Equations (7-14) and (7-15).

It is natural to imagine that if the SIFs are extracted from the total COD data on the crack line, the accuracy will be greatly improved. The global displacement extraction procedure is based on the use of a path independent $J$-integral which is defined as:

$$J = \int_S \left( W n_1 - t_j u_{j,1} \right) dS \tag{7-16}$$

where $S$ is an arbitrary contour surrounding the crack tip, $W$ the strain energy density that can be expressed by $\frac{1}{2}\sigma_{ij}\varepsilon_{ij}$, where $\sigma_{ij}$ and $\varepsilon_{ij}$ are the stress and strain tensors, respectively, $t_j = \sigma_{ij} n_i$, the traction components, where $n_i$ are the components of the unit outwardly normal to the contour path. The relationship between the $J$-integral and SIFs is given by

$$J = \frac{K_{\mathrm{I}}^2 + K_{\mathrm{II}}^2}{E} \tag{7-17}$$

The advantages of using Equations (7-16) and (7-17) to extract SIFs are that the total COD data are used, although indirectly, and the numerical calculations of the $J$-integral are not significantly sensitive to the displacement near fields where the big errors of the displacement fields are encountered. This makes the accuracy of SIFs extracted from Equations (7-16) and (7-17) much better than that of SIFs extracted from Equations (7-14) and (7-15). The disadvantages of this procedure is that (1) more post-processing work is needed; (2) the integral in the right-hand side of Equation (7-16) involves the calculation of stress or strain tensors along a new contour which will introduce more errors than if the integral only involves displacement fields along the crack line itself; and (3) it is not easy to extend this method to 3-D crack problems.

Within the scope of LEFM, the physical meaning of the $J$-integral is the energy release rate $G$ when the crack tip has a virtual crack extension $\Delta a$:

$$J = G = \Delta U / \Delta a \tag{7-18}$$

where $\Delta U$ is the change in strain energy. By using the energy release rate method to extract SIFs, a second analysis of the crack problem should be carried out, and the size of $\Delta a$ is based more on experience, however, the results are as good as those from using the $J$-integral method. Another advantage of the energy release rate method is that it can be used in 3-D crack programs.

Cooper et al. (1995) compared the above-mentioned two procedures on a set of about forty basic test problems by using finite element analysis. They concluded that the continued use of local fitting procedures for SIFs is not recommended, other than possibly for rough hand estimates, and for applications in which reliable estimates of SIFs are sought, path-independent integrals represent a superior choice.

Later, a self-similar crack expansion method (SSCE) was proposed by Xu (1998) and Xu et al. (1997) to evaluate SIFs for the cracks in an infinite medium or semi-infinite medium. By taking advantage of the crack self-similarity, a relation between the SIFs and the crack opening volume defined by the integral of COD over the crack face is established. Since only COD data are needed and all the COD data are used, highly accurate SIF results have been obtained by using the SSCE method. Also, the post-processing is relatively simple, and this method is valid for the analysis of 3-D crack problems. However, if a crack is not self-similar, such as a crack in a finite body, an additional integral, including the traction derivatives around the boundary of the finite body, is needed to evaluate the SIFs.

A reliable, highly accurate and easily implemented procedure to extract the SIFs should have the following features: (1) The procedure should make full use of the COD data to increase the accuracy of the results, and the information of COD fields should also be considered. (2) The numerical calculation should only involve displacement fields to avoid additional errors from the numerical calculation of other field quantities such as stress or strain components. (3) The postprocessing should be as small as possible, i.e. no new contouring and remeshing are needed.

(4) The procedure can easily be applied to mixed mode crack problems with arbitrary crack shapes; and (5) the errors of obtaining SIFs can be directly estimated from the error information of the COD data.

For Mode II, the stress expressions are:

$$\sigma_{xx} = \frac{-K_{\text{II}}}{\sqrt{2\pi r}} \sin\frac{\theta}{2}\left(2+\cos\frac{\theta}{2}\cos\frac{3\theta}{2}\right)$$
$$\sigma_{yy} = \frac{K_{\text{II}}}{\sqrt{2\pi r}} \cos\frac{\theta}{2}\left(\sin\frac{\theta}{2}\sin\frac{3\theta}{2}\right) \tag{7-19}$$
$$\sigma_{xy} = \frac{K_{\text{II}}}{\sqrt{2\pi r}} \cos\frac{\theta}{2}\left(1-\sin\frac{\theta}{2}\sin\frac{3\theta}{2}\right)$$
$$K_{\text{II}} = \sigma_{xy}\sqrt{\pi a} \tag{7-20}$$

For Mode III, the stress expressions are:

$$\sigma_{xz} = \frac{-K_{\text{III}}}{\sqrt{2\pi r}} \sin\frac{\theta}{2}$$
$$\sigma_{yz} = \frac{K_{\text{III}}}{\sqrt{2\pi r}} \cos\frac{\theta}{2} \tag{7-21}$$

The stress intensity factor is one of the most important concepts of fracture mechanics. It is also called fracture toughness. A crack extension will occur when the critical fracture toughness of the material is reached. Therefore, fracture toughness is expected to be a material property.

According to the stress expression, the stress approaches infinity at a crack tip where $r$ tends to zero. This phenomenon is termed the stress singularity at the tip of an elastic crack. Since infinite stress cannot develop in real materials, a certain range of the inelastic zone must exist at the crack tip.

The stress intensity factor can serve as a fracture criterion. A crack propagates whenever $K_{\text{I}}$ is equal to a threshold value, i.e.

$$K_{\text{I}} = K_{\text{Ic}} \tag{7-22}$$

where $K_{\text{Ic}}$ is the critical stress intensity factor for a mode I crack. The value of $K_{\text{Ic}}$ is regarded as a material fracture parameter in linear elastic fracture mechanics.

### 7.2.2 Griffith Strain Energy Release Rate

An energy-based fracture criterion was first developed by Griffith. Crack propagation can also be described by an energy-based criterion. The energy-based fracture criterion was built upon an equilibrium state of a structure with a crack. To demonstrate such an equilibrium state, one might consider a unit thickness plate with a crack subjected to tension, as shown in Figure 7-6. For simplicity, body forces are neglected. The total potential energy in the structure can be expressed as:

$$\Pi = U - F + W \tag{7-23}$$

where $U = U(a, \varepsilon)$ is the strain energy of the structure and a function of crack length and strain, $F$ the work done by the external (applied) load, and $W$ the energy for crack formation. The condition

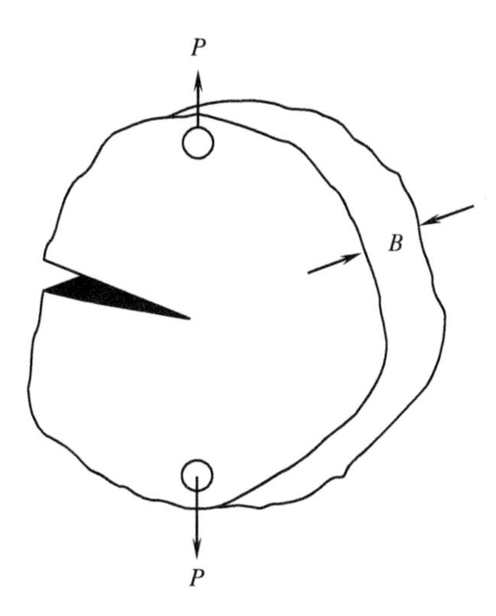


**Figure 7-6** A specimen with an edge crack under loading

for the structure to have an equilibrium state is that the first-order derivative of total potential energy equals zero during an infinitesimal crack extension, *da*, and this results in:

$$\frac{\partial}{\partial a}(F-U)=\frac{\partial W}{\partial a} \tag{7-24}$$

It should be noted that since a change of the equilibrium state results from a change of the crack length, the equation is taken as a derivative with respect to the crack length, *a*. By introducing *G* as the strain energy release rate for the propagation of a unit length of crack in a structure with unit thickness, we have

$$G=\frac{1}{B}\frac{\partial}{\partial a}(F-U) \tag{7-25}$$

Consider a cracked plate of thickness *B* under a load *P*, as shown in Figure 7-6. Under an action of the load, the load-application points undergo a relative displacement *U*. For a fixed load case, when the crack increases in size by an amount *da*, and the displacement will increase by an amount of *dv*. Hence, the work done by the external force is *Pdv*. Thus, we have

$$G=\frac{1}{B}\left(\frac{\partial F}{\partial a}-\frac{\partial U}{\partial a}\right)=\frac{1}{B}\left(P\frac{\partial v}{\partial a}-\frac{\partial U}{\partial a}\right) \tag{7-26}$$

Since $U = 0.5\ Pv = 0.5CP^2$, we can write Equation (7-26) as

$$G=\frac{1}{B}\left(P^2\frac{\partial C}{\partial a}+CP\frac{\partial P}{\partial a}-\frac{1}{2}P^2\frac{\partial C}{\partial a}-CP\frac{\partial P}{\partial a}\right)=\frac{P^2}{2B}\frac{\partial C}{\partial a} \tag{7-27}$$

For the condition of the fixed displacement, we have

$$G=-\frac{P^2}{2B}\frac{\partial C}{\partial a} \tag{7-28}$$

Define $G_c = 1/B(dW/da)$ as the critical strain energy release rate of the materials, then $G = G_c$ represents the condition for the equilibrium state of the structure during crack propagation. Since any propagation of an initial crack means catastrophic failure of a structure made of linearly elastic materials, it can be used as a failure criterion for such materials. Generally, the strain energy release rate, $G$, is a function of the applied load, structural geometry and boundary conditions, whereas the critical strain energy release rate, $G_c$, is a material fracture constant for linearly elastic materials. Since $G$ provides the energy for a crack extension, it is also termed the crack driving force.

The value of $G$ can be evaluated based on a load-deflection curve. It has been shown that for a linearly elastic material under a plane stress condition a relationship between $G$ and $K$ exists as

$$G_{\mathrm{I}} = \frac{K_{\mathrm{I}}^2}{E} \tag{7-29}$$

where $E$ is Young's modulus. For the plane strain case, we have

$$G_{\mathrm{I}} = \frac{K_{\mathrm{I}}^2}{\left(1 - \nu^2\right) E} \tag{7-30}$$

## 7.3 THE CRACK TIP PLASTIC ZONE

According to the elastic stress field solutions, a stress singularity exists at the tip of an elastic crack. However, in practice, materials tend to exhibit a yield stress, above which they deform plastically. There is always a small plastic zone around a crack tip in a metal, and a stress singularity cannot exist. The plastic region is known as the crack tip plastic zone. A rough estimate of the size of the plastic zone is simple to make. For a plane stress case, Figure 7-7 shows the magnitude of stress $\sigma_y$ in the plane $\theta = 0$.

At a distance $r^*_{\mathrm{p}}$ from the crack tip the stress is higher than the yield stress $\sigma_{\mathrm{ys}}$. To a first approximation, this $r^*_{\mathrm{p}}$ is the size of plastic zone. The $r^*_{\mathrm{p}}$ can be calculated by substituting $\sigma_{\mathrm{ys}}$ with $\sigma_{\mathrm{y}}$:

$$\sigma_y = \frac{K_{\mathrm{I}}}{\sqrt{2\pi r_{\mathrm{p}}^*}} = \sigma_{\mathrm{ys}} \tag{7-31}$$

and we have

$$r_{\mathrm{p}}^* = \frac{K_{\mathrm{I}}^2}{2\pi\sigma_{\mathrm{ys}}^2} = \frac{\sigma^2 a}{2\sigma_{\mathrm{ys}}^2} \tag{7-32}$$

However, the size of the actual plastic zone must be larger than $r^*_{\mathrm{p}}$ because it will not satisfy equilibrium by just simply cutting off the stress area above the yield stress. To estimate the actual

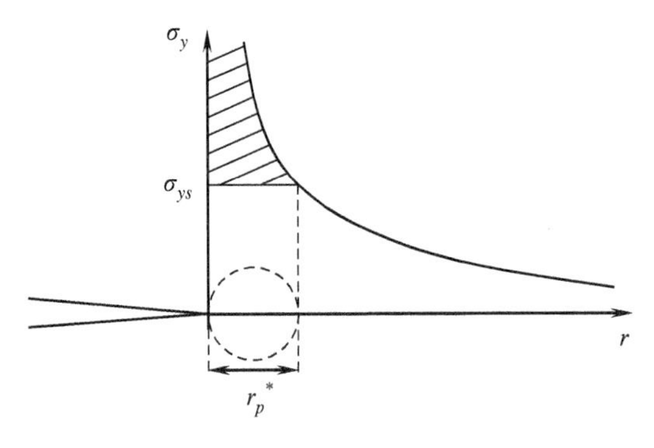


**Figure 7-7** Estimation of the plastic zone in front of a crack tip

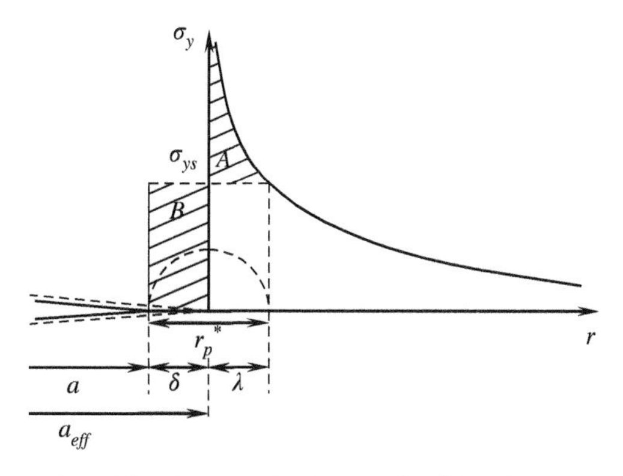


**Figure 7-8** Effective crack size method for the plastic zone

size of the plastic zone, Irwin suggested that an effective crack size method which shifted the stress curve to front of the crack tip a distance of $\delta$, as shown in Figure 7-8. Note that $\delta$ must be large enough to balance the load that is carried by the area of the singularity hat.

$$\delta\sigma_{ys} = \int_0^{\lambda} \sigma_{yy} dr - \lambda\sigma_{ys} \tag{7-33}$$

After substituting the expression for $\sigma_{yy}$ into Equation 7-33, and integral, we get

$$(\delta + \lambda)\,\sigma_{ys} = \sigma\sqrt{2a\lambda} \tag{7-34}$$

It can be further rewritten as:

$$(\delta + \lambda)^2 = \frac{\sigma^2 2a\lambda}{\sigma_{ys}^2} = \frac{2K_I^2\lambda}{\pi\sigma_{ys}^2} = 4\lambda^2 \tag{7-35}$$

Since $\lambda << a$, we have

$$\lambda = \frac{\sigma^2\pi\,(a + \lambda)}{2\pi\sigma_{ys}^2} \approx r_p^* \tag{7-36}$$

Therefore, it can be verified that the actual size of the plastic zone is twice that of the value of $r^*_p$, i.e.,

$$r_p = 2r_p^* = 2\frac{K_I^2}{2\pi\sigma_{ys}^2} = \frac{K_I^2}{\pi\sigma_{ys}^2} \tag{7-37}$$

A different approach to find the size of the plastic zone was proposed by Dugdale (1960). Dugdale assumed the length of the plastic zone to be much greater than the thickness of the sheet and modeled the plastic zone as a yield strip ahead of the crack tip. Dugdale applied the superposition method to calculate the size of the plastic zone. The detailed procedures are stated as follows.

It was assumed that the crack had a crack length of $a + \rho$ by considering the plastic zone. However, the crack edge, a distance of $\rho$ in front of the physical crack tip, was under the action of the yield stress, $\sigma_{ys}$, tending to close the crack in this region. Part of $\rho$ was not really cracked; the material could still bear the yield stress. The size of $\rho$ was chosen such that the stress singularity disappears: $K$ should be zero. This means that the stress intensity factor $K_\sigma$, due to uniform stress $\sigma$, had to be compensated by the stress intensity factor $K_\rho$, due to the edge closing stress, $\sigma_{ys}$.

$$K_\sigma = -K_\rho \tag{7-38}$$

This equation permitted determination of ρ in the following manner. The stress intensity factor at crack tip is due to wedge forces $p$ shown in Figure 7-9, and given as

$$K_A = \frac{p}{\sqrt{\pi a}}\sqrt{\frac{a+x}{a-x}} \tag{7-39}$$

And for the case shown in Figure 7-10, the stress intensity factor was given by

$$K_B = \frac{p}{\sqrt{\pi a}}\sqrt{\frac{a-x}{a+x}} \tag{7-40}$$

If the wedge forces were distributed from $s$ to the crack tip (as in the Dugdale case), the stress intensity becomes

$$K_p = \frac{p}{\sqrt{\pi a}}\int_s^a \left[\sqrt{\frac{a+x}{a-x}} + \sqrt{\frac{a-x}{a+x}}\right] dx \tag{7-41}$$

The integral result is

$$K_p = 2p\sqrt{\frac{a}{\pi}}\arccos\frac{s}{a} \tag{7-42}$$

Applying this result to the Dugdale crack, the integral had to be taken from $s = a$ to $a + \rho$. Hence a had to be substituted for $s$ and $a + \rho$ for $a$, while $p = \sigma_{ys}$.

$$K_p = 2\sigma_y\sqrt{\frac{a+\rho}{\pi}}\arccos\frac{a}{a+\rho} \tag{7-43}$$

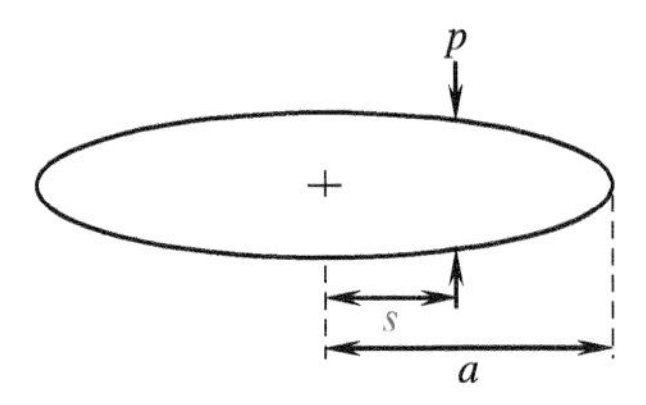


**Figure 7-9** A crack under a pair of wedge force on the right side

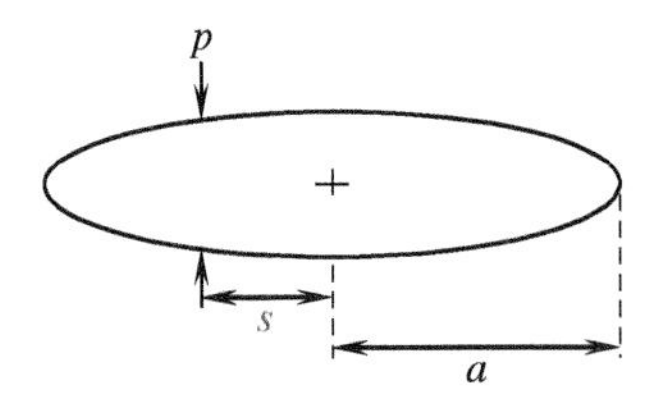


**Figure 7-10** A crack under a pair of wedge forces on the left side

According to Dugdale, this stress intensity should be equal to $K_\sigma$,

$$K_\sigma = \sigma\sqrt{(a+\rho)\,\pi} \tag{7-44}$$

Then it followed that $\rho$ could be determined from:

$$\frac{a}{a+\rho} = \cos\frac{\pi\sigma}{2\sigma_y} \tag{7-45}$$

By expanding the two sides in a Taylor series and neglecting the higher-order terms in the series development of the cosine, $\rho$ was found as:

$$\rho = \frac{\pi^2\sigma^2 a}{8\sigma_{ys}^2} = \frac{\pi K^2}{8\sigma_{ys}^2} \tag{7-46}$$

This result can be compared with

$$r_p = \frac{K_I^2}{\pi\sigma_{ys}^2} \tag{7-47}$$

because the value of π is close to $8/\pi$. When the plastic zone is small compared to the crack size, LEFM can still be used to find the stress and displacement distributions in front of the crack tip. However, if the plastic zone is larger with respect to the crack, the application of LEFM is doubtful, because of the validity of the expressions for $K$ which are based on elastic solutions only.

## 7.4 CRACK TIP-OPENING DISPLACEMENT

A certain size of the inelastic zone always exists at a crack tip in a real material. The presence of the inelastic zone allows the crack to open a small amount. The crack opening displacement (COD) is a general term and can be at any point along a crack. The crack tip opening displacement (CTOD) is the displacement at the point of the crack tip. Therefore, COD is a function but CTOD is a number. In LEFM, the critical value of CTOD is related to the critical value of SIF or energy release rate. Hence, one may use crack tip opening displacement as a fracture criterion. To better understand the CTOD fracture criterion, let us consider an infinite plate with a crack of size $2a$, subjected to tension, as shown in Figure 7-11.

The crack opening displacement (COD) for the structure can be expressed as:

$$\text{COD} = \frac{4\sigma}{E}\sqrt{a^2 - x^2} \tag{7-48}$$

It can be seen that if there is no crack propagation, when $x = a$, COD = 0. When $x = 0$, COD reaches maximum,

$$\text{COD}_{x=0} = \frac{4\sigma}{E}a \tag{7-49}$$

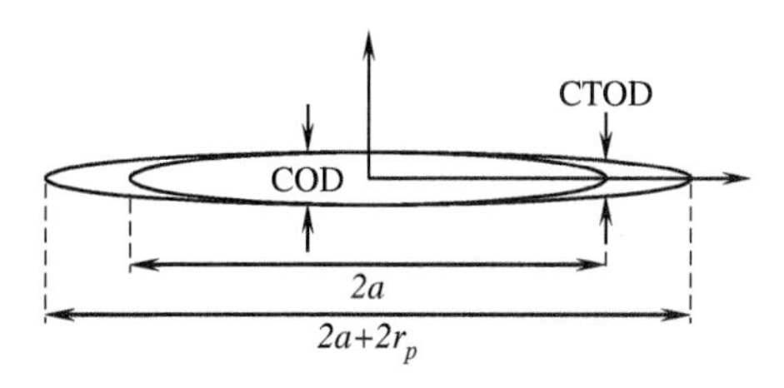


**Figure 7-11** Sketch of crack opening displacement

Since the stress is usually two or three orders of magnitude smaller than Young's modulus, COD is much smaller than the crack length. By introducing the crack length extension, $r_p$, we can write,

$$\text{COD} = \frac{4\sigma}{E}\sqrt{(a + r_p)^2 - x^2} \tag{7-50}$$

where $a + r_p$ is the effective crack size. Since CTOD is the value of COD at the position of $x = a$, it follows:

$$\text{CTOD} = \frac{4\sigma}{E}\sqrt{2ar_p} \tag{7-51}$$

It can be shown that

$$r_p = \frac{K_I^2}{\pi\sigma_{ys}^2} \tag{7-52}$$

where $\sigma_{ys}$ is the yield stress of the material. Substituting $r_p$ into Equation (7-51), we have

$$\text{CTOD} = \frac{4\sqrt{2}K_I^2}{\pi\sigma_{ys}E} = \frac{4\sqrt{2}}{\pi\sigma_{ys}}G_I \tag{7-53}$$

It can be further written as

$$\text{CTOD} = \frac{K_I^2}{m\sigma_{ys}E} = \frac{G_I}{m\sigma_{ys}} \tag{7-54}$$

where $m = \frac{\pi}{4\sqrt{2}}$ and is a constant. When CTOD reaches its critical value, the crack will propagate. Thus, we have one more criterion for the unstable crack as

$$\text{CTOD} = \text{CTOD}_c \tag{7-55}$$

The parameters $K_{Ic}$ and $G_{Ic}$ can be related to $\text{CTOD}_c$ as:

$$K_{Ic} = \sqrt{mE\sigma_{ys}\text{CTOD}_c} \tag{7-56}$$

and

$$G_{Ic} = m\sigma_{ys}\text{CTOD}_c \tag{7-57}$$

Thus, once $\text{CTOD}_c$ is known, $K_{Ic}$ and $G_{Ic}$ can be obtained from the equations.

## 7.5 FRACTURE PROCESS IN CONCRETE

To study fracture progress in concrete, uniaxial tension tests have been carried out with monitoring of the microcrack occurrence using acoustic emission technique. The test set-up is shown in Figure 7-12. As can be seen from Figure 7-12, four LVDTs (linear variable differential transformers) are used to cover the whole test portion of the concrete specimen. An adoptive control method allows the control mode to be switched among the LVDTs and hence the major crack formation can be captured by one of them, and the opening displacement of the major crack can be used as a feedback signal to control the movement of the actuator of the test machine. Six acoustic emission transducers are placed on the surfaces of the specimens. They are used to detect the wave signal generated by the occurrence and the propagation of cracks during the loading process. The acoustic emission information can be used to locate the positions of the microcrack or propagation path of major cracks and to interpret the fracture properties of concrete.

**Figure 7-12** Uniaxial tension test set-up for un-notched specimens

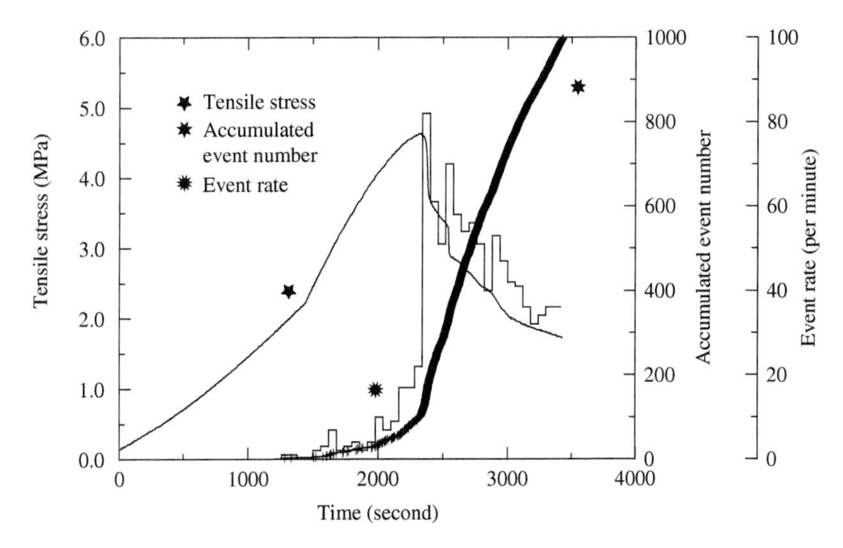


**Figure 7-13** Tensile stress and accumulated acoustic events as function of time

Figure 7-13 shows the tensile stress, accumulated AE event number, and AE event rate as a function of time for a plain concrete tension specimen. The AE data analysis can be divided into five regimes according to the stress level (loading stage) for the convenience of analyses. In the first regime, which is the initial portion of the stress-displacement curve from the start of testing to when the stress reaches about 40% of the peak load, no AE event occurred, implying that there was no microcrack. In the second regime, where the stress level is between 0.4 and $0.8f_t$, AE activity starts to be detectable at around 40% of the peak stress $f_t$, and the AE events gradually increase with the stress increase but with a small slope. At this stress level range, the defects, voids, and other pre-existing flaws in the mortar-aggregate interfaces within the concrete begin to form microcracks and propagate. Later on, with the increment of the external load, AE events gradually increase as more and more damage forms. When the stress level closes to $0.8f_t$, the AE event rate becomes much higher, which implies that AE activity is closely related to the stress level before the peak stress. The interface defects continue to propagate and branch inside the mortar matrix. Matrix cracks (and also voids) begin to link. The consequent propagation and linking of these cracks will lead to the formation of a macrocrack, in the ensuing regime.

A significant increase of the AE event rate occurs when the stress approaches the peak value, $f_t$, which implies that a major crack may have formed, and begins to propagate. The fracture process zone around the crack tip becomes the major emission source. After this loading stage, with the stress level decreased, the AE event rate becomes relatively lower. The map of the microcrack can be obtained by analysis of AE event source locations. Figure 7-14 shows the distribution of the microcracks that occurred during the loading stage between 0.4 to $0.8f_t$, between pre-peak $0.8f_t$ to peak stress, between peak to post-peak stress $0.8f_t$. It can be seen from Figure 7-14(a) that the AE events registered on the source location maps, or the microcracking activities, are randomly and relatively sparsely distributed in the test portion of the specimen. When the stress level proceeds into the regime between pre-peak $0.8f_t$ to peak stress, significant AE events take place in a narrow region of the specimen, clearly showing a strain localization. In the stage from peak stress $f_t$ to post-peak $0.8f_t$, the localization phenomenon become more obvious (see Figure 7-14(c)). During this loading stage, emissions from those defects within the specimen tend to be dominated by a localized area, namely, around a macrocrack (Li and Li, 2000). A similar phenomenon has also been reported by Li and Shah (1994).

One more phenomenon can be observed from the microcrack map is that the area spread by microcracks seems quite large, which is quite different from the case in metal. This phenomenon could be evidenced by other investigations that focused on the fracture process zone, especially its size and its distribution. Castro-Montero et al. (1990) used laser holographic interferometry,

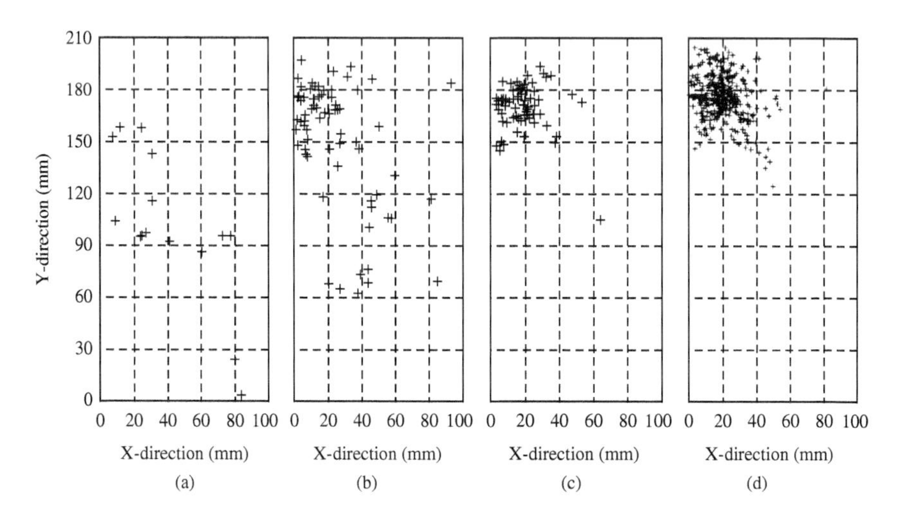


**Figure 7-14** The distribution of microcracks which occurred during the loading stage between (a) 0.4 to $0.8f_t$, (b) pre-peak $0.8f_t$ to peak stress, (c) peak to post-peak $0.8f_t$, and (d) after post-peak $0.8f_t$

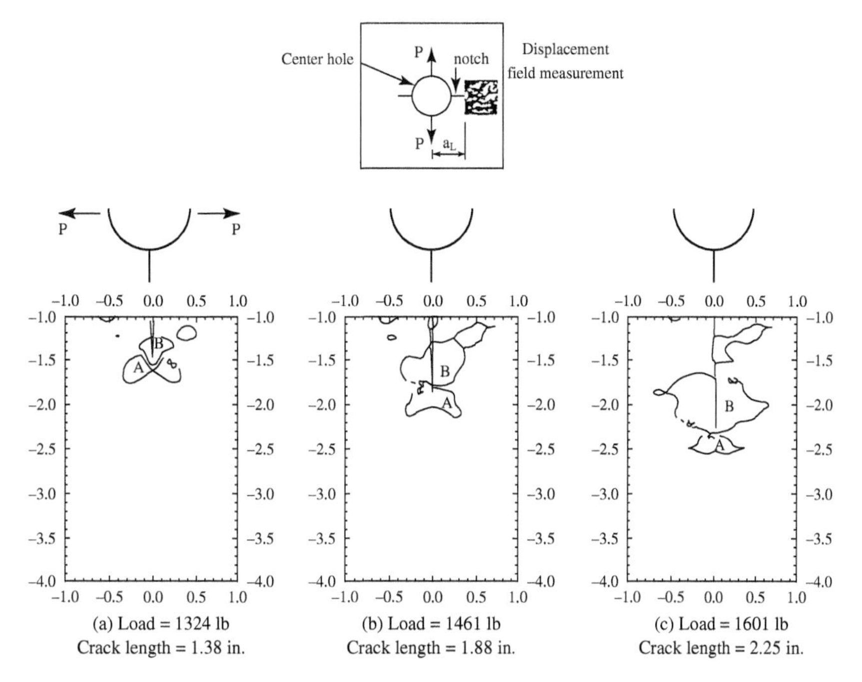


**Figure 7-15** Fracture test with laser holographic interferometry observations

with an accuracy of a quarter of a micron, to study displacement fields at the crack tip. The region where the experimentally determined strain field differed significantly from the LEFM solutions was taken to be the fracture process zone. As shown in Figure 7-15, they reported a large wake zone (labeled B) behind the observed crack-tip. This zone increased with crack extension, whereas the zone in front of the tip (labeled A) remained practically consistent in size. They concluded that most of the toughening in concrete occurs in the wake zone. This fact may suggest that the increasing size of the wake process zone rather than the constant size of the crack-tip process zone should be primarily responsible for the growing R-curve behavior in concrete. Similar observations have been reported for other quasi-brittle materials (Sakai et al., 1988; Homeny and Vaughn, 1990). By using the dye-penetration method, Swartz and Refai (1989) found that the fracture process zone varies along the specimen thickness.

The large area spread by microcracks in concrete is termed microcrack shielding. The reasons for the shielding are that during fracture, the high stress state near the crack tip causes the pre-existing flaws, which resulted from the water-filled pores, air voids acquired during casting, and shrinkage and thermal shock due to the hydration process, to grow into microcracks. Microcrack shielding is a major characteristic of the fracture processing zone in concrete. Other characteristics for the fracture process zone in concrete are further indicated in Figure 7-16. Crack deflection occurs when the path of least resistance is around a relatively strong particle or along with a weak interface (see Figure 7-16b). This mechanism has been studied in detail by Faber and Evans (1983). Other important toughening processes in concrete are grain bridging (Van Mier, 1991), as shown in Figure 7-16c. Bridging occurs when the crack has advanced beyond an aggregate that continues to transmit stresses across the crack until it ruptures or is pulled out. Also, during grain pullout, or the opening of a tortuous crack, there may be some contact (or interlock) between the faces (see Figure 7-16d). This causes energy dissipation through friction, and some bridging across the crack.

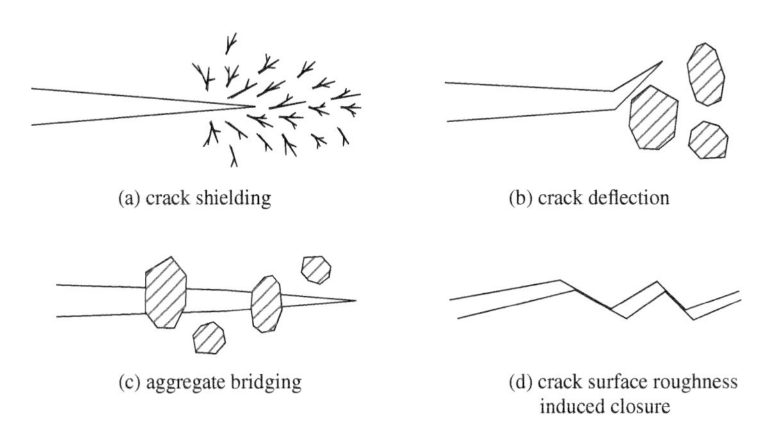


**Figure 7-16** Possible mechanisms for the fracture process zone in concrete

It is obvious that a crack process zone in concrete is not small. In this case, the applicability of LEFM solutions is questionable although LEFM can still be applied when there is a small yield zone, as we indicated earlier.

Shah and McGarry (1971) conducted a three-point bending test on cement paste, mortar and concrete specimens. They found that hardened cement paste was a notch sensitive material. The presence of a notch significantly reduced the tensile or flexural strength of the paste. On the other hand, mortar and concrete, with the normally used amounts and volume of gravel aggregates, were notch insensitive materials. Even for cement paste that showed notch sensitivity there was no brittle behavior, and the direct application of LEFM had to be verified. Higgins and Bailey (1976) tested three-point bending beams with different depths but the same cement paste, and calculated the critical stress intensity factor using linear elastic fracture mechanics. However, they found values of the critical stress intensity factor to be size-dependent, as shown in Figure 7-17. This was due to the fact that linear elastic fracture mechanics did not take the stable crack growth associated

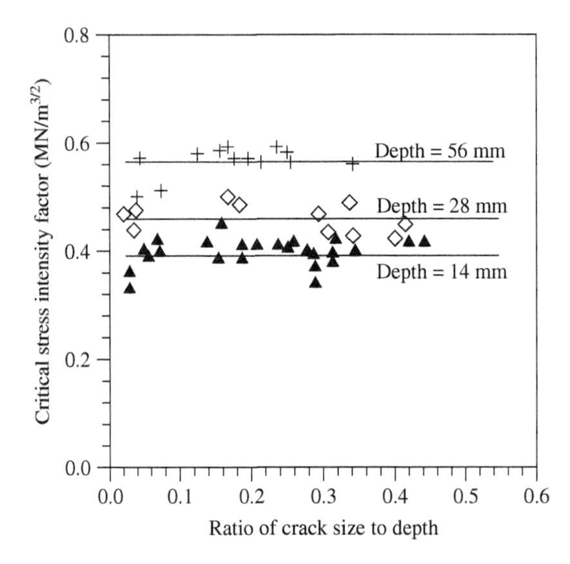


**Figure 7-17** Dependency of the stress intensity factor on the crack size in concrete

with the fracture process zone into account. In other words, in linear elastic fracture mechanics, the initial crack length, rather than the sum of the initial crack length and the crack extension, was used to determine the critical stress intensity factor.

The experimental results have demonstrated that LEFM cannot be directly applied to cement-based composites, especially concrete. For LEFM materials, one parameter is sufficient to describe its failure toughness. For concrete if only one parameter is considered, then one observes that the fracture toughness increases with increasing compressive strength or increasing strain rate. Such single parameter representation is misleading since concrete becomes more brittle as its compressive strength increases. The presence of a sizable fracture process zone and its contribution to the fracture toughness of concrete have to be taken into consideration for concrete fracture mechanics. In other words, nonlinear fracture mechanics should be used for modeling concrete fracture.

## 7.6 NONLINEAR FRACTURE MECHANICS FOR CONCRETE

Fracture behavior of concrete is greatly influenced by its fracture process zone. Since the work of Kaplan (1961), many attempts have been made to apply fracture mechanics to concrete. As elaborated in Section 7.5, linear elastic fracture mechanics cannot be directly applied to concrete due to the sizeable fracture process zone, aggregate bridging, and tortuous, zigzag crack path. It is difficult to determine the exact position of the crack tip in concrete due to particles bridging and variation of the fracture process zone along the thickness direction. An accurate description of concrete fracture needs to specify the torturous three-dimensional crack path, as well as the inelastic material response within the fracture process zone. In order to simplify the application of fracture mechanics to the failure of concrete, most recently, available models attempt to simulate mode I concrete fracture with an effective straight line crack. The variation of the fracture process zone along the structure thickness or width is usually neglected. The inelastic fracture process zone may be taken into account by a cohesive pressure acting on the crack faces.

By analoging Dugdale's model, an effective inelastic crack or effective quasi-brittle crack in a concrete can be drawn. Figure 7-18 shows such a crack, where the initial crack length is noted as $a_o$ and the length of the associated fracture process zone is presented by a crack extension of $\Delta a$. Since the crack extension does have an ability in consuming energy, a continuously distributed cohesive stress is modeled over $\Delta a$.

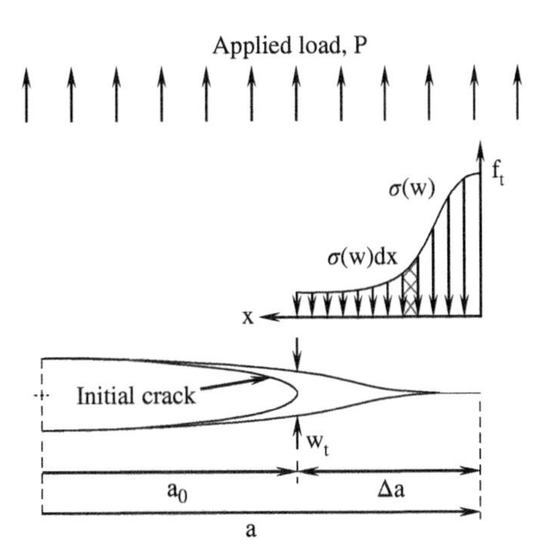


**Figure 7-18** Modeling of cohesive stress for a quasi-brittle crack

Differing from the yield stress in Dugdale's model for metals, the cohesive pressure, $\sigma(w)$, in concrete has to be modeled as a monotonic decreasing function of crack opening displacement, $w$, because of its strain softening behavior. The shape of the cohesive pressure function, $\sigma(w)$, can be determined by referring to the post-peak branch of the stress-displacement curve of concrete under uniaxial tension as a function of the crack opening displacement. The value of $\sigma(w)$ is set equal to the concrete tensile strength $f_t$ at the end of the fracture process zone where $w = 0$ and is gradually reduced towards the original crack tip. The value of $\sigma(w)$ at the original crack tip is usually set to 0 to simplify the computation process. This cohesive pressure tends to close the crack, just like the yield stress in Dugdale's model. It should be noted that the effective crack extension is usually not the same as the length of the real fracture process zone because of the difficulty in determining the variation of the actual fracture process zone along with the thickness or the width direction.

When a concrete structure with an effective quasi-brittle crack is subjected to loading, the applied load will result, at a certain level, in an energy release rate $G$ at the tip of the effective quasi-brittle crack. The energy release rate $G$ is composed of two portions: (1) the energy release rate in creating two new surfaces generated by remote stress during specimen loading (the material surface energy), and (2) the energy release rate in overcoming the local cohesive pressure, $\sigma(w)$ during the separation of the new surfaces. As a result, the total energy release rate, $G_{\mathrm{It}}$ for a mode I quasi-brittle crack can be expressed as,

$$G_{\mathrm{It}} = G_{\mathrm{I}\sigma} + G_{\sigma(w)} = G_{\mathrm{I}\sigma} + \int_0^{\mathrm{CTOD}} \sigma(w)\,dw \tag{7-58}$$

where $G_{\mathrm{I}\sigma}$ is the strain energy rate to create two new crack surfaces due to remote stress for a mode I crack, $\sigma(w)$ the normal traction pressure which is a function of the crack opening displacement $w$, as explained earlier, and CTOD the crack tip opening displacement. The crack shape is assumed to be a line. The two terms, representing two types of energy dissipation mechanisms for the fracture process, have been introduced in the equation. The Griffith-Irwin energy dissipation mechanism is represented by a non-zero stress intensity factor and the Dugdale-Barenblatt energy dissipation mechanism is represented by the traction term. It seems that it is proper to use these two energy dissipation mechanisms to describe the propagation of a quasi-brittle crack. However, one may approximately use models only based on a single fracture energy dissipation mechanism, either the Griffith-Irwin energy dissipation mechanism by assuming $\sigma(w) = 0$, or the Dugdale-Barenblatt energy dissipation mechanism by assuming $K_{\mathrm{I}} = 0$ to simplify the mathematical derivation. Based on the different energy dissipation mechanisms used, nonlinear fracture mechanics models for quasi-brittle materials may be classified as a fictitious crack approach and an equivalent-elastic crack approach (or an effective-elastic crack approach). Fracture models only using the Dugdale-Barenblatt energy dissipation mechanism are usually referred to as the fictitious crack approach, whereas fracture models using only the Griffith-Irwin energy dissipation mechanism are usually referred to as the effective-elastic crack approach.

In the category of the fictitious crack approach, Hillerborg et al. (1976) have proposed a cohesive model with $K_{\mathrm{I}} = 0$. Bažant and Oh (1983) have developed a crack band model using a similar concept. It is noted that by assuming $K_{\mathrm{I}} = 0$ some computational efficiency may be accomplished. However, a single Dugdale-Barenblatt energy dissipation mechanism may only achieve a global energy balance by selecting model parameters. Some actual features associated with crack propagation, such as crack profile, computed based on the pure cohesive model, may not match with those experimentally measured.

In the effective-elastic crack approach category, two representative models are the two-parameter model and the size effect model. In the effective-elastic crack models, instead of the original crack length, an effective-elastic crack length is used. The equivalence between the actual and the corresponding effective crack can be prescribed explicitly. Since a stable crack extension before the peak-load is present in concrete, the critical effective-elastic crack length at the failure of the materials is different from the initial crack length. To predict the failure condition of the materials, two conditions are needed to determine the critical load and the corresponding crack length. Most of these effective crack models use two or more fracture parameters to define the inelastic fracture process.

## 7.7 TWO-PARAMETER FRACTURE MODEL

### 7.7.1 The Model

Using the concept of the effective elastic crack, Jenq and Shah (1985) proposed a two-parameter fracture model. In their model, two parameters, crack tip opening displacement (CTOD) and stress intensity factor ($K_I$), corresponding to the effective-elastic crack, are used to describe the fracture properties of concrete. Jenq and Shah used a beam under three-point bending for the investigation. Figure 7-19 shows the geometrical feature and loading pattern for such a specimen. It can be seen from Figure 7-19 that it is easier to measure the crack mouth opening displacement (CMOD) rather than CTOD, because the crack tip is difficult to capture.

Figure 7-20 shows a typical stress-CMOD curve of loading and unloading. It can clearly be seen that after unloading from the peak stress to zero stress, CMOD cannot return to zero, which means that CMOD contains some plastic deformation, and so does CTOD. Figure 7-21 illustrates the relationship between CMOD and CTOD.

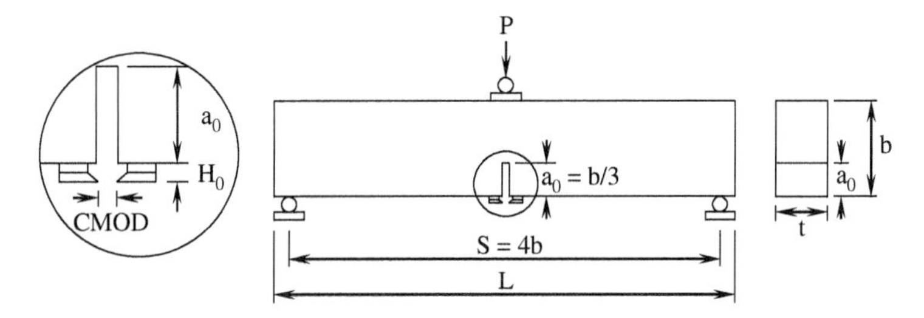


**Figure 7-19** Three-point bending test set-up for the two-parameter model

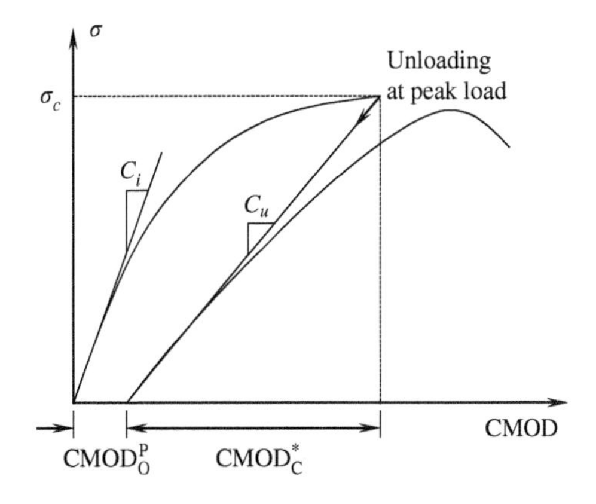


**Figure 7-20** Loading and unloading procedure

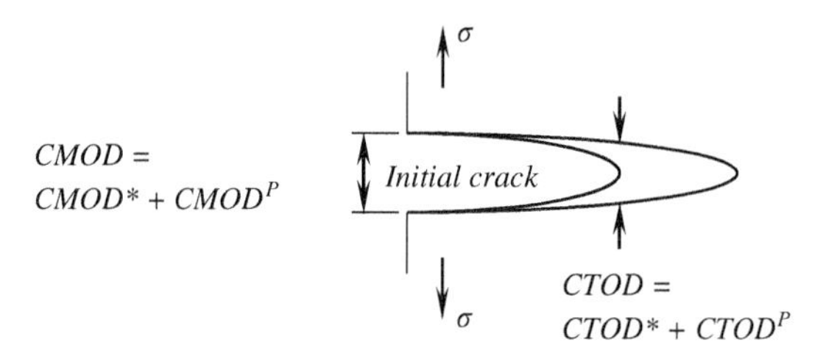


**Figure 7-21** The schematic of CMOD and CTOD

If accounting for the quasi-brittle features of concrete, the $\mathrm{CMOD_c}$, the CMOD value at peak load, can be divided into two parts, the elastic and plastic parts as:

$$\mathrm{CMOD_c} = \mathrm{CMOD_c^e} + \mathrm{CMOD_c^p} \tag{7-59}$$

The procedures to separate these two components involve a loading and unloading process. The measured value of $\mathrm{CTOD_c^e}$ and the maximum stress $\sigma_c$, can be further substituted into the following LEFM equations to calculate the critical effective elastic crack length, $a_c$, and then the critical stress intensity factor, $K_{Ic}^s$, based on maximum stress $\sigma_c$ and critical effective-elastic crack length, $a_c$,

$$\mathrm{CMOD_c^e} = \frac{4\sigma_c a_c}{E_t} g_2\left(\frac{a_c}{b}\right) \tag{7-60}$$

and

$$K_{Ic}^s = \sigma_c \sqrt{\pi a_c} g_1\left(a_c/b\right) \tag{7-61}$$

The value of the critical crack tip opening displacement at peak load $\mathrm{CTOD_c}$ then can be determined based on the obtained values of $\mathrm{CTOD_c^e}$,

$$\mathrm{CTOD_c^e} = \mathrm{CMOD_c^e} g_3\left(\frac{a_c}{b}\right) \tag{7-62}$$

In the above three equations, $g_1$, $g_2$ and $g_3$ are geometrical functions for calculating $K_{Ic}^S$, $\mathrm{CMOD_c^e}$, and $\mathrm{CTOD_c^e}$, respectively. As an example, for a three-point bending beam with a span to height ratio of 4, the expressions are (Murakami et al., 1987):

$$g_1\left(\frac{a_c}{b}\right) = \frac{1.99 - \frac{a_c}{b}\left(1 - \frac{a_c}{b}\right)\left[2.15 - 3.93\frac{a_c}{b} + 2.70\left(\frac{a_c}{b}\right)^2\right]}{\sqrt{\pi}\left(1 + 2\frac{a_c}{b}\right)\left(1 - \frac{a_c}{b}\right)^{3/2}} \tag{7-63}$$

$$g_2\left(\frac{a_c}{b}\right) = \frac{1.73 - 0.85\frac{a_c}{b} + 31.2\left(\frac{a_c}{b}\right)^2 - 46.3\left(\frac{a_c}{b}\right)^3 + 2.70\left(\frac{a_c}{b}\right)^4}{\left(1 - \frac{a_c}{b}\right)^{3/2}} \tag{7-64}$$

$$g_3\left(\frac{a_c}{b}, \frac{a_0}{a_c}\right) = \left\{\left(1 - \frac{a_0}{a_c}\right)^2 + \left(1.081 - 1.149\frac{a_c}{b}\right)\left[\frac{a_0}{a_c} - \left(\frac{a_0}{a_c}\right)^2\right]\right\}^{1/2} \tag{7-65}$$

From the experimental results obtained by using beams with three-point bending, Jenq and Shah (1985) found that for the beams with different sizes but made of the same material, the values of $K_{Ic}^S$ and $\mathrm{CTOD_c^e}$ were basically constant. As a result, they proposed that the critical fracture property of a quasi-brittle material might be characterized by the values of $K_{Ic}^S$ and $\mathrm{CTOD_c^e}$. For a given material, structures with different geometry and sizes when subjected to the critical fracture load (the peak load) would satisfy the following two conditions:

$$K_I = K_{Ic}^S$$
$$CTOD = CTOD_c \tag{7-66}$$

where $K_I$ and CTOD are the stress intensity factor and the crack tip opening displacement, which can be calculated based on LEFM. Also, it is noted that $CTOD_c^e$ is simply denoted as $CTOD_c$. This is why it is called the two-parameter model. It should be emphasized that $K_I$ and CTOD are functions of the applied load, structural geometry and size, as well as the crack length, whereas $K_{Ic}^S$ and $CTOD_c^e$, which are defined in terms of the critical effective elastic crack, are fracture material parameters depending only on the material. Only when the fracture load and crack tip opening displacement meet the two necessary conditions, will a concrete structure start fracture failure. It is seen from the above-mentioned procedure of determining the values of $K_{Ic}^S$ and $CTOD_c^e$ in the two-parameter fracture model that the effective-elastic crack exhibits compliance equal to the unloading compliance of the actual structure. Therefore, the two-parameter model determines the critical fracture state based on its elastic response.

The existence of $CTOD_c$ can be justified by the fact that all materials have some sort of initial flaw. This is especially true for concrete. When the material is subjected to an external load, these flaws will open, propagate and sometimes coalesce with other cracks. The opening displacement can be directly linked to the value of CTOD. At the critical fracture load, the crack opening displacement at the initial crack tip can be defined as $CTOD_c$. For quasi-brittle materials such as concrete, the stable crack extension occurs before the critical fracture. Since $CTOD_c$ is defined at the tip of the initial crack, it may primarily account for the growth in the size of the weak process zone.

### 7.7.2 Determination of Fracture Parameters for the Two-Parameter Model

The RILEM Technical Committee 89-FMT on Fracture Mechanics of Concrete Test Methods has proposed a method for the determination of fracture parameters based on the two-parameter model. It recommended using the three-point bending beam with $s/b = 4$ to obtain $K_{Ic}$ and $CTOD_c$. The size of the beam depends on the maximum size of the aggregate and is listed in Table 7-1.

A typical beam is shown in Figure 7-22. As can be seen from Figure 7-22, the initial notch-to-depth ratio should be 1/3 and the notch width should be less than 5 mm. After casting, the

**Table 7-1** Relationship between maximum size of aggregates and size of the beam

| $D_{max}$ (mm) | Depth, $D$ (mm) | Width, $B$ (mm) | Length, $L$ (mm) | Span, $S$ (mm) |
|---|---|---|---|---|
| 1 to 25 | $150 \pm 5$ | $80 \pm 5$ | $700 \pm 10$ | $600 \pm 5$ |
| 25.1 to 50 | $250 \pm 5$ | $150 \pm 5$ | $1{,}100 \pm 10$ | $1{,}000 \pm 5$ |

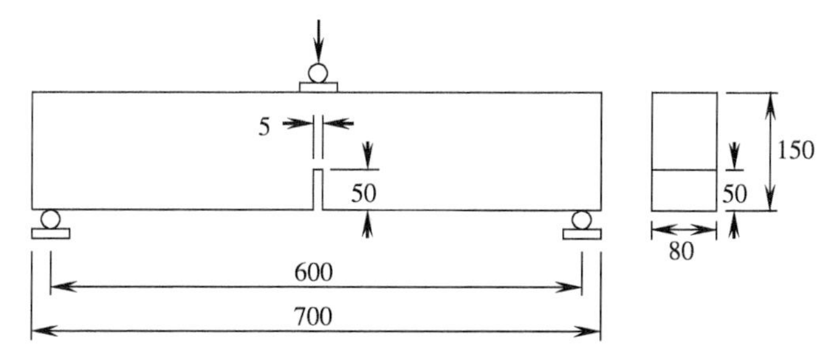


**Figure 7-22** A typical beam geometry for determination of fracture parameters (all dimensions are in mm)

specimens should be cured at 100% RH and at 23 ± 2°C until 4 hours before testing. A minimum of four specimens is required for each batch of concrete.

The test should be conducted with a closed-loop controlled materials testing machine, using CMOD as a feedback signal. The CMOD and the applied load should be recorded continuously during a test. It is preferred to use a clip gage to measure the CMOD. The loading rate should be controlled so that the peak load is reached in 5 minutes. After reaching the peak load in a monotonic increasing mode, an unloading process should be carried out before it passes 95% of the peak load. When the applied load reaches 0, reloading may be applied according to the test need.

As shown in Figure 7-20, a typical load-CMOD curve can provide information concerning initial stiffness (compliance), unloading stiffness (compliance), and peak load. To obtain the critical stress intensity factor $K_{Ic}^{S}$, and the critical crack tip opening displacement $CTOD_c$, the critical effective elastic length has to be determined. It can be done by the following procedures.

By substituting the stress expression into the expression for CMOD and noting that $C_i = CMOD/P$, $E$ can be calculated as:

$$E = \frac{6Sa_0 g_2(\alpha_0)}{C_i b^2 t} \tag{7-67}$$

where $C_i$ is the initial compliance calculated from the load-CMOD curve and $g_2(\alpha_0)$ is,

$$g_2(\alpha_0) = 0.76 - 2.28\alpha_0 + 3.78\alpha_0^2 - 2.04\alpha_0^3 + \frac{0.66}{(1-\alpha_0)^2} \tag{7-68}$$

where $\alpha_0 = (a_0 + \text{HO})/(b + \text{HO})$, in which HO is the height of the holding plates for the clip gage or LVDT and $a_0$ is the initial notch length.

Similarly, $E$ can also be expressed using unloading compliance, $C_u$, as:

$$E = \frac{6Sa_c g_2(\alpha_c)}{C_u b^2 t} \tag{7-69}$$

where $C_u$ is the unloading compliance within 95% of the peak load calculated from the load-CMOD curve, and the geometrical function $g_2(\alpha_c)$ is the same as the expression of $g_2(\alpha_0)$, except that $\alpha_c$ should be used instead of $\alpha_0$ as shown,

$$\alpha_c = \frac{a_c + \text{HO}}{b + \text{HO}} \tag{7-70}$$

Since the modulus of elasticity of concrete should be the same, i.e., $E = E$, the two expressions for $E$ above should be equal, so we have:

$$\frac{6Sa_0 g_2(\alpha_0)}{C_i b^2 t} = \frac{6Sa_c g_2(\alpha_c)}{C_u b^2 t} \tag{7-71}$$

Therefore, the critical crack length can be obtained as:

$$a_c = \frac{C_u a_0 g_2(\alpha_0)}{C_i g_2(\alpha_c)} \tag{7-72}$$

It should be noted that $g_2(\alpha_c)$ is also a function of $a_c$ and hence numerical procedures are usually needed to solve for $a_c$. After knowing $a_c$ and substituting the stress expression into the critical stress intensity factor expression, we have:

$$K_{\mathrm{Ic}}^{\mathrm{S}} = 3\left(P_{\mathrm{c}} + 0.5W_{\mathrm{h}}\right)\frac{S\sqrt{\pi a_{\mathrm{c}}}g_1\left(\alpha_{\mathrm{c}}\right)}{2b^2t} \tag{7-73}$$

where $P_{\mathrm{c}}$ is the peak load, $W_{\mathrm{h}} = W_0S/L$ and $W_0$ is the self-weight of the beam, and $g_1(a_{\mathrm{c}})$ is

$$g_1\left(\alpha_{\mathrm{c}}\right) = \frac{1.99 - \alpha_{\mathrm{c}}\left(1-\alpha_{\mathrm{c}}\right)\left[2.15 - 3.93\left(\alpha_{\mathrm{c}}\right) + 2.70\left(\alpha_{\mathrm{c}}\right)^2\right]}{\sqrt{\pi}\left(1+2\alpha_{\mathrm{c}}\right)\left(1-\alpha_{\mathrm{c}}\right)^{3/2}} \tag{7-74}$$

in which $\alpha_{\mathrm{c}} = a_{\mathrm{c}}/b$.

The $\mathrm{CTOD_c}$ can be calculated using the equation:

$$\begin{aligned}\mathrm{CTOD_c} &= \mathrm{CMOD_c}\sqrt{\left[\left(1-\beta_0\right)^2 + \left(1.081 - 1.149\alpha_{\mathrm{c}}\right)\left(\beta_0 - \beta_0^2\right)\right]}\\ &= \frac{6\left(P_{\mathrm{c}} + 0.5W_h\right)Sa_{\mathrm{c}}g_2\left(\alpha_{\mathrm{c}}\right)}{Eb^2t}\sqrt{\left[\left(1-\beta_0\right)^2 + \left(1.081 - 1.149\alpha_{\mathrm{c}}\right)\left(\beta_0 - \beta_0^2\right)\right]}\end{aligned} \tag{7-75}$$

where $\beta_0 = a_0/a_{\mathrm{c}}$.

It has been shown that the values of $K_{\mathrm{Ic}}^{\mathrm{S}}$ evaluated at the effective elastic crack are independent of the dimensions of the notched beams by a large number of round robin tests. However, $\mathrm{CTOD_c}$ has not been verified, since it is a small quantity and difficult to measure.

### 7.7.3 Some Applications of the Two-Parameter Model

**(a)** *Material length, Q*. When the values of $K_{\mathrm{Ic}}^{\mathrm{S}}$ and $\mathrm{CTOD_c^e}$ are known for a concrete, some material characteristic properties can be identified. For example, a material length, Q, can be defined as:

$$Q = \left(\frac{E \times \mathrm{CTOD_c}}{K_{\mathrm{Ic}}^{\mathrm{S}}}\right)^2 \tag{7-76}$$

The material length is proportional to the size of the fracture process zone for the same material. It is found that the values of $Q$ ranged from 12.5 to 50 mm for hardened cement paste, 50 to 150 mm for mortar, and 150 to 350 mm for concrete. Hence, the material length, $Q$, can be used as a brittleness index for a material. The larger the value of $Q$, the less brittle the material.

**(b)** *Size effect*. Experimental results have indicated that the strength of concrete usually decreases with the increasing size of the structure and then remains constant, as shown in Figure 7-23, where the nominal strength of a three-point bend beam is equal to its failure load (peak load) divided by its cross-sectional area. This size effect on concrete strength may be primarily explained by the fracture process zone. When a concrete structure is loaded, the strain energy produced by the applied load is converted to the energy consumed to create a new fracture surface and the energy absorbed in the fracture process zone. For larger structures, the latter is negligible compared to the former, whereas for small structures, the latter can be comparable to the former. Therefore, the larger the structures' size, the lower the nominal strength. However, the concrete strength approaches a constant when the sizes of concrete structures become very large.

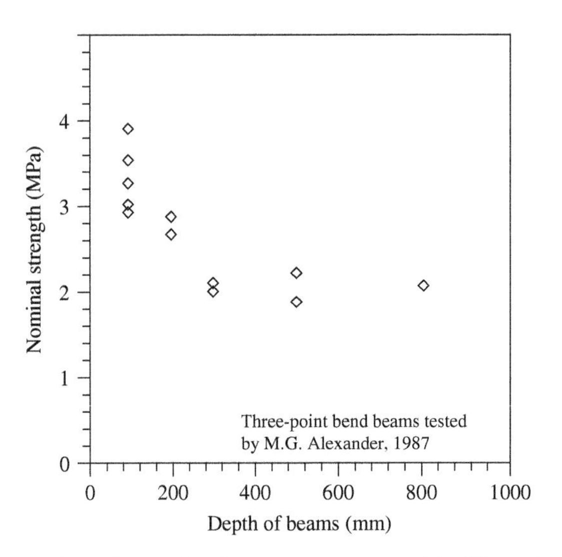


**Figure 7-23** Strength of concrete beams with different sizes

This size effect on concrete tensile strength can be predicted by using the fracture mechanics approach. The nominal stress of a concrete structure can be obtained directly as,

$$\sigma_N = \frac{K_{Ic}}{\sqrt{\pi a_c} f\left(\frac{a_c}{b}\right)} \tag{7-77}$$

If the values of $K_{Ic}$ and $a_c$ are determined from the two-parameter model, the size effect on the nominal tensile strength of concrete can be predicted by Equation (7-77). The results obtained for three-point bending test are shown in Figure 7-24, which shows a similar trend with the experimental values.

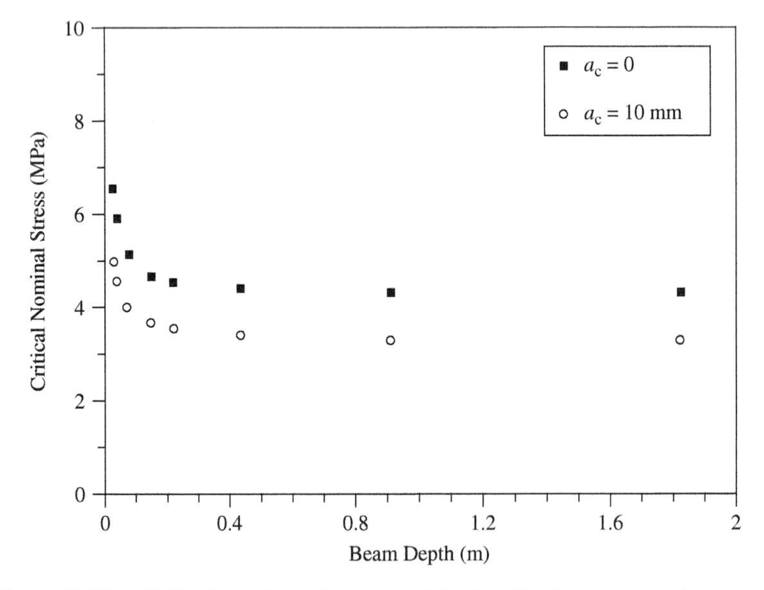


**Figure 7-24** Critical nominal stress predicted by the two-parameter model

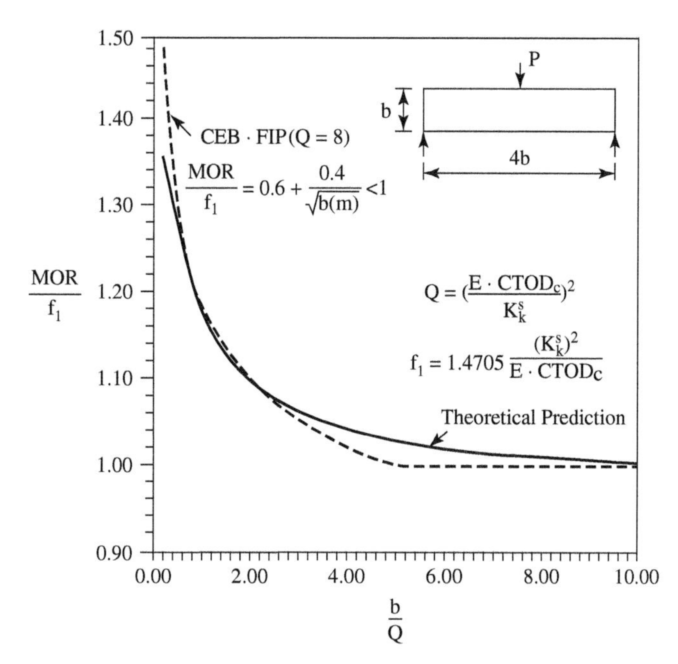


**Figure 7-25** Effect of beam depth on MOR by the two-parameter fracture model and CEB-FIP Model Code

The two-parameter model can also be applied to an unnotched structure. This is illustrated in Figure 7-25. By assuming that the material length $Q$ is equal to 20 mm, which is the average material length of concrete and mortar, the influence of the structural size on the modulus of rupture (MOR) of unnotched beams predicted by the two-parameter model is plotted in Figure 7-25. It is seen that the CEB-FIP size effect MOR compares favorably with the prediction by the two-parameter model.

(c) *High strength concrete*. Typical high strength concrete has a matrix which is very strong and stiff, is compact, and possesses well-bonded aggregate-mortar interfaces. Due to its composition, several of the toughening mechanisms, found in normal concrete, are absent during the fracture process. Microcracking at interfaces, flaws, and voids are infrequent, and cracks propagate through the coarse aggregate instead of being deflected by them. Carrasquillo et al. (1981) observed these differences, and concluded that as its compressive strength $f_c$ increases, concrete behaves more like a homogeneous material. This decrease in toughness leads to an increase in brittleness. Designers have been forced to confine high strength concrete with steel in order to prevent catastrophic failure, especially under seismic loading. Obviously, a less brittle material would make the design safer. Also, as seen earlier, an increase in brittleness implies a decrease in nominal strength. Figure 7-26 shows a plot of critical crack extension against compressive strength. It is seen that the critical crack extension decreases with increasing compressive strength. Thus, the two-parameter model correctly simulates the observed brittle response of high-strength concrete.

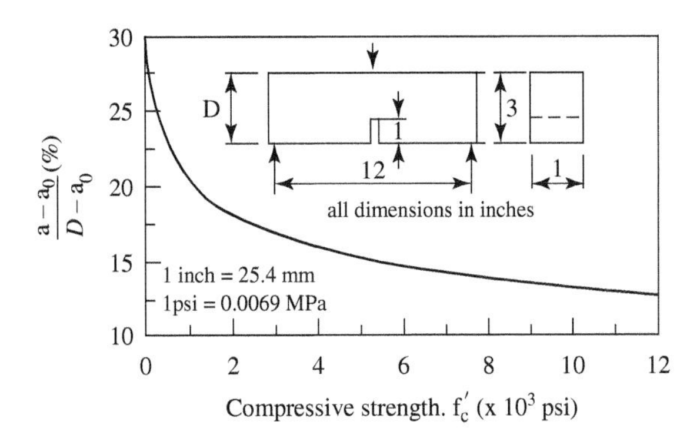


**Figure 7-26** Critical crack extension against compressive strength

## 7.8 SIZE EFFECT MODEL

### 7.8.1 Bažant's Model

Size effect is a characteristic behavior for nonlinear fracture in concrete, in which the nominal strength (stress at maximum load) of a structure is a function of structural size. In 1972, Walsh reported experimental evidence of the size effect from the nominal strength of notched concrete beams under three-point bending (Walsh, 1972). One year later, Leicester (1973) described the size effect on the strength of structures made of metals, timber and concrete in a nominal stress form:

$$\sigma_N = A_1 D^{-s}, \quad s \geq 0 \tag{7-78}$$

where $A_1$ is a constant, $D$ the characteristic dimension of the structure and $s$ the size coefficient that is a measure of the size effect for a structure with different geometries and loading patterns. The so-called size effect law was developed by Bažant (1984). Bažant and Kazemi (1990) have simulated the fracture of quasi-brittle materials by an effective elastic crack. For two-dimensional similarity, the ratio of initial crack length, $a_0$, to the characteristic dimension, $D$, is kept constant for different sized structures. $D$ represents the depth of a beam, or the width of a plate. It is, however, an important requirement that the third dimension (thickness) be kept constant. For these geometrically similar structures, the nominal stress at failure is described as,

$$\sigma_N = \frac{c_n P_c}{tD} \tag{7-79}$$

where $P_c$ is the critical fracture load (or the peak load), $t$ the thickness of a structure, and $c_n$ a coefficient representing different types of structures; $c_n = 1$ for a tensile plate and $c_n = 1.55S/b$ for a beam (**S** is the beam span; $b$ is the beam depth). It is noted that the value of $c_n$ is a constant for a series of geometrically similar structures. $D$ is the characteristic dimension of the structure or specimen. For a simply supported beam, $D$ can be defined as the height (depth) of the beam. When the beam is subjected to three-point bending under elastic conditions, we have:

$$\sigma_N = \frac{M_c \frac{D}{2}}{\frac{tD^3}{12}} = \frac{\frac{P_c S}{4}\frac{D}{2}}{\frac{tD^3}{12}} = \frac{3S}{2D}\frac{P_c}{tD} = \frac{c_n P_c}{tD} \tag{7-80}$$

where $c_n = 3S/2D$ is a constant for a series of geometrically similar structures with $S$ representing the span of the beam. The variable $t$ is the width (thickness of the beam) and is taken as a constant for a series of geometric similar structures to simplify the analysis. Likewise, under a plastic bending condition, we have:

$$\sigma_N = \frac{P_c S}{tD^2} = \frac{S}{D}\frac{P_c}{tD} = \frac{c_n P_c}{tD} \tag{7-81}$$

where $c_n = S/D$ is a constant for a series of geometrically similar structures, with $S$ representing the span of the beam.

It is well known that elastic analysis with an allowable stress criterion or any method of analysis with a failure criterion based on stress or strain, as well as plastic limit analysis, exhibits no size effect. In other words, geometrically similar structures or specimens of different sizes will fail at the same $\sigma_{N.}$ LEFM, however, does show a size effect. From the expression of energy release rate and stress intensity factor, we can write:

$$G_I = \frac{K_I^2}{E'} = \frac{\sigma_N^2 \pi a D}{E' D} \tag{7-82}$$

then the nominal stress can be obtained as:

$$\sigma_N = \sqrt{\frac{G_I E' D}{\pi a D}} = \frac{\text{constant}}{\sqrt{D}} \tag{7-83}$$

after considering that $D_1/a_1 = D_2/a_2 = \cdots = D/a = \text{constant}$. From Equation (7-83), we can see that LEFM has a very simple size effect: $\sigma_N \propto 1/\sqrt{D}$.

For nonlinear concrete fracture mechanics, the size effect is more complicated. Let us consider two geometrically similar specimens as shown in Figure 7-27. Assume that: (1) total energy released at failure, $W_{total}$, depends on fracture length, fracture band width or area of the fracture zone; (2) cracks at failure are similar (i.e., following geometry similarity); and (3) failure does not occur at the crack initiation stage.

According to assumption (1), the total energy can be written as function of nominal stress, specimen geometry and crack geometry:

$$W_{total} = W\left(\sigma_N, t, d, a, w_c\right) \tag{7-84}$$

Let $\alpha_1 = a/d$ and $\alpha_2 = w_c/d$, then Equation (7-84) can be rewritten as:

$$W_{total} = W\left(\sigma_N, t, d, \alpha_1, \alpha_2\right) = F\left(\sigma_N, t, d\right) f\left(\alpha_1, \alpha_2\right) \tag{7-85}$$

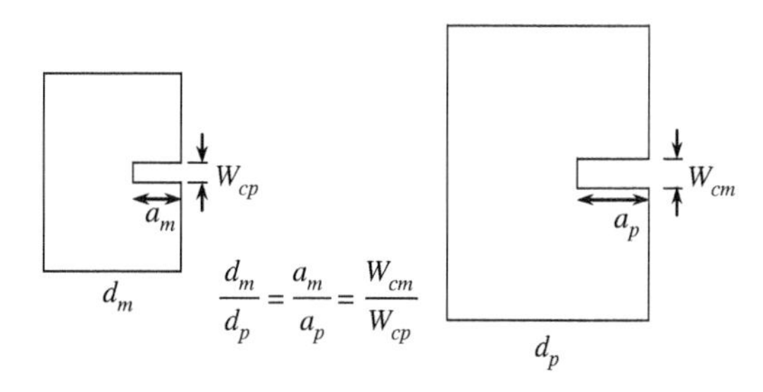


**Figure 7-27** Two geometrically similar specimens with a crack

where $F(\sigma_N, t, d)$ represents the energy consideration in the specimen without a crack and $f(\alpha_1, \alpha_2)$ the crack influence on energy consumption.

For $F(\sigma_N, t, d)$, we have:

$$F(\sigma_N, t, d) = \iint \frac{1}{2}\sigma\, d\varepsilon\, dv = \iint \frac{1}{2}\sigma \frac{d\sigma}{E} dv = \frac{\sigma_N^2}{2E} td^2 \tag{7-86}$$

Hence, we have:

$$W_{total} = \frac{\sigma_N^2}{2E} td^2 f(\alpha_1, \alpha_2) \tag{7-87}$$

Differentiate $W_{total}$ with respect to the crack length, $a$, and we have:

$$\begin{aligned} \frac{\partial W_{total}}{\partial a} &= \frac{\sigma_N^2}{2E} td^2 \frac{\partial f(\alpha_1, \alpha_2)}{\partial a} = \frac{\sigma_N^2}{2E} td^2 \left[ \frac{\partial f(\alpha_1, \alpha_2)}{\partial \alpha_1} \frac{\partial \alpha_1}{\partial a} + \frac{\partial f(\alpha_1, \alpha_2)}{\partial \alpha_2} \frac{\partial \alpha_2}{\partial a} \right] \\ &= \frac{\sigma_N^2}{2E} td^2 \left[ \frac{\partial f(\alpha_1, \alpha_2)}{\partial \alpha_1} \frac{\partial \alpha_1}{\partial a} + \frac{\partial f(\alpha_1, \alpha_2)}{\partial \alpha_2} \frac{\partial \alpha_2}{\partial a} \right] = \frac{\sigma_N^2}{2E} td^2 \left[ \frac{\partial f(\alpha_1, \alpha_2)}{\partial \alpha_1} \frac{1}{d} \right] \\ &= \frac{\sigma_N^2}{2E} td \bullet g(\alpha_1, \alpha_2) \end{aligned} \tag{7-88}$$

Substitute $\partial W_{total}/\partial a = tG_I$, gives:

$$\frac{\sigma_N^2 td}{2E} g(\alpha_1, \alpha_2) = tG_I \tag{7-89}$$

then

$$\sigma_N = \sqrt{\frac{2EG_I}{dg(\alpha_1, \alpha_2)}} \tag{7-90}$$

It may be assumed that the geometrical function $g(\alpha_1, \alpha_2)$ is positive, i.e., $g'(\alpha_1, \alpha_2) > 0$. Then we can expand $g(\alpha_1, \alpha_2)$ with respect to $\alpha_2$, at $\alpha_2 = 0$.

$$g'(\alpha_1, \alpha_2) = g(\alpha_1, 0) + g'(\alpha_1, 0)\alpha_2 + \cdots \tag{7-91}$$

By substituting Equation (7-91) into (7-90), we get

$$\sigma_N = \sqrt{\frac{2EG_I}{d\left[g(\alpha_1, 0) + g'(\alpha_1, 0)\alpha_2\right]}} = \sqrt{\frac{B'EG_I}{1 + d/d_0}} \tag{7-92}$$

where $B' = 2/g'(\alpha_1, 0) w_c$, and $d_0 = g'(\alpha_1, 0) w_c / g(\alpha_1, 0)$.

Furthermore, by considering the area under a simplified stress-strain curve with a post peak response, as shown in Figure 7-28, $G_I$ can be computed as:

$$G_I = \frac{1}{tda} \iint \sigma\, d\varepsilon\, dv = \frac{w_c f_t^2}{2} \left( \frac{1}{E} - \frac{1}{E'} \right) \tag{7-93}$$

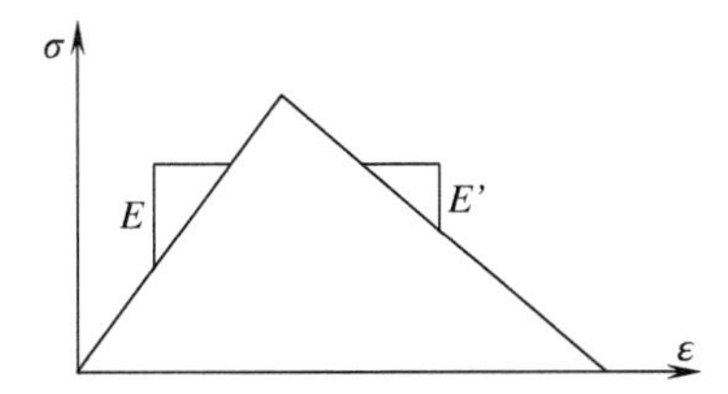


**Figure 7-28** A simplified stress-strain curve with post-peak response

By substituting this expression into Equation (7-92),

$$\sigma_N = \frac{Bf_t}{\sqrt{1+\frac{d}{d_0}}} = \frac{Bf_t}{\sqrt{1+\beta}} \tag{7-94}$$

where $B = \sqrt{B'w_c E\frac{1}{2}\left(\frac{1}{E} - \frac{1}{E'}\right)}$ and $\beta = d/d_0$.

Figure 7-29 shows the size effect law together with the strength criterion and the LEFM criterion. It can be seen that the nominal strength of a series of geometrically similar structures is conventionally described by a strength criterion, which is a constant, regardless of structural size, and may only be used for relatively small size structures. On the other hand, the single parameter failure criterion based on LEFM predicts a linear curve for the nominal strength, and this criterion may only be used for relatively large size structures. The above equation is based on nonlinear fracture and thus provides a smooth curve for the nominal strength of various sizes of structures. It can be seen from Figure 7-29 that for a small size, the size effect law approaches the strength criterion, and for a large size, it approaches the LEFM solution.

The brittleness parameter, $\beta$, may be obtained by calculating the parameter, $D_0$, as the value of $D$ at the intersection of the horizontal strength criterion, $\sigma_N = Bf_t$, with the inclined asymptote, $\sigma_N = c_n\sqrt{G_f E'/g(\alpha_0)D}$ from LEFM, as shown in Figure 7-29. Equating these two expressions yields $D_0$, and then $\beta = D/D_0$ or

$$\beta = \frac{B^2 f_t^2 g(\alpha_0)}{c_n E' G_f} D \tag{7-95}$$

It can be noted that the brittleness number, $\beta$, depends not only on the material fracture parameter $G_f$ but also on the structural geometry function $g(a_0/D)$. Since the value of $B = \sigma_N/f_t$ relates to

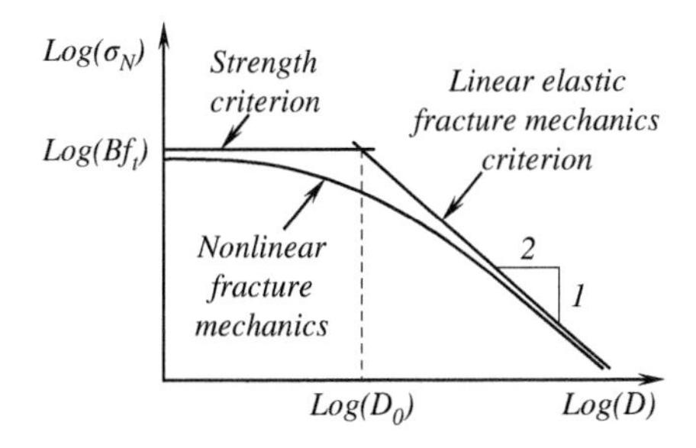


**Figure 7-29** The size effect law, together with the strength criterion and the LEFM criterion

the nominal strength of a very small structure on the basis of the strength criterion, the above equation may be suitable for small size structures. For relatively large size structures, the brittleness parameter, $\beta$, may also be expressed as (Bažant and Pfeiffer, 1987):

$$\beta = \frac{D}{c_f}\left[\frac{g(\alpha_0)}{g'(\alpha_0)}\right] \tag{7-96}$$

Based on these definitions, the brittleness of a structure not only relates to the material but also the geometry and size of the structure. The greater the value of $\beta$, the more brittle the structure. For $\beta < 0.1$, the strength criterion may be used. For $0.1\ \beta < 10$, the size effect model should be used. For $\beta > 10$, the LEFM criterion may be used.

Since an effective-elastic crack is used in the size effect law, the critical energy release rate for a certain size of structure, $G_{\mathrm{Ic}}$, can be written as:

$$G_{\mathrm{Ic}} = \frac{K_{\mathrm{Ic}}^2}{E} = \frac{\sigma_{\mathrm{N}}^2 \pi a_c}{E} g_1^2\left(\frac{a_c}{D}\right) = \frac{\sigma_{\mathrm{N}}^2 D}{E c_{\mathrm{n}}^2} g\left(\frac{a_c}{D}\right) \tag{7-97}$$

where $a_c = a_0 + \Delta a_c$ is the critical crack length which is expressed as the sum of the initial crack length $a_0$ and the effective crack extension $\Delta a_c$. The geometrical function is defined as:

$$g\left(\frac{a_c}{D}\right) = \frac{c_{\mathrm{n}}^2 \pi a_c}{D} g_1^2\left(\frac{a_c}{D}\right) \tag{7-98}$$

Bažant proposed using the critical energy release rate and the critical crack extension for an infinitely large structure, with $G_f$ and $c_f$ as fracture parameters of a quasi-brittle material. The parameters $G_f$ and $c_f$ are defined as:

$$G_f = \lim_{D\to\infty} G_{\mathrm{Ic}} \tag{7-99}$$

and

$$c_f = \lim_{D\to\infty} \Delta a_c \tag{7-100}$$

Substituting $G_{\mathrm{Ic}}$ expression Equation (7-97) into Equation (7-99), we have:

$$G_f = \lim_{D\to\infty} \frac{\sigma_{\mathrm{N}}^2 D}{E c_{\mathrm{n}}^2} g\left(\frac{a_c}{D}\right) = \lim_{D\to\infty} \frac{B^2 f_t^2 D_0}{E c_{\mathrm{n}}^2} \frac{D}{D + D_0} g\left(\frac{a_c}{D}\right) \tag{7-101}$$

Note that

$$\lim_{D\to\infty} \frac{D}{D + D_0} = 1 \tag{7-102}$$

and

$$\lim_{D\to\infty} g\left(\frac{a_c}{D}\right) = g\left(\frac{a_0}{D}\right) \tag{7-103}$$

These two results from $\Delta a_c$ are negligible compared to the initial crack length when the structure size approaches infinity. Thus, we have

$$G_f = \frac{B^2 f_t^2 D_0}{E c_{\mathrm{n}}^2} g\left(\frac{a_0}{D}\right) \tag{7-104}$$

This equation relates the material fracture parameter $G_f$ to an infinite size structure with a characteristic size of $D_0$.

The value of $c_f$ can be determined from Equation (7-96) as:

$$c_f = \frac{g\left(\frac{a_0}{D}\right)}{g'\left(\frac{a_0}{D}\right)} D_0 \tag{7-105}$$

or

$$D_0 = \frac{g'\left(\frac{a_0}{D}\right)}{g\left(\frac{a_0}{D}\right)} c_f \tag{7-106}$$

This equation represents a relationship between $c_f$ and $D_0$. Substituting the expression for $D_0$ as a function of $c_f$ into $G_f = B^2 f_t^2 D_0 / E c_n^2 \, g(a_0/D)$, we get:

$$Bf_t = c_n \sqrt{\left[\frac{EG_f}{c_f g'(a_0/D)}\right]} \tag{7-107}$$

Through these equations, the values of $D_0$ and $Bf_t$ are related to the fracture parameters $G_f$ and $c_f$. Further substituting Equations (7-104) and (7-106) into Equation (7-94), we can get:

$$\sigma_N = c_n \sqrt{\left[\frac{EG_f}{c_f g'\left(\frac{a_0}{D}\right) + g\left(\frac{a_0}{D}\right) D}\right]} \tag{7-108}$$

This equation describes the nominal strengths for a series of geometrically similar structures made of the same quasi-brittle material. Once the fracture parameters, $G_f$ and $c_f$, are known for a given material, the nominal strength can be determined. The influence of geometry is taken into account by the functions $g(a_0/D)$ and $g'(a_0/D)$. Since the functions $g(a_0/D)$ and $g'(a_0/D)$ are constants for geometrically similar structures, the nominal strength, $\sigma_N$, decreases with increase of $D$.

This equation also provides a handy method to determine $G_f$ and $c_f$ through experiments. To explain this, let us rewrite the equation as

$$c_n^2 \sigma_N^2 = \frac{EG_f}{c_f g'\left(\frac{a_0}{D}\right) + g\left(\frac{a_0}{D}\right) D} \tag{7-109}$$

Inversing the equation gives:

$$\frac{1}{c_{nn}^2 \sigma_N^2} = \frac{c_f g'\left(\frac{a_0}{D}\right) + g\left(\frac{a_0}{D}\right) D}{EG_f} = \frac{c_f g'\left(\frac{a_0}{D}\right)}{EG_f} + \frac{g\left(\frac{a_0}{D}\right) D}{EG_f} \tag{7-110}$$

If we take $1/c_{nn}^2 \sigma_N^2$ as $y$ and $D$ as $x$, $c_f g'\left(a_0/D\right)/EG_f$ and $g\left(a_0/D\right)/EG_f$ as constants, it is a linear equation.

### 7.8.2 Method of Bažant et al. to Determine $G_f$ and $c_f$

**(a)** *Specimens and test procedures*. The three-point bending beams to be used, as shown in Figure 7-30, should have a span-to-depth ratio $S/b$ not less than 2.5. The initial notch-to-depth ratio is between 0.15 to 0.5. The notch width should be as small as possible and should not exceed 0.5 times the maximum aggregate size $d_a$. The width and the depth of the beam, $b$, must not be less than $3d_a$.

Specimens of at least three different sizes, characterized by beam depth $b = b_1, b_2, \ldots, b_n$, and spans $S = S_1, S_2, \ldots, S_n$, must be tested. The smallest depth $b_1$ must not be larger than $5d_a$, and the largest depth $b_n$ must not be smaller than $10d_a$. The ratio of $b_n$ and $b_1$ must be at least 4. The ratios of the adjacent sizes should be approximately constant. The choice of $b_i/d_a = 3$, 6, 12 and 24 is preferable. At least three identical specimens should be tested for each beam size. The geometrical ratios should be the same for all specimens. It is important that the thickness $t$ must be kept constant.

Specimens should be loaded at a constant displacement rate. The loading rates should be such that the maximum load is reached in about 5 minutes. Although an ordinary uniaxial testing machine may be used, a machine with high strength or closed-loop control is better. All tests should be performed on the same test machine and in an identical manner.

**(b)** *Test results and calculations*. First, the corrected maximum loads, $P_j^0$ should be obtained by taking the weight of the specimen into account,

$$P_j^0 = P_j + \frac{S_j}{2L_J} W_j \tag{7-111}$$

where $P_j$ is the maximum load, $S_j$ the span, $L_j$ the total length, and $W_j$ the weight of the specimen $j$.

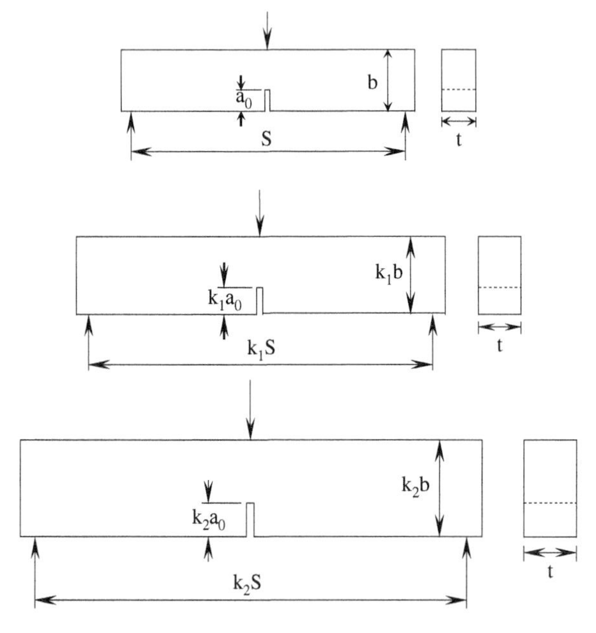


**Figure 7-30** Series of geometrically similar structures

By introducing

$$Y_j = \left(\frac{b_j t}{P_j^0}\right)^2 \tag{7-112}$$

and

$$X_j = b_j \tag{7-113}$$

for $j = 1, 2, 3, \ldots, n$, a linear regression $Y = Ax + C$ can be plotted. The values of the slope $A$ and the intercept $C$ can be obtained as below:

$$A = \frac{\sum_{j=1}^{n}\left(X_j - \overline{X}\right)\left(Y_j - \overline{Y}\right)}{\sum_{j=1}^{n}\left(X_j - \overline{X}\right)^2} \tag{7-114}$$

and

$$C = \overline{Y} - A\overline{X} \tag{7-115}$$

where

$$\overline{X} = \frac{1}{n}\sum_{j=1}^{n}\left(X_j\right) \tag{7-116}$$

and

$$\overline{Y} = \frac{1}{n}\sum_{j=1}^{n}\left(Y_j\right) \tag{7-117}$$

are the centroids of all data points.

The geometrical factor $g(a_0/b)$ or $g(\alpha_0)$ is calculated as

$$g\left(\alpha_0\right) = \left(\frac{S}{b}\right)^2 \pi\alpha_0\left[1.5g_1\left(\alpha_0\right)\right]^2 \tag{7-118}$$

The value of $g_1(\alpha_0)$ is determined according to ratios of $S/b$. For $S/b = 2.5$,

$$g_1\left(\frac{a_0}{b}\right) = \frac{1.0 - 2.5\frac{a_0}{b} + 4.49\left(\frac{a_0}{b}\right)^2 - 3.98\left(\frac{a_0}{b}\right)^3 + 1.33\left(\frac{a_0}{b}\right)^4}{\left(1 - \frac{a_0}{b}\right)^{3/2}} \tag{7-119}$$

For $S/b = 4$,

$$g_1\left(\frac{a_0}{b}\right) = \frac{1.99 - \frac{a_0}{b}\left(1 - \frac{a_0}{b}\right)\left[2.15 - 3.93\left(\frac{a_0}{b}\right) + 2.70\left(\frac{a_0}{b}\right)^2\right]}{\sqrt{\pi}\left(1 + 2\frac{a_0}{b}\right)\left(1 - \frac{a_0}{b}\right)^{3/2}} \tag{7-120}$$

For $S/b = 8$,

$$g_1\left(\frac{a_0}{b}\right) = 1.11 - 1.552\frac{a_0}{b} + 7.71\left(\frac{a_0}{b}\right)^2 - 13.55\left(\frac{a_0}{b}\right)^3 + 14.25\left(\frac{a_0}{b}\right)^4 \tag{7-121}$$

Linear interpolation may be used to obtain $g_1(\alpha_0)$ for the other values of $S/b$. The values of the material fracture energy $G_f$ can be determined from

$$G_f = \frac{g(\alpha_0)}{EA} \tag{7-122}$$

The critical value of the fracture process zone length, $c_f$, can be further obtained from

$$c_f = \frac{g(\alpha_0)}{g'(\alpha_0)} \frac{C}{A} \tag{7-123}$$

where $g'(\alpha_0)$ is

$$g'(\alpha_0) = \frac{dg(\alpha_0)}{d\alpha}\Big|_{\alpha=\alpha_0} \tag{7-124}$$

To verify the linear regression, one needs to calculate the following statistical parameters:

$$s_x^2 = \frac{1}{n-1}\sum_{j=1}^{n}\left(X_j - \overline{X}\right)^2 \tag{7-125}$$

$$s_y^2 = \frac{1}{n-1}\sum_{j=1}^{n}\left(Y_j - \overline{Y}\right)^2 \tag{7-126}$$

$$s_{x|y}^2 = s_{y|x}^2 = \frac{n-1}{n-2}\left(s_y^2 - A^2 s_x^2\right) \tag{7-127}$$

$$\omega_{y|x} = \frac{s_{y|x}}{\overline{Y}} \tag{7-128}$$

$$\omega_x = \frac{s_x}{\overline{X}} \tag{7-129}$$

$$\omega_A = \frac{s_{y|x}}{A s_x \sqrt{n-1}} \tag{7-130}$$

$$\omega_c = \frac{s_{y|x}}{C\sqrt{n-1}}\sqrt{1 + \frac{1}{\omega_x^2}} \tag{7-131}$$

and

$$m = \frac{\omega_{y|x}}{\omega_x} \tag{7-132}$$

where $X_j$ are the specimen sizes, $Y_j$ the measured data points, $s$ the square deviation, $\omega$ the coefficient of variation, and $m$ the relative width of the scatter band. The value of $\omega_A$ should not exceed 0.10 and the values of $\omega_c$ or $m$ should not exceed 0.20; otherwise, the result is highly uncertain. To avoid a large scatter, very broad size ranges should be used.

It should be noted that a slight error in the calculation of A can significantly alter the value of $G_f$. Therefore, extreme care should be exercised in determining the slope $A$ of the regression line.

The value of $c_f$ has been reported as $c_f = 13.55$ mm for concrete with a maximum aggregate size of 13 mm, and $c_f = 1.90$ mm for a mortar with a maximum aggregate size of 5 mm.

## 7.9 THE FICTITIOUS MODEL BY HILLERBORG

### 7.9.1 The Model

The basic premise of the present analysis is that the best available fracture model for concrete is the cohesive crack of the fictitious crack model. It was pioneered and generalized by Hillerborg et al. (1976). Hillerborg proposed a fictitious crack model based on the assumption of strain localization and consideration of the softening curve of the cohesive stress versus the separation of a crack. The assumption has been verified by experiments conducted in uniaxial tension tests by Li and Shah (1994) and Li and Li (2000). Figure 7-31 shows the deformations measured at different portions of a specimen as a function of time. It can be seen from Figure 7-31 that initially, the deformations are almost identical at the different parts of the specimen. Then some deviation occurs, indicating non-uniform strain development occurs. Finally, only one LVDT keeps increasing in tensile strain, while the other LVDTs show compression, which means that the tensile strain concentrates in the region covered by that particular LVDT, showing significant strain localization.

In Hillerborg's model, the elastic part of the stress-deformation curve is neglected and the post-peak response can be characterized by a stress-elongation curve. The area under the entire curve is denoted as $G_F$, and is given by

$$G_F = \int_0^{w_c} \sigma(w)\, dw \tag{7-133}$$

where $w_c$ is the critical crack separation displacement when the softening stress is zero. The crack width $w$ is usually defined as zero at the peak load. The material fracture toughness, $G_F$, represents the energy absorbed per unit area of the crack. The computation of $G_F$ is per definition for the complete separation of the crack surfaces, where the crack area $A$ is the projected area and not the real area. The beauty of the model is that the softening stress-separation displacement curve,

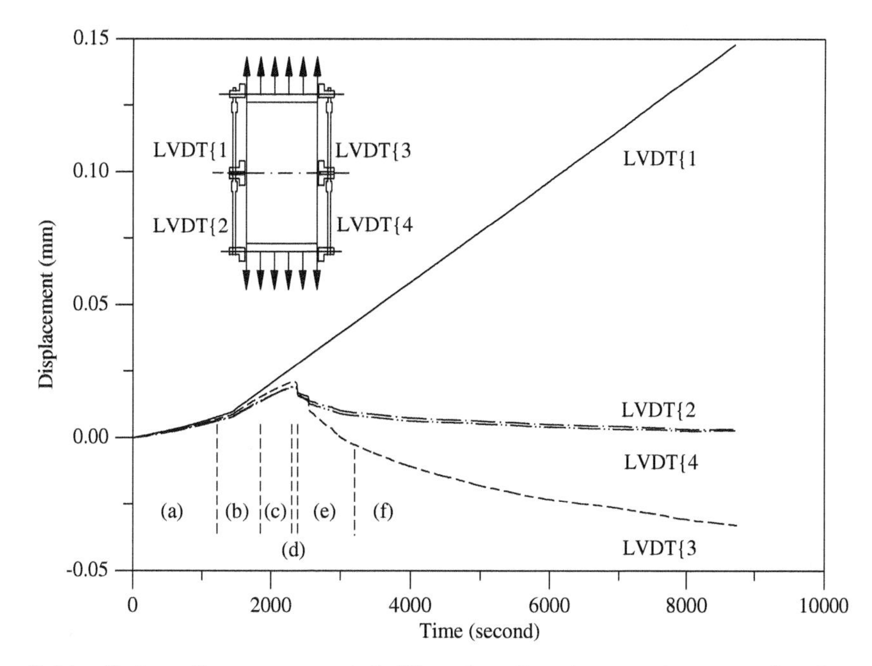


**Figure 7-31** Deformations measured at different portions in a specimen as a function of time

$\sigma = \sigma\left(f_t, w\right)$, represents a type of constitutive relationship and can easily be incorporated into numerical analysis. The expression of $\sigma = \sigma\left(f_t, w\right)$ can be linear, for example, $\sigma = f_t\left(1 - w/w_c\right)$ or exponential, for example, $\sigma = f_t e^{-Aw/w_c}$.

The fictitious crack model requires three parameters, $f_t$, $G_F$, and the shape of $\sigma(w)$. When the shape of $\sigma(w)$ is given, the material fracture property is determined by $f_t$ and $G_F$. Along this line, a new parameter, the characteristic length, is proposed by combining $f_t$ and $G_F$ into:

$$l_{ch} = \frac{EG_F}{f_t^2} \tag{7-134}$$

where $l_{ch}$ is proportional to the length of the fracture process zone. The value of $l_{ch}$ for concrete roughly ranges from 100 mm to 400 mm. The length of the fracture process zone in concrete from the initial crack tip is of the order of $0.3 \sim 0.5\ l_{ch}$.

### 7.9.2 Determination and Influence of $\sigma(w)$ Relationship

The cohesive model requires a unique $\sigma(w)$ curve to qualify the energy dissipation. The choice of the $\sigma(w)$ function influences the prediction of the structural response significantly. The local fracture behavior, for example, the crack opening displacement, is particularly sensitive to the shape of $\sigma(w)$. Many different shapes of $\sigma(w)$ curves, including linear, bi-linear, trilinear, exponential and power functions, have been used previously.

The critical value of the crack opening displacement plays an important role in the energy calculation. Hillerborg et al. (1976) simply used $w_c = 0.01 \sim 0.02$ mm. On the other hand, Cedolin et al. (1987) obtained values of $f_t$ and $w_c$ based on their test results and numerical modeling. They tested two mixes of concrete with a water-to-cement ratio of 0.5. The first mix had a maximum aggregate size of 12 mm, $E = 34.2$ GPa and $f_c = 41.6$ MPa, where $f_c$ is the cylindrical compressive strength of the concrete. For the first concrete mix, Cedolin et al. estimated $f_t = 4$ MPa and $w_c = 0.035$ mm using a least-square data fitting procedure based on the solution of an inverse problem, and assumed the shape of the stress-strained curve. The second mix of concrete had a maximum aggregate size of 8 mm, $E = 39.0$ GPa and $f_c' = 41.6$ MPa. For this second concrete mix, the values of $f_t = 3.6$ MPa and $w_c = 0.037$ mm were estimated.

Roelfstra and Whittmann (1986) proposed a bilinear curve for $\sigma(w)$ with $\sigma_1$ and $w_1$ defined as the coordinates of the intersection point, see Figure 7-32. They tested three-point bend concrete beams to determine the values of $f_t$, $w_c$, $\sigma_1$ and $w_1$. The concrete used had a water-to-cement ratio of 0.58, and maximum aggregate sizes of 3, 8 and 16 mm, respectively. For the concrete mix with a maximum aggregate size of 3 mm, the values of $f_t = 4.7$ MPa, $w_c = 0.077$ mm, $\sigma_1 = 0.57$ Mpa and $w_1 = 0.017$ mm were obtained. For the concrete mix with a maximum aggregate size of 8 mm, the values of $f_t = 4.0$ MPa, $w_c = 0.100$ mm, $\sigma_1 = 0.64$ MPa and $w_1 = 0.022$ mm were obtained. For the concrete mix with a maximum aggregate size of 16 mm, the values of $f_t = 4.4$ MPa, $w_c = 0.123$ mm, $\sigma_1 = 0.82$ MPa and $w_1 = 0.020$ mm were obtained.

Cho et al. (1984) and Liaw et al. (1990) used a trilinear curve for $\sigma(w)$ as shown in Figure 7-33. In this model, the values of $w_1$ and $w_2$ were determined according to the compressive strength, $f_c'$. It was reported that for 23.9 MPa $< f_c' <$ 34.5 MPa, $w_1 = 0.37f_c' - 2.54$, $w_2 = 5w_1$ and $w_c = w_2 + 228.6$, whereas for 34.5 MPa $< f_c' <$ 62.5 Mpa, $w_1 = 19 - 0.26f_c'$, $w_2 = (18.2f_c' - 1.25)$ $w_1$ and $w_c = w_2 + 228.6$. Here all the crack-opening displacements are given in micrometers.

Gopalaratnam and Shah (1985) proposed an exponential curve for $\sigma(w)$ as shown in Figure 7-35, in which $A$ and $B$ are constants depending on the concrete mix. Based on the

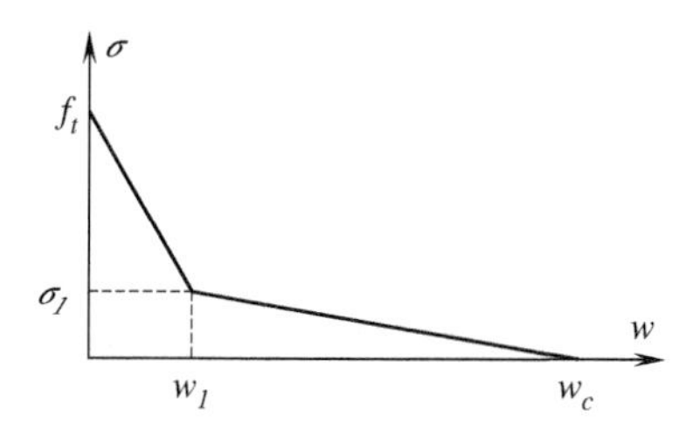


**Figure 7-32** Bilinear curve modeling of the closing pressure. Source: Roelfstra and Whittmann, 1986 / With permission of Elsevier

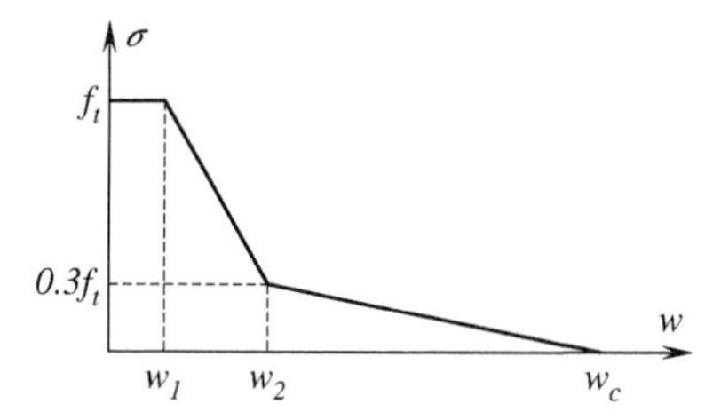


**Figure 7-33** Trilinear curve modeling of the closing pressure. Source: Based on Cho et al., 1984 and Liaw et al., 1990

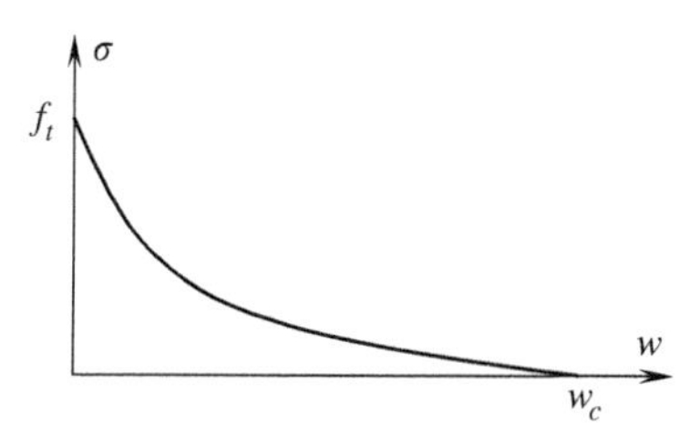


**Figure 7-34** Exponential curve modeling of the closing pressure. Source: Based on Cedolin et al., 1987

experimental data obtained, they reported $A = -0.06163$ and $B = 1.01$ for concrete with $f'_c = 33 - 47$ MPa, with the crack opening displacement also in micrometers. Note that no value of $w_c$ is found using this approach. The value of $w$ corresponding to $\sigma(w) = 0$ depends on the constants $A$ and $B$ selected in Figure 7-34. A similar exponential relationship for $\sigma(w)$ was also suggested by Cedolin et al. (1987).

A power function for $\sigma(w)$ was suggested by Du et al. (1992), as shown in Figure 7-35. It is noted that the $\sigma(w)$ function suggested by Du et al. has a sudden drop after the initial opening of the crack.

It should be emphasized that the assessment of the fracture behavior of a concrete structure is largely influenced by the $\sigma(w)$ function, therefore, reasonable and accurate determinations of the curve and the corresponding parameters become crucial for the cohesive crack approach. Experimental determination directly from tension tests has been suggested as the right method, but it is difficult to proceed and the results may vary with specimen size and shape. Thus, the concept that the shape of the softening curve is a material property is currently not accepted by many researchers in the field.

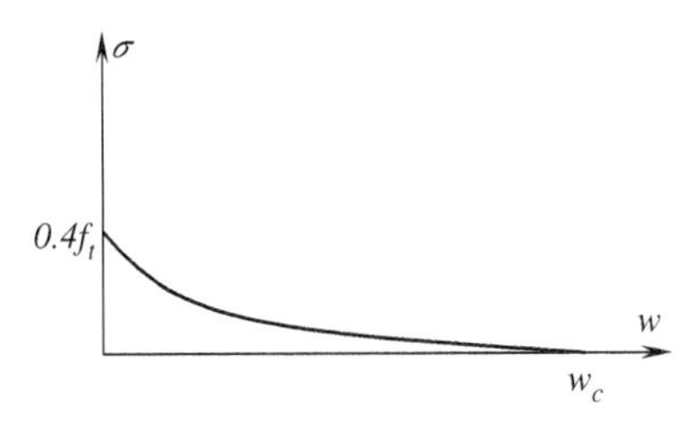


**Figure 7-35** Power curve modeling of the closing pressure. Source: Du et al., 1992 / American Concrete Institute

### 7.9.3 Test Method to Determine $G_F$

A three-point bend beam is recommended to be used for determining $G_F$. The size of the beam depends on the maximum size of aggregate. The notch depth is equal to half the beam depth ±5 mm, and the notch width at the tip should be less than 10 mm. The notch should be sawn under wet conditions at least one day before testing. The curing ambient temperature should be 20 ±2°C. The beam should be stored in lime-saturated water until 30 min. before testing. The peak load should be reached within 30–60 sec with a deformation accuracy of at least 0.01 mm, and a load accuracy of 2% relatively. Capturing the post-peak response is an essential requirement for determining $G_F$. Therefore, a closed-loop control or a very stiff test machine is required.

For the most-used experimental set-up, as shown in Figure 7-36, the weight of the beam needs to be considered using the usual equation,

$$\sigma_w = \frac{btS^2\rho g}{8}\frac{6}{t(b-a_0)^2} \tag{7-135}$$

where $b$ is the height of the beam, $t$ the thickness of the beam, $S$ the span, and $a_0$ the length of the notch. For a normal strength concrete beam, with $a_0/b = 0.5$, we get,

$$\sigma_w \approx \frac{7000S^2}{b} \tag{7-136}$$

where $S$ and $b$ are in meters and $\sigma_w$ in Pa. As a result, $\sigma_w \approx 0.04$ MPa is obtained for $S^2/b = 6.4$.

The total load $P$ is $P = P_w + P_a$, where $P_a$ is the applied load. Since $W = W_0 + W_1 + W_2$, where $W_0$ is the area below the measured $P_a - \delta$ curve, and $W_1 = P_w\delta_0$, in which the $\delta_0$ is the $\delta$ value when $P_a = 0$, $W_2$ approximately equals to $W_1$, thus,

$$W = W_0 + 2P_w\delta_0 \tag{7-137}$$

By assuming that energy absorption takes place only in the fracture zone, the fracture energy per unit projected area can be calculated by

$$G_F = \frac{W_0 + 2P_w\delta_0}{(b-a_0)\,t} \tag{7-138}$$

A rather deep notch has been used in order to keep a low maximum bending moment for the test. Thus, the assumption that energy absorption takes place only in the fracture zone can be considered valid.

In additional to the downward loading pattern, the lateral loading and upward loading are other possible patterns to determine $G_F$, as shown in Figures 7-37 and 7-38, respectively.

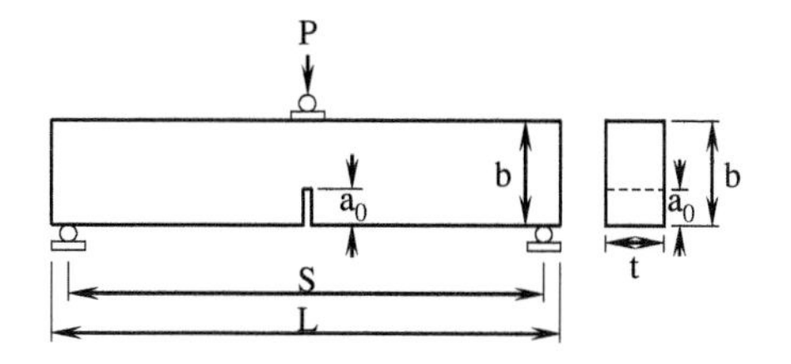


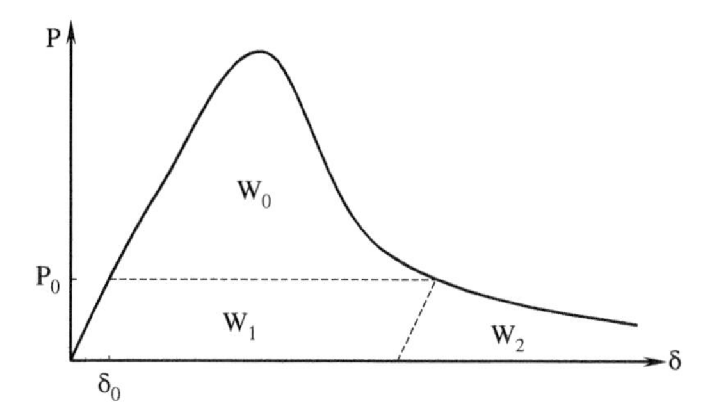


**Figure 7-36** Experimental set-up and load-displacement curve

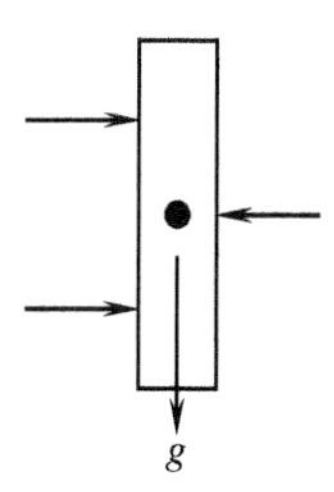


**Figure 7-37** Lateral loading setup

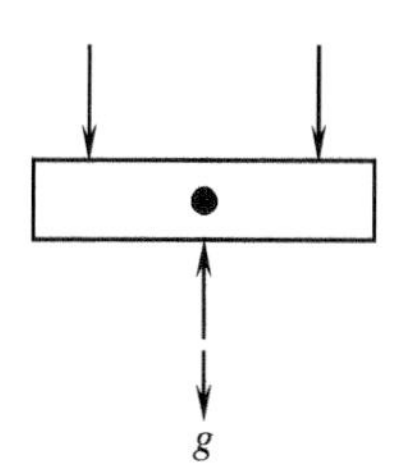


**Figure 7-38** Upward loading setup

By noting the influence of gravity in these two loading patterns, which are different from the downward loading pattern, we can readily write the $G_F$ expressions. For a lateral loading pattern, $G_F$ can be calculated as:

$$G_F = \frac{W_0}{\left(b - a_0\right) t} \tag{7-139}$$

and for an upward loading pattern, $G_F$ can be calculated as:

$$G_F = \frac{W_0 - P_w \delta_0}{\left(b - a_0\right) t} \tag{7-140}$$

A round-robin test among different laboratories was performed to measure the value of $G_F$ using different sizes of beams. The results of the tests showed that the values of $G_F$ are size-dependent. On average, the values of $G_F$ increased by 20% when the beam depth increased by factor of 2, and by 30% when the depth increased by a factor of 3. This size dependency on the $G_F$ values may partially be due to the unwanted energy absorption outside the fracture zone. The value of $G_F$ was found to be 90 N/m from some experiment results.

## 7.10 *R*-CURVE METHOD FOR QUASI-BRITTLE MATERIALS

### 7.10.1 General Description of *R*-curve

According to energy principles, the necessary condition for a structure in an equilibrium state during crack propagation is that the first order of derivative of the potential energy with respect to the crack length be equal to zero,

$$\frac{\partial (F - U + W)}{\partial a} = 0$$
$$\frac{\partial (U - F)}{\partial a} = \frac{\partial (W)}{\partial a} \tag{7-141}$$

where $U$ is the strain energy of the structure, $F$ the work done by the applied force, and $W$ the energy required for crack formation. The second part of the above equation can be rewritten as,

$$G_q = R \tag{7-142}$$

where $G_q = 1/t\, \partial (U - F)/\partial a$ is the energy release rate for the propagation of a unit length of crack in a structure with a unit thickness ($t$ is the thickness of the structure), and,

$$R = \frac{\partial W}{\partial a} \tag{7-143}$$

is the fracture resistance of the material. Taking $R$ as a constant that is independent of the crack length is approximately true only for cracks under the plane strain. In the case of plane stress, $R$ varies with the amount of crack growth. The $R$-curve method was developed based on such observation of the incremental of the resistance with increase of the stable crack extension for a plane stress case. For a crack in a sheet thin enough for plane stress to occur, when it is subjected to a certain stress level, the crack starts propagating. However, the crack growth is stable and fracture does not yet occur. If the stress level is kept constant, the crack propagates only over a small distance, and then stops.

A further increase of the stress is required to maintain the crack growth: although the crack is longer if it can withstand a higher stress. In other words, for a linear elastic material, $R$ is material constant and any extension of the initial crack means catastrophic failure of the structure. However, for a quasi-brittle material, $R$ is a function of crack extension increment, i.e., $R$ increases with crack extension. When the value of $G_q$ increases, due to the increase of applied load, the value of $R$ also increases due to the incremental crack length. Therefore, $G_q = R$ can only serve as a necessary condition for failure. A crack may be stable or unstable depending on the changing trend of the function of $G_q$. Thus, another condition is needed to describe the onset of unstable crack propagation in quasi-brittle materials. This condition is defined as:

$$\frac{\partial G_q}{\partial a} = \frac{\partial R}{\partial a} \tag{7-144}$$

This equation can be regarded as a sufficient condition for catastrophic propagation while $G_q = R$ serves as a necessary condition. In order to use the two equations to describe the failure of a structure, both $G_q$ and $R$ should be known. Generally, $G_q$ is a function of the applied load, structural geometry, and boundary conditions and can be obtained through analytical or numerical stress analysis. However, the fracture resistance $R$ is an unknown function depending on the material fracture properties and structural geometry.

For a simple case, the function of the fracture resistance $R$ can be obtained through a graphical approach. To illustrate this approach, let us take a look at Figure 7-39.

In Figure 7-39, the horizontal axis represents the crack length and the vertical axis represents the $G$ or $R$ function. The curve represents the $G$ function while the straight lines the $R$ function. It can be seen that the lower two straight lines satisfy the condition of $G_q = R$. However, after the intersection point with $G_q = R$, the lines are in the safety region under the $G$ function. Hence, a catastrophic fracture will not occur. Instead, a stable crack increase will stop. For the top straight line, its point intersects with the $G$ function with satisfies both $G_q = R$ and $\partial G_q/\partial a = \partial R/\partial a$, hence signals an unstable crack propagation.

Based on a knowledge of mathematics, the equation of the top line can be written as:

$$G = G_i + \left(\frac{dG}{da}\right)_i (a - \Delta a_c) \tag{7-145}$$

where $G_i$ is the $G$ value at the intersection point, $(dG/da)_i$ the slope of $G$ function at the intersection point and $\Delta a_c$ the a value at the intersection point, which also represents the critical crack extension.

At $a = -a_0$, $G = 0$. We have,

$$G_i = \left(\frac{dG}{da}\right)_i a_0 + \left(\frac{dG}{da}\right)_i \Delta a_c \tag{7-146}$$

Let

$$a_0 + \Delta a_c = \alpha\, a_0 \tag{7-147}$$

Then we have

$$a_0 = \frac{\Delta a_c}{(\alpha - 1)} \tag{7-148}$$

and Equation (7-146) becomes

$$(\alpha - 1)\, G_i = \alpha \left(\Delta a_c\right) \left(\frac{dG}{da}\right)_i \tag{7-149}$$

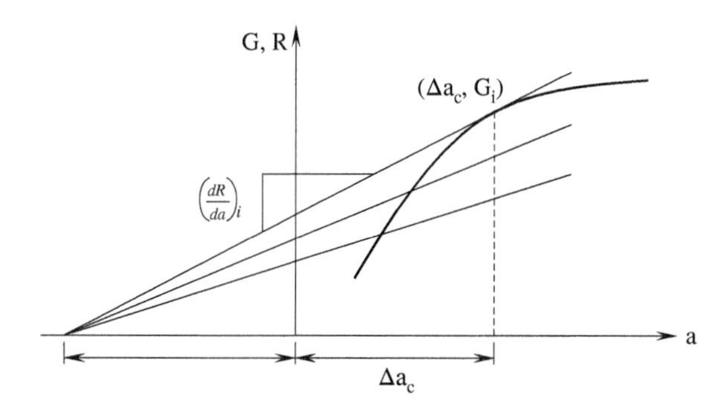


**Figure 7-39** A graphical approach for the determination of fracture resistance

or

$$\frac{\alpha}{\alpha-1}\left(\Delta a_c\right)\frac{dG}{d\left(\Delta a_c\right)} - G = 0 \tag{7-150}$$

It is a typical form of the Euler equation and its solution can be obtained by the variable exchange technique as:

$$G = \beta\left(\Delta a_c\right)^{\alpha-1/\alpha} \tag{7-151}$$

where $\beta$ is a constant.

For a complex case, two methods may be used to measure or construct $R$-curves. One is the experimental method by monitoring positions of the crack length during the loading process and constructing the $R$-curve using $R = G$. It can be done either in a single test from successive values of $G$ during slow crack growth or in a series of tests from the instability point of each test. For a single test, usually, a film record is made during the test to derive the slow growth curve. The film recording method has shown fair accuracy. However, the crack tip observed on the specimen surface may not coincide with the tip of the effective crack. This will greatly influence the determination of $R$, especially for concrete.

Another method is to construct the $R$-curve using semi-analytical and semi-experimental methods. The $R$-curve can be constructed from a purely mathematical derivation first. Then the unknown constants in the mathematical functions of $R$-curve are determined, using the relatively easy measured fracture parameters.

### 7.10.2 *R*-curve Based on Equivalent-Elastic Crack (Ouyang et al.)

Based on the two conditions for unstable crack propagation, a governing differential equation for the $R$-curve can be derived. The $R$-curve obtained by solving this differential equation is interpreted as the envelope of the energy release rate for a series of structures with the same specimen-geometry and initial crack length, but different sizes.

The energy release rate $G_q$, as shown in Figure 7-40, can be expressed in a Taylor's series at a critical point, $c$,

$$G_q = \left(G_q\right)_c + \sum_{n=1}^{\infty}\frac{1}{n!}\left(\frac{d^n G_q}{da^n}\right)\left(a - a_c\right)^n \tag{7-152}$$

where $a_c$ is the critical crack length, $(G_q)_c$ is the value of $G_q$ at the critical point, and the subscript c indicates the critical point.

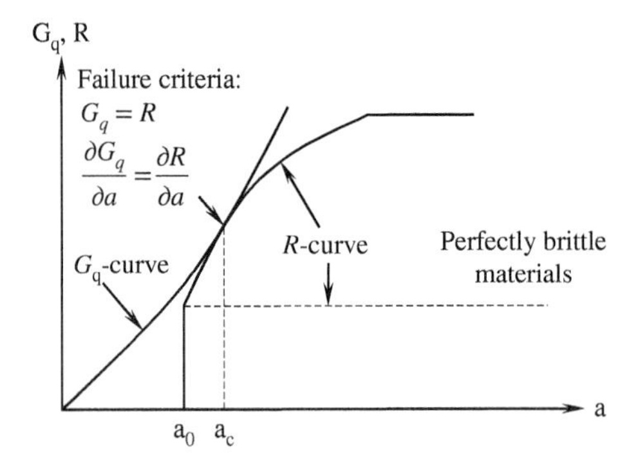


**Figure 7-40** Schematic of the *R*-curve

Using $G_q = 0$ at $a = 0$ results in the following equation,

$$(G_q)_c = -\sum_{n=1}^{\infty} \frac{1}{n!}\left(\frac{d^n G_q}{da^n}\right)(a - a_c)^n \tag{7-153}$$

For materials with a pre-critical stable crack length, $a_c$ is proportional to the initial crack length, $a_0$, i.e.

$$a_c = a_0 + \Delta a_c = \alpha\, a_0 \tag{7-154}$$

where α describes the prescribed pre-critical stable crack growth ($\alpha \geq 1$), and may be regarded as a brittleness index.

Substituting $a_c$ into $(G_q)_c$, one gets,

$$(G_q)_c = -\sum_{n=1}^{\infty} \frac{1}{n!}\left(\frac{d^n G_q}{d\Delta a^n}\right)\left(\frac{\alpha}{\alpha - 1}\right)^n (-c)^n = 0 \tag{7-155}$$

Replacing $(G_q)_c$ with $R$ and rearranging the terms, one gets,

$$R + \sum_{n=1}^{\infty} \frac{1}{n!}\left(\frac{d^n R}{dc^n}\right)\left(\frac{\alpha}{\alpha - 1}\right)^n (-c)^n = 0 \tag{7-156}$$

where $c = \Delta a_c$. If the $R$-curve is an envelope of the $G$-curves for a certain category of structure, every point on the $R$-curve corresponds to a critical point for a particular structure. Therefore, $c$ is a variable. From the mathematical point of view, the above equation can be classified as a Euler equation. The solution of the equation can be written as

$$R = \sum_{n=1}^{\infty} \beta^n (a - a_0)^{d_n} \tag{7-157}$$

For $n = 1$, Broek (1987) obtained the solution as

$$R = \beta_0 (a - a_0)^{d_0} \tag{7-158}$$

with

$$d_0 = \frac{\alpha - 1}{\alpha} \tag{7-159}$$

This $R$-curve can be used for materials with a small crack extension, such as metals. For quasi-brittle materials, more terms have to be considered. Ouyang et al. (1991) developed an expression for $n = 2$ as the follows,

$$\frac{1}{2}\left(\frac{\alpha}{\alpha - 1}\right)^2 c^2 \frac{d^2 R}{dc^2} - \frac{\alpha}{\alpha - 1} c \frac{dR}{dc} + R = 0 \tag{7-160}$$

Letting $t = \ln c$, we have:

$$\frac{dR}{dc} = \frac{dR}{dt}\frac{dt}{dc} = \frac{1}{c}\frac{dy}{dt} \tag{7-161}$$

and

$$\frac{d^2 R}{dc_2} = \frac{1}{c^2}\left(\frac{d^2 R}{dt^2} - \frac{dR}{dt}\right) \tag{7-162}$$

Then, Equation (7-160) can be rewritten as:

$$\frac{1}{2}\left(\frac{\alpha}{\alpha-1}\right)^2\frac{d^2R}{dt^2}-\frac{\alpha}{\alpha-1}\frac{dR}{dt}+R=0 \tag{7-163}$$

The solution of Equation (7-163) can be easily obtained through the characteristic equation. After substituting the original parameters back into the solution, we have:

$$R=\beta_1\left(a-a_0\right)^{d_1}+\beta_2\left(a-a_0\right)^{d_2} \tag{7-164}$$

where

$$d_1=\frac{1}{2}+\frac{\alpha-1}{\alpha}+\left[\frac{1}{4}+\frac{\alpha-1}{\alpha}-\left(\frac{\alpha-1}{\alpha}\right)^2\right]^{1/2} \tag{7-165}$$

and

$$d_2=\frac{1}{2}+\frac{\alpha-1}{\alpha}-\left[\frac{1}{4}+\frac{\alpha-1}{\alpha}-\left(\frac{\alpha-1}{\alpha}\right)^2\right]^{1/2} \tag{7-166}$$

where $\alpha$, $\beta_1$ and $\beta_2$ are constants. Once these constants are determined, the $R$-curve can be plotted as a known function. The values of $\alpha$, $\beta_1$ and $\beta_2$ have to be determined, based on experimental results. They are dependent on the geometrical function and material fracture property of a structure of any given size. As an example, a method using an infinitely sized specimen and fracture parameters from the size effect law is introduced for simplicity.

According to LEFM, for an infinite size specimen, we have:

$$G_q=\frac{\sigma^2\pi a f_1^2}{E} \tag{7-167}$$

where $f_1$ is a geometrical factor of an infinitely sized structure for the stress intensity factor. For the tension mode specimen, its value is 1.0 for a specimen with a center crack, and 1.12 for a specimen with a single edge crack under tension.

Substituting Equations (7-164) and (7-167) into the two basic definitions of the $R$-curve, Equations (7-142) and (7-144), a relationship between $\beta_1$ and $\beta_2$ can be obtained, and the $R$-curve becomes,

$$R=\beta_2\psi\left(a-a_0\right)^{d_2} \tag{7-168}$$

where

$$\psi=1-\left(\frac{d_2\alpha-\alpha+1}{d_1\alpha-\alpha+1}\right)\left(\frac{\alpha a_0-a_0}{a-a_0}\right)^{d_2-d_1} \tag{7-169}$$

Only two parameters, $\alpha$ and $\beta_2$ need to be further determined. By using the critical crack extension, $c_f$, and the fracture toughness, $G_f$, for an infinite-size structure, obtained by the size effect law, the condition of $R=G_f$ when $a=\alpha a_0=a_0+c_f$ yields the following equations.

$$\alpha=1+\frac{c_f}{a_0} \tag{7-170}$$

and

$$\beta_2=\frac{G_f\left(d_1\alpha-\alpha+1\right)}{\alpha\left(d_1-d_2\right)\left(\alpha a_0-a_0\right)^{d_2}} \tag{7-171}$$

After obtaining $\alpha$, $\beta_2$ and $\psi$, the $R$-curve can be finally determined. The $R$-curve obtained in such a way is regarded as an envelope of the $G_q$-curves of a series of structures with the same geometry and initial crack length, but different sizes.

## 7.11 DOUBLE-K CRITERION

### 7.11.1 The Criterion

Distinguishing the fracture process of concrete structures into crack initiation, stable crack propagation and unstable fracture, Xu and Reinhardt (1999a, 1999b, 1999c) proposed the double-*K* fracture criterion. The double-*K* criterion involves two material parameters, both of which are given in terms of stress intensity factor. The first is called initial cracking toughness $K_{Ic}^{ini}$ which defines the onset of cracking. The second is called unstable fracture toughness $K_{Ic}^{un}$ which defines the onset of unstable cracking or failure. The two material parameters, both derived from linear-elastic fracture mechanics (LEFM) principles, well represent the fracture response of quasi-brittle materials like concrete. They can be used to predict the fracture process, including crack initiation and crack propagation. In addition, the crack stability can be determined by comparing the stress intensity factor *K* at the tip of a propagating crack with the initial cracking toughness $K_{Ic}^{ini}$ and the unstable fracture toughness $K_{Ic}^{un}$.

In the process of employing LEFM to determine the double-*K* fracture parameters and to describe the complete fracture process, the evaluation of effective crack length is necessary. A linear asymptotic superposition assumption is introduced to calculate the effective crack length in the double-*K* criterion. The linear asymptotic superposition assumption is based on the following two hypotheses:

- The nonlinear characteristic of the load-CMOD curve is caused by fictitious crack extension in front of a stress-free crack.
- An effective crack consists of an equivalent-elastic stress-free crack and an equivalent-elastic fictitious crack extension.

Figure 7-41 shows a typical load-CMOD curve for a concrete specimen with a preformed notch after the loading-unloading-reloading cycle.

Between point O and point A in Figure 7-41, a preformed crack does not cause any slow extension. Hence, the load-CMOD curve is linear in this region. At point A, the preformed crack starts stable extension and the load-CMOD curve begins to appear nonlinear. In the region between point O and point A, the length of the crack is constant (i.e., equal to $a_0$) and LEFM approaches can be directly employed to calculate the stress intensity factor at the preformed crack tip under external loads. Beyond point A, LEFM is inapplicable.

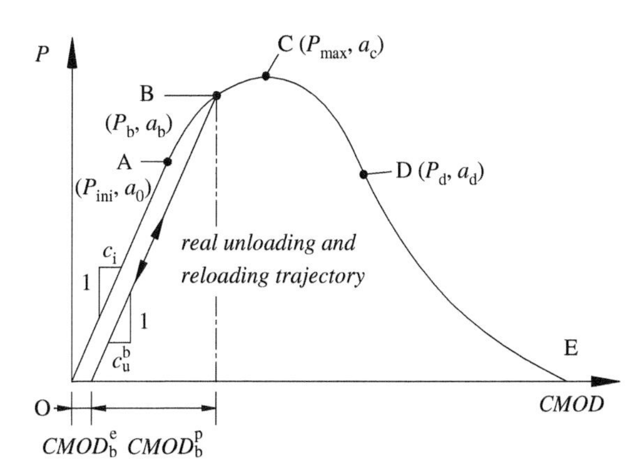


**Figure 7-41** Illustration of real unloading and reloading trajectory at an arbitrary load point

From Figure 7-41, it can also be seen that the total CMOD at point B consists of the elastic deformation $CMOD_b^e$ and the inelastic deformation $CMOD_b^p$. According to the hypotheses taken for the double-$K$ fracture criterion, the progressively extending crack with somewhat inelastic characteristics should be equivalent to an elastic crack taking into account not only the influence of the elastic deformation $CMOD_b^e$ but also the inelastic deformation $CMOD_b^p$ on the crack propagation.

The linear asymptotic superposition assumption takes the permanent deformation to be zero after unloading. Based on this assumption, the fictitious unloading trajectory will go back to the original point O, see Figure 7-42. The line OABCDE is the load-CMOD curve measured from beam A with an initial preformed crack length equal to $a_0$. At point B for beam A, the equivalent-elastic crack has extended its length to $a_b$. If unloading starts at point B, the fictitious unloading trajectory follows the line OB. On reloading, the load vs CMOD will rise up along the same unloading trajectory OB. Corresponding to the extension of the equivalent-elastic crack from $a_0$ to $a_b$, the compliance changes from $c_i$ to $c_s$. Since there is no permanent deformation after unloading, the reloading curve OBCDE could be imagined to be the load-CMOD curve measured from a brand-new beam B with an initially preformed crack length equal to $a_b$. The material and geometrical dimensions of beam B are still the same as beam A.

For beam B, the load-CMOD curve is linear in the region between point O and point B. In this region, the length of the crack stays constant at $a_b$. At point B, the preformed crack starts to propagate. With the increase of load beyond point B, the crack will propagate further and the same load-CMOD curve applies for both beams A and B. Therefore, point B at which beam A already exhibits nonlinear behavior can be treated as within the elastic regime for beam B. The value of the equivalent-elastic crack length $a_b$ can then be determined using LEFM approaches.

As point B is an arbitrary point on the nonlinear load-CMOD curve, the complete load-CMOD curve could be regarded as the envelope of many points, each representing a special linear case for a series of beams with a different initially preformed crack length according to the linear asymptotic superposition assumption, as described in Figure 7-43.

The linear asymptotic superposition assumption has been verified by (Xu and Reinhardt, 1999b) with the experiment results of two series of three-point bending beams provided by (Refai and Swartz, 1987).

### 7.11.2 Determination Method of Fracture Parameters

**(a)** *Determination of initial cracking toughness* $K_{Ic}^{ini}$. The onset of the cracking is the time when the crack is about to initiate. Before this moment, no crack growth occurs. The structure is elastic and the corresponding load-CMOD curve is linear. After this moment, crack extension

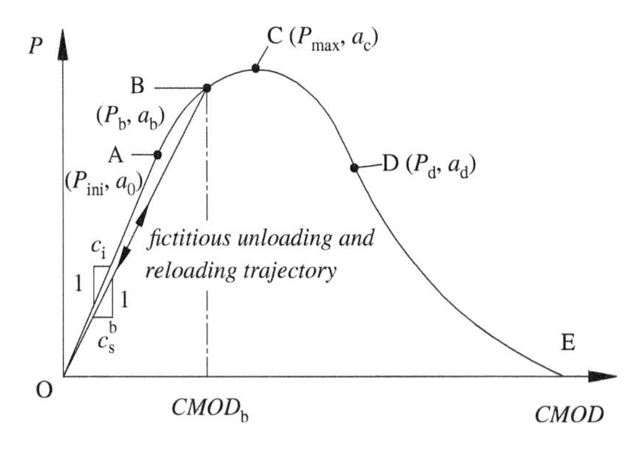


**Figure 7-42** Illustration of fictitious unloading and reloading trajectory at an arbitrary load point

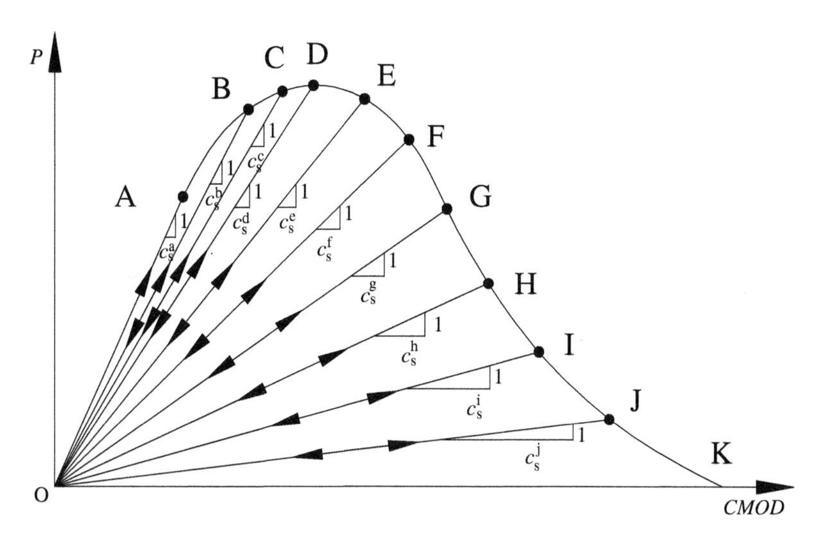


**Figure 7-43** Illustration of asymptotical superposition of initially linear load-CMOD curves

results in inelastic behavior and the corresponding load-CMOD curve becomes nonlinear. The initial cracking load $P_{ini}$ is defined as the corresponding load of the turning point at which the linear ascending part of the load-CMOD curve becomes nonlinear. At the turning point, LEFM is still applicable, so the initial cracking toughness can be determined by inserting the initial cracking load $P_{ini}$ and the preformed crack length $a_0$ into the proper LEFM formula directly.

In theory, the initial cracking load $P_{ini}$ can be determined from the turning point between the linear and nonlinear parts of the ascending load-CMOD curve. Practically, determination of the position of the turning point is very subjective, resulting in uncertainty of the initial cracking load $P_{ini}$. Given this, the strain gauge method, which has been used to determine the initial cracking load in several studies (Gao and Xu, 2007; Xu and Zhang, 2008; Xu et al., 2008), is recommended for identifying the initial cracking load $P_{ini}$. In this method, two electric strain gauges that make up a full-bridge circuit are symmetrically bonded in the horizontal direction at both sides of the tip of the preformed notch with an approximate 10-mm interval. Once a crack appears at the notch tip, the strain will decrease because of the release of the stored elastic energy in the structure body near the crack tip. The point that the strain begins to decrease on the load-strain curve of the strain gauges, which is easy to identify, corresponds to the onset of cracking. The corresponding load at this point is the initial cracking load $P_{ini}$.

**(b)** *Determination of unstable fracture toughness* $K_{Ic}^{un}$. Since extensive investigations have shown that the nonlinearity of a load-CMOD curve is mainly due to crack propagation in front of the preformed crack, the crack extension must be taken into account in computing the unstable fracture toughness $K_{Ic}^{un}$. At the critical situation of unstable crack propagation, the load reaches its maximum value $P_{max}$ and the crack mouth opening displacement CMOD achieves its critical value CMODc. Based on the linear asymptotic superposition assumption, the maximum load can be taken as the initial crack load of an imaginary beam, which has the same material, geometry, and dimensions as the practical beam, but with the preformed crack length equal to the critical effective crack length $a_c$. Therefore, the unstable fracture toughness, $K_{Ic}^{un}$ can be evaluated by the same LEFM formula for $K_{Ic}^{ini}$, by using the critical effective crack length $a_c$ and the maximum load $P_{max}$.

**(c)** *Determination of critical effective crack length $a_c$ and Young's modulus E.* According to the above-mentioned linear asymptotic superposition assumption, the critical effective crack length is the equivalent-elastic crack length $a_c$ at maximum load, which can be obtained from LEFM formulas.

When computing the critical effective crack length $a_c$ with LEFM formulas, Young's modulus $E$ needs to be known first. A method has been proposed by (RILEM Technical Committee 89-FMTT, 1990) that the Young's modulus $E$ can be calculated by the initial compliance $c_i$ measured from the linear part of the load-CMOD curve. However, as reported in the study of Karihaloo and Nallathambi (1991), the average values of Young's modulus $E$ measured from the load-CMOD curve and the cylinder compressive tests are almost the same. Based on this observation, Young's modulus $E$ that is measured from the standard cylinder compressive tests can be used to predict the averaged critical crack length for a group of specimens as well.

### 7.11.3 Specific Calculation Methods

Specific calculation methods to determine the fracture parameters in the double-$K$ criterion for the three-point bending beam test and wedge-splitting test have been recommended by RILEM (RILEM Technical Committee 265-TDK 2011), which are separately illustrated in the following.

**(a)** *Calculation methods for three-point bending beam test,* The geometry for three-point bending beam is illustrated in Figure 7-44.

As described in LEFM approaches, the formula to evaluate the stress intensity factor for a standard three-point bending beam with span-to-depth ratio equal to 4 (which is proposed to be used) is:

$$K_{\mathrm{I}} = \frac{1.5PSa^{1/2}}{BD^2} f(\alpha) \quad \left[\mathrm{Pa} \bullet \mathrm{m}^{1/2}\right] \tag{7-172}$$

where

$$f(\alpha) = \frac{1.99 - \alpha(1-\alpha)\left(2.15 - 3.93\alpha + 2.7\alpha^2\right)}{(1+2\alpha)(1-\alpha)^{3/2}} \tag{7-173}$$

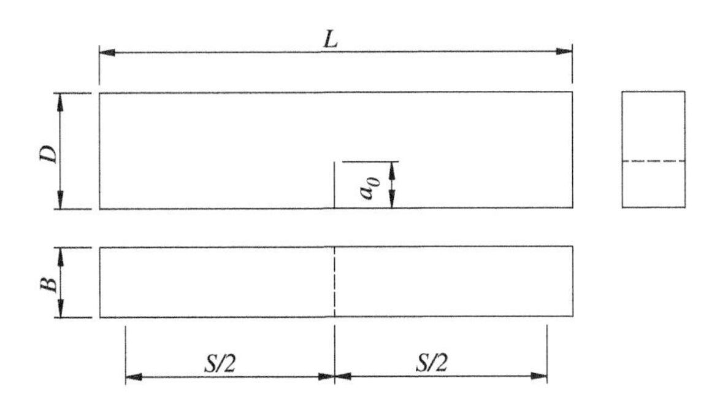


$L$ = specimen length, $D$ = specimen depth,
$B$ = specimen thickness, $S$ = specimen span,
$a_0$ = initial notch depth

**Figure 7-44** Geometry of specimen for three-point bending test

where $\alpha = a/D$; $P$ is the measured load [N]; $S$ is specimen loading span [m]; $B$ is specimen thickness [m]; $D$ is specimen depth, and $a$ *is* crack length [m].

As mentioned above, one can evaluate the values of the initial cracking toughness and unstable fracture toughness for the three-point bending beam by Equation (7-172). Note that the beam is under both the imposed load and its weight during the test unless a special device is adopted to compensate for self-weight. The weight of the beam should therefore be considered as the additional central load when calculating the initial fracture toughness $K_{\mathrm{I\,c}}^{\mathrm{ini}}$ and the unstable fracture toughness $K_{\mathrm{I\,c}}^{\mathrm{un}}$. The specific functions are given by:

$$K_{\mathrm{I\,c}}^{\mathrm{ini}} = \frac{1.5\left(P_{\mathrm{ini}} + P_{\mathrm{w}}\right) S a_0{}^{1/2}}{BD^2} f(\alpha) \quad \left[\mathrm{Pa} \bullet \mathrm{m}^{1/2}\right] \tag{7-174}$$

where

$$f(\alpha) = \frac{1.99 - \alpha(1-\alpha)\left(2.15 - 3.93\alpha + 2.7\alpha^2\right)}{(1+2\alpha)(1-\alpha)^{3/2}} \tag{7-173}$$

in which $\alpha = \frac{a_0 + h_0}{D + h_0}$ and $P_{\mathrm{ini}}$ is initial cracking load [N], which is measured by the strain gauge technique; $h_0$ is the thickness of the holder of the clip gauge [m]. $P_{\mathrm{w}}$ is the additional central load caused by the self-weight [N], which gives rise to the same central bending moment as $(M_1 - M_2)$, is computed by Equation (7-175):

$$P_{\mathrm{w}} = \frac{4\left(M_1 - M_2\right)}{S} \tag{7-175}$$

where $M_2$ is the midspan moment compensated by the support moments [N • m]; $M_1$ is the midspan bending moment, caused by the self-weight of the beam between the supports as well as the weight of the loading arrangements that is not attached to the machine.

$$M_1 = \frac{m_1 g S^2}{8L} + \frac{m_2 g S}{4} \ [\mathrm{N} \bullet \mathrm{m}] \tag{7-176}$$

in which $m_1$ is mass of the beam [kg]; $m_2$ is mass of the loading arrangement which is not attached to the machine [kg]; g is the acceleration due to gravity [$\mathrm{m}^2$/s]; $L$ is specimen length [m]; $a_0$ is initial notch depth [m].

$$K_{\mathrm{I\,c}}^{\mathrm{un}} = \frac{1.5\left(P_{\max} + P_{\mathrm{w}}\right) S a_{\mathrm{c}}{}^{1/2}}{BD^2} f(\alpha) \quad \left[\mathrm{Pa} \bullet \mathrm{m}^{1/2}\right], \tag{7-177}$$

where

$$f(\alpha) = \frac{1.99 - \alpha(1-\alpha)\left(2.15 - 3.93\alpha + 2.7\alpha^2\right)}{(1+2\alpha)(1-\alpha)^{3/2}} \tag{7-173}$$

in which $\alpha = \frac{a_0 + h_0}{D + h_0}$; $P_{\max}$ is measured maximum load at peak point on the measured load-CMOD curve and $a_{\mathrm{c}}$ is critical effective crack length.

To obtain the critical effective crack length $a_{\mathrm{c}}$, the LEFM expression of CMOD is employed.

$$\mathrm{CMOD} = \frac{6PSa}{D^2BE} V_{\mathrm{I}}(\alpha) \quad [\mu\mathrm{m}] \tag{7-178}$$

for $S/D = 4$, $V_{\mathrm{I}}(\alpha)$ is

$$V_{\mathrm{I}}(\alpha) = 0.76 - 2.28\alpha + 3.87\alpha^2 - 2.04\alpha^3 + \frac{0.66}{(1-\alpha)^2} \tag{7-179}$$

where $\alpha = (\alpha + h_0) / (D + h_0)$; CMOD is the crack mouth opening displacement [μm] and $E$ is Young's modulus [MPa].

The critical effective crack length $a_c$ can be obtained from Equation (7-178), in which the measured load $P$ is taken to be the sum of the maximum load $P_{max}$ and $P_w$ while $CMOD_w$ is the crack mouth opening displacement at the maximum load including the influence of additional central load, and determined by the following equation.

$$CMOD_w = CMOD_c + P_w c_i \quad [\mu m], \tag{7-180}$$

where $CMOD_c$ is the critical crack mouth opening displacement collected by the auto data acquisition system at the maximum load; $c_i$ is the initial compliance, $c_i = \frac{CMOD_i}{P_i}$; $P_i$ and $CMOD_i$ are the corresponding coordinates of an arbitrary point on the linear ascending part of the load-CMOD curve.

Young's modulus $E$ is calculated from the measured initial compliance $c_i$ through the method proposed by RILEM (1990), using the following equation:

$$E = \frac{6Sa_0}{c_i D^2 B} V_I(\alpha_0) = \frac{24a_0}{c_i DB} V_I(\alpha_0) \quad [MPa] \tag{7-181}$$

where $\alpha_0 = (a_0 + h_0) / (D + h_0)$.

However, Equation (7-178) is not easy to solve, as it involves $\alpha$ to the 6th power. To simplify the solution, the following empirical formula from (Chen et al., 1977) can be applied:

$$CMOD = \frac{P}{BE}\left[3.70 + 32.60 \tan^2\left(\frac{\pi}{2} \cdot \alpha\right)\right] \quad [\mu m] \tag{7-182}$$

where $\alpha = (a_0 + h_0) / (D + h_0)$.

Within the range of $0.2 \leq \alpha \leq 0.75$, the maximum difference between the results obtained by Equation (7-178) and Equation (7-182) is less than 2.0%, and when $\alpha$ is 0.8, the maximum error is less than 3.5% (Xu and Reinhardt, 2000).

After substituting $CMOD_w$ and $P_{max} + P_w$ for CMOD and $P$ respectively into Equation (7-182), the critical effective crack length $a_c$ is readily available as below.

$$a_c = \frac{2}{\pi}(D + h_0) \arctan\left(\frac{EBCMOD_w}{32.6(P_{max} + P_w)} - 0.1135\right)^{1/2} - h_0 \quad [m] \tag{7-183}$$

Similarly, Young's modulus $E$ can be computed by Equation (7-184) derived from Equation (7-182) as following:

$$E = \frac{1}{Bc_i}\left[3.70 + 32.60 \tan^2\left(\frac{\pi}{2} \cdot \frac{a_0 + h_0}{D + h_0}\right)\right] \quad [MPa] \tag{7-184}$$

**(b)** *Calculation methods for wedge-splitting test*. The geometry for wedge-splitting specimen is illustrated in Figure 7-45.

The expression of the initial cracking toughness $K_{Ic}^{ini}$ is:

$$K_{Ic}^{ini} = \frac{P_{Hini}}{BD^{1/2}} f(\alpha) \quad [Pa \bullet m^{1/2}] \tag{7-185}$$

where

$$f(\alpha) = \frac{3.675[1 - 0.12(\alpha - 0.45)]}{(1 - \alpha)^{3/2}} \tag{7-186}$$

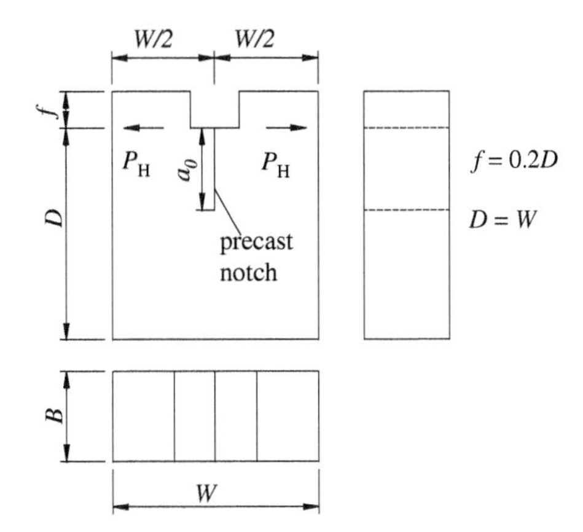


**Figure 7-45** Recommended specimen dimensions for wedge-splitting test

in which $\alpha = \frac{a_0+h_0}{D+h_0}$; $P_{\text{Hini}}$ is initial horizontal splitting force [N], which can be calculated by $P_{\text{ini}}$ through Equation (7-189); $h_0$ is the thickness of the holder of the clip gauge; $B$ is specimen thickness [m]; $D$ is specimen depth [m]; $a_0$ is initial notch depth [m]; $P_{\text{ini}}$ is initial cracking load [N], which is measured by the strain gauge technique.

The expression of the unstable fracture toughness $K_{\text{Ic}}^{\text{un}}$ is:

$$K_{\text{Ic}}^{\text{un}} = \frac{P_{\text{H max}}}{BD^{1/2}} f(\alpha) \left[\text{Pa} \bullet \text{m}^{1/2}\right] \tag{7-187}$$

where

$$f(\alpha) = \frac{3.675\left[1 - 0.12(\alpha - 0.45)\right]}{(1-\alpha)^{3/2}} \tag{7-188}$$

in which $\alpha = \frac{a_c+h_0}{D+h_0}$; $P_{\text{Hmax}}$ is the maximum horizontal splitting force [N], which is also computed by $P_{\text{max}}$ through Equation (7-189); $P_{\text{max}}$ is the measured maximum load [N], which is the load at peak point on the measured load-CMOD curve; $a_c$ is the critical effective crack length [m].

The horizontal splitting force $P_H$ is related to the measured load $P$ through Equation (7-189):

$$P_H = \frac{\left(P + mg \times 10^{-3}\right)}{2\tan\theta} \,[\text{N}] \tag{7-189}$$

where $P$ is the measured load [N]; $\theta$ is the wedge angle of the wedging device; $m$ is the mass of the wedging device [kg], if the wedging device is attached to the testing machine, $m$ is omitted; g is the acceleration due to gravity [m²/s].

Likewise, another empirical formula, Equation (7-190), was proposed by Xu and Reinhardt (1999c) to approximately evaluate the crack mouth opening displacement of wedge-splitting specimen with recommended dimensions described in Figure 7-45. When $0.3 \leq \alpha \leq 0.7$, the empirical formula Equation (7-190) gives result with accuracy of less than 2.9% of the actual values.

$$\text{CMOD} = \frac{P_H}{BE}\left[13.18\left(1 - \frac{a+h_0}{D+h_0}\right)^{-2} - 9.16\right] \quad [\mu\text{m}] \tag{7-190}$$

where $P_H$ is the horizontal splitting force [N].

The critical effective crack length $a_c$ can be derived from Equation (7-190). Two different situations are considered: one is attaching the wedging device to the testing machine while the other not. In these two cases, calculations of the critical effective crack length $a_c$ are different. For the situation that the wedging device is attached, $a_c$ is obtained by:

$$a_c = (D + h_0)\left[1 - \left(\frac{13.18}{\frac{\mathrm{EBCMOD_c}}{P_{\mathrm{Hmax}}} + 9.16}\right)^{1/2}\right] - h_0 \quad [\mathrm{m}] \tag{7-191}$$

in which $\mathrm{CMOD_c}$ is the critical crack mouth opening displacement collected by auto data acquisition system at the maximum load.

Where the situation is that the wedging device is not attached to the testing machine, the critical effective crack length $a_c$ can be calculated as following:

$$a_c = (D + h_0)\left[1 - \left(\frac{13.18}{\frac{\mathrm{EBCMOD_d}}{P_{\mathrm{Hmax}}} + 9.16}\right)^{1/2}\right] - h_0 \tag{7-192}$$

in which $\mathrm{CMOD_d}$ is the crack mouth opening displacement at the maximum load considering the influence of the weight of the wedging device, and can be computed by the following formula.

$$\mathrm{CMOD_d} = \mathrm{CMOD_c} + \frac{mgc_i}{2\tan\theta} \quad [\mu\mathrm{m}] \tag{7-193}$$

where $\mathrm{CMOD_c}$ is the critical crack mouth opening displacement, crack mouth opening displacement collected by the auto data acquisition system at the maximum load; $c_i$ is initial compliance, $c_i = \mathrm{CMOD_i}/P_{\mathrm{Hi}}$.

$P_{\mathrm{Hi}}$ relies on $P_i$ according to the following relationship:

$$P_{\mathrm{Hi}} = \frac{P_i}{2\tan\theta} \quad [\mathrm{N}] \tag{7-194}$$

$P_i$ and $\mathrm{CMOD_i}$ are the corresponding coordinates of an arbitrary point on the linear ascending part of the load-CMOD curve.

Young's modulus $E$ is given by.

$$E = \frac{1}{Bc_i}\left[13.18\left(1 - \frac{a_0 + h_0}{D + h_0}\right)^{-2} - 9.16\right] \tag{7-195}$$

## 7.12 THE APPLICATION OF FRACTURE MECHANICS IN THE DESIGN CODE OF CONCRETE STRUCTURES

In the 1970s, accumulating experimental evidence indicated the existence of a strong size effect in shear failure of concrete beams, as a result, the significance of size effect for concrete for structural design codes has been recognized among the academic community. The Japan Society of Civil Engineering (JSCE), apparently thinking that "better something than nothing," pioneered the size effect for design code in the 1980s (Dönmez and Bažant, 2019). It introduced the Weibull's statistical power-law size effect, expressed by purely empirical equations, into the shear specifications of the design code on the ultimate shear force. At the same time, the size effect on cracking shear force has been taken into account by the Comité Européen du Béton (CEB) in its design code. Accordingly, FIB (Fédération Internationale du Béton), the successor to CEB, has adopted the size effect equations based on the Modified Compression Field Theory (MCFT) for beam shear and the critical shear crack theory (CSCT) for punching shear in its Model Code 2010. Compared with JSCE and CEB introducing the size effect into their design specifications expeditiously, the American Concrete Institute (ACI), insisting that "better nothing than something controversial," did not include the size effect in their design code immediately due to the absence of broad verification and the persistent controversies about alternative theories of size effect (Dönmez and Bažant, 2019). After more than three decades of more detailed theoretical studies, extensive computational modeling, failure probability analyses, and extensive experimental verifications, including the collection of a vast worldwide database of the results of uncoordinated experiments and its filtered statistical interpretation, a consensus was gradually crystallized in the ACI that the size effect law (SEL) should be approved for the one-way shear of beams, two-way shear of slabs, and the compression struts of the strut-and-tie model. Therefore, in 2019, the SEL was incorporated into ACI Standard 318-19, and, consequently, ACI became the first code-making society that modified its design specifications underlying the quasi-brittle fracture mechanics (Carloni et al., 2019).

Although the Weibull statistical theory, MCFT, and CSCT have been incorporated into various codes for considering the existence of size effect in concrete, there are, more or less, apparent problems within these methods. Within a decade after the adoption of Weibull's theory, it was gradually accepted that Weibull's theory is valid only for structures' immediate failure right at the initiation of fracture growth from a smooth surface, while large crack or a cracking zone develops in concrete structures before the maximum load is reached (Bažant, 2004). The CSCT derivation and calculation procedure are prone to obfuscate the mechanics of shear failure due to the lack of support in mechanics. For example, it is evidenced by numerical results that the opening width, which is one of the fundamental factors of the CSCT, cannot be what controls failure. The size effect of MCFT has an incorrect large-size asymptote leading to a misprediction of the size effect in large members. Furthermore, both the CSCT and the MCFT are incompatible with the strut-and-tie method, while SEL is. For SEL, on the contrary, it can capture the transition from ductile behavior in small concrete members to a brittle behavior in large ones and has been extensively verified experimentally and theoretically (Dönmez and Bažant, 2019).

### 7.12.1 One-Way Shear

In ACI Standard 318-19, the one-way shear equations, proposed for the shear design of structural concrete beams, were modified by including the size effect to calculate the shear strength provided by concrete, $V_c$. According to the Standard, $V_c$ for nonprestressed members shall be calculated in accordance with Table 7-2 (ACI 318-19, 2019).

**Table 7-2** $V_c$ for nonprestressed members

| Criteria | | $V_c$ | |
|---|---|---|---|
| $A_v \geq A_{v,min}$ | Either of: | $\left[2\lambda\sqrt{f_c'} + \frac{N_u}{6A_g}\right] b_w d$ | (7-196) |
| | | $\left[8\lambda(\rho_w)^{1/3}\sqrt{f_c'} + \frac{N_u}{6A_g}\right] b_w d$ | (7-197) |
| $A_v < A_{v,min}$ | | $\left[8\lambda\lambda_s(\rho_w)^{1/3}\sqrt{f_c'} + \frac{N_u}{6A_g}\right] b_w d$ | (7-198) |

Source: Based on ACI 318-19, 2019.

where $A_v$ and $A_{v,min}$ are the area of shear reinforcement within spacing and minimum area of shear reinforcement within spacing, respectively, where $b_w$ and $d$ are the web width and the distance from extreme compression fiber to the centroid of longitudinal tension reinforcement, respectively, therefore, $b_w d$ is represented for the effective cross-section, where $N_u$ is the axial load normal to cross-section, $A_g$ is the gross area of concrete section, where $\lambda$ is the modified factor to reflect the reduced mechanical properties of lightweight concrete, $f_c'$ is the specified compressive strength of concrete, $\rho_w$ is the ratio of area of nonprestressed longitudinal tension reinforcement to $b_w d$, where $\lambda_s$ is the size effect modification factor determined by

$$\lambda_s = \sqrt{\frac{2}{1 + \frac{d}{10}}} \leq 1 \tag{7-199}$$

It is noted that the $\lambda_s$ (i.e., the size effect) is considered in the calculation of $V_c$ only if $A_v < A_{v,\min}$. Based on the extensive experiments conducted to access the shear capacity of concrete members, it suffices to prove that the beams without shear reinforcement, which is the most extreme condition in the range of $A_v < A_{v,\min}$, exhibit an apparent size effect, especially whose $d$ is greater than 40 in. (1 m). Are the beams with stirrups, however, safe enough to ignore the size effect? The answer is definitely no. In the test conducted at the University of Toronto (2004), for example, a beam 74.41 in. (1.89 m) with approximately minimum stirrups has had a shear force applied to evaluate the reduction of $V_c$ caused by size effect. It was shown that, after a proper statistical analysis, the non-negligible reduction of $V_c$, approximately 41%, is observed though still less than the 76% reduction measured in a companion beam without stirrups. That is, the minimum shear reinforcement can mitigate the size effect significantly, but not enough to make it negligible (Bažant et al., 2007). Therefore, on the one hand, for most practical members (either with small size $d$ or with minimum shear reinforcement), a size effect modification is not needed. On the other hand, it is also indicated that the failure possibility of large size beams will significantly increase if the size effect is ignored. That is the reason why the size effect should be considered when the insufficient shear reinforcement is provided.

### 7.12.2 Two-Way Shear

Size effect also has been adopted for the two-way shear (or punching shear) equations in ACI Standard 318-19. Compared with one-way shear equations, which are calibrated and verified by analyzing numerous data statistically, relatively inadequate data is available in the database for two-way shear failure. Therefore, the two-way shear equations derived from theoretical support in quasi-brittle fracture mechanics are more reliable than pure empirical ones (Dönmez and Bažant, 2017). Meanwhile, according to the former code, the existence of an excessive safety factor makes it possible to ignore the size effect but leads to the design for small members as uneconomic. With the size effect modification, the economics for smaller structures and the safety for larger members can be achieved simultaneously by the current design equations.

For nonprestressed two-way members without shear reinforcement, the nominal two-way shear strength $v_c$ at critical sections should be calculated in accordance with Table 7-3 (ACI 318-19, 2019) (Equations 7-200–7-202).

**Table 7-3** $v_c$ for two-way members without shear reinforcement

| | $v_c$ | |
|---|---|---|
| Least of (7-200), (7-201), and (7-202): | $4\lambda_s \lambda \sqrt{f_c'}$ | (7-200) |
| | $\left(2 + \frac{4}{\beta}\right) \lambda_s \lambda \sqrt{f_c'}$ | (7-201) |
| | $\left(2 + \frac{\alpha_s d}{b_0}\right) \lambda_s \lambda \sqrt{f_c'}$ | (7-202) |

Source: Based on ACI 318-19, 2019.

where $\beta$ is the ratio of long to short sides of a column, concentrated load, or reaction area, $\alpha_s$ is a constant factor to reflect the side number of the critical section, $b_0$ is the perimeter of critical section for two-way shear.

The size effect is also reflected by the size effect modification factor, $\lambda_s$, which is determined by Equation (7-199). Similarly, with one-way shear, the size effect on two-way shear members must be considered when shear reinforcement is not provided. More importantly, for two-way shear members without a minimum amount of shear reinforcement and with $d > 10$ in. (0.25 m), the size effect reduces the shear strength of two-way members below $4\sqrt{f_c'}b_0d$, which was regarded as a safe value in the former code (ACI 318-14, 2014).

For nonprestressed two-way members with shear reinforcement, the nominal two-way shear strength $v_c$ at critical sections should be calculated in accordance with Table 7-4 (ACI 318-19, 2019) (Equations 7-203–7-207).

For both stirrups and headed shear stud reinforcement, if a minimum amount of shear reinforcement is provided, the size effect in slabs with $d > 10$ in. can be significantly mitigated. Notably, the ability of headed shear stud reinforcement to effectively mitigate the size effect on the two-way shear may be compromised if the studs in the members are longer than 10 in. Thus, according to the ACI Standard 318-19, it will be permitted to take $\lambda_s$ as 1.0, i.e., the size effect can be ignored, if the two requirements are satisfied: (1) the specification of location and spacing of shear reinforcement in ACI Standard 318-19, which is to guarantee its efficiency in resisting the punching shear, and (2) the minimum shear reinforcement, which is reflected by the inequality, $A_v/s \geq 2\sqrt{f_c'}b_0/f_{yt}$. Otherwise, the size effect on the two-way shear strength of slabs must be considered.

### 7.12.3 Strut-and-Tie Method

The strut-and-tie method, which can visualize the flow of stresses in the member, is an effective tool for analysis of the discontinuity region in reinforcement members by modeling the region as an idealized truss. For the idealized truss, struts are the compression members, ties are the tension members, and nodes are the joints. A typical strut-and-tie model of a single-span deep beam with two concentrated loads is shown in Figure 7-46. Based on the truss model, only two types of failure can occur: (1) ductile failure of the steel ties or longitudinal bars, (2) brittle failure of the diagonal compression struts (interior strut in Figure 7-46). The former type of failure is well understood, but the latter possibly induced by the shear failure of beams can involve a size effect, which is of

**Table 7-4** $v_c$ for two-way members with shear reinforcement

| Type of shear reinforcement | Critical sections | | $v_c$ | |
|---|---|---|---|---|
| Stirrups | All | | $2\lambda_s\lambda\sqrt{f_c'}$ | (7-203) |
| Headed shear stud reinforcement | According to 22.4.6.1 in ACI 318-19 | Least of (7-204), (7-205), and (7-206) | $3\lambda_s\lambda\sqrt{f_c'}$ | (7-204) |
| | | | $\left(2+\frac{4}{\beta}\right)\lambda_s\lambda\sqrt{f_c'}$ | (7-205) |
| | | | $\left(2+\frac{\alpha_s d}{b_0}\right)\lambda_s\lambda\sqrt{f_c'}$ | (7-206) |
| | According to 22.4.6.2 in ACI 318-19 | | $2\lambda_s\lambda\sqrt{f_c'}$ | (7-207) |

Source: Based on ACI 318-19, 2019.

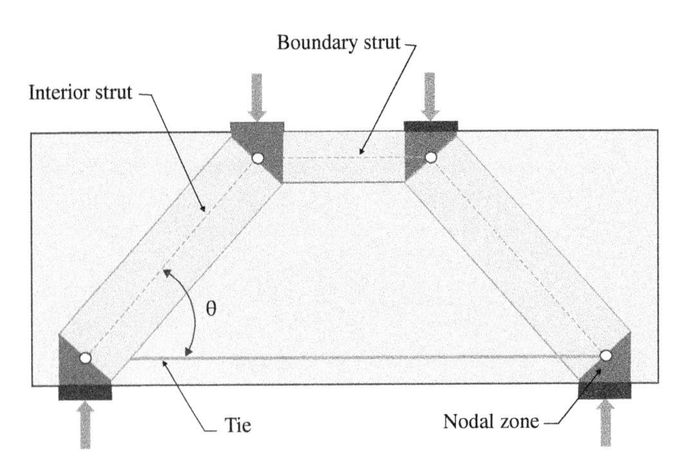


**Figure 7-46** Description of strut-and-tie model. Source: after ACI 318-19

great interest in this section. Much effort has been devoted to investigating such shear failure in concrete by truss model, and, consequently, the existence of the size effect on the nominal strength of shear failure has been proved, which approximately follows the size effect law (Bažant, 1997). In particular, the compression fracture occurs only within a portion of the length of the strut where the fracture zone exists. Intriguingly, the width of the fracture zone only depends on the material characteristics, i.e., it is not proportional to the beam size, but, due to the geometrical similarity of beams, the length of the fracture zone is. Hence, the fixed width but the varying length of the fracture zone causes the size effect.

In ACI Standard 318-19, the size effect has been introduced to calculate the effective compressive strength of concrete in a strut, $f_{ce}$, as (ACI 318-19, 2019):

$$f_{ce} = 0.85\beta_c\beta_s f_c' \tag{7-208}$$

where $\beta_c$ is the strut and node confinement modification factor reflecting the confinement provided by the surrounding concrete, $\beta_s$ is the strut coefficient influenced by cracking and confining reinforcement in a strut, for which the value of $\beta_s$ is related to the size effect on the nominal strength of shear failure.

As mentioned, the size effect must be considered, which is reflected by the value of $\beta_s$, when calculating the $f_{ce}$ in an interior strut. If the member dimensions of an interior strut satisfy (ACI 318-19, 2019):

$$V_u \leq \phi 5 \tan \phi \lambda \lambda_s \sqrt{f_c'} b_w d \tag{7-209}$$

where $V_u$ is factored shear strength at section, $\phi$ is strength reduction factor. Again, the size effect is incorporated in Equation (7-209) by $\lambda_s$. Similarly, with one-way shear and two-way shear, if the requirements of reinforcement or beam size could be satisfied, the size effect can be ignored, i.e., $\lambda_s$ will be taken as 1.0. That is: (1) minimum distributed reinforcement, which can promote redistribution of the internal force in the cracked state of members, is provided, or (2) if there is insufficient distributed reinforcement, the beam depth $d$ must be smaller than 10 in. Otherwise, $\lambda_s$ will be taken in accordance with Equation (7-199). In fact, Equation (7-209) is intended to preclude diagonal tension failure in the discontinuity region, where diagonal tension stress increases as the strut angle increases (ACI 318-19, 2019). Therefore, the existence of the size effect causes a reduction of the shear strength at section and may transfer the shear failure into the diagonal tension failure, thus decreasing the $f_{ce}$ reflected by a smaller value of $\beta_s$.

## DISCUSSION TOPICS

Why did linear fracture mechanics develop?
How to describe the stress criterion and energy criterion for LEFM?
What is a singularity?
Why do we have to consider the plastic zone in front of a crack tip?
What is the superposition method?
What is the Dugdale model?
Why can LEFM not be applied to concrete directly?
Describe the special characteristics of a concrete fracture process.
What is the two-parameter model?
Describe the size effect law.
What is the main difference between the fictitious model and the size effect law?
How to construct a *R*-curve for concrete?
How to determine the necessary parameters for a two-parameter model, the size effect law, and the *R*-curve approach?
What are the differences in fracture behavior for a cement paste and a normal strength concrete?
Do you think fracture-based design will soon become dominant for civil engineers? Why?
For shear and flexural design of reinforced concrete structures, which is more suitable to use in fracture mechanics as a design guideline?

## PROBLEMS

1. A type of steel has a $K_{II}$ value of 66 MPa $\sqrt{m}$ and an ultimate strength, $\sigma_u$, of 840 MPa. If the allowed stress in the material is half of its ultimate strength, what will be the corresponding allowed crack length? For a crack length of 1.25 mm, what will be the allowed stress value?
2. Consider a cylindrical pressure vessel as shown in Figure P7-1, with $R = 1.5$ m, $h = 0.025$ m, $\sigma_y = 1500$ MPa, and $K_{IC} = 60$ MPa $\sqrt{m}$. Assuming that the vessel is designed on the basis of the yield stress, $\sigma_y$, reduced by a safety factor of 4, solve the following problems:
   **(a)** Find the expression for the normal stress, $\sigma$, in the wall of the vessel as a function of the internal pressure *p*.
   **(b)** Assuming that $K_I = 1.12\sigma\sqrt{\pi a}$, calculate the failure pressure for a minimum detectable flaw size of $a = 1.5$ mm.
   **(c)** Calculate the critical crack depth at the design pressure.

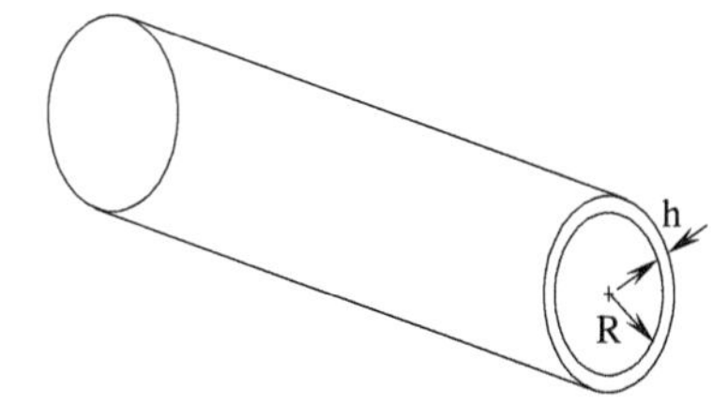


**Figure P7-1** A cylindrical pressure vessel

3. Consider a crack located parallel to the edge of a thin, semi-infinite plate, as shown in Figure P7-2. The edge of the plate is loaded by a concentrated force, *P*, as shown in Figure P7-2, and

the thickness of the plate is $b$. Determine an approximation for the stress intensity factor of the configuration.

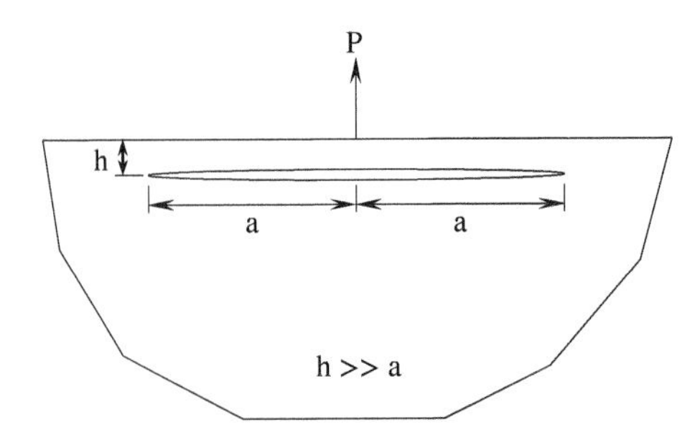


**Figure P7-2** A thin plate with a crack near the edge

4. Derive the following expression for $E$:

$$E = \frac{6Sa_0 g_2(\alpha_0)}{C_i b^2 t} \tag{7-210}$$

from the equation of $\text{CMOD}_c = (4\sigma_c a_0/E)\, g_2(\alpha_0)$, where $t$ is the beam thickness and $b$ the beam height, referring to Figure P7-3 (the self-weight of the beam can be ignored).

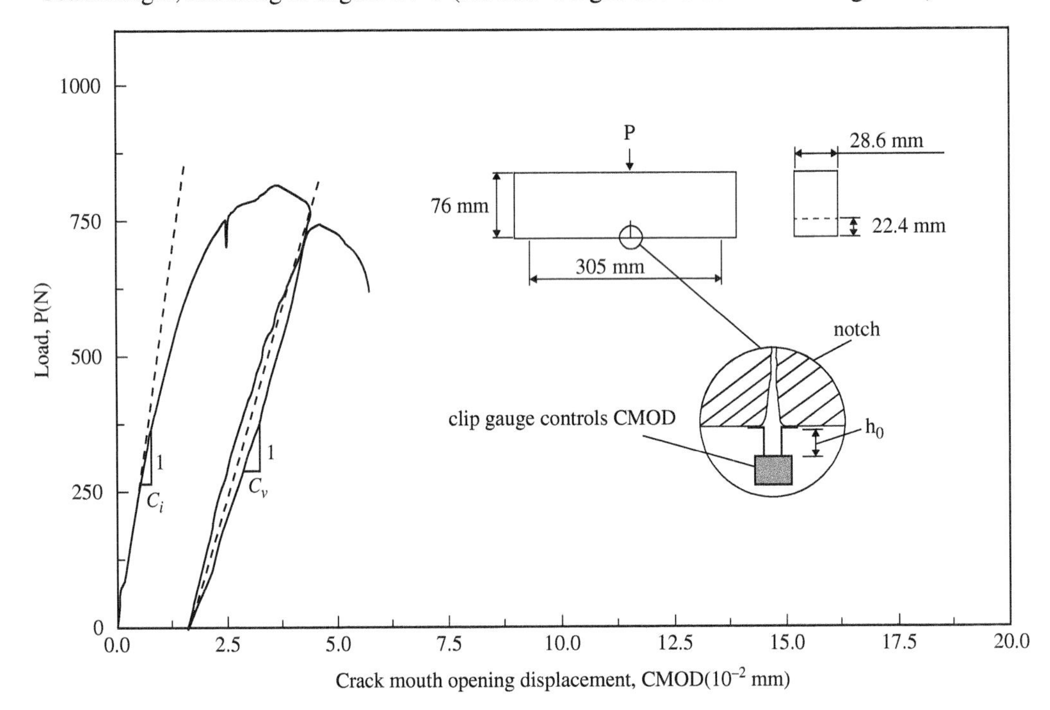


**Figure P7-3** A typical measured load-CMOD curve under three-point bending

5. Determine the value of $K_\text{I}$ for a round crack, as shown in Figure P7-4a, loaded by normal tensile stress $\sigma$ over $e < r < b$ by considering the case shown in Figure P7-4b with $K_\text{I}$ expression below.

$$K_\text{I} = \frac{2\sigma}{\sqrt{\pi a}}\left(a - \sqrt{a^2 - b^2}\right) \tag{7-211}$$

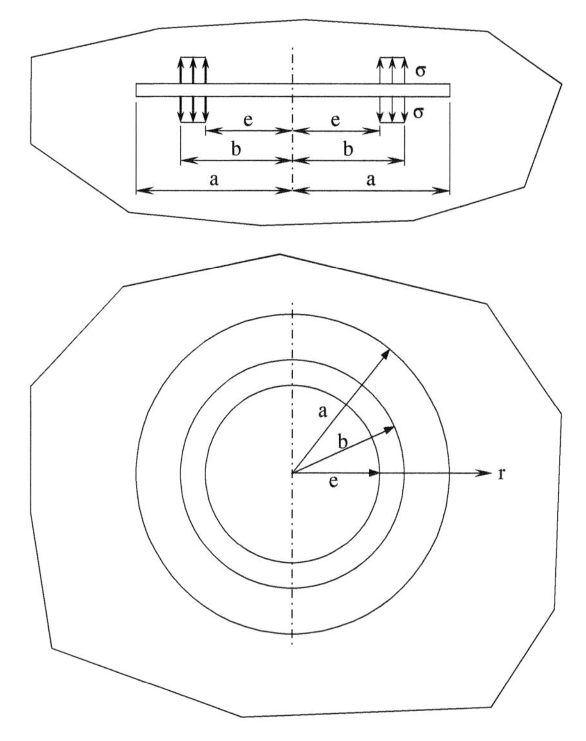


**Figure P7-4a** A penny-shaped crack under symmetrical loading

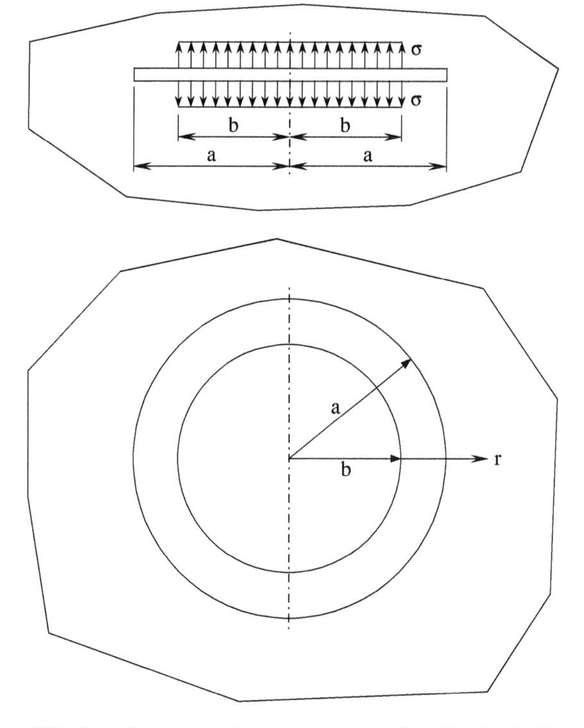


**Figure P7-4b** A penny-shaped crack under distributed loading

6. For $n = 2$ and 3, derive the solution of the $R$-curve (with detailed procedures) given by

$$R + \sum_{n=1}^{\infty} \frac{1}{n!}\left(\frac{d^n R}{dc^n}\right)\left(\frac{\alpha}{\alpha - 1}\right)^n (-c)^n = 0 \tag{7-212}$$

7. Consider a semi-infinite crack in a thin plate as shown in Figure P7-5. The faces of the crack are subjected to normal line loads. The stress intensity factor is

$$K_{\mathrm{I}} = \frac{2p}{\sqrt{2\pi l}} \tag{7-213}$$

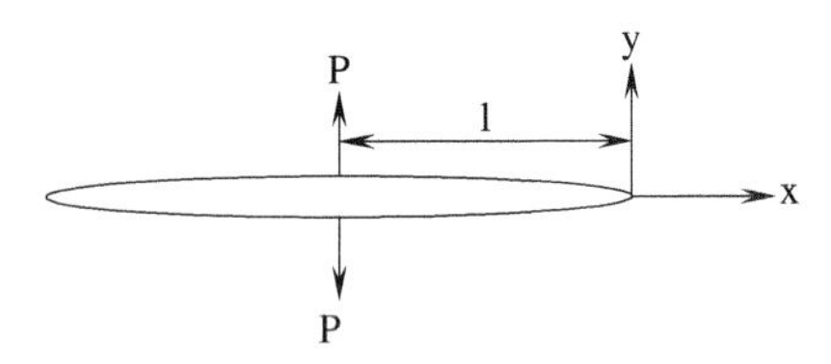


**Figure P7-5** A semi-infinite crack in a thin plate

Suppose the yield stress is $\sigma_{ys}$. Compute the plastic zone on the basis of the Dugdale model.

8. Using the configuration shown in Figure P7-6, show that for plane stress, we have the following equation. (*Hint*: The relationship between $K$ and $G$ for mode I can be applied to mode II.)

$$K_{\mathrm{II}} = \frac{P}{t\sqrt{2h}} \tag{7-214}$$

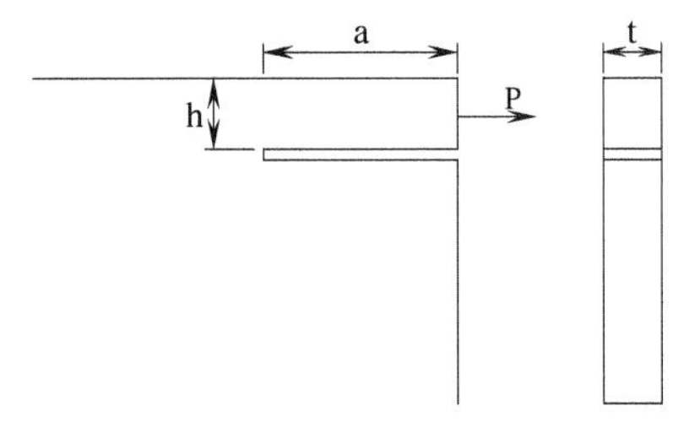


**Figure P7-6** A thin plate with an edge crack under plane stress

9. Consider two series of three-point bending tests, one series on concrete specimens and the other on cement paste specimens. The setup of the test is shown in Figure P7-7 and the test results are provided in Table 7-5. Solve the following problems:
   (a) By using beam theory, calculate the modulus of rupture for all six specimens. (Actually, it is not strictly valid to use beam theory. However, for simplification, some people still use it to get an approximate result.) What can you say about the notch sensitivity of the two materials?

**(b)** Use the compliance method to find $G_{\text{Ic}}$ for all specimens. Can LEFM be applied to either paste or concrete?

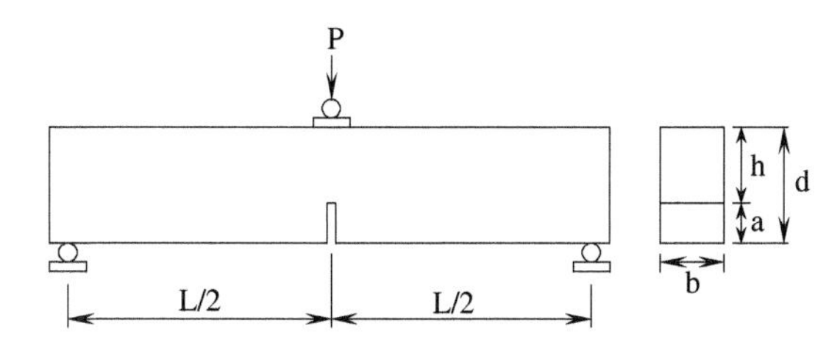


**Figure P7-7** Specimen configuration used in the bending tests

**Table 7-5** Results of two series of 3-point bending tests

| *b* (mm) | *L* (mm) | *a* (mm) | *d* (mm) | *h* (mm) | *P* (paste) (N) | *P* (concrete) (N) |
|---|---|---|---|---|---|---|
| 50 | 450 | 25 | 150 | 125 | 3,685 | 5,148 |
| 75 | 675 | 37.5 | 225 | 187.5 | 6,772 | 11,588 |
| 100 | 900 | 50 | 300 | 250 | 10,422 | 20,560 |

## REFERENCES


ACI Committee 318 (2014) "Building code requirements for structural concrete (ACI 318-14) and commentary (ACI 318R-14)," Farmington Hills, MI: American Concrete Institute.

ACI Committee 318 (2019) "Building code requirements for structural concrete (ACI 318-19) and commentary (ACI 318R-19)," Farmington Hills, MI: American Concrete Institute.

Bažant, Z.P. (1984) "Size effect in blunt fracture: concrete, rock, metal," *Journal of Engineering Mechanics, ASCE*, 110(4): 518–535.

Bažant, Z.P. (1997) "Fracturing truss model: size effect in shear failure of reinforced concrete," *Journal of Engineering Mechanics, ASCE*, 123(12), 1276–1288.

Bažant, Z.P. (2004) "Scaling theory for quasibrittle structural failure," *Proceedings of the National Academy of Sciences*, 101(37), 13397–13399.

Bažant, Z.P. and Kazemi, M.T. (1990) "Determination of fracture energy, process zone length and brittleness number from size effect, with application to rock and concrete," *International Journal of Fracture*, 44, 111–131.

Bažant, Z.P. and Oh, B.H. (1983) "Crack band theory for fracture of concrete," *Materials and Structures*, RILEM, 16, 155–177.

Bažant, Z.P. and Pfeiffer, P.A. (1987) "Determination of fracture energy from size effect and brittleness number," *ACI Materials Journal*, 84(6), 463–480.

Bažant, Z.P., Yu, Q., Gerstle, W., Hanson, J., and Ju, J.W. (2007) "Justification of ACI 446 code provisions for shear design of reinforced concrete beams," *ACI Structural Journal*, 104(5), 601–610.

Broek, D. (1987) *Elementary engineering fracture mechanics*, 4th ed., Hingham: Martinus Nijhoff.

Carloni, C., Cusatis, G., Salviato, M., Le, J., Hoover, C.G., and Bažant, Z.P. (2019) "Critical comparison of the boundary effect model with cohesive crack model and size effect law," *Engineering Fracture Mechanics*, 215, 193–210.

Carrasquillo, R.L., Slate, F.O. and Nilson, A.H. (1981) "Microcracking and behavior of high strength concrete subject to short-term loading," *Journal of the American Concrete Institute* 78(3), 179–186.

Castro-Montero, A., Shah, S.P., and Miller, R.A. (1990) "Strain field measurement in fracture process zone," *Journal of Engineering Mechanics, ASCE*, 116(11), 2463–84.

Cedolin, L., DeiPoli, S., and Iori, I., (1987) "Tensile behavior of concrete," *Journal of Engineering Mechanics, ASCE,* 113(3), 431–449.

Chen, C., Cai, Q. and Wang, R. (1977) *Engineering Fracture Mechanics* Beijing: The Press of National Defense Industry (in Chinese).

Cho, K.Z., Kobayashi, A.S., Hawkins, N.M., Barker, D.B., and Jeang, F.L. (1984) "Fracture process zone of concrete cracks," *Journal of Engineering Mechanics, ASCE*, 110(8), 1174–1184.

Cooper, D.B., Meda, G., and Sinclair, G.B. (1995) "A comparison of crack-flank displacement fitting for estimating K with a path independent integral," *International Journal of Fracture*, 70(3), 237–251.

Dönmez, A. and Bažant, Z.P. (2017) "Size effect on punching strength of reinforced concrete slabs with and without shear reinforcement," *ACI Structural Journal*, 114(4), 875–886.

Dönmez, A. and Bažant, Z.P. (2019) "Critique of critical shear crack theory for fib Model Code articles on shear strength and size effect of reinforced concrete beams," *Structural Concrete*, 2019, 1–13.

Du, J., Yon, J.H., Hawkins, N.M., Arakawa, K., and Kobayashi, A.S., (1992) "Fracture process zone for concrete for dynamic loading," *ACI Materials Journal*, 89(3), 252–258.

Dugdale, D.S. (1960) "Yielding of steel sheets containing slits," *Journal of the Mechanics and Physics of Solids*, 8(2), 100–104.

Faber, K.T. and Evans, A.G. (1983) "Crack deflection processes-I theory," *Acta Metallurgica*, 31(4), 565–576.

Gao, S. and Xu, S. (2007) "Critical concrete crack length determination using strain gauges," *Journal of Tsinghua University (Science and Technology)*, 47(9): 1432–1434 (in Chinese).

Gopalaratnam, V.S., and Shah, S.P. (1985) "Softening response of plain concrete in direct tension," *ACI Journal*, 82(3), 310–323.

Higgins, D.D. and Bailey, J.E. (1976) "Fracture measurements on cement paste," *Journal of Materials Science*, 11, 1995–2003.

Hillerborg, A., Modeer, M., and Petersson, P.E. (1976) "Analysis of crack formation and crack growth in concrete by means of fracture mechanics and finite elements," *Cement and Concrete Research*, 6(6), 773–782.

Homeny, J. and Vaughn, W.L. (1990) "R-curve behavior in a silicon carbide whisker/alumina matrix composite," *Journal of the American Ceramic Society*, 73(7), 2060–2062.

Inglis, C.E. (1913) "Stresses in a plate due to the presence of cracks and sharp corners," *Transactions of the Institute of Naval Architects*, 55: 219–241

Irwin, G.R. (1958) *Fracture: Handbook der Physik*, VI, Flugge Ed., Berlin: Springer, pp 551–590.

Irwin G.R. (1960) "Plastic zone near a crack and fracture toughness," in *Proceedings of the 7th Sagamore Conference*, pp. IV–63.

Jenq, Y.S. and Shah, S.P. (1985) "A two parameter fracture model for concrete," *Journal of Engineering Mechanics, ASCE*, 111(4), 1227–1241.

Kaplan, M.E. (1961) "Crack propagation and fracture of concrete," *Journal of ACI*, 58(5), 591–610.

Karihaloo, B.L. and Nallathambi, P. (1991) "Notched beam test: Mode I fracture toughness," in *Fracture Mechanics Test Methods for Concrete. Report of RILEM Technical Committee 89-FMT*, Shah, S.P. and Carpinteri, A., eds., London: Chapman & Hall, pp. 1–86.

Leicester, R.H. (1973) "Effect of size on the strength of structures," Report paper No. 71, CSIRO Forest Products Laboratory, Division of Building Research Technology, Commonwealth Scientific and Industrial Research Organization, Australia.

Li, F. and Li, Z. (2000) "AE monitoring of fracture of fiber reinforced concrete in tension," *ACI Materials Journal*, 97(6), 629–636.

Li, Z. and Shah, S.P. (1994) "Microcracking localization in concrete under uniaxial tension: AE technique application," *ACI Materials Journal*, 91(4), 372–389.

Liaw, B.M., Jeang, F.L., Du, J.J., Hawkins, N.M., and Kobayashi, A.S. (1990) "Improved non-linear model for concrete fracture," *Journal of Engineering Mechanics, ASCE*, 116(2), 429–445.

Lim, I.L., Johnston, I.W., Choi, S.K., and Murti, V. (1992) "An improved numerical inverse isoparametric mapping technique for 2D mesh rezoning," *Engineering Fracture Mechanics*, 41(3), 417–435.

Murakami, Y., et al. (1987) *Stress intensity factors handbook*, New York: Pergamon.

Ouyang, C., Landis, E., and Shah, S.P. (1991) "Damage assessment in concrete using quantitative acoustic emission," *Journal of Engineering Mechanics, ASCE*, 117(11), 2681–2698.

Pang, H.L.J. (1993) "Linear elastic fracture mechanics benchmarks: 2D finite element test cases," *Engineering Fracture Mechanics*, 44(5), 741–751.

Refai, T.M.E. and Swartz, S.E. (1987) "Fracture behavior of concrete beams in three-point bending considering the influence of size effects," Report No. 190, Engineering Experiments Station, Kansas State University.

Rice, J.R. (1968) "A path independent integral and the approximate analysis of strain concentration by notches and cracks," *Journal of Applied Mechanics, ASME*, 35(6), 379–386.

RILEM Technical Committee 89-FMT (1990) "Determination of fracture parameters (KsIc and CTODc) of plain concrete using three-point bend tests, proposed RILEM draft recommendations," *RILEM, Materials and Structures* 23(138), 457–460.

RILEM Technical Committee 265-TDK (2011) "Testing methods for determination of double-K criterion for crack propagation in concrete," *RILEM, Materials and Structures*

Roelfstra, R.E. and Wittmann, F.H. (1986) "A numerical method of link strain softening with fracture in concrete," in *Fracture Toughness and Fracture Energy in Concrete*, Wittmann F.H., ed., Amstesrdam: Elsevier Science, pp. 163–175.

Sakai, M., Yoshimura, J., Goto, Y. and Inagaki, M. (1988) "R-curve behavior of a polycrystalline graphite: microcracking and grain bridging in the wake region," *Journal of the American Ceramic Society*, 71(8), 609–616.

Shah, S.P. and McGarry, F.J. (1971) "Griffith fracture criterion and concrete," *Journal of Engineering Mechanics Division, ASCE*, 97(6), 1663–1676.

Swartz, S.E. and Refai, T. (1989) "Cracked surface revealed by dye and its utility in determining fracture parameters," in *Fracture toughness and fracture energy: test method for concrete and rock*, Mihashi, H., et al., eds., Brookfield, VT: Balkema, pp. 509–520.

Tada, H., Paris, P.C., and Irwin, G.R. (2000) *The stress analysis of cracks handbook*, New York: ASME Press.

Van Mier, J.G.M. (1991) "Mode I fracture of concrete: Discontinuous crack growth and crack interface grain bridging," *Cement and Concrete Research*, 21(1), 1–15.

Walsh, P.F. (1972) "Fracture of plain concrete," *Indian Concrete Journal*, 46(11), 469–470.

Wecharatana, M. and Shah, S.P. (1982) "Slow crack growth in cement composites," *Journal of Structural Engineering*, 108(6), 1400–1413.

Wecharatana, M. and Shah, S.P. (1983) "Predictions of nonlinear fracture process zone in concrete," *Journal of Engineering Mechanics*, 109(5): 1231–1246.

Xu, S., Bu, D. and Zhang, X. (2008) "A study on double-K fracture parameters by using wedge-splitting test on compact tension specimens of various sizes," *China Civil Engineering Journal* 41(2), 70–76 (in Chinese).

Xu, S. and Reinhardt, H.W. (1999a) "Determination of double-K criterion for crack propagation in quasi-brittle materials, Part I: Experimental investigation of crack propagation," *International Journal of Fracture*, 98(2), 111–149.

Xu, S, and Reinhardt, H.W. (1999b) "Determination of double-K criterion for crack propagation in quasi-brittle materials, Part II: Analytical evaluating and practical measuring methods for three-point bending notched beams," *International Journal of Fracture*, 98(2), 151–177.

Xu, S. and Reinhardt, H.W. (1999c) "Determination of double-K criterion for crack propagation in quasi-brittle materials, Part III: compact tension specimens and wedge splitting specimens," *International Journal of Fracture*, 98(2), 179–193.

Xu, S. and Reinhardt, H.W. (2000) "A simplified method for determining double-K fracture parameters for three-point bending tests," *International Journal of Fracture*, 104(2), 181–209.

Xu, S. and Zhang, X. (2008) "Determination of fracture parameters for crack propagation in concrete using an energy approach," *Engineering Fracture Mechanics*, 75(15), 4292–4308.

Xu, Y. (1998) "Self-similar crack expansion method for two-dimensional cracks under mixed mode loading conditions," *Engineering Fracture Mechanics*, 59(2), 165–182.

Xu, Y., Moran, B., and Belytschko, T., (1997) "Self-similar crack expansion method for three-dimensional crack analysis," *Journal of Applied Mechanics, Transactions ASME* 64(4), 729–736.

CHAPTER 8

# NONDESTRUCTIVE TESTING IN CONCRETE ENGINEERING

## 8.1 INTRODUCTION

### 8.1.1 General Description

There are two kinds of tests, destructive and nondestructive, that can be used to assess the properties of concrete material or structures. Destructive testing obtains the materials' or structural properties or information through actions that destroy the integrity of the materials or structures, while nondestructive testing obtains the information without destroying the integrity of the materials or structures. Hence, nondestructive testing can be defined as the measurement, inspection, or analysis of materials, existing structures, and processes of manufacturing without destroying the integrity of the materials and structures. The common terms used in this field are as follows:

NDT: nondestructive test
NDI: nondestructive inspection
NDE: nondestructive evaluation
QNDE: quantitative nondestructive evaluation

The first three terms are the same and the terms are interchangeable. The last term, QNDE, emphasizes the quantitative nature and is more difficult to apply. The application fields of NDT include the following:

**(a)** *Quality control of concrete materials and structures*, including hydration processing monitoring, strength development evaluation, curing process monitoring of concrete, welding quality evaluation of reinforcing steel, and/or porosity assessment in matured concrete members.

**(b)** *In-service inspections of concrete materials and structures*, including inspection on the corrosion degree and rate, residual strength, fatigue damage, existing crack length, debonding of an interface, and impact damage.

Nondestructive testing in concrete engineering (NDT-CE) is an important tool for assessing the quality and maturity of a concrete structure during construction, which helps decision-making in construction procedures and speed. For new constructions, NDT-CE can act as a quality-control tool to evaluate reliability, assess integrity, and monitor the conditions of construction as a whole or in part. NDT-CE is very useful for condition assessments of serviceability and stability of concrete structures and maintenance and renovation decision making of aged buildings and infrastructures by estimating the properties and performance of materials and structures. NDT-CE can localize and measure the defects or damage inside a structure for repair or removal purposes. It also can be used to find the position of the prestressed elements to avoid them being destroyed by accidental drilling. In general, NDT-CE can make a significant contribution to evaluating and guaranteeing building safety and protecting invaluable monuments at a low cost, and in a rapid and simple manner. The following are some general requirements for a practical NDT-CE technique:

It must be effective for determining the quantities of interest.
It must be accurate, with the ability to determine the shape, size, and depth of the defective areas.
It must be insensitive to the surface condition and the shape of the object to be tested
It must be efficient in terms of labor and equipment, and rapid and simple to use.
It must be economical, with a low cost for equipment and maintenance.
It must be unobtrusive to the surrounding environment.
It must be safe for operators and equipment.
It must be convenient to the users.
It is better to be non-contacting.
It should be able to inspect large areas as well as localized areas.

For a specific NDT-CE technique, it is difficult to satisfy all of the above requirements, but it should at least meet some of these requirements, and the more the better. The performance of concrete in a building is often different from that in the laboratory. Hence, in situ testing directly on a concrete structure becomes important.

It is said that many NDT methods have been tried on concrete structures but none of them has proved satisfactory. This is partially true, because concrete, the most widely used building material, is an inhomogeneous, porous, and highly variable material. The average values of the concrete properties depend on the raw material types and proportions, and they vary with water content, curing conditions, the microstructure of hardened binders, the cement mix, the type, shape, orientation, size, and distribution of the aggregate, as well as the state of the concrete reinforcement. Another problem with the NDT application to concrete structures is that a building or bridge structure is usually very large. All of these factors have created problems in producing an accurate result. In regard to the NDT-CE techniques themselves, even if they can perform well in some aspects, they also have their own limitations.

Generally speaking, to better apply NDT-CE techniques, one needs to thoroughly understand the problem of the concrete structures or materials to be tested, and to select NDT-CE techniques correctly, with a sound knowledge of the basis of the testing techniques and devices to be used. Usually, this requires interdisciplinary cooperation, prior knowledge of the specimen, calibration of the instruments, and experienced operators. Since the 1980s, NDT-CE techniques have been developed into a more advanced stage. Many excellent results have been achieved with these techniques, benefitting from the advanced development of digital electronics and computer technology. Measurements are mostly automatically performed by computer-controlled instruments and the test results are automatically analyzed and displayed on a computer screen in an easily understood way. The analyzing and displaying process is usually called digital signal processing. Significant improvements were achieved in the 1990s in ultrasonic pulse-echo testing, in the application of radar and the impact echo technique, in data analysis, and in the interpretation and simulation of test results.

In NDT-CE, an allowable tolerance philosophy is frequently applied. First, there is always a detection limit in an NDT method and this limit has to be tolerated. Second, even when a flaw or crack is detected, it can be allowed to remain if its size is less than a critical value. Moreover, tolerance can be applied to the inspection interval on different concrete structures, according to their importance and from economic considerations. When conducting NDT, there may not be a clear cutoff in the detection resolution, as shown in Figure 8-1; hence, probability theory is very useful in NDT-CE.

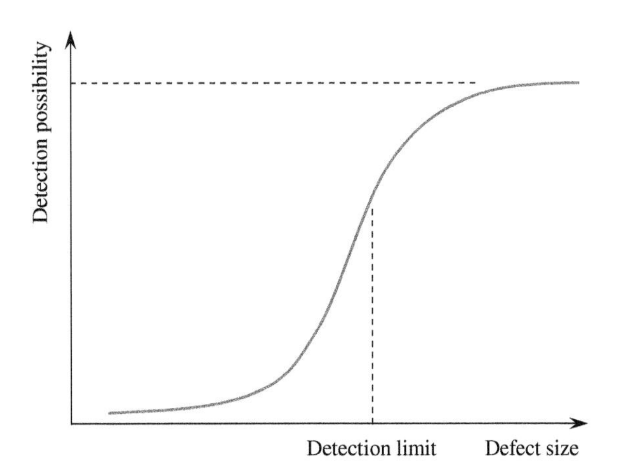


**Figure 8-1** No clear cutoff in detection resolution

### 8.1.2 Principles and Classifications

Generally speaking, NDT is a technique to know the properties or to "see" the structures inside the component to be inspected, especially for optically opaque components, without damaging them. There are many NDT techniques available, such as mechanical techniques, elastic wave techniques, electromagnetic techniques (EMT), and optical techniques (OT). There are four main aspects related to a specific NDT technique. The first is the physical principle on which the technique is based, called the *principle* for simplicity. The second aspect is the NDT techniques employed, termed *method*. The third is the equipment used to perform the function specified, and the fourth is the signal processing scheme adopted. Any advances in these four aspects would improve the performance of the relevant NDT technique.

Among the four aspects, the principle is the most important, because it is the foundation of an NDT technique. The principle controls the way to link the received signals to the properties of the materials and to interpret material or structural conditions and qualities. Different NDT techniques depend on different physical rules. For example, the principle of the elastic wave technique involves the physical rules on elastic wave behavior. The capability and the limitation of an NDE technique essentially depend on the basic physical rules followed.

The NDT method means the manner or way to realize the measurement and is the key aspect of an NDT technique. Following a principle, there are lots of possible methods. For example, there is the transmission method and the reflection method for wave-related NDT techniques. The equipment is a physical medium to undertake the NDT measurements and is designed based on the principle and method. The equipment determines whether an NDT technique can be applied or not.

The signal processing scheme is the way to extract useful information from the gathered signals. The bases of interpreting the measured data are the principles of the method followed, but the signal processing scheme sometimes plays a significant role in the data interpretation procedure. Nowadays, signal processing (SP) has become an essential and necessary procedure in NDT measurements. Signal processing can be analog or digital. Digital signal processing (DSP) has become more and more popular due to the rapid development of advanced computer techniques and mathematical analysis tools.

The difference between terminology words "signal" and "information" should be pointed out. The signal is something that a sensor directly measures. Information is something interpreted from a signal. The signal is the information carrier. Sometimes, there are many signals received, but little information can be interpreted. In this situation, the signal may be noisy or may lack ways to be interpreted. Having fewer signals but carrying more information is more desirable.

The NDT techniques can be classified into different categories according to the above-mentioned four aspects. That is to say, they can be classified as to what kind of physical principle they follow, what methods they adopt, what kind of equipment they use, and what kind of signal processing scheme they use. They can also be classified according to the final goals the technique is intended to reach, or the objectives defined in the technique.

Plenty of NDT-CE techniques can be found in practice. Some are mechanical techniques like building dynamics, impact echo, ultrasound, and acoustic emission techniques. Some are electromagnetic techniques like eddy current, electrochemical methods, microwaves, nuclear magnetic resonance method, magnetic method, and radiography. Some are optical methods, like interferometry, infrared thermography, and fiber optic sensors. Computer tomography and other imaging techniques are also used as NDT-CE techniques.

According to the physical working principle, NDT-CE techniques can be roughly classified into three categories: (1) direct measurement; (2) load-induced reaction measurement; and (3) measurements through an inquiring agency.

#### 8.1.2.1 Direct-Measurement NDT-CE Techniques

Sometimes the information or properties of a material can be directly measured or interpreted through NDT measurement. This kind of testing is referred to as the direct NDT-CE technique. Direct-measurement NDT-CE techniques should be solidly based on a fundamental physical principle. For example, the density can be obtained by weighing the object, providing its volume is known. The weighing technique also can be used to determine the moisture content of a concrete sample. The procedure involves weighing the sample, drying the sample, and reweighing the dried sample. The weight difference is directly related to the moisture content. Another example is the half-cell potential measurement technique, which is the most commonly used electrochemical NDT-CE technique for reinforced concrete structures. It measures the macrocurrent of the corroding steel bars of reinforced concrete, and then the corrosion condition can be easily deduced. The magnetic leakage technique measures the premagnetized magnetic field of the steel-reinforced concrete, and the deformation of the magnetic field can indicate the fracture of the steel rebar.

The direct NDT-CE testing technique is usually simple, but, like the weighing–drying–weighing technique, for example, it is not suitable for in situ applications. It is still useful in practice for testing structures by means of a slightly destructive manner—samples could be taken from the structure, e.g., by coring from a column, beam, slab, or wall.

#### 8.1.2.2 Load-Induced Reaction Measurement NDT-CE Techniques

This kind of technique is used to detect the response of an object to be tested by loading it to a certain level. However, the load should be controlled in a reasonable range without damaging the test specimen. The signals are gathered to reflect the nature of the test object. Building dynamics is an example. In building dynamics, the natural frequency or eigenfrequency of a structure (a building, a bridge, etc.) is excited by a random load and measured. The mode shapes can also be determined when needed. These data can be compared with the theoretical prediction of the model of this structure to extract information about the integrity, stiffness, stability, and some macro defects. Another example is the impact echo method, which measures the dynamic response of a structure to an applied impact load.

#### 8.1.2.3 Measurement Through Inquiring Agent NDT-CE Techniques

These methods use an inquiring agent as a probe, and the agent usually is a wave packet. They can be divided into active and passive techniques. In the active technique, a wave packet is generated by the test instrument and then transmitted into the materials to be inspected. Examples are ultrasonic testing and microwave testing. The probe agency, the wave packet, interacts with the matrix material and embedded objects if they exist. Then, it gives out a signal, which carries the information on the specimen. In passive techniques, the wave packet stems from the tested specimen itself and carries information about the process the specimen is undergoing. An example is the acoustic emission NDT-CE technique.

The most widely used inquiring agents in NDT-CE are mechanical waves (acoustic waves and ultrasonic waves), electromagnetic waves (ultrahigh frequencies (UHF), L-band, and S-band microwaves), infrared lights, lights, and X-ray and gamma-ray radiation. From a physical point of view, infrared light, light, and X-ray and gamma-ray radiation are essentially electromagnetic waves.

### 8.1.3 Components of NDT-CE

As mentioned above, NDT-CE stands for nondestructive testing in concrete engineering. It is a relatively young area in NDT applications, and large-scale experiments only started in the 1980s. NDT-CE activities can be viewed in different categories: (1) the test objects or problem, and (2) the testing method. However, these categories interweave. A problem may require several different NDT techniques, or a technique may be used for attacking several test objects or problems.

#### 8.1.3.1 Testing Objects

The testing objects of NDT-CE may include masonry buildings, usually low, multistory residential or office buildings made of concrete blocks, concrete bridges, tall buildings, dams, highways, and airport runways. Differing from NDT applied in other fields, the test structures in NDT-CE are usually much larger, more complicated in shape, and more complex in properties. Hence, NDT-CE usually requires more carefully designed inspection plans and several different techniques.

#### 8.1.3.2 Testing Problems

The test problems of NDT-CE can be viewed from the load-carrying abilities and environment resistance abilities.

**(a)** *Strength*: The strength of a concrete cast on-site can be very different from the strength measured in a laboratory. Moreover, the strength of concrete can vary with its service life. Hence, the determination of concrete strength on site is a big issue for NDT-CE. In addition, if a concrete structure has gone through a fire, the residual strength of the damaged concrete has to be determined in order to make decisions for renovation work. The true steel yield stress of a concrete structure during service and after a natural disaster has to be determined using NDT-CE, for maintenance purposes.

**(b)** *Cracks and fractures*: Concrete is a quasi-brittle and tension-weak material. In addition, due to the boundary constraints in shrinkage and creep, large tensile stresses can be generated in concrete. When such stress exceeds its tensile strength, concrete will crack. The existence of cracks can induce many durability problems, as they can channel many harmful agents into the concrete. Thus, detecting the properties of cracks in concrete is a big challenge. The detection of a crack includes the length and width of the crack, the position of the crack, and the propagation rate of the crack. From these parameters, it can be judged whether the crack is active or not, or dangerous or not.

**(c)** *Thickness*: The thickness measurement of a pavement, a restraining wall, the cover of reinforcing steel, and slab is an important quality assurance issue in concrete structure construction and maintenance.

**(d)** *Moisture*: Moisture or water in a concrete structure can play an important role in deterioration. It can cause a lot of durability problems, such as corrosion, CH leaching, alkali-aggregate reaction (AAR) expansion, and leakage. Hence, detection of the moisture location and content is a common NDT-CE problem.

**(e)** *Corrosion*: Corrosion is the most serious and dangerous durability problem, and also is the most commonly encountered durability problem. Corrosion can crack the concrete cover and make it spall. Corrosion can also significantly reduce the effective area of the reinforcing steel and put a concrete structure in danger, especially a prestressed concrete structure. It is obvious that the detection of corrosion occurring, corrosion rate, and corrosion degree is one of the most important issues of NDT-CE.

**(f)** *Debonding*: Debonding is the separation of two adjacent materials originally bonded together. Debonding is usually an indication of severe damage or deterioration of a concrete structure. Frequently encountered debonding problems include those caused by corrosion in losing the adhesive effect and interfaces damaged by shear stress.

#### 8.1.3.3 Testing Methods

The classification principle of NDT-CE methods was briefly discussed in Section 8.1.2 and more details of commonly used NDT-CE methods are described in the following sections. Here, only a brief introduction is provided for most NDT-CE methods.

**(a)** *Mechanical wave techniques (MWT)*: These techniques take the mechanical waves as the working agency. The principle of these techniques is the generation and propagation of laws concerning mechanical waves. Ultrasonic waves, acoustic waves, and subacoustic waves are mechanical waves. A special feature of these waves is that they are directly related to the mechanical properties of the media they propagate through, hence, this is a remarkable advantage for determining the mechanical performance of the materials. Mechanical wave NDT techniques are widely used in NDT-CE and are also some of the most popularly used conventional NDT techniques. They can be used in a wide range of objectives, as a tool for measurement, detection, and monitoring. They can meet most of the requirements for NDE in civil engineering. They are accurate in determining the shape, size, and depth of the defective areas, with high sensitivity, deep penetration, low cost, and they are easy and fast to operate and convenient for in situ use. Mechanical waves are not harmful to the human body, which is an important advantage over radiation.

The basic theory involved in MWT methods is elastic wave generation, propagation, and reception. Wave propagation involves reflection, transmission, and scattering, as well as diffraction and interference. The wave packet transmitted into the materials to be tested interacts with the material and changes its own parameters as a result, which carries the information about the properties of the test object. The most frequently measured wave parameters are amplitude, phase, and frequency, as well as the transmission time when the wave passes through the object, which is usually called the time of flight (TOF). It is important to establish some relations between these wave parameters and the properties of the object to be tested.

Mechanical wave techniques can be active or passive. In the active ones, the testing apparatus produces mechanical waves. In the passive techniques, the mechanical waves stem from the test object itself. The main mechanical wave NDT-CE techniques are ultrasound testing, impact echo, and acoustic emission.

**(i)** *Ultrasonic techniques (UT)*: The ultrasonic technique works in the ultrasonic band. Most ultrasonic inspection is conducted at frequencies between 0.1 and 25 MHz (human hearing range is 20 Hz to 20 kHz.). Ultrasonic waves are mechanical waves that consist of oscillations or vibrations of the atomic or molecular particles about the equilibrium position of these particles. This method introduces high-frequency sound waves into materials for the detection of surface and subsurface flaws in the material and hence is an active NDT method. The sound waves travel through the material and are reflected at the interfaces. The reflected sound wave or the first arrival time can be displayed and analyzed to define the presence and location of flaws or dislocations. UT has the following advantages: accurate, convenient to use, cheap to buy and maintain, as well as low labor requirements. In addition, UT can detect almost every kind of defect. Usually, piezoelectric transducers (PZT) or transducer arrays are used to generate ultrasound in UT. The waves used can be continuous (CW) or impulse (PW). PW ultrasound is now the dominant technique. The ultrasonic waves propagate in the material and are reflected or scattered by the materials inside the specimen or at the interface of the structure of the specimen. The reflected or scattered wave pulse, especially the backscattered wave pulse, is usually referred to as an echo. To receive an ultrasound, the same emitting transducer/array could be used, or another one. The receiver can be placed at the same side of the emitter (reflection mode) or opposite the emitter (transmission mode). Both the emitter and the receiver can be individual or in an array. These form a great variation in NDT-CE-UT testing methods. Last but not insignificant, the procedure is signal processing and display. Signal processing, especially digital signal processing, is so important that it has become an essential part of a UT technique.

**(ii)** *Acoustic emission (AE)*: The acoustic emission method is a passive NDT method. It relies on the detection of the energy released due to microcracks, dislocations, and inclusions in concrete. The beauty of the AE technique is that it can provide information about an active crack that could be dangerous to a structure.

**(iii)** *Impact echo (IE)*: The impact-echo method usually generates an impact on a concrete specimen and then measures the response of the specimen under the action of the impact. Through time domain and frequency domain analysis of the received response, the properties of concrete, such as the dynamic modulus and Poisson's ratio, can be interpreted. Moreover, the damage degree of the material or structure can be interpreted from the changes in the stiffness.

**(b)** *Electromagnetic wave technique (EMT)*: Among NDT methods in concrete engineering, the electromagnetic wave technique is one of the most powerful, as most nonmetal materials are amenable to electromagnetic waves. This kind of technology uses a special band of electromagnetic waves as the inquiring agent and is an active method. It can be applied in either a noncontact or a contact mode, and measurements can be taken in a remote as well as in a real-time manner. It possesses a lot of advantages over other techniques.

Electromagnetic waves cover a very wide range of frequencies in the scientific sense. According to the different frequency ranges, from lower to higher, they have different nomenclature, such as radio waves, high-frequency waves (RF), very high-frequency waves (VHF), ultra-high-frequency waves, microwaves (L-band, S-band, etc.), infrared light (IR), and light, X-rays. In NDT engineering, they are usually referred to according to what kind of electromagnetic wave is used as the working agent. The electromagnetic wave techniques are usually related to a relatively lower frequency, namely, the frequency from RF to the S-band usually called microwaves. Microwave testing in the reflection mode is often called the radar technique, whereas the infrared light, light, and X-ray techniques, etc. are usually regarded as other independent NDT techniques. For most cases, the frequency range for EMT is from 90 MHz to 1 GHz, and the electrical techniques are well developed in this band.

EMT is suitable mostly for testing non-metallic materials. It can penetrate much more deeply than ultrasound and is very sensitive in detecting the metal objects inside the nonmetal matrix, like steel bars in concrete. It can detect a flat defect, which is difficult to detect by X-ray techniques. It is not sensitive to aggregate size and type. By means of a microwave technique, radar, a 6-mm-diameter steel bar and 20 by 20-mm voids can be detected when the cover thickness is around 200 mm. It does not require knowledge of the specific state of the testing surface, such as the shape, oil pollution, or roughness. It can make one-sided access to an object under testing, and it can precisely locate the internal faults. It has a wide range of measurements, from microns in paint coatings to meters in concrete.

The operation of the measurement system is fast and simple. Electromagnetic shielding, by the microwave technique, can detect only the surface of a metal object but not the insides. Nor can it give information on the structures beyond a metallic sheet. In addition, the equipment is relatively expensive. EMT is based on the physical laws of microwave generation and propagation; thus, problems similar to those in UT would be encountered. These include electromagnetic wave (EMW) generation, propagation and reception, reflection, transmission, and scattering, as well as diffraction and interference.

The most frequently measured parameters of the wave are amplitude, phase, and frequency, as well as the time of flight. It is important to establish some relations between these wave parameters and the properties of the object to be tested. The most reliable and reproducible results are obtained by determining the wave velocity, reflection coefficient, and attenuation and velocity of the microwave packet to be tested. The wave velocity and the reflection and absorption of the wave depend on the electromagnetic parameters of the material, such as their permittivity, permeability, and conductivity, and the geometrical parameters of the object, such as their thickness, shape, orientation, direction, etc. Usually, in different frequencies, a wave exhibits different characteristics. Hence, by measuring the time of flight of microwaves, the attenuation during passing through the testing object, or the reflection/transmission coefficients, one can deduce the parameters of the test object, and hence deduce information about the structure and the mechanical properties.

The microwave NDT technique can be used to perform building inspections and quality control and to detect steel objects, as well as moisture, water content, and cracks. It is well suited for locating the lateral positions of rebars, the location of metallic ducts, and metallic T-girders and metallic anchors in prefabricated three-layer concrete elements, the location of plastic and metallic leads, measurement of the cover thickness of concrete, and moisture measurement of concrete. The artificial birefringence properties can be found for ceramics and cement with microwaves, of which the wavelength is between 7.5 and 12.5 mm, similar to that of photoelasticity with visible light (Dannheim et al., 1995). Hence, visualizing the inside of the specimen is possible with microwaves.

A very powerful EMT instrument is ground-penetrating radar (GPR). A GPR image can profile a bridge pile in a bridge scour investigation. A maximum frequency should be recommended in a GPR survey design for clutter reduction. It is suggested that the wavelength should be ten times larger than the characteristic dimension for geological materials (Davidson et al., 1995). For most cases, the frequency range is from 90 MHz to 1 GHz and the pulse repetition frequency is 25 Hz. The following typical results could be reached: a scour depth of 1.2 m relative to the general riverbed; a filling consisting of a sandy layer 0.5 m deep overlying gravel 0.8 m thick at the center of the hole; and a water depth of about 1 m. The GPR measurement is fast enough to detect a large area like a 10,000-m$^2$ airport pavement (Weil, 1995).

The EMT technique can make a reliable assessment of the integrity of a structure and locate tendons and metal ducts (diameter 85 mm) to offer information on geometrical dimensions. It can detect defects in concrete members, voids in tendon ducts, compression faults, or honeycombing (voids from 80 to 160 mm) in concrete. The EMT technique is not sensitive to aggregate size. For example, for aggregate size 8 to 32 mm and cover thickness 340 ± 10 mm, the lateral position of reinforcement could be found with an accuracy of ±10 mm (Attoh-Okine, 1995). A microwave moisture content meter offers online water content monitoring in clay (Kalinski, 1995).

**(c)** *Optical techniques*: These techniques use light waves as the inquiring agent, and are also the active methods. The main techniques are based on interferometry and energy transportation. The representative technique of the former is electronic speckle pattern interferometry (Hung, 1982; Jones and Wykes, 1989), and the latter is infrared detection (Reynolds, 1988; Favro et al., 1991).

Interferometric methods use the principle of interferometry of two lights to measure the surface change of a concrete specimen. The examples include holographic, moire, and electronic speckle pattern interferometry methods.

Infrared thermography uses the principles of radiation theory. By measuring the surface temperature difference due to the difference in the radiation of the materials underneath, infrared thermography can detect debonding and moisture content in concrete materials or structures.

**(d)** *Electrical and electrochemical methods*: These methods usually measure the electrical parameters, such as potential, current, or resistivity that can be influenced by property changes of the concrete material or structure, and hence these electrical parameters are used to interpret the concrete properties. Examples include noncontact resistivity measurement, half-cell potential measurement, and corrosion current measurement.

Other electrical methods for detecting reinforcement corrosion are available. One such technique is based on electrical resistance measurements on a thin section of in situ reinforcement (Vassie, 1978). The electrical resistance of the reinforcement bar is inversely proportional to its thickness; therefore, as the thin slice is gradually consumed by corrosion, it becomes thinner, with a corresponding increase in resistance. In this technique, to facilitate measurements, the probe is normally incorporated into a Wheatstone bridge network (Figure 8-2). One of the probes is protected from corrosion, while the other arm is the in-place portion of the reinforcement. The measured resistance ratio can be used to monitor the corrosion rate. The significant disadvantages are the need to position the exposed arm of the probes during construction and the concerns of associated sampling techniques required for locating the probes in a large structure subject to localized corrosion.

**(e)** *Magnetic methods*: Magnetic techniques use magnetic fields as an essential tool to detect or interpret material properties. As far as applications in concrete structures go, currently, there are three different aspects of magnetic field phenomena: (1) alternating current excitation of conducting materials and their magnetic inductance; (2) direct current excitation, resulting in magnetic flux leakage fields around defects in ferromagnetic materials; and (3) nuclear magnetic resonance (Malhotra and Carino, 2004). The magnetic induction method is applicable only to ferromagnetic materials, in which the test equipment circuitry resembles a simple transformer and the test object acts as a core (Figure 8-3). With a piece of metal close to the coil of the transformer, the inductance increases and the change in induced current depends on the magnetization characteristics, location, and geometry of the metal. The inductance of the coil can be used to measure coil-to-place distance if the relationship between mutual inductance and the coil-to-place distance is known. The magnetic induction theory has resulted in the development of equipment for determining the location, sizes, and depth of reinforcement or depth of concrete covers, such as a cover meter to detect the location of embedded steel.

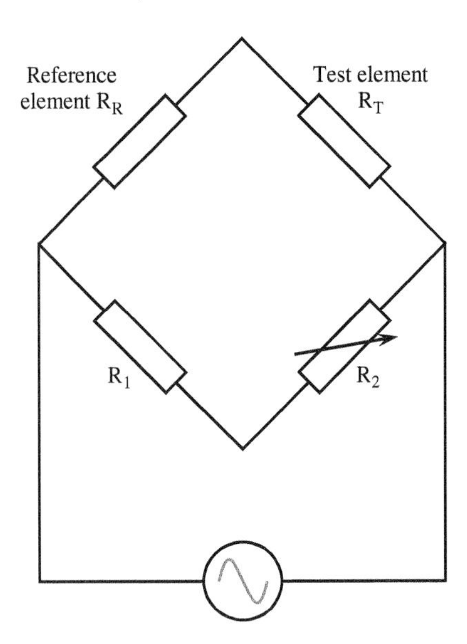


**Figure 8-2** Basic circuit for the electrical resistance probe technique

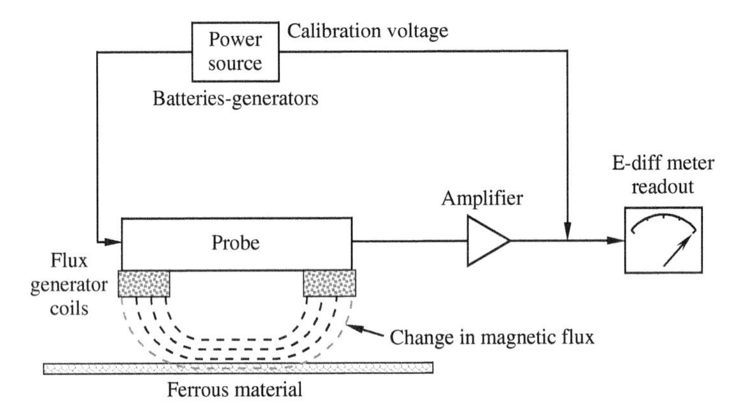


**Figure 8-3** Principle of an induction meter used to locate the reinforcement

Magnetic flux leakage (MFL) nondestructive testing consists of magnetizing a test part, generally a ferromagnetic material, and scanning its surface with some form of the flux-sensitive sensor for the leakage field (Bray and Stanley, 1997). The fundamental theory of MFL has been explained elsewhere (Bray and McBride, 1992; Bray and Stanley, 1997). When ferromagnetic materials are magnetized, magnetic lines of force (or flux) pass through the material and complete a magnetic path between the poles, which increases from zero at the center of the specimens to increased density and strength toward the outer surface. When cracks or defects exist in materials, they can lead to perturbations of the magnetic flux. Its magnetic permeability is drastically changed and the leakage flux provides a basis for non-destructive identification of such discontinuities. The amount of leakage flux produced also depends on the defect geometry. Broad, shallow defects will not produce a large outward component of leakage flux. What's more, a defect whose long axis is parallel to the lines of flux will not produce leakage flux either. By sensing the magnetic flux along with a test member, damage location can be deduced. MFL techniques are suitable for the detection of surface or near-surface anomalies in ferromagnetic materials, such as detection area reduction in reinforcements and breakage of tendons in prestressed cables. Internal defects in thick parts may not be detected because the magnetic lines of flux nearly bypass the defect with little leakage. These techniques generally do not require mechanical contact with the testing object and are very amenable to automatic signal recognition schemes, both of which are of great benefit for automated, high-speed inspection.

Nuclear magnetic resonance (NMR) is based on the interaction between nuclear magnetic dipole moments and a magnetic field. Magnetic resonance occurs in electrons, atoms, molecules, and nuclei in response to excitation by certain discrete radiation frequencies as a result of a space quantization in a magnetic field. The NMR technique can be used as a basis for determining the amount of moisture content in concrete by the detection of a signal from the hydrogen nuclei in the water molecules. Careful surface preparation is required and the object for testing must be clean and originally demagnetized. The test requires a source of high-current electric power to magnetize the object. The operation is relatively messy and good operator skill is needed to interpret the results.

Magnetic particle testing is another nondestructive magnetic method, mainly for steel structures. In this method, the object is magnetized and covered with magnetic powder. Surface and/or near-surface discontinuities in the magnetized materials may create leakage in the magnetic field, which, consequently, may affect the orientation of the particles above those areas. A variation of this test involves using wet fluorescent particles visible in backlight through a borescope. The magnetic particle technique can be used for locating surface cracks, laps, voids, seams, and other irregularities. Some subsurface defects can also be detected to a depth of about 0.635 mm (0.25 in.) (Newman, 2001). This method is relatively fast, simple to administer, and inexpensive, but there is a limited depth penetration for these methods. Magnetic particle testing has a long history and has been implemented into many standards and specifications. This method is the most effective way to nondestructively detect surface and near-surface discontinuities in ferromagnetic material.

**(f)** *Building dynamics*: Building dynamics measures the vibration of a concrete building under random dynamic loading with a very small amplitude. The measured waveform is analyzed in both the time and frequency domains to induce the fundamental frequencies and modal modes. The analysis is extended into the change of the stiffness of the structure to access the damage degree of the building under inspection.

Building dynamics techniques are widely used in structural engineering. Simple methods are useful for assessing localized integrity, such as delamination, while more complex methods are used in pile integrity testing, determination of member thickness, and examination of the change of stiffness of members affected by cracking or other deterioration. In general, there are three categories of dynamic response methods for NDT purposes: modal analysis, resonant, and damping techniques. The test equipment for the building dynamics techniques is roughly composed of two parts, one generating mechanical vibrations and the other sensing these vibrations. The pulse-echo method is simple and easy to apply to the building dynamics techniques. It involves measuring the reflected shock waves caused by a surface hammer and analyzing them in the time and/or frequency domains. Dynamic response testing of large structures may similarly involve hammer impacts or the application of vibrating loads. The vibration response is recorded by carefully located accelerometers. Through measuring the natural frequencies and/or the rate of attenuation (or damping) of the vibrations of the building structure, information about the dynamic properties, defects, and damage of the building structure, even individual member stiffness, can be obtained.

The resonant frequency technique is also widely used in deriving the dynamic properties of a building structure. Since every elastic object has many resonant frequencies, which are related to its stiffness and mass distribution, many physical characteristics of the object may be determined from the characteristics of the induced vibration. Typical vibration mode shapes include flexural, longitudinal, and torsional, as well as fundamental and higher orders. Usually, the fundamental flexural and extensional modes are most easily excited and are important for NDT inspection of building structures. When the test object is caused to vibrate in one of its natural or resonant modes by an applied external force, its mode shape can reveal the configuration and composition of the test object. This method is mainly used to determine the dynamic modulus of elasticity, stiffness and Poisson's ratio. Several factors may influence the results of the resonant frequency method in the inspection of concrete structures, including mix proportions and properties of aggregates (Jones, 1962); specimen-size (Obert and Duvall, 1941; Kesler and Higuchi, 1954); and curing conditions (Obert and Duvall, 1941; Kesler and Higuchi, 1953).

The damping test is another method for testing building dynamics. Damping is closely related to the dynamic motion of an object. When a solid object is subjected to dynamic forces, the amplitude of its free vibration will decrease with time after the exciting forces are removed. This is because some of the internal energy of the vibrating object is converted into heat. This phenomenon is called damping. What's more, solid objects exhibit a hysteresis loop, i.e., the downward stress-strain curve due to unloading does not exactly retrace its upward path. In addition, engineering materials always exhibit mechanical relaxation by an asymptotic increase in strain resulting from the sudden application of fixed stress, and, conversely, by an asymptotic relaxation in stress whenever they are suddenly strained (Bray and McBride, 1992). This mechanical relaxation has an associated relaxation time, the direct result of which is the significant attenuation of vibrations whenever the imposed frequency has a period that approximates to the relaxation time. Normally, the damping effect is characterized by the specific damping capacity, $Y$, which is given by

$$Y = \frac{\Delta W}{W} \tag{8-1}$$

where $\Delta W$ is the energy dissipated in one cycle; and $W$ the total energy of the cycle. Damping is a relaxation process, which is governed by a characteristic time that corresponds to the peak frequency and is referred to as the relaxation time. The specific damping capacity and associated dynamic response of the material are characterized by the damping coefficient, which can be expressed by

$$\alpha = \frac{1}{N} \ln\left(\frac{A_0}{A_n}\right) \tag{8-2}$$

where $\alpha$ is the damping ratio; $A_0$ the vibration amplitude of the reference cycle; and $A_n$ the vibration amplitude after $N$ cycles. The specific damping capacity ($\Delta W/W$) for the material is calculated as

$$\frac{\Delta W}{W} = 1 - e^{-2\alpha} \tag{8-3}$$

The damping test methods require an input vibration pulse and an associated output signal. The test body is first caused to vibrate in one of its natural vibration modes, the input signal is then interrupted, and the vibration decay of the test object is measured.

**(g)** *Radiography or radiometry*: These types of NDT methods use X-rays or $\gamma$-rays as the agency for detecting the internal defects or microstructure of a concrete material or structure. The techniques of radiometry and radiography are based on radioactive sources, such as X-rays,

$\gamma$-rays, or neutrons. Both X-rays and $\gamma$-rays have a very small wavelength. X-rays have a wavelength range from 3 to 0.03 nm, while $\gamma$-rays' wavelength is much smaller. As the wavelengths of light decrease, they increase in energy. Thus, X-rays and $\gamma$-rays are both at the high-energy end of the electromagnetic spectrum and can penetrate matter with some attenuation. The attenuation of radiation passing through matter is exponential and may be expressed as

$$I = I_0 \exp(-\mu X) \tag{8-4}$$

where $I$ is the energy intensity of the beam at a particular location; $I_0$ the incident energy intensity; $\mu$ the linear absorption coefficient; and $X$ the distance from member surface.

The absorption coefficient depends on the composition of the material. Thus, the measured intensity (with a detector or a radiation-sensitive film) can provide information on the material. In radiographic methods, a radiation source and photographic film are placed on opposite sides of a test object. The result, after exposing the film, is a photographic image of the member's interior. Defects inside the member can thus be identified. These techniques are generally fast and reliable, and can provide information unavailable by other means. However, they involve complex technology, high initial costs, and specific training and licensing requirements. X-rays can go through about 1 m of concrete. Applications include determining the distribution of aggregate particles, the three-dimensional configuration of air voids inside concrete, segregation, and the presence of cracks. However, X-ray equipment is very expensive and operated with high voltage. As the use of $\gamma$-rays does not require electricity, $\gamma$-ray equipment has gained a considerable market. However, $\gamma$-ray equipment has to be properly shielded and additional safety blocks are required to prevent exposure to radioactive materials. The $\gamma$-ray technique is especially valuable for determining the position and condition of reinforcements, voids in the concrete, or the grouting quality of post-tensioned structures.

In radiometric methods, a radiation source and a detector are placed on the same or opposite side of a concrete member. The number of electric pulses produced at the detector is a measure of the dimensions or physical characteristics (e.g., density or composition of the concrete member). $\gamma$-rays are most commonly used in radiometry systems for concrete, although neutron radiometry has been used for asphalt concrete and soil. For the detection of $\gamma$-rays, the Geiger-Müller tube is most commonly used. When high-energy radiation passes through concrete, some energy is absorbed, some energy passes through, and a significant amount is scattered by collisions with electrons in the concrete. So, when employing $\gamma$-rays in examining concrete, there are basically two modes of transmission: the direct transmission mode and the backscatter mode. For the former, depending on the source, a $\gamma$-ray can go through 50–300 mm of concrete. The latter essentially measures concrete within 100 mm of the surface. Some examples of applications include: (1) measurement in the transmission mode with internally embedded probes in fresh concrete to monitor the density and hence the degree of consolidation; and (2) noncontact backscatter measurement allows the monitoring of the density of a relatively thin pavement.

**(h)** *Computer tomography*: This technique is powerful post-processing for X-ray or $\gamma$-ray measurement. It can slice the object under inspection into three dimensions with very thin segments and recompose them into a whole image of the internal microstructure.

Since the most widely used inquiring agent in NDT-CE is mechanical waves (acoustic waves and ultrasonic waves) and electromagnetic waves, understanding the basic wave theory is essential to managing NDT-CE. Thus, the following sections concentrate on a review of wave theories, including wave reflection and transmission.

## 8.2 REVIEW OF WAVE THEORY FOR A 1D CASE

A wave is a physical phenomenon, a disturbance or variation that transfers energy progressively from point to point in a medium. It may take the form of elastic deformation or a variation of pressure, electric or magnetic intensity, electric potential, or temperature. The most important part of this definition is that a wave is a disturbance or variation that travels through a medium. The medium through which the wave travels may experience some local oscillations as the wave passes by, but the particles in the medium do not travel with the wave. The disturbance may take any of a number of shapes, from a finite width pulse to an infinitely long sine wave. Waves can be divided into mechanical waves and electromagnetic waves. The main difference between mechanical and electromagnetic waves is that a mechanical wave needs a medium to travel through while an electromagnetic wave can travel in a vacuum. Mechanical waves can be divided into body waves that travel through the interior of a solid body, and surface waves that travel along a surface of a solid body. The body wave can be further divided into P-waves and S-waves. P-waves are also known as primary, compressional, or longitudinal waves. In a longitudinal wave, the particle vibration direction is the same as the wave propagation direction. P-waves can travel through both solids and liquids. S-waves are also known as secondary, shear, or transverse waves. In shear waves, the particle vibration direction is perpendicular to the wave propagation direction. The most important surface waves for engineering purposes include Rayleigh waves and Love waves. In a homogeneous elastic half-space, only P-waves, S-waves, and Rayleigh waves can exist. Love waves can be developed only when a half-space is overlain by a layer of material with a lower-body wave velocity.

A wave is a disturbance that propagates through a transmission medium, usually with the transference of energy. While a mechanical wave exists in a medium (on which deformation is capable of producing elastic restoring forces), waves of electromagnetic radiation (and probably gravitational radiation) can travel through a vacuum, that is, without a medium. Waves travel and transfer energy from one point to another, often with little or no permanent displacement of the particles of the medium (that is, with little or no associated mass transport); instead, there are oscillations around almost fixed locations.

### 8.2.1 Derivation of the 1D Wave Equation

For an elastic solid, that is homogeneous, isotropic, and linear elastic, we have the following equations:

By Hooke's law

$$\begin{aligned}
\varepsilon_{xx} &= \frac{1}{E}\left[\sigma_{xx} - \nu\left(\sigma_{yy} + \sigma_{zz}\right)\right] \\
\varepsilon_{yy} &= \frac{1}{E}\left[\sigma_{yy} - \nu\left(\sigma_{xx} + \sigma_{zz}\right)\right] \\
\varepsilon_{zz} &= \frac{1}{E}\left[\sigma_{zz} - \nu\left(\sigma_{yy} + \sigma_{xx}\right)\right]
\end{aligned} \tag{8-5a}$$

$$\begin{aligned}
\varepsilon_{xy} &= \frac{\sigma_{xy}}{2\mu} \\
\varepsilon_{yz} &= \frac{\sigma_{yz}}{2\mu} \\
\varepsilon_{zx} &= \frac{\sigma_{zx}}{2\mu}
\end{aligned} \tag{8-5b}$$

and for the displacement-strain relationship

$$\begin{aligned}\varepsilon_{xx} &= \frac{\partial U_x}{\partial x}\\ \varepsilon_{yy} &= \frac{\partial U_y}{\partial y}\\ \varepsilon_{zz} &= \frac{\partial U_z}{\partial z}\end{aligned} \tag{8-6}$$

and

$$\begin{aligned}\varepsilon_{xy} &= \frac{1}{2}\left(\frac{\partial U_x}{\partial y} + \frac{\partial U_y}{\partial x}\right)\\ \varepsilon_{yz} &= \frac{1}{2}\left(\frac{\partial U_z}{\partial y} + \frac{\partial U_y}{\partial z}\right)\\ \varepsilon_{zx} &= \frac{1}{2}\left(\frac{\partial U_x}{\partial z} + \frac{\partial U_z}{\partial x}\right)\end{aligned} \tag{8-7}$$

For a one-dimensional strain case, we have the displacement expression

$$\begin{aligned}U_x &= U(x, t)\\ U_y &= U_z = 0\end{aligned} \tag{8-8}$$

and strain expressions

$$\begin{aligned}\varepsilon_{xx} &= \frac{\partial U_x}{\partial x}\\ \varepsilon_{yy} &= \varepsilon_{zz} = \varepsilon_{xy} = \varepsilon_{yz} = \varepsilon_{zx} = 0\end{aligned} \tag{8-9}$$

For the stress expressions, we have

$$\begin{aligned}\sigma_{xx} &\neq 0\\ \sigma_{yy} &= \frac{\nu}{1-\nu}\sigma_{xx}\\ \sigma_{zz} &= \frac{\nu}{1-\nu}\sigma_{xx}\end{aligned} \tag{8-10}$$

Substituting into Equation 8-5, we have

$$\begin{aligned}\varepsilon_{xx} &= \frac{1}{E}\left[\sigma_{xx} - \nu\left(\sigma_{yy} + \sigma_{zz}\right)\right]\\ &= \frac{1}{E}\frac{(1+\nu)(1-2\nu)}{1-\nu}\sigma_{xx}\end{aligned} \tag{8-11}$$

Or in term of stress, we have

$$\begin{aligned}\sigma_{xx} &= E\frac{1-\nu}{(1+\nu)(1-2\nu)}\varepsilon_{xx}\\ &= (\lambda + 2\mu)\,\varepsilon_{xx}\end{aligned} \tag{8-12}$$

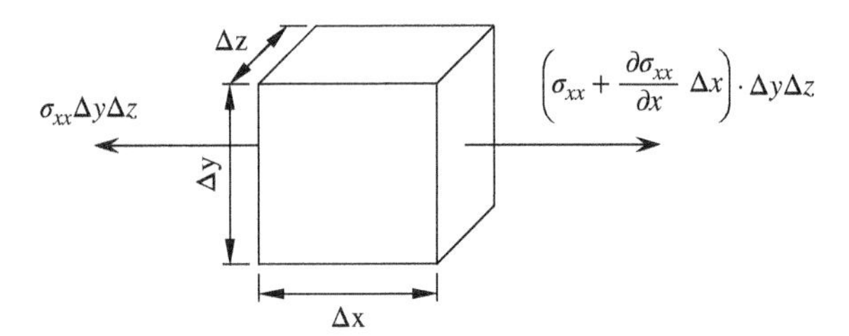


**Figure 8-4** A unit element for stress consideration

Considering an element, as shown in Figure 8-4, we can write the equilibrium equation as

$$\frac{\partial \sigma_{xx}}{\partial x}\Delta x\Delta y\Delta z = \rho\Delta x\Delta y\Delta z\frac{\partial^2 U}{\partial t^2} \tag{8-13}$$

$$\text{or} \quad E\frac{1-\nu}{(1+\nu)(1-2\nu)}\frac{\partial^2 U}{\partial x^2} = \rho\frac{\partial^2 U}{\partial t^2} \tag{8-14}$$

Furthermore, we can derive the one-dimensional wave equation as follows:

$$\frac{\partial^2 U}{\partial x^2} = \frac{1}{C_{L^2}}\frac{\partial^2 U}{\partial t^2} \tag{8-15}$$

where $C_L$ is called the longitudinal wave velocity. For the current case, i.e., a one-dimensional strain case, $C_L$ equals

$$C_L^2 = \frac{E}{\rho}\frac{1-\nu}{(1+\nu)(1-2\nu)} = \frac{\lambda + 2\mu}{\rho} \tag{8-16}$$

For the one-dimensional stress case, we have

$$C_L^2 = \frac{E}{\rho} \tag{8-17}$$

### 8.2.2 Solution for a 1D Wave Equation

#### 8.2.2.1 Longitudinal Wave Case

The wave equation for the longitudinal case is

$$\frac{\partial^2 U}{\partial x^2} = \frac{1}{C_L^2}\frac{\partial^2 U}{\partial t^2} \tag{8-15}$$

where $C_L$ represents the longitudinal wave velocity, and

$$C_L^2 = \frac{E}{\rho} \quad \text{for plane stress} \tag{8-18}$$

$$C_L^2 = \frac{\lambda + 2\mu}{\rho} \quad \text{for plane strain} \tag{8-19}$$

It is clear that the wave equation is a partial differential equation and a function of both space coordinates and time. To get the solution for the wave equation, let us introduce two new parameters, $\alpha$ and $\beta$, and assume that

$$\alpha = t - \frac{x}{C_L} \qquad \beta = t + \frac{x}{C_L} \tag{8-20}$$

Furthermore, we can get:

$$\frac{\partial}{\partial t} = \frac{\partial}{\partial \alpha} + \frac{\partial}{\partial \beta} \tag{8-21}$$

$$\text{and} \quad \frac{\partial^2}{\partial t^2} = \frac{\partial^2}{\partial \alpha^2} + 2\frac{\partial^2}{\partial \alpha \partial \beta} + \frac{\partial^2}{\partial \beta^2} \tag{8-22}$$

$$\text{and} \quad \frac{\partial}{\partial x} = \frac{1}{C_L}\left(-\frac{\partial}{\partial \alpha} + \frac{\partial}{\partial \beta}\right) \tag{8-23}$$

$$\text{and} \quad \frac{\partial^2}{\partial x^2} = \frac{1}{C_L^2}\left(\frac{\partial^2}{\partial \alpha^2} - 2\frac{\partial^2}{\partial \alpha \partial \beta} + \frac{\partial^2}{\partial \beta^2}\right) \tag{8-24}$$

Substituting Equations 8-20 through 8-24 into wave Equation 8-15, one gets

$$\frac{\partial^2 U}{\partial \alpha \partial \beta} = 0 \tag{8-25}$$

Now we can solve this equation easily through an integral process. The first integral is made with respect to β, and we have

$$\frac{\partial U}{\partial \alpha} = \bar{f}(\alpha) \tag{8-26}$$

The second integral is made with respect to $\alpha$, giving

$$U = f(\alpha) + g(\beta) \tag{8-27}$$

$$\text{or} \quad U = f\left(t - \frac{x}{C_L}\right) + g\left(t + \frac{x}{C_L}\right) \tag{8-28}$$

This is the so-called D'Alembert solution. The term $f\left(t - x/C_L\right)$ represents a pulse propagated with the velocity of $C_L$ in the positive $x$-direction while $g\left(t + x/C_L\right)$ represents a pulse propagated with the velocity of $C_L$ in the negative $x$-direction.

#### 8.2.2.2 Transverse Wave Case

For a transverse wave, the particle vibrates in the $y$-direction while the wave moves in the $x$-direction, as shown in Figure 8-5. Thus, we have

$$U(x,t) = \omega(x,t) = 0 \qquad v(x,t) \neq 0 \tag{8-29}$$

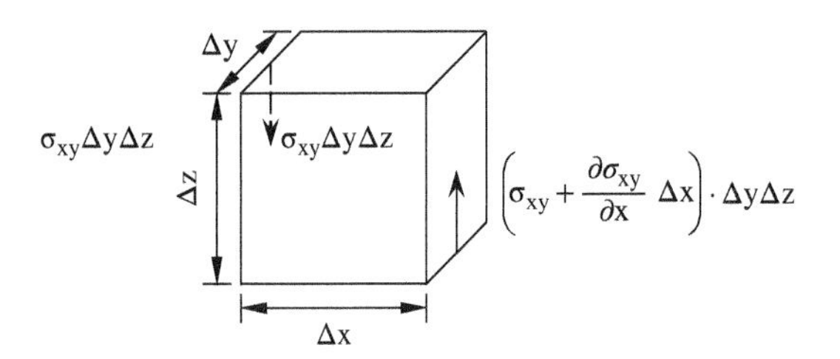


**Figure 8-5** A unit element for the transverse wave case

Consequently, we have

$$\frac{\partial v}{\partial x} \neq 0$$
$$\gamma_{xy} = \frac{\partial v}{\partial x} \tag{8-30}$$

Furthermore, we can write

$$\sigma_{xy} = \mu \frac{\partial v}{\partial x} \tag{8-31}$$

where $\mu$ is the shear modulus of materials. The equilibrium equation can be written as

$$\frac{\partial \sigma_{xy}}{\partial x} \Delta x \Delta y \Delta z = \rho \Delta x \Delta y \Delta z \frac{\partial^2 v}{\partial t^2} \tag{8-32}$$

$$\text{or} \quad \frac{\partial^2 v}{\partial x^2} = \frac{1}{C_{\mathrm{T}}^2} \frac{\partial^2 v}{\partial t^2} \tag{8-33}$$

$$\text{where} \quad C_{\mathrm{T}}^2 = \frac{\mu}{\rho} \tag{8-34}$$

Note that

$$C_{\mathrm{L}}^2 = \frac{E}{\rho} \quad \text{for plane stress} \tag{8-18}$$

$$C_{\mathrm{L}}^2 = \frac{E}{\rho} \quad \text{for plane strain} \tag{8-19}$$

Thus, for the plane strain case, we have

$$\frac{C_{\mathrm{L}}^2}{C_{\mathrm{T}}^2} = \frac{2(1-v)}{1-2v} \tag{8-35}$$

For the plane stress case, we have

$$\frac{C_{\mathrm{L}}^2}{C_{\mathrm{T}}^2} = 2(1+v) \tag{8-36}$$

The solution of the S-wave equations can be obtained by using the same methods and procedures as the longitudinal wave case. The solution of the S-wave equation is

$$U(x,t) = f\left(t - \frac{x}{C_{\mathrm{T}}}\right) + g\left(t + \frac{x}{C_{\mathrm{T}}}\right) \tag{8-37}$$

This is exactly the same as the longitudinal case, Equation 8-28, with a substitution of $C_{\mathrm{T}}$ to $C_{\mathrm{L}}$.

#### 8.2.2.3 Observations on the D'Alembert Solution

Since the D'Alembert solution for the longitudinal case and the transverse case has the same form, here we use the longitudinal case as an example. For the expression of

$$U(x,t) = f\left(t - \frac{x}{C_{\mathrm{L}}}\right) + g\left(t + \frac{x}{C_{\mathrm{L}}}\right) \tag{8-28}$$

the first differential with respect to $x$ yields

$$\frac{\partial U}{\partial x} = -\frac{1}{C_{\mathrm{L}}} f'\left(t - \frac{x}{C_{\mathrm{L}}}\right) + \frac{1}{C_{\mathrm{L}}} g'\left(t - \frac{x}{C_{\mathrm{L}}}\right) \tag{8-29}$$

and the first differential with respect to $t$ leads to

$$\frac{\partial U}{\partial t} = f'\left(t - \frac{x}{C_L}\right) + g'\left(t - \frac{x}{C_L}\right) \tag{8-30}$$

Thus, we have

$$\begin{aligned}\frac{\partial U}{\partial x} &= -\frac{1}{C_{\mathrm{L}}}\left[\frac{\partial U}{\partial t} - 2g'\left(t + \frac{x}{C_{\mathrm{L}}}\right)\right] \\ &= \frac{1}{C_{\mathrm{L}}}\left[\frac{\partial U}{\partial t} - 2f'\left(t + \frac{x}{C_{\mathrm{L}}}\right)\right]\end{aligned} \tag{8-31}$$

Hence, for $g' = 0$, we have

$$\begin{aligned}\sigma_{xx} &= (\lambda + 2\mu)\frac{\partial U}{\partial x} \\ &= -\rho C_{\mathrm{L}}\frac{\partial U}{\partial t}\end{aligned} \tag{8-32}$$

and for $f' = 0$, we have

$$\sigma_{xx} = \rho C_L \frac{\partial U}{\partial t} \tag{8-33}$$

This means that the stress can be expressed as a function of the rate of wave propagation and different directions give different signs. The term $\rho C_{\mathrm{L}}$ is also called acoustic impedance.

#### 8.2.2.4 Specific Solution of a Wave Equation

It should be noted that although the D'Alembert solution has a very clear physical meaning for a moving wave along either the positive $x$-direction or the negative $x$-direction, and sometimes its specific wave form can be obtained by using the initial condition, it does not give a specific solution for most wave cases. Here, we introduce a generally applicable method to give the explicit solution of the wave equation, the variable separation method. It is a commonly used simple method. In this section, we use this method to show how to obtain an explicit solution for the wave equation. For the wave equation with a form of

$$\frac{\partial^2 U}{\partial x^2} = \frac{1}{C_{\mathrm{L}}^2}\frac{\partial^2 U}{\partial t^2} \tag{8-15}$$

the solution of $U$ must be a function of both $x$ and $t$. The variable separation method assumes that $x$ and $t$ can be completely separated in the expression for $U$, i.e.

$$U = T \cdot X \tag{8-43}$$

where $T$ is a function of $t$ only when $X$ is a function of $x$ only. Then, substitution of the expression into the wave equation leads to two normal differential equations of $T$ and $X$

$$X'' + \lambda X = 0 \tag{8-44}$$

$$T'' + \lambda C_{\mathrm{L}}^2 T = 0 \tag{8-45}$$

It is much easier to solve the two normal differential equations using characteristic equations. The constants accompanying the solution can be determined by the boundary conditions and the initial conditions. To further examine the features of the wave equation solution, let us take a look at the solution in the form of

$$U(x,t) = A\cos\left[k\left(x - C_{\mathrm{L}}t\right)\right] = A\cos(kx - \omega t) \tag{8-46}$$

where $A$ is the amplitude of a wave, a measure of the maximum disturbance in the medium during one wave cycle or the maximum distance from the highest point of the crest to the mean value. The units of the amplitude depend on the type of wave: waves on a string have an amplitude expressed as a distance (meter), sound waves as pressure (pascal), and electromagnetic waves as the amplitude of the electric field (volt/meter). The amplitude may be constant (in which case, the wave is a continuous wave), or may vary with time and/or position. The form of the variation of amplitude is called the envelope of the wave.

The variable $k$ is called the wave number, which represents how many complete waves occur in a given period. The wave number is related to the wavelength by the equation

$$k = \frac{2\pi}{\lambda} \tag{8-47}$$

The physical meaning of the wavelength (denoted as $\lambda$) is the distance between two sequential crests (or troughs). This generally is measured in meters; it is also commonly measured in nanometers for the optical part of the electromagnetic spectrum.

The variable $\omega$ is called angular frequency, representing the frequency in terms of radians per second. It is related to the frequency by

$$\omega = 2\pi f = \frac{2\pi}{T} \tag{8-48}$$

where $T$ is the period of the wave motion and is the time required for one complete cycle in the oscillation of a wave. The frequency $f$ defines the number of periods per unit of time (for example, one second) and is measured in hertz. The frequency and period are related by

$$f = \frac{1}{T} \tag{8-49}$$

In other words, the frequency and period of a wave are reciprocals of each other. There are two velocities that are associated with waves. The first is the phase velocity, which gives the rate at which the wave propagates, and is given by

$$v_{\mathrm{p}} = \frac{\omega}{k} = \lambda f \tag{8-50}$$

The second is the group velocity, which gives the velocity at which variations in the shape of the wave's amplitude propagate through space. This is the rate at which information can be transmitted by the wave. It is given by

$$v_g = \frac{\partial \omega}{\partial k} \tag{8-51}$$

For a fixed $t(t = t_i)$, $U(x, t)$ becomes a periodic function in $x$ as shown in Figure 8-6.

The length between two positive peaks or two negative peaks is called the wavelength:

$$\lambda = \frac{2\pi}{k} \tag{8-52}$$

where $k$ is the wave number. For a fixed $x$, we have a traveling harmonic wave

$$\begin{aligned} U(x_i, t) &= A \cos\left[k\left(x_i - C_L t\right)\right] \\ &= A \cos\left[kx_i - kC_L t\right] \\ &= A \cos\left[kx_i - \omega t\right] \end{aligned} \tag{8-53}$$

We can derive the following relationships:

$$\begin{aligned} T &= \frac{2\pi}{\omega} \\ f &= \frac{1}{T} = \frac{\omega}{2\pi} \\ \omega &= 2\pi f \\ k &= \frac{2\pi f}{C_L} \end{aligned} \tag{8-54}$$

*Example of the Variable Separation Method*

Solve the following 1D wave equation using the variable separation method.

$$\begin{aligned} U_{xx} &= \frac{1}{C_L^2} U_{tt} \\ U\big|_{x=0} &= 0 \\ U\big|_{x=l} &= 0 \end{aligned}$$

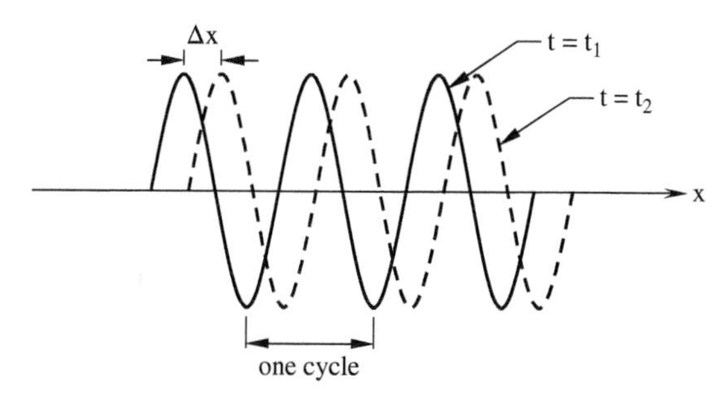


**Figure 8-6** A representative periodic function

Solution:

$$\text{Let} \quad U = XT$$

$$\frac{\partial^2 U}{\partial x^2} = X''T$$

$$\frac{\partial^2 U}{\partial t^2} = T''X$$

$$X''T = \frac{1}{C_L^2} T''X$$

$$\Rightarrow \frac{X''}{X} = C_L \frac{T''}{T} = -\lambda$$

Then, we have

$$X'' + \lambda X = 0$$

$$T'' + \lambda C_L^2 T = 0$$

$$\text{For} \quad X'' + \lambda X = 0,$$

$$r^2 + \lambda = 0$$

$$r = \pm\sqrt{-\lambda}$$

The first case is $\lambda < 0$:

$$X(x) = C_1 e^{\sqrt{-\lambda}x} + C_1 e^{-\sqrt{-\lambda}x}$$

From boundary conditions $X(0) = 0$ and $X(l) = 0$,

$$\begin{cases} C_1 + C_2 = 0 \\ C_1 e^{\sqrt{-\lambda}l} + C_2 e^{-\sqrt{-\lambda}l} = 0 \end{cases}$$

and $C_1 = C_2 = 0$ (no meaning).
The second case is $\lambda = 0$:

$$X(x) = C_1 x + C_2$$

From boundary conditions $X(0) = 0$ and $X(l) = 0$,

$$\begin{cases} C_2 = 0 \\ C_1 l + C_2 = 0 \end{cases}$$

Again, $X(x) \equiv 0$ (No meaning)
The third case is $\lambda > 0$:

$$X(x) = C_1 \cos\sqrt{\lambda}x + C_2 \sin\sqrt{\lambda}x$$

From boundary conditions $X(0) = 0$ and $X(l) = 0$,

$$\begin{cases} C_1 = 0 \\ C_2 \sin\sqrt{\lambda}l = 0 \end{cases}$$

Since $C_2 \neq 0$, $\sin\sqrt{\lambda}l = 0$

$$\sqrt{\lambda}l = n\pi,\ \sqrt{\lambda} = \frac{n\pi}{l},\ \lambda = \frac{n^2\pi^2}{l^2}$$

$$\Rightarrow X(x) = C_2 sin\frac{n\pi}{l}x$$

From $T'' + \lambda C_L^2 T = 0$

Let $T(t) = Acos\frac{n\pi C_L}{l}t + Bsin\frac{n\pi C_L}{l}t$

Thus,

$$U(x,t) = \left(A'\cos\frac{n\pi C_L}{l}t + B'\sin\frac{n\pi C_L}{l}t\right)\sin\frac{n\pi x}{l}$$

## 8.3 REFLECTED AND TRANSMITTED WAVES

As mentioned above, the major types of body waves are longitudinal (P-) waves and transverse (S-) waves. For longitudinal waves, the propagation and particle motion directions are the same. Longitudinal waves can propagate in solids, liquids, and gases and are the most widely used wave modes for nondestructive testing of materials and structures. On the other hand, shear waves have particle motion transverse to the direction of the propagation of the wave. Shear waves cannot pass through liquids and are thus limited to solid inspection only. The reflection and transmission behavior of the wave front at an interface plays an important role in ultrasonic investigations. The wavefront defines the leading edge of a stress wave as it propagates through a medium. The reflection and transmission describe the behavior of the wavefront at an interface of two different materials. The acoustic impedance ratio of two materials plays an important role in determining the reflection and transmission parameters. The acoustic impedance is defined as a product of the density and wave velocity of a material.

The reflection and transmission behavior of the wavefront at an interface plays an important role in ultrasonic investigations and provides the basis for determining the presence of a flaw and other anomalies. Here, let us consider a one-dimensional wave propagation case, as in Figure 8-7. Figure 8-7 shows that two media have different material properties with a boundary, in which $\rho$ and $C_L$ are the density and longitudinal wave velocity for the medium $I$, while $\rho^A$ and $C^A{}_L$ are for medium II. A plane wave traveling in the medium $I$ approaches the boundary from the left. This wavefront is parallel to the boundary and is designed as the incident wave. Upon striking the boundary, part of the energy is reflected back to medium I and part of it is transmitted into medium II through the boundary. Their portions of energy transmitted and reflected are a function of the properties of media I and II. The displacements for the incident (i), reflected (r), and transmitted (t) waves are as follows:

$$\begin{aligned} U^i &= f\left(t - xC_L\right) \\ U^r &= g\left(t - aC_L + \frac{x-a}{C_L}\right) \\ U^t &= h\left(t - aC_L - \frac{x-a}{C_L^A}\right) \end{aligned} \tag{8-55}$$

where $U^i$ is the displacement of incident wave; $U^r$ the displacement of reflected wave; $U^t$ the displacement of transmitted wave; $C_L$ the longitudinal wave velocity for medium I; and $C^A{}_L$ for medium II.

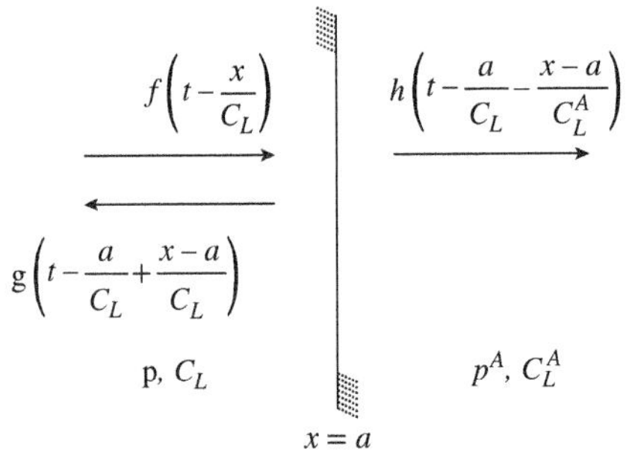

**Figure 8-7** One-dimensional wave propagation at an interface between two mediums

At the interface ($x = a$), stress continuity and displacement compatibility have to be satisfied. From these conditions, we can obtain the relationships

$$A_{\mathrm{R}} = \frac{1 - \frac{\rho^{\mathrm{A}} C_{\mathrm{L}}^{\mathrm{A}}}{\rho C_{\mathrm{L}}}}{\frac{\rho^{\mathrm{A}} C_{\mathrm{L}}^{\mathrm{A}}}{\rho C_{\mathrm{L}}+1} + 1} A_{\mathrm{I}} = \frac{1 - z}{z + 1} A_{\mathrm{I}}$$

$$A_{\mathrm{T}} = \frac{2\rho C_{\mathrm{L}}}{\rho^{\mathrm{A}} C_{\mathrm{L}}^{\mathrm{A}} + \rho C_{\mathrm{L}}} A_{\mathrm{I}} = \frac{2}{z + 1} A_{\mathrm{I}} \tag{8-56}$$

where $A_{\mathrm{R}}$ is the displacement amplitude of reflected wave; $A_{\mathrm{I}}$ the displacement amplitude of incident wave; and $A_{\mathrm{T}}$ the displacement amplitude of transmitted wave; and

$$\sigma_{\mathrm{R}} = \frac{\frac{\rho^{\mathrm{A}} C_{\mathrm{L}}^{\mathrm{A}}}{\rho C_{\mathrm{L}}} - 1}{\frac{\rho^{\mathrm{A}} C_{\mathrm{L}}^{\mathrm{A}}}{\rho C_{\mathrm{L}}+1} + 1} \sigma_{\mathrm{I}} = \frac{z - 1}{z + 1} \sigma_{\mathrm{I}} = R\sigma_{\mathrm{I}}$$

$$\sigma_{\mathrm{T}} = \frac{2\rho^{\mathrm{A}} C_{\mathrm{L}}^{\mathrm{A}}}{\rho^{\mathrm{A}} C_{\mathrm{L}}^{\mathrm{A}} + \rho C_{\mathrm{L}}} \sigma_{\mathrm{I}} = \frac{2z}{z + 1} \sigma_{\mathrm{I}} = T\sigma_{\mathrm{I}} \tag{8-57}$$

where $R$ and $T$ are the reflection and transmission parameters. These expressions show that the ratio of the acoustic impedances completely determines the nature of the reflection and transmission at the interface. They can be plotted as a function of the ratio of acoustic impedance of the two materials, as shown in Figure 8-8. Let's look at a few extreme cases.

For $\rho^{\mathrm{A}} C_{\mathrm{L}}^{\mathrm{A}} / \rho C_{\mathrm{L}} = 0$, $T = 0$ and $R = -1$, which means that the incident wave is reaching a free surface. No stress can be transmitted. To satisfy the zero-stress boundary condition, the displacement must be twice the displacement of the incident wave. The reflected wave has the same amplitude as the incident wave but opposite polarity. It implies that a free end will reflect a compression wave to a tension wave with identical amplitude and shape and vice versa.

For $\rho^{\mathrm{A}} C_{\mathrm{L}}^{\mathrm{A}} / \rho C_{\mathrm{L}} = 1$, it means the same materials, $T = 1$ and $R = 0$. The wave is completely transmitted.

For $\rho^{\mathrm{A}} C_{\mathrm{L}}^{\mathrm{A}} / \rho C_{\mathrm{L}} = \infty$, it goes to infinity, which means that an incident wave is approaching a fixed end. No displacement can occur, $U^{\mathrm{t}} = 0$. The stress at the boundary is twice that of the incident wave and the reflected wave has the same amplitude and polarity as the incident wave.

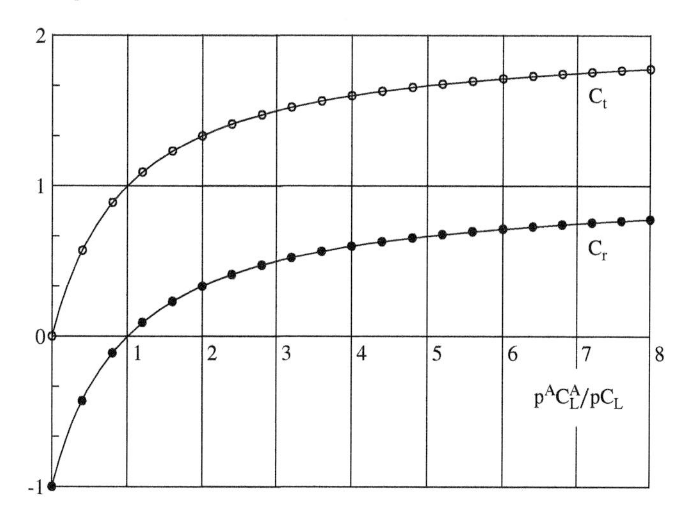


**Figure 8-8** The reflection and transmission coefficients vs. acoustic impedance

When the wavefronts propagate on the boundary between two materials with different properties, with an angle not normal to the interface, the reflection and transmission will depend on the angle. Let $\theta$ be the wave incidence angle. The reflection angle of the wave is also $\theta$. However, the angle of transmission, $\beta$, is a function of the angle of incidence, $\theta$, and the ratio of wave velocities in the two media as given by Snell's law:

$$\sin\beta = \frac{C_L^A}{C_L}\sin\theta \tag{8-58}$$

Stress waves can change their mode of propagation when striking a boundary at an oblique angle. Depending on the angle of incidence, a P-wave can be partially reflected as both P- and S-waves and can be transmitted as both P- and S-waves. An S-wave reflects and transmits at angles, determined using Snell's law, which are less than the angles of reflection and transmission for a P-wave.

## 8.4 ATTENUATION AND SCATTERING

An acoustic wave traveling through engineering materials will lose energy for a variety of reasons. This behavior can lead to a loss in amplitude and is sometimes called attenuation. Attenuation is generally expressed in the form of

$$P = P_0 e^{aL} \tag{8-59}$$

where $P_0$ is original pressure level at a source or a reference point; $P$ the pressure level at measured place; $a$ the attenuation coefficient (in Nepers per meter, i.e. Np/m); and $L$ the distance between the two points $P_0$ and $P$.

Ultrasonic pulse attenuation is typically expressed in units of decibels (dB). Decibels are based on a logarithmic scale and are convenient to use over a large range. The relative sound pressure level (SPL) is defined as

$$\text{SPL} = 20\log\frac{P}{P_0} \tag{8-60}$$

Considering two points,

$$\mathrm{SPL}_1 - \mathrm{SPL}_2 = 20 \log \frac{P_1}{P_2} \tag{8-61}$$

$$\text{or} \quad \alpha \mathrm{L} = 20 \log \frac{P_1}{P_2} \tag{8-62}$$

As an example, let us look at an attenuation test conducted on an aluminum specimen. The test results showed that a pulse traveling 200 mm had an amplitude of 70% as great as that of a pulse, which had traveled 100 mm. For this case, the α is

$$\alpha = \frac{20}{L} \log \frac{P_1}{P_2} = \frac{20}{0.1} \log \frac{1}{0.7} = 31\,\mathrm{dB/m} \tag{8-63}$$

It should be pointed out that $a$ and α have a relationship of $\alpha = 8.686\,a$.

## 8.5 MAIN COMMONLY USED NDT-CE TECHNIQUES

### 8.5.1 Ultrasonic Technique

#### 8.5.1.1 Principle of Ultrasound

As mentioned earlier, the ultrasonic technique (UT) is based on time-varying deformations or vibrations in materials, which are generally referred to as acoustics. All material substances are composed of atoms, which may be forced into vibrational motion about their equilibrium positions. Many different patterns of vibrational motion exist at the atomic level, but most are irrelevant to acoustics and ultrasonic testing. Acoustics is focused on particles that contain many atoms that move in unison to produce a mechanical wave. When a material is not stressed in tension or compression beyond its elastic limit, its individual particles perform elastic oscillations. When the particles of a medium are displaced from their equilibrium positions, internal (electrostatic) restoration forces arise. It is these elastic restoring forces between particles, combined with the inertia of the particles, which lead to the oscillatory motions of the medium. There are many different UT methods, such as pulse velocity, pulse-echo, frequency shift, ultrasonic resonance spectroscopy, 3D images, time of flight diffraction (TOFD), and synthetic aperture focusing technique (SAFT). Among all the existing NDT methods, the pulse velocity and pulse-echo methods (PE) are the dominant ones. All the UT methods obey the laws of elastic waves, but different methods follow different physical principles.

The following are the basic characteristics of elastic waves. Two kinds of elastic waves propagate in a homogeneous isotropic solid medium: the longitudinal wave and the transverse wave. The velocity of the longitudinal wave and the transverse wave has been given in Equations 8-28 and 8-37. It can be seen from the expressions that the wave velocities are functions of Young's modulus, Poisson's ratio, and the density of the material only. Hence, when the ultrasonic velocities and density of the material are known, then the elastic modulus can easily be deduced. This forms the basis of the pulse velocity method.

There are another two important wave types in solids. One is the surface wave and the other is the plate wave. Surface waves, also called Rayleigh waves, travel from the surface of a relatively thick, solid material penetrating to a depth of one wavelength. Surface waves combine both longitudinal and transverse motion to create an elliptical orbit motion. The major axis of the ellipse is perpendicular to the surface of the solid. As the depth of an individual atom from the surface increases, the width of its elliptical motion decreases. Surface waves are generated when a longitudinal wave intersects a surface near the second critical angle and they travel at a velocity between 0.87 and 0.95 of a shear wave. Rayleigh waves are useful because they are very sensitive to surface defects (and other surface features) and they follow the curvature of a surface. Because of this, Rayleigh waves can be used to inspect areas that other waves might have difficulty in reaching. Plate waves are similar to surface waves except they can be generated only in a thin, plate-like specimen, a few wavelengths thick. Lamb waves are the most commonly used plate waves in NDT. Lamb waves are complex vibrational waves that propagate parallel to the test surface throughout the thickness of the material. Propagation of Lamb waves depends on the density and the elastic properties of the material. They are also influenced a great deal by the test frequency and material thickness. Lamb waves are generated at an incident angle in which the parallel component of the velocity of the wave in the source is equal to the velocity of the wave in the test material. Lamb waves can travel several meters in steel and so are useful to scan plates, wires, and tubes.

With Lamb waves, a number of modes of particle vibration are possible, but the two most common are symmetrical and asymmetrical, as shown in Figure 8-9. The complex motion of the particles is similar to the elliptical orbits for surface waves. Symmetrical Lamb waves move in a symmetrical fashion about the median plane of the plate. This is sometimes called the extensional mode because the wave is "stretching and compressing" the plate in the wave motion direction. Wave motion in the symmetrical mode is most efficiently produced when the exciting force is parallel to the plate. The asymmetrical Lamb wave mode is often called the *flexural mode* because a large portion of the motion is in a normal direction to the plate, and little motion occurs in the direction parallel to the plate. In this mode, the body of the plate bends as the two surfaces move in the same direction.

The ultrasonic waves will be attenuated during propagation. There are three main mechanisms: beam spreading, absorption by the medium, and scattering by inhomogeneities, as discussed in the previous section. The ultrasonic wave is also subjected to reflection, refraction, diffraction, and scattering when it encounters an interface between two media with different acoustic impedances. The elastic wave behavior at the interface obeys Snell's law and is complex. The longitudinal wave and transverse wave can transform each other, which is called mode conversion. The reflection factor depends on the difference of the acoustic impedance $Z_i$ of the two materials and the incidence angle, as mentioned earlier. In the simplest case, the wave is of normal incidence, there is no mode conversion, and the reflection factor can be obtained by using Equation 8-57. The reflected wave and the scattered wave are usually called echoes. The time of flight of an echo is determined by the distance it is propagated and the wave velocity

$$t = \frac{2h}{c} \tag{8-64}$$

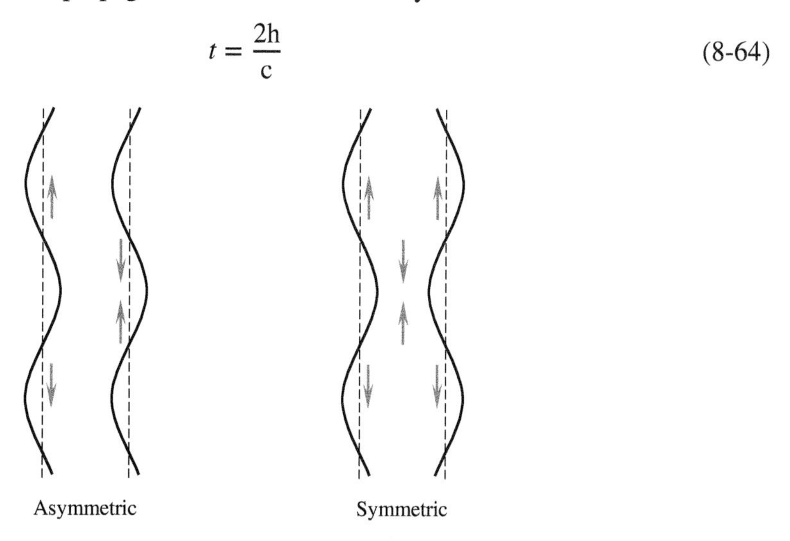


**Figure 8-9** Asymmetric and symmetric Lamb waves

where $h$ is the thickness of the specimen. Combining the data of the reflection ratio and the TOF, one can deduce the thickness and material type of the layers. This forms the theoretical basis of the pulse-echo method. All the velocity, TOF, and reflection ratio data can be used to form a 2D or 3D image of the specimen by means of computer tomography (CT), synthetic aperture focusing technique, C-scan, or B-scan methods.

#### 8.5.1.2 Technical Features and Advances

When applying UT to concrete or other cement-based materials and structures, some difficulties can be encountered due to the material features. First, concrete is an inhomogeneous, porous, multiscale, and heterogeneous building material. Reinforced concrete is also anisotropic. The elastic wave behavior becomes more complicated when it propagates in these kinds of materials. Second, the surface condition of a cement-based material such as concrete is usually very rough as compared to metals, which makes the coupling between the UT transducer and the concrete surface difficult to realize. Finally, concrete constructions, such as buildings and bridges, are usually very large and have very complicated shapes that bring complexity to the carrying out of inspection and analysis of the complicated data collected. All of these require attention to the special technique features of conventional NDT-UT techniques.

**(a)** *Working frequency*: To overcome the difficulties mentioned above, first of all, relative lower frequencies, say, hundreds of kHz, are usually used for UT inspection in concrete rather than the higher frequencies used for UT applications in the metallurgical field. Ultrasonic wave scattering is significantly dependent on the frequency, and the higher the frequency, the stronger the scattering. Under UT frequency ranges, the scattering is proportional to the fourth-order of the frequency. Hence, using lower frequencies in concrete can greatly reduce the clutter and noise generated by the scattered waves at the boundaries of aggregates. Specifically, a low-frequency wave, with a wavelength as large as 4–5 times the nominal size of the maximum aggregate, is preferred for UT inspection in concrete. Low-frequency waves can also penetrate deep into the specimen, as the attenuation of elastic waves is reduced as the working frequency goes down. However, since the resolution of UT inspection is limited to half the wavelength, the working frequency cannot go too low, otherwise, there is insufficient resolution.

**(b)** *Acoustic coupling*: Mechanical waves may lose significant strength at an interface between air and a solid material because the difference in the acoustic impedance between air and solid materials is very large. As more than 95% of ultrasound energy will be reflected from a smooth air–solid interface, UT-NDT is very effective for detecting cracks or voids inside solids. Hence, coupling between a UT transducer and solid material to be inspected is a necessary measure to ensure a meaningful measurement. In the case of a concrete structure, the surfaces of the specimens are usually in a rough condition from the UT point of view, and it is difficult to couple the transducers well. To make good contact for a UT transducer with concrete, the concrete surface is usually smoothed with sandpaper before coupling materials are applied. The commonly used coupling materials include fast-set glue, grease, and oil. Some kinds of fast-solidifying cement have also been developed as a couplet for UT inspection in concrete (Krause et al., 1995).

To test large structures like bridge piles, the cross-hole sonic logging technique (CSL) is very useful (Sack and Olson, 1995). The coupling problem can be partly solved by the CSL method. A set of parallel tubes is built into the structure. The tubes are filled with water, and the ultrasonic transducers (probes) are hung inside the tubes. The probes are worked in the transmission mode without coupling problems. It is easy to take the measurements of the ultrasonic pulse velocity of the B-scan, etc. CSL also can work in the tomogram configuration. By means of the new digital technique, the states of the object tested at different periods can be compared. That is to say, it can be used as both instant and long-term monitor. The CSL technique can undertake quality assurance of drilled shafts, slurry walls, steel footings, piers and dams, etc.

**(c)** *High-power transducer*: Another way to overcome the coupling problem, is to develop a stronger emitter and a more sensitive receiver. In the 1990s, an airborne ultrasonic flaw detector was developed (Curlin Air Tech Note, 1998). A pair of powerful 50-kHz transducers produce an ultrasound beam to punch through and test brick or concrete samples without contact. The powerful transducers matched the high-performance electrical circuits, which greatly enhanced the ability of the NDT-UT instruments.

**(d)** *Signal processing and displaying techniques*: Recently, a fast sampling frequency and low-cost analogue to digital transform device (ADT) has been developed and is readily available. It allows the digital signal processing of high-frequency signals to become possible. Relying on the DSP, such as by applying spectrum analysis, digital filtering, self/co-relations, and artificial neural networks, the application of traditional UT has become broader. Now signal processing is so important that it is an essential part of a UT technique.

**(e)** *Wave propagation and modeling*: In concrete, ultrasonic wave propagation is complicated. To understand the phenomenon computer modeling and simulation can be performed by modern powerful computers. The production parameters, such as the maximum aggregate size, cement mix, moisture content, and pore size are selected to meet practical conditions.

#### 8.5.1.3 Applications

The applications are wide for UT. For example, it can be used for quality assurance, condition assessment, and reliability evaluation, can locate and identify defects, cracks, fractures, voids, and inhomogeneities, can measure the thickness and size of defects, and can estimate the strength of concrete for bridges, drilled shafts, slurry walls, and dams.

**(a)** *Scanning technique for flaw detection*: The pulse-echo method is the simplest and most common ultrasonic application method. In this method, a single piezoelectric transducer is used to excite an acoustic wave in the object being examined and it moves around to detect the inside flaws by receiving the reflected signal and by using an imaging technique. There are three kinds of scan methods for flaw detection—A-scan, B-scan, and C-scan—in the presentation of pulse-echo data.

The A-scan is known as an amplitude scan. It can be done by placing a transducer directly against the solid material to be examined and making contact between the transducer and the sample with grease. It displays the signal received in amplitude to the time base. An adjustable threshold of the amplitude level is used to suppress the noise. Suppose that the transducer emits an acoustic pulse, the generated acoustic pulse passes into the specimen and is reflected by the acoustic impedance discontinuities caused by the presence of flaws. The returned echo signal is received at the transducer and is amplified and displayed as a function of time on an oscilloscope. Since the amplitude of the returned echo depends on the size and shape of the flaw, a rough estimate of the size of the flaw can be obtained by measuring the amplitude of returned echo. Figure 8-10 shows the flowchart of the A-scan technique.

As there are usually many signals stemming from the multireflection and multiscattering from the inside structure boundaries or the inhomogeneities, the A-scan is difficult to understand and it is easy to lose a lot of information at times, especially for applications in civil engineering. An alternative display format is the B-scan, which is widely used in

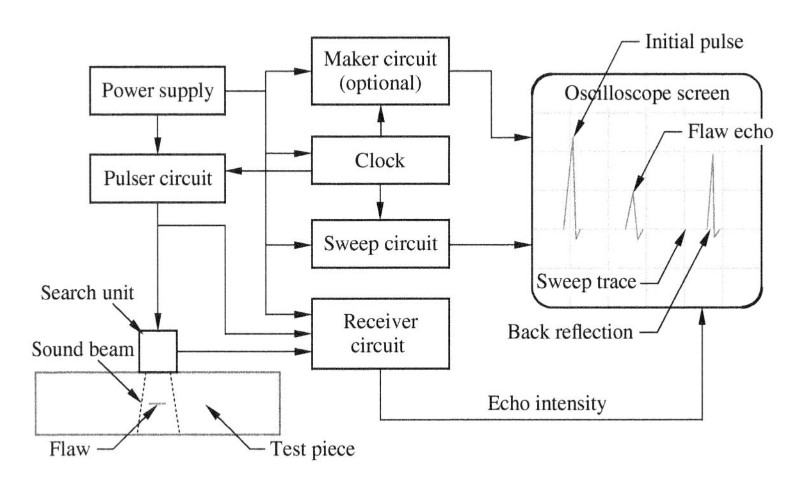


**Figure 8-10** Sketch of A-scan technique

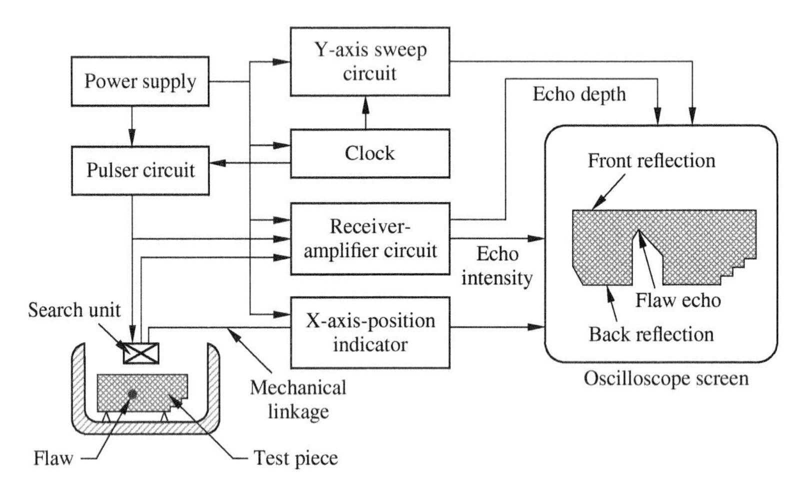


**Figure 8-11** Schematics of the B-scan technique

medical diagnostics. In the B-scan, the echoes are displayed as bright points, of which the brightness is proportional to the echo amplitude, the point is located in a line, and the distance from the starting point of the line represents the time of flight of the echo. The line direction is along the ultrasonic wave direction of propagation. By scanning the ultrasonic wave source along a line, a two-dimensional plot is formed. Hence, a B-scan display forms a longitudinal cross-sectional plot of the specimen tested. In other words, one can use a set of A-scan data to form a B-scan data set.

The B-scan is also called a brightness scan, in which the returned echo signal is used to modulate the intensity of the spot on an oscilloscope. This format provides a quantitative display of time-of-flight data obtained along a line on the test piece. The B-scan display shows the relative depth of reflectors and is used mainly to determine the size (length in one direction), location (both position and depth), and, to a certain degree, the shape and orientation of large flaws. By this means, a crude picture of the structure within a material can be presented, as shown in Figure 8-11. B-scans are very popular in the medical field.

The A-scan and B-scan systems are almost identical, but there are some differences between the two methods. B-scan allows the imaginary cross-section to be viewed; in a B-scan, echoes are indicated by bright spots on the screen rather than by deflections of the time trace. The position of a bright spot along the axis orthogonal to the search unit position axis indicates the depth of the echo within the test piece.

Another widely used display format is called a C-scan. The ultrasound source is moved on a two-dimensional surface. Every source point opens a time window at a fixed distance. Only the echoes inside the window are displayed in a bright point format, of which the brightness is in proportion to the amplitude of the echo. The coordinates of the point displayed are just those of the source. Thus, a lateral cross-sectional plot is formed. The depth in the specimen corresponds to the time window distance selected. B-scan and C-scan methods give more information than A-scan. For both B-scan and C-scan, digital image processing (DIP) techniques can be applied to extract more information. C-scan provides a semi-quantitative or quantitative display of signal amplitudes obtained over an area of the test piece surface. This information can be used to map out the position of flaws on a plan view of the test piece, as shown in Figure 8-12. A C-scan format also records time-of-flight data, which can be converted and displayed by image-processing equipment to provide an indication of the flaw depth. C-scan has an electronic depth gate that can allow echo signals to be received within a limited range to exclude the display of the front and back reflections.

**(b)** *S-wave reflection method for fresh concrete monitoring*: This ultrasonic technique has been proven to be very effective for nondestructive measurement of fresh and hardened cementitious composites. In the 1990s, a shear wave reflection method was developed at Northwestern University for hydration process monitoring of fresh concrete.

The method uses high-frequency (2.25-MHz) ultrasonic waves continuously monitoring the setting and hardening of concrete. The test setup is schematically shown in Figure 8-13. The transducer is placed on the top surface of a steel plate in contact with fresh concrete. The principle of the shear wave reflection measurement consists of monitoring the reflection coefficient of the ultrasonic waves at an interface formed between the steel plate and the concrete to be tested. Since a shear wave cannot pass through a liquid, the reflection factor at the beginning of the hydration is very high. However, with the hydration

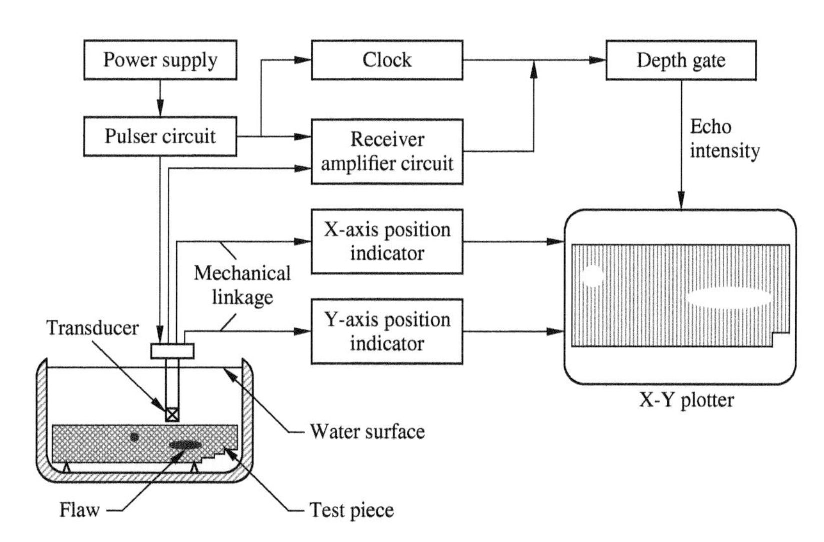


**Figure 8-12** General description of C-scan technique

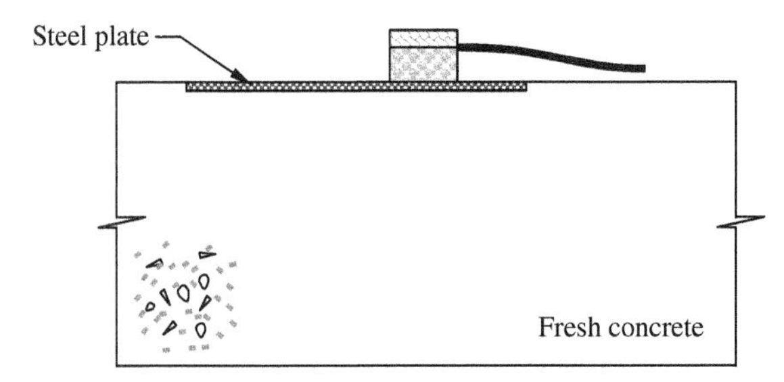


**Figure 8-13** S-wave reflection method for monitoring concrete setting and hardening

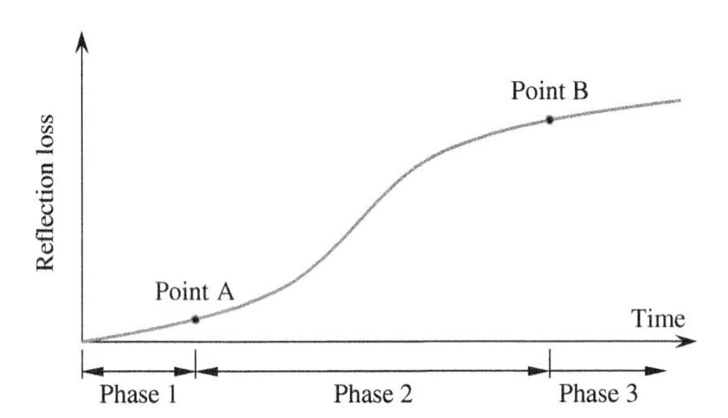


**Figure 8-14** A typical reflection loss versus time plot

process, the concrete changes into a more solid condition, and the reflection of the shear wave becomes weak due to the partial wave transmission into the concrete. Hence, the change of the reflection factor at the interface between steel plate and the concrete is related to the hydration process. The reflections of the shear waves from the steel-concrete interface are received from the transducer in the time domain. To calculate the reflection coefficient, the received signals in the time domain are transformed into the frequency domain by using the Fast Fourier transform algorithm. The reflection loss, $L$, is then defined as

$$L = 1 - \frac{F_2(f)}{F_1(f)} \tag{8-65}$$

where the $F_2(f)$ is the amplitude of the reflection from steel–concrete interface in the frequency domain and $F_1(f)$ the amplitude of the emitted signal in the frequency domain. Typical reflection loss versus time is plotted in Figure 8-14. It can be seen that the curve can be divided into three stages. Up to point $A$, the reflection loss is almost 0, i.e., the shear wave is almost completely reflected. This implies that the concrete is in a liquid state. From point $A$ to point $B$, the reflection loss is gradually increased, which implies that concrete is under a fast hardening process. After point $B$, the reflection loss curve becomes very flat, which signals that the concrete has reached the maturity stage.

The relationship between the reflection loss and the setting time, as well as the compressive strength of the concrete, can be established through a careful calibration process. Figure 8-15 shows the relationship between the setting time and the time at which point $A$ in the reflection loss curve occurs. It can be seen that there is a linear relationship between the two parameters. Hence, there is a potential to use point A as an index to predict the setting time for concrete. Figure 8-16 shows the relationship between compressive strength

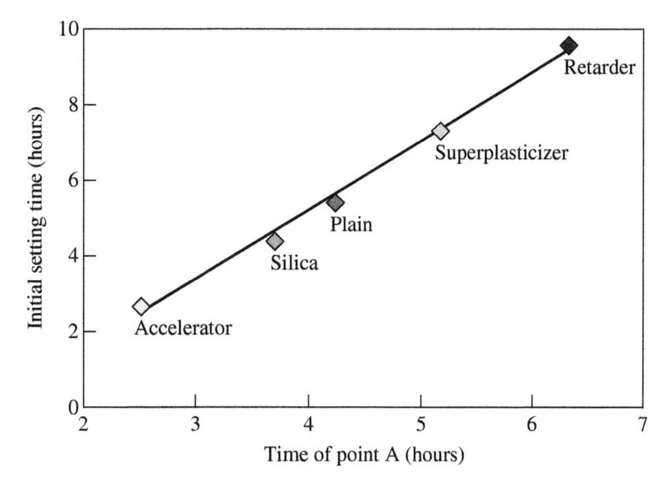


**Figure 8-15** Relationship between setting time and reflection loss

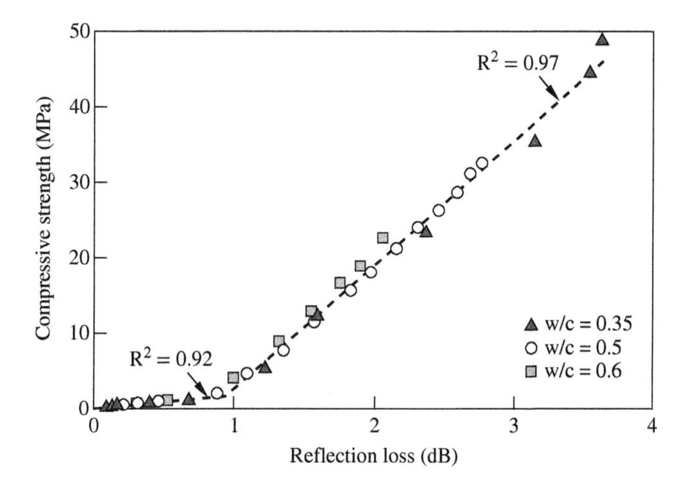


**Figure 8-16** Relationship between compressive strength and reflection loss

and reflection loss factors. It can be seen that a bilinear relationship exists between the two parameters. The transition point is between 6 and 15 hours after casting. It can be seen from Figure 8-16 that the compressive strength of a cement-based composite can be estimated from the reflection loss factor of a shear wave.

(c) *Longitudinal transmission method for fresh concrete*: This method uses ultrasonic waves to investigate the setting and hardening of cement-based materials, and was developed by Grosse and Reinhardt in 1994. It measures the flight time of an ultrasonic pulse traveling through a known distance and then calculates the wave velocity. During the setting and hardening process, fresh concrete changes from a liquid state to a plastic state and then to a solid state. The wave velocity keeps increasing with the process. By interpreting the characteristic points on the wave velocity versus time curve, the behavior of fresh concrete can be identified. Furthermore, since the wave velocity is directly related to the modulus of elasticity of the concrete, the development of the modulus of elasticity can be interpreted by this method.

A typical apparatus for this technology is shown in Figure 8-17. The walls of the container are made of Plexiglas. At the center of the apparatus, a transmitter and a receiver are fixed on opposite sides of the container, and fresh concrete is placed between the two transducers. Once the container is fully filled with fresh concrete, the transmitter starts to emit a P-wave and the receiver receives the signal. The measurement can be continuously conducted and recorded. Since the distance between the two transducers is known, it is easy to calculate the velocity. A typical plot of the wave velocity versus time is shown in Figure 8-18. One limitation of this method is that it is difficult to perform on site.

**(d)** *Embedded transmission method for fresh concrete*: To make the transmission method feasible for on-site hydration monitoring of fresh concrete, an embedded ultrasonic system has been developed at the Hong Kong University of Science and Technology (Qin and Li, 2008). The measurement system includes a functional generator, a power amplifier, a pre-amplifier, and an oscilloscope. An electrical pulse is first generated by an Agilent 33120A functional generator. Before being applied to the transmitter, the pulse is amplified by a power amplifier.

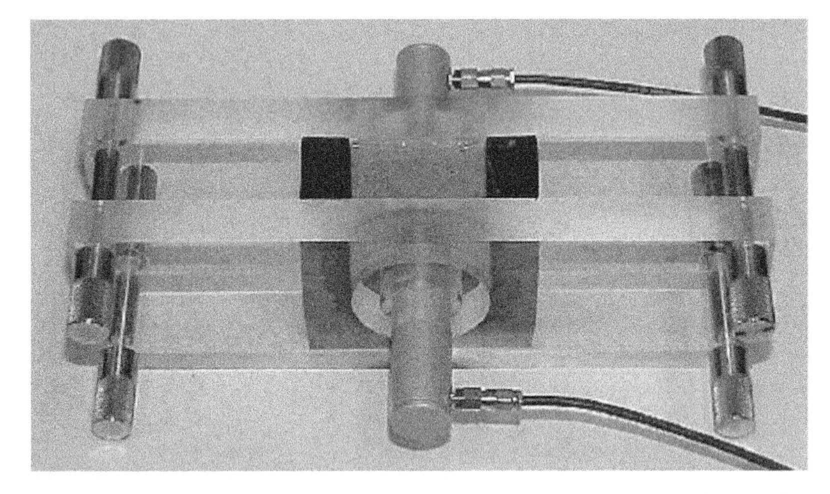

**Figure 8-17** A typical apparatus for using longitudinal transmission method

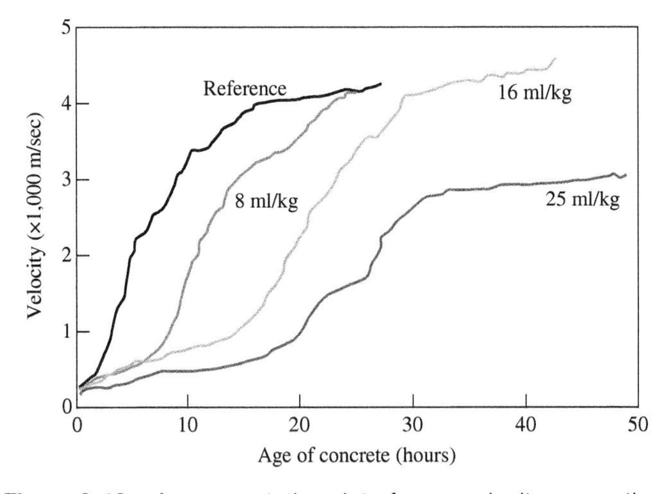


**Figure 8-18** A representative plot of wave velocity versus time

The transmitter is excited by the input electrical pulse and then vibrated in its resonance mode, emitting ultrasonic waves. An electrical field is applied along the polarization direction, and, therefore, the ultrasonic waves are of the longitudinal type. The ultrasonic waves can be received by the receiver through changing the mechanical energy into electrical energy. The voltage of the receiver can be measured and recorded by the oscilloscope through a pre-amplifier. The oscilloscope is a 12-bit Agilent 5462A digital oscilloscope with a sampling frequency of 1 MHz.

This embedded ultrasonic system has many advantages. First, it has a good coupling between the transducer and the matrix. Second, the method is very effective for hydration measurement of large-scale or underground concrete structures. Last, but not least, it can also be used to continuously monitor the structure during its entire service life. For a concrete structure, early-age performance can be monitored by the new system to provide guidance for construction quality control. After the concrete is cured and the strength is fully developed, the system can be used to detect damage accumulation or even impending disasters. Compared with traditional ultrasonic nondestructive methods, this method is inexpensive and effective for any-scale concrete structures. It meets all the health monitoring requirements of concrete structures from the fresh stage to the hardened stage.

The key element in the system is the homemade PZT. The transducer fabrication process includes the following steps: cutting PZT rods, welding coaxial wires to the positive and negative electrodes, coating the epoxy layer for insulation, coating the conductive layer as a shielding layer, and grounding. First, small PZT rods about 2 cm long were cut from a big PZT plate (from Hong Kong Piezo). The length direction was set along the poling direction. After welding a coaxial cable to the two electrodes of the piezo-rod, the patch and the coaxial cable were insulated using a thin layer of epoxy. Due to the embedment requirement, the insulation layer was laid very carefully. The insulation between the power line and the instrument circuit has little effect on the capacitive coupling. Capacitive coupled interference was mainly reduced by shielding the instrument circuit, using an envelope of high conductive material such as aluminum or copper that physically surrounds the instrument. Here, silver paint was used as the shielding layer. The shield was connected at one point to the ground, thereby, encasing the circuit in a zero-voltage surface. Figure 8-19 shows the fabrication of the transducer.

To measure the hydration process of fresh cement-based materials, the transducers should be fixed in a predetermined position in a formwork before the casting of fresh concrete or other cement-based materials. As an example, Figure 8-20 shows a placement of the transducers on the bottom of a plastic mold before casting. For this experiment, cement paste with a *w/c* ratio of 0.4, was mixed thoroughly for 5 minutes, and then was poured into the mold followed by a proper compaction to remove the entrapped air bubbles. After casting the fresh sample in the plastic box, the measurements were immediately started. During the experiment, one transducer, which was excited by an electrical pulse, acted as the signal source, while the other transducer acted as the receiver for ultrasonic signals. Figure 8-21 shows the input and output signals recorded by the two sensors. The input electrical signal is a step function of several hundred volts. The recorded signal from the receiver has very low amplitude, from several millivolts to several volts. The travel time is the interval between the onset time of the input pulse and the onset time of the received signal. A threshold is preset based on the noise level. The time when the signal exceeds the threshold is the onset time. In this study, because of the good coupling between the sensor and the specimen, the signal-to-noise level is quite high. It is easy to determine the onset time accurately according to the threshold. Knowing the travel distance, the velocity can be calculated.

Figure 8-22 shows the development of the velocity in fresh concrete paste. Based on the characteristic points on the experimental curves, evolution of ultrasonic pulse velocity (UPV) can

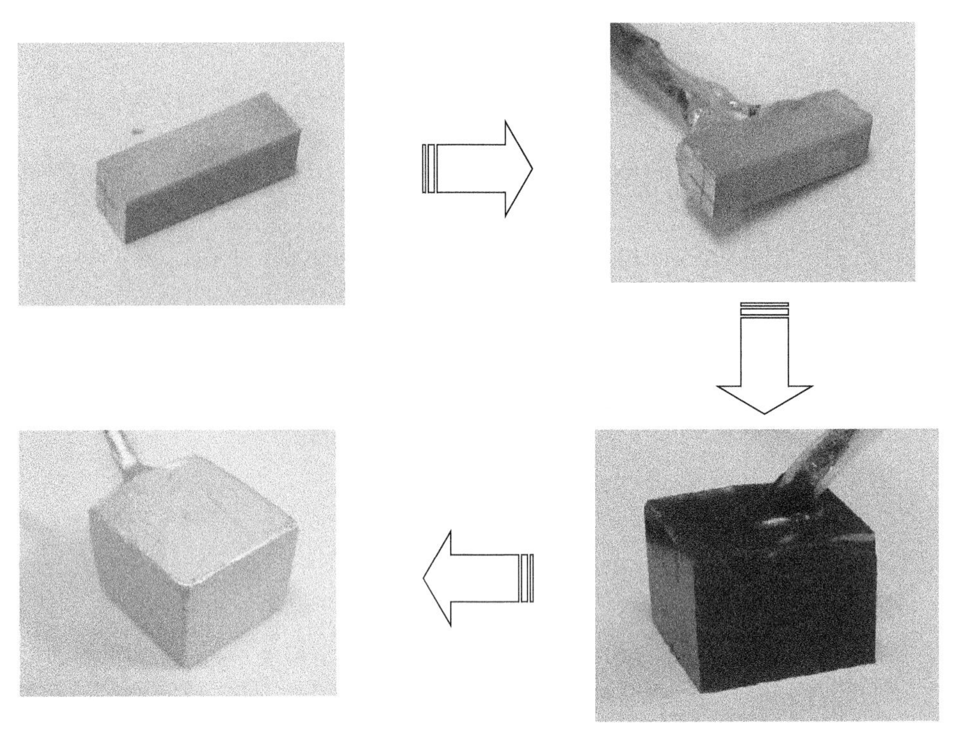

**Figure 8-19** Fabrication process of the transducer

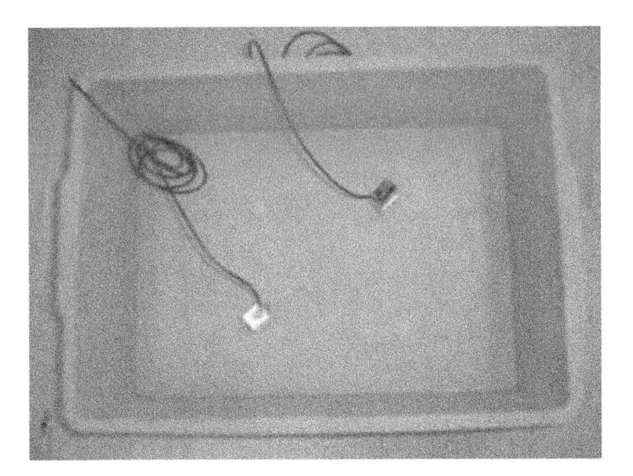

**Figure 8-20** Placement of the transducers before casting

be divided into four stages. In the first stage, the UPV remains nearly constant at around 1500 m/s. This period is related to the dissolution and dynamic balance stages of cement hydration. The wave velocity is very close to that of water. A period of rapid increase in the UPV appears after the dissolution and dynamic balance period, which corresponds to the rapid development of hydration products. When a critical quantity of hydration products is reached, percolation of solid phase seems to occur, and the UPV starts to increase. In stage 2, more and more hydration products are produced and intersected, and the stiffness or the modulus of the material increases rapidly.

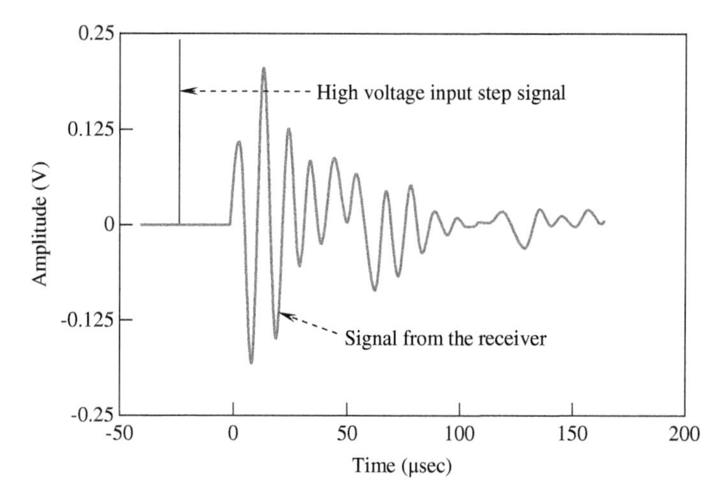


**Figure 8-21** Input and output signals recorded by two sensors

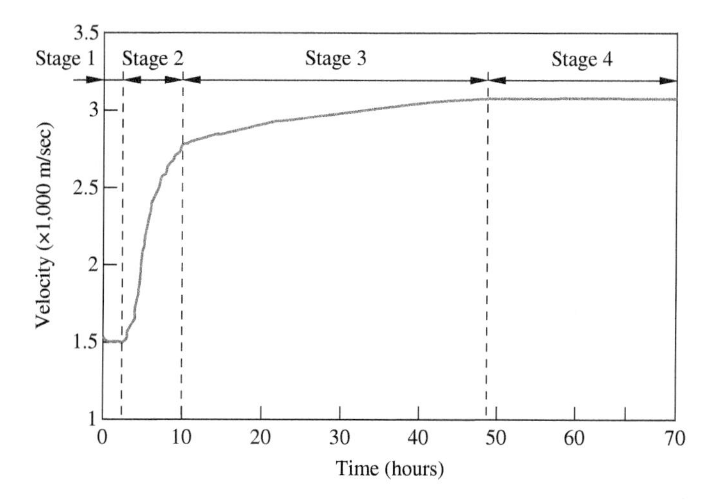


**Figure 8-22** Development of the acoustic velocity in fresh concrete

As a result, the UPV increases notably. In stage 3, the cement hydration process proceeds into diffusion control, and the growth rate of the hydration products slows down. The UPV thereafter increases slightly. In stage 4, a high hydration degree is reached and further hydration becomes minimal. Thus, the velocity development reaches a plateau. After the wave velocities are obtained, the wavelength, $\lambda$, and the dynamic modulus of elasticity, $E$, of the cement paste specimen at different ages can be calculated using the following equations:

$$\lambda = \frac{C_L}{f} \tag{8-66}$$

$$E = \rho C_L^2 \tag{8-67}$$

where $C_L$ is the longitudinal wave velocity, $f$ the resonant frequency, and $\rho$ the density of the cement paste.

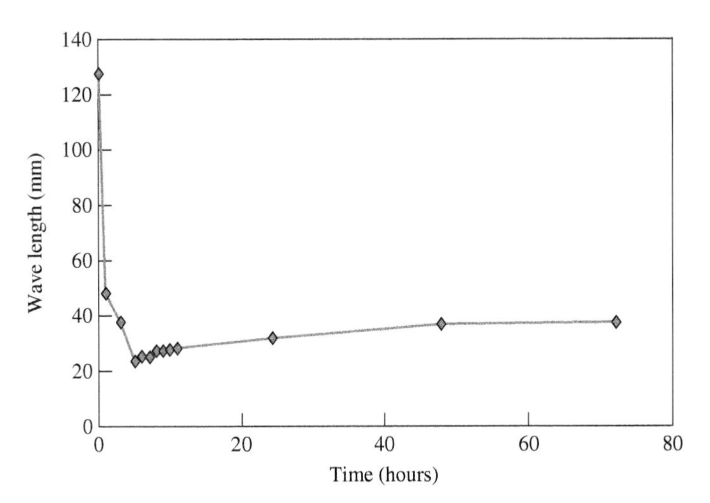


**Figure 8-23** The calculated values of wavelength

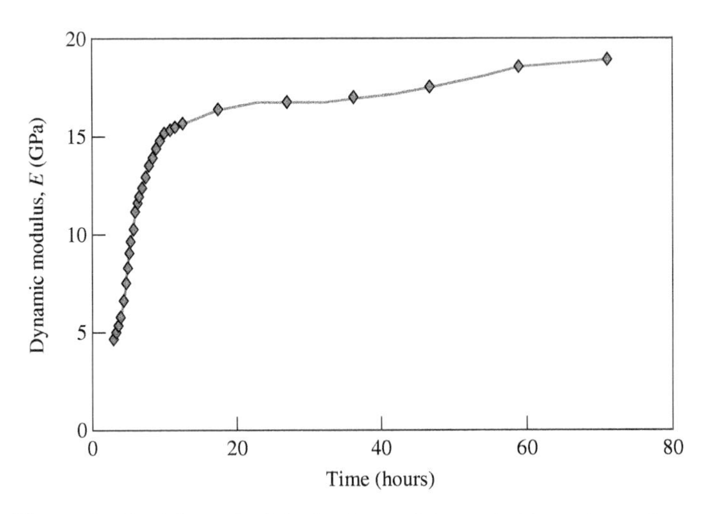


**Figure 8-24** The calculated values of dynamic Young's modulus

The calculated results are shown in Figure 8-23 for wavelength and in Figure 8-24 for dynamic modulus. It can be seen that at an early age, the wavelength decreases sharply due to the fast increase of the resonant frequency, although the velocity increases considerably in the same period. The dynamic modulus of elasticity increases with the hydration process. The mechanical properties and the microstructure development of the material can be assessed by using the calculated modulus.

### 8.5.2 Acoustic Emission Technique

The acoustic emission (AE) technique is one of the most important passive NDT techniques—it simply listens to "sounds" generated within materials. Acoustic emission is the physical process of an object quickly releasing its inner elastic energy in the form of stress waves when the object is subjected to external loads and/or environmental action. These stress waves, referred to as AE signals, are generated from localized sources, such as dislocations, microcracking, debonding, and other inside micro- as well as macrochanges in a stressed material. In concrete, AE sources may also include the local stress distributions associated with chemical action, such as alkali-aggregate reaction and corrosion of steel inside concrete. The AE signals carry the source information and follow the elastic wave propagation rules in concrete materials. AE activities, thus, can reflect the damage process in a stressed specimen. The locations where AE events occur and the material properties inside the object can be deduced from the received AE signals.

These form the basis of the NDT-AE as an acoustic passive NDT technique. In other words, the NDT-AE is a technique to characterize the material properties and to identify the structure of the test specimen, or even occurring inside the specimen, by means of receiving these AE signals and analyzing their characteristics. The AE technique can be used to monitor the change of material conditions in real time and to determine the location of these emission events as well. It cannot detect a crack that already exists but is not propagated. Since AE is a result of local stress redistribution, any micromechanisms that will cause stress release can be the sources of AE. Some typical AE sources in concrete are summarized as follows:

Microcrack and macrocrack initiation and propagation due to loading
Separation of reinforcing members
Mechanical rubbing of separated surfaces
Corrosion of reinforced steel inside concrete
Debonding of different interfaces, such as the interfaces between steel bar and concrete and between ceramic tile and concrete substrate
Matrix cracking due to the formation of the aggressive chemical product, such as ASR gel
Fiber debonding, fracture, and pull-out in fiber-reinforced concrete
Damage generated by environmental attack

The AE signal could be affected by several factors simultaneously, such as the structure and shape of the specimen, the quality and state of the specimen material, the strength of the loading, the position and direction of the loading, and the rate and process of the loading. For concrete, the *w/c*, cement type and property, the size and type of aggregates, the technology of processing, the age of the concrete, and so on all affect the AE process. Hence, generally, AE signals are very complex as well as random and blurred. The positive result of this is that an AE signal carries abundant information. The negative result is that the AE signal can be too complex to extract useful information.

Due to the nature of AE, the most important aspect of NDT-AE is the physical basis. Compared to the UT technique, where the source property of the ultrasonic wave is a known factor, the NDT-AE just wants to find the properties of the ultrasound source, that is to say, how to establish the roles on the relation between the quantities to be tested and the behavior of AE signals. Because of the complexity of the AE phenomena, most of these roles are empirical. A great deal of research to reveal these relations has been performed, on the mechanism of acoustic emission, the properties of ultrasonic transducers, the acoustic wave propagation theory, etc. This has led the AE technique onto a more solid and wider physical basis, and the application of NDT-AE has rapidly expanded.

The application of the AE principle relies strongly on the techniques of data acquisition and signal processing. It is difficult to make a prediction when and where AE events will emerge. Hence, it is difficult to record AE events precisely. To characterize the AE source, unlike the NDT-CE-UT, multichannel receiving techniques are necessary for NDT-CE-AE. From an analysis point of view, the more channels, the better. However, from the technical point of view, adding channels requires great effort due to the increasing interaction among channels and the fast-expanding data and cost of equipment.

The early AE testing instruments used analogue electrical devices. The ability to capture instantaneous AE events and to process AE signals is low with these instruments. Early successful applications of NDT-AE as a practical industrial applied technique started around the middle of the 1960s in the United States, where it was used to monitor and to test a rocket shell. The development of NDT-AE techniques was slow until the numerical data acquisition AE instrument was developed by the middle of the 1980s. Thanks to the rapid development of the numerical electrical techniques and computer technology, with the assistance of solid electrical circles, high-speed analog-to-digital converters, DSP technologies, and modern computer data acquisition, the capture and storage of an instant AE signal that is precise and synchronized became possible and easy. More and more fast and reliable new numerical data processing methods have become available. These led the NDT-AE technique to have a reliable basis in both of hardware and software.

After about three decades of research and practice, the AE technique has made great progress and has become a very powerful nondestructive technique. Successful applications are to be extended over more and more areas.

When an AE event occurs at a source within the material, the stress wave travels directly from the source to the receiver in the form of body waves. Surface waves may then arise from mode conversion. When the stress waves arrive at the receiver, the transducer responds to the surface motions that occur. AE signals cover a wide range of energy levels and frequencies. Modern instrumentation can record AE signals in concrete in the range of 50 kHz to about 1 MHz. At lower frequencies, background noise from the test equipment becomes a problem. At very high frequencies, the attenuation of the signals is severe. AE signals are usually divided into two basic types, the burst type and the continuous type. The former corresponds to individual emission events, while the latter refers to an apparently sustained signal level from rapidly occurring emission events (Bray and Stanley, 1997). The most important role for a data analysis method in the AE technique is to acquire an integrated and meaningful AE waveform.

There are several ways to characterize the material behavior based on acoustic emissions, including event counting, rise time, spectrum analysis, AE source location (defect location), energy analysis, signal processing, and signal duration. For example, AE activities (event rate, accumulated event numbers, count rate, and count summation, etc.) have been used to predict the onset and extent of damage in concrete and relate to the imposed stress level. Frequency components, rise time, and amplitude of AE signals have been used to distinguish different sources at different stress levels. The source location of an AE event can be determined based on the time differences among the recorded signals by an array of transducers, and has been used to monitor microcrack localization in concrete and fiber-reinforced concrete under uniaxial tension, and the fracture process zone in concrete under uniaxial tension.

The first representative investigation of AE from metals was carried out by Kaiser, who established the so-called Kaiser effect: "the absence of detectable acoustic emissions at fixed sensitivity level, until previously applied stress levels are exceeded." Concrete and fiber-reinforced concrete are multiphase and flaw-rich composite materials. When subjected to a load, microcracks tend to form along flaws as the load increases, which makes it very suitable for AE monitoring. Unlike other NDE techniques, this technique indicates only active flaws and cannot determine the presence of other kinds of flaws. It responds to changes in flaw size instead of the total size. Therefore, continuous monitoring is required to detect a flaw extension whenever it occurs.

The application of AE in concrete has been studied for the past 30 years. Almost all the applications of the AE technique for concrete are focused on the fracture process zone. An extensive series of investigations has been carried out by Maji and Shah (1988), Li and Shah (1994), Maji et al. (1990), Ouyang et al. (1991), and Landis et al. (1993). In these tests, concrete, mortar, and low-volume-fraction FRC specimens were loaded, in either direct tension or bending, to study the damage initiation and propagation within the materials. Both notched (Maji and Shah, 1989) and unnotched (Li and Shah, 1994) specimens were investigated. Deconvolution techniques were used by Maji et al. (1990), Ouyang et al. (1991), Li (1996), and Suaris and van Mier (1995) to study the orientation and the mode of microcracking. The relative amplitudes of AE signals at different transducer positions were used to distinguish tension and shear microcracks in mortar and aggregate–matrix interfaces.

For fiber-reinforced concrete, Li and Shah (1994) attempted to use the AE technique to study the tensile fracture of short steel fiber-reinforced concrete. Li et al. (1998) improved this AE measuring system for unnotched fiber-reinforced concrete specimens with relatively high volume fractions of fibers. An adaptive AE trigger signal identifier was developed for an adaptive trigger AE measurement system. An automated P-wave arrival time determination method was developed for pinpointing the location of thousands of AE events by using an adaptive low-pass filter (Li et al., 2000).

The AE technique has also been explored for detecting reinforcing steel corrosion in concrete. Li et al. (1998) examined the correlation between the characteristics of an AE event and the behavior of rebar corrosion in HCl solution, and the possibility of the corrosion detection of rebars inside concrete through an accelerated corrosion experimental method. The theoretical prediction and experimental results have shown that the AE technique is able to detect rebar corrosion at an early corrosion stage.

#### 8.5.2.1 AE Measurement System

The main elements of a modern AE instrumentation system are schematically shown in Figure 8-25. A complete AE measurement system includes AE transducers, pre-amplifiers, triggers, digitizers, and computers. The AE transducers are mainly made of piezoelectric composites, preferably wideband, with a linear phase response and small size. These transducers are used to convert the surface displacements into electric signals. The voltage output from the transducer is directly proportional to the strain in the PZT, which depends, in turn, on the amplitude of the surface displacement. Pre-amplifiers are normally necessary because the outputs from the transducers are usually low. The pre-amplifiers have two functions. One is to amplify the signal received by the AE transducers, usually using two magnification scales, 40 and 60 dB. The other function is to act as a filter.

Usually, a bandpass filter is installed in the pre-amplifier with a bandwidth of 30 kHz to 1 MHz. Only signals within this range can get through. A trigger is useful and the triggering is

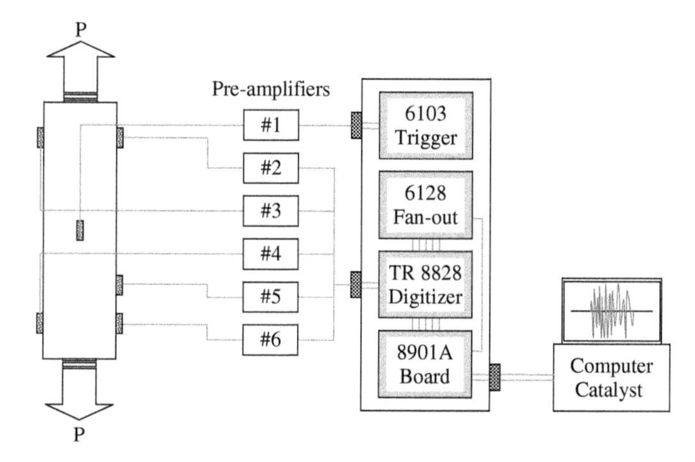


**Figure 8-25** Acoustic emission measurement system

initiated when the output voltage of the transducer exceeds a preset reference voltage (threshold), such that the lower level background noises can be rejected. An analog-digital converter is used to digitize a signal in real time and store the digitized data in memory. A computer is usually utilized to control the AE system and record the AE signals. The radiation patterns of acoustic emission sources are formed according to the same principle described earlier for ultrasonic transducers. In industrial applications, the AE sensors usually are located relatively farther from the source.

#### 8.5.2.2 AE Source Location Method

The AE source location can be deduced by using the time differences among the first-wave arrival times of the AE transducers. Figure 8-26 shows a typical set of AE signals. There are four individual signals in the figure, representing four channels. Each signal is obtained by a specific transducer located at a specific position. It can be seen from Figure 8-26 that the first arrival time of the wave is different for different signals because the distances from the transducer to the source of AE are different. The differences in these first arrival times provide a base for locating the AE sources, as discussed in next section.

For the difference in these arrival times, for a two-dimension case, we can write the error between the distance from transducer 1 to the AE source and the distance from transducer i to the AE source as

$$e_{1i} = \sqrt{(x-x_1)^2 + (y-y_1)^2} - \sqrt{(x-x_i)^2 + (y-y_i)^2} - \Delta t_{1i}C \tag{8-68}$$

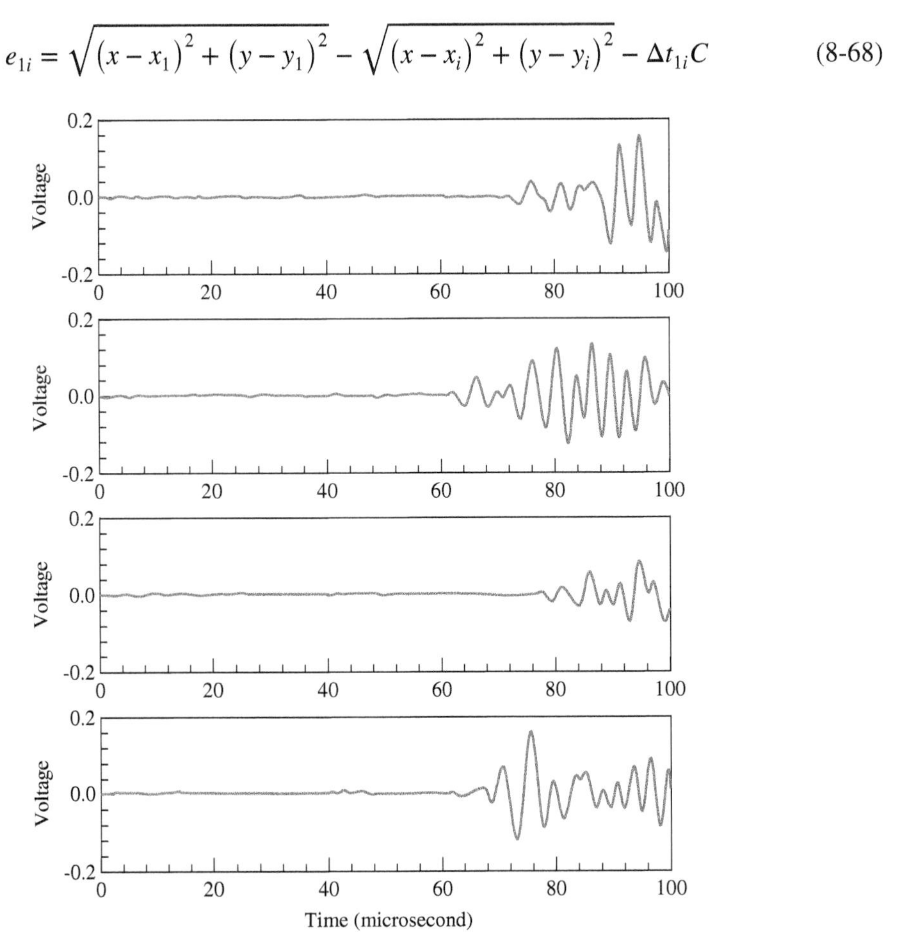


**Figure 8-26** Typical set of AE signals

where $x$ and $y$ are coordinates of the AE source to be determined and $x_1(x_i)$ and $y_1(y_i)$ are the coordinates of the transducers. The term $\Delta t_{1i}$ represents the difference in arrival time between transducer 1 and $i$ ($i = 2, 3, \ldots, n$) and $C$ is the wave velocity. For an ideal situation, the error expression should be zero and the AE source location should be easily determined by an intersection of two sets of equations. However, for a real experiment, error always exists. To minimize the test error, the following method is developed. First, the square of individual error is summed:

$$\begin{aligned} e &= \sum_{i=2}^{n} (e_{1i})^2 \\ &= \sum_{i=2}^{n} (d_1 - d_i - \Delta t_{1i} C)^2 \end{aligned} \tag{8-69}$$

where

$$\begin{aligned} d_1 &= \sqrt{(x - x_1)^2 + (y - y_1)^2} \\ d_i &= \sqrt{(x - x_i)^2 + (y - y_i)^2} \end{aligned} \tag{8-70}$$

Then, the total error is differentiated with respect to $x$ and $y$, giving

$$\begin{aligned} f_1(x, y, C) &= \frac{\partial e}{\partial x} \\ &= 2\sum_{i=2}^{n} \left( \frac{x - x_1}{d_1} - \frac{x - x_i}{d_i} \right) (d_1 - d_i - \Delta t_{1i} C) \end{aligned} \tag{8-71}$$

$$\begin{aligned} f_2(x, y, C) &= \frac{\partial e}{\partial y} \\ &= 2\sum_{i=2}^{n} \left( \frac{y - y_1}{d_1} - \frac{y - y_i}{d_i} \right) (d_1 - d_i - \Delta t_{1i} C) \end{aligned} \tag{8-72}$$

By setting $f_1(x, y, C)$ and $f_2(x, y, C)$ to zero, the values of $x$ and $y$ can be solved by employing a numerical method, provided that the value of the wave velocity is known. Moreover, if the error equation is differentiated with respect to wave velocity, $C$, one gets,

$$\begin{aligned} f_3(x, y, C) &= \frac{\partial e}{\partial C} \\ &= 2\sum_{i=2}^{n} \Delta t_{1i} (\Delta t_{1i} C + d_i - d_1) \end{aligned} \tag{8-73}$$

It should be pointed out that solving Equations 8-71 through 8-73 together will provide not only the source locat ion but also the wave velocity.

The method is easy to expand to a 3D case. In this case, Equation 8-68 becomes

$$e_{1i} = \sqrt{(x - x_1)^2 + (y - y_1)^2 + (z - z_1)^2} - \sqrt{(x - x_i)^2 + (y - y_i)^2 + (z - z_i)^2} - \Delta t_{1i} C \tag{8-74}$$

The sum of the square of the errors for a 3D case will take the same expression as shown in Equation 8-70. However, the expressions for $d_1$ and $d_i$ will take the form

$$
\begin{aligned}
d_1 &= \sqrt{(x-x_1)^2 + (y-y_1)^2 + (z-z_1)^2} \\
d_i &= \sqrt{(x-x_i)^2 + (y-y_i)^2 + (z-z_i)^2}
\end{aligned}
\tag{8-75}
$$

Consequently, the differential with respect to $x$, $y$, $z$ and $C$ are in the forms of

$$
\begin{aligned}
f_x(x, y, z, C) &= \frac{\partial e}{\partial x} \\
&= 2\sum_{i=2}^{n} \left( \frac{x-x_1}{d_1} - \frac{x-x_i}{d_i} \right) (d_1 - d_i - \Delta t_{1i} C)
\end{aligned}
\tag{8-76}
$$

$$
\begin{aligned}
f_y(x, y, z, C) &= \frac{\partial e}{\partial y} \\
&= 2\sum_{i=2}^{n} \left( \frac{y-y_1}{d_1} - \frac{y-y_i}{d_i} \right) (d_1 - d_i - \Delta t_{1i} C)
\end{aligned}
\tag{8-77}
$$

$$
\begin{aligned}
f_z(x, y, z, C) &= \frac{\partial e}{\partial z} \\
&= 2\sum_{i=2}^{n} \left( \frac{z-z_1}{d_1} - \frac{z-z_i}{d_i} \right) (d_1 - d_i - \Delta t_{1i} C)
\end{aligned}
\tag{8-78}
$$

$$
\begin{aligned}
f_C(x, y, z, C) &= \frac{\partial e}{\partial C} \\
&= 2\sum_{i=2}^{n} \Delta t_{1i} (\Delta t_{1i} C + d_i - d_1)
\end{aligned}
\tag{8-79}
$$

Setting the four equations equal to zero and solving them will provide the coordinates of the AE source location, $x$, $y$, and $z$, and the wave velocity, $C$, for a 3D problem.

### 8.5.2.3 Characterization of AE Signals

As mentioned earlier, traditionally the AE signals have been characterized by the so-called wave-shape parameters such as the number of ring echoes, the total energy, the number of AE events, the rate of AE events, the rise time of the wave pulse, the wave pulse length, and the spectrum. An AE test records the AE signal wave-shape parameters first, then by means of some pre-established relationships estimates the material properties under inspection.

As an update, the Physical Acoustics Company (PAC) has developed software, PAC-PARS, using an artificial neural network (ANN) analysis system to characterize AE signals in a more advanced way. It uses twelve basic parameters in an AE signal to form a tensor space. The components of the tensor are load, number of channels, rise time of the pulse, the number of ring echoes, total energy, amplitude, average signal level, the ratio of pulse length over rise time, the product and pulse length, the ratio of the number of ring echoes over pulse length, the product of the number of ring echoes, and the pulse length. Over the tensor space, detailed analysis, such the analysis of eigenvalues and eigenvectors, mode recognition and ANN, can be performed to extract useful information fast and accurately.

Besides the wave-shape analysis, wave-form analysis has been developed as a new scheme to extract information from AE signals. It is based on the AE signal waveform in the time domain and the well-developed analysis techniques, such as frequency spectrum, autorelation function, and corelation function, to extract information from a wave packet. To realize the waveform analysis, the equipment must have the ability to capture the transient wave impulse and the ability of real-time processing. The high-speed numerical AE system (NAES) instrument makes this possible, in an easy and reliable manner.

No NDT technique, other than the AE technique, is so strongly reliant on the instrument used, and only good digital AE instruments can give satisfactory results. The specification of the transient wave record greatly affects the performance of the AE system, of which the most important are the sampling rate, precision, and memory length for one record (a hit). In practice, the sampling rate should be at least 4 or 5 times as fast as the highest frequency of the AE signal to receive the details of the AE signals and hence more information. The maximum dynamic range of the amplitude is determined by the precision of the analog-digital converter (ADC)

$$\text{Dynamic range} = 6 \times \text{bits of ADC (dB)} \tag{8-80}$$

That is to say, the dynamic range is limited to below 48 dB for an 8-bit ADC theoretically. The AE pulse duration is determined by the product of the memory length for one sampling and the sampling period, which is the inverse of the sampling rate. For a certain expected length of AE signal, enough memory length should be kept. High sampling rate and long hitting of the AE signal and multichannel receiver require the very fast post-processing of data.

#### 8.5.2.4 Laboratory Applications

NDT-AE is widely applied in fracture mechanics for metals and composite materials. In civil engineering, it has also been widely used to detect crack formation and propagation in a concrete specimen under load action, to detect corrosion of reinforcing steel under severe environment conditions, or to monitor a debonding process of external decoration attached to concrete structures.

**(a)** *Detection crack occurrence in concrete under uniaxial tension*: In this example of AE application, a method developed to study crack occurrence and propagation in concrete is introduced. The setup for the test is shown in Figure 8-27. Except for four LVDT transducers, five piezoelectric transducers are mounted on the side surfaces of the specimen to pick up the acoustic emission signals originating from the materials, mainly the working portion of the specimen. The mix proportion for the plain concrete specimen is 1:0.45:2:0.65 (binder:water:sand:coarse aggregate, by weight). The specimens were tested about 30 days after casting. The test portion in the center, after gluing the loading plates onto each end, had a length of 210 mm, a width of 100 mm, and a depth of 20 mm.

The acoustic emission measurement system shown in Figure 8-27 was used to acquire AE data during the test. The system was much the same as the one that was described by Li and Shah (1994). There is a major difference, however, in that an eight-channel adaptive AE-trigger-signal identifier was employed in the system. The adaptive triggering was actualized by comparing the signal level to a reference voltage (adjustable) with the help of voltage comparators, and then the identifier can "identify" the first signal, whose amplitude exceeds the threshold, among all signals. By using this trigger identifier, all channels can be used as a trigger. Hence, the requirement of presetting a trigger channel is eliminated. In addition, possible omission of AE events due to the geographical arrangement of a preset trigger can be eliminated.

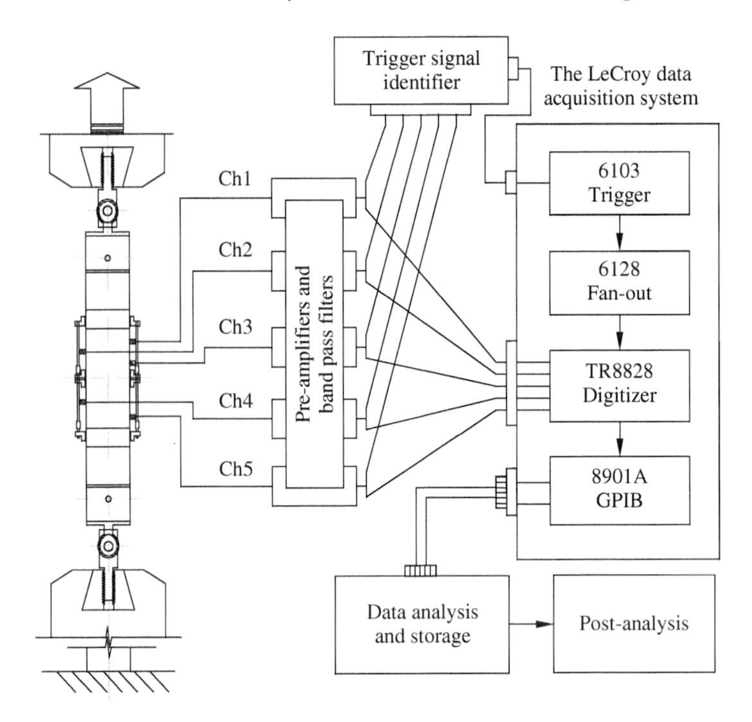


**Figure 8-27** Experimental setup for studying crack occurrence

During the test, the pre-amplifier gain was set at 60 dB and the sampling frequency was set at 6.25 MHz. Considering the maximum frequency of AE signals from concrete to be normally within 1 MHz, this frequency was much higher than the one required by the Nyquist theorem (Oppenheim and Schafer, 1975). Each digitized signal from a channel is set at 4 K (4096) sample points and the threshold was selected as 49 dB. When the external load was increased to near the concrete capacity, diversified fracture events could occur, such as preload microcracks, voids, or other defects, which will propagate due to stress concentrations in the vicinities. During this process, energy is released and an acoustic wave is generated. The level of energy released or the amplitude of the waves depends on the types of fracture events. Thus, in principle, it is possible to detect these fracture events by monitoring the acoustic emission activity. The occurrences of AE events were somewhat related to the imposed stress level. The activity can be characterized by an AE event count and event rate. The source locations can be determined if enough transducers are used for monitoring. Bearing this view in mind, five identical PZT transducers are mounted on the unnotched specimen to monitor the damage and the fracture during the uniaxial tensile testing.

A plain concrete specimen was tested in direct tension and monitored by AE measurement in the meantime. Figure 7-31 (see Chapter 7) shows its deformation–time curves. It can be seen from Figure 7-31 that the deformations in the specimen vary a lot at different loading stages. The tensile stress, accumulated event number, and event rate as a function of time for the plain concrete are depicted in Figure 8-28. In Figure 8-28, it can be seen that the AE data is closely related to the stress level (loading stage). In Figure 8-28, stress is the nominal stress calculated from the imposed tensile load divided by the original cross-sectional area. Displacement is the deformation measured by the control LVDT that catches the formation of major cracks.

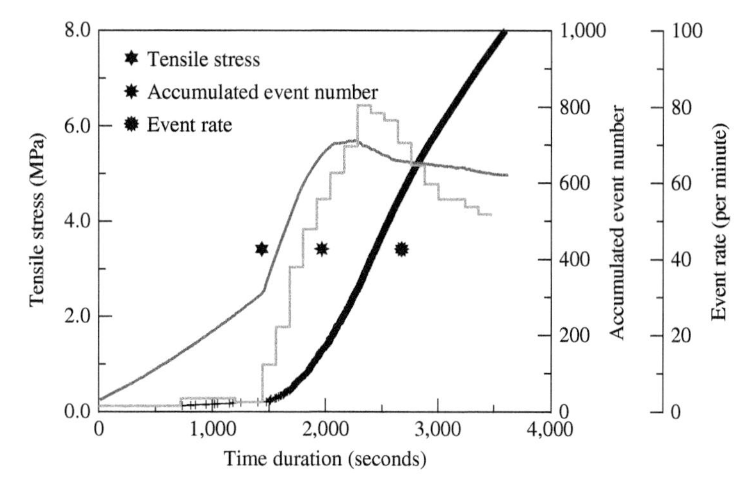


**Figure 8-28** The tensile stress, accumulated even number, and event rate for plain concrete

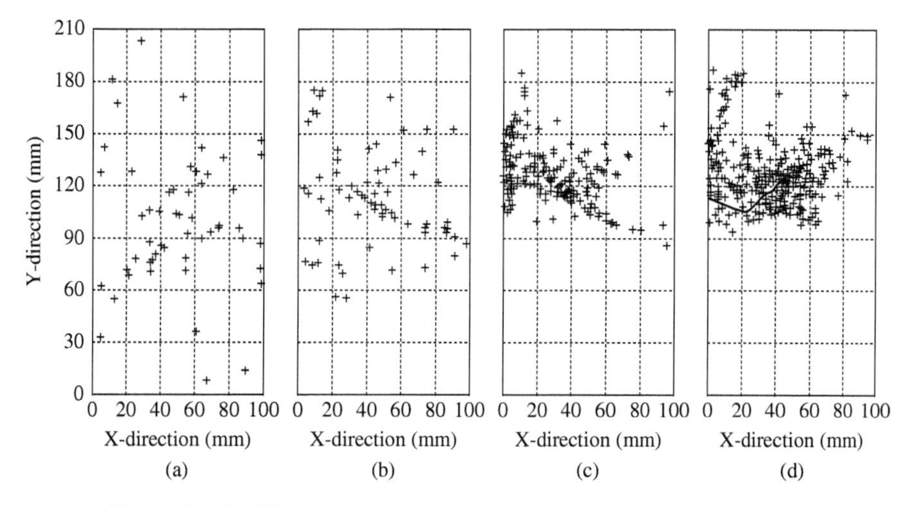


**Figure 8-29** The source location map of the measured AE events

The source locations of the AE events or microcracks have been calculated using the method introduced in the previous section and the source location map has been plotted in Figure 8-29. It should be pointed out that identification of damage localization can be clearly observed in Figure 8-29 for regime (c), which corresponds to a stress level from $0.8f_t$ to $f_t$. A major crack has developed from the localization zone and is propagated across the width of the specimen. The phenomenon agrees with the findings of Li and Shah (1994) very well.

**(b)** *Corrosion detection*: The feasibility of the application of the AE technique to detect corrosion of reinforcing bars in concrete has been studied by Li et al. (1998). The physical background of this application is that under the expansion of the corroded parts, microcracks will be developed in the interface and bulk matrix, and energy will be released. Subsequently, a stress wave will be generated, which propagates along the medium and will reach the outer surface. By placing the AE transducer on the surface, the occurrence of microcracks can be detected. Since these microcracks are caused by corrosion products, the signals detected can be used to interpret the corrosion activity.

The mathematical model for calculating the stress caused by rebar corrosion at the rebar–concrete interface can be simplified as a shrink-fit model. First, let us assume that the rebar can freely expand due to corrosion with an increase of $r$ in radius direction. Then, we try to put the rebar back into the hole that it occupied before. Due to corrosion expansion, the rebar now is too big to fit freely in the concrete hole. To allow the rebar to fit back into concrete, pressure has to be applied to both the surrounding concrete and the rebar. Let us consider the surrounding concrete first. This situation can be treated as a hole under internal pressure in an unbounded medium. According to the elasticity solution, the displacement caused by internal pressure $p$ along radius direction is

$$U_{\mathrm{r}}^{\mathrm{c}}\,(r = a) = \frac{pa}{2\mu_{\mathrm{c}}} \tag{8-81}$$

where $p$ is the pressure, $a$ is the radius of hole, and $\mu_{\mathrm{c}}$ is the shear modulus of the surrounding concrete (interface). For the rebar, it can be treated as an inclusion with pressure on the outside. The displacement along the radial direction can be written as

$$U_{\mathrm{r}}^{\mathrm{s}} = \frac{-p\,(k-1)\,a}{4\mu} \tag{8-82}$$

where $k$ is the Kolosou constant, with values

$$k = \frac{3-v}{1+v} \quad \text{for plane stress} \tag{8-83a}$$

$$k = 3 - 4v \quad \text{for plain strain} \tag{8-83b}$$

Compatibility requires that

$$|U_{\mathrm{r}}^{\mathrm{c}}| + |U_{\mathrm{r}}^{\mathrm{s}}| = \Delta a \tag{8-84}$$

Thus, we can obtain the pressure expression as

$$p = \frac{4\mu_{\mathrm{c}}\mu_{\mathrm{s}}}{2\mu_{\mathrm{s}} + \left(k_{\mathrm{s}} - 1\right)} \frac{\Delta a}{a} \tag{8-85}$$

The stress produced in the surrounding concrete interface is then derived as

$$\begin{aligned} \sigma_{\theta\theta} &= \frac{4\mu_{\mathrm{s}}\mu_{\mathrm{c}}}{2\mu_{\mathrm{s}} + \left(k_{\mathrm{s}} - 1\right)\mu_{\mathrm{c}}} \frac{\Delta a}{a} \\ &= C\frac{\Delta a}{a} \end{aligned} \tag{8-86}$$

For steel, the shear modulus is about 81 GPa and the Kolosou constant is around 2. The shear modulus for concrete is about 12 GPa. Thus, the value of C is $2.23 \times 10^{10}$. For $\Delta a/a$ equaling 0.0001, the stress produced is 2.23 MPa. Note that the stress is, in fact, the shear stress in the interface and this value is large enough to create a microcrack. The stress wave generated by this microcrack can be detected by an acoustic emission transducer. This proves that the detection sensitivity of the AE technique to rebar corrosion is very high (0.0001 for a radius of 10 mm is only 1 μm!).

An experimental setup for corrosion testing is shown in Figure 8-30. The specimen used in the experiment was a concrete block with dimensions of 300 × 300 × 175 mm. The mix proportion of the concrete was 1:0.6:2.5:2.5 (B:W:S:A) by weight. Three deformed rebars, 25 mm in diameter, were placed in the concrete block. One was positioned about 25 mm

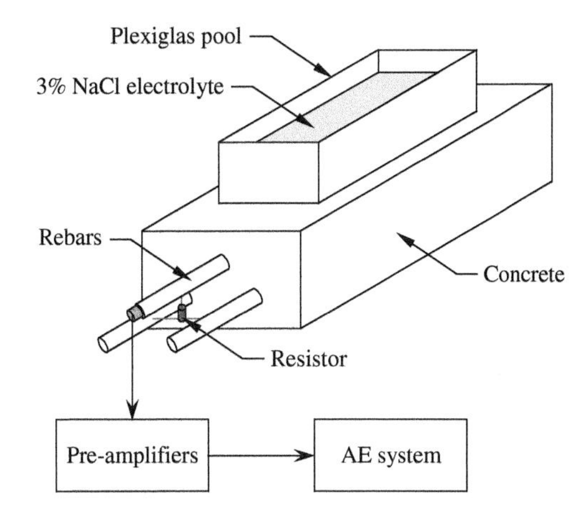


**Figure 8-30** Experimental setup for a corrosion test of reinforcement

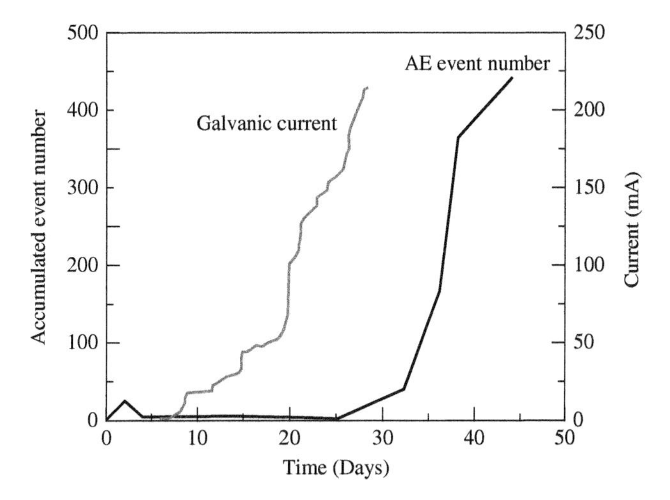


**Figure 8-31** Comparison of the accumulated number of AE signals with galvanic current

from the top surface of the concrete and the other two 250 mm below the top surface. All the reinforcing bars were coated with epoxy resin, 25 mm in length, at the edge surface where they entered the concrete, to prevent edge effect. The two rebars placed at the bottom of the concrete specimen were electrically connected to the top rebar by a shunt resistor, which made galvanic current measurement possible. An acrylic tank was attached to the top surface of the concrete specimen. The reinforced concrete specimen underwent a cyclic exposure of salt solution with 15% of solid NaCl electrolyte. The exposure cycle consisted of 3 days wet and 4 days dry to simulate the condition of wet/dry cycling of seawater on concrete bridge decks and substructures. The AE transducers were mounted on the two ends of the top rebar that was to be corroded. The output of the AE transducers was amplified and filtered by a pre-amplifier and transferred into digital signals by an A/D module. The digital signals were fed into a computer via shielded coaxial cables.

Half-cell and galvanic current were also measured to monitor the corrosion activities of the reinforcing steel. Figure 8-31 shows the comparison of the accumulated number of AE signals with the measured values of galvanic current. It can be seen that there is a significant

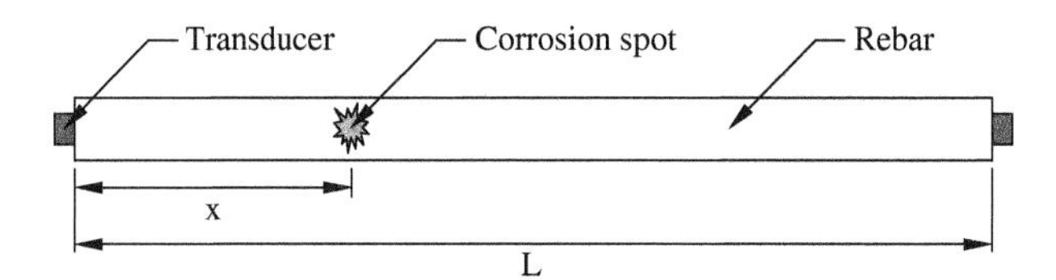


**Figure 8-32** Measurement of the location of corrosion

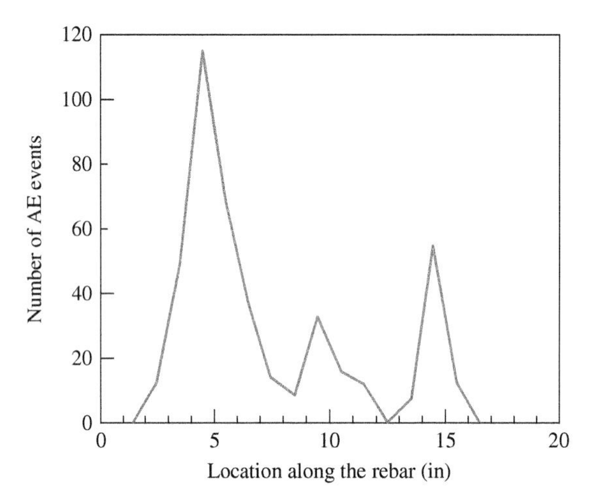


**Figure 8-33** Calculated source locations of corrosion

increase in AE signals at about 20 days into exposure that is most likely due to microcracking caused by the build-up of corrosion products on the rebar. However, the measurement of galvanic current shows no obvious change at this time. A sudden increase in galvanic current is found at approximately 32 days of exposure time. Also, the half-cell potential became more negative, to about −420 mV, at the same time. It is thus verified that AE monitoring can detect corrosion earlier than the other two corrosion-detecting techniques.

For the arrangement of the AE transducers in this way, the AE source location becomes a one-dimensional problem. Referring to Figure 8-32, suppose that * is the position of the corrosion, and the length of reinforcing bar is $L$. Then the distance from corrosion source to the AE transducer on the left is $x$ and that on the right is $L - x$. If the time difference between the first arrival times at the two transducers is $\Delta t$ and the wave velocity of steel is $C$, we can write the following equation:

$$(L - x) - x = \Delta t \cdot C \tag{8-87}$$

and solving the equation for $x$ gives

$$x = \frac{L - \Delta t C}{2} \tag{8-88}$$

Based on this equation, the source locations of corrosion have been calculated for the test specimen. The results are shown in Figure 8-33. To verify the calculated results, the specimen was broken and the reinforcing steel was taken out, as shown in Figure 8-34. Direct comparison shows a good agreement between the predicted corrosion locations and real ones on the rebar.

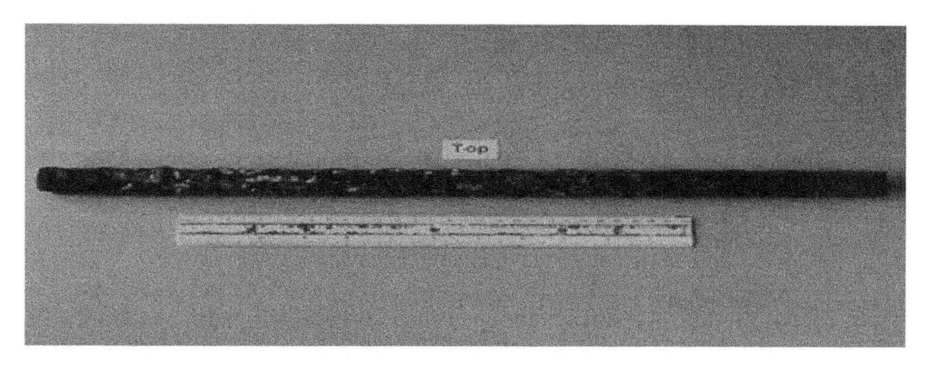


**Figure 8-34** The actual corrosion conditions of the reinforcement

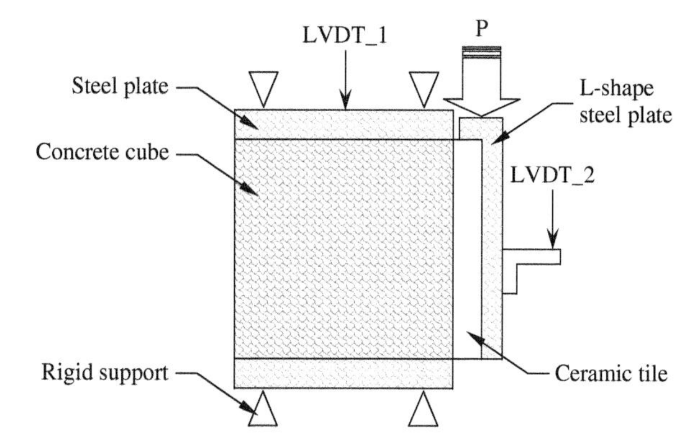


**Figure 8-35** Test setup for debonding detection of ceramic tiles on concrete

(c) *Debonding detection of ceramic tiles on concrete*: Decorating concrete buildings using ceramic tile finishes is common practice throughout the world. The external finishes not only improve the stark appearance of concrete, but also offer some degree of protection from carbonation attack to the concrete surface beneath. Unfortunately, almost every building experiences some degree of failure on the ceramic tile finishes. Most of this failure manifests itself as hollowing, cracks, discoloration, or disintegration of the finishing materials, or by the finishing material separating from the substrate. Thus, debonding is a common and an inevitable occurrence of tile finishes. To prevent or avoid the danger of tiles falling off, damage monitoring during the service period is very important. Its ability to detect weak signals makes the AE technique a strong candidate for this task.

To investigate the feasibility of the AE technique in monitoring the debonding process between the ceramic tiles and concrete, a push-off test was conducted. To prepare the push-off specimens, concrete cubes of size 150 × 150 × 150 mm were cast. The specimens were made by fixing ceramic tiles of 150 × 150 mm to one surface of the cubes with adhesive material. Two loading conditions were applied in this test, monotonic and cyclic loading. The specimens were cured in water at 20°C at room temperature for 26 days after tile fixing. One day before testing, the specimens were taken out of the curing chamber. An L-shaped steel plate was attached to the tile with epoxy resin. The steel plate was used to transfer the load to the tile–adhesive interface.

The test setup is shown in Figure 8-35. As can be seen from Figure 8-35, the specimen was tied to a flat rectangular plate that was connected to a servo hydraulic actuator of the testing machine, through two C-clamps. The entire specimen/fixture could move up with the actuator. A steel rod, connected to the load cell, made contact with the top surface of the L-shaped steel plate with the help of a ball socket joint. A restraining force was provided by resisting the upward movement of the loading fixture, pushed by the L-shaped steel plate, which in turn pushed the ceramic tile downward. Two LVDTs, which were fixed between the top surface of the specimen and the rigid wing of the steel rod, were used to measure the displacement of the top of the L-shaped steel plate relative to the surface of the concrete cube. To study the damage accumulation process of the interface during push-off testing, an AE measurement system was utilized, as shown in Figure 8-36. The system consisted of transducers, pre-amplifiers, and A/D modules. Six piezoelectric transducers were glued on the two opposite surfaces of the specimen to monitor the acoustic emission activities. These transducers had an essentially flat amplitude response, approximately in the frequency range from 0.1 – 1.2 MHz. The pre-amplifiers had a bandwidth of 0.02 – 1.2 MHz and a gain of 40 – 60 dB. In the system, channel 1 was used as a trigger channel and was connected to a LeCroy 6103 amplifier trigger module. The other five channels were used as working channels and were connected to five TR8828D digitizer modules. A LeCroy 6128 fan-out module was used to receive the signal from the LeCroy 6103 amplifier and to trigger all working channels simultaneously. A LeCroy 8901A interface board was used to communicate with a personal computer via a GPIB board. The operation of the LeCroy system was controlled by a Physical Acoustics Catalyst program, which also stored the digitized data on the computer's hard disk. Three thousand data points were stored for each channel at a digitizing rate of 16 MHz.

The electrical signal from each sensor was pre-amplified with a gain of 60 dB. A band-pass filter with a range from 30 to 500 kHz was used to eliminate undesirable low and high frequencies. A threshold of 46 dB was selected to ensure a high signal/noise ratio. A complete set of the acoustic signals obtained from the push-off test is plotted in Figure 8-37. It can be seen that the first arrival times are different for different channels. This provides, as mentioned earlier, the basis for the interpretation of the damaged or debonded areas. The 3D-source locations of AE events are deduced using the algorithm mentioned earlier.

Figure 8-38 shows the AE event occurrence under cyclic (a) and monotonic (b) loading. It can clearly be seen that prior to 30% of the peak load, there was no AE activity. From 30 – 70% of the peak load, AE activities increased slowly. After 70% of the peak load, the rate of the occurrence of the AE events shows an obvious increase. For the specimens tested under cyclic loading with 20 repeated cycles prior to 50% of the peak load, the Kaiser effect can be observed, as shown in Figure 8-39. Figure 8-39 shows the loading history with superposition of the AE events from the test. In the first cycle (*OABC*), the AE events started at point *A*, and

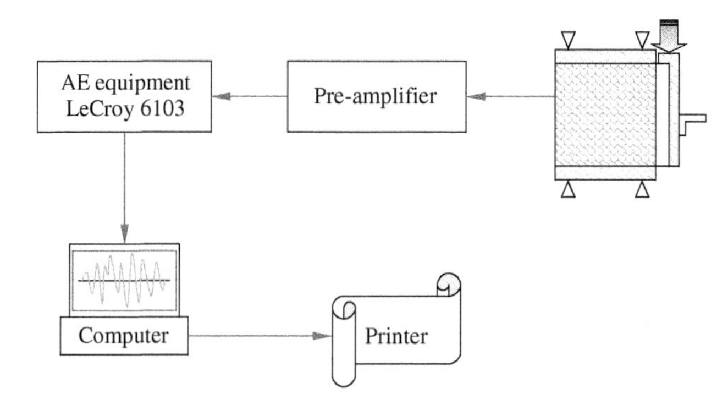


**Figure 8-36** The AE measurement system for debonding detection

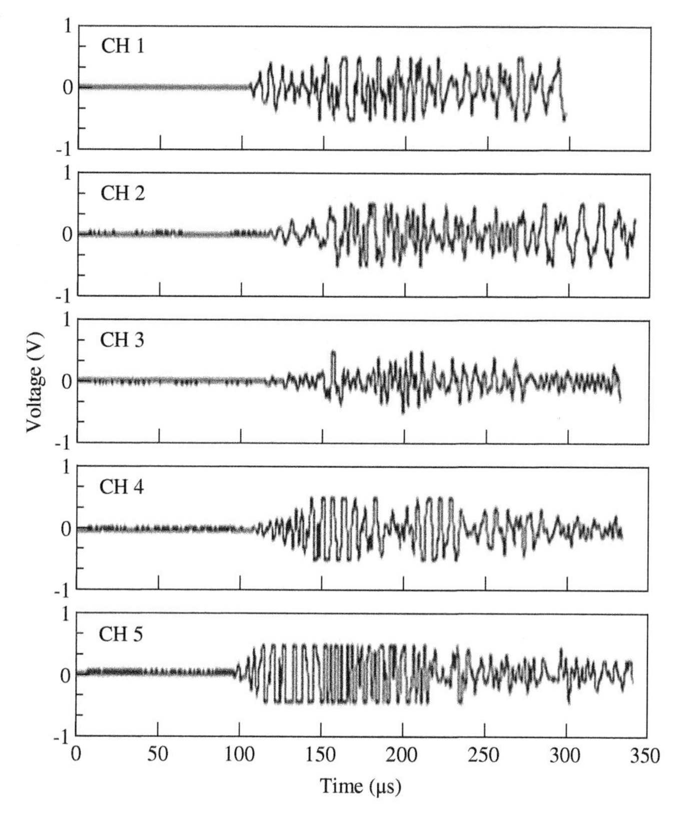


**Figure 8-37** A typical set of acoustic emission signals on ceramic tiles

ended at point $B$, at the maximum load for the cycle. No events occurred on the unloading part. In the second loading cycle, AE events did not occur until point $D$, close to the maximum load of the previous cycle. Similar phenomena were observed for the other 18 loading cycles.

The interpreted AE sources are plotted in Figure 8-40, where (a) and (b) show the source locations for AE events that occurred prior to 70% of peak load for the case of monotonic loading. We can see that the events were randomly distributed in the horizontal and depth directions of the specimen, while in vertical direction AE events were concentrated in the top of specimen, which means that the shear stress in this region was bigger and the main macrocrack may have started from the top of the specimen. The majority of AE events registered after 70% of the peak load showed a trend to localize on a narrow band in the specimen, as can be seen from Figure 8-40 (c) and (d). A clustering of event locations is concentrated around the final failure plane.

**(d)** *Further research*: Due to the limitations of current AE technology, further research is needed to improve its performance. From the theoretical prediction point of view, the inverse of AE source properties is one of the most challenging problems. Only with a well-established reverse theory, can the nature of microcracks in the process zone be clearly understood, the relations between AE characteristics and the microcrack formation, localization, and propagation be resolved, and the relationship between the unknown quantities and the measured AE signal wave-type parameters and waveforms be established. From the hardware point of view, a more advanced acquisition system with sufficient channels needs to be developed.

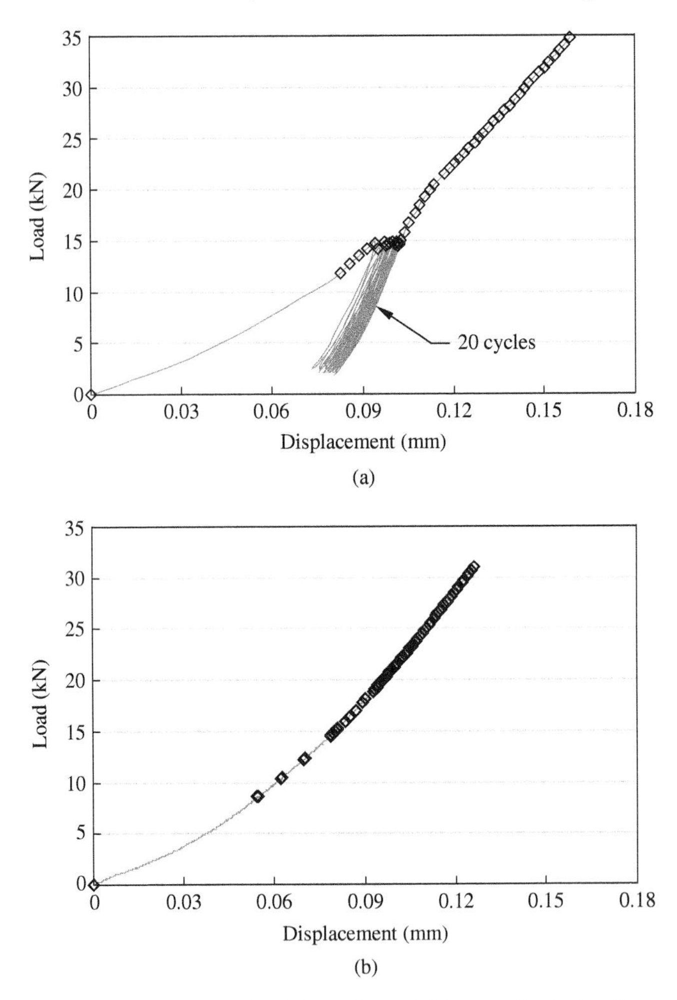


**Figure 8-38** Acoustic emission under (a) cyclic and (b) monotonic loading of push-off test

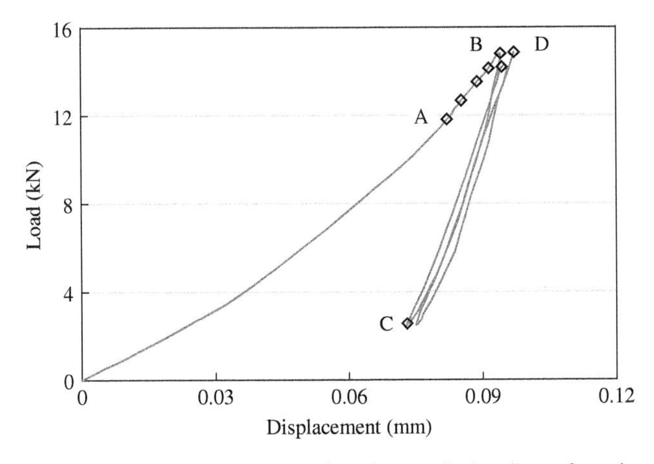


**Figure 8-39** Kaiser effect observed under cyclic loading of push-off test

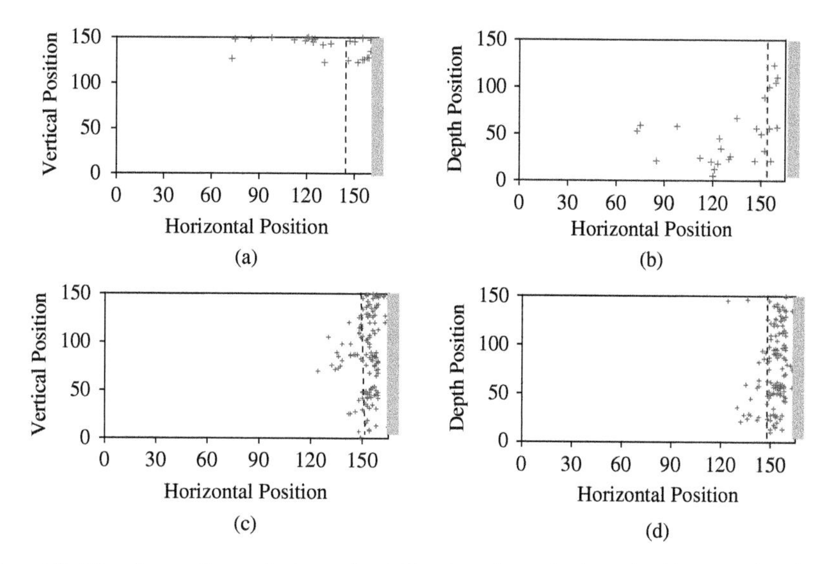


**Figure 8-40** Acoustic emission of relative locations before (a, b) and after (c, d) 70% of peak load under push-off test

The hardware development should absorb the advances in computer technology so the information can be obtained much faster and with real-time processing. From the software point of view, DSP techniques should be adopted for in situ displaying of the crack sources and propagation pattern, damage degree, and safety evaluation.

AE is well suited for laboratory experiments but can be very time-consuming and difficult to monitor on site. However, the AE technique can serve as a dynamic method of flaw detection and its use is actively growing (Brandes et al., 1998). A potentially important application of quantitative AE is to monitor bridges and buildings to assess the degree of damage and service life. To know the history and the present situation, AE might be a suitable tool, but powerful analysis software must be provided. This must also rely on the wave-form analysis and combined multidisciplinary NDT techniques, and new data fusion techniques are needed. Data fusion is the systematic integration of multisensor information (Gros, 1997) and can be defined as the overall understanding of a phenomenon.

### 8.5.3 Impact Echo

Hammers are used in the evaluation of piles (Steinbach and Vey, 1975; Brendenberg, 1980; Olson and Wright, 1990), and produce energetic impacts with long contact times, which are acceptable for testing long, slender structures, but are not suitable for detecting flaws within thin structures such as slabs and walls. Impact sources with shorter-duration impacts, such as small steel spheres and spring-loaded, spherically tipped impactors, have been used for detecting flaws within slab and wall structures (Carino et al., 1986; Sansalone and Carino, 1988). The use of small ball bearings as impact sources was regarded as one of several key breakthroughs in impact-echo (IE) research in the mid-1980s (Sansalone, 1997). Such a source produces a well-defined and mathematically simple input, which in turn generates waves with characteristics that facilitate signal interpretation. Ball bearing impactors are easy to use and the frequency content of the resulting stress waves can be tuned to fit the size of the structure and the sizes of flaws to be detected. The data acquisition system should have a sampling frequency of at least 500 kHz. The optimal sampling frequency depends on the thickness of the test object, but for testing relatively thin members, a high sampling rate is more effective. Nowadays, the impact-echo (IE) test equipment can be portable and has a wide range of applications in the inspection of concrete, both in laboratory and field conditions. The impact-echo technique is more precise than the rebound hammer method.

In these studies, the partial or complete discontinuities of piles, such as voids, abrupt changes in cross-section, weak concrete, and soil intrusions, and the location of these irregularities can be detected by the impact-echo technique. Carino and Sansalone initiated experimental and theoretical studies to develop an impact-echo method for testing structures other than piles (Carino et al., 1986; Sansalone and Carino, 1988). They used the impact-echo technique to detect interfaces and defects in concrete slab and wall structures, including cracks and voids in concrete, depths of surface-opening cracks, voids in prestressing tendon ducts, thicknesses of slabs and overlays, and delaminations in slabs. The impact-echo technique has also been used in layered-plate structures, including concrete pavements with asphalt overlays (Sansalone and Carino, 1988; Sansalone and Carino, 1990). Lin and Sansalone (1997; Lin et al., 1997) developed a new method for determining the P-wave velocity in concrete on the basis of the impact-echo technique.

#### 8.5.3.1 Principle

The impact-echo technique is an important and widely used mechanical wave NDT technique in civil engineering. Impact echo is actually a very old nondestructive testing method. In ancient times, people used this method to evaluate the quality of stone by striking the stone and listening to the ringing sound by ear. This is an active NDT technique. A hammer, a dropped ball, or a layered beam can be used to generate a mechanical acoustic vibration of the specimen and emit acoustic waves into the air. Defects and anomalies change the sounds. The sounds could be received by means of a microphone in air or a transducer on the specimen surface. IE works in the acoustic and low ultrasonic frequency bands. The received signals are analyzed mainly by the Fast Fourier Transform (FFT). The multireflection between the flaw surface and the top surface of the specimen forms a periodic vibration source to produce a peak in the frequency spectrum of the echo. The multireflection process can also occur inside the flaw, say, in the void, which gives out the specific peaks in the frequency spectrum. The geometry of the flaw, such as the depth $h$, can be deduced from the measured frequency peak $f$, provided the velocity $C$ of the ultrasonic waves in the specimen is known:

$$h = \frac{C_i}{2f_i} \tag{8-89}$$

where the subscript represents the wave type of the corresponding frequency peak. In most cases, the wave type is a longitudinal one. It is useful to detect cracks, delaminations, and voids beneath the surface of concrete.

The IE signal actually is the response of the specimen under the impulse loading. This is the main difference of IE from UT and also causes the difference in the wave-emitting technique. In IE, the sound is produced by mechanical impulse by the use of a hammer or in another similar manner; in UT the sound is usually produced by a PZT, an ultrasonic probe. Another difference between UT and IE is the frequency ranges they cover. The IE usually works in the acoustic frequency band, and the UT in the ultrasonic band. As the IE signals are the response of an object, they might carry the fingerprints of this object. By choosing proper advanced DSP methods, one might find out more information from IE testing. Compared to the NDT-CE-UT, the coupling problem of IE is trivial if the receiver uses a microphone, because it uses the air as a couplet. Thus, it is easy to assess surfaces that are rough and complicated in shape, which are found more frequently in NDT-CE. On the other hand, when the PZT is used, all types of waves, including longitudinal waves, transverse waves, and surfaces waves, can be received. As the first arrival wave must be the longitudinal one, the longitudinal wave is dominant.

An impact-echo test system normally consists of three components: an impact source, a receiving transducer, and a digital processing oscilloscope or wave-form analyzer, which is used to capture the transient output of the transducer, store the digitized waveforms, and perform signal analysis. The force-time history of an impact may be approximately a half-sine curve, and the duration of the impact is the contact time, which determines the frequency content of the stress pulse generated by the impact—the shorter the contact time, the broader the range of frequencies contained in the pulse. As the contact time decreases, the higher-frequency components can be generated and smaller defects can be detected. In using the impact-echo method to determine the locations of flaws within an object, tests are performed at regularly spaced points along "scan" lines marked on the surface. Examination of the amplitude spectra from these scans reveals the depth and approximate size of defects that may be present. However, interpretation of test data requires in-depth understanding of the technique (Mailvaganam, 1992). The selection of the impact source is a critical aspect of a successful impact-echo test system.

#### 8.5.3.2 Application

IE can be used to detect and locate flaws, fractures, voids, and laminations in ground decks, concrete pavements, and walls. The depth of the flaws can also be measured. The general non-homogeneous forms might be identified by the method suggested above. IE can also be used to interpret the dynamic modulus of a material.

The development of a standard test method for flaw detection using the impact-echo technique is difficult because of the many variables and conditions that may be encountered in field testing, in which the type of defects and shapes of structures are frequently met. ASTM C1383 (2004) has proposed a standard test method based on the use of the impact-echo method to measure the thickness of plate-like concrete members. This method includes two procedures. Procedure A is used to measure the P-wave velocity in concrete based on the travel time of the P-wave between two transducers at a known distance. In procedure B, the plate thickness is determined using the P-wave velocity measured in procedure A. The data analysis procedure in ASTM C1383 considers the systematic errors associated with the digital nature of the data on procedures A and B.

#### 8.5.3.3 Dynamic Modulus Measurement

The dynamic modulus of concrete can be measured by an impact-echo technique. Jin and Li have developed a new device to conduct such an experiment (Jin and Li, 2001). The device is schematically shown in Figure 8-41. The system involves a box made of sponge, an 8-mm steel sphere,

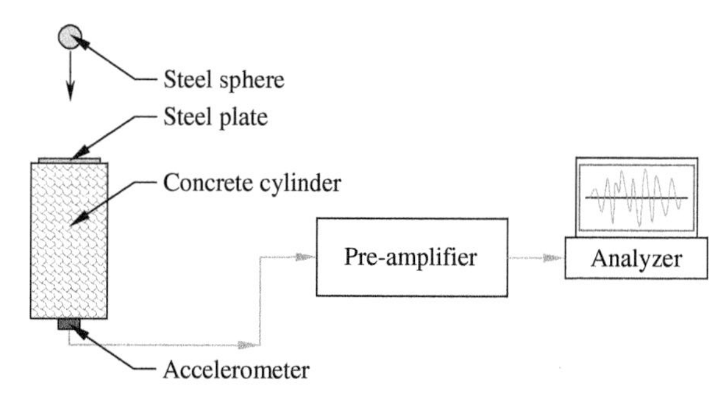


**Figure 8-41** Experimental setup for resonance measurements. Source: Du et al., 1992 / American Concrete Institute

an accelerometer, and a digitized oscilloscope with a Fast Fourier Transform function. To measure the dynamic modulus, a cylindrical specimen is put into the hole in the sponge box with the accelerometer fixed at the bottom of the specimen. Then the steel ball is dropped onto the top surface of the specimen. The vibration generated by such an impact will be detected by the accelerometer and recorded by the oscilloscope. The recorded wave form will be transferred into a frequency spectrum later, for dynamic modulus determination.

For concretes, the dynamic modulus measurement can be carried out on the same specimen at different ages to investigate the influence of age on the development of dynamic modulus. To interpret the dynamic modulus of concrete, the waveforms recorded by the oscilloscope have to be transformed into a frequency spectrum. Figure 8-42 shows a typical spectrum for a concrete specimen at different ages. It can be seen from Figure 8-42 that the first two resonance frequencies do shift from lower values to high values regularly with age all the time. Thus, the resonance frequencies do reflect the process and degree of hydration, setting and hardening.

After determining $f_1$ and $f_2$, the dynamic Poisson's ratio can be calculated using the method developed by Subramaniam et al. (1999). The formula used to calculate Poisson's ratio is

$$\nu = A_1\left(\frac{f_2}{f_1}\right)^2 + B_1\left(\frac{f_2}{f_1}\right) + C_1 \tag{8-90}$$

where $f_1$ and $f_2$ are the measured first and second longitudinal resonance frequency, respectively; and

$$A_1 = -8.6457\left(\frac{L}{D}\right)^2 + 24.443\left(\frac{L}{D}\right) - 12.478 \tag{8-91}$$

$$B_1 = 34.599\left(\frac{L}{D}\right)^2 - 101.72\left(\frac{L}{D}\right) + 56.172 \tag{8-92}$$

$$C_1 = -34.681\left(\frac{L}{D}\right)^2 + 105.98\left(\frac{L}{D}\right) - 62.731 \tag{8-93}$$

where $L/D$ is length to diameter ratio.

The calculated Poisson's ratio from the experiments is plotted as a function of age in Figure 8-43. Generally speaking, Poisson's ratio values show a very small variation within the range of 0.18 to 0.21 and can be regarded as constant. This is consistent with the findings of Boumiz et al. (1996), who observed that the Poisson's ratio of cement paste and mortar were more or less constant after curing for 12 hours.

After Poisson's ratio is obtained, the dynamic modulus can be calculated using its value and the longitudinal resonance frequencies (Subramaniam et al., 1999). The formula used to calculate the dynamic modulus is

$$E_d = 2(1+\nu)\rho\left(\frac{2\pi f_1 R_0}{f_n^1}\right)^2 \tag{8-94}$$

$$\text{where } f_n^1 = A_2(\nu)^2 + B_2(\nu) + C_2 \tag{8-95}$$

$$A_2 = -0.2792\left(\frac{L}{D}\right)^2 + 1.4585\left(\frac{L}{D}\right) - 2.1093 \tag{8-96}$$

$$B_2 = 0.0846\left(\frac{L}{D}\right)^2 - 0.5868\left(\frac{L}{D}\right) + 1.3791 \tag{8-97}$$

$$C_2 = 0.285\left(\frac{L}{D}\right)^2 - 1.7026\left(\frac{L}{D}\right) + 3.3769 \tag{8-98}$$

and $R_0$ is the radius of the solid cylinder and $\rho$ the mass density.

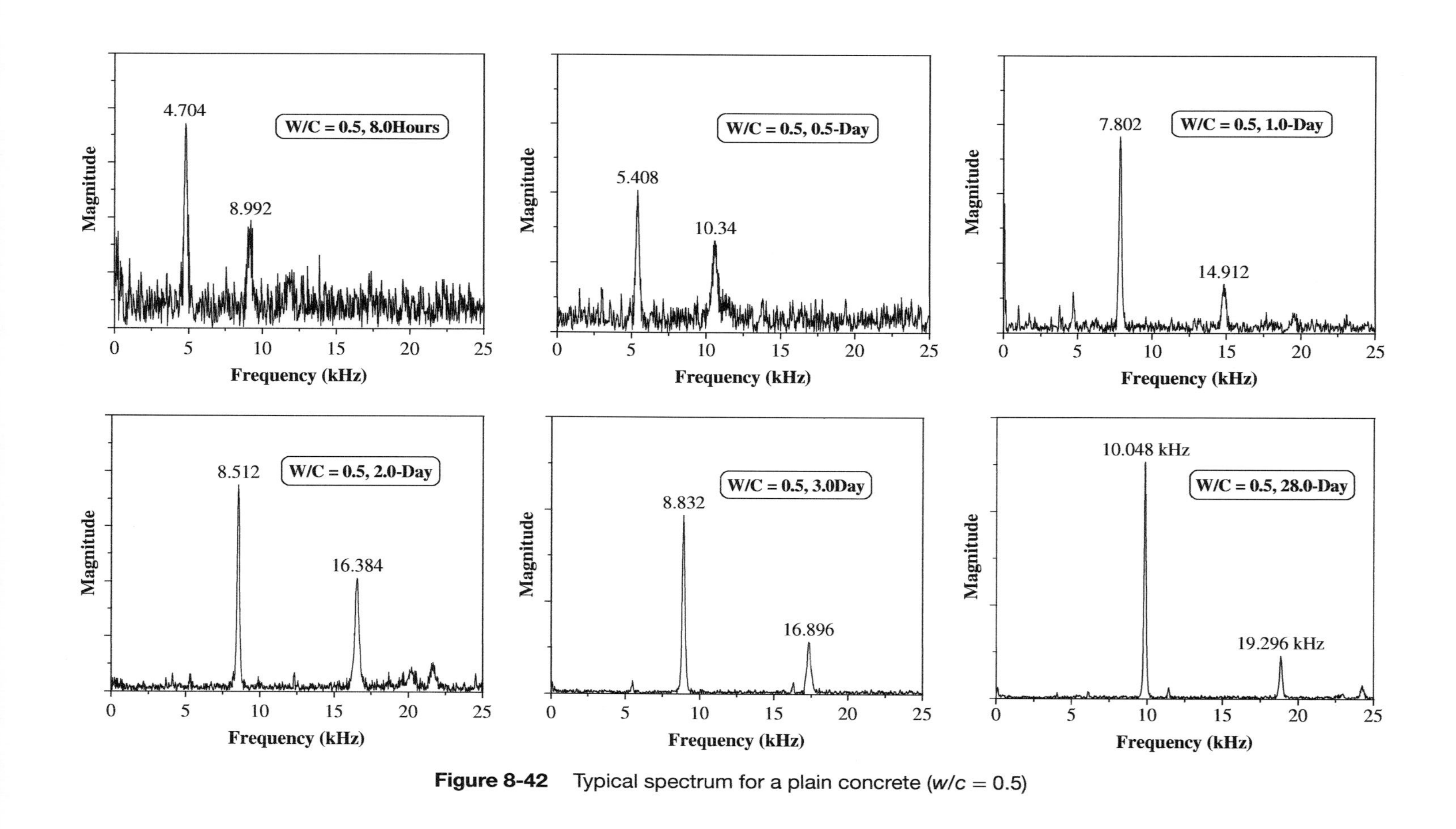


**Figure 8-42** Typical spectrum for a plain concrete ($w/c = 0.5$)

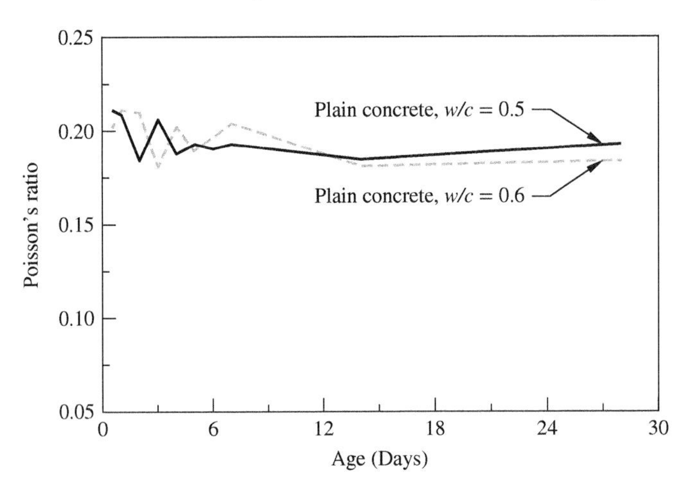


**Figure 8-43** Poisson's ratio versus age

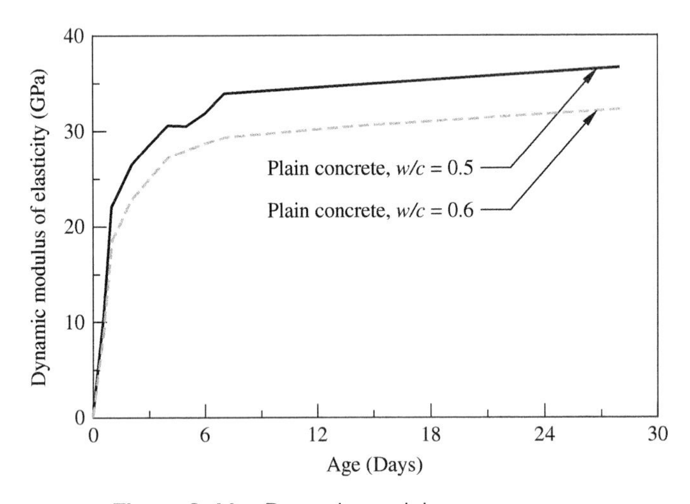


**Figure 8-44** Dynamic modulus versus ages

The calculated results for the dynamic modulus are plotted as a function of age in Figure 8-44. It can be seen from Figure 8-44 that the dynamic modulus increases rapidly, up to the age of 4 days, then the rate decreases and the curve becomes flat. The influence of the *w/c* ratio can also be clearly seen from Figure 8-44, and a higher *w/c* ratio leads to a lower dynamic modulus at all ages.

### 8.5.3.4 Surface Cracking Measurement by Impact Echo

To detect the finer crack on a concrete surface that cannot easily be observed by the naked eye, the impact echo technique can be used. The impact-echo setup for surface crack measurement is shown in Figure 8-45. As can be seen from Figure 8-45, it consists of one impact source and two receivers. When the surface wave is generated by an impact source, usually a steel ball, the wave propagates along the surface of the material or specimen. The stress wave can be picked up by two transducers. If there is no crack between the two transducers, the responses of the two transducers are almost identical. However, if there is a crack between the two transducers, the response of the second transducer will show a significant difference from the first one; hence, the existence

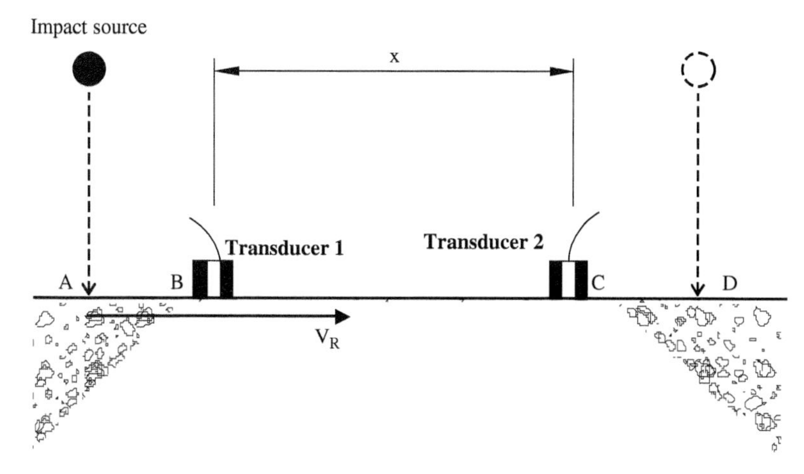


**Figure 8-45** Surface wave velocity measurement for concrete strength assessment

of a crack can be determined. The detection of a surface crack using the impact-echo method usually uses *self-calibration*, developed by Achenbach et al. (1992). The self-calibration method can cancel uncertain parameters of the source characteristics term and the receiver response term, by using the FFT form of the captured time domain signal, $V_{AB}$, $V_{AC}$, $V_{DB}$, and $V_{DC}$. In this, $V_{AB}$ and $V_{AC}$ represent a stress wave signal, generated by an impact (wave source) at location A and received by the near accelerometer at locations B and C, in the frequency domain, respectively. $V_{DC}$ and $V_{DB}$ represent stress wave signals, generated by an impact (wave source) at location D and received by the accelerometer at locations C and B, in the frequency domain, respectively.

The signal transmission function between locations B and C, $d_{BC}$, can be obtained using the following equation:

$$d_{BC} = \sqrt{\frac{V_{AC}V_{DB}}{V_{AB}V_{DC}}} \tag{8-99}$$

where $d_{BC}$ is a function of frequency and can be visualized as the ratio of the amplitude of the signal from the far accelerometer to that of the near accelerometer. Thus, a transmission value of unity indicates no amplitude loss (completely transmission) as the wave propagates from B to C, whereas a value of zero indicates complete signal amplitude loss (no transmission). In the case of a point source, $d_{BC}$ values should be less than 1, even for perfect transmission materials due to some loss from attenuation.

Usually, an impact event will generate waves propagated in all directions. Some wave components propagate along the surface while others reflect at free boundaries. By setting an appropriate window, the surface wave or the first P-wave reflection from the opposing side of the specimen can be distinguished. The $d_{BC}$ measurement can be used to detect a surface crack in concrete. Clearly, if there is a crack between two transducers, the $d_{BC}$ values will suffer a severe reduction in almost all the frequency ranges.

### 8.5.4 Penetrative Radar Technique

The penetrative radar technique was developed based on the theory of electromagnetic waves. When electromagnetic waves propagate in a medium, the wave velocity $v$ is

$$v = \frac{c}{n} \tag{8-100}$$

where $c$ is the velocity of light, and $n$ the index of refraction. For a homogeneous medium,

$$n = \sqrt{\varepsilon\mu} \tag{8-101}$$

where $\varepsilon$ and $\mu$ are the relative permittivity and relative magnetic permeability of the material, respectively. The relative magnetic permeability almost equals unity for most materials. For a mixture consisting of $j$ constituents with volume $V_j$, the relative permittivity can be regarded as the linear combination of its constituents,

$$\varepsilon = \sum_{i=1}^{j} V_i \varepsilon_i V \tag{8-102}$$

where $V$ is the total volume of the mixture. Hence, one could deduce the volume component from the measured velocity provide the materials of the constituents are known.

From the point of view of concrete quality control, water content is one of the most important parameters. Fortunately, the relative permittivity of water is about 81, whereas that of cement paste and normal aggregates is about 5. Hence, the measurement is very sensitive to the water content. The accuracy is about 3–5% for a normal concrete mixture. Moreover, since the relative permittivity of air is 1, it is feasible to use EM wave propagation to detect the porosity for dry concrete. The time of flight can be used to measure the thickness by the formula

$$\text{Thickness} = \frac{150t}{\sqrt{\varepsilon}} \tag{8-103}$$

The scattered wave is usually called an echo, and its amplitude is related to several material parameters. By means of some inverse procedures, one can deduce the characteristic of a layered structure (Attoh-Okine, 1995). By scanning, the echo information, amplitudes, phases, TOF echo data can be used to form an image, or can be used as raw data for tomography.

As concrete is a composite mixture, the scattered wave from the aggregate will form a cluster. The scattering at an interface is strongly dependent on the difference of the complex permittivity of the two materials. As the permittivity of a normal aggregate and a cement matrix is similar, the scattered EM waves from the interface are usually much weaker than ultrasonic waves. This is another advantage of the EM wave to detect water or air defects in concrete that have very different permittivities. For the same reason, the test results are also not sensitive to the aggregate size.

One should be aware of the limitation by the nature of the physical laws. For example, the power reflection coefficient $R$ from air to a board made of lossless material with a perpendicular incidence is

$$R = 4r^2\sin^2(\beta d) \Big/ \left[\left(1 - r^2\right)^2 + 4r^2\sin^2(\beta d)\right] \tag{8-104}$$

where $\beta = 2\pi/\lambda$, $\lambda$ is the wavelength in the medium, $d$ the thickness of the board, and $r$ the amplitude reflection coefficient, which can be expressed as

$$r = (1 - n) / (1 + n) \tag{8-105}$$

in which $n$ is the relative refractive index. Hence, the penetrating ability of a microwave strongly depends on the frequency when the thickness of a plate is fixed. When the thickness is equal to the integer times of the wavelength, then $R = 0$, which means that the technique cannot realize the existence of the board by reflection.

In the case of testing involving waves, the wavelength is the most critical parameter. The resolution and sensitivity of the measurements are all related to the wavelength used. The lateral resolution is mainly limited by the wavelength. The depth resolution depends mostly on the width of the wave pocket. Electromagnetic waves totally reflect from a metal sheet even if the thickness of the sheet is only a few tenths of a millimeter. Hence, it is hardly possible to detect the parts inside or behind the metal components.

Ground penetrative radar (GPR) is a very powerful EMT instrument. A basic radar system consists of a control unit, antennas (one is used for transmitting and one receiving pulses), an oscillographic recorder, and a power converter for DC operation. In the inspection of concrete, it is desirable to use a radar antenna with relatively high resolution so that the small layer of a concrete member can be detected. Radar can be employed in the rapid investigation of concrete structures, such as for measuring the thickness of structural members, for determining the spacing and depth of reinforcement, and for detecting the position and extent of voids and other types of defects in bare or overlaid reinforced concrete decks. In the testing of concrete, normally short-pulse radar is used, which is the electromagnetic analog of sonic and ultrasonic pulse-echo methods. In this method, the equipment generates electromagnetic pulses, which are transmitted to the member under investigation by an antenna close to its surface. The pulse travels through the member and its propagation velocity is determined by the electrical permittivity. The relative permittivity of the concrete is determined predominantly by the moisture content of the concrete. Typical relative permittivity values for concrete range between 5 (oven-dry concrete) and 12 (wet concrete) (Bungey and Millard, 1996).

The radar technique uses the EM wave as an enquiring agent to get information on concrete materials or structures, such as the thickness and defects inside. A GPR can image the bridge pile profile in a bridge scour investigation. A maximum frequency should be recommended in a GPR survey design for clutter reduction. It is suggested that the wavelength $\lambda$ be ten times larger than the characteristic dimension for geological materials (Davidson et al., 1995). For most cases, the frequency range is from 90 MHz to 1 GHz and the pulse repetition frequency is 25 Hz. The following typical results can be achieved: a scour depth of 1.2 m relative to the general riverbed; a filling consisting of a sandy layer 0.5 m deep with overlying gravel 0.8 m thick at the center of the hole; and a water depth of about 1 m (Davidson et al., 1995). The GPR measurement is fast enough to be suitable to detect a large area like a 10,000 $m^2$ pavement at an airport (Weil, 1995).

#### 8.5.4.1 Application

Microwave NDT techniques can be used to perform building inspection and quality control and to detect steel objects as well as moisture, water content, locating and cracks. It is well suited for locating the lateral position of rebars, metallic ducts, metallic T-girders, and metallic anchors in prefabricated three-layer concrete elements; locating plastic and metallic leads, and measuring the cover thickness and moisture content of concrete. Artificial birefringence properties can be found for ceramics and cement with microwaves, of which the wavelength is between 7.5 and 12.5 mm. This is similar to that of photoelasticity with visible light (Dannheim et al., 1995). Hence, visualizing inside the specimen is possible with microwaves.

EMT techniques can make a reliable assessment of integrity. It can locate tendons and metal ducts (diameter 85 mm) and offer information on geometrical dimensions. It can detect concrete member defects, voids in tendon ducts, compression faults or honeycombing (voids from 80 to 160 mm) in concrete. However, using the EMT technique to measure concrete is not sensitive to aggregate size. For example, for aggregate size of 8 to 32 mm, and cover thickness 340 $\pm$10 mm, the lateral position for reinforcement could be within an accuracy of $\pm$10 mm (Attoh-Okine, 1995). A microwave moisture content meter offers online water content monitoring in clay (Kalinski, 1995).

#### 8.5.4.2 Digital Image Processing (DIP) and Modeling

As the microwave frequency is usually very high, the AD device for the microwave signal itself is either very expensive or unavailable with the present technique. The DSP in NDE-CE-EMT is different from other NDT-CE techniques. In NDE-CE-EMT, the AD converter acts on the radar image, which corresponds to the B-scan plot in NDT-CE-UT. Then the general digital image processing technique can be used normally. In this case, the DSP routines, e.g., artificial neural networks, frequency analysis, and digital filtering, can be used to effectively draw the information from the image.

Interpreting the radar image is not so easy sometimes. If the simulation technique is used together with the analysis of the radar response image, the void as well as the shape of the void can be discovered, while it is very difficult to discover the shape of a void only from direct test results.

### 8.5.5 Optical Techniques

These kinds of techniques use light waves as the inquiring agent. They are also active methods. The main techniques are based on interferometry and energy transportation. A representative technique of the former is electronic speckle pattern interferometry (Hung, 1982; Jones and Wykes, 1989), and of the latter is infrared thermography (Milne and Carter, 1988; Favro et al., 1991).

#### 8.5.5.1 Electronic Speckle Pattern Interferometry

Electronic speckle pattern interferometry (ESPI) is a powerful laser-based measurement technique (Butter and Leendertz, 1971a, 1971b; Hung and Liang, 1979). The essential principle is the same as holographic interferometry (HI). ESPI realizes HI by means of a video technique instead of by photography. The reflected light is superposed to a reference laser beam. The resultant interference pattern is recorded by a CCD camera. The interference patterns before and after loading are recorded and the resultant fringe pattern reflects the displacement fields of the illuminated surface. A phase-shift technique resolution of the order of 20 nm can be reached. ESPI is a noncontacting, full-field technique to measure and monitor the 3D-displacement field of an optical diffusely scattering object (Jones and Wykes, 1989). The surface area to be monitored can be more than $70 \times 80\,\text{cm}^2$, depending on the geometrical arrangement. Because ESPI needs a reference beam, isolation of the vibration ion from the environment is important, but difficult to realize.

Speckle pattern shearing interferometry (SPSI), also called shearography (Hung, 1982) is an advanced technique stemming from ESPI. In contrast to HI or ESPI, SPSI measures the derivatives of the surface displacement directory. It does not respond to the rigid body motions. By a special set of optical elements, the light beam scattered from the object surface splits into two beams separated by a small distance in the lateral direction, which is what the shearing really means, which interferes with each other. As the two beams come from the same source, this method is very stable against environmental vibrations. The advantage of this method is that it measures the surface strain directly instead of displacements. Hence, SPSI can work in larger displacement cases than ESPI.

Both ESPI and SPSI can be used as a real-time micro-crack monitor. By means of the phase-shifting technique, quantitative evaluation of fringes is possible.

#### 8.5.5.2 Infrared Thermography

Sunlight is an electromagnetic wave. It travels at $3 \times 10^8$ m/sec in a vacuum. Sunlight can be visible or invisible according to the wavelength. Table 8.1 shows the range of the wavelength of visible light. If wavelength $\lambda < 0.4\,\mu\text{m}$, it is called ultraviolet; if the wavelength falls between 700 and 1500 nm ($0.7\,\mu\text{m} < \lambda < 1.5\,\mu\text{m}$), it is called near-infrared. If the wavelength falls between 1,500 nm to 20,000 nm ($1.5\,\mu\text{m} < \lambda < 20\,\mu\text{m}$), it is called mid-infrared; and if the wavelength falls between 20,000 and 100,000 nm ($20\,\mu\text{m} < \lambda < 100\,\mu\text{m}$), it is called far infrared.

As mentioned above, infrared is a waveband in the electromagnetic spectrum that lies just beyond red and is invisible to the naked eye. According to the physical laws, all matter with a

**Table 8-1** Wavelength ranges of visible light

| Wavelength (nm) | Color |
|---|---|
| 400~450 | Violet |
| 450~480 | Blue |
| 480~510 | Blue-green |
| 510~550 | Green |
| 550~570 | Yellow-green |
| 570~590 | Yellow |
| 590~630 | Orange |
| 630~700 | Red |

temperature above absolute zero, i.e., −273°C, radiates heat. Heat radiation is synonymous with infrared radiation, or we can say that all objects emit infrared radiation. The intensity of the emitted radiation depends on two factors: the temperature of the object and the ability of the object to radiate or its emissivity. The amount of infrared radiation of an object increases with the rise of temperature. When a material body is conducting heat flow, the effects of the material thermal properties and internal structure means that the heat distribution within the body can be the presented in the surface temperature field, and the infrared radiance emitted from the body can be related to its surface temperature. Based on this principle, infrared thermography can be used to inspect the surface temperature field of materials, and further to interpret its internal structural states.

According to Planck's law, the thermal radiation power of a blackbody per unit of area, unit of solid angle, and unit of frequency, $f$, can be written as

$$u(f,T) = \frac{2hf^3}{c^2} \cdot \frac{1}{e^{hf/kT} - 1} \tag{8-106}$$

where $f$ is the frequency, $T$ the temperature on the Kelvin scale, $h$ Planck's constant ($h = 6.626 \times 10^{-34}$ J/sec), $c$ the speed of the light ($c = 299{,}792{,}458$ m/sec), and $k$ Boltzmann's constant ($k = 1.380 \times 10^{-23}$ J/K). This formula mathematically follows from calculation of the spectral energy distribution in a quantized electromagnetic field that is in complete thermal equilibrium with the radiating object.

By integrating the above equation over the power output, Stefan's law is obtained:

$$W = \sigma \cdot A \cdot T^4 \tag{8-107}$$

where $\sigma$ is the Stefan-Boltzmann constant with a value of $5.670 \times 10^{-8}$ Wm$^{-2}$K$^{-2}$. The wavelength $\lambda$, for which the emission intensity is highest, can be obtained by Wien's law:

$$\lambda_{max} = \frac{b}{T} \tag{8-108}$$

where $b$ is Wien's displacement constant, of value $2.897 \times 10^{-3}$ mK.

For surfaces that are not blackbodies, one has to consider the (generally frequency-dependent) emissivity correction factor $\varepsilon(f)$. This correction factor has to be multiplied with the radiation spectrum formula before integration. The resulting formula for the power output can be written in a way that contains a temperature-dependent correction factor, which is (somewhat confusingly) often called $\varepsilon$ as well:

$$W = \varepsilon \cdot \sigma \cdot A \cdot T^4 \tag{8-109}$$

When an analysis of the surface temperature field is required, the surface temperature profile of an object will have to be recorded and an infrared camera is required for scanning and recording. An infrared camera is a radiometer that measures the radiated electromagnetic energy. Once the parameters are defined, the radiometric values registered by the camera can be converted into temperature values. The fundamental equation of thermography is introduced as follows. The radiance $N_{\mathrm{CAM}}$ received by the camera is expressed by

$$N_{\mathrm{CAM}} = \varepsilon N_{\mathrm{obj}} + (1 - \varepsilon) N_{\mathrm{env}} \tag{8-110}$$

where $\varepsilon$ is the object emissivity, $N_{\mathrm{obj}}$ the radiance from the surface of the object, and $N_{\mathrm{env}}$ the radiance of the surrounding environment. If an object has a very high emissivity coefficient, e.g., $\varepsilon > 0.9$, we can roughly write

$$N_{\mathrm{obj}} \approx N_{\mathrm{CAM}} \tag{8-111}$$

In this case, a direct relationship can be derived between the radiometric signal and the temperature of the object using Equation 8-111. However, in many cases, $\varepsilon$ is not big enough and a correction is needed to accurately interpret the properties of the object under inspection. One such correction method is introduced in a later section.

A system used for thermographic inspection and based on the principles presented above must be composed of many subsystems, such as a scanner, an infrared camera, an image acquisition and analysis system, a portable display monitor, and a thermal stimulation system. The mechanism of a thermographic inspection system is described as follows: the scanner contains a gas-cooled multi-element detector. The gas cooling is used as a reference and to eliminate electronic noise. The camera converts electromagnetic energy into electronic video signals. These signals are then amplified and transmitted to a high-resolution portable display monitor. Parts of the scanning system are very sensitive and can identify temperature differences as small as 0.1°C or less. A range of lenses can be used so that surfaces can be inspected from as close as 1 meter or up to several hundred meters from the structure. The portable display monitor portrays the image in monochrome or color, and, generally, the hotter the area, the lighter the image, and the cooler the area, the darker the image. This portrayed image can be video-recorded, in real time, and the system provides an instant thermal picture. The displayed image can be stored in a computer, where further analysis of the captured data can be carried out. The complete thermographic inspection system is illustrated in Figure 8-46.

There are two different ways in which infrared equipment can be deployed for thermography nondestructive evaluation, depending on the specific application envisaged. In general, a thermograph can be deployed either in a passive or an active fashion. In the passive configuration, the

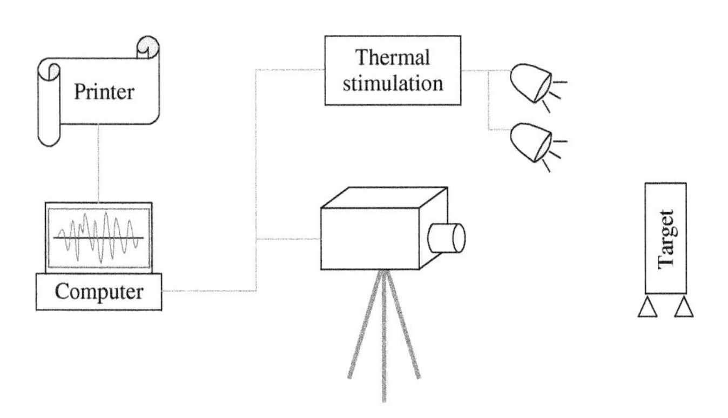


**Figure 8-46** Typical active infrared NDT experimental setup

infrared camera is pointed at the scene, focusing on the object to be inspected, while no external thermal perturbation is applied. Applications such as building inspection make use of this mode of operation, where the surface temperature distribution, as is, contains relevant information concerning the possible presence or absence of any defects. In the active configuration, an externally applied thermal stimulation is needed to generate meaningful contrasts that will yield the detection of subsurface abnormalities. In general, the heat will be conducted away by the mass of the structure. If there are defect areas, the heat will not dissipate there at the same rate as in the sound areas. With this phenomenon, information can be drawn from the inspection, since the surface temperature distribution of a certain object is related to the subsurface structure.

The first attempt to detect anomalies and air leaks in buildings using IR techniques was reported by an AGA salesman in 1965. From 1975 to 1985, there was rapid progress and the potential of nondestructive evaluation of materials by infrared thermograph was increasingly exploited, especially in the 1980s with the availability of commercial infrared cameras whose video signals were compatible with black-and-white television standards. Results have been reported at different international conferences in the United States (SPIE: thermo-sensing) and in Canada (symposium on remote sensing). Most research work on infrared thermography for civil engineering applications has been carried out during the last 10 years. This method has now gained widespread popularity because it is a nondestructive technique that is quick, safe, and easy to implement.

Under certain ambient conditions, a building with surface defects in the form of debonding mosaic tiles or delaminating render emits differing amounts of infrared radiation. Using a suitably modified infrared camera, the surface of a building can be scanned rapidly and areas that are defective will show up as a result of the differential transmittance of infrared radiation. A surface with an even texture, free of cracks and delaminations, will appear quite uniform when viewed by the infrared camera. If there is a defect, the ambient temperature conditions and solar gain will make the surface heat up and cool down at different rates. Different buildings can be situated in different locations that may be surrounded by different complex environmental conditions, which can lead to abnormal distributions of surface temperature fields. Some subsurface defects may be filled with air; some may be partially filled with moisture. In the sunshine or on a cloudy day, these defects present different behaviors, which could impose some potential difficulties for interpreting the results obtained from an infrared thermograph.

Consider a thin tile, in which one surface is in contact with air and is exposed to the sun. The other surface is in contact with the wall. The equation for the temperature of an element of the tile is

$$B\frac{dT}{dt} = H - K\left(T - T_0\right) \tag{8-112}$$

where $B$ is the heat capacitance per unit area of the tile, $H$ the heat flow onto the tile/air surface due to the sunlight, and $K$ the thermal conductivity between the tile and the wall. In the most simplified model, $T_0$ is the wall temperature, which remains constant because of its large thermal capacitance. In a more sophisticated model, the temperature of the thick wall varies across the thickness section of the wall, which is in contact with tile on one surface and with the room air on the other. The temperature variation of the wall is such that its surface on the tile side will follow the change of the tile temperature, while the temperature of its indoor surface remains constant. Since the heat flow is determined by the temperature gradient, an equivalent temperature $T_0$ can be found such that $K(T - T_0)$ is equal to the actual heat flow from the tiles to the wall.

The values of $B$ and $K$ may vary across the plane of the tile due to the variation of the tile/wall bound conditions, thus causing a temperature variation, which can be detected by a thermal camera. Here, the lateral thermal conduction, the thermal conduction of the surface/air, is ignored. These factors can be ignored only within a short period of time when the tile temperature is varying. Over a period (several hours), the thermal conduction to air and the lateral conduction will result in a uniform temperature distribution all over the tiles, and the defects behind the tile will not be easily detected. Therefore, it is important that the thermograph measurements should be carried out within the transient period.

**(a)** *Heating process*: Supposed a tile is at $T_0$ at $t = 0$. This would be the condition in the morning before sunrise, where $H = 0$. A heat flow is quickly imposed on the tile, and could be due to the sunrise, building shadow movement, or the clearing up of clouds. The time-dependent temperature is then followed, by solving the equation below with the proper initial conditions:

$$T = T_0 + \frac{H}{K}\left(1 - e^{-K_t/B}\right) \tag{8-113}$$

It is seen that the time constant in the exponential expression depends on $K/B$, and the steady state is $T_0 + H/K$. In practice, within $t = 5B/K$, $T(t)$ will reach 99% of its steady value. This steady value, which is actually only quasi-steady because the lateral thermal conduction and air conduction are not considered, will gradually be taken over by these much slower processes. A quasi-steady-state thermograph should therefore be taken at times after $5B/K$ but before the lateral conduction takes over.

A debonding defect is a hollow space between the tile and wall. The hollow space could be filled with water, caused by rain or moisture condensation due to the high humidity in Hong Kong, or caused simply by air. However, the two filling media, air and water, have different effects on the thermal properties, $B$ and $K$, of the tiles. An air pocket (void) with its poor thermal conductivity will have a reduced $K$ but little influence on $B$. The surface temperature next to an air hole will then be higher than in intact areas during both the transient and quasi-steady-state stages. A water-filled cavity will probably have a larger $K$ because of the improved thermal contact, even though water probably has about the same heat conductivity as cement, and will increase the thermal capacitance of the tile if the water is considered as part of the tile. Its surface temperature will therefore be lower than an intact tile. Laboratory tests (see next section) are in complete agreement with the above predictions.

**(b)** *Cooling process*: For a cooling process, the initial temperature is then $T_0 + H/K$ where the tiles are still in a quasi-steady state. The solution to the equation is then

$$T = T_0 - \frac{H}{K}e^{-K_t/B} \tag{8-114}$$

Due to lower thermal conductivity, an air-filled defect will cause a faster decrease, or a lower local temperature, while a water-filled hole will have a higher temperature.

**(c)** *Laboratory calibration*: To establish a link between the abnormal isotherms recorded on the inspected target surface and the presence of a subsurface defect, it is necessary to establish a baseline for comparison by calibrating the known defects in the laboratory. The basic principle of (transient) infrared thermograph consists of flashing heat to a specimen and subsequently recording the IR images (also called thermograms). Subsurface defects having different thermal properties with respect to the bulk of material will have an effect on the diffusion of the thermal front inside the material.

Two procedures for calibration are used: (i) reflection, where both the thermal source and the IR camera are located on the same side of the tile wall; and (ii) transmission, where the thermal source and the IR camera are located on opposite sides of the tile wall.

*Case 1. Debonded area filled with water*: Filling the debonded area with water would lead to a lower surface temperature during heat inflow and a higher temperature during heat outflowing. An example of such an analysis was calibrated in the laboratory, as shown in Figure 8-47.

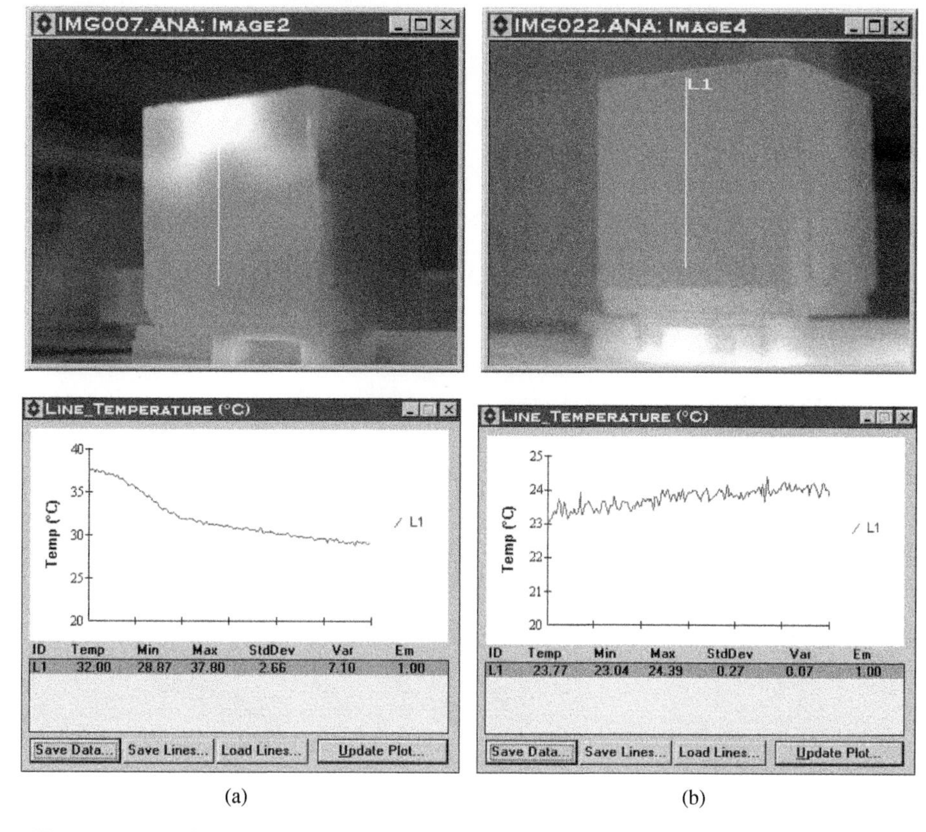


(a) (b)

**Figure 8-47** Surface temperature distribution of a debonded tile sample with upper half dry and lower half filled with water: (a) uniform heating of the inspected face, and (b) after cooling for half an hour

Before testing, the upper half of the tile sample was kept dry and the lower half was soaked with water, and an external thermal source was applied to one side of the tile sample, with the isotherm pattern on the same side being recorded. It was noticed that the surface temperature of the bottom half (wet) was lower than that of top portion of the tile, see Figure 8.47a. Once the thermal source was withdrawn, the tile sample would start to cool off. After half an hour, it was noted that the surface temperature of the bottom half (wet) in the IR image was higher, see Figure 8.47b. This was due to the fact that the moist areas retained more heat energy than the sound dry areas because of the higher thermal capacity of water.

*Case 2. Debonded area filled with air*: Another common phenomenon is the filling of the debonded area with air, which will lead to a higher surface temperature during heat flowing in and a lower surface temperature during heat outflow. Figure 8-48 shows thermograms obtained on a tile sample with air defects under an external thermal stimulation. Figure 8-48a shows the areas of the bonded and debonded tile samples. Under an external heat stimulation, the thermal image presented different temperature distributions, as shown in Figure 8-48b. The high-temperature region was in the debonded area and the line profile also shows that the higher temperature was on the front surface of the debonded area, see Figure 8-48c.

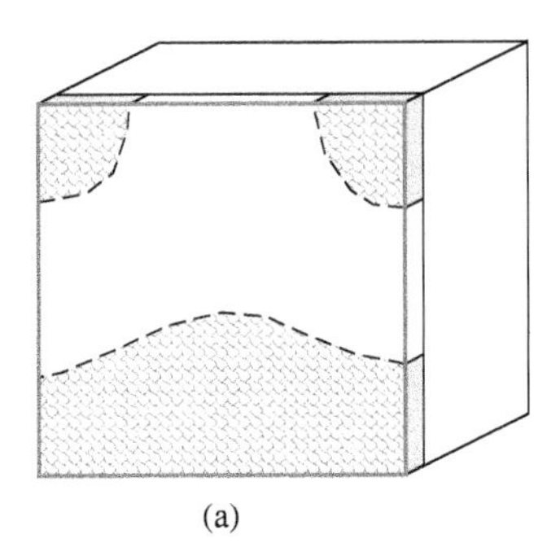

(a)

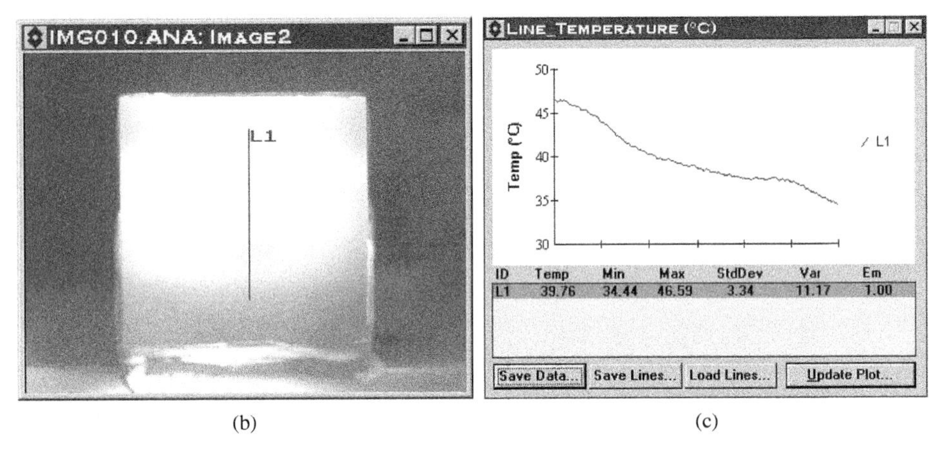


(b) (c)

**Figure 8-48** Tile with debonded area filled with air

### 8.5.5.3 Buildings Inspection

As described in the previous section, a favorable condition for field inspection is a steep thermal gradient between the environment and the target object. This usually occurs in the morning after sunrise and in the afternoon before sunset, or due to the slow movement of shadows or clouds. The steep gradient stimulates the transient behavior of the tiles. When the finishing material has a perfect bond with the substrate, the incidence of heat on the surface is conducted away by the mass of the substrate (concrete wall). However, when there are debonding areas filled with moisture in the finishing materials, their thermal properties will be different. Figure 8-49 shows a corresponding thermogram of such a case. The survey was performed in the morning during the sunrise. Before the survey was conducted, it had been rainy for several days and it was very likely that condensed moisture or water was trapped in the defect areas (indexed by arrows). Due to water having infiltrated the porous material, thereby increasing its specific heat, the surface temperature of the defect area was lower than the sound areas.

For the cases of air-filled debonding, the thermograms behave differently. When the sun shines on the surface, it begins to heat the tile-clad wall, and the surface over a debonding or potential void filled with air will heat up faster, because the fracture plane of the defect acts as a small insulator, trapping the heat in the tile. These "hot spots" on the surface are generally 2–3°C higher than the surrounding areas and can be easily detected by an IR camera. Figure 8-50 identifies this type of pattern. The higher-temperature regions in Figure 8-50 are suspected to be severely debonded areas on the walls, which is later confirmed by the tapping technique.

**Figure 8-49** The lower-temperature areas (indexed by upward arrows) indicating the defect areas filled with water or moisture

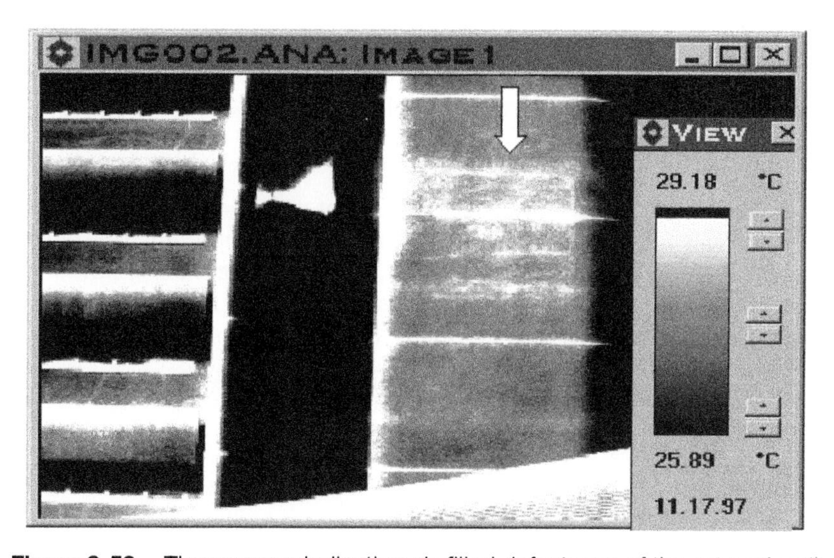


**Figure 8-50** Thermogram indicating air-filled defect area of the external wall

Moreover, even fairly small defects can be detected by the thermography technique. In Figure 8-51, the smallest "hot spot" was only about 50 × 50 mm in size. Debonded areas like this are generally held in place by the interlocking of the surrounding materials. However, any small defect has a potential to propagate into a large detachment. Hence, it should be marked for future surveys to monitor its size stability.

**(a)** *Reflection correction*: As mentioned in the earlier sections, an IR camera in front of an object detects not only the emitted radiance but also a part of the radiance due to reflection of the ambient fluxes by the object surface, whose reflectivity is nonzero. It is necessary to take into account this phenomenon for objects with reflective surfaces, such as ceramic tiles, when

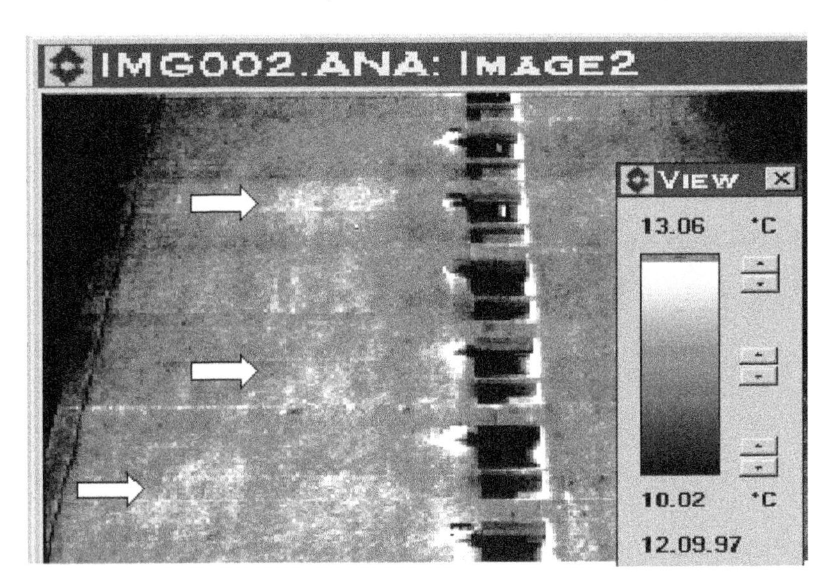


**Figure 8-51** Thermogram indicating several small air-filled defect areas on the external wall of building

interpreting the survey results. As we know, in the fundamental thermograph equation, the radiance $N_{\mathrm{CAM}}$ received by the IR camera is expressed by

$$N_{\mathrm{CAM}} = \tau_{\mathrm{atm}} \varepsilon N_{\mathrm{obj}} + \tau_{\mathrm{atm}} \rho N_{\mathrm{env}} + \left(1 - \tau_{\mathrm{atm}}\right) N_{\mathrm{atm}} \tag{8-115}$$

where $\tau_{\mathrm{atm}}$ is the transmission coefficient of the atmosphere in the spectral window of interest, $\varepsilon$ the object emissivity (the object is considered opaque), $\rho$ the object reflectivity, $N_{\mathrm{obj}}$ the radiance from the surface of the object, $N_{\mathrm{env}}$ the radiance of the surrounding environment considered as a blackbody, and $N_{\mathrm{atm}}$ the radiance of the atmosphere (supposed constant). If the transmission coefficient of the atmosphere is considered to be close to unity, Equation 8-115 can be simplified as follows:

$$N_{\mathrm{CAM}} = \varepsilon N_{\mathrm{obj}} + \rho N_{\mathrm{env}} \tag{8-116}$$

If the emissivity is very high (more than 0.9), according to the Kirchhoff laws, $\varepsilon = 1 - \rho$, Equation 8-116 can be further reduced to $N_{\mathrm{obj}} \approx N_{\mathrm{CAM}}$. According to some relationships, camera signals can be converted to radiance $N_{\mathrm{obj}}$ and temperature $T_{\mathrm{obj}}$ values. If the emissivity $\varepsilon$ is not high enough, but with perfectly diffuse reflection, the temperature of an object can also be qualitatively obtained from the radiance $N_{\mathrm{CAM}}$ received by the IR camera.

For ceramic tiles, the emissivity is about 0.6–0.8, not high enough to ignore the effect of the surface reflection. Moreover, due to the smooth surface, more specular than diffuse, the intensity of the reflected radiation is not constant in all directions. In an actual survey, the specular reflection toward the camera will yield apparent hot spots or strips in the IR images. Hot spots or strips would be falsely interpreted as damage zones following the active thermography principle discussed previously. Consequently, extreme care is required and an adequate approach should be undertaken when IR thermography is to be employed for materials having specular reflective surfaces. We here take the reflectivity to be the sum of a specular component and a diffuse component:

$$\rho = \rho_{\mathrm{s}} + \rho_{\mathrm{D}} \tag{8-117}$$

We define the diffuse radiance $N_D$ as the total diffuse energy leaving the surface per unit area and per unit time, so that

$$N_D = \varepsilon N_{obj} + \rho_D N_{env} \quad (8\text{-}118)$$

In Equations 8-116 and 8-118, it is known that $N_{obj}$ is related to the surface temperature of unit area, $N_{CAM}$ and $N_D$ are related to the detected temperatures by the IR camera for the unit area with and without specular reflection toward the camera, respectively. The temperature detected by the IR camera can be overestimated due to reflectivity. Moreover, the specular reflection will create parasitic hot spots that further complicate the thermogram interpretation.

The most effective way to solve the problem is to remove the influence of specular reflection. This can be done by grabbing the images continually at the same zone (for example, mounting the camera on a tripod, when the images are recorded and neither the object nor camera is moved), and recording the thermography images during a dual transient period, i.e., right after a thermal source is applied and then withdrawn abruptly. During the daytime, a favorable weather condition for this purpose is when the sky is partially cloudy. The movement of a cloud leads to a change from sunshine to shadow and back to sunshine again, providing an opportunity to record the transient behavior with or without specular reflection for the wall tiles under inspection. In sunshine, the object is thermally stimulated but with a reflective disturbance; under cloud cover, the effect of surface reflection is minimized and the detected temperature by the IR camera corresponds to the actual surface temperature of the object. Due to the rapid change of temperature in the transient period, a grab rate of 30 sec/frame is suggested.

When the images are taken in this way, the points of interest will always be in exactly (or very near) the same place in all the images. Figure 8-52 shows some results of the images taken using this approach, and each image was grabbed at a different time (but for the same position). In image 1, we marked down the *X*/*Y* coordinates of three points, P1, P2, and P3. By using the "Trending Analysis" program, the three points (that we placed in image 1) were automatically placed at exactly the same pixel locations in all of the images (see Figure 8.52a), and the graph showing the temperature variation of the spots over the series of images is shown in Figure 8.52b.

In Figure 8-52, it is found that in about 15 minutes, the detected temperatures by the IR camera for the three fixed points are different due to the motion of the clouds. The temperature of P3 is higher than that of other two points both with clouds and without clouds, indicating the zone near this point is a damaged area with a subsurface filled with air. The temperature of P1 is lower at all times, indicating the region near P1 is a sound area. It should be noted that the temperature change of P2 is somewhat different as compared to these two points. Without clouds, its temperature is higher than that of P1, and only slightly lower than that of P3. But with clouds (stage AB, C, DE, and FG in Figure 8.52b), the detected temperature of P2 is nearly equal to that of P1, indicating the influence of specular reflection on P2. Hence, we can conclude that the actual surface temperatures of points P1 and P2 are almost the same, and the region near point P2 is also a sound area. The higher temperature of P2 in the case without clouds, caused by the effect of specular reflection, might be falsely interpreted as a damaged zone if the object was stimulated by sunshine all the time. After all, the proposed approach is very useful to eliminate the effect of reflection and make possible interpretation and analysis, with a high degree of accuracy.

**(b)** *Space resolution*: Before discussing the space resolution of an infrared thermograph camera, let us first talk about the field of view. The field of view (FOV) is defined as the area of an object to be inspected that can be viewed by an infrared thermograph camera. It depends

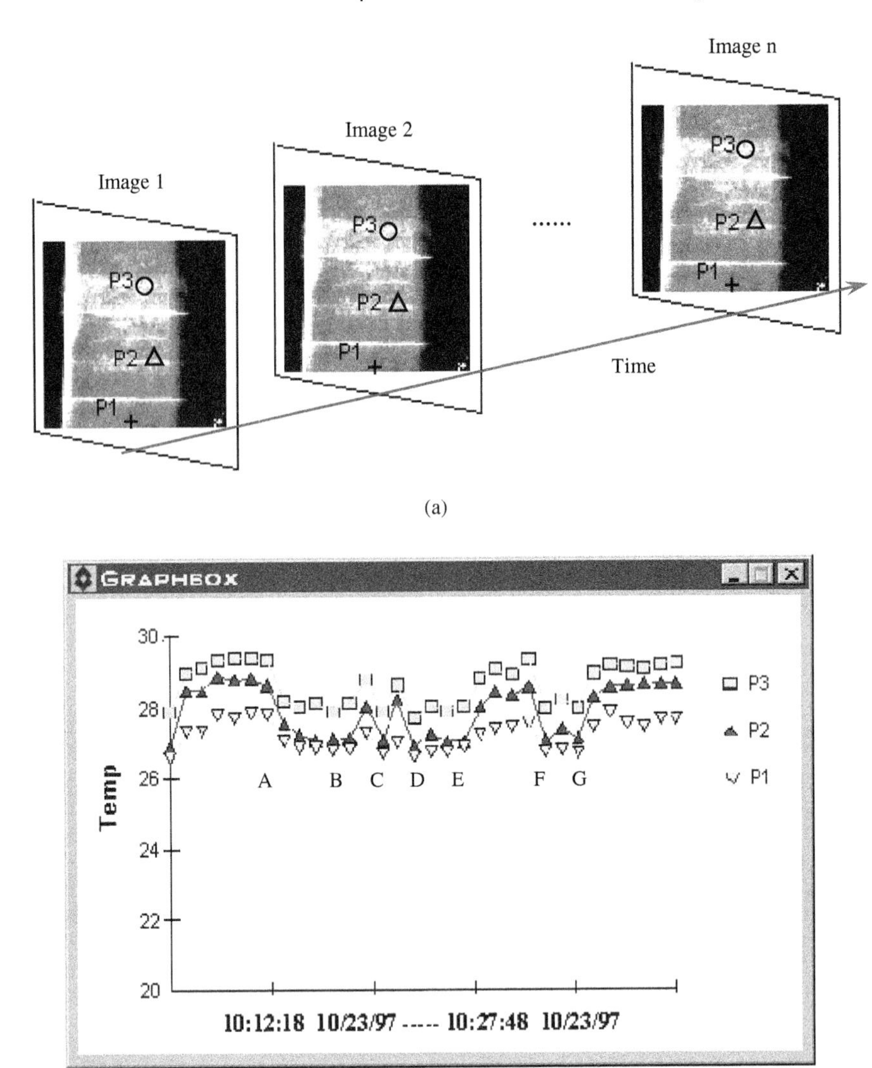


**Figure 8-52** (a) Thermal trend analysis approach; and (b) the development of the spots over the series of images at different times

on the lens of the system and the distance between the camera and the object. Another term frequently encountered is the instantaneous field of view (IFOV). A camera usually consists of many detectors, arranged in an array. The smallest cell in the array is called a picture element or pixel. The area covered by each pixel is the IFOV. For instance, if a camera has 320 × 240 detectors, it has 76,800 cells. The field of view will be distributed on these cells. As an illustration, Table 8.2 shows the field of view and instantaneous field of view for a typical camera (with a 20 by 15-degree lens) at different distances.

It is fair to say that the IFOV is a measure for the spatial resolution from cameras with focus parallel array detectors. The spatial resolution, also called geometrical resolution, is one of the most important parameters for viewing, as well as for temperature measurement of an object. It determines how well an object under inspection can be measured. For the case

**Table 8-2** The field of view and instantaneous field of view for a typical camera

| Distance to object | Field of view | IFOV |
|---|---|---|
| 1 m | 0.35 × 0.26 m | 1.1 × 1.1 mm |
| 5 m | 1.76 × 1.32 m | 5.5 × 5.5 mm |
| 10 m | 3.52 × 2.63 m | 11 × 11 mm |
| 50 m | 17.6 × 13.2 m | 55 × 55 mm |

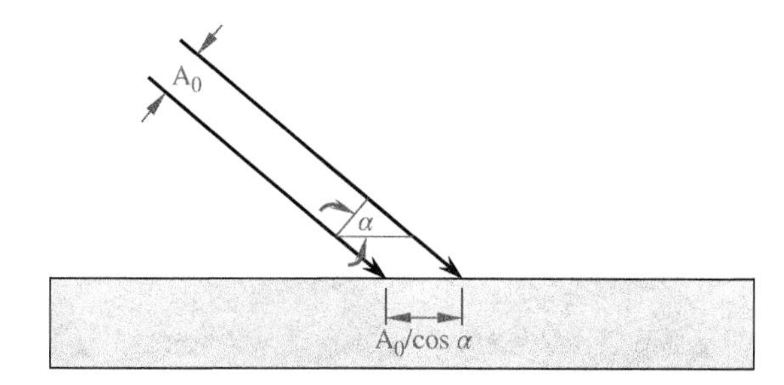


**Figure 8-53** The projection area of an angled detection beam

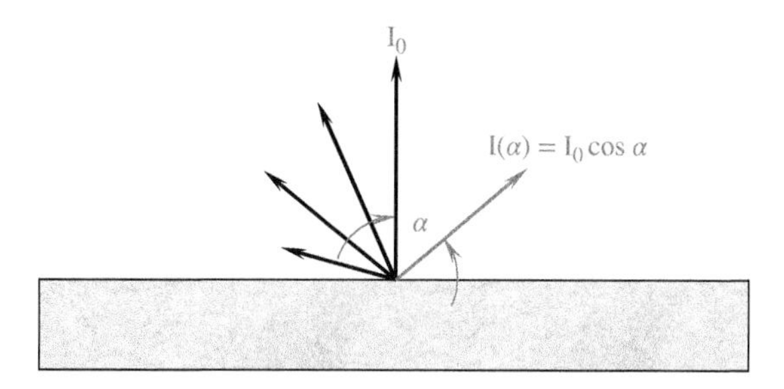


**Figure 8-54** The radiation from an object with the angle of view

where the object to be measured covers IFOV completely, the pixels will receive radiation from the object only and the temperature can be correctly measured. However, if the object to be measured cannot cover IFOV completely, i.e., IFOV is larger than the object, the detector will pick up radiation from other objects surrounding or behind the object to be inspected and hence an inaccurate result will be obtained. Moreover, the optics distort the image, e.g., chromatic aberration, spherical aberration, and a number of other deficiencies, which implies that radiation coming from objects outside the IFOV will fall onto the detector. Hence the real size of the measurement spot is substantially bigger than the IFOV. The IFOV is therefore a measure of the absolute lower limit for measurement.

**(c)** *Angle correction*: The IFOV is usually defined perpendicular to the object. In other words, it is the area seen from a camera when the camera looks at the object at a right angle. However, in many cases, the projection of the detector on the object does not always hit the object at a right angle. When the detector is projected onto the object with an angle $\alpha$, the surface that the detector hits is bigger than that it hits at a right angle, as shown in Figure 8-53. Considering that the radiation from an object varies with the angle of view as shown in Figure 8-54, if we combine the two effects sketched above, i.e., the radiator being looked at by a thermal camera at an angle of $\alpha$, the expression for the received radiation can be obtained:

$$\frac{A_0}{\cos\alpha} I_0 \cos\alpha = A_0 I_0 \tag{8-119}$$

It can be seen that the camera sees the same radiation from a diffuse radiator, no matter what the angle of incidence. However, in reality, no perfectly diffuse bodies exist. A good rule of thumb is that most bodies are Lambertian radiators up about 50 degrees from the normal. After that, the emissivity usually goes down.

## 8.6 NONCONTACTING RESISTIVITY MEASUREMENT METHOD

Many nondestructive methods are based on measuring the changes in electrical properties, including electrical resistance, dielectric constant, and polarization resistance, of concrete. These methods can be classified into electrical and electrochemical nondestructive methods. The electrical resistance of an electrolyte is directly proportional to the length and inversely proportional to the cross-sectional area and is expressed by

$$R = \rho \frac{L}{A} \tag{8-120}$$

where $R$ is the resistance in ohms, $\rho$ the resistivity in $\Omega \cdot$ m, $L$ the length in m, and $A$ the cross-sectional area in $m^2$. The changes in electrical properties of concrete are closely related to the evaporable water content in concrete, which varies with water/cement ratio, degree of hydration, and degree of saturation. The ion concentration in water varies with time, too. The conduction of electricity by moist concrete is essentially electrolytic and can be used to interpret the properties of concrete, despite the complex relationship between the concrete's moisture content and its dielectric constant. The electrical resistance depends on the size of the concrete specimen, while the resistivity is essentially a material property, so that electrical resistivity is more widely used for characterizing concrete property in electrical methods. Resistivity measurement can provide a rapid nondestructive assessment of concrete surface areas, crack size, reinforcing steel corrosion, and cement hydration.

To detect steel corrosion, the ability of the corrosion current to flow through the concrete is assessed in terms of the electrolytic resistivity of the material. An in situ resistivity measurement setup, in conjunction with half-cell measurement, is shown in Figure 8-55. In this test technique, normally low-frequency alternating current is applied and the current flowing between the outer probes and the voltage between the inner probes are measured. Since the ability of the corrosion current to flow through concrete increases with decreasing resistivity, the measured resistivity $\rho$ can be used together with potential measurements to assess the likelihood of corrosion. Classification of the likelihood of corrosion can be obtained following the values in Table 8.3 when half-cell potential measurements show that corrosion is possible. As far as the test results are interpreted, highly negative corrosion potential or a high gradient in a potential map imply that corrosion is thermodynamically favorable. If the resistivity is also low, there is a high chance to have significant corrosion actually occurring in the reinforcing steel.

The resistivity can also be used as a measure of the degree of hydration of cement-based materials. In the past, such measurement always employed contacting electrical resistivity apparatus, which has many drawbacks. Recently, Li and Li (2002) and Li et al. (2003) have invented a noncontacting electrical resistivity apparatus to measure the resistivity of cement-based materials. The measuring system of this nondestructive testing method is shown in Figure 8-56, which adopts the transformer principle. The electrical circuit contains a primary coil with wound wires

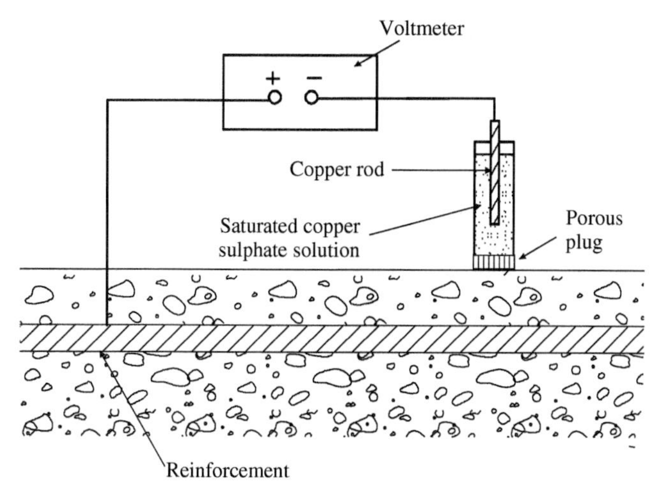


**Figure 8-55** Reinforcement potential measurement by the half-cell method

**Table 8-3** Classification of the likelihood of corrosion by half-cell potential

| Potentials over an area | > −0.20V CSE[a] | −0.20V ~ −0.35V CSE | < −0.35 V CSE |
|---|---|---|---|
| Probability of steel corrosion | > 90% no corrosion | Uncertain | > 90% corrosion |

[a]copper sulfate electrode.

**Figure 8-56** A noncontact resistivity test apparatus for cement-based materials

and a ring-type cementitious specimen acting as the secondary coil of the transformer. The primary coil is the input coil of the transformer and the secondary coil is the output coil. When AC is applied to the primary coil, mutual induction causes current to be induced in the secondary coil. With the measurement of the induction current in a cement-based specimen, the curve of resistivity versus time can be obtained. By interpreting the behavior of the resistivity curve, the hydration characteristics of cement-based materials can be obtained. The characteristics include the hydration stages, setting times, and strength development. It has been found that the noncontact electrical resistivity measurement provides a good nondestructive way to determine and assess the setting time and mechanical properties of the cement-based materials during the entire setting process. In general, fresh concrete behaves essentially as an electrolyte with a resistivity of the order of 1 Ω·m, a value in the range of semiconductors, while hardened concrete has a resistivity of the order of 10 Ω·m, a reasonably good insulator.

The Wenner four-probe technique has been used for soil testing for many years and it has recently been adopted in applications for testing in situ concrete. In this method, four electrodes are placed in a straight line, on, or just below, the concrete surface at equal spacing, as shown in Figure 8-57. A low-frequency alternating electrical current is passed between the two outer electrodes while the voltage drop between the inner electrodes is measured. The apparent resistivity, $\rho$, is calculated as

$$\rho = \frac{2\pi sV}{I} \tag{8-121}$$

where $s$ is the electrode spacing, $V$ the voltage drop, and $I$ the current. This method can also be used in applications for detecting pavement thickness. Since concrete and subgrade pavement materials have different electrical characteristics, the change in slope of the resistivity vs. pavement depth curve will indicate the pavement thickness (Vassie, 1978). When testing a concrete pavement, the electrode system may be adjusted at 25- or 50-mm spacings for the initial readings, and then expanded to 25-mm increments for successive readings, thereby extending to a spacing equal to the pavement depth plus 75–150 mm (Malhotra and Carino, 2004).

The corrosion of steel in concrete is an electrochemical process that requires a flow of electrical current for the chemical corrosion reactions to proceed. The electrochemical method can be used to detect signs of rebar corrosion, based on an electrochemical process, through

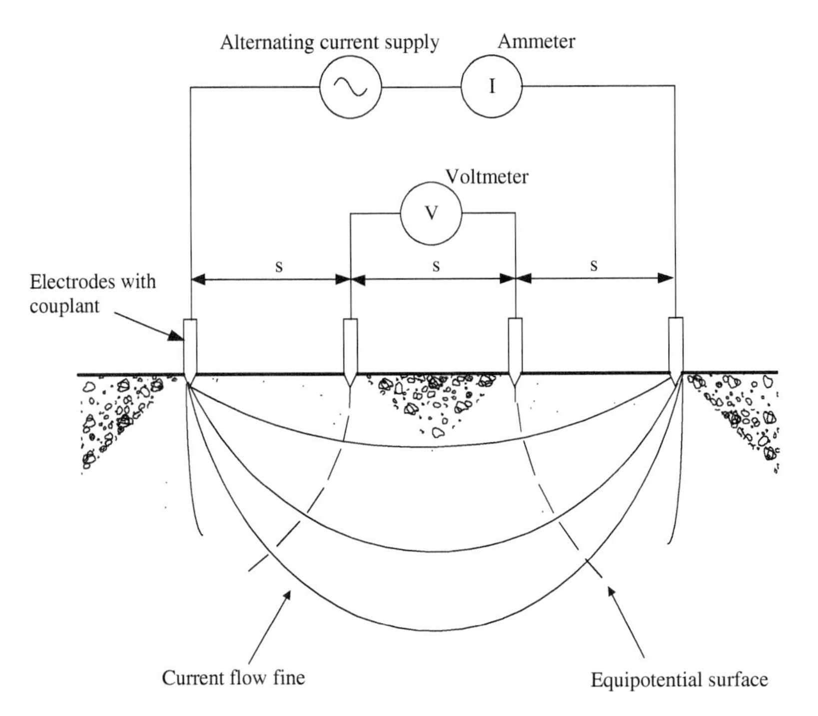


**Figure 8-57** Four-probe resistivity test of concrete

measuring the electric potential of the reinforcing bars (Newman, 2001). Some tests use the half-cell method of measurement described in Chapter 2, in which a copper/copper sulfate half-cell is used. In this technique, an electrical connection is made to the reinforcement at a convenient position. The electric potential difference between the anode and the cathode is measured by a voltmeter; and if the values are more negative than 350 mV, the probability of corrosion is in excess of 90 percent. In practice, a series of measurements can be taken at grid points to map the probable corrosion activity. This technique has been standardized by ASTM C876. According to this standard, the difference between two half-cell readings taken at the same location with the same cell should not exceed 10 mV when the cell is disconnected and reconnected. In addition, the difference between two half-cell readings taken at the same location with two different cells should not exceed 20 mV. This method is readily used in the field, but trained operators are required. What's more, the actual rate of corrosion (such as the percent loss of section) is not provided by this method. The interpretations of these potentials vary with investigator and agency.

The variations of electrical resistivity are closely related to the physical and chemical properties of cement-based materials at early ages. Resistivity is used to study and analyze the hydration process of young cement-based materials (EL-Enein et al., 1995; Khalaf and Wilson, 1999; Levita et al., 2000). Early studies of concrete resistivity applied a DC voltage on the concrete specimen, then obtained electrical resistance according to Ohm's law. There are several disadvantages to DC voltage probing, such as the polarization effect, the release gas effect, and the change of ion distribution (Banthia et al., 1992; Lakshminarayanan et al., 1992). These factors will cause the measured results to be inaccurate, hence, using AC voltage as the probing signal is more reasonable. In recent years, there have been many papers concerning this subject (Hughes et al., 1985). These conventional methods for resistance measurements (consequently, resistivity) of cement-based materials are usually conducted by measuring the AC current and the AC voltage between two plate electrodes placed at opposite sides of a prism specimen (McCarter et al., 1981). Obviously, the contact between the electrodes and specimen plays an important role in determining the accuracy of the measurement. Loose contacts will lead to inaccurate and sometimes ridiculous results. To improve the contact, some researchers inserted fresh cement paste or colloidal graphite between the electrodes and concrete (McCarter et al., 1981; Whittington et al., 1981), and others used external force to fasten the electrodes and concrete specimen from two ends.

These measures are effective only at the beginning of experiments. As the hydration of cement proceeds, shrinkage will occur, which can lead to some fissures appearing in the electrode–concrete boundary. If this phenomenon occurs, the experiment cannot be continued. In addition, the cement paste is highly alkaline (pH 12.5–13.5), so the electrodes are gradually corroded within several days. Due to the poor connection problem, the data obtained by this kind of method are suspect. Besides, fresh cement paste and colloidal graphite can diffuse inside the specimen being measured, which affects the measured results.

### 8.6.1 Principle of the Novel Method

The schematic of the resistance measurement for the novel method is shown in Figure 2-15 (Chapter 2), in which there are no electrodes, and the cement-based specimen is a ring with a rectangular section. The specimen ring acts as a secondary transformer coil.

When an AC voltage is applied on the primary coil of the transformer, a toroidal voltage ($V$) will be inducted in the secondary, i.e., in the specimen ring. Subsequently, a toroidal current ($I$) will be incurred inside the specimen ring. Supposing the impedance of specimen ring is $Z$:

$$Z = VI \tag{8-122}$$

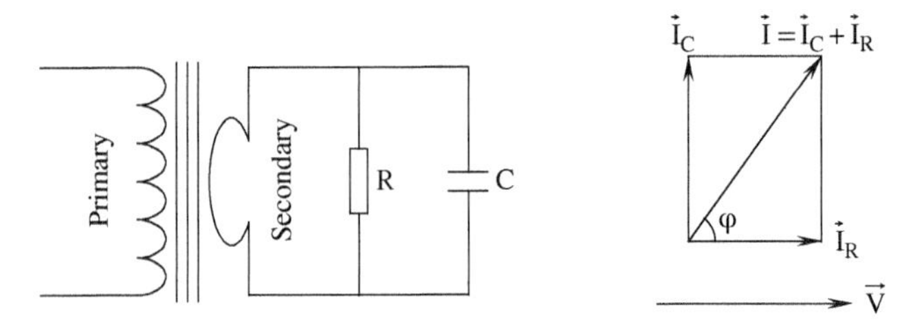


**Figure 8-58** An equivalent circuit for the resistivity measurement

This principle can be represented by an equivalent circuit as shown in Figure 8-58. In Figure 8-58, $R$ is resistance, and $C$ capacitance; then

$$\begin{aligned} V &= R \cdot I_R \\ \mathbf{I} &= \mathbf{I}_C + \mathbf{I}_R \end{aligned} \tag{8-123}$$

The impedance of the specimen is the parallel connection of $R$ and $C$:

$$Z = R\left[1 + (\omega CR)^2\right]^{-12} = R\left[1 + (2\pi fCR)^2\right]^{-12} \tag{8-124}$$

where $Z$ is the impedance of specimen ring, $R$ the resistance of specimen ring, $C$ the capacitance of specimen ring, and $f$ the probing frequency. According to the phase difference between the toroidal voltage $V$ and the toroidal current $I$, $R$ can be obtained:

$$R = V\,(I \cos \varphi) \tag{8-125}$$

where $V$ is the toroidal voltage, $I$ the current passed through specimen ring, and $\varphi$ the phase difference between $V$ and $I$.

### 8.6.2 Formulation of Resistivity Calculation

The resistance can be found in Equation 8-125. Since the resistance of a specimen depends on the geometric parameters, resistance is therefore not significant for material research. The resistivity is a necessary parameter for comparison between different specimens. Due to the geometric form of the ring, the resistivity cannot be directly calculated from $R$. In these circumstances, a formula for resistivity calculation was derived from integral equations. As shown in Figure 8-59, the cross-section of the specimen is rectangular. In fact, the cross-section of a specimen ring can be either rectangular or circular. However, a rectangular section is recommended due to the ease of mold making. On the other hand, this shape can meet drill sampling for hardened cement-based materials. The procedures of formula derivation for the resistivity calculation are as follows. For a conductor, such as a metal wire, the resistance $R$ can be represented by

$$R = \rho LS \tag{8-126}$$

where $\rho$ is the electrical resistivity of conductor, $L$ the length of conductor, and $S$ the cross-section of conductor. For a conducting ring with rectangle section, as shown in Figure 8-59, the rectangle section can be divided into many area elements:

$\Delta R_1, \Delta R_2, \Delta R_3, \ldots$ are the resistance elements

$r_{\text{in}}$ the internal radius

$r_{\text{ex}}$ the external radius

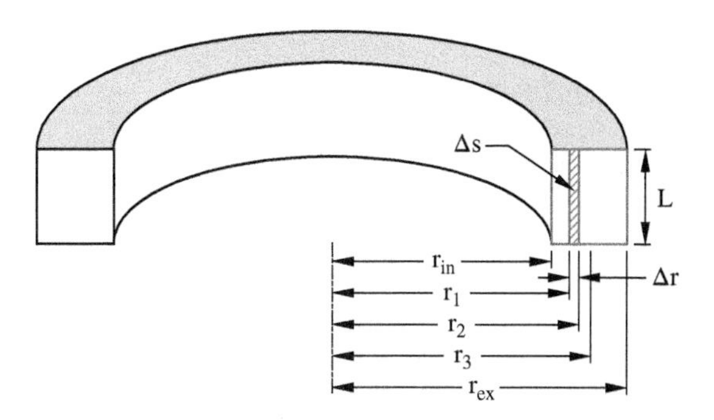


**Figure 8-59** Cross-sections of the resistivity measurement ring

$\Delta S$ the area element

$\Delta S = h\Delta r$

The circumferences for different radii are

$$L_i = 2\pi r_i, \qquad i = 1 \text{ to } n, \qquad n = \left(r_{\text{ex}} - r_{\text{in}}\right)/\Delta r$$

$$\begin{aligned}
\Delta R_1 &= \rho L_1/\Delta S_1 = \rho 2\pi r_1/\Delta S_1 = \rho 2\pi r_1/h\Delta r, &\quad r_1 &= r_{in} + \Delta r \\
\Delta R_2 &= \rho L_2/\Delta S_2 = \rho 2\pi r_2/\Delta S_2 = \rho 2\pi r_2/h\Delta r, &\quad r_2 &= r_1 + \Delta r \\
\Delta R_3 &= \rho L_3/\Delta S_3 = \rho 2\pi r_3/\Delta S_3 = \rho 2\pi r_3 h/\Delta r, &\quad r_3 &= r_2 + \Delta r \\
&\vdots \\
\Delta R_n &= \rho L_n/\Delta S_n = \rho 2\pi r_n/\Delta S_n = \rho 2\pi r_n/h\Delta r, &\quad r_n &= r_{n-1} + \Delta r
\end{aligned} \tag{8-127}$$

Supposing the whole resistance is $R$, which equals all resistance elements connected in parallel. Then

$$\begin{aligned}
1R &= 1\Delta R_1 + 1\Delta R_2 + 1\Delta R_3 + \cdots \\
&= \sum_{i=1}^{\infty} 1\Delta R_i \\
&= (h2\pi\rho)\int_{r_{\text{in}}}^{r_{\text{ex}}} (1r)\, dr \\
&= (h2\pi\rho)\, ln\left(r_{\text{ex}} r_{\text{in}}\right)
\end{aligned} \tag{8-128}$$

$$\text{hence,} \quad R = \frac{2\pi\rho}{h\ln\left(r_{\text{ex}} r_{\text{in}}\right)} \tag{8-129}$$

The resistivity $\rho$ can be obtained as

$$\rho = \frac{Rh\ln\left(r_{\text{ex}} r_{\text{in}}\right)}{2\pi} \tag{8-130}$$

The rectangular section ring can be made of cement-based materials, because the resistivities of cement-based material are larger than those of metals (metal conductor $\rho$ $10^{-6}$–$10^{-4}$ Ω·cm). In addition, the electrical current is produced by ion movement and the frequency used in the experiments is lower (1–20 kHz). Thus, there is no skin effect inside the concrete. The current density in the specimen section is proportional to $1/r$. These initial conditions can satisfy the requirements of the equations used in the derivation process.

When the resistance $R$ is obtained from Equation 8-125, concurrently, the relative resistivity can be found by Equation 8-130, which can be used to accurately calculate the resistivities of cement-based products.

### 8.6.3 Measuring System

The measuring system of the novel method is shown in Figure 8-60, which consists of a generator, amplifier, oscilloscope, transformer core, current sensor, data acquisition interface, and computer. The generator produces a sine wave at a given frequency. The amplifier magnifies the sine wave to match the transformer primary, and the data acquisition interface samples the data needed and transfers them to computer. The sampling period is 5 minutes. An oscilloscope is used to observe the phase between the toroidal voltage and current. This system can operate continuously. To understand this system clearly, it is necessary to give the relevant parameters of this system. The signal applied on the primary of transformer is a sine wave of 1 kHz frequency, the turn number of the transformer primary is 36, and the secondary toroidal voltage is 1.2 V. The measuring range of the leakage current clamp is 0–100 mA, and the resolution is 0.1 mA.

Before measuring the resistances of cement-based specimens, a circuit of parallel connection by a resistor and a capacitor is used to assess the accuracy of the measuring system. The values measured by this method, such as resistance, and capacitance, are perfectly coincident with the nominal values. The phase relations between $V$ and $I$ are correct, when using a pure resistance to simulate the cement specimen. The phase measured is zero. These results proved that the system is credible.

#### 8.6.3.1 Resistivity Measurements of Cement Specimens

When using this system to monitor the resistivity of a cement-based paste for studying the hydration process, the paste has to be cast in a mold. The geometric shape of the mold is shown in Figure 2-16 (see Chapter 2), and the mold is cast in the position shown in Figure 2-16. If using the system to measure the resistivity of hardened cement-based materials, it is necessary to make a specimen of this material, because the specimen has to be a ring with a rectangular section (as demold shape). This shape can be taken using drill sampling. After such a specimen is finished, it is placed in the window of transformer core, and the current leakage clamp and transformer core are opened.

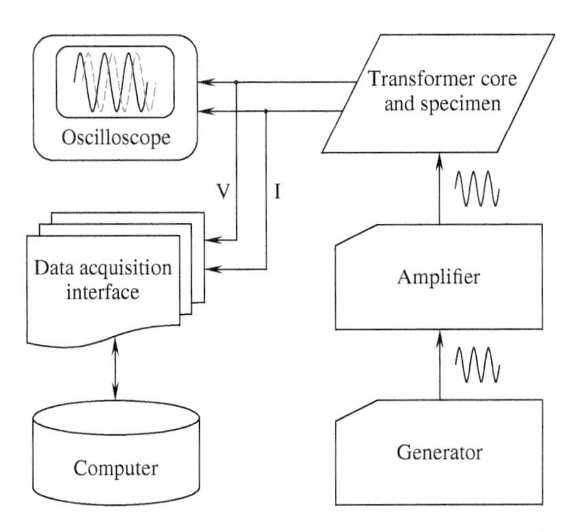


**Figure 8-60** Measurement system for the resistivity test

Some results of resistivity measurements for OPC pastes and OPC mortars in different *w/c* ratios have been measured. The relevant curves are given in Figures 8-61 and 8-62. Since the purpose of this discussion is to present the novel method of electrical resistivity measurement for cement-based materials, analyses concerning the measured curves have not be given.

#### 8.6.3.2 Repeatability

To observe the repeatability of this system, three specimens of cement paste with the same water-cement ratio ($w/c = 0.5$) have been measured. The repeatability is excellent. The measured results are shown in Figure 8-63.

#### 8.6.3.3 Calibration

The simplest calibration is to measure the electrical resistivity and phase of a closed circuit with a resistor or with a resistor plus capacitor, and to compare the theoretical values from an equivalent

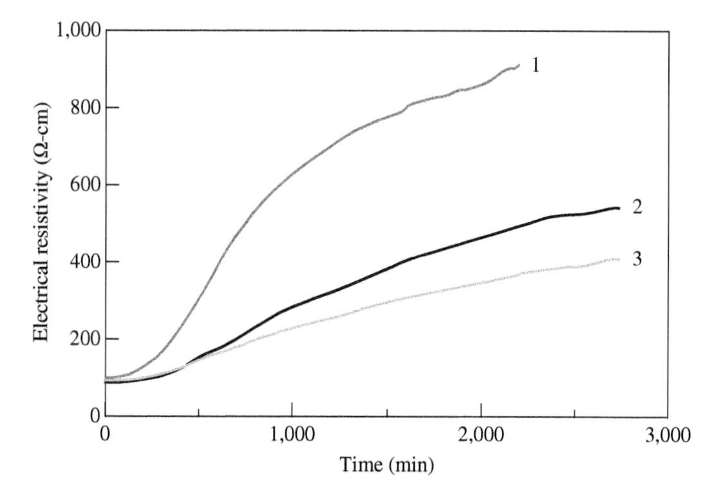


**Figure 8-61** Resistivity of cement pastes (*w/c*: 1. 0.30, 2. 040, 3. 0.50)

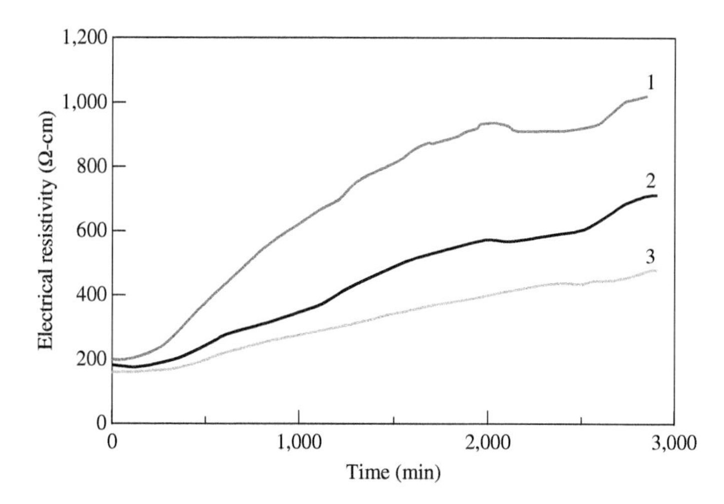


**Figure 8-62** Resistivity of mortars (cement/sand = 1, *w/c* = 1. 0.35, 2. 0.50, 3. 070)

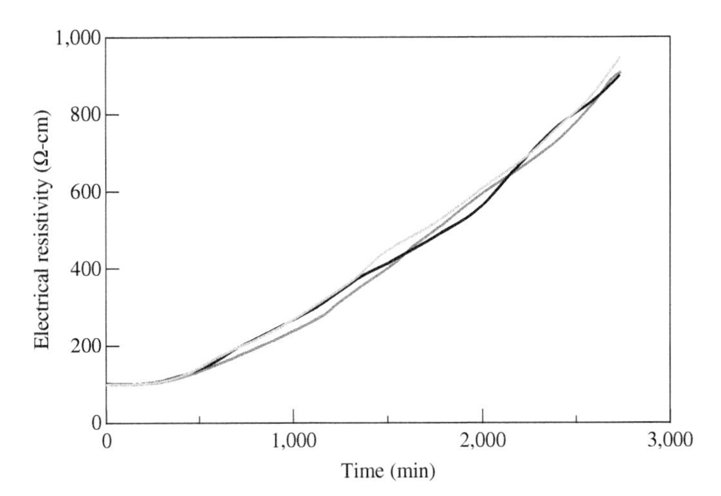


**Figure 8-63** Repeatability of the measuring system

circuit. For this purpose, a sinusoidal waveform with frequency 1 kHz is again used. The primary turns of the transformer are 36 and the secondary toroidal voltage generated is 2.5 V. In the first calibration case, only one resistor of 600 Ω is used. The measured current by the sensor in this case is 4.1 mA. The theoretically calculated result is

$$2.5\text{V}/600 = 4.17\,\text{mA} \tag{8-131}$$

In the second calibration case, one 600-Ω resistor and one 2.2-μF capacitor are used. The measured output of the current sensor is 35 mA and measured phase shift from the oscilloscope is 83 degrees. The theoretically calculated results are 34.5 mA and 83.1 degrees, respectively.

To further prove the reliability of this method, a standard aqueous solution of potassium chloride has been used in the experiments to verify the measurement accuracy. The parameters of the standard KCl solution are as follows: 7.455 g (0.1 mol) KCl per kg of distilled water. The KCl is extra pure (max. 0.0001% Al), and the temperature of the solution is 25°C. The conductivity of such a solution is given in David (1994), and the resistivity is $\rho = 0.77993$. The result measured by this method is almost coincident with the value given in the reference, the difference between both values is less than 0.4%. So it can undoubtedly be said that the principle and the accuracy of the novel method are accurate and credible.

### 8.6.4 Application

#### 8.6.4.1 Hydration Dynamics

Noncontact electrical resistivity measurement can be used to study the dynamics of the cement hydration process, which includes the stages of dissolution, dynamic balance, setting, hardening, and hardening deceleration. The corresponding application details of this method have already been included and discussed in the Section 2.2.4 (Chapter 2); therefore, they are not reiterated here.

#### 8.6.4.2 Setting Time

As mentioned in Section 3.2.5.1 (Chapter 3), the measurement of the concrete setting time can be conducted according to ASTM 403, using the experimental setup shown in Figure 3-8. Resistivity measurement can also be used to determine concrete setting time. Introduction of this innovative method in determining concrete setting time is presented in detail in Section 3.2.5.3.

#### 8.6.4.3 Porosity

A new method using resistivity measurement to determine the porosity in cement-based materials has been developed by Zhang and Li (2009). They developed a general effective media (GEM) model to predict the porosity using the measured resistivity values. The equation is as follows:

$$\phi = \left[\left(1 - \phi_c\right) F^{-1/t} + \phi_c\right] \cdot \left(\frac{M^{1/t} - F^{1/t}}{M^{1/t} - 1}\right) \tag{8-132}$$

where $\phi$ is the capillary porosity in cement-based materials; $F$ the resistivity formation factor determined by $\rho$ and $\rho_0$; $\rho$ the overall resistivity of a cement-based system, which can be provided by noncontact resistivity measurement; $\rho_0$ the resistivity of the capillary pore phase, which can be determined experimentally by measuring the pore solution extracted from the specimen; $M$ the magnification coefficient between $\rho_1$ and $\rho_0$; $\rho_1$ the resistivity of the phase consisting of C[bond]S[bond]H gel, CH, and unhydrated cement particles; $\phi_c$ the value of the percolation threshold, below which the low-resistivity capillary pore phase is isolated by the other phase; $t$ a free parameter that is the critical exponent for electrical conductivity.

When we consider the capillary porosity $\phi$ as a function of the formation factor $F$, then $\rho_1$ (or $M$), $\phi_c$, and $t$ are the three parameters in the equation. By applying the parameter values in traditional research, the equation can be rewritten as

$$\phi = \left[(1 - 0.18) F^{-1/2} + 0.18\right] \left(\frac{400^{1/2} - F^{1/2}}{400^{1/2} - 1}\right) \tag{8-133}$$

Once the $F$ values are determined by resistivity measurement, the porosity can be predicted by this equation. Figure 8-64 shows a comparison between the porosity predicted by the equation and that measured by the MIP technique. The results show a good agreement.

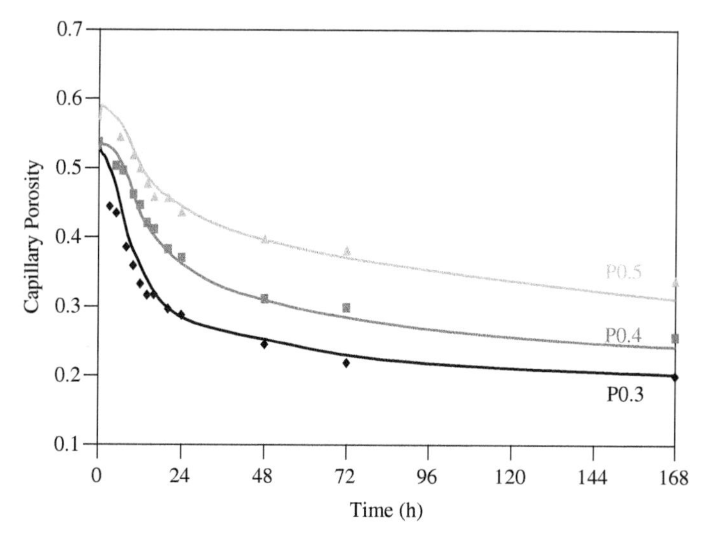


**Figure 8-64** Comparison of theoretical predicted and measured porosity (solid lines are the theoretical porosity developments from the GEM-based equation; dots are the experimental results from the MIP test)

#### 8.6.4.4 Other Applications

The noncontact resistivity measurement method has also been applied to study the influences of mineral and chemical admixtures on the hydration process of cement-based materials and the relationship between compressive strength and resistivity values. Very promising results have been obtained (Li et al., 2003; Wei and Li, 2005; He et al., 2006; Xiao and Li, 2006; Xiao et al., 2007; Wei et al., 2008).

## 8.7 AN INNOVATIVE MAGNETIC CORROSION DETECTION TRANSDUCERS

Corrosion monitoring can provide significant advantages when integrated into both preventative maintenance and the processes inherent to safety management programs. Corrosion detection by the magnetic method is an innovation in the NDT area. As Figure 8-65 shows, the EM sensor and the measuring system consist of a primary coil or solenoid that magnetizes the steel samples; the secondary coil wound on the test sample senses changes in magnetic flux. Voltage is induced by these flux changes, which is processed by an electronic integrator, which takes into consideration a fixed cross-section area of the as-received 12.5 mm diameter rod (122.7 $mm^2$). The secondary coil is placed concentrically with the primary coil. The tangential component of the intensity of the magnetic field is measured by an array of Hall sensors conveniently located close to the surface of the specimen. A magnetizing yoke and non-surrounding sensing element sensor configuration are currently under investigation to develop a more realistic tool for field applications. The Hall sensors were initially calibrated to reliably relate the magnetic intensity to the output voltage. The array of Hall sensors can be treated as a line arrangement, provided that the configuration is placed in a magnetic field with cylindrical symmetry. Then, the value of the magnetic field at different sensor positions corresponds to the magnetic field at different distances from the test sample, and the magnetic intensity at the surface of the sample can be found by extrapolation. The output voltages from the Hall sensors and integrator are read by an analog/digital (A/D) converter implemented on a data acquisition board CYDAS1600, which is controlled by a graphical programming language called DASYLAB, suitable for the development of applications such as the monitoring of the magnetizing process and generation of hysteresis curves. Magnetic properties are determined from the hysteresis curves (Polar et al., 2010).

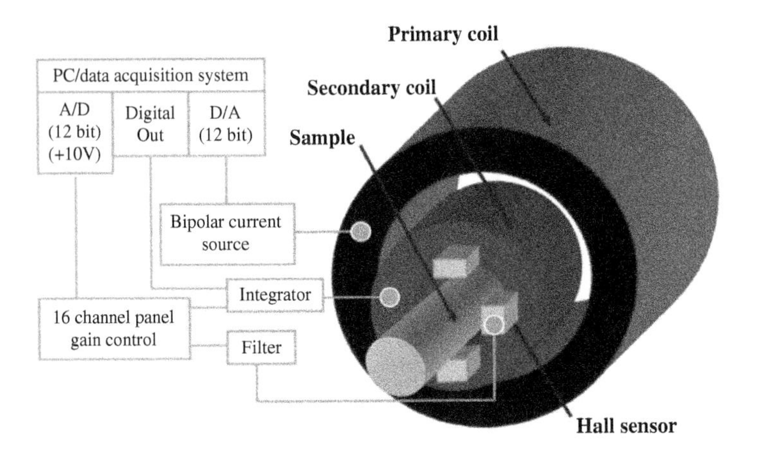


**Figure 8-65** Schematics of the electromagnetic sensor and measuring system

The EM sensor has the potential to be used as a reliable NDE tool to detect corrosion based on the variation in magnetic properties. The sensor can detect effectively, within 95% confidence limits, early stages of corrosion in structural steels with mass losses as small as 0.01 *g/cm²* or 0.5% in reduction in a cross-section area. The reductions in the cross-section area of the steel rods cause a decrease in the induced saturated flux, as well as in the magnetic retentivity. Both quantities had a linear correlation with the mass loss. Also, the sensor is most sensitive when the corrosion of the structural steel is uniform (Rumiche et al., 2008).

### 8.7.1 Useful Equations

*Permeability*: Permeability is the magnetization degree of the response of a medium in a magnetic field. Magnetic permeability is represented by $\mu$. In the International System of Units, permeability is measured in H/m or N/A$^2$.

*Relative permeability*: Relative permeability is the ratio between the permeability of a medium and the permeability of free space, $\mu_0$. $\mu_0 = 4\pi \times 10^7$ H/m $\approx 1.25664\ldots \times 10^{-6}$ N/A$^2$ Relative permeability is represented by $\mu$.

$$\mu_r = \frac{\mu}{\mu_0} \tag{8-134}$$

In this equation $\mu_0 = 4\pi \times 10^7$ H/m $\approx 1.25664\ldots \times 10^{-6}$ N/A$^2$.

*Relative equations*:

$$H = \frac{B}{\mu_0} - M \tag{8-135}$$

where $H$ is magnetic field intensity, $B$ is magnetic induction; $M$ is magnetization.

$$M = \chi_m H \tag{8-136}$$

where $\chi_m$is the magnetic susceptibility.

$$\mu_r = \chi_m + 1 \tag{8-137}$$

From Equations 8-134–8-137, the magnetic induction can be derived as:

$$B = \mu_0 (H + M) = \mu_0 \left(1 + \chi_m\right) H = \mu_0 \mu_r H = \mu H \tag{8-138}$$

If the material does not respond to the magnetic field by magnetizing, then the field in the material will be only the applied field and the relative permeability $\mu_r = 1$. A positive relative permeability greater than 1 implies that the material magnetizes in response to the applied magnetic field. It is only the permeability minus 1 if the material does not respond with any magnetization. So both quantities give the same information, and both are dimensionless quantities.

For ordinary solids and liquids at room temperature, the relative permeability $\mu_r$ is typically in the range 1.00001 to 1.003. Their great contrast to the magnetic response of ferromagnetic materials was recognized. More precisely, they are either paramagnetic or diamagnetic, but that represents a very small magnetic response compared to ferromagnetic materials.

Table 8-4 shows the permeability and relative permeability of some common materials. It can be seen that as a ferromagnetic material, the permeability of carbon steel is about 100 times larger than the permeability of concrete. Compared to ferromagnetic materials, the permeability of ferric oxide is much closer to 1. It means that corrosion of reinforcing steel in concrete can be detected by using magnetic or electromagnetic methods.

**Table 8-4** Permeability and relative permeability of common materials

| Medium | Permeability (H/m) | Relative permeability |
|---|---|---|
| Iron (99.8% pure) | $6.3 \times 10^{-3}$ | 5000 |
| Electrical steel | $5 \times 10^{-3}$ | 4000 |
| Carbon steel | $1.26 \times 10^{-3}$ | 100 |
| Platinum | $1.256980 \times 10^{-6}$ | 1.000265 |
| Aluminum | $1.256665 \times 10^{-6}$ | 1.000022 |
| Copper | $1.256629 \times 10^{-6}$ | 0.999994 |
| Air | $1.25663753 \times 10^{-6}$ | 1.00000037 |
| Vacuum | $4\pi \times 10^{-7} (\mu_0)$ | 1, exactly |
| Water | $1.256627 \times 10^{-6}$ | 0.999992 |
| Ferric oxide | N/A | 1.0072 |
| Concrete | N/A | 1 |

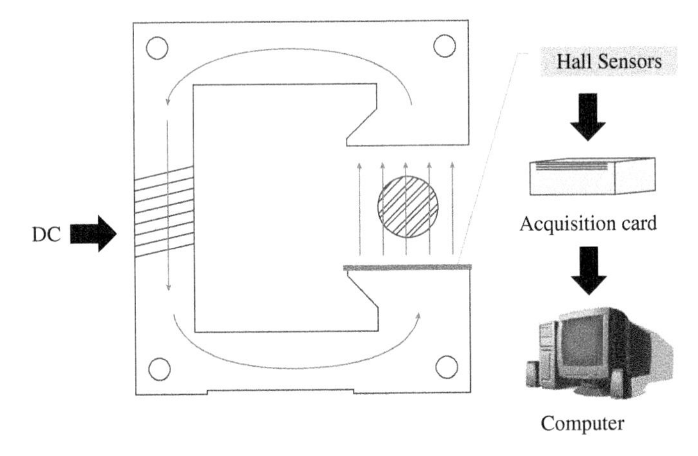


**Figure 8-66** Apparatus conception

### 8.7.2 Electromagnetic Testing Device

The test device is illustrated in Figure 8-66. The main part is made of electrical steel, which is tailored to produce certain magnetic properties, such as a small hysteresis area and high permeability.

When the direct current accesses the coil, which is wound on the other side of this part, a magnetic field is generated in the testing area. The magnetic field change will be measured by sensors on the top or bottom surface of the testing area. They send the data to the acquisition card and to the computer for analysis. On the left side is the coil and specimens to go through the interspace, which will be detected by sensors pasted on the bottom surface. Also, the demagnetizing coil is set in the bottom of the part to prevent magnetization when it is powered on for a long time. The electromagnetic testing device has only one coil to generate a magnetic field and there is no need to place specimens concentrically with the coil compared to the EM sensor in Figure 8-65. It means that the main difference is the angle between the specimen and the magnetic induction line direction.

From the previous study, it is predicted that the direction of the magnetic field in the testing area can be split into three situations as shown in Figure 8-67. Where there is no specimen between

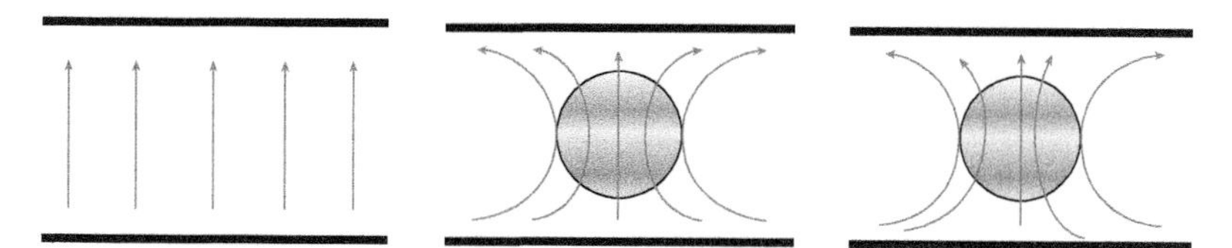

**Figure 8-67** Prediction of magnetic field changes in testing area

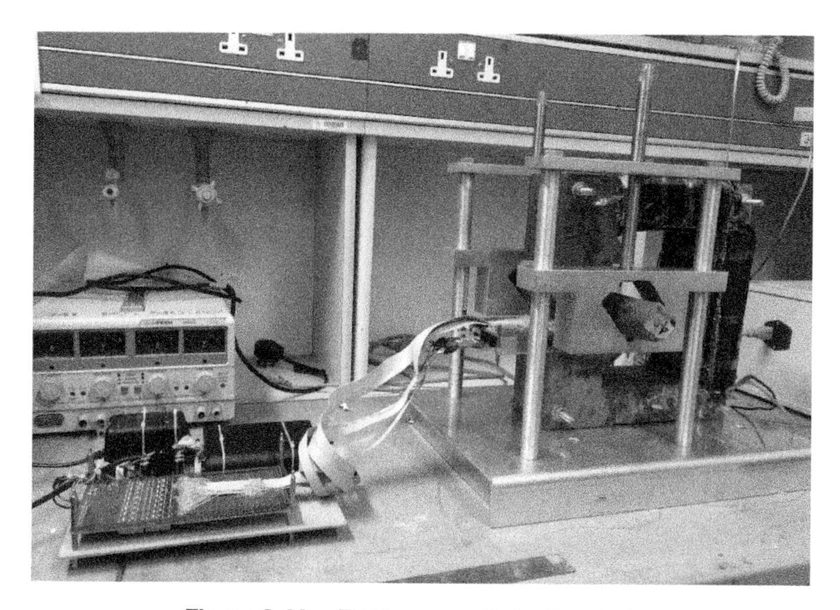

**Figure 8-68** Electromagnetic testing device

two surfaces, the direction of the magnetic field is vertical to the surface uniformity. If a rebar without corrosion is in the magnetic field, the direction will converge to the specimen first and then disperse in uniformity. However, if a corroded rebar is put there instead, the direction will be led irregularly as well as the magnetic field intensity will change during the different magnetic induction.

The first generation electromagnetic testing device is shown in Figure 8-68. The specimen is put on the specimen holder, marking the relative position between the specimen and specimen holder, and then one begins to measure the output voltage. Average voltage output of the steel sheet in the experiment is shown in Figure 8-69. It can be seen that during the corrosion, the degree is increasing and the absolute value of the electromagnetic induction rises higher and higher. The prediction has been proved that the corrosion process of steel in the electromagnetic field will affect the electromagnetic field intensity around it. From experiments using the electromagnetic testing device above, several advantages were found during the process. One of the most important advantages is that the output result of the device is stable, which indicates the degree and distribution of corrosion on specimens. On the other hand, the electromagnetic testing device also has some problems. The major issue is that the device has to have a period of operating temperature. The coil and the silicon steel part have to be preheated for several minutes or an error result will be produced in the experiments. For the same reason, if the device remains on for a long time, the coil will become overheated and lead to an operation error.

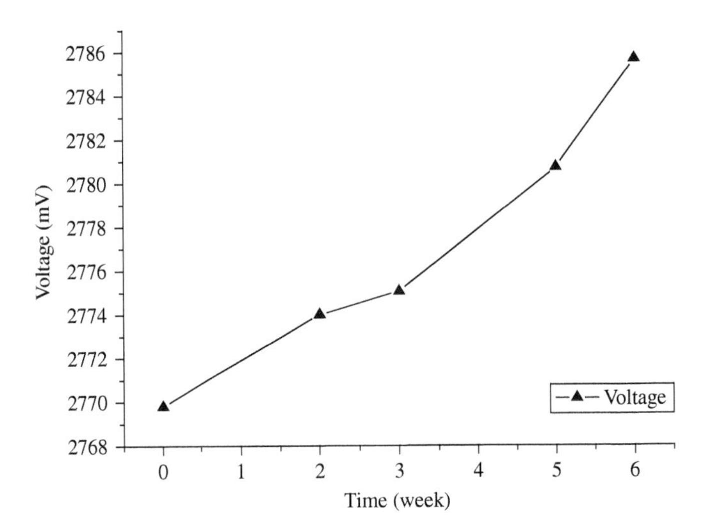


**Figure 8-69** Relationship between average output voltage and time

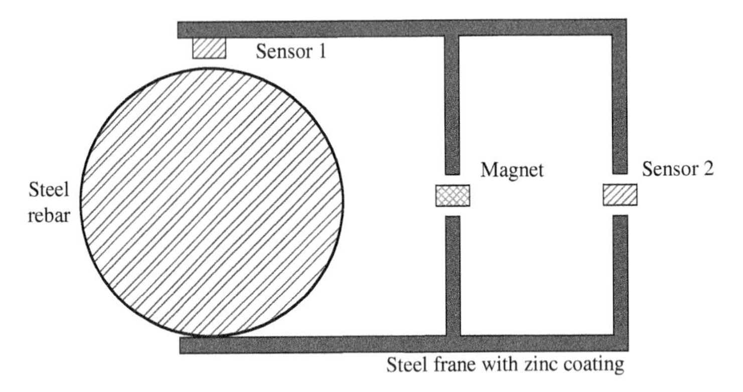


**Figure 8-70** Conception of corrosion monitoring apparatus

### 8.7.3 Magnetic Corrosion Monitoring Device

The electromagnetic corrosion detection device possesses high accuracy and is nondestructive to specimens. In order to detect and monitor the corrosion process of reinforcing steel in concrete, the apparatus, which can work in reinforced concrete structures, has been developed based on an electromagnetic testing device.

In this apparatus, a magnetic field is generated by a group of permanent magnets. Magnetic circuits are formed in the steel frame structures and specimens. The acquisition card collects the strength of the magnetic field detects through Hall effect sensors, which are placed in different locations in the apparatus. The whole testing part of the device, which is shown in Figure 8-70 will be cast in concrete structures.

Monitoring shows the variation of the reluctance of the magnetic circuit to estimate if the steel rusts. Between the reinforcing steel and the permanent magnet is a detection circuit in the device, and to indicate machining error, self-demagnetization, or different installing positions and other uncontrollable elements, a magnetic correction circuit is added from the permanent magnet to the sensor on the other side of the apparatus.

#### 8.7.3.1 Analysis of Equivalent Model

The equivalent model is based on three circuit laws. The first is the sum of magnetic fluxes $\Phi_1$, $\Phi_2$ ... into any node is always zero:

$$\Phi_1 + \Phi_2 + \ldots = 0 \tag{8-139}$$

This law follows Gauss's law and is analogous to Kirchhoff's current law for analyzing electrical circuits (Balabanian et al., 1969).

The second law is that the voltage excitation applied to a loop is equal to the sum of the voltage drops around the loop; the magnetic analog states that the magnetomotive force is equal to the sum of the magnetomotive force drop across the rest of the loop. This also follows Ampère's law and is analogous to Kirchhoff's voltage law.

The third law is Hopkinson's law, which is a counterpart to Ohm's law used in magnetic circuits. It states that:

$$F = \Phi R_m \tag{8-140}$$

where $F$ is the magnetomotive force across a magnetic element, $\Phi$ is the magnetic flux through the magnetic element, and $R_m$ is the magnetic reluctance of that element. Figure 8-71 shows the equivalent model of a magnetic corrosion device.

Because the relative permeability of steel materials is far greater than air and concrete's permeability, it can be considered that $R_{m1}$ and $R_{m2}$ mainly depend upon the size of the interspace in magnetic circuits.

#### 8.7.3.2 Corrosion Gradients Detection

Based on a conception of corrosion monitoring apparatus, which is shown in Figure 8-70, two more channels are added to the frame structure to implement gradients detection. As shown in Figure 8-72, the end surfaces of the steel frame part were close to sensor 3 and sensor 4. It means these end surfaces become corroded during environment humidity, chloride penetration, and other factors through concrete surfaces. The end surfaces of the extra channels can be corroded. The corrosion condition of the end surfaces can be monitored to predict the corrosion rate and direction of reinforcing steel in the concrete. The numbers of the channels are flexible, considering the size

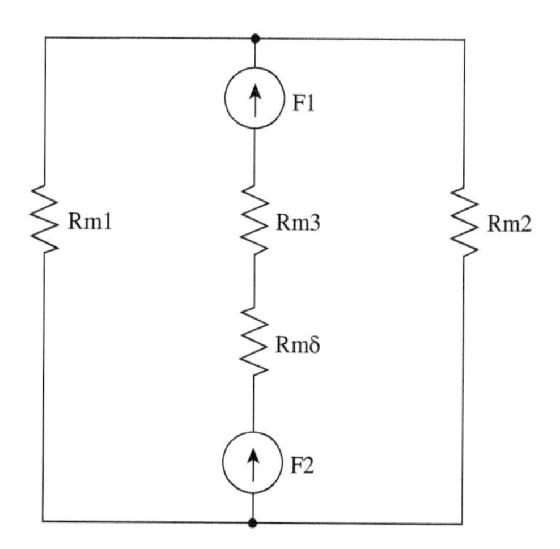


**Figure 8-71** Equivalent model of a magnetic corrosion device

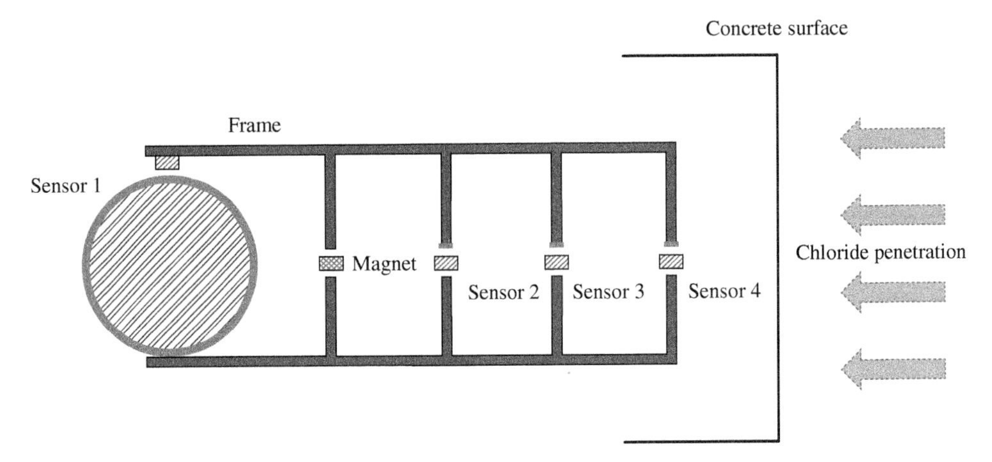


**Figure 8-72** Corrosion gradients detection apparatus

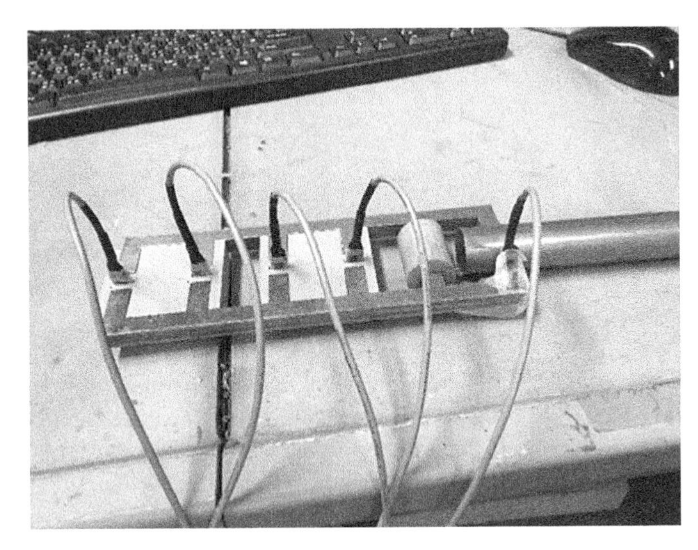

**Figure 8-73** Corrosion testing section of apparatus

of the apparatus in normal concrete structures, three channels were set in the final version of the apparatus.

### 8.7.3.3 Relationship of Mass Loss and the Output Voltage of Specimens

An experimental setup for corrosion testing is shown in Figure 8-73. Shield wires and 3-pin terminal blocks were used in the data transmission. More quality connection structures were printed by the 3D printer. The three channels version of the steel frames was used in the following experiments of calibration.

As shown in Figure 8-74 and Figure 8-75, the results of a 25-mm smooth surface rebar experiment and a 16-mm reinforced steel calibration experiment show that the mass loss of specimens is linear with changes of magnetic induction. After the fitting process, the mass loss of the specimen and the changes of output voltage were found in a manner of linear regularity. The different diameter of the specimen corresponds to their own slope in linear regularity.

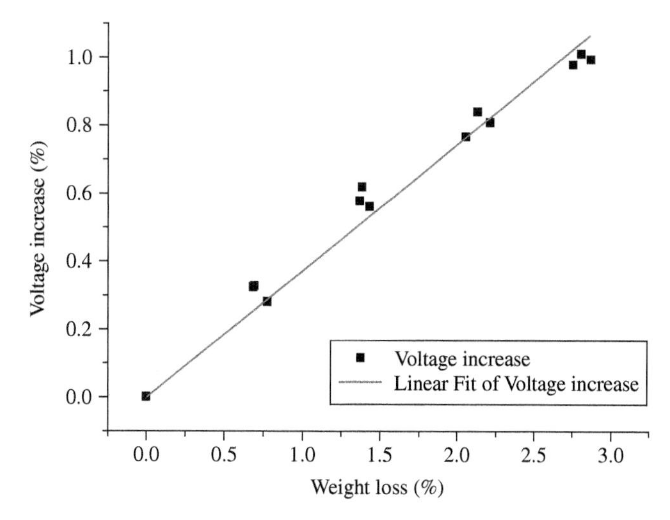


**Figure 8-74** Linear fit of experiment result of 25-mm smooth surface steel rebar

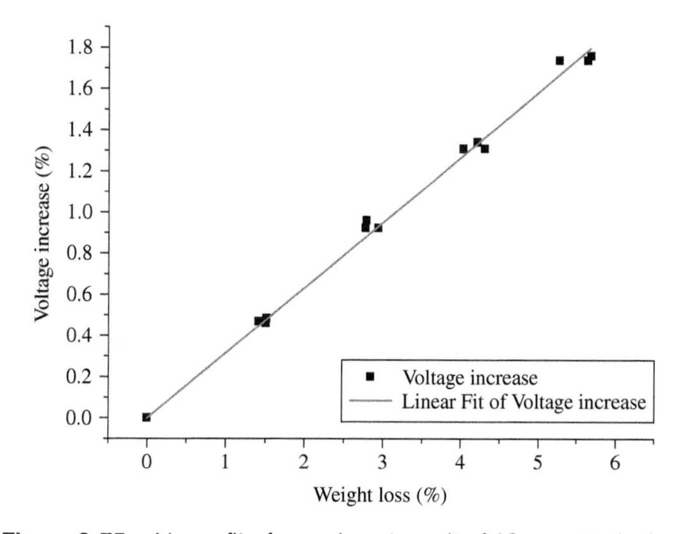


**Figure 8-75** Linear fit of experiment result of 16-mm steel rebar

### 8.7.4 Multi-Techniques Corrosion Monitoring Experiment

A steel-reinforced OPC beam was cast (180 mm × 120 mm × 2000 mm), which was reinforced by three steel rebars, one of them in the tensile zone with 16-mm diameter and the other two in the compressive zone with 12-mm diameter. The thickness of the concrete cover is about 25 mm. After 28 days of the curing period, the concrete beam was subjected to the following procedure.

Various types of non-destructive techniques, including the magnetic monitoring technique, the magnetic corrosion monitoring transducer, the acoustic emission technique and half-cell technique were used to measure the corrosion condition in the following accelerated corrosion test, as shown in Figure 8-76. A rectangular plastic pool with dimensions of 400 × 100 × 80 mm was built on the upper surface of the concrete beam, 3.5% sodium chloride solution soaked the tensile zone of the concrete beam in the pool. 3-day wet (soaked in NaCl solution) and 4-day dry (23 ± 1 °C, 50 ± 2% r.h.) cycles were applied to the reinforced concrete beam to accelerate the

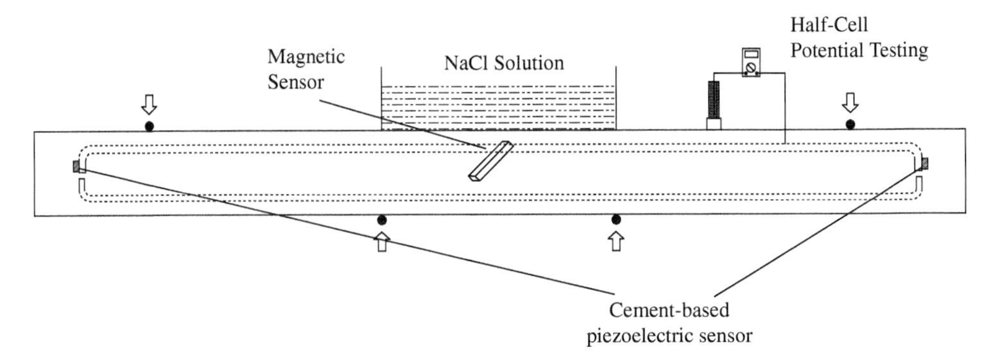


**Figure 8-76** Diagram of corrosion monitoring experiment

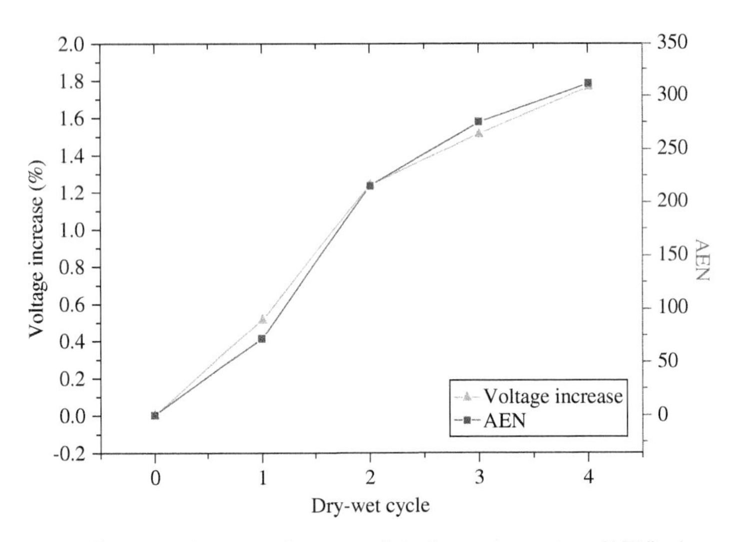


**Figure 8-77** Output voltage and accumulated event number (AEN) changes percentage in four dry-wet cycles

corrosion condition. The magnetic corrosion monitoring transducer was fixed in the middle of the tensile reinforced rebar. Two cement-based piezoelectric sensors were fixed on both sides of the straight line on the rebar by epoxy resin and hot melt adhesive. In addition, a wire was soldered to the rebar and at the other side, it emerged for half-cell potential testing. The reinforced concrete beam was subjected to four-point upwards bending coupled with wet-dry cycles. The applied bending moment was set as 4 kN·m, i.e., 40% of its ultimate bearing capacity, to avoid a crack width larger than 0.3 mm according to Eurocode 2 (2004).

In the concrete beam experiment, a magnetic monitoring device measures the output signals by magnetic transducer every dry-wet cycle. The collected data will be processed to determine the corrosion condition of the reinforced steel. Half-cell potential testing was conducted simultaneously for comparison, according to ASTM C876-91 (1999). For AE measurement, the trigger threshold was set just beyond the background noise level. Cement-based piezoelectric sensors detected signals continuously and transmitted them to the data acquisition card through a pre-amplifier (gain of 40 dB, the bandwidth of 30 kHz~1 MHz). AE signals are processed by frequency analysis to characterize the corrosion-induced deterioration of steel-reinforced OPC beams.

Figure 8-77 shows the percentage change of output voltages and accumulated event number (AEN) over four dry-wet cycles. It can be seen that signals continuously go upward, which indicates that the reinforced steel is becoming corroded and the cross-section area is decreasing. In other words, the interspace between the rebar and the Hall effect sensor is expanded. The AEN detected by the AE technique is used in this study to reflect the degree of corrosion in the reinforced steel during wet-dry cycles. The growth rate of AEN is closely related to the corrosion activity of steel reinforcement. A rapid rise of AEN shows in the first two wet-dry cycles, which indicates an active and continuous corrosion process. With the accumulation of corrosion products, the corrosion rate slows down due to the lack of sufficient oxygen to the corrosion surface of the rebar and becomes flat and stable in the growth of AEN after the second wet-dry cycle.

The half-cell potential testing result in Figure 8-78 reveals the trend of corrosion probability of the rebar over four dry-wet cycles. The result from the two methods is verified in this experiment.

For the acoustic emission detection system, AE signals of the reinforced concrete beams were detected continuously. After four dry-wet cycles experiments, only a few signals were detected. Figures 8-79–8-82 show the selected typical AE waveform (left side) and corresponding power spectral density (PSD) curve in each cycle of the experiment. It can be concluded that the frequency

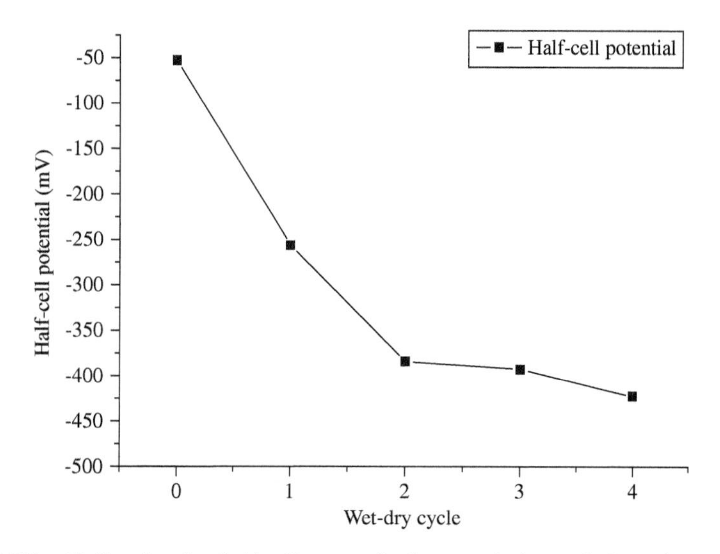


**Figure 8-78** Half-cell potential testing result of concrete beam in four dry-wet cycles

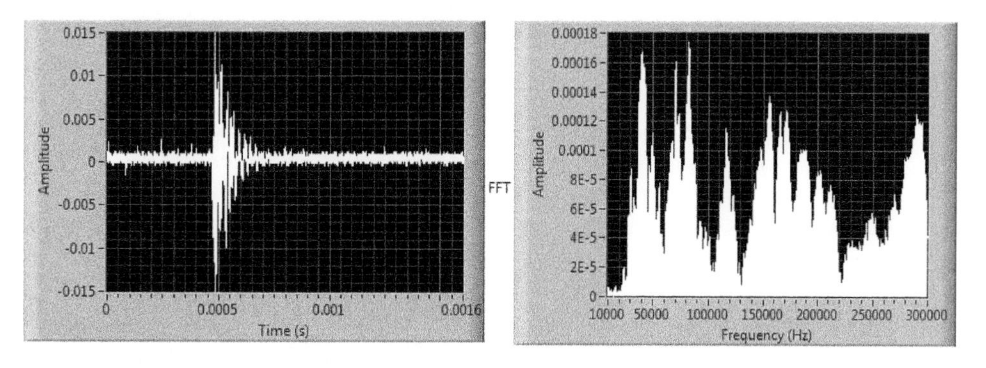


**Figure 8-79** Typical AE waveform and its corresponding PSD curve (1st cycle)

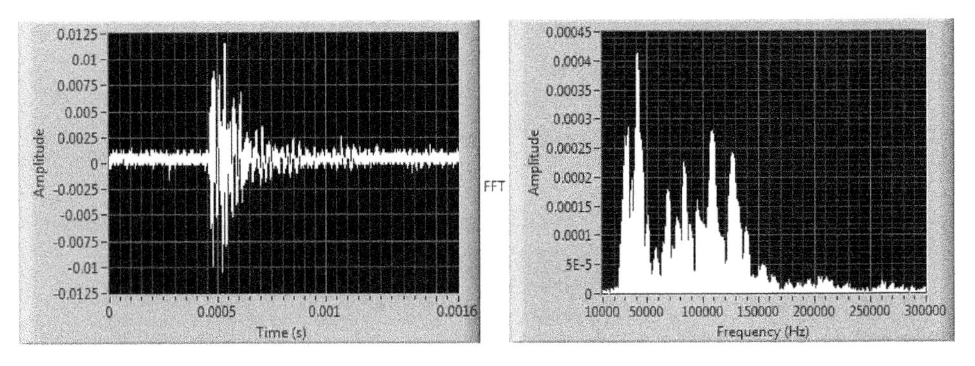


**Figure 8-80** Typical AE waveform and its corresponding PSD curve (2nd cycle)

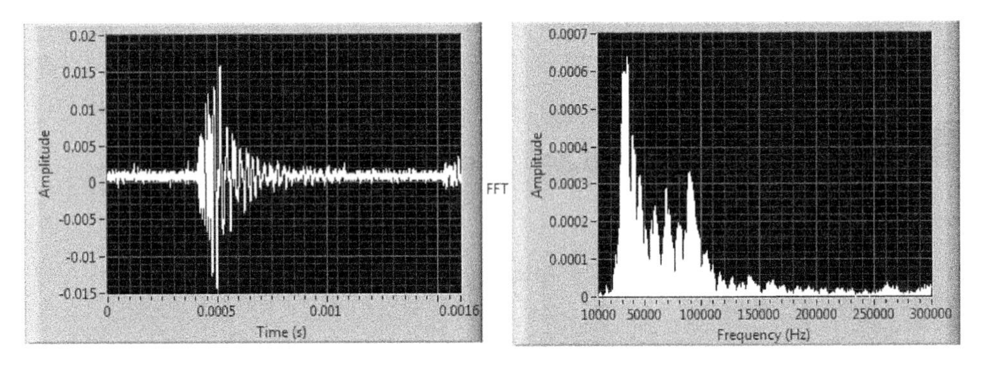


**Figure 8-81** Typical AE waveform and its corresponding PSD curve (3rd cycle)

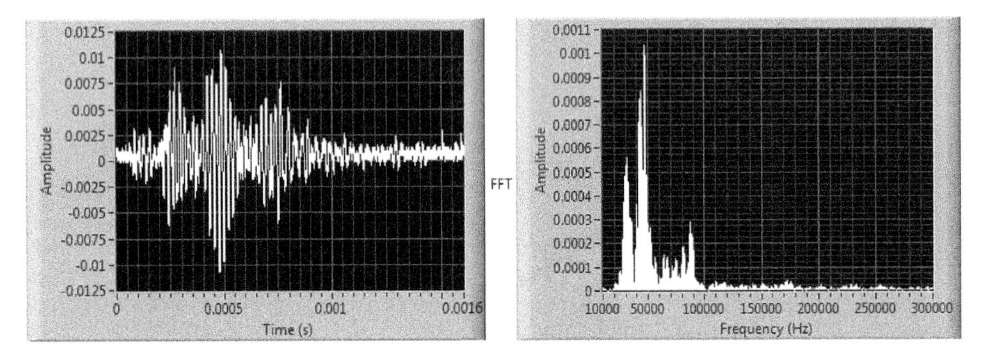


**Figure 8-82** Typical AE waveform and its corresponding PSD curve (4th cycle)

shift downward is a sign that the corrosion condition increased, and the structure damage level is increasing. This result from the acoustic emission detection system has proved the testing result of magnetic corrosion monitoring transducer by the concrete crack study.

The magnetic corrosion monitoring device shows that it possesses good capability in monitoring the corrosion process, and the result indicates the corrosion condition quantitatively of reinforced steel after comparing with the calibration experiment. The acoustic emission detection system and the half-cell potential testing in this experiment strengthen the effectiveness and credibility of the conclusions of the experiment result by mutual authentication. Significantly, the Hall effect sensors worked well in the concrete beam for 56 days in the experiment. As corrosion monitoring is a long-term job, high-quality package technology will sustain sensors in the structure for decades.

The experimental results of the half-cell potential and the acoustic emission measurement demonstrate good consistency with the magnetic monitoring apparatus result. In particular, the comparative study of three corrosion monitoring methods verifies that the magnetic device proposed in this study has a good capacity for quantitative analysis of the corrosion rate.

## DISCUSSION TOPICS

Why is NDT needed?
Which wave will travel faster, a P-wave or an S-wave?
Why is acoustic impedance an important index?
Where are sources for attenuation?
What are the differences between A-scan, B-scan, and C-scan?
In what way is the acoustic emission method unique?
How does infrared thermography detect defects below the surface of a material?
Can you give some examples of NDT in concrete engineering?
How are the durability problem and NDT related?
Do you trust the results of a single NDT method? How can you improve the accuracy?

## PROBLEMS

1. Consider an elastic rod with Young's modulus $E$, mass density of $\rho$, and a length of $L$, which is rigidly clamped at $x = L$, as shown in the Figure P8-1. The rod is initially at rest. At time $t = 0$, the end $x = 0$ is subjected to a pressure $P(t)$.

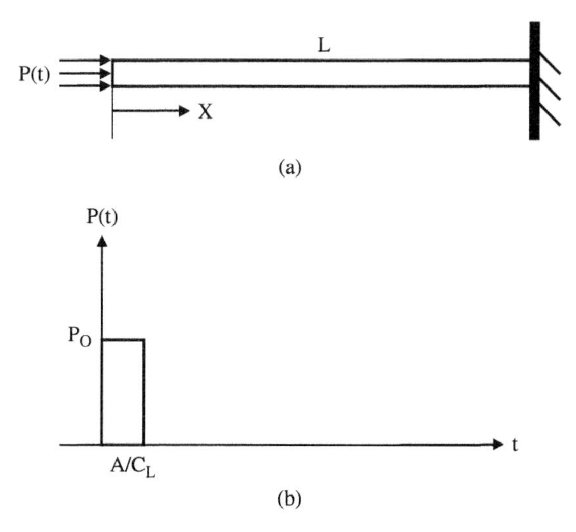


**Figure P8-1** A fixed rod is subjected to (a) pressure pulse; (b) rectangular pressure pulse

(a) If the material is brittle, and much weaker in tension than in compression, and if $p_0$ exceeds the tensile strength of the rod, at what time would you expect the rod to fail ($A$ is not negligible)?

(b) Is it possible that the material in case (a) will not fail if the length of a square wave can be adjusted? What is the minimum requirement for the length, $A$.

2. Solve the following 1D wave equation using the variable separation method:

$$U_{xx} = \frac{1}{C_{\mathrm{L}}^2} U_{tt}$$

$$U\,|_{x=0} = 0$$

$$U_x\,|_{x=l} = 0$$

3. To monitor the hydration process of fresh concrete, an experimental setup has been developed as shown in Figure P8-3. An ultrasonic transducer is attached to the surface of a steel mold. It sends a shear wave signal to the specimen and receives the reflected signal from the steel-concrete interface. The material properties of steel are Young's modulus, $E = 200\,\text{GPa}$; Poisson's ratio, $\nu = 0.3$; and relative density, $\rho = 7.8$. The properties of fresh concrete at one day are Poisson's ratio, $\nu = 0.2$; and relative density, $\rho = 2.3$. If the reflection factor measured from the first interface at 1 day is $-0.77$, calculate the modulus of the concrete at that time. What will be the reflection factor at the age of 1 day at the second interface?

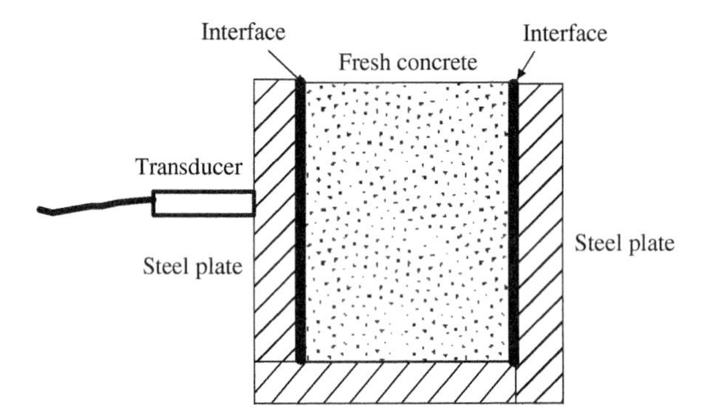


**Figure P8-3** Test setup for fresh concrete hydration process using ultrasonic shear wave

4. Consider an elastic rod with Young's modulus $E$, mass density of $\rho$, and length of $L$, which is rigidly clamped at $x = L$ ($L = 2.4\,\text{m}$), as shown in Figure P8-4. The rod is initially at rest. At time $t = 0$, the end $x = 0$ is subjected to a pressure $P(t)$. $p_0 = 5\,\text{MPa}$. The tensile strength of the material is 6 MPa (compressive strength is much higher).

(a) Will the rod fail? If so, where and when? (A = 0.2 m)

(b) Find the value of attenuation coefficient at which the material will not fail.

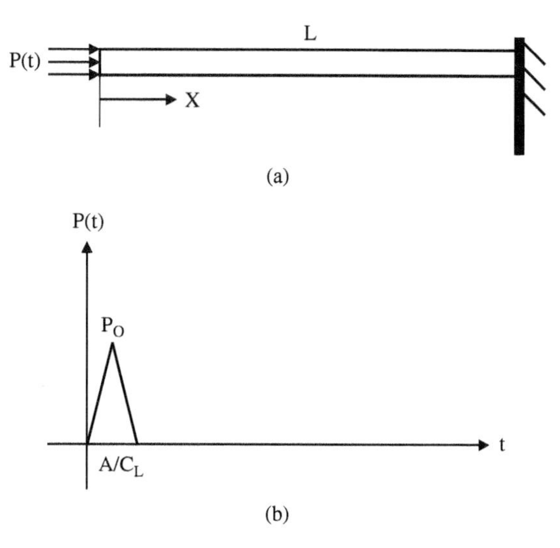


**Figure P8-4** A fixed rod is subjected to (a) pressure pulse; (b) triangle pressure pulse

5. For the infrared thermograph technique, find the solution of the heat transfer differential equation of

$$BdTdt = H - K\left(T - T_0\right)$$

$$\text{Initial condition}: t = 0,\ T = T_0.$$

(Detailed procedures are required.)

6. The resistivity of a fresh concrete can be measured using the transformer principle. The cement-based ring specimen acts as a secondary in the transformer. For a concrete ring with the shape and dimension shown in Figure P8-6, show that the resistivity can be calculated by using the following equation:

$$\rho = R_{\text{total}} \frac{h}{2\pi} \left( \left[ \ln \frac{r_3}{r_2} + \frac{r_4}{r_4 - r_3} \ln \frac{r_4}{r_3} - \frac{r_1}{r_2 - r_1} \ln \frac{r_2}{r_1} \right] \right)$$

where $R_{total}$ is the total resistance of concrete ring.

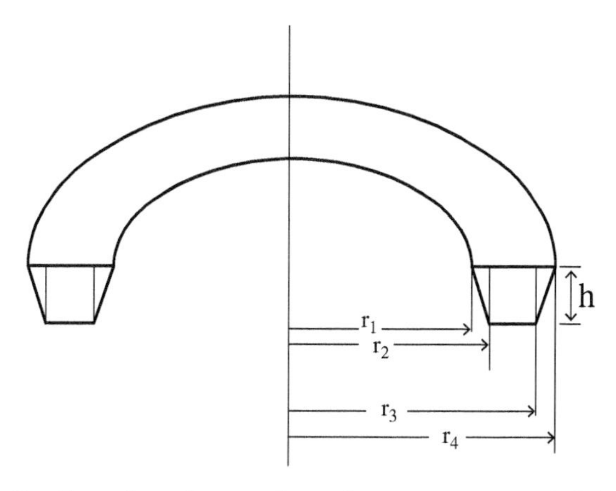


**Figure P8-6** Configuration of a specimen for noncontact resistivity measurement

**7.** To detect the corrosion in a reinforcing steel bar, three acoustic emission transducers were used. The arrangement of the transducers is shown in the Figure P8-7. The first arrive-time differences between transducers T1 and T2 is 105 μsec and between transducers T1 and T3 is 156 μsec. Find out the position of the corrosion.

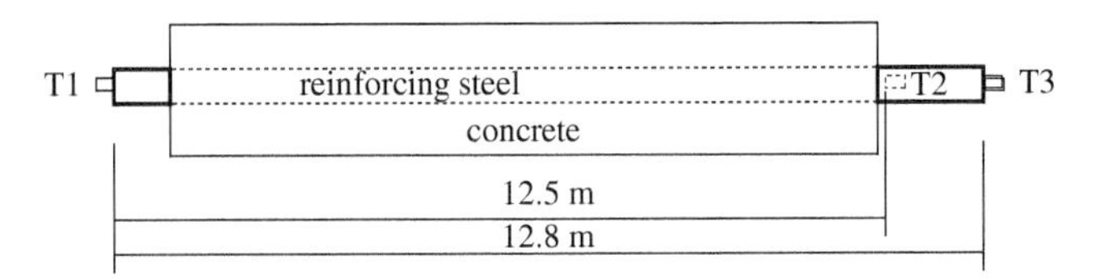


**Figure P8-7** Test setup for corrosion detection of reinforcement using acoustic emission

**8.** A concrete beam was repaired for the cracks using the injection method. To ensure the repair quality, the ultrasonic pulse method with a shear wave was used to evaluate the resin injection. The recorded pulse amplitude from the front interface of repair material and repaired concrete is −0.48 (see Figure P8-8). The properties of concrete are $E = 30$ GPa, $v = 0.2$, and $r = 2.4$. If the repair material has a Poisson's ratio ($v$) of 0.33 and density ($r$) of 1.6,

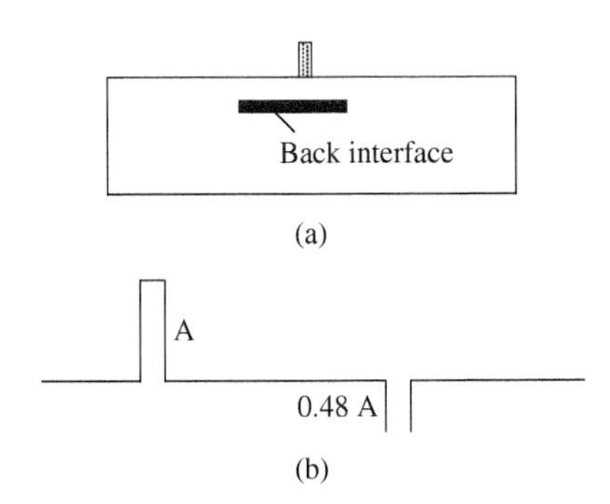


**Figure P8-8** (a) Test setup for detection of an internal crack in a concrete specimen; (b) the initial pulse amplitude and pulse amplitude reflected from the front interface of repair material and repaired concrete

**(a)** What will be the modulus of elasticity of the repair material?

**(b)** What will be the amplitude of the reflected pulse from the back interface of the repair material and repaired concrete if there is an air gap at the back interface?

**9.** A material has the following properties: for plane strain case, CL (P-wave velocity) = 5300 m/s, CT (shear wave velocity) = 3300 m/s, and density $\rho = 7800\ \text{kg/m}^3$. Find Poisson's ratio, the shear modulus, and Young's modulus for the material. Calculate $C_L$ for the plane stress case.

## REFERENCES

Achenbach, J. D., Komsky, I.N., Lee, Y.C., and Angel, Y.C. (1992) "Self-calibrating ultrasonic technique for crack depth measurement," *Journal of Nondestructive Evaluation*, 11(2), 103–108.

American Standards for Testing and Materials (1999) ASTM C876-91. "Standard test method for half-cell potentials of uncoated reinforcing steel in concrete."

American Standards for Testing and Materials (2004) ASTM C1383-04. "Standard test method for measuring the P-wave speed and the thickness of concrete plates using the impact-echo method."

Attoh-Okine, N. (1995) "Use of artificial neural networks in ground penetrating radar applications in pavement evaluation and assessment." In Schickert, G. and Wiggerhanser, H., eds., *International Symposium Non-Destructive Testing in Civil Engineering(NDT-C, E)*, Sept. 26–28, Berlin, pp. 93–100.

Balabanian, N., Bickart, T.A., and Seshu, S. (1969). Electrical network theory: Wiley New York.

Banthia, N., Djeridane, S., and Pigeon, M. (1992) "Electrical resistivity of carbon and steel micro-fiber reinforced cements," *Cement and Concrete Research*, 22(5), 804–814.

Boumiz, A., Vernet, C., and Cohen Tenoudji, F. (1996) "Mechanical properties of cement pastes and mortars at early ages," *Advanced Cement Based Materials*, 3(4), 94–106.

Brandes K., Herter J., and Helmerich R. (1998) "Assessment of remaining fatigue life of steel bridges assisted by adaptive NDT," *ISNDT-CE* [P77], 1233–1240.

Bray, D. E. and McBride, D. (1992) *Nondestructive testing techniques*, New York: Wiley.

Bray, D.E. and Stanley, R.K. (1997) *Nondestructive evaluation: A tool in design, manufacturing and service*, Boca Raton, FL: CRC Press.

Brendenberg, H., ed. (1980) *Proceedings of international seminar on the application of stress-wave theory on piles*, Stockholm, June.

Bungey, J.H. and Millard, S.G. (1996) *Testing of concrete in structures*, 3rd ed., Glasgow, U.K.: Blackie Academic & Professional.

Butters, J.N. and Leendertz, J.A. (1971a) "A double exposure technique for speckle pattern interferometry," *Journal of Physics E: Scientific Instruments*, 4(4), 277-279.

Butters, J.N. and Leendertz, J.A. (1971b) "Holographic and video techniques applied to engineering measurement," *Journal of Measurement and Control*, 4, 349–354.

Carino, N.J., Sansalone, M., and Hsu, N.N. (1986) "A point source-point receiver technique for flaw detection in concrete," *Journal of the American Concrete Institute*, 83(2), 199.

Curlin Air Tech Note (1998) NDTnet, www.ndt.net/article/1298/ndts/ndts.htm ; 3(12)

Dannheim, H., Haedrich, H., and Ruckdaeschel, R. (1995) "Measurements of stresses in ceramics and cement with microwaves." In Schickert, G. and Wiggerhanser, H., eds., *International Symposium Non-Destructive Testing in Civil Engineering(NDT-C, E)*, Sept. 26–28, Berlin, p. 1041.

Davidson, N., Padaratz, I., and Forde, M. (1995) "Quantification of bridge scour using impulse radar." In Schickert, G. and Wiggerhanser, H., eds., *International Symposium Non-Destructive Testing in Civil Engineering(NDT-C, E)*, Sept. 26–28, Berlin, pp. 61–68.

El-Enein, S. A.A., Kotkata, M.F., Hanna, G.B., Saad, M., and El Razek, M.M.A. (1995) "Electrical conductivity of concrete containing silica fume," *Cement and Concrete Research*, 25(8), 1615–1620.

Eurocode 2 (2004) "Design of concrete structures." EN 92-1-1- www.phd.eng.br/.../2015/12/en.1992.1.1 .2004.pdf

Favro, L.D., Ahmed, T., Crowther, D., Jin, H.J., Kuo, P.K., and Thomas, R.I. (1991) "Infrared thermal-wave studies of coatings and composites," in *Proceedings SPIE, 1467*, Thermosense VIII, Orlando, FL, pp. 290–294.

Gros, X. E. (1997) *NDT data fusion*, London: Arnold.

He, Z., Li, Z. J., Chen, M.Z., and Liang, W.Q. (2006) "Properties of shrinkage-reducing admixture-modified pastes and mortar," *Materials and Structures RILEM*, 39(4), 445–453.

Hughes, B.P., Soleit, A.K.O. and Brierley, R.W. (1985) "New technique for determining the electrical resistivity of concrete," *Magazine of Concrete Research*, 37(133), 243–248.

Hung, Y. Y. (1982) "Shearography: A new optical method for strain measurement and non-destructive testing," *Optical Engineering*, 21(3), 391–395.

Hung, Y.Y. and Liang, C.Y. (1979) "Image-shearing camera for direct measurement of surface strains," *Applied Optics*, 18(7), 1046–1051.

Jin, X.Y. and Li, Z. J. (2001) "Dynamic property determination for early-age concrete," *ACI Materials Journal*, 98(5), 365–370.

Jones, R. (1962) *Non-destructive testing of concrete*, London: Cambridge University Press.

Jones, R. and Wykes, C., (1989) *Holographic and speckle interferometry*, 2nd ed., London: Cambridge University Press.

Kalinski, J. (1995) "On-line water content monitoring in clay by means of microwave method and instrumentation." In Schickert, G. and Wiggerhanser, H., eds., *International Symposium Non-Destructive Testing in Civil Engineering(NDT-C, E)*, Sept. 26–28, Berlin, p. 1005.

Kesler, C.E. and Higuchi, Y. (1953) "Determination of compressive strength of concrete by using its sonic properties," *Proceedings of the ASTM*, 53, 1044.

Kesler, C.E. and Higuchi, Y. (1954) "Problems in the sonic testing of plain concrete," *Proceedings of the International Symposium on Nondestructive Testing of Materials and Structures*, Vol. 1, Paris: RILEM, p. 45.

Khalaf, F.M. and Wilson, J.G. (1999) "Electrical properties of freshly mixed concrete," *Journal of Materials in Civil Engineering*, 11(3), 242–248.

Krause, M. et al. (1995) "Comparison of pulse-echo-methods for testing concrete." In Schickert, G. and Wiggerhanser, H., eds., *International Symposium Non-Destructive Testing in Civil Engineering (NDT-C, E)*, Sept. 26–28, Berlin, pp. 281–296.

Lakshminarayanan, V., Ramesh, P.S., and Rajagopalan, S.R. (1992) "A new technique for the measurement of the electrical resistivity of concrete," *Magazine of Concrete Research*, 44(158), 47–52.

Landis, E.N., and Shah, S.P. (1993) "Recovery of microcrack parameters in mortar using quantitative acoustic emission," *Journal of Nondestructive Evaluation*, 12(4), 219–232.

Levita, G., Marchetti, A., Gallone, G., Princigallo, A., and Guerrini, G.L. (2000) "Electrical properties of fluidified Portland cement mixes in the early stage of hydration," *Cement and Concrete Research*, 30(6), 923–930.

Li, Z. (1996) "Microcrack characterization in concrete under uniaxial tension," *Magazine of Concrete Research*, 48(176), 219–228.

Li, Z., Li, F.M., Li, X.S., and Yang W.L. (2000) "P-wave arrival determination and AE characterization of concrete," *Journal of Engineering Mechanics*, 126(2), 194–200.

Li, Z., Li, F., Zdunek, A., Landis, E., and Shah, S.P. (1998) "Application of acoustic emission to detection of rebar corrosion in concrete," *ACI Materials Journal*, 95(1): 68–76.

Li, Z. and Li, W. (2002) "Contactless, transformer-based measurement of the resistivity of materials," United States Patent 6639401.

Li, Z. and Shah, S.P. (1994) "Microcracking localization in concrete under uniaxial tension: AE technique application," *ACI Materials Journal*, 91(4): 372–389.

Li, Z. and Wei, X. (2003) "The electrical resistivity of cement paste incorporated with retarder," *Journal of Wuhan University of Technology-Materials Science Edition*, 18(3), 76–78.

Li, Z., Wei, X., and Li, W. (2003) "Preliminary interpretation of Portland cement hydration process using resistivity measurements," *ACI Materials Journal*, 100(3), 253–257.

Lin, J. M. and Sansalone, M. (1997) "A procedure for determining P-wave speed in concrete for use in impact-echo testing using a Rayleigh wave speed measurement technique," *Innovations in Nondestructive Testing*, SP-168, Detroit, MI: American Concrete Institute.

Lin, J.M., Sansalone, M., and Streett, W.B. (1997) "A procedure for determining P-wave speed in concrete for use in impact-echo testing using a P-wave speed measurement technique," *ACI Materials Journal*, 94(6), 531–539.

Mailvaganam, N.P. (1992) *Repair and protection of concrete structures*, Boca Raton, FL: CRC Press.

Maji, A.K., Ouyang, C., and Shah, S.P. (1990) "Fracture mechanism of quasi-brittle material based on acoustic emission," *Journal of Material Research*, 5(1), 206–217.

Maji, A. K., and Shah, S.P. (1988) "Process zone and acoustic emission measurement in concrete," *Experimental Mechanics*, 28, 27–33.

Maji, A. K., and Shah, S.P. (1989) "Application of acoustic emission and laser holography to study micro-fracture in concrete," in *Nondestructive Testing of Concrete,* SP-112, Detroit, MI: American Concrete Institute, pp. 83–109.

Malhotra, V.M. and Carino, N.J. (2004) *Handbook of nondestructive testing of concrete*, 2nd ed., Boca Raton, FL: CRC Press.

McCarter, W.J., Forde, M.C., and Whittington, H.W. (1981) "Resistivity characteristics of concrete," *Proceedings of the Institution of Civil Engineers (London), Part 1: Design & Construction*, 71(2), pp. 107–117.

Milne, J.M. and Carter, P. (1988) "A transient thermal method of measuring the depths of sub-surface flaws in metals," *British Journal of Nondestructive Testing*, 30(5), 333–336.

Newman, A. (2001) *Structural renovation of buildings: Methods, details, and design examples*, New York: McGraw-Hill.

Obert, L. and Duvall, W.I. (1941) "Discussions of dynamic methods of testing concrete with suggestions for standardization," *Proceedings of the ASTM*, 41, 1053.

Olson, L.D. and Wright, C.C. (1990) "Seismic, sonic, and vibration methods for quality assurance and forensic investigation of geotechnical, pavement, and structural systems." In dos Reis, H.L.M., ed., *Proceedings of Conference on Nondestructive Testing and Evaluation for Manufacturing and Construction, Hemisphere*, p. 263.

Oppenheim, Alan V. and Schafer, Ronald W. (1975) *Digital signal processing*, Englewood Cliffs, NJ: Prentice Hall.

Ouyang, C.S., Landis, E., and Shah, S.P. (1991) "Damage assessment in concrete using quantitative acoustic emission," *ASCE Journal of Engineering Mechanics*, 117(11), 2681–2698.

Polar, A., Indacochea J.E., and Wang, M.L. (2010) "Sensing creep evaluation in 410 stainless steel by magnetic measurements." *Journal of Engineering Materials and Technology*, 132(4), 041004.

Qin, L. and Li, Z.J. (2008) "Monitoring of cement hydration using embedded piezoelectric transducers," *Smart Materials and Structures*, 17(5), Article Number 055005.

Reynolds, W.N. (1988) "Inspection of laminates and adhesive bonds by pulse-video thermograph," *NDT International*, 21(4), 229–232.

Rumiche, F., Indacochea J.E., and Wang M.L. (2008) "Detection and monitoring of corrosion in structural carbon steels using electromagnetic sensors." *Journal of Engineering Materials and Technology*, 130(3), 031008.

Sack, D. and Olson, L. (1995) "High speed testing technologies for NDT of structures." In: Schickert, G. and Wiggerhanser, H., eds., *International Symposium Non-Destructive Testing in Civil Engineering (NDT-C, E)*, Sept. 26–28, Berlin, pp. 43–50.

Sansalone, M. (1997) "Impact-echo: The complete story," *ACI Structural Journal*, 94(6), 777–786.

Sansalone, M. and Carino, N.J. (1988) "Impact-echo method: Detecting honeycombing, the depth of surface-opening cracks, and ungrouted ducts," *Concrete International*, 10(4), 38–46.

Sansalone, M. and Carino, N.J. (1990) "Finite element studies of the impact-echo response of layered plates containing flaws." In McGonnagle, W., ed., *International advances in nondestructive testing*, 15th ed., New York: Gordon & Breach, pp. 313–336.

Steinbach, J. and Vey, E. (1975) "Caisson evaluation by stress wave propagation method," *ASCE Journal of Geotechnical Engineering, ASCE*, 101(4), 361–378.

Suaris, W. and van Mier, J.G.M. (1995) "Acoustic emission source characterization in concrete under biaxial loading," *Materials and Structures*, 28(182), 444–449.

Subramaniam, K.V., Popovics, J.S., and Shah, S.R. (1999) "Fatigue response of concrete subjected to biaxial stresses in the compression-tension region," *ACI Materials Journal*, 96(6), 663–669.

Vassie, P.R. (1978) "Evaluation of techniques for investigating the corrosion of steel in concrete," Department of Transport, TRRL Report SR397, Crowthorne, U.K: Department of the Environment.

Wei, X.S. and Li, Z.J. (2005) "Study on hydration of Portland cement with fly ash using electrical measurement," *Materials and Structures, RILEM*, 38(5), 411–417.

Wei, X. S., Xiao, L.Z. and Li, Z.J. (2008) "Hyperbolic method to analyze the electrical resistivity curve of Portland cements with superplasticizer," *Journal of Wuhan University of Technology-Materials Science Edition*, 23(2), 245–248.

Weil, S. (1995) "Non-destructive testing of bridge, highway and airport pavements." In Schickert, G. and Wiggerhanser, H., eds., *International Symposium Non-Destructive Testing in Civil Engineering (NDT-C, E)*, Sept. 26–28, Berlin, pp. 467–474

Whittington, H.W., McCarter, J., and Forde, M.C. (1981) "Conduction of electricity through concrete," *Magazine of Concrete Research*, 33(114), 48–60.

Xiao, L.Z. and Li, Z.J. (2006) "Hydration monitoring of cementitious materials by using electrical resistivity measurement." In Reinhardt, H.W., ed., *Proceedings of Advanced Testing of Fresh Cementitious Materials*, Aug. 3–4, Stuttgart, Germany, pp. 167–176.

Xiao, L.Z., Li, Z.J., and Wei, X.S, (2007) "Selection of superplasticizer in concrete mix design by measuring the early electrical resistivities of pastes," *Cement Concrete and Composites*, 29, 350–356.

Zhang, J. and Li, Z.J. (2009) "Application of GEM equation in microstructure characterization of cement-based materials," *Journal of Materials in Civil Engineering*, 21(11), 648–656.

CHAPTER 9

# THE FUTURE AND DEVELOPMENT TRENDS OF CONCRETE

Due to the unique advantages of concrete, it will continue to be the most popular and most widely used man-made material in the foreseeable future. The global demand for concrete, especially in developing countries and regions, will keep increasing. Subsequently, the research and development of concrete have to be advanced to meet the need and new requests from end users and to face new challenges. Hence, it is important to correctly predict the future and development trends of concrete in the twenty-first century. Here, the issues closely related to this topic are briefly discussed.

## 9.1 SUSTAINABILITY OF CONCRETE

Of all the future development trends, how to make concrete more sustainable is the most important issue. Sustainability can be defined as the "development that meets the needs of the present without compromising the ability of future generations to meet their own needs," according to the Brundtland Commission Report (Brundtland, 1987). Sustainability covers all aspects of society, such as civil infrastructural systems, energy, environment, health, safety, and so on. Concrete is the most widely used material and consumes a great amount of resources. Contemporary concrete, typically, contains about 16–20% cementitious materials, 6–8% mixing water, 60–70% aggregate, and 2–3% of admixtures by mass. The most reliable way to estimate world production of concrete is to determine the amount of cement produced, calculate the amount of other composite using the mass ratio of the common practice, and finally add them together. According to the amount of cement produced in 2020, it can be estimated that concrete consumes about 4.1 billion tonnes of cement, 1.6 billion tonnes of mixing water, 16 billion tonnes of aggregates together with 600 million tonnes of chemical admixtures annually. It shows that concrete is a 22-billion-tonnes-a-year industry, the largest user of natural resources in the world. The mining, processing, and transport of huge quantities of aggregate, in addition to the billions of tonnes of raw materials needed for cement manufacturing, use considerable energy and adversely affect the ecology of virgin lands. Moreover, a large amount of $CO_2$ is produced. Obviously, such a huge consumption of natural resources has a great impact on the environment, and if not well controlled, will definitely compromise "the ability of future generations to meet their own needs." The considerations regarding the sustainability of concrete have three aspects: (1) how to reduce greenhouse gas emissions, especially $CO_2$ emissions; (2) how to reduce energy consumption; and (3) how to reduce the consumption of raw materials from natural resources during the production, construction, and application of concrete. The ways to achieve the sustainability of concrete include reduction (e.g., $CO_2$ emission and energy consumption), reusing (e.g., valuable resources), recycling (e.g., demolished waste), and reinventing (e.g., breakthrough technologies to replace the current $CO_2$- and energy-intensive practices).

A multi-stakeholder working group initiated by the United Nations Environment Program Sustainable Building and Climate Initiative (UNEP-SBCI) has identified three categories of effective approaches to improve the eco-efficiency of cement: (1) increasing use of $CO_2$- and energy-efficient supplementary cementitious materials (SCMs); (2) developing sustainable alternative cements; and (3) improving cement efficiency (including production efficiency and utilization efficiency, such as longer service life, higher strength-to-density ratio, and lower cement content in unit volume of concrete) (United Nations et al., 2018). Similar strategies also apply to aggregate. To be effective, any approaches in such categories have to be based on abundant raw material resources and to achieve the desired performance (e.g., comparable to regular concrete to ensure market acceptance).

### 9.1.1 Scientific Utilization of More Eco-Efficient Resources

One way to make concrete more sustainable is to use low-carbon-embodiment industrial wastes/by-products or natural resources to replace the raw materials (i.e., cement and aggregate) for making concrete.

Such materials for replacing cement are usually known as SCMs. Currently, ground-granulated blast-furnace slag (GGBS), coal-combustion fly ash, limestone powders, silica fume, and metakaolin are the most commonly used SCMs. These SCMs can be obtained in large and regular amounts with a relatively consistent composition. They can be added into cement during the final grinding process of cement production to reduce the amount of clinker used, and the products are called blended cements. They can also be added into concrete mixtures during concrete production to reduce the amount of cement. No matter which way it is done, the use of SCMs can reduce the amount of cement, which is the most $CO_2$- and energy-intensive component of concrete. The production of one tonne of cement consumes 4 GJ energy, and releases 0.8 to 1 tonnes of $CO_2$ into the atmosphere, as indicated by the following calculations.

In cement manufacture, limestone has to be decomposed as

$$\underset{(100)}{\text{Limestone}(CaCO_3)(1000^\circ C)} \rightarrow \underset{(56)}{CaO} + \underset{(44)}{CO_2} \tag{9-1}$$

One tonne of cement contains 620 kg CaO on average, and hence $CO_2 = 620 \times 44/56 = 487$ kg. $CO_2$ is also produced from fuel burning during the cement production. The amount ranges from 320 kg to about 450 kg, depending on the advances of the burning technique. The world's yearly cement output accounts for nearly 10% of the global anthropogenic $CO_2$ emissions. Thus, use of SCMs in large volumes can greatly reduce the environmental impact of the concrete industry. Moreover, the SCMs could benefit the performance of concrete too. For instance, SCMs containing reactive silica (e.g., fly ash and silica fume) can react with CH (from cement hydration) in concrete to form secondary C–S–H, densify the microstructure, and improve the concrete properties; and SCMs containing reactive alumina (e.g., GGBS) could improve the chloride binding capacity, and, thus, enhance the durability of concrete structures in marine environments. Due to the incorporation of SCMs into concrete, the structure of concrete at the nanometer and micrometer levels becomes more complex. Thus, deeper studies have to be carried out to reveal the influence of SCMs to discover to what extent and how they can modify the structure of hydration products and subsequently the macroscopic properties of concrete. The optimized amount of cement replacement at which concrete can most benefit should be investigated and proposed. In this way, the scientific use of current SCMs can be achieved.

It should be pointed out that many countries and regions have been fully using their regular SCMs in concrete and even facing severe shortages of quality SCMs. To further reduce the tonnage of Portland cement in concrete and improve the eco-efficiency, new resources of SCMs have to be explored. Such resources should include other industrial by-products that also contain reactive silica and/or reactive alumina, such as Pb/Zn mine tailing, phosphorous slag, coal gangue, and copper slag. It may be possible for them to achieve effects similar to those of the currently used SCMs. The search for industrial wastes as SCMs may be extended to the resources that have been historically ignored or considered unreactive, hazardous or deleterious, such as steel slag, air-cooled iron slag, off-specification coal ash (not conforming to ASTM C618; including bottom ash and reclaimed fly ash), and municipal solid waste incineration (MSWI) ashes. Taking the US as an example: ~90 million tonnes of cement are currently produced every year, and around 40 million tonnes of SCMs are in demand. However, the dominant regular SCMs (i.e., GGBS and on-specification fly ash) are far from sufficient. Meanwhile, the US generates ~7 million tonnes of steel/air-cooled iron slags (USGS, 2020), ~20 million tonnes of off-specification coal ashes (ACAA, 2019), and ~10 million tonnes of MSWI ashes (EPA, 2017), plus a more than 2.5 billion tonnes of these wastes in repository—stockpiles, landfills, and/or ash ponds. Simply from the view of mass, these wastes can provide sufficient SCMs for the domestic concrete industry. The same situation may apply to other major economies, such as China, where the annual cement production is more than 20 times higher than that of the US and the generation of the underused wastes is also at least one order of magnitude larger. However, the new resources of wastes are typically characterized by high compositional variation and high content of deleterious materials (with respect to concrete performance, e.g., chloride, sulfate, and unburned carbon) and toxic materials (with respect to human health and environmental safety, e.g., heavy metals). Extensive studies are needed on each of them before they can be readily used. Only with a better understanding of their reaction mechanisms and effective approaches to suppress their adverse effects on concrete and the environment can we open up them for SCM resources so they can contribute to the sustainable development of concrete.

Natural mineral deposits also provide promising resources for the SCMs. Apart from the limestone powder mentioned above, these also include natural pozzolans, such as volcano ash, which is ready to be used, and calcined clays (e.g., metakaolin, which is thermally activated kaolinite-rich clay) (Scrivener and Kirkpatrick, 2007). Metakaolin has similar effect to that of silica fume, but is less expensive. Hence, it has great potential for future applications in concrete. More details on metakaolin, calcined clay, and limestone calcined clay cement can be found in Chapter 2. Because of the vast natural repositories and relatively eco-efficient processing, limestone calcined clay cement has been considered an eco-efficient solution for the cement and concrete industry, and launched commercially as FutureCem™ by Cementir Holding in 2021 (Cementir Holding, 2020).

To support the discovery and design of a broader range of SCMs, ASTM has launched the *Standard Guide for Evaluation of Alternative Supplementary Cementitious Materials (ASCM) for Use in Concrete* (ASTM C1709). This guide is intended to specify a detailed protocol to evaluate potential SCM resources that are beyond the scope of ASTM C618 (Class F/C coal fly ash and Class N natural pozzolan), C989 (GGNS/slag cement), and C1240 (silica fume). If some potential resources fail to pass ASTM C1709, future technologies may be developed to process them, even into carbon-efficient multi-phase SCMs similar to limestone-calcined clay.

Aggregate (fine and coarse) represents the most extracted group of materials, even more than fossil fuels (Torres et al., 2017). Its usage exceeds the rate of natural renewal of these minerals, which entails that the demand will outstrip supply before 2050 (Sverdrup et al., 2017). To make the extraction of natural resources more sustainable, some industrial wastes can be used to replace aggregates in making concrete. These wastes include, but are not limited to, waste glass, demolished concrete, bottom ash, and large-size slag (e.g., steel slag and copper slag). Bottom ash and slags can normally be accepted as aggregate (e.g., to partially replace natural aggregate) if potential deleterious substances (i.e., free lime and unburned carbon) can be effectively controlled to avoid premature degradations of concrete.

Figure 9-1 shows crushed waste glass as a replacement for fine aggregate. Grading of waste glass particles is usually limited to 75 μm to 5 mm. The main concern about glass incorporation into concrete as aggregate is the possibility of the alkali–silica reaction (ASR, detailed in Chapter 5). It has been reported that the risk of ASR reduces with the decreasing particle size of crushed waste glass. When the glass is ground into a fine powder comparable to cement, it can even serve as a pozzolanic material. The ASR risk of waste glass incorporated in concrete can also be controlled by using dry-mixing (versus wet-mixing) and partially replacing cement by pozzolanic SCMs (Lee et al., 2011).

Since sand and gravel are used mainly in concrete, reusing demolished concrete has been listed as a prerequisite for sustainable extraction of these natural resources (Bendixen et al., 2019). It should be emphasized that as more and more concrete structures age, more and more demolished concrete will be produced. The U.S. Environmental Protection Agency (EPA) estimated that concrete constituted more than 70% of the 548 million tons of construction and demolition waste generated in the US in 2015 (a 69% increase since 2003) (EPA, 2004; 2016). For a long time, waste concrete was only disposed of in landfills or used as backfilling material (Gastaldi et al., 2015). Notwithstanding, since the 1990s, recycling of concrete has gained increasing importance, because of the above-mentioned concerns pertaining to sustainability. Currently, the construction and demolition recycling industry is worth over US$17 billion, and the landfill avoided by recycling the construction and demolition waste is equivalent to an area of 17,401,498 $m^2$ (4,300 acres) at a depth of 15.24 m (50 ft) per year (CDRA, 2014). In current practice, concrete waste is typically processed into recycled concrete aggregates (RCAs) to fulfill the enormous need for aggregate. However, the majority of RCAs are used in road base/subbase constructions and engineering fills, while only a small fraction is used in making new structural concrete. These primary applications qualify as low-value uses, because the cost of transporting, crushing, and processing concrete waste into RCAs can exceed the sale price of normal sub-base aggregate. The recycling industry can only be economically stimulated to produce RCAs if they are used in high-value applications

**Figure 9-1** Recycled waste glass as an aggregate for concrete

(i.e., upcycling), such as making high-strength, structural concrete (Katz, 2003; Topcu and Sengel, 2004). In spite of these environmental and economic merits, RCAs have not been widely employed in "high-value" applications, due to a series of technical concerns about RCAs.

The main difference between RCAs and natural aggregates is that hydrated cement paste is attached to RCAs (de Juan and Gutiérrez, 2009). The paste is generally more porous and weaker than natural sand and rocks, making the quality of RCAs poorer as compared to natural aggregates. The higher water absorption ratio and the rougher surface result in a higher water demand for RCA concrete to achieve a specified workability (Xiao et al., 2012). The incorporation of RCAs can negatively affect the compressive, tensile, and flexural strength, as well as the modulus of elasticity of concrete, mainly due to the presence of micro-cracks, damaged interfacial transition zone (ITZ), and weaker cement paste components (González-Taboada et al., 2016; Waseem and Singh, 2017). Drying shrinkage is another challenge that limits large-scale use of RCAs in concrete. RCAs can increase drying shrinkage because of their higher water absorption ratio and poorer water restraining capacities (Sagoe-Crentsil et al., 2011; Silva et al., 2015). When used at high replacement level of natural aggregate, RCAs could also lead to higher chloride ion permeability (Andreu and Miren, 2014; Duan and Poon, 2014), larger carbonation depth (Silva et al., 2016), and poorer freeze-thaw damage resistance if the parent concrete was not properly air-entrained (Gokce et al., 2004). In summary, the incorporation of RCAs tends to be detrimental to concrete properties. However, researchers do not agree to what extent these detriments could be. The highly variable and discrete experimental results can be attributed to the variability of RCAs' properties, including disparate porosity/water absorption, and mechanical properties (e.g., Los Angeles abrasion mass loss), apart from other minor factors like chemicals and organic residues in RCAs or the parent concrete (Sadati, 2017). According to the U.S. Federal Highway Administration, RCAs used in pavement can have specific gravities from 2.0 to 2.5, water absorption ratios from 2% to over 10%, and Los Angeles abrasion mass loss from 20% to 45% (Shi et al., 2016). The highly variable properties, resulting from the quality of the parent concrete and the crushing method, make the design and application of high-value RCA concrete quite difficult.

To summarize the above discussion, two issues must be solved to facilitate high-efficiency use of RCAs in making new concrete: (1) the properties of RCAs must be enhanced; and (2) the variability in RCAs' properties must be reduced. To enhance the properties of RCAs, removing and strengthening the adhered paste/mortar are the two general strategies. Removal of paste is typically accomplished by mechanical grinding or pre-soaking. Grinding, sometimes assisted by heating, is the simplest procedure to remove the mortar adhering to coarse RCA (Bru et al., 2014). However, this procedure can create new cracks due to collision. Pre-soaking in water (assisted by ultrasound) and acid solutions have been used to remove the weakly adhered mortar and cement hydrates from RCA grains (Katz, 2004; Tam et al., 2007). The shortcomings of the "removal" strategy are obvious—the need for large amounts of energy or chemicals, and potential harm to the produced RCAs. Strengthening can be achieved through different surface treatments or impregnations, such as using latexes, silane/siloxane solution, paraffin, or pozzolan slurries containing fly ash, silica fume, or nano-silica (Katz, 2004; Tsujino et al., 2007; Kou and Poon, 2012). Carbonation (strengthening RCAs by taking up $CO_2$) has been considered as the most sustainable choice (Thiery et al., 2013; Zhang et al., 2015), but, the benefits might be limited because the amount of CH in the surface layer of RCAs is fairly low. In conclusion, many methods may enhance the properties of RCAs to some extent, but the benefits could be limited due to the underlying mechanisms or detrimental side effects. Furthermore, these methods are incapable of reducing the variability of RCAs. In the future, while methodologies are improved to enhance the properties of RCA and RCA concrete, more attentions may be paid to reducing the variability. After all, it is difficult to produce concrete with relatively consistent properties using extremely varying raw material feedstock.

### 9.1.2 Low Energy and Low $CO_2$ Emission Binders

To reduce the environmental impact of Portland cement, attempts have been made to search for other type of binders with low energy demand and less emission of carbon dioxide. One system along this line is alkali-activated cementitious materials, of which geopolymer and alkali-activated slag are two good examples. Geopolymer was discussed in detail in Chapter 2. Activated slag also uses water glass and/or other alkaline solutions to trigger the chemical reaction with slag to form bonding. However, although the feasibility of alkali-activated materials has been frequently demonstrated in the laboratory, their commercial exploitation has not been widespread due to variability of their performance, especially with changes in temperature; the high cost of the most effective alkaline activators (e.g., water glass); and the lack of robustness with cheaper activators. Alkali-activated systems are generally fast-setting and the use of superplasticizers is not effective. More research is needed to promote the practical use of alkali-activated cementitious materials.

Another system uses the MgO-based binder, such as magnesium phosphate cement, as introduced in detail in Chapter 2. As pointed out earlier, how to control the setting time is a big issue for such material to be applied on a large scale. Other systems include new clinker types that contain higher amounts of alumina and sulfate than Portland cement. These clinkers include high aluminate cement, sulfoaluminate cement, and high-belite cement. A new high-belite cement was developed recently in China (Sui et al., 2009). This type of belite-based Portland cement (high-belite cement, HBC) contains 45–60% $C_2S$, 20–30% $C_3S$, 3–7% $C_3A$, and 10–15% $C_4AF$. Laboratory research, industrial production, and field application of the resultant HBC concrete demonstrate that HBC compared with normal PC is a kind of low energy consumption, low $CO_2$ emission Portland cement with low hydration heat evolution, high later-age strength, and high performance. For example, the clinkering temperature of HBC is 1350°C, which is 100°C lower than normal PC. The $CO_2$ emission for clinkering HBC is reduced by 10% due to the low calcium design in the clinker mineral composition and low consumption of coal for clinker burning. The resultant HBC concrete shows excellent performance not only in better workability, higher mechanical strength, and excellent durability, but also in excellent thermal properties and crack resistance. Table 9-1 compares the energy consumption and $CO_2$ emission for $C_2S$ and $C_3S$. Obviously, the high $C_2S$ content in the new HBC has led to energy savings and $CO_2$ emission reduction.

The concrete produced using HBC shows a better workability compared to the one made of normal Portland cement when low *w/c* ratios are used, as shown in Table 9-2. Although the strength development of the concrete made from HBC is still slower at the early stage, up to an age of 7 days, it catches up to the strength of concrete made with ordinary Portland cement at the age of 28 days and exceeds it at 90 days, as shown in Table 9-3 for flexural, splitting, and compressive strengths. Compared to the existing low-heat Portland cements (e.g., ASTM C150 Type IV), which also contains a large amount of belite, the new binder has achieved a much higher strength at early ages, as shown in Table 9-4.

Concrete made of the new HBC has also demonstrated a better freeze–thaw resistance, smaller drying shrinkage, and lower heat release. The new HBC concrete has been applied in

**Table 9-1** Comparison of energy consumption and $CO_2$ emission for $C_2S$ and $C_3S$

| Mineral | Formation enthalpy kJ/kg | Formation temperature °C | CaO % | $CO_2$ emission coefficient per unit mass of mineral |
|---|---|---|---|---|
| $C_3S$ | 1848 | 1450 | 73.7 | 0.578 |
| $C_2S$ | 1336 | 1300 | 65.1 | 0.511 |

**Table 9-2** Workability of HBC high-strength, high-performance concrete in comparison with OPC concrete

| Concrete type | Cement amount (kg/m³) | *W/B* | Fly ash (kg/m³) | Water (kg/m³) | Superplasticizer (%) | Initial slump (cm) | Slump at 90 min (cm) |
|---|---|---|---|---|---|---|---|
| C60 HBC | 414 | 0.32 | 104 | 165 | 1.0 | 23.2 | 22.5 |
| C60 OPC | 414 | 0.32 | 104 | 165 | 1.0 | 22.4 | 20.3 |
| C80 HBC | 510 | 0.25 | 90 | 150 | 1.7 | 24.4 | 20.6 |
| C80 OPC | 510 | 0.25 | 90 | 150 | 1.7 | 0 | 0 |

**Table 9-3** Strength comparison between HBC and OPC concretes

| | Flexural strength (MPa) | | | Splitting strength (MPa) | | | Compressive strength (MPa) | | |
|---|---|---|---|---|---|---|---|---|---|
| Type | 7 d | 28 d | 90 d | 7 d | 28 d | 90 d | 7 d | 28 d | 90 d |
| HPC-60 | **3.5** | 7.2 | 8.9 | 3.81 | 5.37 | 7.44 | 54.6 | 77.6 | 98.5 |
| OPC-60 | 5.2 | 6.7 | 8.2 | 4.45 | 5.16 | 6.36 | 60.3 | 78.5 | 90.2 |
| HPC-80 | 4.5 | 8.4 | 10.3 | 4.76 | 6.72 | 8.25 | 68.3 | 95.4 | 116.4 |
| OPC-80 | 6.2 | 7.9 | 9.1 | 5.28 | 5.94 | 7.35 | 72.5 | 91.0 | 107.5 |

**Table 9-4** Comparison of the strength requirements for low-heat Portland cement in different standards (the one in China is represented by HBC)

| Standard | Compressive, 7 days (MPa) | Compressive, 28 days (MPa) |
|---|---|---|
| China GB200-2003 | ≥13.0 | ≥42.5 |
| USA ASTM C150 | ≥7.0 | ≥17.0 |
| Japan JIS R 521 | ≥7.5 | ≥22.5 |

several major infrastructure projects in China, such as the third phase of the Three Gorges Dam in Yichang, Zipingpu and Shenxigou dams in Sichuan Province, Beijing International Airport, and the 5th Ring Road in Beijing. Figure 9-2 shows the application of the new HBC concrete in Three Gorges Dam.

Studies on hydraulic alternative cements, similar to what was discussed above, have been ongoing. Some relatively new developments include belite–ye'elimite–ferrite cements (e.g., belite-sulfoaluminate cement) and the incorporation of a ternesite phase (which may potentially reduce the clinkering temperature to ~1100°C) (Gartner and Hirao, 2015; Shen et al., 2015). However, although these alternative cements could reduce the carbon emission to some extent, the reductions are not significant. As pointed out in UNEP-SBCI's report (United Nations et al., 2018), the International Energy Agency's 2050 goal of reducing $CO_2$ emissions from cement manufacturing by 24% (IEA et al., 2018) could be too rigorous to be addressed by existing and emerging technologies; not to mention the carbon-neutral goals of the US by 2050 and China by 2060. These goals must rely on new, breakthrough technologies, such as carbon capture, utilization and storage (CCUS) technologies and innovative alternative cements that are fundamentally different from the existing ones.

**Figure 9-2** Application of new HBC concrete in the Three Gorges Dam

**Table 9-5** Comparison of pathways of direct use of $CO_2$ in cement and concrete

| Pathways | Market | Tech nature | Cost | $CO_2$ source |
|---|---|---|---|---|
| CarbonBuilt | Prefabrication | $CaO+CO_2 \rightarrow CaCO_3$ | Depends on Ca source | Prefers high concentration; with elevated temperature and pressure |
| Solidia | Prefabrication | $CaSiO_3+CO_2 \rightarrow CaCO_3 + SiO_2$ | ~$100/t Solidia cement | |
| Novacem | Prefabrication | $MgO+CO_2 \rightarrow MgCO_3$ | ~$100/t Novacem | |
| CarbonCure | Concrete | $CaO+CO_2 \rightarrow CaCO_3$ | Not changing much price of concrete | Currently food grade; may allow low-concentration |
| Blue Planet | Aggregate | $CaO+CO_2 \rightarrow CaCO_3$ | <$25/t | >5%, e.g., flue gas |

Stimulated by the urgent need of CCUS for eco-efficient cement and concrete, several technologies have been developed. Ones that have been commercialized include CarbonBuilt (www.carbonbuilt.com), Solidia (www.solidiatech.com), Novacem (Walling and Provis, 2016), CarbonCure (https://www.carboncure.com), and Blue Planet (www.blueplanet-ltd.com). In this list, CarbonCure (injecting $CO_2$ into fresh concrete) and Blue Planet (carbonating geo-mass to produce aggregate) are not binder technologies. They are compared here because of their ability to decarbonize concrete. It is worth noting that CarbonBuilt and CarbonCure are the two final winners of the NRG COSIA Carbon XPRIZE in 2021, beating many other carbon reduction concepts (e.g., $CO_2$ to chemicals or fuels) because of their large total $CO_2$ utilization in the implementation phase and permanent carbon storage. These technologies are discussed below and compared in Table 9-5.

CarbonBuilt, formerly known as $CO_2$Concrete, carbonates calcium hydroxide, high-calcium fly ash, blast-furnace slag, etc. in well-controlled (temperature and pressure) $CO_2$ curing conditions, to make prefabricated construction elements (e.g., concrete blocks). Though the concept claimed to use flue gas, such technologies prefer using high-concentration $CO_2$ since the production duration and product quality are determined by $CO_2$ diffusion and interior carbonation reaction. If industrial wastes are used as the calcium source, the product cost can be relatively low (depending on the cost of the "waste") and the $CO_2$ uptake can be up to 0.3 ton/ton calcium source (assuming 40% CaO content in the source, e.g., slags). However, if calcium hydroxide or Portland cement (instead of solid wastes) is used as the calcium source, the net $CO_2$ uptake will be negative since calcium hydroxide needs to be produced from limestone decomposition. The biggest limitations of the technology are that it cannot be used for cast-in-place practices which are still dominating concrete construction; and the size of the prefabricated products based on this technology may not be large because the hardening process is determined by the diffusion of the $CO_2$.

Solidia is technically similar to CarbonBuilt, but the calcium source adopted is wollastonite ($CaSiO_3$). Since natural wollastonite is not widely available, it needs to be manufactured by the calcination of limestone and quartz. In the best scenario, carbonation of Solidia can absorb all the $CO_2$ released from the limestone decomposition. So Solidia cannot achieve carbon-neutral state because grinding and calcination in the process of manufacturing release $CO_2$. Due to the similarity to cement production, the cost of manufacturing of Solidia cement should be comparable to Portland cement (marked as ~$100/t in Table 9-5). Also similar to CarbonBuild, Solidia cannot be used in cast-in-place applications and should be limited to carbonation-based prefabrications.

Novacem is another technology similar to CarbonBuilt and Solida, but developed in the UK. It starts from magnesium silicate, and produces $MgCO_3$ from enhanced carbonation of magnesium silicate (forming $MgCO_3$ and $SiO_2$, under elevated temperature and pressure). The $MgCO_3$ is then thermally decomposed to produce MgO, which is used as a cement to absorb $CO_2$. Since the $CO_2$ released in magnesite decomposition can be looped back to carbonate silicate, the process of MgO production is chemically carbon-neutral. The equivalent emissions due to grinding and calcination can be offset by the hardening process of Novacem ($MgO+CO_2 \rightarrow MgCO_3$, in which the $CO_2$ uptake is as high as 1.1 ton/ton MgO). Therefore, it is theoretically possible for Novacem to achieve carbon-neutral or even negative emission. Apart from the common limitations of carbonation-hardening cements, Novacem may also be limited by easily processable magnesium silicate minerals.

CarbonCure is currently injecting food-grade or purified $CO_2$ (sourced and stored at concrete plants) into fresh-state concrete before it is delivered to the construction site. The advantages include: (1) easy to deploy (i.e., just needs a $CO_2$ tank for each mixing truck); (2) inducing negligible cost (so concrete can be sold as usual); and (3) it does not limit the applications of concrete (i.e., either cast-in-place or pre-fabrication). However, this technology has a limited $CO_2$ uptake rate. According to a report published by the company (Thomas, 2019), the $CO_2$ uptake rate is typically less than 0.2% by weight of cement, to avoid strength compromise and steel corrosion. That is to say, even if the technology is applied to all cement produced in the world (assuming 4.1 billion tons), the total $CO_2$ utilization is just 8.2 million tons. Of course, this may have underestimated the carbon credit of this technology if it can effectively improve the strength and durability of concrete.

Blue Planet is dedicated to producing calcium carbonate coated aggregate. The current demonstration plant of Blue Planet uses recycled concrete as substrate and a calcium source (e.g., lime) to produce aggregate, and the carbonation can be implemented using flue gas or captured $CO_2$ at >5% concentration. This technology is easy to use. However, it could suffer from the two limitations of recycled concrete aggregate: (1) low quality (i.e., low strength, high adsorption, and low modulus because of the adhered paste/mortar); and (2) extreme quality variation. The calcite coating may overcome the first limitation to some extent but can do little to fix the second. Furthermore, composition of recycled concrete is dominated by aggregate (i.e., quartz and silicates) and the calcium content is low. Taking a typical concrete (200 kg water, 400 kg cement, and 1800 kg aggregate in 1 $m^3$ of concrete) as an example, the content of CaO in concrete that can be carbonated is only roughly $400 \times 60\%/2400 = 10\%$. So, one ton of carbonated aggregate from recycled concrete may sequester up to 7.3% $CO_2$, if no supplemental lime is used to coat the surface. Nevertheless, using lime to compensate CaO source in this process cannot improve the eco-efficiency of this technology since production of lime is energy- and $CO_2$-intensive.

So far, no perfect solutions have been found to significantly improve the eco-efficiency of cement and concrete. To be effective, the limitations of the existing and emerging technologies must be solved. Innovators should be encouraged to think beyond the scope of the existing cement and concrete technologies. For instance, since the $CO_2$ and energy burdens of contemporary cement and concrete are mainly attributed to the decomposition of carbonate minerals and calcination, is it possible to develop calcination-free alternative cements based on non-carbonate minerals and greenhouse gases-derived materials (e.g., activators)?

### 9.1.3 Prolonging the Service Life of Concrete

Of all the measures to make concrete sustainable, how to prolong the service life of a concrete structure is one of the most important. Prolonging the service life of a concrete structure not only saves the resources of raw materials for new buildings, but also reduces construction waste due to the demolition of the existing buildings and infrastructure. Since durable concrete can reduce the frequency of replacement, improving the durability of concrete contributes directly to its service life. Durability is especially important to contemporary concrete because the contemporary Portland cement concrete mixtures, which are usually designed to obtain high strength at an early age, are very prone to cracking. The interconnections between surface and interior cracks, microcracks, and voids in concrete provide the pathway for the penetration of water and harmful ions that are implicated in all kinds of durability problems.

To improve the durability of concrete, much research has been conducted on developing concrete mixes with the incorporation of chemical and mineral admixtures. Along this line, concrete durability-enhancing admixtures have been developed, and some of them could be derived from natural polymers (Li et al., 2000). Due to the natural renewability, low cost, ease of manufacture, and excellent improvement in durability, these admixtures can readily be used to develop more durable concrete. Based on one of such durability-enhancing admixtures, several concrete mixes with different formulae have been prepared to evaluate the effects on concrete properties.

The cement used for the test was type I Portland cement (OPC) with a specific gravity of 3.15 and a fineness of 385 $m^2/kg$. River sand with a fineness modulus of 2.3 was employed as fine aggregate and a crushed limestone aggregate with a nominal maximum size of 10 mm was used as coarse aggregate. These concrete mix proportions are listed in Table 9-6.

The freeze–thaw test was performed to evaluate the effect of the admixture on concrete properties. In the test, specimens were exposed to repeated cycles of freezing in air at −18°C, and thawing in water at 6°C. After the freeze–thaw test, compression tests were conducted on the specimens.

**Table 9-6** Mixture proportions ($kg/m^3$) of the concrete series used for evaluation of the natural polymer-derived durability enhancing admixture

| Code | Cement | Water | Sand | Aggregate |
|---|---|---|---|---|
| C1 | 415 | 250 | 625 | 1035 |
| C2 | Mixed with 0.1% effective solid content of natural polymer by weight of cement | | | |
| C3 | Mixed with 0.5% effective solid content of natural polymer | | | |
| C4 | Mixed with 1.0% effective solid content of natural polymer | | | |
| C5 | With DAREX (75 mL/100 kg cement) | | | |

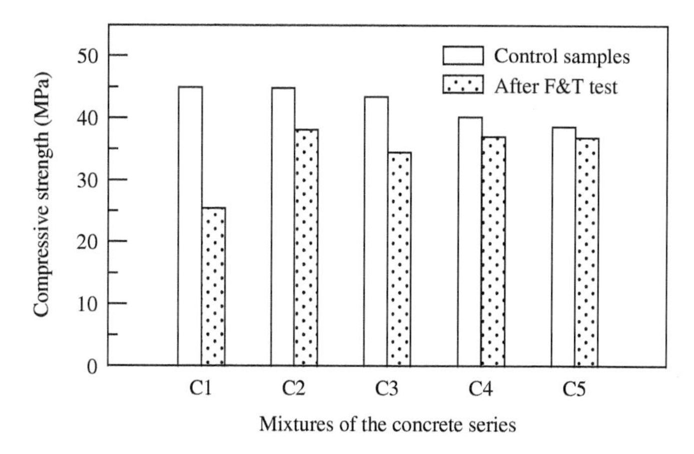


**Figure 9-3** Comparison of compressive strengths of control samples and samples that underwent the freezing and thawing test for the concrete series

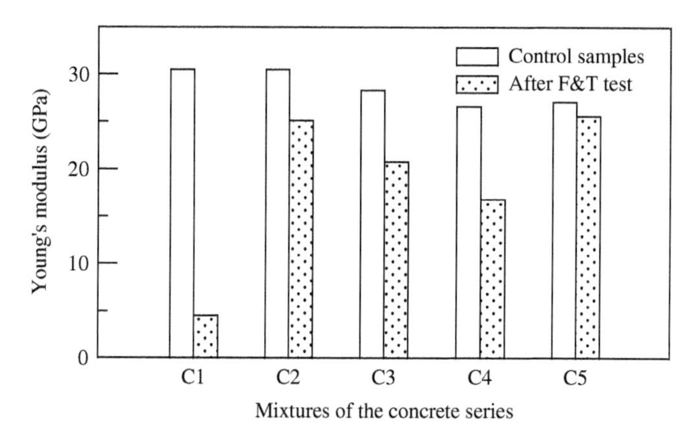


**Figure 9-4** Comparison of the elastic modulus of control samples and samples that underwent the freezing and thawing test for the concrete series

Figures 9-3 and 9-4 show the strength and Young's modulus of the concrete mixtures following the severe freezing and thawing cycles, and comparisons with the control specimens. The results show that the application of the durability-enhancing admixture in the concrete mixture leads to a significant improvement in the frost resistance of concrete. Even with only 0.1% effective solid content of the natural polymer, an obvious enhancement of concrete frost resistance in terms of less deficiency of compressive strength and elastic modulus was observed.

For the plain concrete mixture C1, the compressive strength and elastic modulus after a severe freezing and thawing process have largely deteriorated, by 43% and 85%, respectively, compared to control samples. For the air-entrained concrete mixture C5, there was a deficiency of only a few percent in both strength and modulus after the frost resistance test. This demonstrated the capability of improving the frost resistance of concrete by applying an air-entraining agent; however, the entrained air void system itself caused strength deficiency to a certain extent.

For the modified concrete C2, the reductions in compressive strength and elastic modulus after freezing and thawing tests were only 15% and 18%, respectively. This can be seen more clearly from Figure 9-5 which shows the representative stress–strain curves for the

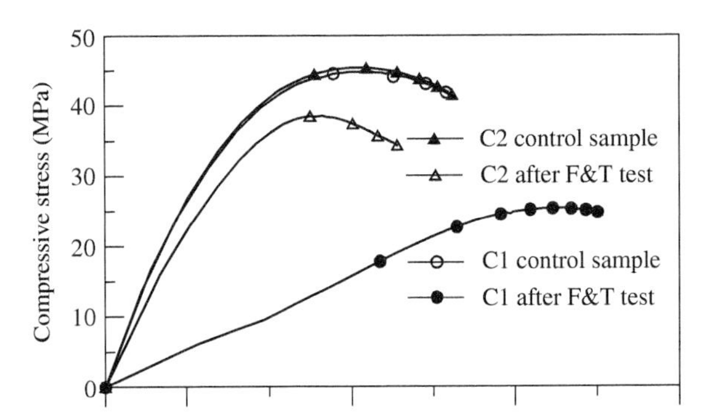


**Figure 9-5** Representative stress–strain curves of the concrete series

concrete series. It was shown that the incorporation of the durability-enhancing admixture can significantly improve the frost resistance of modified concrete in its unique ways.

The degradation of concrete materials under frost attack begins with the initiation of microcracks on the exposed concrete surface. Therefore, an inspection of the surface quality and cracking of concrete that has undergone freeze–thaw cycles also offers evidence of the improvement in the frost resistance of concrete employing the new admixture. In this study, wide and continuous cracks were observed on the surfaces of the plain concrete mixture, while only a few thin cracks were found on the surfaces of the natural polymer-modified concrete mixture. From the frost-resistance test results, it can be concluded that the optimal usage of the new admixture is about 0.1% of the effective solid content of the natural polymer by weight of cement usage in the concrete mixture.

Although only one example is discussed in detail here, it serves as an excellent indicator that the durability of concrete can be improved effectively and eco-efficiently. In this direction, future studies may be conducted to discover a wider range of eco-efficient materials and/or technologies that improve durability. These may include natural polymer-based admixtures that modify the microstructure, surface coatings that control the penetration of moisture and harmful substances, and fibers that improve the cracking-resistance of concrete.

### 9.1.4 Efficiently Utilizing Materials: HSC and UHSC Applications

Improving the usage efficiency of materials is another way to make concrete sustainable. Using high-strength concrete (HSC) and ultra-high-strength concrete (UHSC) and a pre-stress technique during design and construction provides a good way of increasing the efficiency of concrete usage. It may be argued that HSC and UHSC use more cementitious materials and thus imply higher energy consumption and $CO_2$ emissions. However, when HSC and UHSC are used in design and construction, the cross-section of structural members such as columns and beams can be largely reduced. Subsequently, the total amount of materials used for a concrete structure can be reduced compared to the case where normal-strength concrete is used. As a result, the cement amount is reduced and so are energy consumption and $CO_2$ emissions. Similarly, by using a pre-stressing technique, steel tendons with higher tensile strength can be effectively used and the total amount of steel can be reduced. The San Francisco Public Utilities Commission Building is a good example of this concept (Schokker, 2010). It is a concrete building of 14 stories above grade and 1 story below grade with a total area of 26,000 $m^2$. The building design and construction adopt a high-strength concrete that is vertically post-tensioned throughout the building core, which provides a 30% reduction in concrete and reinforcement over a normal-strength concrete system.

Future developments will certainly yield more methodologies of improved material utilization efficiency. Apart from materials innovations, new methods of construction (e.g., additive manufacturing and prefabrication) will also make relevant contributions.

## 9.2 DEEP UNDERSTANDING OF THE NATURE OF HYDRATION

Enhancement of concrete performance (e.g., its mechanical properties and durability) requires a deep understanding of the nature of hydration products. One of the most important, long-standing needs in cement science (Powers and Brownyard, 1948) is a quantitative understanding of C–S–H on the atomic to 100 nm scale and how the structures on this scale control the mechanical, transport, and chemical properties of hydrated cement paste. Numerical methods are commonly used to simulate the C–S–H structures. As mentioned earlier, the structure of C–S–H, the major hydration product of cement, has not yet been revealed. Moreover, as concrete is becoming more complex with the increased incorporation of admixtures and SCMs, the hydration process and products become more complex. The phase assemblages formed from the reaction of the cement and various SCMs may be different from the hydration products of pure cement. The pore structure may also be changed along with the cohesive forces between phases and the resulting mechanical properties. Adoption of alternative binders (e.g., geopolymers and C-A-S-H/N-A-S-H gels) will further increase the complexity. Understanding the structure of the hydration products of contemporary concrete is essential to improving the microstructure and, thus, the performance of concrete.

Fortunately, with the advances in technology, a wide range of experimental and computational tools have become available in recent decades to discover the hydration of cementitious materials. Experimental methods such as atomic force microscopy (AFM), small-angle neutron and X-ray scattering, nuclear magnetic resonance (NMR), nanoindentation, and high-resolution scanning and transmission electron microscopies can be used to reveal at least part of the natures of C–S–H at different scales. For example, AFM can be used to examine the aggregation shape of C–S–H at a tens of nanometer scale and to determine the cohesive force nature of the hydration products (Plassard et al., 2005). Small-angle neutron and X-ray scattering can be applied to probe the "gel" porosity of C–S–H (Allen et al., 2004, 2007; Faraone et al., 2004; Fratini et al., 2006). NMR can be used to determine the C–S–H structure and the pore structure of cementitious materials (Richardson, 2004; McDonald et al., 2005). Nanoindentation can measure the modulus of C–S–H and other hydrates (Ulm et al., 2007). High-resolution scanning and transmission electron microscopies coupled with chemical microanalysis can determine microstructural development and microchemistry of hydration phases (Richardson, 2002; Scrivener et al., 2004).

The data obtained from these experiments provide a predictive framework that spans the atomic to the macroscopic level to some extent. However, using only these experimental methods, it is still not possible to completely reveal the structure of concrete at the atomic and molecular level. Computational modeling and numerical simulation are needed in concrete science.

The numerical simulation methods at the atomic and molecular scale include methods based on quantum chemical and molecular potential. The quantum (first-principles) approaches involve the solution of the Schrödinger equation under the Born-Oppenheimer approximation, mean field approximation, and periodic potential approximation, describing the interaction of multi-particle systems. The quantum chemical methods can be classified into wave-function-based and density-functional-theory-based (DFT) methods according to the quantity that is being calculated. The Hartree-Fock (HF) approach is the theoretical basis of the wave-function-based method. It is believed that the wave function can fully describe the properties of electrons. The molecular orbital of the electron wave function of the system can be obtained through the self-consistent field method in the HF approach. However, this method does not consider the electronic correlation effect, which limits its accuracy. The post-Hartree-Fork method, including configuration interaction (CI), coupled cluster (CC), Møller-Plesset perturbation theory (MP), and the quadratic configuration interaction (QCI) method improves the precision considerably, but the computational cost also increases exponentially. While the DFT method uses electron density as the basic quantity of research, making the question reduced from 3N to 3 dimensions. The theoretical basis of the DFT method is the Kohn-Sham (KS) equation, which considers both the electronic exchange and the correlation interaction. Nonetheless, this method contains a self-interaction effect and is not robust for the calculation of weak interactions. Using the hybrid functional (i.e. B3LYP and PWPB95-D3) is a good way to moderate these negative impacts.

Potential-based methods such as the classical molecular dynamics (MD) approach, are based on empirical or semi-empirical potentials between or among atomic or molecular entities. They involve treating the atoms or molecules as classical (nonquantum) entities and computing their positions, motion, and energies as they interact with each other under the influence of potential functions. These functions can describe short-range atomic repulsion, van der Waals forces, and attractive and repulsive coulombic interactions. MD methods follow the time evolution of the structure and energy of the computed system, thus allowing calculation of dynamical properties, such as vibrational spectra. Kalinichev et al. (2002) provided a more detailed discussion of the applications of potential-based, molecular dynamics simulations to cement systems. With the development of high-performance computing technology, quantum (*ab initio*) MD (AIMD) becomes more accessible. In an AIMD calculation, finite-temperature dynamical trajectories are generated by using forces obtained directly from electronic structure calculations performed "on the fly" as the simulation proceeds. Therefore, AIMD can show the breakage and formation of chemical bonds and account for electronic polarization effects. AIMD can be categorized into Born-Oppenheimer MD (BOMD) and Car–Parrinello MD (CPMD) (Car and Parrinello, 1985; Remler and Madden, 1990; Marx and Hutter, 2000). AIMD has been successfully applied to a wide variety of important issues in physics, chemistry and biology, and has also begun to influence cement science. AIMD can reveal some new physical phenomena and microscopic mechanisms that cannot be uncovered by empirical methods, and even leads to new interpretations of experimental data and suggests new experiments to perform.

Each of these methods has pros and cons. The quantum methods are normally more accurate than the potential-based methods in describing atomic positions, interaction energies, and spectroscopic properties, but even with the most recent generation of supercomputers, they are limited to a few hundred of atoms. The hybrid quantum mechanics/molecular mechanics (QM/MM) approach may partly solve this drawback. In contrast, MD and Monte Carlo (MC) simulations trade off accuracy for increased system size and reaction time. They can now be used for systems of the order of $10^6$ atoms for a few nanoseconds.

To build up an appropriate C–S–H model using simulation, the parameters have to be considered and used as objective functions to achieve a reasonable result. The first criterion is the Ca/Si ratio, which is the key factor that the simulation has to satisfy. Theoretically, as indicated in Chapter 5 by Equation 5-62, the stoichiometric equation for fully hydrated $C_3S$ results in a Ca/Si ratio of 1.75. Experimentally, as summarized by Richardson (1999), the results of small-angle X-ray scattering (SAXS) and small-angle neutron scattering (SANS) have demonstrated that the values of the Ca/Si ratio range from 1.2 to 2.3, with a mean of 1.7–1.75. These values were also echoed by Pellenq et al. (2009).

Another important property for determining C–S–H structure is its density. In reality, the density depends on the water content in C–S–H. Recently, Allen et al. (2007) obtained a value of 2604 kg/m$^3$ as the density of C–S–H by utilizing SANS and SAXS simultaneously with consideration of combined water. Such an estimation is thought to be closer to the intrinsic property of C–S–H gel and hence can be adopted in simulations.

Silicate connection is another important factor that has to be considered in interpreting the structure of C–S–H. Such information can be obtained utilizing $^{29}Si$ NMR. The basic unit of the silicates is the $SiO_4$ tetrahedron. When referring to $^{29}Si$ NMR, the peak in the spectrum for silicate can be classified by the number of bridging oxygen atoms in a specific $SiO_4$ unit. Theoretically, five categories of $Q_n$ sites along with a chemical shift (centered) can be observed in NMR: $Q_0$ (−70 ppm), $Q_1$ (−80 ppm), $Q_2$ (−88 ppm), $Q_3$ (−98 ppm), and $Q_4$ (−110 ppm), where the subscript $n$ represents the number of additional tetrahedra linked by sharing an oxygen atom with the base tetrahedron (Johansson, 1999). For C–S–H, it has been found that $Q_1$ and $Q_2$ Si sites are predominant; $Q_3$ and $Q_4$ Si sites are not expected to be seen (Kirkpatrick, 1996; Hewlett; 2004). As for $Q_0$, it is conventionally considered to be a surface hydroxylation of $C_3S$; however, it has been found that such monomers persist in forming from the induction period to the late stages of the hydration reaction. The amount of $Q_0$ may range from 2% to 10%. Therefore, the role of $Q_0$ silicate sites should also be taken into consideration for numerical simulation. It is believed that only with the correct selection of a combination of Ca/Si ratio, density, and silicate connection projected as functions in a numerical simulation, can meaningful and realistic results be obtained.

A more realistic route to construct a C–S–H structure at the atomistic level is to understand its formation mechanism from the $C_3S$ hydration process. When the $C_3S$ encounters water, a large amount of Ca ions and small amount of silicate groups dissolve (Garrault et al., 2005), then the silicate and oxygen ions on the surface would be hydrolyzed. Because of the hydrophobic nature and small diffusion coefficient for silicate groups, there is a Si-rich region formed around the $C_3S$ surface, which may be an intermediate phase for C–S–H. The formation of synthetic C–S–H would go through a phase transformation from the metastable into the thermodynamically more favored structure (Kumar et al., 2017), but whether this mechanism is also applied for the C–S–H formed through $C_3S$ hydration should be further studied.

To gain a deeper understanding of the above processes, it is necessary to use the state-of-the-art simulation method at the atomistic level to give an insight into the nature of $C_3S$ hydration and the C–S–H formation mechanism. Simulating this large time-span hydration process embodying a series of chemical reactions is inaccessible to the traditional MD method due to its small simulation time scale and large energy barrier between the two stable states. To solve this conundrum, the enhancing sampling methods, such as umbrella sampling (US), metadynamics (MetaD) and variationally enhancing sampling (VES), need to be implemented to explore the possible reaction pathways. For example, the process of dissolution of the Ca ion from the $C_3S$ surface has been explored through the *ab initio* metadynamics simulations with two collective variables CNs (the coordination number of Ca with O on the $C_3S$ surface) and CNw (the coordination number of Ca with O from water molecules). As shown in Figure 9-6a, the breakage of the last Ca-Os bond requires to absorb six water molecules. After the Ca ion is detached from the original position, it will coordinate with five, six or seven water molecules. Figure 9-6b shows the configurations of the free energy minimums and we can find the previous Ca site is replaced by one water molecule and one hydron ion bonded with the O ion in the second surface layer. The diffusion of water molecules into the bottom of the Ca ion may need a relatively high activation energy and this water molecule provides an electrostatic repulsion on the Ca ion, which promotes the vertical displacement of this Ca ion.

Recently, the machine learning-based (ML) or deep learning-based (DL) interatomic potential energy surface (PES) models derived from the quantum chemical method have been proposed to perform MD for a large-scale system with quantum accuracy and we believe this cutting-edge technology would have an astonishing influence on the fundamental understanding of cement hydration in the future.

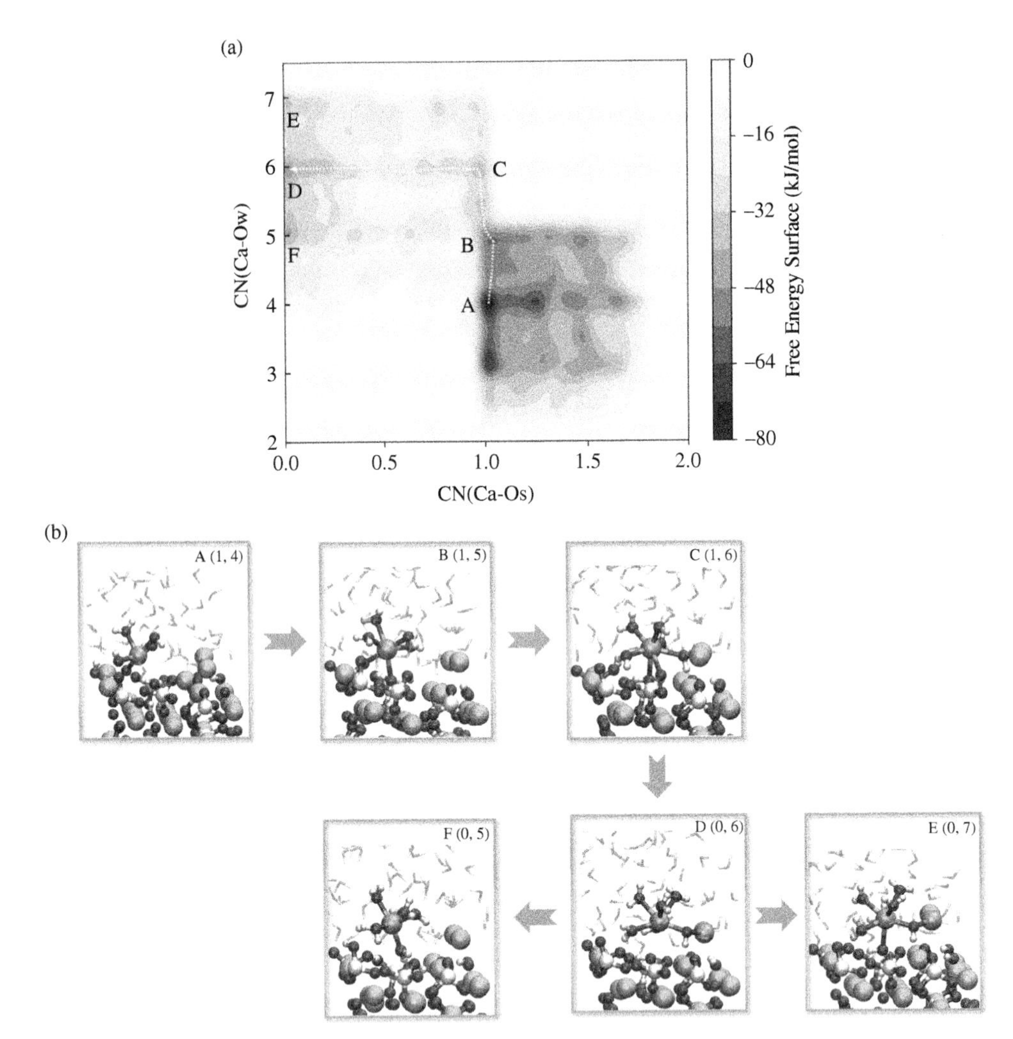


**Figure 9-6** Dissolution of the Ca ion from the $C_3S$ surface through *ab initio* metadynamics simulations: (a) the two-dimensional free energy surface (FES) of dissolution of the Ca ion with variables of CNs (the coordination number of Ca with O on the $C_3S$ surface) and CNw (the coordination number of Ca with O from water molecules); (b) the configurations of the free energy minimum states on the FES.

## 9.3 INTEGRATED MATERIALS AND STRUCTURAL DESIGN

### 9.3.1 Introduction

Design is a decision-making exploration with goal-oriented planning activity for creating products, processes, and systems with desired functions through specifications (Ming et al., 2021). In concrete engineering, structural design is used to create buildings and infrastructures. Integrated materials and structural design is one type of design philosophy that takes an interdisciplinary and holistic approach. Recently, structural design in many countries has emphasized a performance-based philosophy that considers the safety, durability, serviceability, and sustainability of a structure simultaneously. Integrated design combines materials engineering and structural engineering at the design stage: it fully uses the strengths of materials by selecting the most suitable structural forms.

The intertwined relationship between materials and structures is illustrated in Figure 9-7. The base of the bottom pyramid represents materials engineering, with four corners each representing synthesis, microstructures/composition, properties, and performance. The top of the inverted pyramid represents four structural engineering aspects: safety, durability, serviceability, and sustainability. Clearly, material and structural engineering both have to meet the end-use needs and constraints of the final structure. As shown in Figure 9-7, integrated design includes two approaches: (1) from bottom to top: this is a bottom-up design approach. In this approach, the properties of existing materials determine the possible structural forms. This approach is the traditional method employed by builders and engineers in the past centuries; and (2) from top to bottom: this is a top-down approach. In this approach, structural engineers need to translate targeted structural performance to desired materials properties that may include mechanical properties, microstructure, or macrostructure; and material engineers need to invent new composite materials or customize the existing materials to meet the desired material properties. In other words, the desired performance of structures guides the material's design and development. This top-down approach is a relatively new engineering paradigm in concrete engineering, which is made possible by the advances in material engineering in recent decades. Advanced design methods—e.g., the materials-oriented integrated design and construction of structures (MIDCS) method (Ming et al., 2021)—can bridge the "top to bottom" or "bottom to top" paths between materials and structures as shown in Figure 9-8.

### 9.3.2 Load-Carrying Capability: Durability Unified Service Life Design Theory

Concrete structures are usually designed to bear mechanical loads as well as be tolerant of various environmental conditions. The mechanical loads and environmental factors degrade the properties of concrete structures with time. However, traditional codes and standards usually treat the mechanical parameters of materials and structures as constant values, which will overestimate the service lives of concrete structures under harsh environmental conditions. Since the 1990s,

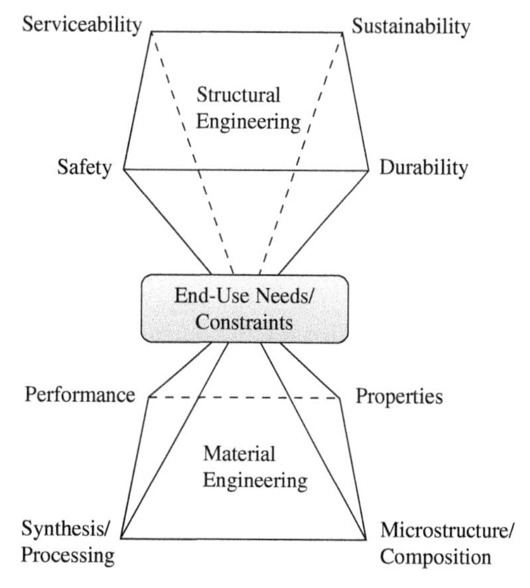


**Figure 9-7** The intertwined relationship between materials and structures

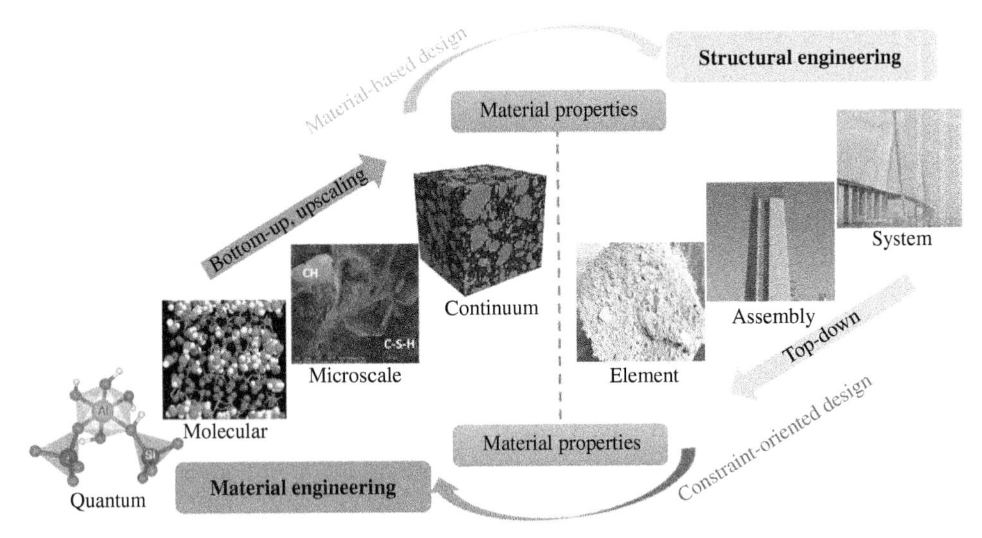


**Figure 9-8** Schematic illustration of the concepts of MIDCS in civil engineering

durability issues have attracted increasing attention because of more and more durability-related damages to concrete structures. In their current practice, engineers simplify the durability issues by selecting key parameters (e.g., the maximum water-to-cement ratio of concrete and the minimum concrete cover thickness) for different environmental conditions, according to developed guidelines, codes or standards. Thus, there is no scientific or quantitative formula for measuring the environmental factors, and the dynamically changed mechanical properties of materials and structures induced by environmental conditions are not considered in codes and standards.

It has long been realized that a concrete structure carries the load while being exposed to various environmental conditions. The mechanical load and environmental condition influence each other during the service life of concrete. Property degradation of concrete is caused by the coupling effect of loading and the environment. However, traditionally, the design of concrete structure has considered only the load-carrying capability of the structure. Moreover, the design code treats the mechanical properties of concrete at both the material and structural levels as a constant having no variation with time. Since the 1990s, the durability issue has received more attention and some design codes concerning concrete durability have been developed. However, in these preliminary attempts, the durability of the concrete structure is taken care of only by the details described in the code, such as the cover thickness of a structure under a certain environmental condition. There is no scientific formulation to quantify the effect of environmental conditions. In addition, the codes do not consider the dynamic changes of the properties and performance of concrete as either a material or a structure with time. Thus, the codes cannot reflect the true service conditions of a concrete structure.

In a national basic research project started in 2009 in China, "Basic study on environmentally friendly contemporary concrete," the load-carrying capability-durability unified service life design theory was proposed for the first time to develop a new and scientific design philosophy where the safety, durability, serviceability, and sustainability of concrete structures can be considered in a unified way. The new theory must resolve two fundamental issues that the current codes and standards do not address. The first issue is to quantitatively measure the environmental factors and combine them with the mechanical loads. The second is to carefully consider the time-dependent properties of materials and structures.

In detail, to address the first issue, this theory has developed a feasible method that can convert the environmental factor to an equivalent stress based on thermodynamics and porous media theory. Take an example of the volume elements of cement-based materials, as shown in Figure 9-9: these elements bear a mechanical load in the cement pore solution environment. Based on the thermodynamics and virtual energy calculation, the chemical reactions happening in the pore solutions can be converted into strain or stress. Thus, this theory can quantitatively couple the environmental factors with the mechanical loads by superposition of the two stresses (environment factors-induced stress and mechanical loads-induced stress). In this way, the design processes can still follow the existing stress analysis-based codes and standards.

For the second issue, one approach is to figure out the deterioration mechanisms of materials and structures under the combined effects of environmental factors and mechanical loads through advanced experimental methods and/or multi-scale simulations (Ma, 2013; Hu, 2014). Then, we can get the time-dependent properties of materials and structures, and predict the performance of concrete structures as a function of time as shown in Figure 9-10. With this function, the performance of a concrete structure at different service periods can be predicted and incorporated into the design/evaluation accordingly.

Such a unified design method must consider material properties and structural performance all the way through design, construction, service, and maintenance. It is a typical designer-centered, multi-disciplinary, and holistic approach and takes the entire life cycle of buildings and civil infrastructures into consideration. This design method meets the requirement of the Vision 2030 of the American Concrete Institute (ACI, 2001) as well as the future development of the concrete industry.

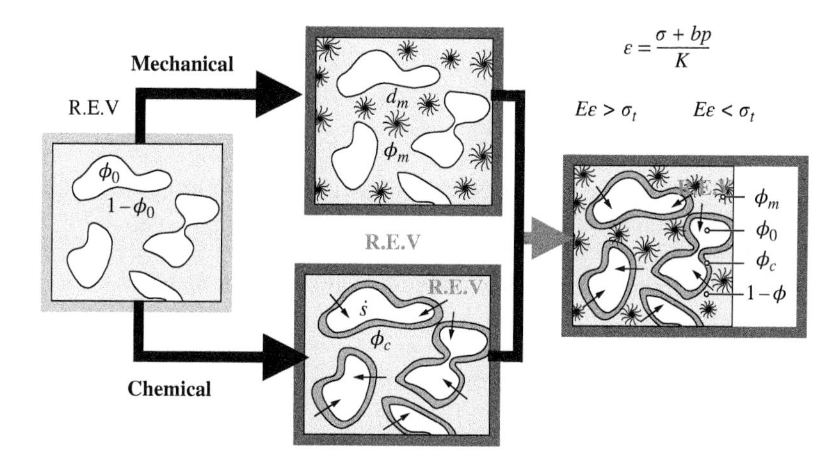


**Figure 9-9** Conversion of chemical/environmental influence to mechanical effect

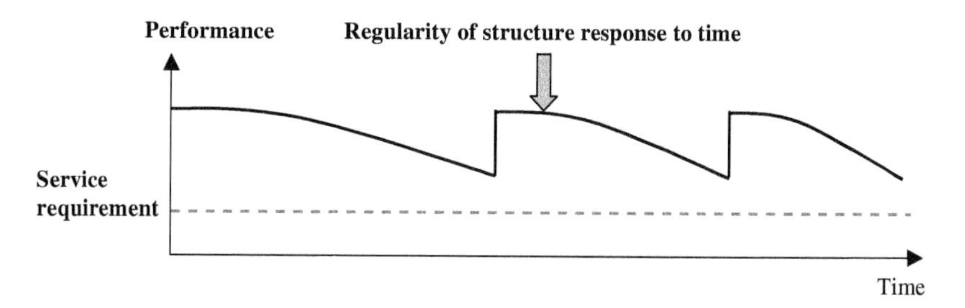


**Figure 9-10** Regularity of concrete structure performance as a function of time

## 9.4 HIGH-TENSILE-STRENGTH AND HIGH-TOUGHNESS CEMENT-BASED MATERIALS

Concrete has fatal disadvantages—low tensile strength, a quasi-brittle nature, tendency to crack, low toughness, and low ductility. Since Portland cement was invented in the nineteenth century, many attempts have been made to overcome these disadvantages and improve the toughness of Portland cement concrete. The development of reinforced concrete using steel bars to carry out tension has advanced the application of concrete greatly. The invention of pre-stressed concrete has further improved the crack resistance of concrete. Later, incorporating high-strength, small-diameter fibers into concrete was developed to improve the toughness and ductility of concrete. Two kinds of fibers are commonly used: continuous and short fibers. The use of short fibers seems to be preferable due to the simple and economical nature of the fabrication process. Experiments have verified that the response of the short fiber reinforced cement-based composites can be largely influenced by dispersion of the fiber. By adding the fibers to concrete, the toughness can be improved. However, due to increased cost and dispersion difficulties, short fibers cannot be added in a large amount. Thus, the improvement to tensile strength and toughness of concrete is limited.

Another attempt at toughness improvement is to incorporate polymer into concrete. There are three ways of doing this nowadays. One is to use polymer as the binder in concrete. This method uses a two-part polymer system, such as epoxy resin and hardener, premixed first and then mixed with aggregate to produce a hardened plastic material with aggregate as filler. The concrete thus obtained is called polymer concrete (PC). The second way is polymer impregnated concrete (PIC). This is usually done by impregnating a monomer and catalyst/initiator into a hardened concrete followed by *in situ* polymerization facilitated using steam or infrared heat. The third way uses a polymer (usually latex) that is dispersed in water and then mixed with Portland cement and aggregate. This is identified as polymer-modified or latex-modified concrete (LMC). PC and PIC are very expensive and thus their use is limited to emergency or essential concreting jobs in mines, tunnels, and highways as well as the production of high-strength, precast products. LMC is economically available and the most commonly used in the practice of pavement overlay. Reactive latexes (e.g., polyacrylics) can interact with the hydration products to form an organic-inorganic hybrid matrix, and inert latexes (e.g., styrene-butadiene rubber) can result in an organic-inorganic interpenetrated structure (Ma et al., 2011; Ma and Li, 2013). In both cases, the tensile strength may be improved, and the modulus of elasticity and impermeability may be reduced. However, LMC does not offer very significant improvement in tensile strength and toughness of concrete. It cannot change the quasi-brittle nature of concrete.

There is a burgeoning interest in both academia and industry to develop a technology that can improve the flexural strength and toughness of concrete by eliminating its quasi-brittle nature. Breakthroughs in this direction may result in a new generation of concrete. Recently, many attempts have been made toward this goal.

At the Hong Kong University of Science and Technology and the University of Macau, research studies have been conducted on flexural strength enhancement through on-site polymerization (Liang, 2018; Liu, 2020). This is realized by incorporating organic monomer, initiator, and accelerator into the cement-based composite mix during the mixing process and letting the polymerization happen during the cement hydration period. Liang (2018) used acrylamide (AM) as the monomer, ammonium persulfate (APS) as the initiator, and N, N, N', N'-tetramethyl-ethylenediamine (TEMED) as the accelerator. It was found that the polyacrylamide (PAM) network formed in the composite and there was a strong chemical bond formed between PAM and $Ca^{2+}$ in hydration products of cement by Time-of-Flight Secondary Ion Mass Spectrometry (ToF-SIMS), as shown in Figure 9-11. The PAM network can greatly improve the flexural strength of the composite. With only 1% polymer content, the flexural

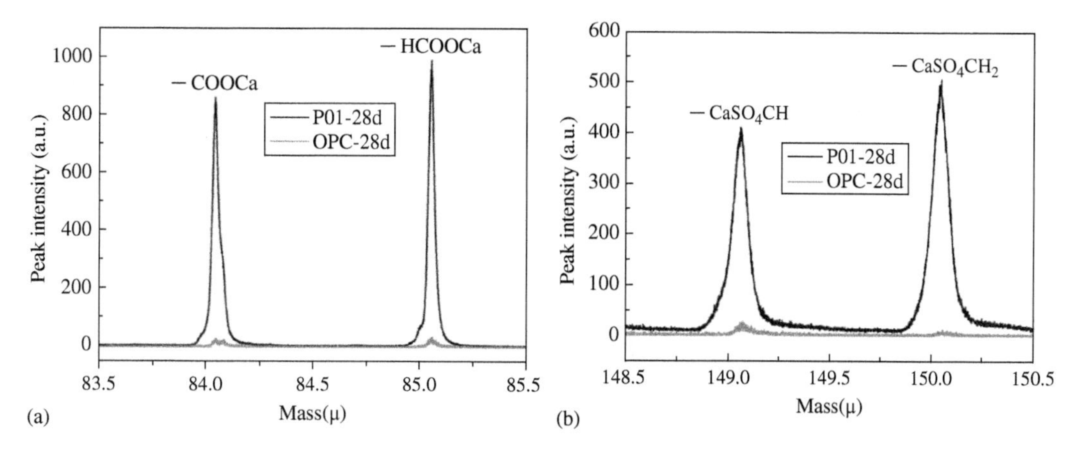


**Figure 9-11** ToF-SIMS tests of OPC and PAM/cement composite with AM/cement weight ratios of 1% (P01): (a) peaks of –COOCa (mass 84.0) and –HCOOCa (mass 85.0) appear in P01; (b) peaks of $-CaSO_4CH$ (mass 149.0) and $-CaSO_4CH_2$ (mass 150.0) appear in P01

strength increased almost three times and at the same time the compressive strength can still be maintained with almost the same value as the reference specimens. Liu (2020) used acrylic acid (AA) and 2-acrylanmido-2-methylpropanesulfonic acid (AMPS) as the monomer, sodium hydroxide as the neutralizer for the monomer solution, ammonium persulfate (APS) as the initiator, and N,N,N',N'-tetramethyl-ethylenediamine (TEMED) as the accelerator in the concrete mix. The experiment results showed that the flexural strength increased 61.2% for concrete with 7% AA-AMPS incorporation as compared to the reference concrete. The toughness of the 7% AA-AMPS modified concrete is also significantly increased, almost tripled as compared to the reference specimen.

In Europe, under the Nanocem project "Organo Mineral Composites," led by Sika (www.nanocem.org), a new method has been proposed: a mineral organic composite having two poles, so that one end can chemically bond with the hydrate of the Portland cement, and the other end can polymerize. The hybrid organic–inorganic polymer dispersion is based on the emulsion polymerization of acrylates, urethane, or vinyl acetate and siloxane molecular precursors. Hybrid organic–inorganic particles are functionalized with siloxane molecules $R'n$Si(OR)4-$n$, where $R'$ can polymerize or copolymerize. Terminal functional groups undergo hydrolysis and bond with the mineral matrix of concrete (cement or aggregates). With this new technology, the concrete was expected to have high tensile strength and be very ductile, which will be completely different from traditional concrete.

In China, under a national key research project on concrete, a study has been conducted to increase the toughness of concrete through incorporation of a new admixture. The new admixture is called a concrete toughness enhancer. Its chemical composition is schematically shown in Figure 9-12. To verify the effect of the newly developed admixture, two types of concrete specimens have been prepared for a three-point bending fracture test. One is made of regular concrete as reference and the other incorporates the new admixture. The test results of two specimens are listed in Table 9-7. As shown in Table 9-7, with the incorporation of this admixture into concrete, the fracture toughness of the modified concrete is significantly increased.

Although preliminary investigation shows some promising results, it will take a long time to reach the final goal. However, with our understanding of the nature of the concrete hydrates, especially the C–S–H, it is possible to develop a method and a material that can react with C–S–H

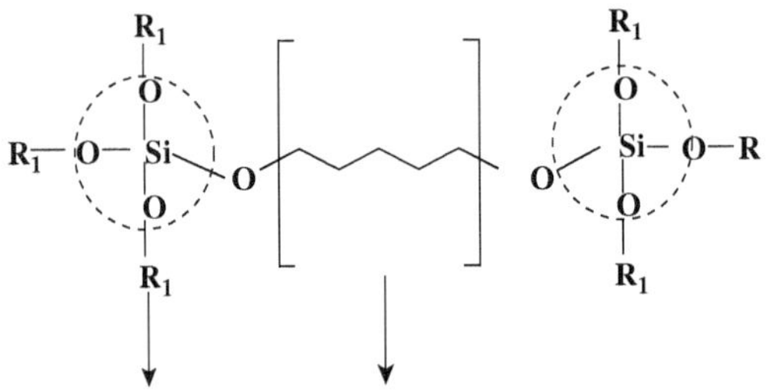

**Figure 9-12** Schematic drawing of the toughness enhancer structure

**Table 9-7** Fracture test results for two types of specimens

| Type (7 d) | Dosage (%) | $P_{max}$ (kN) | $K_{IC}$ (MPa·m$^{0.5}$) | $W_1$ (N·m) | $W_2$ (N·m) | $G_F$ (N·m$^{-1}$) | Ratio (%) |
|---|---|---|---|---|---|---|---|
| Reference | 0 | 4.12 | 1.03 | 0.54 | 0.23 | 91 | 100 |
| Modified | 1.5 | 5.53 | 1.38 | 0.89 | 0.99 | 220 | 242 |

and self-polymerize at the molecular scale to eliminate concrete's quasi-brittle nature and open an era for a new generation of concrete, i.e., the fourth generation of concrete.

## 9.5 APPLICATION OF NANOTECHNOLOGY IN CONCRETE

Normally, nanotechnology means the understanding, control, and restructuring of materials on the scale of less than 100 nm to enable fundamentally new properties and/or functions. There are two main approaches for nanotechnology, that is, the "top-down" and the "bottom-up" approaches. The former reduces the dimension of larger structures to the nanoscale while maintaining their original properties without atomic level control, and the latter engineers materials from the atomic or molecular scale through a process of assembly or self-assembly. Although both approaches have been proposed for concrete technology and much progress has been made in the nanoscale science of cement-based materials (e.g., the origins of cohesion and creep and the mechanisms of hydration and deterioration), real-world applications of nanotechnologies in concrete have been limited (Pellenq et al., 2009; Vandamme and Ulm, 2009; Sanchez and Sobolev, 2010).

To date, various nanomaterials have been proposed and used to modify cement-based materials. These materials include nano-particles, nano-reinforcements, and organic molecules. Relevant technologies based on such materials can effectively improve the properties and performance of concrete (e.g., mechanical strength and resistance to various deterioration mechanisms) or enable new functions that concrete does not originally possess. Applications like this belong to the bottom-up approach. Upon finding intrinsic defects of concrete in the solid phase (e.g., hydration products susceptible to deterioration), the liquid/pore phase (e.g., a well-connected pore network), or interfaces (e.g., an interfacial transition zone) can be modified by nanomaterials, so the performance of concrete will be improved (Garboczi, 2009).

Typical nano-particles that have been used in cement-based materials include nano-silica ($SiO_2$), nano-alumina ($Al_2O_3$), nano-iron oxide ($Fe_2O_3$), nano-clay, nano-$CaCO_3$, and nano-$TiO_2$. These nano-particles can act as nuclei for cement hydration products, facilitating cement hydration; and participate in cement hydration, leading to the formation of extra hydration products and densifying the microstructure. These effects could modify both the bulk cement paste and the interfacial transition zones between the paste and the aggregate, which is the weakest link in concrete, and, thus, improve the mechanical and durability properties. In addition to these effects, some of the particles can impart special functions to concrete, such as self-cleaning and pollutant removal by incorporating nano-$TiO_2$ (Jin et al., 2019), and self-sensing by adding nano-$Fe_2O_3$ (Li et al., 2004). Apart from adding nano-particles to cement-based materials, there is also limited research on producing nano-sized cementitious materials, through a top-down approach (i.e., reducing the particle size of the cement to the nano-scale by high-energy grinding, sometimes co-grinding with mineral additives) or a bottom-up approach (i.e., synthesize nano-cement chemically) (Lee and Kriven, 2005; Sobolev, 2005). The nano-size cementitious materials could also achieve higher strength than regular Portland cement.

Since the late 1990s, carbon nanotubes (CNT), nanofibers (CNF), and graphitic nano-plates (GNP) have been used as nano-reinforcements of cement-based materials to improve compressive and flexural strength as well as the modulus of elasticity. Since the early 2010s, graphene and graphene oxide (GO) have also been added to the list of nano-reinforcements. These nano-additives are frequently advocated as 1D (e.g., CNT) or 2D (e.g., GO) nano-reinforcements that could bridge weak hydration products, interfaces, or initial cracks, and, thus, improve the mechanical properties (especially tensile strength). However, it has been pointed out that bare, non-modified CNT, GNP, and graphene may not have beneficial interactions with hydration products of cement (Sanchez and Zhang, 2008; Hou et al., 2017). To be effective, the surface of the nano-reinforcements needs to be modified, such as by grafting functional groups (e.g., hydroxyl, carboxylate, carbonyl, and amine groups).

Development of nano-science of cement in 2000s—based on characterization and simulation—has found that grafting small organic groups or polymeric chains into growing C–S–H gel may be able to achieve covalent hybridization to improve the fracture energy and strain capacity of cement-based materials (Pellenq et al., 2008). In this line, organic matters that have been used include poly(methacrylic acid) (PMA), poly(acrylic acid) (PAA), poly(vinyl sulfonic acid) (PSA), poly(vinyl alcohol) (PVA), poly(ethylene glycol) (PEG), poly(allylamine) (PAM), polysaccharides, celluloses, etc. These molecules were selected because their structures contain large amount of functional groups (e.g., anionic: carboxyl ($-COO^-$) and sulfonate ($-SO_3^-$); or nonionic: hydroxyl ($-OH$), alkoxy ($-O-$), carbonyl ($=O$), and amino ($-NH_2$)) which can interact with C–S–H. It was found that such molecules can be intercalated into C–S–H to a certain extent, but the majority of them could stay in gel pores interlinking C–S–H colloidal particles. Both functional groups of the polymer molecules and features of C–S–H play important roles in the polymer/C–S–H interactions. The functional groups not only improve the H-bonds and the electrostatic forces at the polymer/C–S–H interfaces, but also fill the defects/damage of the silicate chains, repairing defective silicate chains that are supposedly broken during the fracture process (e.g., by forming Si-O-C bonds) (Hou et al., 2019). Such interaction mechanisms are expected to modify C–S–H by increasing or maintaining a degree of silicate chain polymerization and alter its mechanical properties and even long-term stability (e.g., calcium leaching and creep) (Khoshnazar et al., 2014, 2016).

So far, the enhancement in compressive and flexural strengths of cement-based materials achieved by nano-particles and nano-reinforcement could only be up to 30%, though improvement in the modulus of elasticity may reach 50%. Most of the studies relevant to organic-hybridization have been limited to synthetic polymer/C–S–H composites (based on sol-gel methods) and/or MD simulations. In realistic cement-based materials, the hybridizations of polymeric molecules sometimes even reduce the bulk mechanical properties, such as indentation modulus, hardness, and strength (Zhou et al., 2019; Chi et al., 2020). The ineffectiveness of nano-materials and nano-modifications in realistic cement systems could be attributed to the agglomeration of nano-particles and reinforcements, or the steric, entropic, and/or electrostatic interactions which tend to hinder the entry of organic molecules into C–S–H. In the future development and applications of

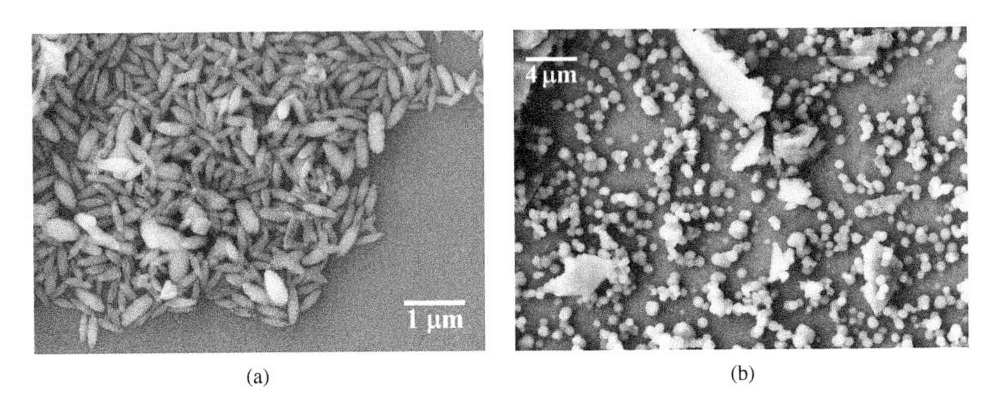


**Figure 9-13** Nano-particles formed *in situ* in fresh concrete: (a) calcium carbonate nano-particles; and (b) organic-inorganic nano-particles

nanotechnology in concrete, some challenges will need to be solved, such as proper dispersion, compatibility between nano-materials and cement, public safety concerns, and cost. An ideal way to solve these issues could be to generate nano-materials *in situ* in fresh concrete economically and eco-efficiently. On-going studies at Missouri University of Science and Technology have recently generated *in situ* calcium carbonate nano-particles (by carbonating calcium-bearing mixing water) and organic-inorganic nano-particles (via chelation of metal cations using selected organic molecules), as shown in Figure 9-13, which have been proven to significantly enhance engineering properties of concrete.

## 9.6 DATA SCIENCE AND ARTIFICIAL INTELLIGENCE IN CONCRETE TECHNOLOGY

Concrete is perhaps the most complex engineering material, mainly due to the substantial diversity in the characteristics of its raw materials and service environments. The future of studies on concrete may rely on a large amount of data, or "Big Data." The accumulated data in concrete industry is big enough to train artificial intelligence-based (AI, e.g., supervised machine learning algorithms) models to predict properties with high accuracy and reliability, to design concrete mixtures based on particular targets (i.e., performance, eco-efficiency, and cost) and service environment, or to fully validate numerical models (e.g., kinetic and thermodynamic models).

In recent years, there have been a burgeoning number of studies on using machine learning methods in concrete technology. Most efforts have been given to predictions of the mechanical properties of concrete, such as compressive strength, tensile strength, shear strength, and modulus of elasticity (Chaabene et al., 2020). Machine learning models can be classified into supervised learning and unsupervised learning. Most machine learning models that have been adopted in concrete technology belong to the supervised learning, because these models can generate patterns and hypotheses through a provided dataset to predict target values explicitly—which is more consistent with traditional engineering philosophy. The most widely adopted algorithms have included: artificial neural networks (ANN), such as multilayer perceptron ANN; support vector machines (SVM); decision trees, such as random forest (RF); and evolutionary algorithms, such as genetic algorithms. Besides, some algorithms like the firefly algorithm (FFA) have been used to determine the optimum values of the hyper-parameters of machine learning models, such as the number of trees and number of leaves per tree in the forest of a RF model (Cook et al., 2019). It has been shown that, when a dataset is big enough, standalone machine learning methods (e.g., SVM) may give satisfactory predictions. However, in most cases, ensemble learning (including RF), developed by integrating multiple models to solve a particular computational intelligence problem, can increase the prediction result (Cai et al., 2020). Apart from mechanical properties of concrete, machine learning has also been adopted in predictions of hydration behaviors of novel or multi-phase cementitious materials (Cook et al., 2021; Lapeyre et al., 2021), the design of chemical admixtures (Washburn et al., 2018), and improving the prediction accuracy of durability properties of concrete materials and structures (Taffese and Sistonen, 2017; Cai et al., 2020). In the field of health monitoring of concrete structures, deep learning algorithms, such as convolutional neural networks, have been implemented to detect concrete cracks (Cha et al., 2017).

Society is entering the era of Industry 4.0, where data-driven AI, among other emerging tools, has been fusing the world composed of traditional art, science and technology (e.g., concrete technology). While adopting data-driven AI in concrete technology, cement and concrete researchers may face with several challenges: (1) these researchers normally do not invent algorithms, so how to select the most suitable algorithms will become the key to using the new tool in a traditional field; and (2) a database is normally not big enough to ensure reliable outcomes. More and more researchers are sharing their data through open access protocols, which is a good trend. In the future, some general protocols will also be needed to screen reliable, high-fidelity datasets and fuse them together to achieve Big Data. Then, machine learning methods, including physics-based machine learning and deep learning, will truly change the rules of the game of concrete design and evaluation.

## REFERENCES


ACAA (American Coal Ash Association) (2019) "Coal Combustion Products Production & Use Survey Report," https://acaa-usa.org/publications/production-use-reports/.

Allen A. J., et al. (2004) "In-situ quasi-elastic scattering characterization of particle size effects on the hydration of tricalcium silicate," *Journal of Materials Research*, 19, 3242–3254.

Allen, A. J., Thomas, J.J., and Jennings, H.M. (2007) "Composition and density of nanoscale calcium–silicate–hydrate in cement," *Nature Materials*, 6, 311–316.

American Concrete Institute (2001) *Vision 2030.*

Andreu, G., and Miren, E. (2014) "Experimental analysis of properties of high performance recycled aggregate concrete," *Construction and Building Materials*, 52, 227–235.

Bendixen, M., Best, J., Hackney, J., and Iverson, L. L. (2019) "Time is running out for sand," *Nature*, 571, 29–31.

Bru, K., Touze, S., Bourgeois, F., Lippiatt, N., and Ménard, Y. (2014) "Assessment of a microwave-assisted recycling process for the recovery of high-quality aggregates from concrete waste," *International Journal of Mineral Processing*, 126, 90–98.

Bruntland, G., ed. (1987) *Our common future: The world commission on environment and development*, Oxford: Oxford University Press.

Cai, R., Han, T., Liao, W., Huang, J., Li, D., Kumar, A., and Ma, H. (2020) "Prediction of surface chloride concentration of marine concrete using ensemble machine learning," *Cement and Concrete Research*, 163, 106164.

Car, R. and Parrinello, M. (1985) "Unified approach for molecular dynamics and density-functional theory," *Physical Review Letters*, 55, 2471–2474.

CDRA, Construction & Demolition Recycling Association (2014) *The benefits of construction and demolition materials recycling in the United States*. CDRA.

Cementir Holding (2020) "The cement of the future is now here: Cementir launches FUTURECEM™ With up to 30 percent lower carbon emissions," https://www.cementirholding.com/en/media/whats-new/cement-future-now-here-cementir-launches-futurecemtm-30-percent-lower-carbon.

Cha, Y., Choi, W., and Büyüköztürk, O. (2017) "Deep learning-based crack damage detection using convolutional neural networks," *Computer-Aided Civil and Infrastructure Engineering*, 32(5), 361–378.

Chaabene, W., Flah, M., and Nehdi, M. (2020) "Machine learning prediction of mechanical properties of concrete: Critical review," *Construction and Building Materials*, 260, 119889.

Chi, Y., Huang, B., Saafi, M., Ye, J., and Lambert, C. (2020) "Carrot-based covalently bonded saccharides as a new 2D material for healing defective calcium-silicate-hydrate in cement: Integrating atomistic computational simulation with experimental studies," *Composites Part B: Engineering*, 199, 108235.

Cook, R., Han, T., Childers, A. Ryckman, C., Khayat, K., Ma, H., Huang, J., Kumar, A. (2021) "Machine learning for high-fidelity prediction of cement hydration kinetics in blended systems," *Materials & Design*, 208, 109920.

Cook, R., Lapeyre, J., Ma, H., and Kumar, A. (2019) "Prediction of compressive strength of concrete: A critical comparison of performance of a hybrid machine learning model with standalone models," *ASCE Journal of Materials in Civil Engineering,* 31, 11, 04019255.

De Juan, M. S. and Gutiérrez, P. A. (2009) "Study on the influence of attached mortar content on the properties of recycled concrete aggregate," *Construction and Building Materials*, 23(2), 872–877.

Duan, Z. and Poon, C. (2014) "Properties of recycled aggregate concrete made with recycled aggregates with different amounts of old adhered mortars," *Materials & Design*, 58, 19–29.

EPA (United States Environmental Protection Agency) (2004) "RCRA in focus: construction, demolition, and renovation," https://www.epa.gov/hwgenerators/resource-conservation-and-recovery-act-rcra-focus-hazardous-waste-generator-guidance.

EPA (United States Environmental Protection Agency) (2016) "Advancing sustainable materials management: 2014 fact sheet," https://www.epa.gov/sites/production/files/2016-11/documents/2014_smmfact sheet_508.pdf.

EPA (United States Environmental Protection Agency) (2017) "National overview: Facts and figures on materials, wastes and recycling," https://www.epa.gov/facts-and-figures-about-materials-waste-and-recycling/national-overview-facts-and-figures-materials.

Faraone, A., et al. (2004) "Quasielastic and inelastic neutron scattering on hydrated calcium silicate pastes," *Journal of Chemical Physics*, 121, 3212–3220.

Fratini, E., et al. (2006) "Hydration water and microstructure in calcium silicate and aluminium hydrates," *Journal of Physics: Condensed Matter*, 18, 2467–2483.

Garboczi, E. J. (2009) "Concrete nanoscience and nanotechnology: Definitions and applications," In: *Nanotechnology in Construction: Proceedings of the NICOM3* (3rd International Symposium on Nanotechnology in Construction), 81–88, Prague, Czech Republic.

Garrault, S., Finot, E., Lesniewska, E., and Nonat, A. (2005) "Study of CSH growth on C 3 S surface during its early hydration," *Materials and Structures*, 38, 435–442.

Gartner, E., and Hirao, H. (2015) "A review of alternative approaches to the reduction of $CO_2$ emissions associated with the manufacture of the binder phase in concrete," *Cement and Concrete Research*, 78 (Part A), 126–142.

Gastaldi, D., Canonico, F., Capelli, L., Buzzi, L., Boccaleri, E., and Irico, S. (2015) "An investigation on the recycling of hydrated cement from concrete demolition waste," *Cement and Concrete Composites*, 61, 29–35.

Gokce, A., Nagataki, S., Saeki, T., and Hisada, M. (2004) "Freezing and thawing resistance of air-entrained concrete incorporating recycled coarse aggregate: The role of air content in demolished concrete," *Cement and Concrete Research*, 34(5), 799–806.

González-Taboada, I., González-Fonteboa, B., Martínez-Abella, F., and Carro-López, D. (2016) "Study of recycled concrete aggregate quality and its relationship with recycled concrete compressive strength using database analysis," *Materiales de Construcción*, 66(323), e089 (18 pages).

Hewlett, P. (2004) *Lea's chemistry of cement and concrete*. Oxford: Elsevier Science & Technology Books.

Hou, D., Lu, Z., Li, X., Ma, H., and Li, Z. (2017) "Reactive molecular dynamics and experimental study of graphene-cement composites: Structure, dynamics and reinforcement mechanisms," *Carbon*, 115, 188–208.

Hou, D., Yu, J., and Wang, P. (2019) "Molecular dynamics modeling of the structure, dynamics, energetics and mechanical properties of cement-polymer nanocomposite," *Composites Part B: Engineering*, 162, 433–444.

Hu, C. (2014) "Multi-scale characterization of concrete," PhD thesis, the Hong Kong University of Science and Technology.

IEA, CSI, and WBCSD (2018) *Technology roadmap: Low-carbon transition in the cement industry*, Geneva: World Business Council for Sustainable Development.

Jin, Q., Saad, E., Zhang, W., Tang, Y., and Kurtis, K. (2019) "Quantification of NOx uptake in plain and $TiO_2$-doped cementitious materials," *Cement and Concrete Research*, 122, 251–256.

Johansson K.C.L. (1999) "Kinetics of the hydration reactions in the cement paste with mechanochemically modified cement 29Si magic-angle-spinning NMR study." *Cement and Concrete Research*, 29(10), 1575–1581.

Kalinichev, A. G., et al. (2002) "Molecular dynamics modeling of chloride binding to the surfaces of Ca hydroxide, hydrated Ca-aluminate and Ca-silicate phases," *Chemistry of Materials*, 14, 3539–3549.

Katz, A. (2003) "Properties of concrete made with recycled aggregate from partially hydrated old concrete," *Cement and Concrete Research*, 33(5), 703–711.

Katz, A. (2004) "Treatments for the improvement of recycled aggregate," *Journal of Materials in Civil Engineering*, 16(6), 597–603.

Khoshnazar, R., Beaudoin, J. J., Raki, L., and Alizadeh, R. (2014) "Volume stability of calcium-silicate-hydrate/polyaniline nanocomposites in aqueous salt solutions," *ACI Materials Journal*, 111(6), 623–632.

Khoshnazar, R., Beaudoin, J. J., Raki, L., and Alizadeh, R. (2016) "Interaction of 2-, 3- and 4-nitrobenzoic acid with the structure of calcium-silicate-hydrate," *Materials and Structures*, 49, 499–506.

Kirkpatrick, X. C. (1996) "29Si and 17O NMR investigation of the structure of some crystalline calcium silicate hydrates." *Advances in Cement Based Materials* 3, 133–143.

Kou, S., and Poon, C. (2012) "Enhancing the durability properties of concrete prepared with coarse recycled aggregate," *Construction and Building Materials*, 35, 69–76.

Kumar, A., Walder, B. J., Mohamed, A. K., Hofstetter, A., Srinivasan, B., Rossini, A. J., Scrivener, K., Emsley, L., and Bowen, P. (2017) "The atomic-level structure of cementitious calcium silicate hydrate," *The Journal of Physical Chemistry C*, 12117188-17196.

Lapeyre, J., Han, T., Wiles, B., Ma, H., Huang, J., Sant, G., and Kumar, A. (2021) "Machine learning enables prompt prediction of hydration kinetics of multicomponent cementitious systems," *Scientific Reports*, 11, 3922.

Lee, G., Ling, T., Wong, Y., and Poon, C. (2011) "Effects of crushed glass cullet sizes, casting methods and pozzolanic materials on ASR of concrete blocks," *Construction and Building Materials*, 25(5), 2611–2618.

Lee, S. J., and Kriven, W. M. (2005) "Synthesis and hydration study of Portland cement components prepared by the organic steric entrapment method," *Materials and Structures*, 38(1), 87–92.

Li, H., Xiao, H., and Ou, J. (2004) "A study on mechanical and pressure-sensitive properties of cement mortar with nanophase materials," *Cement and Concrete Research*, 34(3), 435–438.

Li, Z., Chau, C. K., Ma, B., and Li, F. (2000) "New natural polymer based durability enhancement admixture and corresponding concretes made with the admixture," US Patent No. 6,153,006.

Liang, R. (2018) "Mechanism and performance study of cement-organic integrated materials," PhD thesis, The Hong Kong University of Science and Technology.

Liu, Q. (2020) "Flexural strength and durability enhancements for cement-based materials by in situ polymerization of monomers," PhD thesis, University of Macao.

Ma, H. (2013) "Multi-Scale modeling of the microstructure and transport properties of contemporary concrete," PhD thesis, the Hong Kong University of Science and Technology.

Ma, H. and Li, Z. (2013) "Microstructures and mechanical properties of polymer modified mortars under distinct mechanisms," *Construction and Building Materials*, 47, 579–587.

Ma, H., Tian, Y., and Li, Z. (2011) "Interactions between organic and inorganic phases in PA-and PU/PA-modified-cement-based materials," *Journal of Materials in Civil Engineering*, 23(1), 1412–1421.

Marx, D. and Hutter, J. (2000) In: Grotendorst, J., ed., *Modern methods and algorithms of quantum chemistry*. NIC, FZ Julich, pp. 301–449.

McDonald, P. J., Korb, J-P., Mitchell, J., and Monteilhet, L. (2005) "Surface relaxation and chemical exchange in hydrating cement pastes: A two-dimensional NMR relaxation study," *Physical Review E*, 72, 011409.

Ming, X., Huang, J. and Li, Z. (2021) "Materials-oriented integrated design and construction of structures in civil engineering – a review," Manuscript submitted to *Frontiers of Structural and Civil Engineering*.

Pellenq, R., Kushima, A., Shahsavari, R., Van Vliet, K. J., Buehler, M., Yip, S., and Ulm, F.-J. (2009) "A realistic molecular model of cement hydrates," *Proceedings of the National Academy of Sciences on the United States of America*, 106(38), 16102–16107.

Pellenq, R., Lequeux, N., and van Damme, H. (2008) "Engineering the bonding scheme in C–S–H: The iono-covalent framework," *Cement and Concrete Research*, 38(2), 159–174.

Plassard, C., Lesniewska, E., Pochard, I., and Nonat, A. (2005) "Nanoscale experimental investigation of particle interactions at the origin of the cohesion of cement," *Langmuir*, 21, 7263–7270.

Powers, T. C. and Brownyard, T. L. (1948) "Studies of the physical properties of hardened Portland cement paste," *Bulletin of the Portland Cement Association*, p. 22.

Remler, D. K. and Madden, P. A. (1990) "Molecular-dynamics without effective potentials via the Car-Parrinello approach," *Molecular Physics*, 70, 921–966.

Richardson, I. (1999) "The nature of C–S–H in hardened cements," *Cement and Concrete Research*, 29, 1131–1147.

Richardson, I. G. (2002) "Electron microscopy of cement," in: Bensted, J. and Barnes, P., eds, *Structure and performance of cements*. London: Spon Press.

Richardson, I. G. (2004) "Tobermorite/jennite- and tonermorite/calcium hydroxide-based models for the structure of C–S–H: applicability to hardened pastes of tricalcium silicates, dicalcium silicate, Portland cement, and blends of Portland cement with blast-furnace slag, metakaolin, or silica fume," *Cement and Concrete Research*, 34, 1733–1777.

Sadati, S. (2017) "High-volume recycled materials for sustainable transportation infrastructure," PhD thesis, Missouri University of Science and Technology.

Sagoe-Crentsil, K. K., Brown, T., and Taylor, A. (2011) "Performance of concrete made with commercially produced coarse recycled concrete aggregate," *Cement and Concrete Research* 31(5), 707–712.

Sanchez, F. and Sobolev, K. (2010) "Nanotechnology in concrete – A review," *Construction and Building Materials*, 24, 2060–2071.

Sanchez, F. and Zhang, L. (2008) "Molecular dynamics modeling of the interface between surface functionalized graphitic structures and calcium–silicate–hydrate: interaction energies, structure, and dynamics," *Journal of Colloid and Interface Science*, 323(2), 349–358.

Schokker, A. J. (2010) *The sustainable concrete guide: Strategies and examples*, U.S. Green Concrete Council.

Scrivener, K. L. et al. (2004) "Quantitative study of Portland cement hydration by X-ray diffraction/Rietveld analysis and independent methods," *Cement and Concrete Research*, 34(9) 1541–1547.

Scrivener, K. L. and Kirkpatrick, R. J. (2007) "Innovation in use and research on cementitious material," in: Beaudoin, J.J., Makar, J.M., and Raki, L., eds., *Proceedings of the 12th international congress on the chemistry of cement*, July 8–13, 2007, Montreal, Canada. On CD.

Shen, Y., Qian, J., Huang, Y., and Yang, D. (2015) "Synthesis of belite sulfoaluminate-ternesite cements with phosphogypsum," *Cement and Concrete Composites*, 63, 67–75.

Shi, C., Li, Y., Zhang, J., Li, W., Chong, L., and Xie, Z. (2016) "Performance enhancement of recycled concrete aggregate - A review," *Journal of Cleaner Production*, 112, 466–472.

Silva, R. V., de Brito, J., and Dhir, R. K. (2015) "Prediction of the shrinkage behavior of recycled aggregate concrete: a review," *Construction and Building Materials*, 77, 327–339.

Silva, R. V., Silva, A., Neves R., and de Brito, J. (2016) "Statistical modeling of carbonation in concrete incorporating recycled aggregates," *Journal of Materials in Civil Engineering*, 28(1), 04015082.

Sobolev, K. (2005) "Mechano-chemical modification of cement with high volumes of blast furnace slag," *Cement and Concrete Composites*, 27(7–8), 848–853.

Sui, T.B., Fan, L., Wen, Z. J., and Wang, J. (2009) "Low energy and low emission cement with high performance and low hydration heat and its concrete application," *Materials China*, 28(11), 46–52.

Sverdrup, H. U., Koca D., and Schlyter, P. (2017) "A simple system dynamics model for the global production rate of sand, gravel, crushed rock and stone, market prices and long-term supply embedded into the WORLD6 model," *BioPhysical Economics and Resource Quality*, 2, 8 (20 pages).

Taffese, W., and Sistonen, E. (2017) "Machine learning for durability and service-life assessment of reinforced concrete structures: Recent advances and future directions," *Automation in Construction*, 77, 1–14.

Tam, V., Tam, C., Le, K. (2007) "Removal of cement mortar remains from recycled aggregate using pre-soaking approaches," *Resources, Conservation and Recycling*, 50(1), 82–101.

Thiery, M., Dangla, P., Belin, P., Habert, G., and Roussel, N. (2013) "Carbonation kinetics of a bed of recycled concrete aggregates: a laboratory study on model materials," *Cement and Concrete Research*, 46, 50–65.

Thomas, M. (2019) "Impact of $CO_2$ utilization in fresh concrete on corrosion of steel reinforcement," CarbonCure technical report.

Topcu, I. B., and Sengel, S. (2004) "Properties of concrete produced with waste concrete aggregate," *Cement and Concrete Research*, 34(8), 1307–1312.

Torres, A., Brandt, J., Lear, K., and Liu, J. (2017) "A looming tragedy of the sand commons," *Science*, 357(6355), 970–971.

Tsujino, M., Noguchi, T., Tamura, M., Kanematsu, M., and Maruyama, I. (2007) "Application of conventionally recycled coarse aggregate to concrete structure by surface modification treatment," *Journal of Advanced Concrete Technology*, 5(1), 13–25.

Ulm, F.J., et al. (2007) "Statistical indentation techniques for hydrated nanocomposites: concrete, bone, and shale," *Journal of the American Ceramic Society*, 90(9), 2677–2692.

United Nations, Scrivener, K. L., John, V. M., and Gartner, E. M. (2018) "Eco-efficient cements: Potential economically viable solutions for a low-$CO_2$ cement-based materials industry," *Cement and Concrete Research*, 114, 2–26.

USGS, United States Geological Survey (2020) "Iron and steel slag." Available online: https://pubs.usgs.gov/periodicals/mcs2020/mcs2020-iron-steel-slag.pdf.

Vandamme, M. and Ulm, F.-J. (2009) "Nanogranular origin of concrete creep," *Proceedings of the National Academy of Sciences on the United States of America*, 106(26), 10552–10557.

Walling, S. A. and Provis, J. L. (2016) "Magnesia-based cements: a journey of 150 years, and cements for the future?" *Chemical Reviews*, 116(7), 4170–4204.

Waseem, S. A. and Singh, B. (2017) "Shear strength of interfaces in natural and in recycled aggregate concrete," *Canadian Journal of Civil Engineering*, 44(3), 212–222.

Washburn, N., Menon, A., Childs, C., Poczos, B., and Kurtis, K. (2018) "Machine learning approaches to admixture design for clay-based cements," *Calcined Clays for Sustainable Concrete*, 488–493.

Xiao, J., Li, W., Fan, Y., and Huang, X. (2012) "An overview of study on recycled aggregate concrete in China (1996–2011)," *Construction and Building Materials*, 31, 364–383.

Zhang, J., Shi, C., Li, Y., Pan, X., Poon, C., and Xie, Z. (2015) "Performance enhancement of recycled concrete aggregates through carbonation," *Journal of Materials in Civil Engineering*, 27(11), 04015029.

Zhou, Y., Tang, L., Liu, J., and Miao, C. (2019) "Interaction mechanisms between organic and inorganic phases in calcium silicate hydrates/poly(vinyl alcohol) composites," *Cement and Concrete Research*, 125, 105891.

# INDEX

## G



## H

**I**

**N**

## O



## P

## T

## W